RSMeans

Square Foot Costs

19th Annual Edition

D1264747

1998

Senior Editor

John H. Ferguson, PE

Contributing Editors

Thomas J. Akins
Barbara Balboni
Howard M. Chandler
John H. Chiang, PE
Paul C. Crosscup
Jennifer L. Fordyce
Robert W. Mewis
Melville J. Mossman, PE
John J. Moylan
Jeannene D. Murphy
Peter T. Nightingale
Jesse R. Page
Stephen C. Plotner
Michael J. Regan
Kornelis Smit
William R. Tennyson, II
Phillip R. Waier, PE
James N. Wills
Rory Woolsey

Manager, Engineering Operations

John H. Ferguson, PE

President

Durwood S. Snead

Vice President and General Manager

Roger J. Grant

Vice President, Sales and Marketing

John M. Shea

Vice President, Operations

Andrew J. Centauro

Production Manager

Karen L. O'Brien

Production Coordinator

Marion E. Schofield

Technical Support

Michele S. Able
Wayne D. Anderson
Thomas J. Dion
Michael H. Donelan
Paul C. Hebert
Gary L. Hoitt
Mark H. Kaplan, Jr.
Marla A. Marek
Paula Reale-Camelio
Kathryn S. Rodriguez
Ali Vaghar

Art Director

Helen A. Marcella

Book & Cover Design

Norman R. Forgit

First Printing

Foreword

R.S. Means Co., Inc. is a member of the **Construction Market Data Group**, a leading provider of construction information in North America. The group's other companies, besides **R.S. Means,** include **Architects' First Source,** the most extensive and widely circulated product selection resource for specifiers; **Construction Market Data (CMD),** the source for construction project information for sales leads, market analysis and marketing; **Manufacturers' Survey Associates,** a leading estimating and quantity survey company in the U.S.; and **BIMSA/Mexico,** the dominant distributor of information on building projects and construction costs throughout Mexico.

Our Mission

Since 1942, R.S. Means Company, Inc. has been actively engaged in construction cost publishing and consulting throughout North America.

Today, over fifty years after the company began, our primary objective remains the same: to provide you, the construction and facilities professional, with the most current and comprehensive construction cost data possible.

Whether you are a contractor, an owner, an architect, an engineer, a facilities manager, or anyone else who needs a quick construction cost estimate, you'll find this publication to be a highly useful and necessary tool.

Today, with the constant flow of new construction methods and materials, it's difficult to find the time to look at and evaluate all the different construction cost possibilities. In addition, because labor and material costs keep changing, last year's cost information is not a reliable basis for today's estimate or budget.

That's why so many construction professionals turn to R.S. Means. We keep track of the costs for you, along with a wide range of other key information, from city cost indexes . . . to productivity rates . . . to crew composition . . . to contractor's overhead and profit rates.

R.S. Means performs these functions by collecting data from all facets of the industry, and organizing it in a format that is instantly accessible to you. From the preliminary budget to the detailed unit price estimate, you'll find the data in this book useful for all phases of construction cost determination.

The Staff, the Organization, and Our Services

When you purchase one of R.S. Means' publications, you are in effect hiring the services of a full-time staff of construction and engineering professionals.

Our thoroughly experienced and highly qualified staff works daily at collecting, analyzing, and disseminating comprehensive cost information for your needs. These staff members have years of practical construction experience and engineering training prior to joining the firm. As a result, you can count on them not only for the cost figures, but also for additional background reference information that will help you create a realistic estimate.

The Means organization is always prepared to help you solve construction problems through its five major divisions: Construction and Cost Data Publishing, Electronic Products and Services, Consulting Services, Insurance Division, and Educational Services.

Besides a full array of construction cost estimating books, Means also publishes a number of other reference works for the construction industry. Subjects include construction estimating and project and business management; special topics such as HVAC, roofing, plumbing, and hazardous waste remediation; and a library of facility management references.

In addition, you can access all of our construction cost data through your computer with Means CostWorks '98 CD-ROM, an electronic tool that offers over 50,000 lines of Means construction cost data.

What's more, you can increase your knowledge and improve your construction estimating and management performance with a Means Construction Seminar or In-House Training Program. These two-day seminar programs offer unparalleled opportunities for everyone in your organization to get updated on a wide variety of construction-related issues.

Means also is a worldwide provider of construction cost management and analysis services for commercial and government owners and of claims and valuation services for insurers.

In short, R.S. Means can provide you with the tools and expertise for constructing accurate and dependable construction estimates and budgets in a variety of ways.

Robert Snow Means Established a Tradition of Quality That Continues Today

Robert Snow Means spent years building his company, making certain he always delivered a quality product.

Today, at R.S. Means, we do more than talk about the quality of our data and the usefulness of our books. We stand behind all of our data, from historical cost indexes... to construction materials and techniques... to current costs.

If you have any questions about our products or services, please call us toll-free at 1-800-334-3509. Our customer service representatives will be happy to assist you.

Table of Contents

RESIDENTIAL
- INTRODUCTION
- WORKSHEET
- ECONOMY
- AVERAGE
- CUSTOM
- LUXURY
- ADJUSTMENTS

COMM./INDUST./INSTIT.
- INTRODUCTION
- EXAMPLES
- BUILDING TYPES
- DEPRECIATION

ASSEMBLIES
- FOUNDATIONS — 1
- SUBSTRUCTURES — 2
- SUPERSTRUCTURES — 3
- EXTERIOR CLOSURE — 4
- ROOFING — 5
- INTERIOR CONSTRUCTION — 6
- CONVEYING SYSTEMS — 7
- MECHANICAL — 8
- ELECTRICAL — 9
- SPECIAL CONSTRUCTION — 11
- SITE WORK — 12

REFERENCE INFORMATION
- GENERAL CONDITIONS
- LOCATION FACTORS
- COST INDEXES
- GLOSSARY

How the Book Is Built: An Overview

A Powerful Construction Tool

You have in your hands one of the most powerful construction tools available today. A successful project is built on the foundation of an accurate and dependable estimate. This book will enable you to construct just such an estimate.

For the casual user the book is designed to be:

- quickly and easily understood so you can get right to your estimate
- filled with valuable information so you can understand the necessary factors that go into the cost estimate

For the regular user, the book is designed to be:

- a handy desk reference that can be quickly referred to for key costs
- a comprehensive, fully reliable source of current construction costs for typical building structures when only minimum dimensions, specifications and technical data are available. It is specifically useful in the conceptual or planning stage for preparing preliminary estimates which can be supported and developed into complete engineering estimates for construction.
- a source book for rapid estimating of replacement costs and preparing capital expenditure budgets

To meet all of these requirements we have organized the book into the following clearly defined sections.

How To Use the Book: The Details

This section contains an in-depth explanation of how the book is arranged . . . and how you can use it to determine a reliable construction cost estimate. It includes information about how we develop our cost figures and how to completely prepare your estimate.

Residential Section

Model buildings for four classes of construction—economy, average, custom and luxury—are developed and shown with complete costs per square foot.

Commercial, Industrial, Institutional Section

This section contains complete costs for 70 typical model buildings expressed as costs per square foot.

Assemblies Section

The cost data in this section has been organized in an "Assemblies" format. These assemblies are the functional elements of a building and are arranged according to the 12 divisions of the UniFormat classification system. The costs in this section include standard mark-ups for the overhead and profit of the installing contractor. The assemblies costs are those used to calculate the total building costs in the Commercial, Industrial, Institutional section.

Reference Section

This section includes information on General Conditions, Location Factors, Historical Cost Indexes, a Glossary, and a listing of Abbreviations used in this book.

Location Factors: Costs vary depending upon regional economy. You can adjust the "national average" costs in this book to 930 locations throughout the U.S. and Canada by using the data in this section.

Historical Cost Indexes: These indexes provide you with data to adjust construction costs over time. If you know costs for a project completed in the past, you can use these indexes to calculate a rough estimate of what it would cost to construct the same project today.

Glossary: A listing of common terms related to construction and their definitions is included.

Abbreviations: A listing of the abbreviations used throughout this book, along with the terms they represent, is included.

The Scope of This Book

This book is designed to be as comprehensive and as easy to use as possible. To that end we have made certain assumptions and limited its scope in two key ways:

1. We have established material prices based on a "national average."
2. We have computed labor costs based on a 30-city "national average" of union wage rates.

All costs and adjustments in this manual were developed from the *Means Building Construction Cost Data, Means Assemblies Cost Data, Means Residential Cost Data,* and *Means Light Commercial Cost Data.* Costs include labor, material, overhead, and profit, and general conditions and architect's fees where specified. No allowance is made for sales tax, financing, premiums for material and labor, or unusual business and weather conditions.

What's Behind the Numbers? The Development of Cost Data

The staff at R.S. Means continuously monitors developments in the construction industry in order to ensure reliable, thorough and up-to-date cost information.

While *overall* construction costs may vary relative to general economic conditions, price fluctuations within the industry are dependent upon many factors. Individual price variations may, in fact, be opposite to overall economic trends. Therefore, costs are continually monitored and complete updates are published yearly. Also, new items are frequently added in response to changes in materials and methods.

Costs—$ (U.S.)

All costs represent U.S. national averages and are given in U.S. dollars. The Means Location Factors can be used to adjust costs to a particular location. The Location Factors for Canada can be used to adjust U.S. national averages to local costs in Canadian dollars.

Factors Affecting Costs

Due to the almost limitless variation of building designs and combinations of construction methods and materials, costs in this book must be used with discretion. The editors have made every effort to develop costs for "typical" building types. By definition, a square foot estimate, due to the short time and limited amount of detail required, involves a relatively lower degree of accuracy than a detailed unit cost estimate, for example. The accuracy of an estimate will increase as the project is defined in more detailed components. The user should examine closely and compare the specific requirements of the project being estimated with the components included in the models in this book. Any differences should be accounted for in the estimate.

In many building projects, there may be factors that increase or decrease the cost beyond the range shown in this manual. Some of these factors are:

- Substitution of building materials or systems for those used in the model.
- Custom designed finishes, fixtures and/or equipment.
- Special structural qualities (allowance for earthquake, future expansion, high winds, long spans, unusual shape).
- Special foundations and substructures to compensate for unusual soil conditions.
- Inclusion of typical "owner supplied" equipment in the building cost (special testing and monitoring equipment for hospitals, kitchen equipment for hotels, offices that have cafeteria accommodations for employees, etc.).
- An abnormal shortage of labor or materials.

- Isolated building site or rough terrain that would affect the transportation of personnel, material or equipment.
- Unusual climatic conditions during the construction process.

We strongly urge that, for maximum accuracy, these factors be considered each time a structure is being evaluated.

If users require greater accuracy than this manual can provide, the editors recommend that the *Means Building Construction Cost Data* or the *Means Assemblies Cost Data* be consulted.

Final Checklist

Estimating can be a straightforward process provided you remember the basics. Here's a checklist of some of the items you should remember to do before completing your estimate.

Did you remember to . . .

- include the Location Factor for your city
- mark up the entire estimate sufficiently for your purposes
- include all components of your project in the final estimate
- double check your figures to be sure of your accuracy
- call R.S. Means if you have any questions about your estimate or the data you've found in our publications

Remember, R.S. Means stands behind its publications. If you have any questions about your estimate . . . about the costs you've used from our books . . . or even about the technical aspects of the job that may affect your estimate, feel free to call the R.S. Means editors at 1-781-585-7898 or 1-800-448-8182.

Residential Section

Table of Contents

Introduction to Residential Section

The residential section of this manual contains costs per square foot for four classes of construction in seven building types. Costs are listed for various exterior wall materials which are typical of the class and building type. There are cost tables for Wings and Ells with modification tables to adjust the base cost of each class of building. Non-standard items can easily be added to the standard structures.

Cost estimating for a residence is a three-step process:
  (1) Identification
  (2) Listing dimensions
  (3) Calculations

Guidelines and a sample cost estimating procedure are shown on the following pages.

Identification

To properly identify a residential building, the class of construction, type, and exterior wall material must be determined. Page 6 has drawings and guidelines for determining the class of construction. There are also detailed specifications and additional drawings at the beginning of each set of tables to further aid in proper building class and type identification.

Sketches for eight types of residential buildings and their configurations are shown on pages 8 and 9. Definitions of living area are next to each sketch. Sketches and definitions of garage types are on page 10.

Living Area

Base cost tables are prepared as costs per square foot of living area. The living area of a residence is that area which is suitable and normally designed for full time living. It does not include basement recreation rooms or finished attics, although these areas are often considered full time living areas by the owners.

Living area is calculated from the exterior dimensions without the need to adjust for exterior wall thickness. When calculating the living area of a 1-1/2 story, two story, three story or tri-level residence, overhangs and other differences in size and shape between floors must be considered.

Only the floor area with a ceiling height of six feet or more in a 1-1/2 story residence is considered living area. In bi-levels and tri-levels, the areas that are below grade are considered living area, even when these areas may not be completely finished.

Base Tables and Modifications

Base cost tables show the base cost per square foot without a basement, with one full bath and one full kitchen for economy and average homes and an additional half bath for custom and luxury models. Adjustments for finished and unfinished basements are part of the base cost tables. Adjustments for multi family residences, additional bathrooms, townhouses, alternate roofs, and air conditioning and heating systems are listed in modification tables below the base cost tables.

Costs for other modifications, including garages, breezeways and site improvements, are in the modification tables on pages 58 to 61.

Listing of Dimensions

To use this section of the manual only the dimensions used to calculate the horizontal area of the building and additions and modifications are needed. The dimensions, normally the length and width, can come from drawings or field measurements. For ease in calculation, consider measuring in tenths of feet, i.e., 9 ft. 6 in. = 9.5 ft., 9 ft. 4 in. = 9.3 ft.

In all cases, make a sketch of the building. Any protrusions or other variations in shape should be noted on the sketch with dimensions.

Calculations

The calculations portion of the estimate is a two-step activity:
  (1) The selection of appropriate costs from the tables
  (2) Computations

Selection of Appropriate Costs

To select the appropriate cost from the base tables, the following information is needed:
  (1) Class of construction
       (a) Economy
       (b) Average
       (c) Custom
       (d) Luxury
  (2) Type of residence
       (a) One story
       (b) 1-1/2 story
       (c) 2 story
       (d) 3 story
       (e) Bi-level
       (f) Tri-level
  (3) Occupancy
       (a) One family
       (b) Two family
       (c) Three family
  (4) Building configuration
       (a) Detached
       (b) Town/Row house
       (c) Semi-detached
  (5) Exterior wall construction
       (a) Wood frame
       (b) Brick veneer
       (c) Solid masonry
  (6) Living areas

Modifications are classified by class, type and size.

Computations

The computation process should take the following sequence:
  (1) Multiply the base cost by the area
  (2) Add or subtract the modifications
  (3) Apply the location modifier

When selecting costs, interpolate or use the cost that most nearly matches the structure under study. This applies to size, exterior wall construction and class.

How to Use the Residential Square Foot Cost Pages

The following is a detailed explanation of a sample entry in the Residential Square Foot Cost Section. Each bold number below corresponds to the item being described on the facing page with the appropriate component of the sample entry following in parenthesis.

Prices listed are costs that include overhead and profit of the installing contractor. Total model costs include an additional mark-up for General Contractors' overhead and profit, and fees specific to class of construction.

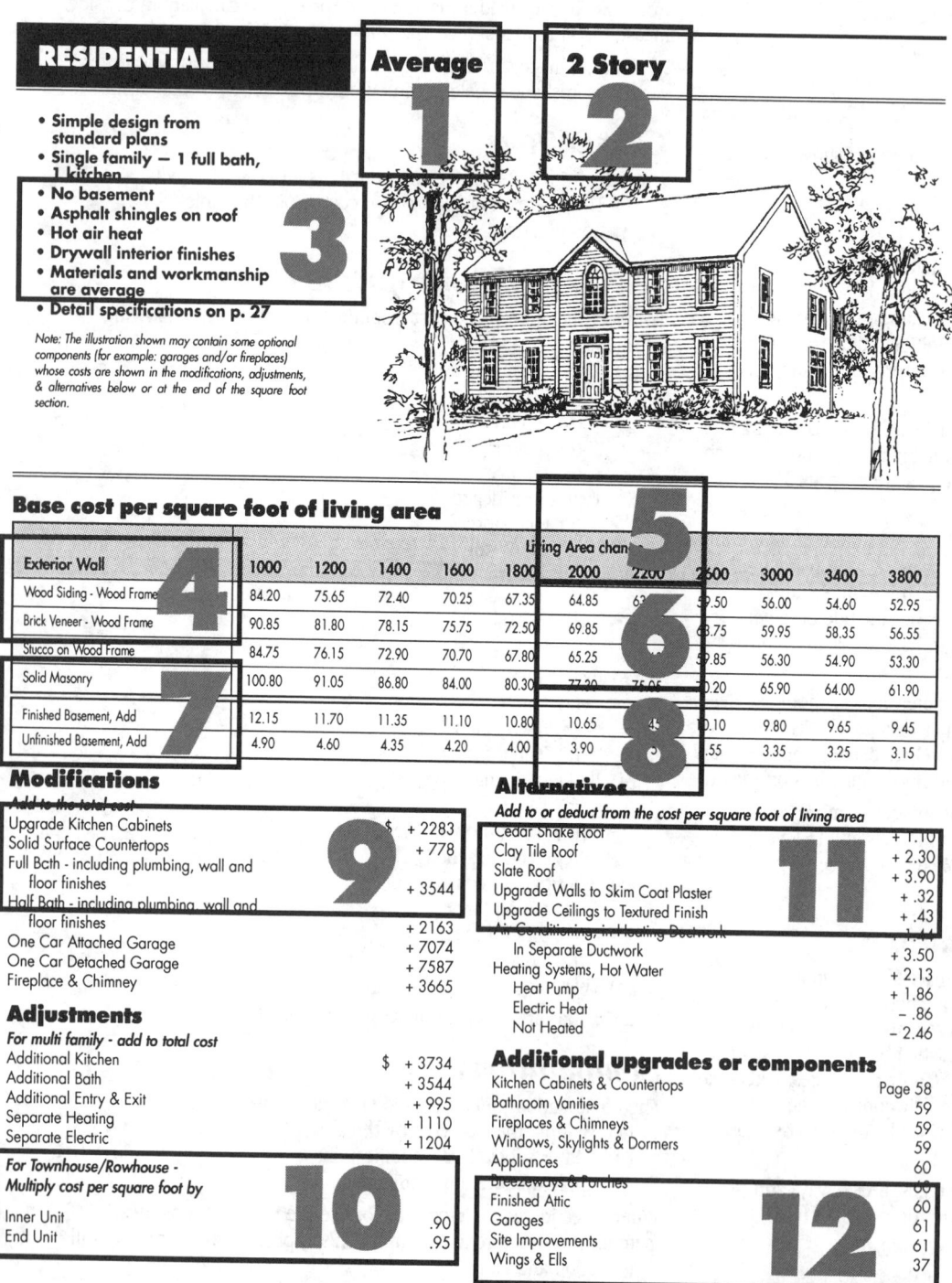

RESIDENTIAL — **Average** **1** — **2 Story** **2**

- Simple design from standard plans
- Single family — 1 full bath, 1 kitchen
- No basement
- Asphalt shingles on roof **3**
- Hot air heat
- Drywall interior finishes
- Materials and workmanship are average
- Detail specifications on p. 27

Note: The illustration shown may contain some optional components (for example: garages and/or fireplaces) whose costs are shown in the modifications, adjustments, & alternatives below or at the end of the square foot section.

Base cost per square foot of living area

Exterior Wall **4**	1000	1200	1400	1600	1800	2000 **5**	2200	2600	3000	3400	3800
Wood Siding - Wood Frame	84.20	75.65	72.40	70.25	67.35	64.85	63.	59.50	56.00	54.60	52.95
Brick Veneer - Wood Frame	90.85	81.80	78.15	75.75	72.50	69.85	**6** 63.75	59.95	58.35	56.55
Stucco on Wood Frame	84.75	76.15	72.90	70.70	67.80	65.25	59.85	56.30	54.90	53.30
Solid Masonry **7**	100.80	91.05	86.80	84.00	80.30	77.30	75.05	70.20	65.90	64.00	61.90
Finished Basement, Add	12.15	11.70	11.35	11.10	10.80	10.65	**8** 10.10	9.80	9.65	9.45
Unfinished Basement, Add	4.90	4.60	4.35	4.20	4.00	3.90	.55	3.35	3.25	3.15

Living Area change **5**

Modifications

Add to the total cost

Upgrade Kitchen Cabinets	$ + 2283
Solid Surface Countertops	+ 778
Full Bath - including plumbing, wall and floor finishes	+ 3544
Half Bath - including plumbing, wall and floor finishes	+ 2163
One Car Attached Garage	+ 7074
One Car Detached Garage	+ 7587
Fireplace & Chimney	+ 3665

9

Adjustments

For multi family - add to total cost

Additional Kitchen	$ + 3734
Additional Bath	+ 3544
Additional Entry & Exit	+ 995
Separate Heating	+ 1110
Separate Electric	+ 1204

For Townhouse/Rowhouse - Multiply cost per square foot by **10**

Inner Unit	.90
End Unit	.95

Alternatives

Add to or deduct from the cost per square foot of living area

Cedar Shake Roof	+ 1.10
Clay Tile Roof	+ 2.30
Slate Roof	+ 3.90
Upgrade Walls to Skim Coat Plaster	+ .32
Upgrade Ceilings to Textured Finish	+ .43
Air Conditioning, in Heating Ductwork	1.44
In Separate Ductwork	+ 3.50
Heating Systems, Hot Water	+ 2.13
Heat Pump	+ 1.86
Electric Heat	– .86
Not Heated	– 2.46

11

Additional upgrades or components

Kitchen Cabinets & Countertops	Page 58
Bathroom Vanities	59
Fireplaces & Chimneys	59
Windows, Skylights & Dormers	59
Appliances	60
Breezeways & Porches	60
Finished Attic	60
Garages	61
Site Improvements	61
Wings & Ells	37

12

Important: See the Reference Section for Location Factors (to adjust for your city) and Estimating Forms

1 Class of Construction (Average)

The class of construction depends upon the design and specifications of the plan. The four classes are economy, average, custom and luxury.

2 Type of Residence (2 Story)

The building type describes the number of stories or levels in the model. The seven building types are 1 story, 1-1/2 story, 2 story, 2-1/2 story, 3 story, bi-level and tri-level.

3 Specification Highlights (Hot Air Heat)

These specifications include information concerning the components of the model, including the number of baths, roofing types, HVAC systems, and materials and workmanship. If the components listed are not appropriate, modifications can be made by consulting the information shown lower on the page or the Assemblies Section.

4 Exterior Wall System (Wood Siding - Wood Frame)

This section includes the types of exterior wall systems and the structural frame used. The exterior wall systems shown are typical of the class of construction and the building type shown.

5 Living Areas (2000 SF)

The living area is that area of the residence which is suitable and normally designed for full time living. It does not include basement recreation rooms or finished attics. Living area is calculated from the exterior dimensions without the need to adjust for exterior wall thickness. When calculating the living area of a 1-1/2 story, 2 story, 3 story or tri-level residence, overhangs and other differences in size and shape between floors must be considered. Only the floor area with a ceiling height of six feet or more in a 1-1/2 story residence is considered living area. In bi-levels and tri-levels, the areas that are below grade are considered living area, even when these areas may not be completely finished. A range of various living areas for the residential model are shown to aid in selection of values from the matrix.

6 Base Costs per Square Foot of Living Area ($64.85)

Base cost tables show the cost per square foot of living area without a basement, with one full bath and one full kitchen for economy and average homes and an additional half bath for custom and luxury models. When selecting costs, interpolate or use the cost that most nearly matches the residence under consideration for size, exterior wall system and class of construction. Prices listed are costs that include overhead and profit of the installing contractor, a general contractor markup and an allowance for plans that vary by class of construction. For additional information on contractor overhead and architectural fees, see the Reference Section.

7 Basement Types (Finished)

The two types of basements are finished or unfinished. The specifications and components for both are shown on the Building Classes page in the Introduction to this section.

8 Additional Costs for Basements ($10.65)

These values indicate the additional cost per square foot of living area for either a finished or an unfinished basement.

9 Modifications and Adjustments (Solid Surface Countertops $778)

Modifications and Adjustments are costs added to or subtracted from the total cost of the residence. The total cost of the residence is equal to the cost per square foot of living area times the living area. Typical modifications and adjustments include kitchens, baths, garages and fireplaces.

10 Multiplier for Townhouse/ Rowhouse (Inner Unit .90)

The multipliers shown adjust the base costs per square foot of living area for the common wall condition encountered in townhouses or rowhouses.

11 Alternatives (Skim Coat Plaster $.32)

Alternatives are costs added to or subtracted from the base cost per square foot of living area. Typical alternatives include variations in kitchens, baths, roofing, air conditioning and heating systems.

12 Additional Upgrades or Components (Wings & Ells page 37)

Costs for additional upgrades or components, including wings or ells, breezeways, porches, finished attics and site improvements, are shown in other locations in the Residential Square Foot Cost Section.

Building Classes

Economy Class

An economy class residence is usually built from stock plans. The materials and workmanship are sufficient to satisfy building codes. Low construction cost is more important than distinctive features. The overall shape of the foundation and structure is seldom other than square or rectangular.

An unfinished basement includes 7' high 8" thick foundation wall composed of either concrete block or cast-in-place concrete.

Included in the finished basement cost are inexpensive paneling or drywall as the interior finish on the foundation walls, a low cost sponge backed carpeting adhered to the concrete floor, a drywall ceiling, and overhead lighting.

Custom Class

A custom class residence is usually built from plans and specifications with enough features to give the building a distinction of design. Materials and workmanship are generally above average with obvious attention given to construction details. Construction normally exceeds building code requirements.

An unfinished basement includes a 7'-6" high 10" thick cast-in-place concrete foundation wall or a 7'-6" high 12" thick concrete block foundation wall.

A finished basement includes painted drywall on insulated 2" x 4" wood furring as the interior finish to the concrete walls, a suspended ceiling, carpeting adhered to the concrete floor, overhead lighting and heating.

Average Class

An average class residence is a simple design and built from standard plans. Materials and workmanship are average, but often exceed minimum building codes. There are frequently special features that give the residence some distinctive characteristics.

An unfinished basement includes 7'-6" high 8" thick foundation wall composed of either cast-in-place concrete or concrete block.

Included in the finished basement are plywood paneling or drywall on furring that is fastened to the foundation walls, sponge backed carpeting adhered to the concrete floor, a suspended ceiling, overhead lighting and heating.

Luxury Class

A luxury class residence is built from an architect's plan for a specific owner. It is unique both in design and workmanship. There are many special features, and construction usually exceeds all building codes. It is obvious that primary attention is placed on the owner's comfort and pleasure. Construction is supervised by an architect.

An unfinished basement includes 8' high 12" thick foundation wall that is composed of cast-in-place concrete or concrete block.

A finished basement includes painted drywall on 2" x 4" wood furring as the interior finish, suspended ceiling, tackless carpet on wood subfloor with sleepers, overhead lighting and heating.

Configurations

Detached House

This category of residence is a free-standing separate building with or without an attached garage. It has four complete walls.

Semi-Detached House

This category of residence has two living units side-by-side. The common wall is a fireproof wall. Semi-detached residences can be treated as a row house with two end units. Semi-detached residences can be any of the five types.

Town/Row House

This category of residence has a number of attached units made up of inner units and end units. The units are joined by common walls. The inner units have only two exterior walls. The common walls are fireproof. The end units have three walls and a common wall. Town houses/row houses can be any of the five types.

Building Types

One Story

This is an example of a one-story dwelling. The living area of this type of residence is confined to the ground floor. The headroom in the attic is usually too low for use as a living area.

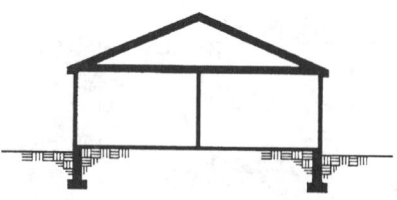

One-and-a-half Story

The living area in the upper level of this type of residence is 50% to 90% of the ground floor. This is made possible by a combination of this design's high-peaked roof and/or dormers. Only the upper level area with a ceiling height of 6 feet or more is considered living area. The living area of this residence is the sum of the ground floor area plus the area on the second level with a ceiling height of 6 feet or more.

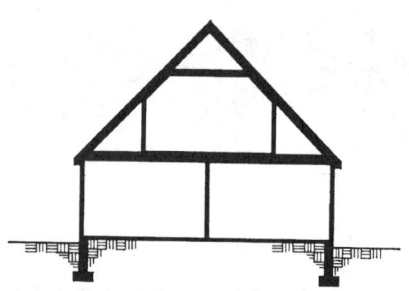

One Story with Finished Attic

The main living area in this type of residence is the ground floor. The upper level or attic area has sufficient headroom for comfortable use as a living area. This is made possible by a high peaked roof. The living area in the attic is less than 50% of the ground floor. The living area of this type of residence is the ground floor area only. The finished attic is considered an adjustment.

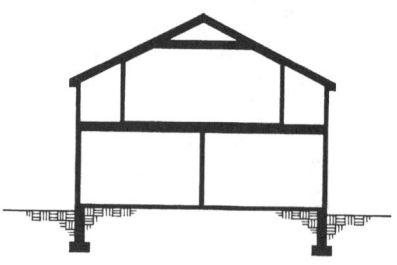

Two Story

This type of residence has a second floor or upper level area which is equal or nearly equal to the ground floor area. The upper level of this type of residence can range from 90% to 110% of the ground floor area, depending on setbacks or overhangs. The living area is the sum of the ground floor area and the upper level floor area.

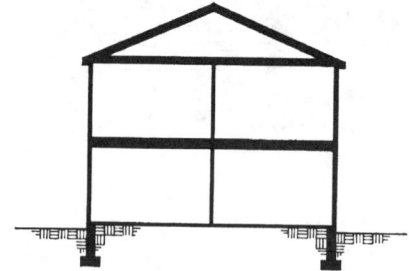

Two-and-one-half Story

This type of residence has two levels of equal or nearly equal area and a third level which has a living area that is 50% to 90% of the ground floor. This is made possible by a high peaked roof, extended wall heights and/or dormers. Only the upper level area with a ceiling height of 6 feet or more is considered living area. The living area of this residence is the sum of the ground floor area, the second floor area and the area on the third level with a ceiling height of 6 feet or more.

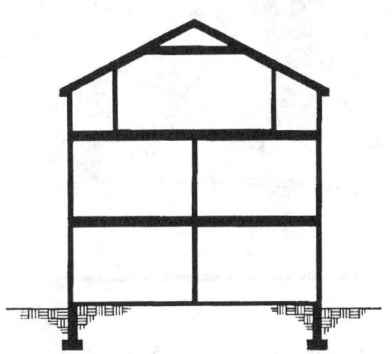

Three Story

This type of residence has three levels which are equal or nearly equal. As in the 2 story residence, the second and third floor areas may vary slightly depending on setbacks or overhangs. The living area is the sum of the ground floor area and the two upper level floor areas.

Bi-level

This type of residence has two living areas, one above the other. One area is about 4 feet below grade and the second is about 4 feet above grade. Both areas are equal in size. The lower level in this type of residence is originally designed and built to serve as a living area and not as a basement. Both levels have full ceiling heights. The living area is the sum of the lower level area and the upper level area.

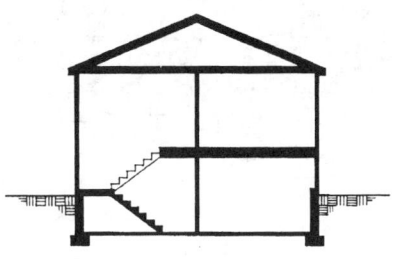

Tri-level

This type of residence has three levels of living area, one at grade level, one about 4 feet below grade and one about 4 feet above grade. All levels are originally designed to serve as living areas. All levels have full ceiling heights. The living area is a sum of the areas of each of the three levels.

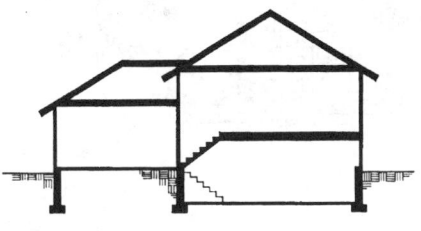

Garage Types

Attached Garage

Shares a common wall with the dwelling. Access is typically through a door between dwelling and garage.

Basement Garage

Constructed under the roof of the dwelling but below the living area.

Built-In Garage

Constructed under the second floor living space and above basement level of dwelling. Reduces gross square feet of living area.

Detached Garage

Constructed apart from the main dwelling. Shares no common area or wall with the dwelling.

Building Components

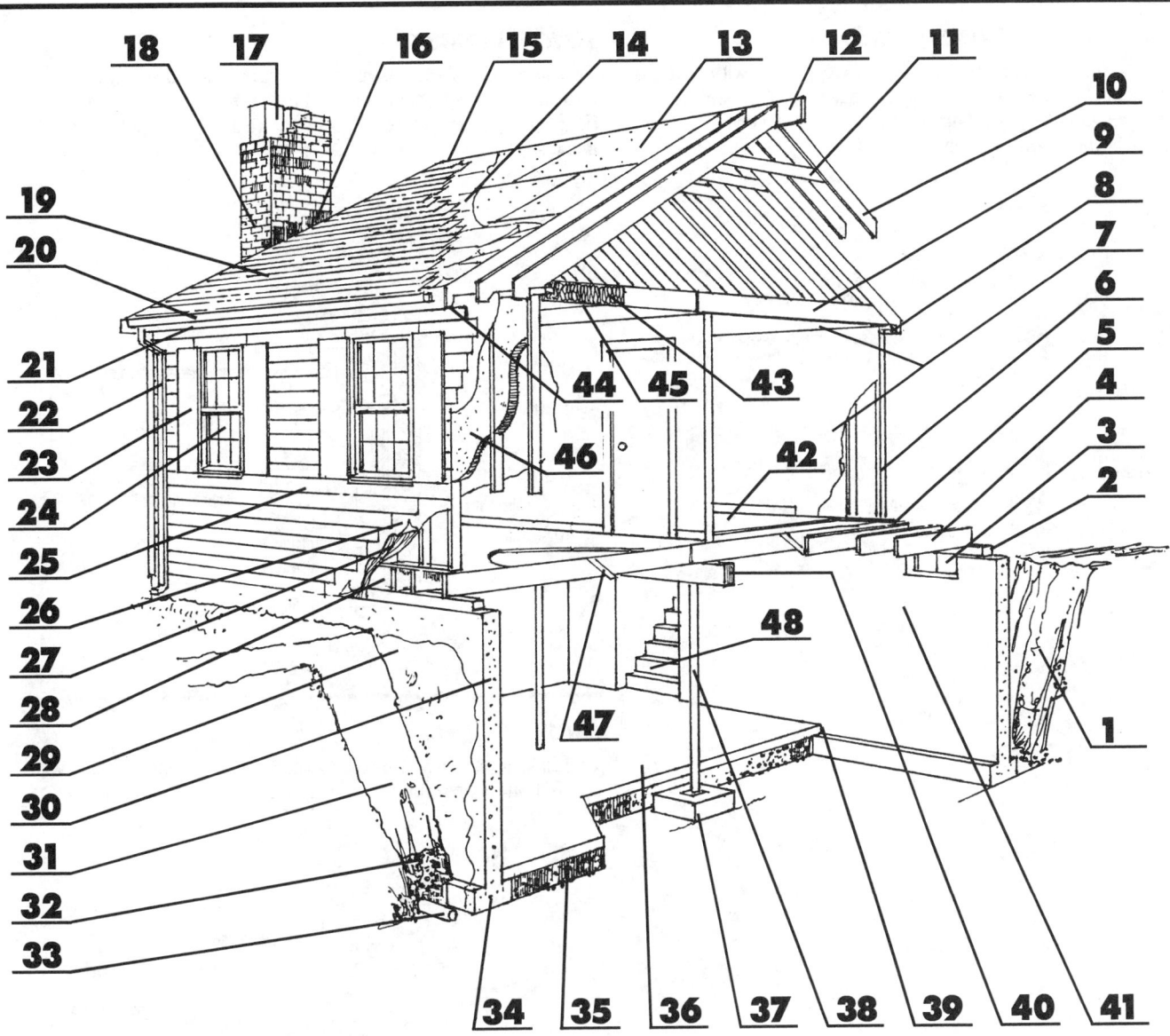

1. Excavation
2. Sill Plate
3. Basement Window
4. Floor Joist
5. Shoe Plate
6. Studs
7. Drywall
8. Plate
9. Ceiling Joists
10. Rafters

11. Collar Ties
12. Ridge Rafter
13. Roof Sheathing
14. Roof Felt
15. Roof Shingles
16. Flashing
17. Flue Lining
18. Chimney
19. Roof Shingles
20. Gutter

21. Fascia
22. Downspout
23. Shutter
24. Window
25. Wall Shingles
26. Building Paper
27. Wall Sheathing
28. Fire Stop
29. Dampproofing
30. Foundation Wall

31. Backfill
32. Drainage Stone
33. Drainage Tile
34. Wall Footing
35. Gravel
36. Concrete Slab
37. Column Footing
38. Pipe Column
39. Expansion Joint
40. Girder

41. Sub-floor
42. Finish Floor
43. Attic Insulation
44. Soffit
45. Ceiling Strapping
46. Wall Insulation
47. Cross Bridging
48. Bulkhead Stairs

Exterior Wall Construction

Typical Frame Construction

Typical wood frame construction consists of wood studs with insulation between them. A typical exterior surface is made up of sheathing, building paper and exterior siding consisting of wood, vinyl, aluminum or stucco over the wood sheathing.

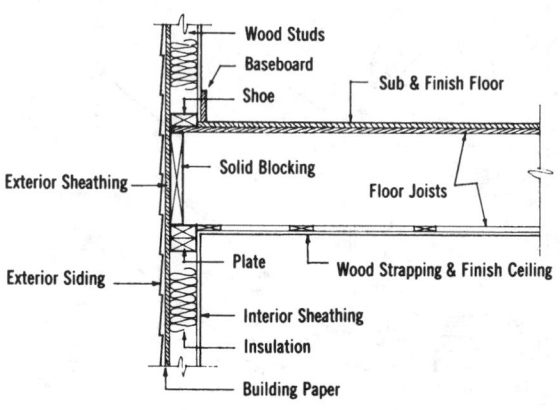

Brick Veneer

Typical brick veneer construction consists of wood studs with insulation between them. A typical exterior surface is sheathing, building paper and an exterior of brick tied to the sheathing, with metal strips.

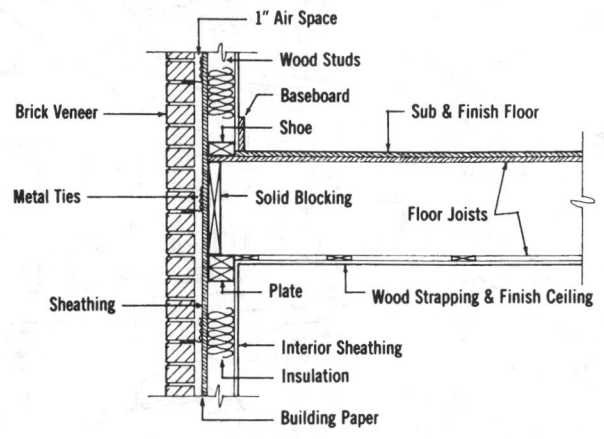

Stone

Typical solid masonry construction consists of a stone or block wall covered on the exterior with brick, stone or other masonry.

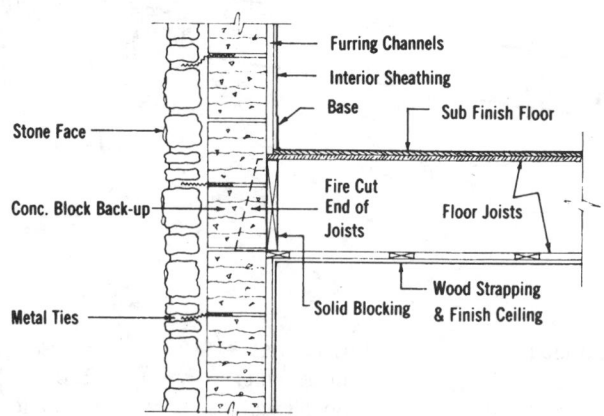

Residential Cost Estimate Worksheets

Worksheet Instructions

The residential cost estimate worksheet can be used as an outline for developing a residential construction or replacement cost. It is also useful for insurance appraisals. The design of the worksheet helps eliminate errors and omissions. To use the worksheet, follow the example below.

1. Fill out the owner's name, residence address, the estimator or appraiser's name, some type of project identifying number or code, and the date.

2. Determine from the plans, specifications, owner's description, photographs or any other means possible the class of construction. The models in this book use economy, average, custom and luxury as classes. Fill in the appropriate box.

3. Fill in the appropriate box for the residence type, configuration, occupancy, and exterior wall. If you require clarification, the pages preceding this worksheet describe each of these.

4. Next, the living area of the residence must be established. The heated or air conditioned space of the residence, not including the basement, should be measured. It is easiest to break the structure up into separate components as shown in the example. The main house (A), a one-and-one-half story wing (B), and a one story wing (C). The breezeway (D), garage (E) and open covered porch (F) will be treated differently. Data entry blocks for the living area are included on the worksheet for your use. Keep each level of each component separate, and fill out the blocks as shown.

5. By using the information on the worksheet, find the model, wing or ell in the following square foot cost pages that best matches the class, type, exterior finish and size of the residence being estimated. Use the *modifications, adjustments, and alternatives* to determine the adjusted cost per square foot of living area for each component.

6. For each component, multiply the cost per square foot by the living area square footage. If the residence is a townhouse/rowhouse, a multiplier should be applied based upon the configuration.

7. The second page of the residential cost estimate worksheet has space for the additional components of a house. The cost for additional bathrooms, finished attic space, breezeways, porches, fireplaces, appliance or cabinet upgrades, and garages should be added on this page. The information for each of these components is found with the model being used, or in the *modifications, adjustments, and alternatives* pages.

8. Add the total from page one of the estimate worksheet and the items listed on page two. The sum is the adjusted total building cost.

9. Depending on the use of the final estimated cost, one of the remaining two boxes should be filled out. Any additional items or exclusions should be added or subtracted at this time. The data contained in this book is a national average. Construction costs are different throughout the country. To allow for this difference, a location factor based upon the first three digits of the residence's zip code must be applied. The location factor is a multiplier that increases or decreases the adjusted total building cost. Find the appropriate location factor and calculate the local cost. If depreciation is a concern, a dollar figure should be subtracted at this point.

10. No residence will match a model exactly. Many differences will be found. At this level of estimating, a variation of plus or minus 10% should be expected.

Adjustments Instructions

No residence matches a model exactly in shape, material, or specifications.

The common differences are:

1. Two or more exterior wall systems
   a. Partial basement
   b. Partly finished basement
2. Specifications or features that are between two classes
3. Crawl space instead of a basement

EXAMPLES

Below are quick examples. See pages 15-17 for complete examples of cost adjustments for these differences:

1. Residence "A" is an average one-story structure with 1,600 S.F. of living area and no basement. Three walls are wood siding on wood frame, and the fourth wall is brick veneer on wood frame. The brick veneer wall is 35% of the exterior wall area.

   Use page 28 to calculate the Base Cost per S.F. of Living Area.
   Wood Siding for 1,600 S.F.   = $65.35 per S.F.
   Brick Veneer for 1,600 S.F.   = $74.75 per S.F.
   .65 ($65.35) + .35 ($74.75) = $68.64 per S.F. of Living Area.

2. a. Residence "B" is the same as Residence "A"; However, it has an unfinished basement under 50% of the building. To adjust the $68.64 per S.F. of living area for this partial basement, use page 28.
   $68.64 + .50 ($6.25) = $71.77 per S.F. of Living Area.

   b. Residence "C" is the same as Residence "A"; However, it has a full basement under the entire building. 640 S.F. or 40% of the basement area is finished.

   Using Page 30:
   $68.64 + .40 ($16.65) + .60 ($6.25) = $79.05 per S.F. of Living Area.

3. When specifications or features of a building are between classes, estimate the percent deviation, and use two tables to calculate the cost per S.F.

   A two-story residence with wood siding and 1,800 S.F. of living area has features 30% better than Average, but 70% less than Custom.

   From pages 30 and 42:

Custom 1,800 S.F. Base Cost	=	$88.85 per S.F.
Average 1,800 S.F. Base Cost	=	$67.35 per S.F.
DIFFERENCE	=	$21.50 per S.F.

   Cost is $67.35 + .30 ($21.50) = $73.80 per S.F. of Living Area.

4. To add the cost of a crawl space, use the cost of an unfinished basement as a maximum. For specific costs of components to be added or deducted, such as vapor barrier, underdrain, and floor, see the "Assemblies" section, pages 233 to 425.

Model Residence Example

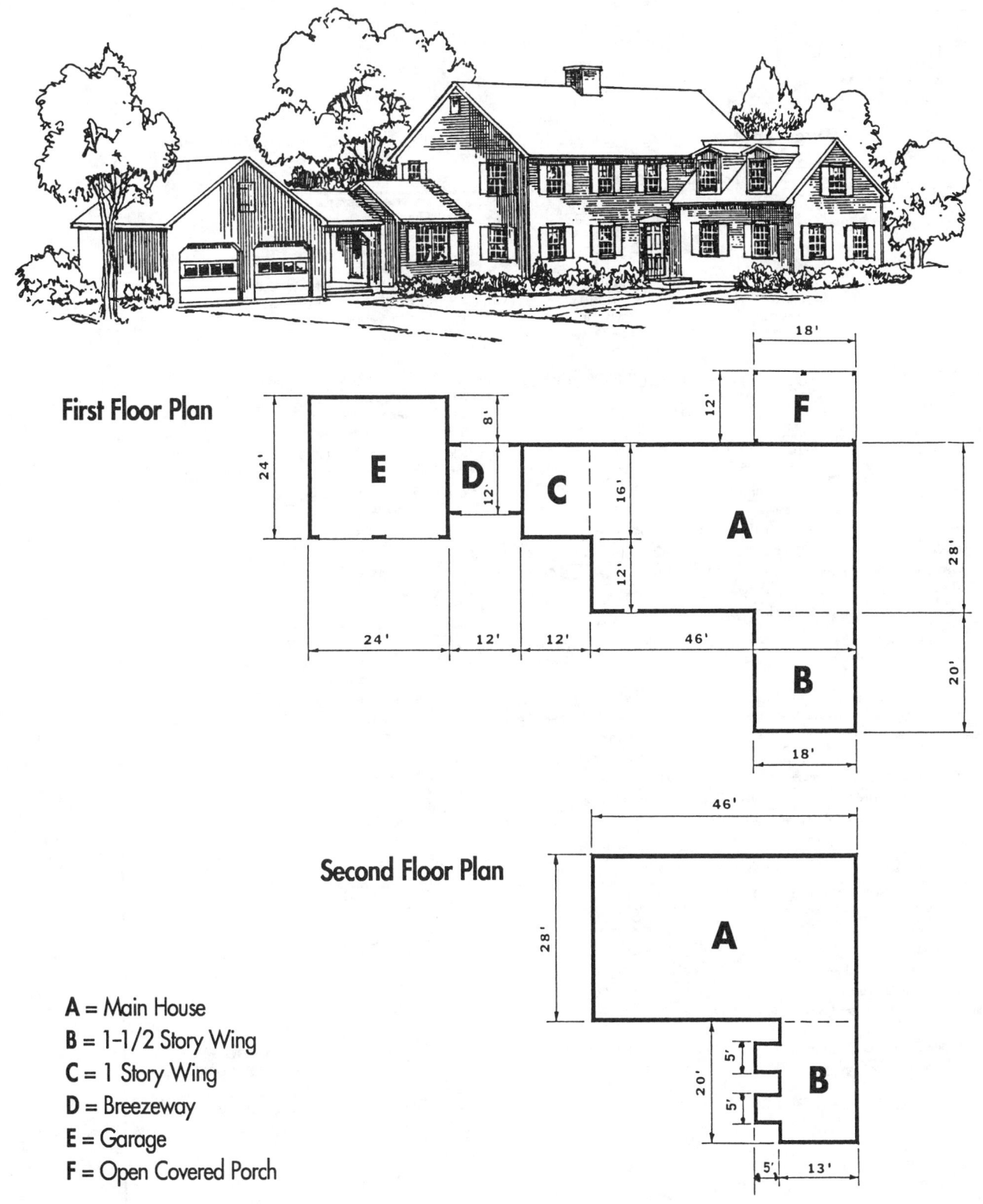

First Floor Plan

Second Floor Plan

A = Main House
B = 1-1/2 Story Wing
C = 1 Story Wing
D = Breezeway
E = Garage
F = Open Covered Porch

Means Forms

RESIDENTIAL
COST ESTIMATE

OWNERS NAME:	Albert Westenberg	APPRAISER:	Nicole Wojtowicz
RESIDENCE ADDRESS:	300 Sygiel Road	PROJECT:	# 55
CITY, STATE, ZIP CODE:	Three Rivers, MA 01080	DATE:	January 2, 1998

CLASS OF CONSTRUCTION	RESIDENCE TYPE	CONFIGURATION	EXTERIOR WALL SYSTEM
☐ ECONOMY	☐ 1 STORY	☑ DETACHED	☑ WOOD SIDING - WOOD FRAME
☑ AVERAGE	☐ 1½ STORY	☐ TOWN/ROW HOUSE	☐ BRICK VENEER - WOOD FRAME
☐ CUSTOM	☑ 2 STORY	☐ SEMI-DETACHED	☐ STUCCO ON WOOD FRAME
☐ LUXURY	☐ 2 1/2 STORY		☐ PAINTED CONCRETE BLOCK
	☐ 3 STORY	OCCUPANCY	☐ SOLID MASONRY (AVERAGE & CUSTOM)
	☐ BI-LEVEL	☑ ONE FAMILY	☐ STONE VENEER - WOOD FRAME
	☐ TRI-LEVEL	☐ TWO FAMILY	☐ SOLID BRICK (LUXURY)
		☐ THREE FAMILY	☐ SOLID STONE (LUXURY)
		☐ OTHER	

* LIVING AREA (Main Building)		* LIVING AREA (Wing or Ell) (B)		* LIVING AREA (Wing or Ell) (C)	
First Level	1288 S.F.	First Level	360 S.F.	First Level	192 S.F.
Second level	1288 S.F.	Second level	310 S.F.	Second level	S.F.
Third Level	S.F.	Third Level	S.F.	Third Level	S.F.
Total	2576 S.F.	Total	670 S.F.	Total	192 S.F.

* Basement Area is not part of living area.

MAIN BUILDING	COSTS PER S.F. LIVING AREA	
Cost per Square Foot of Living Area, from Page ___30___	$	59.50
Basement Addition _____ % Finished, __100__ % Unfinished	+	3.55
Roof Cover Adjustment: **Cedar Shake** Type, Page __30__ (Add or Deduct)	( )	1.10
Central Air Conditioning: ☐ Separate Ducts ☑ Heating Ducts, Page __30__	+	1.38
Heating System Adjustment: _____ Type, Page _____ (Add or Deduct)	( )	—
Main Building: Adjusted Cost per S.F. of Living Area	$	65.53

MAIN BUILDING TOTAL COST	$ 65.53 /S.F. x	2,576 S.F. x	_____	= $ 168,805
	Cost per S.F. Living Area	Living Area	Town/Row House Multiplier (Use 1 for Detached)	TOTAL COST

WING OR ELL (B) __1-1/2__ STORY	COSTS PER S.F. LIVING AREA	
Cost per Square Foot of Living Area, from Page __37__ (WOOD SIDING)	$	52.15
Basement Addition __100__ % Finished, _____ % Unfinished	+	15.00
Roof Cover Adjustment: _____ Type, Page _____ (Add or Deduct)	( )	—
Central Air Conditioning: ☐ Separate Ducts ☑ Heating Ducts, Page __29__	+	1.73
Heating System Adjustment: _____ Type, Page _____ (Add or Deduct)	( )	—
Wing or Ell (B): Adjusted Cost per S.F. of Living Area	$	68.88

WING OR ELL (B) TOTAL COST	$ 68.88 /S.F. x	670 S.F.	= $ 46,150
	Cost per S.F. Living Area	Living Area	TOTAL COST

WING OR ELL (C) __1__ STORY	COSTS PER S.F. LIVING AREA	
Cost per Square Foot of Living Area, from Page __37__ (WOOD SIDING)	$	77.95
Basement Addition _____ % Finished, _____ % Unfinished	+	—
Roof Cover Adjustment: _____ Type, Page _____ (Add or Deduct)	( )	—
Central Air Conditioning: ☐ Separate Ducts ☐ Heating Ducts, Page _____	+	—
Heating System Adjustment: _____ Type, Page _____ (Add or Deduct)	( )	—
Wing or Ell ( ): Adjusted Cost per S.F. of Living Area	$	77.95

WING OR ELL (C) TOTAL COST	$ 77.95 /S.F. x	192 S.F.	= $ 14,966
	Cost per S.F. Living Area	Living Area	TOTAL COST

TOTAL THIS PAGE 229,921

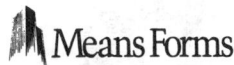 Means Forms

**RESIDENTIAL
COST ESTIMATE**

Total Page 1				$	229,921
		QUANTITY	UNIT COST		
Additional Bathrooms: __2__ Full, __1__ Half 2 @ 3477 1 @ 2122					9,076
Finished Attic: __N/A__ Ft. x _____ Ft.			S.F.	+	
Breezeway: ☑ Open ☐ Enclosed __12__ Ft. x __12__ Ft.		144 S.F.	14.15	+	2,038
Covered Porch: ☑ Open ☐ Enclosed __18__ Ft. x __12__ Ft.		216 S.F.	21.25	+	4,590
Fireplace: ☑ Interior Chimney ☐ Exterior Chimney					
☑ No. of Flues (2) ☑ Additional Fireplaces 1 - 2nd Story				+	6,180
Appliances:				+	—
Kitchen Cabinets Adjustments: (±)					—
☑ Garage ☐ Carport: __2__ Car(s) Description _Wood, Attached_ (±)					10,039
Miscellaneous:				+	
		ADJUSTED TOTAL BUILDING COST		$	262,114

REPLACEMENT COST		
ADJUSTED TOTAL BUILDING COST	$	262,114
Site Improvements		
(A) Paving & Sidewalks	$	
(B) Landscaping	$	
(C) Fences	$	
(D) Swimming Pools	$	
(E) Miscellaneous	$	
TOTAL	$	262,114
Location Factor	x	1.07
Location Replacement Cost	$	280,462
Depreciation - 10 %	-$	28,046
LOCAL DEPRECIATED COST	$	252,416

INSURANCE COST		
ADJUSTED TOTAL BUILDING COST	$	
Insurance Exclusions		
(A) Footings, site work, Underground Piping	-$	
(B) Architects Fees	-$	
Total Building Cost Less Exclusion	$	
Location Factor	x	
LOCAL INSURABLE REPLACEMENT COST	$	

SKETCH AND ADDITIONAL CALCULATIONS

ECONOMY

1 Story

© Home Planners, Inc.

1 - 1/2 Story

2 Story

Bi-Level

Tri-Level

		Components
1	**Site Work**	Site preparation for slab or excavation for lower level; 4' deep trench excavation for foundation wall.
2	**Foundations**	Continuous reinforced concrete footing, 8" deep x 18" wide; dampproofed and insulated 8" thick reinforced concrete block foundation wall, 4' deep; 4" concrete slab on 4" crushed stone base and polyethylene vapor barrier, trowel finish.
3	**Framing**	Exterior walls—2" x 4" wood studs, 16" O.C.; 1/2" insulation board sheathing; wood truss roof 24" O.C. or 2" x 6" rafters 16" O.C. with 1/2" plywood sheathing, 4 in 12 or 8 in 12 roof pitch.
4	**Exterior Walls**	Beveled wood siding and #15 felt building paper on insulated wood frame walls; sliding sash or double hung wood windows; 2 flush solid core wood exterior doors. **Alternates:** • Brick veneer on wood frame, has 4" veneer of common brick. • Stucco on wood frame has 1" thick colored stucco finish. • Painted concrete block has 8" concrete block sealed and painted on exterior and furring on the interior for the drywall.
5	**Roofing**	20 year asphalt shingles; #15 felt building paper; aluminum gutters, downspouts, drip edge and flashings.
6	**Interiors**	Walls and ceilings—1/2" taped and finished drywall, primed and painted with 2 coats; painted baseboard and trim; rubber backed carpeting 80%, asphalt tile 20%; hollow core wood interior doors.
7	**Specialties**	Economy grade kitchen cabinets—6 L.F. wall and base with plastic laminate counter top and kitchen sink; 30 gallon electric water heater.
8	**Mechanical**	1 lavatory, white, wall hung; 1 water closet, white, 1 bathtub, enameled steel, white; gas fired warm air heat.
9	**Electrical**	100 Amp. service; romex wiring; incandescent lighting fixtures, switches, receptacles.
10	**Overhead and Profit**	General Contractor overhead and profit.

Adjustments

Unfinished Basement:
7' high 8" concrete block or cast-in-place concrete walls.

Finished Basement:
Includes inexpensive paneling or drywall on foundation walls. Inexpensive sponge backed carpeting on concrete floor, drywall ceiling, and lighting.

ECONOMY

- Mass produced from stock plans
- Single family — 1 full bath, 1 kitchen
- No basement
- Asphalt shingles on roof
- Hot air heat
- Drywall interior finishes
- Materials and workmanship are sufficient to meet codes
- Detail specifications on page 19

Note: The illustration shown may contain some optional components (for example: garages and/or fireplaces) whose costs are shown in the modifications, adjustments, & alternatives below or at the end of the square foot section.

© Home Planners, Inc.

Base cost per square foot of living area

Exterior Wall	Living Area change										
	600	800	1000	1200	1400	1600	1800	2000	2400	2800	3200
Wood Siding - Wood Frame	72.95	66.45	61.15	56.75	53.15	50.65	49.35	47.80	44.25	41.85	40.20
Brick Veneer - Wood Frame	78.50	71.40	65.65	60.80	56.85	54.10	52.65	50.90	47.10	44.50	42.65
Stucco on Wood Frame	71.45	65.15	59.95	55.70	52.20	49.70	48.45	46.95	43.50	41.15	39.55
Painted Concrete Block	74.35	67.70	62.25	57.80	54.10	51.50	50.20	48.60	45.00	42.55	40.85
Finished Basement, Add	16.35	15.45	14.75	14.15	13.65	13.35	13.15	12.85	12.50	12.25	12.00
Unfinished Basement, Add	8.15	7.35	6.80	6.25	5.80	5.55	5.40	5.15	4.85	4.60	4.40

Modifications

Add to the total cost

Upgrade Kitchen Cabinets	$ + 233
Solid Surface Countertops	+ 448
Full Bath - including plumbing, wall and floor finishes	+ 2824
Half Bath - including plumbing, wall and floor finishes	+ 1724
One Car Attached Garage	+ 6364
One Car Detached Garage	+ 6890
Fireplace & Chimney	+ 3280

Adjustments

For multi family - add to total cost

Additional Kitchen	$ + 2012
Additional Bath	+ 2824
Additional Entry & Exit	+ 995
Separate Heating	+ 1111
Separate Electric	+ 624

For Townhouse/Rowhouse -
Multiply cost per square foot by

Inner Unit	.95
End Unit	.97

Alternatives

Add to or deduct from the cost per square foot of living area

Composition Roll Roofing	– .50
Cedar Shake Roof	+ 2.45
Upgrade Walls and Ceilings to Skim Coat Plaster	+ .41
Upgrade Ceilings to Textured Finish	+ .43
Air Conditioning, in Heating Ductwork	+ 2.28
In Separate Ductwork	+ 4.33
Heating Systems, Hot Water	+ 2.58
Heat Pump	+ 1.53
Electric Heat	– 1.26
Not Heated	– 2.87

Additional upgrades or components

Kitchen Cabinets & Countertops	Page 58
Bathroom Vanities	59
Fireplaces & Chimneys	59
Windows, Skylights & Dormers	59
Appliances	60
Breezeways & Porches	60
Finished Attic	60
Garages	61
Site Improvements	61
Wings & Ells	25

Important: See the Reference Section for Location Factors (to adjust for your city) and Estimating Forms

- Mass produced from stock plans
- Single family — 1 full bath, 1 kitchen
- No basement
- Asphalt shingles on roof
- Hot air heat
- Drywall interior finishes
- Materials and workmanship are sufficient to meet codes
- Detail specifications on page 19

Note: The illustration shown may contain some optional components (for example: garages and/or fireplaces) whose costs are shown in the modifications, adjustments, & alternatives below or at the end of the square foot section.

ECONOMY

Base cost per square foot of living area

Exterior Wall	Living Area change										
	600	800	1000	1200	1400	1600	1800	2000	2400	2800	3200
Wood Siding - Wood Frame	84.15	70.40	62.45	59.35	57.05	52.95	51.25	49.10	44.90	43.50	41.70
Brick Veneer - Wood Frame	91.90	76.05	67.70	64.20	61.65	57.20	55.25	52.90	48.25	46.65	44.65
Stucco on Wood Frame	82.10	68.90	61.10	58.05	55.80	51.85	50.20	48.10	44.00	42.65	40.90
Painted Concrete Block	86.10	71.85	63.80	60.55	58.20	54.05	52.25	50.05	45.75	44.30	42.45
Finished Basement, Add	12.55	10.70	10.25	9.90	9.65	9.25	9.05	8.85	8.45	8.30	8.10
Unfinished Basement, Add	7.10	5.50	5.10	4.80	4.55	4.25	4.05	3.90	3.55	3.40	3.25

Modifications

Add to the total cost

Upgrade Kitchen Cabinets	$	+ 233
Solid Surface Countertops		+ 448
Full Bath - including plumbing, wall and floor finishes		+ 2824
Half Bath - including plumbing, wall and floor finishes		+ 1724
One Car Attached Garage		+ 6364
One Car Detached Garage		+ 6890
Fireplace & Chimney		+ 3280

Adjustments

For multi family - add to total cost

Additional Kitchen	$	+ 2012
Additional Bath		+ 2824
Additional Entry & Exit		+ 995
Separate Heating		+ 1111
Separate Electric		+ 624

For Townhouse/Rowhouse - Multiply cost per square foot by

Inner Unit	.95
End Unit	.97

Alternatives

Add to or deduct from the cost per square foot of living area

Composition Roll Roofing	– .35
Cedar Shake Roof	+ 1.75
Upgrade Walls and Ceilings to Skim Coat Plaster	+ .41
Upgrade Ceilings to Textured Finish	+ .43
Air Conditioning, in Heating Ductwork	+ 1.72
In Separate Ductwork	+ 3.75
Heating Systems, Hot Water	+ 2.25
Heat Pump	+ 1.70
Electric Heat	– 1.02
Not Heated	– 2.63

Additional upgrades or components

Kitchen Cabinets & Countertops	Page 58
Bathroom Vanities	59
Fireplaces & Chimneys	59
Windows, Skylights & Dormers	59
Appliances	60
Breezeways & Porches	60
Finished Attic	60
Garages	61
Site Improvements	61
Wings & Ells	25

Important: See the Reference Section for Location Factors (to adjust for your city) and Estimating Forms

ECONOMY

- Mass produced from stock plans
- Single family — 1 full bath, 1 kitchen
- No basement
- Asphalt shingles on roof
- Hot air heat
- Drywall interior finishes
- Materials and workmanship are sufficient to meet codes
- Detail specifications on page 19

Note: The illustration shown may contain some optional components (for example: garages and/or fireplaces) whose costs are shown in the modifications, adjustments, & alternatives below or at the end of the square foot section.

Base cost per square foot of living area

Exterior Wall	Living Area change										
	1000	1200	1400	1600	1800	2000	2200	2600	3000	3400	3800
Wood Siding - Wood Frame	68.20	61.15	58.35	56.45	54.15	51.90	50.45	47.30	44.30	43.05	41.75
Brick Veneer - Wood Frame	74.20	66.70	63.55	61.40	58.80	56.45	54.70	51.20	47.85	46.45	45.00
Stucco on Wood Frame	66.60	59.70	57.00	55.15	52.95	50.70	49.30	46.30	43.35	42.15	40.90
Painted Concrete Block	69.70	62.60	59.70	57.70	55.35	53.05	51.55	48.30	45.20	43.95	42.55
Finished Basement, Add	8.55	8.20	7.95	7.75	7.55	7.40	7.25	7.00	6.75	6.65	6.55
Unfinished Basement, Add	4.40	4.10	3.85	3.70	3.50	3.40	3.25	3.00	2.85	2.75	2.65

Modifications

Add to the total cost

Upgrade Kitchen Cabinets	$ + 233
Solid Surface Countertops	+ 448
Full Bath - including plumbing, wall and floor finishes	+ 2824
Half Bath - including plumbing, wall and floor finishes	+ 1724
One Car Attached Garage	+ 6364
One Car Detached Garage	+ 6890
Fireplace & Chimney	+ 3625

Adjustments

For multi family - add to total cost

Additional Kitchen	$ + 2012
Additional Bath	+ 2824
Additional Entry & Exit	+ 995
Separate Heating	+ 1111
Separate Electric	+ 624

For Townhouse/Rowhouse -
Multiply cost per square foot by

Inner Unit	.93
End Unit	.96

Alternatives

Add to or deduct from the cost per square foot of living area

Composition Roll Roofing	– .25
Cedar Shake Roof	+ 1.25
Upgrade Walls and Ceilings to Skim Coat Plaster	+ .41
Upgrade Ceilings to Textured Finish	+ .43
Air Conditioning, in Heating Ductwork	+ 1.37
In Separate Ductwork	+ 3.41
Heating Systems, Hot Water	+ 2.06
Heat Pump	+ 1.83
Electric Heat	– .86
Not Heated	– 2.47

Additional upgrades or components

Kitchen Cabinets & Countertops	Page 58
Bathroom Vanities	59
Fireplaces & Chimneys	59
Windows, Skylights & Dormers	59
Appliances	60
Breezeways & Porches	60
Finished Attic	60
Garages	61
Site Improvements	61
Wings & Ells	25

- **Mass produced from stock plans**
- **Single family — 1 full bath, 1 kitchen**
- **No basement**
- **Asphalt shingles on roof**
- **Hot air heat**
- **Drywall interior finishes**
- **Materials and workmanship are sufficient to meet codes**
- **Detail specifications on page 19**

Note: The illustration shown may contain some optional components (for example: garages and/or fireplaces) whose costs are shown in the modifications, adjustments, & alternatives below or at the end of the square foot section.

ECONOMY

Base cost per square foot of living area

Exterior Wall	Living Area change										
	1000	1200	1400	1600	1800	2000	2200	2600	3000	3400	3800
Wood Siding - Wood Frame	63.05	56.40	53.90	52.20	50.15	48.05	46.80	44.00	41.20	40.15	38.95
Brick Veneer - Wood Frame	67.55	60.55	57.80	55.90	53.60	51.40	50.00	46.90	43.90	42.70	41.40
Stucco on Wood Frame	61.85	55.25	52.85	51.20	49.20	47.10	45.95	43.25	40.50	39.50	38.30
Painted Concrete Block	64.20	57.40	54.90	53.10	51.00	48.90	47.60	44.75	41.90	40.80	39.60
Finished Basement, Add	8.55	8.20	7.95	7.75	7.55	7.40	7.25	7.00	6.75	6.65	6.55
Unfinished Basement, Add	4.40	4.10	3.85	3.70	3.50	3.40	3.25	3.00	2.85	2.75	2.65

Modifications

Add to the total cost

Upgrade Kitchen Cabinets	$ + 233
Solid Surface Countertops	+ 448
Full Bath - including plumbing, wall and floor finishes	+ 2824
Half Bath - including plumbing, wall and floor finishes	+ 1724
One Car Attached Garage	+ 6364
One Car Detached Garage	+ 6890
Fireplace & Chimney	+ 3625

Adjustments

For multi family - add to total cost

Additional Kitchen	$ + 2012
Additional Bath	+ 2824
Additional Entry & Exit	+ 995
Separate Heating	+ 1111
Separate Electric	+ 624

For Townhouse/Rowhouse -
Multiply cost per square foot by

Inner Unit	.94
End Unit	.97

Alternatives

Add to or deduct from the cost per square foot of living area

Composition Roll Roofing	– .25
Cedar Shake Roof	+ 1.25
Upgrade Walls and Ceilings to Skim Coat Plaster	+ .41
Upgrade Ceilings to Textured Finish	+ .43
Air Conditioning, in Heating Ductwork	+ 1.37
In Separate Ductwork	+ 3.41
Heating Systems, Hot Water	+ 2.06
Heat Pump	+ 1.83
Electric Heat	– .86
Not Heated	– 2.47

Additional upgrades or components

Kitchen Cabinets & Countertops	Page 58
Bathroom Vanities	59
Fireplaces & Chimneys	59
Windows, Skylights & Dormers	59
Appliances	60
Breezeways & Porches	60
Finished Attic	60
Garages	61
Site Improvements	61
Wings & Ells	25

ECONOMY

- **Mass produced from stock plans**
- **Single family — 1 full bath, 1 kitchen**
- **No basement**
- **Asphalt shingles on roof**
- **Hot air heat**
- **Drywall interior finishes**
- **Materials and workmanship are sufficient to meet codes**
- **Detail specifications on page 19**

Note: The illustration shown may contain some optional components (for example: garages and/or fireplaces) whose costs are shown in the modifications, adjustments, & alternatives below or at the end of the square foot section.

© Design Basics, Inc.

Base cost per square foot of living area

Exterior Wall	Living Area change										
	1200	1500	1800	2000	2200	2400	2800	3200	3600	4000	4400
Wood Siding - Wood Frame	57.40	52.70	49.05	47.75	45.80	43.95	42.55	40.70	38.55	37.90	36.20
Brick Veneer - Wood Frame	61.55	56.40	52.45	51.00	48.90	46.80	45.35	43.25	40.95	40.20	38.35
Stucco on Wood Frame	56.30	51.70	48.15	46.90	45.00	43.20	41.85	40.05	37.90	37.25	35.60
Solid Masonry	58.45	53.60	49.90	48.55	46.60	44.70	43.30	41.35	39.20	38.45	36.75
Finished Basement, Add*	10.30	9.85	9.45	9.25	9.10	8.90	8.75	8.55	8.35	8.25	8.10
Unfinished Basement, Add*	4.90	4.50	4.15	4.00	3.85	3.70	3.60	3.40	3.25	3.15	3.05

*Basement under middle level only.

Modifications

Add to the total cost

Upgrade Kitchen Cabinets	$ + 233
Solid Surface Countertops	+ 448
Full Bath - including plumbing, wall and floor finishes	+ 2824
Half Bath - including plumbing, wall and floor finishes	+ 1724
One Car Attached Garage	+ 6364
One Car Detached Garage	+ 6890
Fireplace & Chimney	+ 3625

Adjustments

For multi family - add to total cost

Additional Kitchen	$ + 2012
Additional Bath	+ 2824
Additional Entry & Exit	+ 995
Separate Heating	+ 1111
Separate Electric	+ 624

For Townhouse/Rowhouse -
Multiply cost per square foot by

Inner Unit	.93
End Unit	.96

Alternatives

Add to or deduct from the cost per square foot of living area

Composition Roll Roofing	– .35
Cedar Shake Roof	+ 1.75
Upgrade Walls and Ceilings to Skim Coat Plaster	+ .41
Upgrade Ceilings to Textured Finish	+ .43
Air Conditioning, in Heating Ductwork	+ 1.14
In Separate Ductwork	+ 3.19
Heating Systems, Hot Water	+ 1.94
Heat Pump	+ 1.89
Electric Heat	– .77
Not Heated	– 2.38

Additional upgrades or components

Kitchen Cabinets & Countertops	Page 58
Bathroom Vanities	59
Fireplaces & Chimneys	59
Windows, Skylights & Dormers	59
Appliances	60
Breezeways & Porches	60
Finished Attic	60
Garages	61
Site Improvements	61
Wings & Ells	25

LUXURY

1 Story — Base cost per square foot of living area

Exterior Wall	Living Area							
	50	100	200	300	400	500	600	700
Wood Siding - Wood Frame	99.95	75.70	65.15	53.40	49.95	47.90	46.50	46.75
Brick Veneer - Wood Frame	114.00	85.70	73.50	58.95	54.95	52.55	50.95	51.05
Stucco on Wood Frame	96.25	73.05	62.95	51.90	48.60	46.65	45.35	45.65
Painted Concrete Block	103.55	78.25	67.30	54.80	51.25	49.10	47.65	47.85
Finished Basement, Add	25.25	20.60	18.65	15.45	14.75	14.40	14.15	13.90
Unfinished Basement, Add	15.50	11.60	10.00	7.25	6.70	6.40	6.15	6.00

1-1/2 Story — Base cost per square foot of living area

Exterior Wall	Living Area							
	100	200	300	400	500	600	700	800
Wood Siding - Wood Frame	79.45	63.65	53.95	47.70	44.75	43.25	41.50	41.00
Brick Veneer - Wood Frame	91.95	73.65	62.30	54.20	50.75	48.95	46.85	46.30
Stucco on Wood Frame	76.15	61.00	51.75	45.95	43.15	41.80	40.10	39.60
Painted Concrete Block	82.65	66.20	56.10	49.35	46.25	44.70	42.85	42.35
Finished Basement, Add	16.85	14.90	13.65	12.20	11.80	11.55	11.35	11.30
Unfinished Basement, Add	9.60	8.00	6.90	5.70	5.35	5.15	4.95	4.90

2 Story — Base cost per square foot of living area

Exterior Wall	Living Area							
	100	200	400	600	800	1000	1200	1400
Wood Siding - Wood Frame	81.90	60.50	51.15	41.40	38.35	36.55	35.30	35.70
Brick Veneer - Wood Frame	95.90	70.50	59.50	46.95	43.35	41.20	39.75	40.00
Stucco on Wood Frame	78.20	57.85	49.00	39.95	37.00	35.30	34.15	34.55
Painted Concrete Block	85.50	63.05	53.30	42.85	39.65	37.75	36.45	36.75
Finished Basement, Add	12.65	10.30	9.35	7.75	7.40	7.25	7.10	7.00
Unfinished Basement, Add	7.75	5.80	5.00	3.65	3.35	3.20	3.10	3.00

Base costs do not include bathroom or kitchen facilities. Use Modifications/Adjustments/Alternatives on pages 58-61 where appropriate.

1 Story

© Home Planners, Inc.

1 - 1/2 Story

© By Designer

2 Story

2-1/2 Story

Bi-Level

Tri-Level

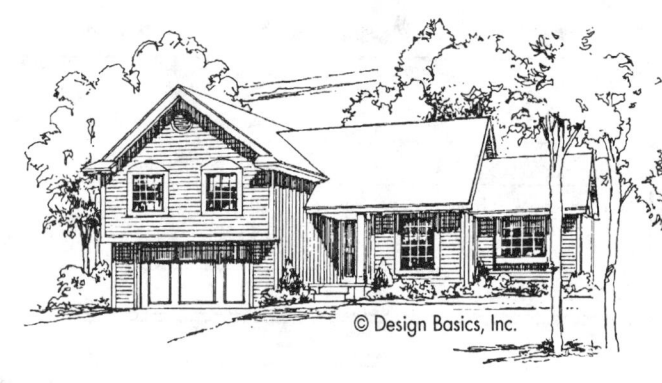

© Design Basics, Inc.

	Components
1 Site Work	Site preparation for slab or excavation for lower level; 4' deep trench excavation for foundation wall.
2 Foundations	Continuous reinforced concrete footing, 8" deep x 18" wide; dampproofed and insulated 8" thick reinforced concrete foundation wall, 4' deep; 4" concrete slab on 4" crushed stone base and polyethylene vapor barrier, trowel finish.
3 Framing	Exterior walls—2" x 4" wood studs, 16" O.C.; 1/2" plywood sheathing; 2" x 6" rafters 16" O.C. with 1/2" plywood sheathing, 4 in 12 or 8 in 12 roof pitch, 2" x 6" ceiling joists; 1/2" plywood subfloor on 1" x 2" wood sleepers 16" O.C.; 2" x 8" floor joists with 5/8" plywood subfloor on models with more than one level.
4 Exterior Walls	Beveled wood siding and #15 felt building paper on insulated wood frame walls; double hung windows; 3 flush solid core wood exterior doors with storms. **Alternates:** • Brick veneer on wood frame, has 4" veneer of common brick. • Stucco on wood frame has 1" thick colored stucco finish. • Solid masonry has a 6" concrete block load bearing wall with insulation and a brick or stone exterior. Also solid brick or stone walls.
5 Roofing	25 year asphalt shingles; #15 felt building paper; aluminum gutters, downspouts, drip edge and flashings.
6 Interiors	Walls and ceilings—1/2" taped and finished drywall, primed and painted with 2 coats; painted baseboard and trim, finished hardwood floor 40%, carpet with 1/2" underlayment 40%, vinyl tile with 1/2" underlayment 15%, ceramic tile with 1/2" underlayment 5%; hollow core and louvered interior doors.
7 Specialties	Average grade kitchen cabinets—14 L.F. wall and base with plastic laminate counter top and kitchen sink; 40 gallon electric water heater.
8 Mechanical	1 lavatory, white, wall hung; 1 water closet, white; 1 bathtub, enameled steel, white; gas fired warm air heat.
9 Electrical	100 Amp. service; romex wiring; incandescent lighting fixtures, switches, receptacles.
10 Overhead and Profit	General Contractor overhead and profit.

Adjustments

Unfinished Basement:
7'-6" high cast-in-place concrete walls 8" thick or 8" concrete block.

Finished Basement:
Includes plywood paneling or drywall on furring on foundation walls, sponge backed carpeting on concrete floor, suspended ceiling, lighting and heating.

- **Simple design from standard plans**
- **Single family — 1 full bath, 1 kitchen**
- **No basement**
- **Asphalt shingles on roof**
- **Hot air heat**
- **Drywall interior finishes**
- **Materials and workmanship are average**
- **Detail specifications on p. 27**

Note: The illustration shown may contain some optional components (for example: garages and/or fireplaces) whose costs are shown in the modifications, adjustments, & alternatives below or at the end of the square foot section.

© Home Planners, Inc.

AVERAGE

Base cost per square foot of living area

Exterior Wall	Living Area change										
	600	800	1000	1200	1400	1600	1800	2000	2400	2800	3200
Wood Siding - Wood Frame	91.75	84.05	77.65	72.45	68.35	65.35	63.65	61.90	57.75	54.95	52.95
Brick Veneer - Wood Frame	103.45	95.10	88.20	82.50	78.05	74.75	72.95	70.90	66.50	63.45	61.25
Stucco on Wood Frame	97.85	90.05	83.65	78.40	74.30	71.25	69.55	67.75	63.60	60.75	58.75
Solid Masonry	111.95	102.65	95.10	88.70	83.65	80.00	78.00	75.70	70.85	67.50	64.95
Finished Basement, Add	20.90	19.70	18.85	18.00	17.35	16.90	16.65	16.30	15.75	15.45	15.05
Unfinished Basement, Add	9.00	8.20	7.65	7.10	6.65	6.40	6.25	6.00	5.70	5.45	5.25

Modifications

Add to the total cost

Upgrade Kitchen Cabinets	$ + 2283
Solid Surface Countertops	+ 778
Full Bath - including plumbing, wall and floor finishes	+ 3544
Half Bath - including plumbing, wall and floor finishes	+ 2163
One Car Attached Garage	+ 7074
One Car Detached Garage	+ 7587
Fireplace & Chimney	+ 3285

Adjustments

For multi family - add to total cost

Additional Kitchen	$ + 3734
Additional Bath	+ 3544
Additional Entry & Exit	+ 995
Separate Heating	+ 1110
Separate Electric	+ 1204

For Townhouse/Rowhouse - Multiply cost per square foot by

Inner Unit	.92
End Unit	.96

Alternatives

Add to or deduct from the cost per square foot of living area

Cedar Shake Roof	+ 2.20
Clay Tile Roof	+ 4.60
Slate Roof	+ 7.75
Upgrade Walls to Skim Coat Plaster	+ .27
Upgrade Ceilings to Textured Finish	+ .43
Air Conditioning, in Heating Ductwork	+ 2.35
In Separate Ductwork	+ 4.44
Heating Systems, Hot Water	+ 2.65
Heat Pump	+ 1.56
Electric Heat	− 1.02
Not Heated	− 2.63

Additional upgrades or components

Kitchen Cabinets & Countertops	Page 58
Bathroom Vanities	59
Fireplaces & Chimneys	59
Windows, Skylights & Dormers	59
Appliances	60
Breezeways & Porches	60
Finished Attic	60
Garages	61
Site Improvements	61
Wings & Ells	37

- **Simple design from standard plans**
- **Single family — 1 full bath, 1 kitchen**
- **No basement**
- **Asphalt shingles on roof**
- **Hot air heat**
- **Drywall interior finishes**
- **Materials and workmanship are average**
- **Detail specifications on p. 27**

Note: The illustration shown may contain some optional components (for example: garages and/or fireplaces) whose costs are shown in the modifications, adjustments, & alternatives below or at the end of the square foot section.

© By Designer

AVERAGE

Base cost per square foot of living area

Exterior Wall	Living Area change										
	600	800	1000	1200	1400	1600	1800	2000	2400	2800	3200
Wood Siding - Wood Frame	103.50	87.90	78.35	74.65	72.00	67.20	65.20	62.60	57.75	56.20	53.95
Brick Veneer - Wood Frame	112.05	94.15	84.10	80.00	77.10	71.85	69.65	66.80	61.50	59.70	57.20
Stucco on Wood Frame	104.20	88.45	78.80	75.10	72.40	67.60	65.60	62.90	58.10	56.45	54.25
Solid Masonry	123.90	102.75	92.00	87.45	84.20	78.30	75.75	72.55	66.60	64.55	61.70
Finished Basement, Add	17.15	14.60	14.00	13.55	13.20	12.70	12.40	12.15	11.60	11.40	11.10
Unfinished Basement, Add	7.70	6.10	5.70	5.40	5.15	4.85	4.65	4.50	4.15	4.00	3.85

Modifications

Add to the total cost

Upgrade Kitchen Cabinets	$ + 2283
Solid Surface Countertops	+ 778
Full Bath - including plumbing, wall and floor finishes	+ 3544
Half Bath - including plumbing, wall and floor finishes	+ 2163
One Car Attached Garage	+ 7074
One Car Detached Garage	+ 7587
Fireplace & Chimney	+ 3285

Adjustments

For multi family - add to total cost

Additional Kitchen	$ + 3734
Additional Bath	+ 3544
Additional Entry & Exit	+ 995
Separate Heating	+ 1110
Separate Electric	+ 1204

For Townhouse/Rowhouse - Multiply cost per square foot by

Inner Unit	.92
End Unit	.96

Alternatives

Add to or deduct from the cost per square foot of living area

Cedar Shake Roof	+ 1.60
Clay Tile Roof	+ 3.30
Slate Roof	+ 5.60
Upgrade Walls to Skim Coat Plaster	+ .31
Upgrade Ceilings to Textured Finish	+ .43
Air Conditioning, in Heating Ductwork	+ 1.76
In Separate Ductwork	+ 3.86
Heating Systems, Hot Water	+ 2.31
Heat Pump	+ 1.74
Electric Heat	– .94
Not Heated	– 2.55

Additional upgrades or components

Kitchen Cabinets & Countertops	Page 58
Bathroom Vanities	59
Fireplaces & Chimneys	59
Windows, Skylights & Dormers	59
Appliances	60
Breezeways & Porches	60
Finished Attic	60
Garages	61
Site Improvements	61
Wings & Ells	37

- **Simple design from standard plans**
- **Single family — 1 full bath, 1 kitchen**
- **No basement**
- **Asphalt shingles on roof**
- **Hot air heat**
- **Drywall interior finishes**
- **Materials and workmanship are average**
- **Detail specifications on p. 27**

Note: The illustration shown may contain some optional components (for example: garages and/or fireplaces) whose costs are shown in the modifications, adjustments, & alternatives below or at the end of the square foot section.

Base cost per square foot of living area

Exterior Wall	Living Area change										
	1000	1200	1400	1600	1800	2000	2200	2600	3000	3400	3800
Wood Siding - Wood Frame	84.20	75.65	72.40	70.25	67.35	64.85	63.20	59.50	56.00	54.60	52.95
Brick Veneer - Wood Frame	90.85	81.80	78.15	75.75	72.50	69.85	67.90	63.75	59.95	58.35	56.55
Stucco on Wood Frame	84.75	76.15	72.90	70.70	67.80	65.25	63.60	59.85	56.30	54.90	53.30
Solid Masonry	100.80	91.05	86.80	84.00	80.30	77.30	75.05	70.20	65.90	64.00	61.90
Finished Basement, Add	12.15	11.70	11.35	11.10	10.80	10.65	10.45	10.10	9.80	9.65	9.45
Unfinished Basement, Add	4.90	4.60	4.35	4.20	4.00	3.90	3.75	3.55	3.35	3.25	3.15

Modifications

Add to the total cost

Upgrade Kitchen Cabinets	$ + 2283
Solid Surface Countertops	+ 778
Full Bath - including plumbing, wall and floor finishes	+ 3544
Half Bath - including plumbing, wall and floor finishes	+ 2163
One Car Attached Garage	+ 7074
One Car Detached Garage	+ 7587
Fireplace & Chimney	+ 3665

Adjustments

For multi family - add to total cost

Additional Kitchen	$ + 3734
Additional Bath	+ 3544
Additional Entry & Exit	+ 995
Separate Heating	+ 1110
Separate Electric	+ 1204

For Townhouse/Rowhouse -
Multiply cost per square foot by

Inner Unit	.90
End Unit	.95

Alternatives

Add to or deduct from the cost per square foot of living area

Cedar Shake Roof	+ 1.10
Clay Tile Roof	+ 2.30
Slate Roof	+ 3.90
Upgrade Walls to Skim Coat Plaster	+ .32
Upgrade Ceilings to Textured Finish	+ .43
Air Conditioning, in Heating Ductwork	+ 1.44
In Separate Ductwork	+ 3.50
Heating Systems, Hot Water	+ 2.13
Heat Pump	+ 1.86
Electric Heat	− .86
Not Heated	− 2.46

Additional upgrades or components

Kitchen Cabinets & Countertops	Page 58
Bathroom Vanities	59
Fireplaces & Chimneys	59
Windows, Skylights & Dormers	59
Appliances	60
Breezeways & Porches	60
Finished Attic	60
Garages	61
Site Improvements	61
Wings & Ells	37

- **Simple design from standard plans**
- **Single family — 1 full bath, 1 kitchen**
- **No basement**
- **Asphalt shingles on roof**
- **Hot air heat**
- **Drywall interior finishes**
- **Materials and workmanship are average**
- **Detail specifications on p. 27**

Note: The illustration shown may contain some optional components (for example: garages and/or fireplaces) whose costs are shown in the modifications, adjustments, & alternatives below or at the end of the square foot section.

AVERAGE

Base cost per square foot of living area

Exterior Wall	Living Area change										
	1200	1400	1600	1800	2000	2400	2800	3200	3600	4000	4400
Wood Siding - Wood Frame	83.65	79.20	71.45	70.45	68.20	64.00	61.30	57.85	56.45	53.30	52.50
Brick Veneer - Wood Frame	90.75	85.55	77.30	76.35	73.70	69.00	66.10	62.15	60.55	57.10	56.20
Stucco on Wood Frame	84.25	79.70	71.95	70.95	68.65	64.45	61.70	58.25	56.80	53.65	52.80
Solid Masonry	100.50	94.35	85.45	84.50	81.35	75.90	72.70	68.15	66.20	62.35	61.30
Finished Basement, Add	10.35	9.90	9.50	9.50	9.25	8.90	8.70	8.40	8.25	8.10	8.00
Unfinished Basement, Add	4.10	3.80	3.55	3.50	3.35	3.10	3.00	2.80	2.75	2.60	2.55

Modifications

Add to the total cost

Upgrade Kitchen Cabinets	$ + 2283
Solid Surface Countertops	+ 778
Full Bath - including plumbing, wall and floor finishes	+ 3544
Half Bath - including plumbing, wall and floor finishes	+ 2163
One Car Attached Garage	+ 7074
One Car Detached Garage	+ 7587
Fireplace & Chimney	+ 4170

Adjustments

For multi family - add to total cost

Additional Kitchen	$ + 3734
Additional Bath	+ 3544
Additional Entry & Exit	+ 995
Separate Heating	+ 1110
Separate Electric	+ 1204

For Townhouse/Rowhouse - Multiply cost per square foot by

Inner Unit	.90
End Unit	.95

Alternatives

Add to or deduct from the cost per square foot of living area

Cedar Shake Roof	+ .95
Clay Tile Roof	+ 2.00
Slate Roof	+ 3.35
Upgrade Walls to Skim Coat Plaster	+ .31
Upgrade Ceilings to Textured Finish	+ .43
Air Conditioning, in Heating Ductwork	+ 1.29
In Separate Ductwork	+ 3.38
Heating Systems, Hot Water	+ 2.05
Heat Pump	+ 1.89
Electric Heat	− 1.26
Not Heated	− 2.87

Additional upgrades or components

Kitchen Cabinets & Countertops	Page 58
Bathroom Vanities	59
Fireplaces & Chimneys	59
Windows, Skylights & Dormers	59
Appliances	60
Breezeways & Porches	60
Finished Attic	60
Garages	61
Site Improvements	61
Wings & Ells	37

- **Simple design from standard plans**
- **Single family — 1 full bath, 1 kitchen**
- **No basement**
- **Asphalt shingles on roof**
- **Hot air heat**
- **Drywall interior finishes**
- **Materials and workmanship are average**
- **Detail specifications on p. 27**

Note: The illustration shown may contain some optional components (for example: garages and/or fireplaces) whose costs are shown in the modifications, adjustments, & alternatives below or at the end of the square foot section.

Base cost per square foot of living area

Exterior Wall	Living Area change										
	1500	1800	2100	2500	3000	3500	4000	4500	5000	5500	6000
Wood Siding - Wood Frame	78.50	70.60	67.80	65.75	60.95	59.15	56.00	52.80	51.95	50.60	49.50
Brick Veneer - Wood Frame	85.15	76.75	73.55	71.25	65.90	63.85	60.25	56.75	55.75	54.25	52.95
Stucco on Wood Frame	79.05	71.10	68.30	66.20	61.35	59.55	56.35	53.15	52.25	50.90	49.80
Solid Masonry	94.30	85.20	81.45	78.80	72.75	70.40	66.20	62.20	61.05	59.30	57.75
Finished Basement, Add	8.90	8.60	8.35	8.20	7.90	7.75	7.50	7.30	7.25	7.20	7.05
Unfinished Basement, Add	3.30	3.10	2.95	2.85	2.65	2.55	2.40	2.30	2.25	2.20	2.10

Modifications

Add to the total cost

Upgrade Kitchen Cabinets	$ + 2283
Solid Surface Countertops	+ 778
Full Bath - including plumbing, wall and floor finishes	+ 3544
Half Bath - including plumbing, wall and floor finishes	+ 2163
One Car Attached Garage	+ 7074
One Car Detached Garage	+ 7587
Fireplace & Chimney	+ 4170

Adjustments

For multi family - add to total cost

Additional Kitchen	$ + 3734
Additional Bath	+ 3544
Additional Entry & Exit	+ 995
Separate Heating	+ 1110
Separate Electric	+ 1204

For Townhouse/Rowhouse -
Multiply cost per square foot by

Inner Unit	.88
End Unit	.94

Alternatives

Add to or deduct from the cost per square foot of living area

Cedar Shake Roof	+ .75
Clay Tile Roof	+ 1.55
Slate Roof	+ 2.60
Upgrade Walls to Skim Coat Plaster	+ .31
Upgrade Ceilings to Textured Finish	+ .43
Air Conditioning, in Heating Ductwork	+ 1.29
In Separate Ductwork	+ 3.38
Heating Systems, Hot Water	+ 2.05
Heat Pump	+ 1.89
Electric Heat	− 1.06
Not Heated	− 2.67

Additional upgrades or components

Kitchen Cabinets & Countertops	Page 58
Bathroom Vanities	59
Fireplaces & Chimneys	59
Windows, Skylights & Dormers	59
Appliances	60
Breezeways & Porches	60
Finished Attic	60
Garages	61
Site Improvements	61
Wings & Ells	37

- **Simple design from standard plans**
- **Single family — 1 full bath, 1 kitchen**
- **No basement**
- **Asphalt shingles on roof**
- **Hot air heat**
- **Drywall interior finishes**
- **Materials and workmanship are average**
- **Detail specifications on p. 27**

Note: The illustration shown may contain some optional components (for example: garages and/or fireplaces) whose costs are shown in the modifications, adjustments, & alternatives below or at the end of the square foot section.

AVERAGE

Base cost per square foot of living area

Exterior Wall	Living Area change										
	1000	1200	1400	1600	1800	2000	2200	2600	3000	3400	3800
Wood Siding - Wood Frame	78.50	70.35	67.55	65.55	62.90	60.60	59.15	55.80	52.60	51.40	49.90
Brick Veneer - Wood Frame	83.45	75.00	71.85	69.65	66.80	64.30	62.70	59.05	55.55	54.20	52.60
Stucco on Wood Frame	78.90	70.75	67.90	65.90	63.25	60.90	59.45	56.10	52.80	51.65	50.10
Solid Masonry	90.35	81.35	77.75	75.30	72.10	69.45	67.55	63.45	59.65	58.05	56.30
Finished Basement, Add	12.15	11.70	11.35	11.10	10.80	10.65	10.45	10.10	9.80	9.65	9.45
Unfinished Basement, Add	4.90	4.60	4.35	4.20	4.00	3.90	3.75	3.55	3.35	3.25	3.15

Modifications

Add to the total cost

Upgrade Kitchen Cabinets	$ + 2283
Solid Surface Countertops	+ 778
Full Bath - including plumbing, wall and floor finishes	+ 3544
Half Bath - including plumbing, wall and floor finishes	+ 2163
One Car Attached Garage	+ 7074
One Car Detached Garage	+ 7587
Fireplace & Chimney	+ 3285

Adjustments

For multi family - add to total cost

Additional Kitchen	$ + 3734
Additional Bath	+ 3544
Additional Entry & Exit	+ 995
Separate Heating	+ 1110
Separate Electric	+ 1204

For Townhouse/Rowhouse -
Multiply cost per square foot by

Inner Unit	.91
End Unit	.96

Alternatives

Add to or deduct from the cost per square foot of living area

Cedar Shake Roof	+ 1.10
Clay Tile Roof	+ 2.30
Slate Roof	+ 3.90
Upgrade Walls to Skim Coat Plaster	+ .30
Upgrade Ceilings to Textured Finish	+ .43
Air Conditioning, in Heating Ductwork	+ 1.44
In Separate Ductwork	+ 3.50
Heating Systems, Hot Water	+ 2.13
Heat Pump	+ 1.86
Electric Heat	– .86
Not Heated	– 2.46

Additional upgrades or components

Kitchen Cabinets & Countertops	Page 58
Bathroom Vanities	59
Fireplaces & Chimneys	59
Windows, Skylights & Dormers	59
Appliances	60
Breezeways & Porches	60
Finished Attic	60
Garages	61
Site Improvements	61
Wings & Ells	37

Important: See the Reference Section for Location Factors (to adjust for your city) and Estimating Forms

AVERAGE

- **Simple design from standard plans**
- **Single family — 1 full bath, 1 kitchen**
- **No basement**
- **Asphalt shingles on roof**
- **Hot air heat**
- **Drywall interior finishes**
- **Materials and workmanship are average**
- **Detail specifications on p. 27**

Note: The illustration shown may contain some optional components (for example: garages and/or fireplaces) whose costs are shown in the modifications, adjustments, & alternatives below or at the end of the square foot section.

© Design Basics, Inc.

Base cost per square foot of living area

Exterior Wall	Living Area change										
	1200	1500	1800	2100	2400	2700	3000	3400	3800	4200	4600
Wood Siding - Wood Frame	73.30	67.75	63.45	60.10	57.55	56.05	54.70	53.15	50.80	48.75	47.70
Brick Veneer - Wood Frame	77.90	71.90	67.20	63.50	60.75	59.15	57.60	55.95	53.35	51.20	50.10
Stucco on Wood Frame	73.70	68.10	63.80	60.40	57.85	56.35	55.00	53.40	51.00	48.95	47.95
Solid Masonry	84.20	77.65	72.35	68.20	65.15	63.40	61.60	59.80	56.95	54.55	53.35
Finished Basement, Add*	13.95	13.35	12.80	12.35	12.10	11.90	11.65	11.55	11.30	11.10	10.95
Unfinished Basement, Add*	5.55	5.15	4.80	4.50	4.30	4.20	4.05	4.00	3.80	3.70	3.60

*Basement under middle level only.

Modifications

Add to the total cost

Upgrade Kitchen Cabinets	$ + 2283
Solid Surface Countertops	+ 778
Full Bath - including plumbing, wall and floor finishes	+ 3544
Half Bath - including plumbing, wall and floor finishes	+ 2163
One Car Attached Garage	+ 7074
One Car Detached Garage	+ 7587
Fireplace & Chimney	+ 3665

Adjustments

For multi family - add to total cost

Additional Kitchen	$ + 3734
Additional Bath	+ 3544
Additional Entry & Exit	+ 995
Separate Heating	+ 1110
Separate Electric	+ 1204

For Townhouse/Rowhouse - Multiply cost per square foot by

Inner Unit	.90
End Unit	.95

Alternatives

Add to or deduct from the cost per square foot of living area

Cedar Shake Roof	+ 1.60
Clay Tile Roof	+ 3.30
Slate Roof	+ 5.60
Upgrade Walls to Skim Coat Plaster	+ .26
Upgrade Ceilings to Textured Finish	+ .43
Air Conditioning, in Heating Ductwork	+ 1.17
In Separate Ductwork	+ 3.27
Heating Systems, Hot Water	+ 1.98
Heat Pump	+ 1.94
Electric Heat	– .72
Not Heated	– 2.32

Additional upgrades or components

Kitchen Cabinets & Countertops	Page 58
Bathroom Vanities	59
Fireplaces & Chimneys	59
Windows, Skylights & Dormers	59
Appliances	60
Breezeways & Porches	60
Finished Attic	60
Garages	61
Site Improvements	61
Wings & Ells	37

Important: See the Reference Section for Location Factors (to adjust for your city) and Estimating Forms

- Post and beam frame
- Log exterior walls
- Simple design from standard plans
- Single family — 1 full bath, 1 kitchen
- No basement
- Asphalt shingles on roof
- Hot air heat
- Drywall interior finishes
- Materials and workmanship are average
- Detail specification on page 27

Note: The illustration shown may contain some optional components (for example: garages and/or fireplaces) whose costs are shown in the modifications, adjustments, & alternatives below or at the end of the square foot section.

SOLID WALL

Base cost per square foot of living area

Exterior Wall	Living Area										
	600	800	1000	1200	1400	1600	1800	2000	2400	2800	3200
6" Log - Solid Wall	89.65	81.05	74.30	69.00	64.70	61.05	59.30	58.05	54.10	51.00	49.30
8" Log - Solid Wall	88.95	80.35	73.70	68.45	64.25	60.60	58.90	57.65	53.75	50.45	48.90
Finished Basement, Add	20.60	19.35	18.50	17.65	17.05	16.65	16.40	16.05	15.50	15.15	14.85
Unfinished Basement, Add	8.80	8.00	7.45	6.90	6.55	6.25	6.10	5.90	5.55	5.30	5.10

Modifications

Add to the total cost

Upgrade Kitchen Cabinets	$ + 2283
Solid Surface Countertops	+ 778
Full Bath - including plumbing, wall and floor finishes	+ 3544
Half Bath - including plumbing, wall and floor finishes	+ 2163
One Car Attached Garage	+ 7074
One Car Detached Garage	+ 7587
Fireplace & Chimney	+ 3285

Adjustments

For multi family - add to total cost

Additional Kitchen	$ + 3734
Additional Bath	+ 3544
Additional Entry & Exit	+ 995
Separate Heating	+ 1110
Separate Electric	+ 1204

For Townhouse/Rowhouse - Multiply cost per square foot by

Inner Unit	.92
End Unit	.96

Alternatives

Add to or deduct from the cost per square foot of living area

Cedar Shake Roof	+ 2.20
Air Conditioning, in Heating Ductwork	+ 2.35
In Separate Ductwork	+ 4.44
Heating Systems, Hot Water	+ 2.65
Heat Pump	+ 1.56
Electric Heat	– 1.02
Not Heated	– 2.63

Additional upgrades or components

Kitchen Cabinets & Countertops	Page 58
Bathroom Vanities	59
Fireplaces & Chimneys	59
Windows, Skylights & Dormers	59
Appliances	60
Breezeways & Porches	60
Finished Attic	60
Garages	61
Site Improvements	61
Wings & Ells	37

Important: See the Reference Section for Location Factors (to adjust for your city) and Estimating Forms

- Post and beam frame
- Log exterior walls
- Simple design from standard plans
- Single family — 1 full bath, 1 kitchen
- No basement
- Asphalt shingles on roof
- Hot air heat
- Drywall interior finishes
- Materials and workmanship are average
- Detail specification on page 27

Note: The illustration shown may contain some optional components (for example: garages and/or fireplaces) whose costs are shown in the modifications, adjustments, & alternatives below or at the end of the square foot section.

Base cost per square foot of living area

Exterior Wall	Living Area										
	1000	1200	1400	1600	1800	2000	2200	2600	3000	3400	3800
6" Log-Solid	89.75	74.00	70.60	68.10	65.35	62.70	60.90	57.30	53.10	51.70	50.35
8" Log-Solid	86.20	77.30	73.50	71.20	68.30	65.30	63.50	59.65	55.60	54.10	52.45
Finished Basement, Add	11.95	11.55	11.15	10.95	10.60	10.50	10.25	9.90	9.65	9.50	9.35
Unfinished Basement, Add	4.80	4.50	4.25	4.10	3.95	3.70	3.70	3.40	3.25	3.20	3.05

Modifications

Add to the total cost

Upgrade Kitchen Cabinets	$ + 2283
Solid Surface Countertops	+ 778
Full Bath - including plumbing, wall and floor finishes	+ 3544
Half Bath - including plumbing, wall and floor finishes	+ 2163
One Car Attached Garage	+ 7074
One Car Detached Garage	+ 7587
Fireplace & Chimney	+ 3665

Adjustments

For multi family - add to total cost

Additional Kitchen	$ + 3734
Additional Bath	+ 3544
Additional Entry & Exit	+ 995
Separate Heating	+ 1110
Separate Electric	+ 1204

For Townhouse/Rowhouse -
Multiply cost per square foot by

Inner Unit	.92
End Unit	.96

Alternatives

Add to or deduct from the cost per square foot of living area

Cedar Shake Roof	+ 1.60
Air Conditioning, in Heating Ductwork	+ 1.76
In Separate Ductwork	+ 3.50
Heating Systems, Hot Water	+ 2.13
Heat Pump	+ 1.86
Electric Heat	– .86
Not Heated	– 2.47

Additional upgrades or components

Kitchen Cabinets & Countertops	Page 58
Bathroom Vanities	59
Fireplaces & Chimneys	59
Windows, Skylights & Dormers	59
Appliances	60
Breezeways & Porches	60
Finished Attic	60
Garages	61
Site Improvements	61
Wings & Ells	37

1 Story — Base cost per square foot of living area

Exterior Wall	Living Area							
	50	100	200	300	400	500	600	700
Wood Siding - Wood Frame	117.80	90.10	77.95	65.40	61.40	59.05	57.45	58.05
Brick Veneer - Wood Frame	127.75	95.65	81.65	65.95	61.35	58.65	56.80	57.25
Stucco on Wood Frame	118.95	90.90	78.60	65.75	61.75	59.35	57.70	58.30
Solid Masonry	154.65	116.45	99.95	80.00	74.60	71.35	70.05	70.10
Finished Basement, Add	33.65	27.15	24.45	19.95	19.00	18.50	18.10	17.85
Unfinished Basement, Add	16.35	12.45	10.85	8.10	7.55	7.25	7.00	6.85

1-1/2 Story — Base cost per square foot of living area

Exterior Wall	Living Area							
	100	200	300	400	500	600	700	800
Wood Siding - Wood Frame	96.90	77.70	66.15	59.40	55.90	54.25	52.15	51.50
Brick Veneer - Wood Frame	129.20	98.00	81.55	71.20	66.25	63.60	60.70	59.65
Stucco on Wood Frame	116.55	87.85	73.10	64.65	60.15	57.85	55.25	54.30
Solid Masonry	148.25	113.25	94.25	81.15	75.40	72.25	68.90	67.80
Finished Basement, Add	22.75	20.05	18.25	16.25	15.70	15.35	15.00	14.95
Unfinished Basement, Add	10.35	8.70	7.60	6.40	6.10	5.85	5.65	5.65

2 Story — Base cost per square foot of living area

Exterior Wall	Living Area							
	100	200	400	600	800	1000	1200	1400
Wood Siding - Wood Frame	97.15	72.45	61.55	51.10	47.45	45.40	43.95	44.70
Brick Veneer - Wood Frame	132.45	93.45	75.75	60.50	55.50	52.55	50.50	50.85
Stucco on Wood Frame	118.25	83.30	67.30	54.90	50.45	47.80	46.00	46.50
Solid Masonry	153.85	108.70	88.50	69.00	63.15	59.65	57.30	57.40
Finished Basement, Add	18.10	14.80	13.45	11.20	10.75	10.50	10.30	10.20
Unfinished Basement, Add	8.30	6.30	5.50	4.15	3.90	3.70	3.60	3.55

Base costs do not include bathroom or kitchen facilities. Use Modifications/Adjustments/Alternatives on pages 58-61 where appropriate.

Important: See the Reference Section for Location Factors (to adjust for your city) and Estimating Forms.

1 Story

© Design Basics, Inc.

1 - 1/2 Story

© Donald A. Gardner

2 Story

2 - 1/2 Story

Bi-Level

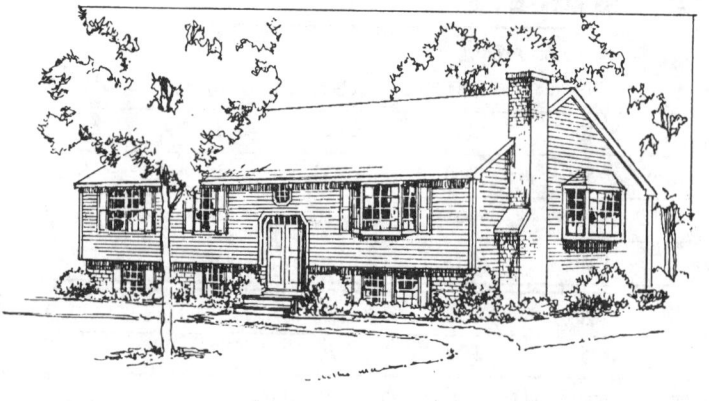

Tri-Level

© Design Basics, Inc.

38

	Components
1 Site Work	Site preparation for slab or excavation for lower level; 4' deep trench excavation for foundation wall.
2 Foundations	Continuous reinforced concrete footing, 8" deep x 18" wide; dampproofed and insulated 8" thick reinforced concrete foundation wall, 4' deep; 4" concrete slab on 4" crushed stone base and polyethylene vapor barrier, trowel finish.
3 Framing	Exterior walls—2" x 6" wood studs, 16" O.C.; 1/2" plywood sheathing; 2" x 8" rafters 16" O.C. with 1/2" plywood sheathing, 4 in 12, 6 in 12 or 8 in 12 roof pitch, 2" x 6" or 2" x 8" ceiling joists; 1/2" plywood subfloor on 1" x 3" wood sleepers 16" O.C.; 2" X 10" floor joists with 5/8" plywood subfloor on models with more than one level.
4 Exterior Walls	Horizontal beveled wood siding and #15 felt building paper on insulated wood frame walls; double hung windows; 3 flush solid core wood exterior doors with storms. **Alternates:** • Brick veneer on wood frame, has 4" veneer high quality face brick or select common brick. • Stone veneer on wood frame has exterior veneer of field stone or 2" thick limestone. • Solid masonry has a 8" concrete block wall with insulation and a brick or stone facing. It may be a solid brick or stone structure.
5 Roofing	30 year asphalt shingles; #15 felt building paper; aluminum gutters, downspouts and drip edge; copper flashings.
6 Interiors	Walls and ceilings—5/8" drywall, skim coat plaster, primed and painted with 2 coats; hardwood baseboard and trim; sanded and finished, hardwood floor 70%, ceramic tile with 1/2" underlayment 20%, vinyl tile with 1/2" underlayment 10%; wood panel interior doors, primed and painted with 2 coats.
7 Specialties	Custom grade kitchen cabinets—20 L.F. wall and base with plastic laminate counter top and kitchen sink; 4 L.F. bathroom vanity; 75 gallon electric water heater; medicine cabinet.
8 Mechanical	Gas fired warm air heat/air conditioning; one full bath including bathtub, corner shower, built in lavatory and water closet; one 1/2 bath including built in lavatory and water closet.
9 Electrical	100 Amp. service; romex wiring; incandescent lighting fixtures, switches, receptacles.
10 Overhead and Profit	General Contractor overhead and profit.

Adjustments

Unfinished Basement:
7'-6" high cast-in-place concrete walls 10" thick or 12" concrete block.

Finished Basement:
Includes painted drywall on 2" x 4" wood furring with insulation, suspended ceiling, carpeting on concrete floor, heating and lighting.

- A distinct residence from designer's plans
- Single family — 1 full bath, 1 half bath, 1 kitchen
- No basement
- Asphalt shingles on roof
- Forced hot air heat/air conditioning
- Drywall interior finishes
- Materials and workmanship are above average
- Detail specifications on page 39

Note: The illustration shown may contain some optional components (for example: garages and/or fireplaces) whose costs are shown in the modifications, adjustments, & alternatives below or at the end of the square foot section.

CUSTOM

© Design Basics, Inc.

Base cost per square foot of living area

Exterior Wall	Living Area change										
	800	1000	1200	1400	1600	1800	2000	2400	2800	3200	3600
Wood Siding - Wood Frame	115.90	105.65	97.35	91.05	86.25	83.55	80.65	74.65	70.60	67.60	64.25
Brick Veneer - Wood Frame	125.00	114.45	105.85	99.30	94.35	91.60	88.55	82.40	78.20	75.05	71.60
Stone Veneer - Wood Frame	130.00	118.95	109.90	103.00	97.80	94.95	91.70	85.25	80.80	77.50	73.85
Solid Masonry	130.10	119.05	110.00	103.10	97.90	95.00	91.80	85.35	80.90	77.55	73.95
Finished Basement, Add	28.95	27.55	26.25	25.30	24.60	24.25	23.70	22.95	22.40	21.90	21.45
Unfinished Basement, Add	12.15	11.45	10.80	10.30	9.95	9.75	9.50	9.10	8.85	8.55	8.35

Modifications

Add to the total cost

Upgrade Kitchen Cabinets	$ + 614
Solid Surface Countertops	+ 1111
Full Bath - including plumbing, wall and floor finishes	+ 4200
Half Bath - including plumbing, wall and floor finishes	+ 2563
Two Car Attached Garage	+ 11,601
Two Car Detached Garage	+ 12,064
Fireplace & Chimney	+ 3490

Adjustments

For multi family - add to total cost

Additional Kitchen	$ + 8100
Additional Full Bath & Half Bath	+ 6763
Additional Entry & Exit	+ 995
Separate Heating & Air Conditioning	+ 4054
Separate Electric	+ 1204

For Townhouse/Rowhouse -
Multiply cost per square foot by

Inner Unit	.90
End Unit	.95

Alternatives

Add to or deduct from the cost per square foot of living area

Cedar Shake Roof	+ 1.85
Clay Tile Roof	+ 4.25
Slate Roof	+ 7.45
Upgrade Ceilings to Textured Finish	+ .43
Air Conditioning, in Heating Ductwork	Base System
In Separate Ductwork	+ 2.19
Heating Systems, Hot Water	+ 2.69
Heat Pump	+ 1.57
Electric Heat	+ .35
Not Heated	– 3.00

Additional upgrades or components

Kitchen Cabinets & Countertops	Page 58
Bathroom Vanities	59
Fireplaces & Chimneys	59
Windows, Skylights & Dormers	59
Appliances	60
Breezeways & Porches	60
Finished Attic	60
Garages	61
Site Improvements	61
Wings & Ells	47

Important: See the Reference Section for Location Factors (to adjust for your city) and Estimating Forms

- A distinct residence from designer's plans
- Single family — 1 full bath, 1 half bath, 1 kitchen
- No basement
- Asphalt shingles on roof
- Forced hot air heat/air conditioning
- Drywall interior finishes
- Materials and workmanship are above average
- Detail specifications on page 39

Note: The illustration shown may contain some optional components (for example: garages and/or fireplaces) whose costs are shown in the modifications, adjustments, & alternatives below or at the end of the square foot section.

© Donald A. Gardner, Inc.

CUSTOM

Base cost per square foot of living area

Exterior Wall	Living Area change										
	1000	1200	1400	1600	1800	2000	2400	2800	3200	3600	4000
Wood Siding - Wood Frame	106.50	100.30	95.80	89.25	86.05	82.20	75.40	72.75	69.55	67.25	64.00
Brick Veneer - Wood Frame	110.00	103.55	98.90	92.05	88.70	84.70	77.65	74.85	71.55	69.10	65.75
Stone Veneer - Wood Frame	115.20	108.45	103.55	96.30	92.70	88.50	81.00	78.05	74.45	71.95	68.40
Solid Masonry	115.35	108.55	103.65	96.40	92.80	88.60	81.10	78.15	74.55	72.05	68.45
Finished Basement, Add	19.25	18.50	18.00	17.20	16.80	16.40	15.60	15.25	14.85	14.60	14.25
Unfinished Basement, Add	8.15	7.80	7.55	7.15	6.95	6.75	6.35	6.15	5.95	5.85	5.65

Modifications

Add to the total cost

Upgrade Kitchen Cabinets	$ + 614
Solid Surface Countertops	+ 1111
Full Bath - including plumbing, wall and floor finishes	+ 4200
Half Bath - including plumbing, wall and floor finishes	+ 2563
Two Car Attached Garage	+ 11,601
Two Car Detached Garage	+ 12,064
Fireplace & Chimney	+ 3490

Adjustments

For multi family - add to total cost

Additional Kitchen	$ + 8100
Additional Full Bath & Half Bath	+ 6763
Additional Entry & Exit	+ 995
Separate Heating & Air Conditioning	+ 4054
Separate Electric	+ 1204

For Townhouse/Rowhouse -
Multiply cost per square foot by

Inner Unit	.90
End Unit	.95

Alternatives

Add to or deduct from the cost per square foot of living area

Cedar Shake Roof	+ 1.35
Clay Tile Roof	+ 3.05
Slate Roof	+ 5.35
Upgrade Ceilings to Textured Finish	+ .43
Air Conditioning, in Heating Ductwork	Base System
In Separate Ductwork	+ 2.19
Heating Systems, Hot Water	+ 2.34
Heat Pump	+ 1.78
Electric Heat	+ .85
Not Heated	– 2.75

Additional upgrades or components

Kitchen Cabinets & Countertops	Page 58
Bathroom Vanities	59
Fireplaces & Chimneys	59
Windows, Skylights & Dormers	59
Appliances	60
Breezeways & Porches	60
Finished Attic	60
Garages	61
Site Improvements	61
Wings & Ells	47

- A distinct residence from designer's plans
- Single family — 1 full bath, 1 half bath, 1 kitchen
- No basement
- Asphalt shingles on roof
- Forced hot air heat/air conditioning
- Drywall interior finishes
- Materials and workmanship are above average
- Detail specifications on page 39

Note: The illustration shown may contain some optional components (for example: garages and/or fireplaces) whose costs are shown in the modifications, adjustments, & alternatives below or at the end of the square foot section.

Base cost per square foot of living area

Exterior Wall	Living Area change										
	1200	1400	1600	1800	2000	2400	2800	3200	3600	4000	4400
Wood Siding - Wood Frame	101.90	96.70	93.05	88.85	85.30	79.30	74.70	71.15	69.00	67.00	65.05
Brick Veneer - Wood Frame	105.60	100.15	96.35	91.95	88.30	82.00	77.15	73.45	71.25	69.10	67.05
Stone Veneer - Wood Frame	111.15	105.35	101.30	96.60	92.80	86.10	80.85	76.90	74.55	72.25	70.10
Solid Masonry	111.35	105.50	101.45	96.75	92.95	86.20	80.95	77.00	74.65	72.35	70.20
Finished Basement, Add	15.50	14.90	14.50	14.10	13.80	13.20	12.70	12.35	12.15	11.90	11.75
Unfinished Basement, Add	6.55	6.25	6.10	5.85	5.75	5.40	5.15	5.00	4.90	4.75	4.70

Modifications

Add to the total cost

Upgrade Kitchen Cabinets	$ + 614
Solid Surface Countertops	+ 1111
Full Bath - including plumbing, wall and floor finishes	+ 4200
Half Bath - including plumbing, wall and floor finishes	+ 2563
Two Car Attached Garage	+ 11,601
Two Car Detached Garage	+ 12,064
Fireplace & Chimney	+ 3940

Adjustments

For multi family - add to total cost

Additional Kitchen	$ + 8100
Additional Full Bath & Half Bath	+ 6763
Additional Entry & Exit	+ 995
Separate Heating & Air Conditioning	+ 4054
Separate Electric	+ 1204

For Townhouse/Rowhouse -
Multiply cost per square foot by

Inner Unit	.87
End Unit	.93

Alternatives

Add to or deduct from the cost per square foot of living area

Cedar Shake Roof	+ .95
Clay Tile Roof	+ 2.15
Slate Roof	+ 3.70
Upgrade Ceilings to Textured Finish	+ .43
Air Conditioning, in Heating Ductwork	Base System
In Separate Ductwork	+ 2.19
Heating Systems, Hot Water	+ 2.16
Heat Pump	+ 1.89
Electric Heat	+ .85
Not Heated	– 2.57

Additional upgrades or components

Kitchen Cabinets & Countertops	Page 58
Bathroom Vanities	59
Fireplaces & Chimneys	59
Windows, Skylights & Dormers	59
Appliances	60
Breezeways & Porches	60
Finished Attic	60
Garages	61
Site Improvements	61
Wings & Ells	47

- **A distinct residence from designer's plans**
- **Single family — 1 full bath, 1 half bath, 1 kitchen**
- **No basement**
- **Asphalt shingles on roof**
- **Forced hot air heat/air conditioning**
- **Drywall interior finishes**
- **Materials and workmanship are above average**
- **Detail specifications on page 39**

Note: The illustration shown may contain some optional components (for example: garages and/or fireplaces) whose costs are shown in the modifications, adjustments, & alternatives below or at the end of the square foot section.

Base cost per square foot of living area

Exterior Wall	Living Area change										
	1500	1800	2100	2400	2800	3200	3600	4000	4500	5000	5500
Wood Siding - Wood Frame	103.30	92.55	87.00	83.10	79.15	74.35	72.20	68.10	66.25	64.15	62.00
Brick Veneer - Wood Frame	107.15	96.05	90.20	86.10	82.05	76.95	74.65	70.40	68.40	66.20	63.95
Stone Veneer - Wood Frame	112.95	101.40	94.95	90.65	86.35	80.90	78.40	73.85	71.65	69.30	66.90
Solid Masonry	113.05	101.55	95.05	90.75	86.45	81.00	78.45	73.95	71.75	69.40	67.00
Finished Basement, Add	12.30	11.75	11.15	10.80	10.55	10.10	9.85	9.60	9.40	9.20	9.05
Unfinished Basement, Add	5.25	5.00	4.65	4.50	4.40	4.15	4.05	3.90	3.80	3.70	3.65

Modifications

Add to the total cost

Upgrade Kitchen Cabinets	$ + 614
Solid Surface Countertops	+ 1111
Full Bath - including plumbing, wall and floor finishes	+ 4200
Half Bath - including plumbing, wall and floor finishes	+ 2563
Two Car Attached Garage	+ 11,601
Two Car Detached Garage	+ 12,064
Fireplace & Chimney	+ 4450

Adjustments

For multi family - add to total cost

Additional Kitchen	$ + 8100
Additional Full Bath & Half Bath	+ 6763
Additional Entry & Exit	+ 995
Separate Heating & Air Conditioning	+ 4054
Separate Electric	+ 1204

For Townhouse/Rowhouse -
Multiply cost per square foot by

Inner Unit	.87
End Unit	.94

Alternatives

Add to or deduct from the cost per square foot of living area

Cedar Shake Roof	+ .80
Clay Tile Roof	+ 1.85
Slate Roof	+ 3.20
Upgrade Ceilings to Textured Finish	+ .43
Air Conditioning, in Heating Ductwork	Base System
In Separate Ductwork	+ 2.19
Heating Systems, Hot Water	+ 2.09
Heat Pump	+ 1.94
Electric Heat	+ .95
Not Heated	– 2.57

Additional upgrades or components

Kitchen Cabinets & Countertops	Page 58
Bathroom Vanities	59
Fireplaces & Chimneys	59
Windows, Skylights & Dormers	59
Appliances	60
Breezeways & Porches	60
Finished Attic	60
Garages	61
Site Improvements	61
Wings & Ells	47

- **A distinct residence from designer's plans**
- **Single family — 1 full bath, 1 half bath, 1 kitchen**
- **No basement**
- **Asphalt shingles on roof**
- **Forced hot air heat/air conditioning**
- **Drywall interior finishes**
- **Materials and workmanship are above average**
- **Detail specifications on page 39**

Note: The illustration shown may contain some optional components (for example: garages and/or fireplaces) whose costs are shown in the modifications, adjustments, & alternatives below or at the end of the square foot section.

Base cost per square foot of living area

Exterior Wall	Living Area change										
	1500	1800	2100	2500	3000	3500	4000	4500	5000	5500	6000
Wood Siding - Wood Frame	103.35	92.65	88.40	85.00	78.50	75.75	71.45	67.15	65.85	63.95	62.40
Brick Veneer - Wood Frame	107.30	96.35	91.85	88.25	81.45	78.60	74.00	69.55	68.15	66.20	64.50
Stone Veneer - Wood Frame	113.30	101.90	97.00	93.25	86.00	82.85	77.90	73.10	71.65	69.50	67.65
Solid Masonry	113.45	102.05	97.15	93.35	86.10	83.00	78.00	73.20	71.70	69.60	67.75
Finished Basement, Add	10.80	10.30	9.90	9.60	9.15	8.90	8.55	8.30	8.20	8.05	7.90
Unfinished Basement, Add	4.60	4.35	4.15	4.05	3.80	3.70	3.50	3.35	3.30	3.25	3.15

Modifications

Add to the total cost

Upgrade Kitchen Cabinets	$ + 614
Solid Surface Countertops	+ 1111
Full Bath - including plumbing, wall and floor finishes	+ 4200
Half Bath - including plumbing, wall and floor finishes	+ 2563
Two Car Attached Garage	+ 11,601
Two Car Detached Garage	+ 12,064
Fireplace & Chimney	+ 4450

Adjustments

For multi family - add to total cost

Additional Kitchen	$ + 8100
Additional Full Bath & Half Bath	+ 6763
Additional Entry & Exit	+ 995
Separate Heating & Air Conditioning	+ 4054
Separate Electric	+ 1204

For Townhouse/Rowhouse -
Multiply cost per square foot by

Inner Unit	.85
End Unit	.93

Alternatives

Add to or deduct from the cost per square foot of living area

Cedar Shake Roof	+ .60
Clay Tile Roof	+ 1.40
Slate Roof	+ 2.50
Upgrade Ceilings to Textured Finish	+ .43
Air Conditioning, in Heating Ductwork	Base System
In Separate Ductwork	+ 2.19
Heating Systems, Hot Water	+ 2.09
Heat Pump	+ 1.94
Electric Heat	+ .95
Not Heated	– 2.48

Additional upgrades or components

Kitchen Cabinets & Countertops	Page 58
Bathroom Vanities	59
Fireplaces & Chimneys	59
Windows, Skylights & Dormers	59
Appliances	60
Breezeways & Porches	60
Finished Attic	60
Garages	61
Site Improvements	61
Wings & Ells	47

- **A distinct residence from designer's plans**
- **Single family — 1 full bath, 1 half bath, 1 kitchen**
- **No basement**
- **Asphalt shingles on roof**
- **Forced hot air heat/air conditioning**
- **Drywall interior finishes**
- **Materials and workmanship are above average**
- **Detail specifications on page 39**

Note: The illustration shown may contain some optional components (for example: garages and/or fireplaces) whose costs are shown in the modifications, adjustments, & alternatives below or at the end of the square foot section.

CUSTOM

Base cost per square foot of living area

Exterior Wall	Living Area change										
	1200	1400	1600	1800	2000	2400	2800	3200	3600	4000	4400
Wood Siding - Wood Frame	94.90	90.15	86.80	82.90	79.60	74.15	70.05	66.75	64.80	63.00	61.20
Brick Veneer - Wood Frame	97.70	92.75	89.25	85.25	81.85	76.15	71.90	68.50	66.45	64.55	62.70
Stone Veneer - Wood Frame	101.85	96.60	93.00	88.75	85.20	79.25	74.65	71.10	68.95	66.90	64.95
Solid Masonry	101.95	96.75	93.05	88.85	85.30	79.30	74.75	71.20	69.05	66.95	65.05
Finished Basement, Add	15.50	14.90	14.50	14.10	13.80	13.20	12.70	12.35	12.15	11.90	11.75
Unfinished Basement, Add	6.55	6.25	6.10	5.85	5.75	5.40	5.15	5.00	4.90	4.75	4.70

Modifications

Add to the total cost

Upgrade Kitchen Cabinets	$ + 614
Solid Surface Countertops	+ 1111
Full Bath - including plumbing, wall and floor finishes	+ 4200
Half Bath - including plumbing, wall and floor finishes	+ 2563
Two Car Attached Garage	+ 11,601
Two Car Detached Garage	+ 12,064
Fireplace & Chimney	+ 3490

Adjustments

For multi family - add to total cost

Additional Kitchen	$ + 8100
Additional Full Bath & Half Bath	+ 6763
Additional Entry & Exit	+ 995
Separate Heating & Air Conditioning	+ 4054
Separate Electric	+ 1204

For Townhouse/Rowhouse -
Multiply cost per square foot by

Inner Unit	.89
End Unit	.95

Alternatives

Add to or deduct from the cost per square foot of living area

Cedar Shake Roof	+ .95
Clay Tile Roof	+ 2.15
Slate Roof	+ 3.70
Upgrade Ceilings to Textured Finish	+ .43
Air Conditioning, in Heating Ductwork	Base System
In Separate Ductwork	+ 2.19
Heating Systems, Hot Water	+ 2.16
Heat Pump	+ 1.89
Electric Heat	+ .85
Not Heated	− 2.57

Additional upgrades or components

Kitchen Cabinets & Countertops	Page 58
Bathroom Vanities	59
Fireplaces & Chimneys	59
Windows, Skylights & Dormers	59
Appliances	60
Breezeways & Porches	60
Finished Attic	60
Garages	61
Site Improvements	61
Wings & Ells	47

- **A distinct residence from designer's plans**
- **Single family — 1 full bath, 1 half bath, 1 kitchen**
- **No basement**
- **Asphalt shingles on roof**
- **Forced hot air heat/air conditioning**
- **Drywall interior finishes**
- **Materials and workmanship are above average**
- **Detail specifications on page 39**

Note: The illustration shown may contain some optional components (for example: garages and/or fireplaces) whose costs are shown in the modifications, adjustments, & alternatives below or at the end of the square foot section.

© Design Basics, Inc.

Base cost per square foot of living area

Exterior Wall	Living Area change										
	1200	1500	1800	2100	2400	2800	3200	3600	4000	4500	5000
Wood Siding - Wood Frame	97.30	88.90	82.45	77.45	73.70	70.95	67.60	64.25	62.95	59.50	57.65
Brick Veneer - Wood Frame	100.05	91.40	84.70	79.50	75.60	72.80	69.35	65.85	64.50	60.90	59.00
Stone Veneer - Wood Frame	104.15	95.15	88.05	82.55	78.50	75.55	71.90	68.20	66.80	63.05	61.00
Solid Masonry	104.30	95.25	88.15	82.65	78.55	75.65	71.95	68.30	66.90	63.05	61.10
Finished Basement, Add*	19.30	18.40	17.55	16.85	16.45	16.15	15.70	15.35	15.15	14.80	14.50
Unfinished Basement, Add*	8.10	7.65	7.20	6.90	6.65	6.50	6.30	6.10	6.00	5.80	5.70

*Basement under middle level only.

Modifications

Add to the total cost

Upgrade Kitchen Cabinets	$ + 614
Solid Surface Countertops	+ 1111
Full Bath - including plumbing, wall and floor finishes	+ 4200
Half Bath - including plumbing, wall and floor finishes	+ 2563
Two Car Attached Garage	+ 11,601
Two Car Detached Garage	+ 12,064
Fireplace & Chimney	+ 3940

Adjustments

For multi family - add to total cost

Additional Kitchen	$ + 8100
Additional Full Bath & Half Bath	+ 6763
Additional Entry & Exit	+ 995
Separate Heating & Air Conditioning	+ 4054
Separate Electric	+ 1204

For Townhouse/Rowhouse -
Multiply cost per square foot by

Inner Unit	.87
End Unit	.94

Alternatives

Add to or deduct from the cost per square foot of living area

Cedar Shake Roof	+ 1.35
Clay Tile Roof	+ 3.05
Slate Roof	+ 5.35
Upgrade Ceilings to Textured Finish	+ .43
Air Conditioning, in Heating Ductwork	Base System
In Separate Ductwork	+ 2.19
Heating Systems, Hot Water	+ 2.03
Heat Pump	+ 1.97
Electric Heat	+ 1.12
Not Heated	− 2.48

Additional upgrades or components

Kitchen Cabinets & Countertops	Page 58
Bathroom Vanities	59
Fireplaces & Chimneys	59
Windows, Skylights & Dormers	59
Appliances	60
Breezeways & Porches	60
Finished Attic	60
Garages	61
Site Improvements	61
Wings & Ells	47

CUSTOM

1 Story — Base cost per square foot of living area

Exterior Wall	Living Area							
	50	100	200	300	400	500	600	700
Wood Siding - Wood Frame	148.90	112.65	96.90	79.60	74.40	71.30	69.25	69.90
Brick Veneer - Wood Frame	158.25	119.35	102.45	83.30	77.70	74.45	72.20	72.70
Stone Veneer - Wood Frame	172.25	129.30	110.80	88.85	82.75	79.10	76.65	77.00
Solid Masonry	172.65	129.60	111.05	89.00	82.85	79.25	76.80	77.10
Finished Basement, Add	47.15	38.25	34.55	28.35	27.10	26.35	25.85	25.50
Unfinished Basement, Add	33.35	23.05	18.30	14.00	12.85	12.20	11.75	11.40

1-1/2 Story — Base cost per square foot of living area

Exterior Wall	Living Area							
	100	200	300	400	500	600	700	800
Wood Siding - Wood Frame	122.15	98.20	83.55	74.25	69.80	67.65	64.95	64.20
Brick Veneer - Wood Frame	130.45	104.85	89.15	78.55	73.80	71.40	68.50	67.70
Stone Veneer - Wood Frame	142.95	114.85	97.45	85.10	79.80	77.05	73.85	73.05
Solid Masonry	143.30	115.15	97.70	85.25	79.95	77.25	74.00	73.15
Finished Basement, Add	31.30	27.60	25.15	22.40	21.65	21.15	20.70	20.60
Unfinished Basement, Add	19.90	15.20	12.90	10.95	10.30	9.80	9.45	9.30

2 Story — Base cost per square foot of living area

Exterior Wall	Living Area							
	100	200	400	600	800	1000	1200	1400
Wood Siding - Wood Frame	124.20	92.05	78.00	63.50	58.90	56.15	54.25	55.05
Brick Veneer - Wood Frame	133.55	98.70	83.55	67.20	62.20	59.25	57.25	57.90
Stone Veneer - Wood Frame	147.55	108.70	91.85	72.75	67.20	63.90	61.70	62.20
Solid Masonry	147.95	109.00	92.10	72.90	67.35	64.05	61.80	62.30
Finished Basement, Add	23.65	19.15	17.30	14.20	13.60	13.20	12.95	12.80
Unfinished Basement, Add	16.70	11.55	9.15	7.00	6.45	6.10	5.90	5.70

Base costs do not include bathroom or kitchen facilities. Use Modifications/Adjustments/Alternatives on pages 58-61 where appropriate.

Important: See the Reference Section for Location Factors (to adjust for your city) and Estimating Forms.

1 Story

© Home Planners, Inc.

1-1/2 Story

© Larry E. Belk Designs

2 Story

2-1/2 Story

© Larry W. Garnett & Associates, In

Bi-Level

Tri-Level

© Home Planners, Inc.

LUXURY

		Components
1	**Site Work**	Site preparation for slab or excavation for lower level; 4' deep trench excavation for foundation wall.
2	**Foundations**	Continuous reinforced concrete footing, 8" deep x 18" wide; dampproofed and insulated 12" thick reinforced concrete foundation wall, 4' deep; 4" concrete slab on 4" crushed stone base and polyethylene vapor barrier, trowel finish.
3	**Framing**	Exterior walls—2" x 6" wood studs, 16" O.C.; 1/2" plywood sheathing; 2" x 8" rafters 16" O.C. with 1/2" plywood sheathing, 4 in 12, 6 in 12 or 8 in 12 roof pitch, 2" x 6" or 2" x 8" ceiling joists; 1/2" plywood subfloor on 1" x 3" wood sleepers 16" O.C.; 2" X 10" floor joists with 5/8" plywood subfloor on models with more than one level.
4	**Exterior Walls**	Face brick veneer and #15 felt building paper on insulated wood frame walls; double hung windows; 3 flush solid core wood exterior doors with storms. **Alternates:** • Wood siding on wood frame, has top quality cedar or redwood siding or hand split cedar shingles or shakes. • Solid brick may have solid brick exterior wall or brick on concrete block. • Solid stone concrete block with selected fieldstone or limestone exterior.
5	**Roofing**	Red cedar shingles; #15 felt building paper; aluminum gutters, downspouts and drip edge; copper flashings.
6	**Interiors**	Walls and ceilings—5/8" drywall, skim coat plaster, primed and painted with 2 coats; hardwood baseboard and trim; sanded and finished, hardwood floor 70%, ceramic tile with 1/2" underlayment 20%, vinyl tile with 1/2" underlayment 10%; wood panel interior doors, primed and painted with 2 coats.
7	**Specialties**	Luxury grade kitchen cabinets—25 L.F. wall and base with plastic laminate counter top and kitchen sink; 6 L.F. bathroom vanity; 75 gallon electric water heater; medicine cabinet.
8	**Mechanical**	Gas fired warm air heat/air conditioning; one full bath including bathtub, corner shower, built in lavatory and water closet; one 1/2 bath including built in lavatory and water closet.
9	**Electrical**	100 Amp. service; romex wiring; incandescent lighting fixtures, switches, receptacles.
10	**Overhead and Profit**	General Contractor overhead and profit.

Adjustments

Unfinished Basement:
8" high cast-in-place concrete walls 12" thick or 12" concrete block.

Finished Basement:
Includes painted drywall on 2" x 4" wood furring with insulation, suspended ceiling, carpeting on subfloor with sleepers, heating and lighting.

- Unique residence built from an architect's plan
- Single family — 1 full bath, 1 half bath, 1 kitchen
- No basement
- Cedar shakes on roof
- Forced hot air heat/air conditioning
- Double drywall interior
- Many special features
- Extraordinary materials and workmanship
- Detail specifications on page 49

Note: The illustration shown may contain some optional components (for example: garages and/or fireplaces) whose costs are shown in the modifications, adjustments, & alternatives below or at the end of the square foot section.

LUXURY

© Home Planners, Inc.

Base cost per square foot of living area

Exterior Wall	Living Area change										
	1000	1200	1400	1600	1800	2000	2400	2800	3200	3600	4000
Wood Siding - Wood Frame	130.15	120.50	113.20	107.65	104.45	101.10	94.25	89.60	86.10	82.35	79.25
Brick Veneer - Wood Frame	133.05	123.15	115.65	109.95	106.65	103.20	96.10	91.35	87.70	83.85	80.65
Solid Brick	141.45	130.75	122.55	116.35	112.85	109.00	101.45	96.25	92.30	88.10	84.65
Solid Stone	143.00	132.10	123.80	117.55	114.00	110.10	102.45	97.15	93.10	88.90	85.35
Finished Basement, Add	33.65	32.05	30.80	29.95	29.50	28.80	27.90	27.15	26.50	26.00	25.50
Unfinished Basement, Add	12.80	12.10	11.55	11.15	10.95	10.65	10.20	9.90	9.60	9.35	9.10

Modifications

Add to the total cost

Upgrade Kitchen Cabinets	$ + 810
Solid Surface Countertops	+ 1245
Full Bath - including plumbing, wall and floor finishes	+ 4872
Half Bath - including plumbing, wall and floor finishes	+ 2974
Two Car Attached Garage	+ 17,149
Two Car Detached Garage	+ 17,273
Fireplace & Chimney	+ 4935

Adjustments

For multi family - add to total cost

Additional Kitchen	$ + 10,371
Additional Full Bath & Half Bath	+ 7846
Additional Entry & Exit	+ 1486
Separate Heating & Air Conditioning	+ 4054
Separate Electric	+ 1204

For Townhouse/Rowhouse -
Multiply cost per square foot by

Inner Unit	.90
End Unit	.95

Alternatives

Add to or deduct from the cost per square foot of living area

Heavyweight Asphalt Shingles	– 1.85
Clay Tile Roof	+ 2.40
Slate Roof	+ 5.60
Upgrade Ceilings to Textured Finish	+ .43
Air Conditioning, in Heating Ductwork	Base System
In Separate Ductwork	+ 2.39
Heating Systems, Hot Water	+ 2.90
Heat Pump	+ 1.72
Electric Heat	+ .90
Not Heated	– 3.25

Additional upgrades or components

Kitchen Cabinets & Countertops	Page 58
Bathroom Vanities	59
Fireplaces & Chimneys	59
Windows, Skylights & Dormers	59
Appliances	60
Breezeways & Porches	60
Finished Attic	60
Garages	61
Site Improvements	61
Wings & Ells	57

- Unique residence built from an architect's plan
- Single family — 1 full bath, 1 half bath, 1 kitchen
- No basement
- Cedar shakes on roof
- Forced hot air heat/air conditioning
- Double drywall interior
- Many special features
- Extraordinary materials and workmanship
- Detail specifications on page 49

Note: The illustration shown may contain some optional components (for example: garages and/or fireplaces) whose costs are shown in the modifications, adjustments, & alternatives below or at the end of the square foot section.

© Larry E. Belk Designs

Base cost per square foot of living area

Exterior Wall	Living Area change										
	1000	1200	1400	1600	1800	2000	2400	2800	3200	3600	4000
Wood Siding - Wood Frame	122.50	115.10	109.75	102.25	98.45	94.05	86.30	83.20	79.50	76.85	73.15
Brick Veneer - Wood Frame	125.95	118.30	112.80	105.00	101.05	96.55	88.50	85.25	81.45	78.70	74.90
Solid Brick	135.65	127.40	121.50	112.85	108.55	103.60	94.75	91.20	86.90	84.00	79.80
Solid Stone	137.45	129.10	123.05	114.30	109.90	104.95	95.90	92.30	87.95	85.00	80.65
Finished Basement, Add	23.60	22.65	22.00	21.05	20.50	20.00	19.00	18.55	18.00	17.75	17.30
Unfinished Basement, Add	9.15	8.75	8.45	8.00	7.80	7.55	7.10	6.90	6.65	6.55	6.35

Modifications

Add to the total cost

Upgrade Kitchen Cabinets	$ + 810
Solid Surface Countertops	+ 1245
Full Bath - including plumbing, wall and floor finishes	+ 4872
Half Bath - including plumbing, wall and floor finishes	+ 2974
Two Car Attached Garage	+ 17,149
Two Car Detached Garage	+ 17,273
Fireplace & Chimney	+ 4935

Adjustments

For multi family - add to total cost

Additional Kitchen	$ + 10,371
Additional Full Bath & Half Bath	+ 7846
Additional Entry & Exit	+ 1486
Separate Heating & Air Conditioning	+ 4054
Separate Electric	+ 1204

For Townhouse/Rowhouse -
Multiply cost per square foot by

Inner Unit	.90
End Unit	.95

Alternatives

Add to or deduct from the cost per square foot of living area

Heavyweight Asphalt Shingles	– 1.35
Clay Tile Roof	+ 1.70
Slate Roof	+ 4.00
Upgrade Ceilings to Textured Finish	+ .43
Air Conditioning, in Heating Ductwork	Base System
In Separate Ductwork	+ 2.39
Heating Systems, Hot Water	+ 2.53
Heat Pump	+ 1.92
Electric Heat	+ .90
Not Heated	– 2.98

Additional upgrades or components

Kitchen Cabinets & Countertops	Page 58
Bathroom Vanities	59
Fireplaces & Chimneys	59
Windows, Skylights & Dormers	59
Appliances	60
Breezeways & Porches	60
Finished Attic	60
Garages	61
Site Improvements	61
Wings & Ells	57

- **Unique residence built from an architect's plan**
- **Single family — 1 full bath, 1 half bath, 1 kitchen**
- **No basement**
- **Cedar shakes on roof**
- **Forced hot air heat/air conditioning**
- **Double drywall interior**
- **Many special features**
- **Extraordinary materials and workmanship**
- **Detail specifications on page 49**

Note: The illustration shown may contain some optional components (for example: garages and/or fireplaces) whose costs are shown in the modifications, adjustments, & alternatives below or at the end of the square foot section.

Base cost per square foot of living area

Exterior Wall	Living Area change										
	1200	1400	1600	1800	2000	2400	2800	3200	3600	4000	4400
Wood Siding - Wood Frame	115.80	109.75	105.45	100.60	96.55	89.75	84.50	80.45	78.00	75.70	73.55
Brick Veneer - Wood Frame	119.45	113.15	108.70	103.70	99.50	92.40	86.95	82.70	80.20	77.80	75.50
Solid Brick	129.85	122.80	117.95	112.40	107.90	99.95	93.85	89.15	86.45	83.60	81.15
Solid Stone	131.75	124.60	119.65	114.00	109.45	101.35	95.10	90.30	87.55	84.75	82.20
Finished Basement, Add	18.95	18.20	17.70	17.15	16.85	16.05	15.40	15.05	14.75	14.45	14.25
Unfinished Basement, Add	7.35	7.00	6.80	6.55	6.40	6.05	5.75	5.60	5.50	5.30	5.25

Modifications

Add to the total cost

Upgrade Kitchen Cabinets	$ + 810
Solid Surface Countertops	+ 1245
Full Bath - including plumbing, wall and floor finishes	+ 4872
Half Bath - including plumbing, wall and floor finishes	+ 2974
Two Car Attached Garage	+ 17,149
Two Car Detached Garage	+ 17,273
Fireplace & Chimney	+ 5400

Adjustments

For multi family - add to total cost

Additional Kitchen	$ + 10,371
Additional Full Bath & Half Bath	+ 7846
Additional Entry & Exit	+ 1486
Separate Heating & Air Conditioning	+ 4054
Separate Electric	+ 1204

For Townhouse/Rowhouse -
Multiply cost per square foot by

Inner Unit	.86
End Unit	.93

Alternatives

Add to or deduct from the cost per square foot of living area

Heavyweight Asphalt Shingles	– .95
Clay Tile Roof	+ 1.20
Slate Roof	+ 2.80
Upgrade Ceilings to Textured Finish	+ .43
Air Conditioning, in Heating Ductwork	Base System
In Separate Ductwork	+ 2.39
Heating Systems, Hot Water	+ 2.33
Heat Pump	+ 2.06
Electric Heat	+ 1.22
Not Heated	– 2.79

Additional upgrades or components

Kitchen Cabinets & Countertops	Page 58
Bathroom Vanities	59
Fireplaces & Chimneys	59
Windows, Skylights & Dormers	59
Appliances	60
Breezeways & Porches	60
Finished Attic	60
Garages	61
Site Improvements	61
Wings & Ells	57

- **Unique residence built from an architect's plan**
- **Single family — 1 full bath, 1 half bath, 1 kitchen**
- **No basement**
- **Cedar shakes on roof**
- **Forced hot air heat/air conditioning**
- **Double drywall interior**
- **Many special features**
- **Extraordinary materials and workmanship**
- **Detail specifications on page 49**

Note: The illustration shown may contain some optional components (for example: garages and/or fireplaces) whose costs are shown in the modifications, adjustments, & alternatives below or at the end of the square foot section.

© Larry W. Garnett & Associates, Inc.

LUXURY

Base cost per square foot of living area

Exterior Wall	Living Area change										
	1500	1800	2100	2500	3000	3500	4000	4500	5000	5500	6000
Wood Siding - Wood Frame	116.35	104.05	97.85	92.65	86.55	81.35	76.40	74.30	71.90	69.50	67.50
Brick Veneer - Wood Frame	120.15	107.55	100.95	95.60	89.25	83.80	78.65	76.45	73.95	71.45	69.40
Solid Brick	130.95	117.50	109.85	104.00	96.85	90.70	85.05	82.55	79.75	76.95	74.65
Solid Stone	132.90	119.35	111.45	105.55	98.25	91.95	86.25	83.65	80.80	77.95	75.70
Finished Basement, Add	15.05	14.40	13.60	13.20	12.55	12.10	11.70	11.45	11.20	11.05	10.85
Unfinished Basement, Add	5.90	5.60	5.25	5.05	4.75	4.55	4.35	4.25	4.15	4.05	4.00

Modifications

Add to the total cost

Upgrade Kitchen Cabinets	$ + 810
Solid Surface Countertops	+ 1245
Full Bath - including plumbing, wall and floor finishes	+ 4872
Half Bath - including plumbing, wall and floor finishes	+ 2974
Two Car Attached Garage	+ 17,149
Two Car Detached Garage	+ 17,273
Fireplace & Chimney	+ 5915

Adjustments

For multi family - add to total cost

Additional Kitchen	$ + 10,371
Additional Full Bath & Half Bath	+ 7846
Additional Entry & Exit	+ 1486
Separate Heating & Air Conditioning	+ 4054
Separate Electric	+ 1204

For Townhouse/Rowhouse - Multiply cost per square foot by

Inner Unit	.86
End Unit	.93

Alternatives

Add to or deduct from the cost per square foot of living area

Heavyweight Asphalt Shingles	– .80
Clay Tile Roof	+ 1.05
Slate Roof	+ 2.40
Upgrade Ceilings to Textured Finish	+ .43
Air Conditioning, in Heating Ductwork	Base System
In Separate Ductwork	+ 2.39
Heating Systems, Hot Water	+ 2.25
Heat Pump	+ 2.09
Electric Heat	+ 1.32
Not Heated	– 2.79

Additional upgrades or components

Kitchen Cabinets & Countertops	Page 58
Bathroom Vanities	59
Fireplaces & Chimneys	59
Windows, Skylights & Dormers	59
Appliances	60
Breezeways & Porches	60
Finished Attic	60
Garages	61
Site Improvements	61
Wings & Ells	57

- **Unique residence built from an architect's plan**
- **Single family — 1 full bath, 1 half bath, 1 kitchen**
- **No basement**
- **Cedar shakes on roof**
- **Forced hot air heat/air conditioning**
- **Double drywall interior**
- **Many special features**
- **Extraordinary materials and workmanship**
- **Detail specifications on page 49**

Note: The illustration shown may contain some optional components (for example: garages and/or fireplaces) whose costs are shown in the modifications, adjustments, & alternatives below or at the end of the square foot section.

LUXURY

Base cost per square foot of living area

Exterior Wall	Living Area change										
	1500	1800	2100	2500	3000	3500	4000	4500	5000	5500	6000
Wood Siding - Wood Frame	116.10	103.85	99.05	95.05	87.75	84.60	79.80	75.05	73.55	71.35	69.65
Brick Veneer - Wood Frame	120.05	107.55	102.45	98.30	90.70	87.40	82.30	77.35	75.80	73.55	71.70
Solid Brick	131.25	117.90	112.10	107.55	99.10	95.35	89.55	84.05	82.25	79.75	77.55
Solid Stone	133.30	119.80	113.90	109.25	100.60	96.80	90.90	85.30	83.40	80.90	78.65
Finished Basement, Add	13.25	12.65	12.15	11.80	11.20	10.90	10.45	10.15	10.00	9.85	9.60
Unfinished Basement, Add	5.15	4.90	4.65	4.50	4.25	4.15	3.90	3.80	3.70	3.65	3.55

Modifications

Add to the total cost

Upgrade Kitchen Cabinets	$ + 810
Solid Surface Countertops	+ 1245
Full Bath - including plumbing, wall and floor finishes	+ 4872
Half Bath - including plumbing, wall and floor finishes	+ 2974
Two Car Attached Garage	+ 17,149
Two Car Detached Garage	+ 17,273
Fireplace & Chimney	+ 5915

Adjustments

For multi family - add to total cost

Additional Kitchen	$ + 10,371
Additional Full Bath & Half Bath	+ 7846
Additional Entry & Exit	+ 1486
Separate Heating & Air Conditioning	+ 4054
Separate Electric	+ 1204

For Townhouse/Rowhouse - Multiply cost per square foot by

Inner Unit	.84
End Unit	.92

Alternatives

Add to or deduct from the cost per square foot of living area

Heavyweight Asphalt Shingles	– .60
Clay Tile Roof	+ .80
Slate Roof	+ 1.85
Upgrade Ceilings to Textured Finish	+ .43
Air Conditioning, in Heating Ductwork	Base System
In Separate Ductwork	+ 2.39
Heating Systems, Hot Water	+ 2.25
Heat Pump	+ 2.09
Electric Heat	+ 1.32
Not Heated	– 2.69

Additional upgrades or components

Kitchen Cabinets & Countertops	Page 58
Bathroom Vanities	59
Fireplaces & Chimneys	59
Windows, Skylights & Dormers	59
Appliances	60
Breezeways & Porches	60
Finished Attic	60
Garages	61
Site Improvements	61
Wings & Ells	57

- **Unique residence built from an architect's plan**
- **Single family — 1 full bath, 1 half bath, 1 kitchen**
- **No basement**
- **Cedar shakes on roof**
- **Forced hot air heat/air conditioning**
- **Double drywall interior**
- **Many special features**
- **Extraordinary materials and workmanship**
- **Detail specifications on page 49**

Note: The illustration shown may contain some optional components (for example: garages and/or fireplaces) whose costs are shown in the modifications, adjustments, & alternatives below or at the end of the square foot section.

Base cost per square foot of living area

Exterior Wall	Living Area change										
	1200	1400	1600	1800	2000	2400	2800	3200	3600	4000	4400
Wood Siding - Wood Frame	108.40	102.80	98.85	94.35	90.55	84.30	79.60	75.80	73.55	71.55	69.45
Brick Veneer - Wood Frame	111.15	105.35	101.25	96.65	92.75	86.25	81.40	77.50	75.20	73.10	70.95
Solid Brick	118.90	112.65	108.20	103.20	99.05	91.95	86.55	82.35	79.85	77.45	75.20
Solid Stone	120.35	113.95	109.45	104.40	100.20	93.00	87.55	83.25	80.75	78.25	75.95
Finished Basement, Add	18.95	18.20	17.70	17.15	16.85	16.05	15.40	15.05	14.75	14.45	14.25
Unfinished Basement, Add	7.35	7.00	6.80	6.55	6.40	6.05	5.75	5.60	5.50	5.30	5.25

Modifications

Add to the total cost

Upgrade Kitchen Cabinets	$ + 810
Solid Surface Countertops	+ 1245
Full Bath - including plumbing, wall and floor finishes	+ 4872
Half Bath - including plumbing, wall and floor finishes	+ 2974
Two Car Attached Garage	+ 17,149
Two Car Detached Garage	+ 17,273
Fireplace & Chimney	+ 4935

Adjustments

For multi family - add to total cost

Additional Kitchen	$ + 10,371
Additional Full Bath & Half Bath	+ 7846
Additional Entry & Exit	+ 1486
Separate Heating & Air Conditioning	+ 4054
Separate Electric	+ 1204

For Townhouse/Rowhouse -
Multiply cost per square foot by

Inner Unit	.89
End Unit	.94

Alternatives

Add to or deduct from the cost per square foot of living area

Heavyweight Asphalt Shingles	– .95
Clay Tile Roof	+ 1.20
Slate Roof	+ 2.80
Upgrade Ceilings to Textured Finish	+ .43
Air Conditioning, in Heating Ductwork	Base System
In Separate Ductwork	+ 2.39
Heating Systems, Hot Water	+ 2.33
Heat Pump	+ 2.05
Electric Heat	+ 1.22
Not Heated	– 2.79

Additional upgrades or components

Kitchen Cabinets & Countertops	Page 58
Bathroom Vanities	59
Fireplaces & Chimneys	59
Windows, Skylights & Dormers	59
Appliances	60
Breezeways & Porches	60
Finished Attic	60
Garages	61
Site Improvements	61
Wings & Ells	57

- **Unique residence built from an architect's plan**
- **Single family — 1 full bath, 1 half bath, 1 kitchen**
- **No basement**
- **Cedar shakes on roof**
- **Forced hot air heat/air conditioning**
- **Double drywall interior**
- **Many special features**
- **Extraordinary materials and workmanship**
- **Detail specifications on page 49**

Note: The illustration shown may contain some optional components (for example: garages and/or fireplaces) whose costs are shown in the modifications, adjustments, & alternatives below or at the end of the square foot section.

LUXURY

© Home Planners, Inc.

Base cost per square foot of living area

Exterior Wall	Living Area change										
	1500	1800	2100	2400	2800	3200	3600	4000	4500	5000	5500
Wood Siding - Wood Frame	102.00	94.50	88.80	84.45	81.25	77.45	73.65	72.15	68.25	66.15	63.80
Brick Veneer - Wood Frame	104.45	96.75	90.85	86.35	83.10	79.15	75.20	73.65	69.60	67.45	65.05
Solid Brick	111.45	103.05	96.60	91.75	88.25	83.95	79.70	77.95	73.60	71.20	68.60
Solid Stone	112.75	104.20	97.65	92.75	89.20	84.80	80.50	78.75	74.35	71.90	69.25
Finished Basement, Add*	22.45	21.35	20.55	19.95	19.65	19.05	18.60	18.40	17.85	17.60	17.30
Unfinished Basement, Add*	8.55	8.10	7.70	7.45	7.30	7.05	6.85	6.70	6.50	6.35	6.25

*Basement under middle level only.

Modifications

Add to the total cost

Upgrade Kitchen Cabinets	$ + 810
Solid Surface Countertops	+ 1245
Full Bath - including plumbing, wall and floor finishes	+ 4872
Half Bath - including plumbing, wall and floor finishes	+ 2974
Two Car Attached Garage	+ 17,149
Two Car Detached Garage	+ 17,273
Fireplace & Chimney	+ 5400

Adjustments

For multi family - add to total cost

Additional Kitchen	$ + 10,371
Additional Full Bath & Half Bath	+ 7846
Additional Entry & Exit	+ 1486
Separate Heating & Air Conditioning	+ 4054
Separate Electric	+ 1204

*For Townhouse/Rowhouse -
Multiply cost per square foot by*

Inner Unit	.86
End Unit	.93

Alternatives

Add to or deduct from the cost per square foot of living area

Heavyweight Asphalt Shingles	− 1.35
Clay Tile Roof	+ 1.70
Slate Roof	+ 4.00
Upgrade Ceilings to Textured Finish	+ .43
Air Conditioning, in Heating Ductwork	Base System
In Separate Ductwork	+ 2.39
Heating Systems, Hot Water	+ 2.18
Heat Pump	+ 2.13
Electric Heat	+ 1.53
Not Heated	− 2.69

Additional upgrades or components

Kitchen Cabinets & Countertops	Page 58
Bathroom Vanities	59
Fireplaces & Chimneys	59
Windows, Skylights & Dormers	59
Appliances	60
Breezeways & Porches	60
Finished Attic	60
Garages	61
Site Improvements	61
Wings & Ells	57

1 Story — Base cost per square foot of living area

Exterior Wall	Living Area							
	50	100	200	300	400	500	600	700
Wood Siding - Wood Frame	167.95	128.50	111.35	92.80	87.15	83.80	81.55	82.30
Brick Veneer - Wood Frame	177.15	135.10	116.80	96.45	90.45	86.90	84.50	85.10
Solid Brick	203.25	153.75	132.40	106.85	99.80	95.60	92.75	93.10
Solid Stone	208.10	157.15	135.25	108.75	101.50	97.20	94.30	94.60
Finished Basement, Add	59.15	47.70	42.90	34.90	33.30	32.30	31.70	31.25
Unfinished Basement, Add	25.55	19.95	17.65	13.75	12.95	12.50	12.20	12.00

1-1/2 Story — Base cost per square foot of living area

Exterior Wall	Living Area							
	100	200	300	400	500	600	700	800
Wood Siding - Wood Frame	136.70	110.60	94.70	84.85	80.00	77.65	74.75	73.90
Brick Veneer - Wood Frame	144.90	117.15	100.15	89.10	83.95	81.40	78.25	77.35
Solid Brick	168.20	135.80	115.70	101.25	95.15	91.95	88.25	87.30
Solid Stone	172.50	139.25	118.60	103.50	97.15	93.90	90.10	89.15
Finished Basement, Add	39.10	34.30	31.10	27.60	26.65	26.00	25.40	25.30
Unfinished Basement, Add	16.45	14.10	12.55	10.85	10.35	10.05	9.80	9.75

2 Story — Base cost per square foot of living area

Exterior Wall	Living Area							
	100	200	400	600	800	1000	1200	1400
Wood Siding - Wood Frame	137.30	102.45	87.20	71.90	66.85	63.90	61.90	62.80
Brick Veneer - Wood Frame	146.50	109.05	92.70	75.55	70.15	67.00	64.80	65.65
Solid Brick	172.60	127.70	108.25	85.90	79.50	75.70	73.05	73.65
Solid Stone	177.45	131.10	111.10	87.80	81.20	77.25	74.60	75.10
Finished Basement, Add	29.60	23.85	21.45	17.45	16.65	16.20	15.90	15.60
Unfinished Basement, Add	12.80	10.00	8.80	6.90	6.50	6.25	6.10	6.00

Base costs do not include bathroom or kitchen facilities. Use Modifications/Adjustments/Alternatives on pages 58-61 where appropriate.

Important: See the Reference Section for Location Factors (to adjust for your city) and Estimating Forms.

Kitchen cabinets - Base units, hardwood *(Cost per Unit)*

	Economy	Average	Custom	Luxury
24" deep, 35" high,				
One top drawer,				
one door below				
12" wide	$107	$143	$190	$250
15" wide	108	144	190	250
18" wide	122	162	215	285
21" wide	129	172	230	300
24" wide	131	174	230	305
four drawers				
12" wide	107	142	190	250
15" wide	133	177	235	310
18" wide	143	191	255	335
24" wide	160	213	285	375
Two top drawers,				
two doors below				
27" wide	159	212	280	370
30" wide	173	231	305	405
33" wide	175	233	310	410
36" wide	182	243	325	425
42" wide	198	264	350	460
48" wide	215	286	380	500
Range or sink base				
Two doors below				
30" wide	137	182	240	320
33" wide	143	190	255	335
36" wide	148	197	260	345
42" wide	161	214	285	375
48" wide	173	231	305	405
Corner Base Cabinet				
36" wide	152	202	270	355
Lazy Susan				
With revolving door	186	248	330	435

Kitchen cabinets - Wall cabinets, hardwood *(Cost per Unit)*

	Economy	Average	Custom	Luxury
12" deep, 2 doors				
12" high				
30" wide	$ 88	$117	$155	$205
36" wide	96	128	170	225
15" high				
30" wide	91	121	160	210
33" wide	92	122	160	215
36" wide	99	132	175	230
24" high				
30" wide	112	149	200	260
36" wide	120	160	215	280
42" wide	127	169	225	295
30" high, 1 door				
12" wide	86	114	150	200
15" wide	84	112	150	195
18" wide	95	126	170	220
24" wide	107	143	190	250
30" high, 2 doors				
27" wide	116	155	205	270
30" wide	128	171	225	300
36" wide	140	186	245	325
42" wide	151	201	265	350
48" wide	162	216	285	380
Corner wall, 30" high				
24" wide	122	162	215	285
30" wide	152	203	270	355
36" wide	173	231	305	405
Broom closet				
84" high, 24" deep				
18" wide	223	297	395	520
Oven Cabinet				
84" high, 24" deep				
27" wide	285	380	505	665

Kitchen countertops *(Cost per L.F.)*

	Economy	Average	Custom	Luxury
Solid Surface				
24" wide, no backsplash	62	83	110	145
with backsplash	69	93	125	160
Stock plastic laminate, 24" wide				
with backsplash	10	14	20	25
Custom plastic laminate, no splash				
7/8" thick, alum. molding	19	26	35	45
1-1/4" thick, no splash	22	30	40	50
Marble				
1/2" - 3/4" thick w/splash	36	48	65	85
Maple, laminated				
1-1/2" thick w/splash	36	48	65	85
Stainless steel				
(per S.F.)	68	91	120	160
Cutting blocks, recessed				
16" x 20" x 1" (each)	58	77	100	135

Vanity bases (Cost per Unit)

2 door, 30" high, 21" deep	Economy	Average	Custom	Luxury
24" wide	118	157	210	275
30" wide	128	170	225	300
36" wide	144	192	255	335
48" wide	180	240	320	420

Solid surface vanity tops (Cost Each)

Center bowl	Economy	Average	Custom	Luxury
22" x 25"	$270	$365	$511	$552
22" x 31"	310	419	587	634
22" x 37"	355	479	671	725
22" x 49"	445	601	841	908

Fireplaces & Chimneys (Cost per Unit)

	1-1/2 Story	2 Story	3 Story
Economy (prefab metal)			
Exterior chimney & 1 fireplace	$3280	$3625	$3965
Interior chimney & 1 fireplace	3140	3490	3660
Average (masonry)			
Exterior chimney & 1 fireplace	3285	3665	4170
Interior chimney & 1 fireplace	3080	3450	3760
For more than 1 flue, add	245	400	675
For more than 1 fireplace, add	2330	2330	2330
Custom (masonry)			
Exterior chimney & 1 fireplace	3490	3940	4450
Interior chimney & 1 fireplace	3275	3705	3995
For more than 1 flue, add	275	470	645
For more than 1 fireplace, add	2505	2505	2505
Luxury (masonry)			
Exterior chimney & 1 fireplace	4935	5400	5915
Interior chimney & 1 fireplace	4710	5150	5445
For more than 1 flue, add	405	680	950
For more than 1 fireplace, add	3885	3885	3885

Windows and Skylights (Cost Each)

	Economy	Average	Custom	Luxury
Fixed Picture Windows				
3'-6" x 4'-0"	$ 183	$ 229	$ 265	$ 380
4'-0" x 6'-0"	318	398	460	660
5'-0" x 6'-0"	398	497	575	845
6'-0" x 6'-0"	480	601	695	1175
Bay/Bow Windows				
8'-0" x 5'-0"	880	1000	1125	1475
10'-0" x 5'-0"	1150	1225	1775	1950
10'-0" x 6'-0"	1328	1415	2050	2225
12'-0" x 6'-0"	1506	1605	2325	2800
Palladian Windows				
3'-2" x 6'-4"		1238	1375	1602
4'-0" x 6'-0"		1260	1400	1631
5'-5" x 6'-10"		1575	1750	2039
8'-0" x 6'-0"		1958	2175	2534
Skylights				
46" x 21-1/2"	391	434	564	818
46" x 28"	416	462	597	866
57" x 44"	504	560	715	1037

Dormers (Cost/S.F. of plan area)

	Economy	Average	Custom	Luxury
Framing and Roofing Only				
Gable dormer, 2" x 6" roof frame	$17	$19	$22	$35
2" x 8" roof frame	18	20	23	37
Shed dormer, 2" x 6" roof frame	11	12	14	22
2" x 8" roof frame	12	13	15	23
2" x 10" roof frame	13	14	16	24

ADJUSTMENTS

ADJUSTMENTS

Appliances *(Cost per Unit)*

	Economy	Average	Custom	Luxury
Range				
30" free standing, 1 oven	$ 365	$ 724	$1134	$1250
2 oven	770	1060	1465	1675
30" built-in, 1 oven	465	649	844	1400
2 oven	1000	1230	1485	1900
21" free standing				
1 oven	365	421	487	540
Counter Top Ranges				
4 burner standard	244	335	454	595
As above with griddle	405	509	623	610
Microwave Oven	175	525	1199	1800
Combination Range,				
Refrigerator, Sink				
30" wide	865	1222	1487	1725
60" wide	2500	2744	3081	3168
72" wide	2825	2983	3464	3272
Comb. Range, Refrig., Sink,				
Microwave Oven & Ice Maker	3960	4566	4935	5847
Compactor				
4 to 1 compaction	380	462	544	545
Deep Freeze				
15 to 23 C.F.	470	500	662	780
30 C.F.	825	912	1005	1025
Dehumidifier, portable, auto.				
15 pint	198	209	235	246
30 pint	248	261	275	287
Washing Machine, automatic	455	747	1045	1250
Water Heater				
Electric, glass lined				
30 gal.	223	294	370	475
80 gal.	440	660	857	950
Water Heater, Gas, glass lined				
30 gal.	315	400	600	695
50 gal.	460	676	817	905
Water Softener, automatic				
30 grains/gal.	605	634	672	847
100 grains/gal.	745	796	820	824
Dishwasher, built-in				
2 cycles	325	445	530	645
4 or more cycles	360	513	662	880
Dryer, automatic	445	558	783	895
Garage Door Opener	245	296	332	350
Garbage Disposal	75	122	173	204
Heater, Electric, built-in				
1250 watt ceiling type	117	159	195	200
1250 watt wall type	112	125	141	189
Wall type w/blower				
1500 watt	167	179	187	205
3000 watt	233	257	269	296
Hood For Range, 2 speed				
30" wide	105	204	321	375
42" wide	242	314	374	380
Humidifier, portable				
7 gal. per day	99	103	109	120
15 gal. per day	160	170	179	197
Ice Maker, automatic				
13 lb. per day	580	615	651	691
51 lb. per day	1075	1182	1223	1271
Refrigerator, no frost				
10-12 C.F.	490	642	814	745
14-16 C.F.	560	653	847	820
18-20 C.F.	645	885	1141	1400
21-29 C.F.	875	1593	2402	2625
Sump Pump, 1/3 H.P.	169	303	414	480

Breezeway *(Cost per S.F.)*

Class	Type	Area (S.F.)			
		50	100	150	200
Economy	Open	$ 14.45	$12.35	$10.35	$10.15
	Enclosed	69.45	53.70	44.55	39.05
Average	Open	18.40	16.20	14.15	12.85
	Enclosed	78.05	58.10	47.55	41.85
Custom	Open	26.35	23.15	20.20	18.50
	Enclosed	108.20	80.50	65.80	57.85
Luxury	Open	28.35	24.85	22.50	21.70
	Enclosed	114.20	84.70	68.50	61.35

Porches *(Cost per S.F.)*

Class	Type	Area (S.F.)				
		25	50	100	200	300
Economy	Open	$ 42.75	$ 28.65	$22.35	$18.95	$16.15
	Enclosed	85.50	59.55	45.05	35.10	30.10
Average	Open	52.20	33.30	25.50	21.25	21.25
	Enclosed	103.15	69.85	52.90	40.95	34.65
Custom	Open	69.00	45.80	34.50	30.10	27.00
	Enclosed	137.20	93.75	71.35	55.50	47.90
Luxury	Open	74.45	48.65	35.95	32.30	28.70
	Enclosed	145.70	102.35	75.95	58.95	50.80

Finished attic *(Cost per S.F.)*

Class	Area (S.F.)				
	400	500	600	800	1000
Economy	$11.45	$11.00	$10.55	$10.40	$10.10
Average	17.60	17.10	16.75	16.50	16.10
Custom	21.65	21.15	20.75	20.40	20.00
Luxury	28.00	27.40	26.75	26.15	25.65

Alarm system *(Cost per System)*

	Burglar Alarm	Smoke Detector
Economy	$ 340	$ 51
Average	395	63
Custom	663	126
Luxury	1025	156

Sauna, prefabricated
(Cost per unit, including heater and controls—7' high)

Size	Cost
6' x 4'	$3550
6' x 5'	3875
6' x 6'	4175
6' x 9'	5600
8' x 10'	6500
8' x 12'	7200
10' x 12'	8500

Garages*

(Costs include exterior wall systems comparable with the quality of the residence. Included in the cost is an allowance for one door, manual overhead door(s) and electrical fixture.)

Class	Type									
	Detached			Attached			Built-in		Basement	
	One Car	Two Car	Three Car	One Car	Two Car	Three Car	One Car	Two Car	One Car	Two Car
Economy										
Wood	$ 6890	$ 9391	$12,889	$ 6364	$ 9203	$13,353	$-1247	$-2123	$ 612	$1018
Masonry	9072	12,061	15,989	7045	10,342	14,246	-1439	-2786		
Average										
Wood	7587	10,241	14,459	7074	10,039	14,194	-1287	-2412	664	1258
Masonry	9573	12,701	15,869	8285	10,970	14,794	-1471	-3254		
Custom										
Wood	8909	12,064	16,406	8167	11,601	15,931	-1437	-2662	822	1571
Masonry	11,495	14,347	19,213	9289	12,309	17,332	-1810	-3566		
Luxury										
Wood	14,084	17,273	21,540	13,442	17,149	21,409	-1630	-3023	3701	5457
Masonry	20,602	25,975	31,123	18,227	23,771	29,203	-2144	-4093		

*See the Introduction to this section for definitions of garage types.

Swimming pools (Cost per S.F.)

Residential (includes equipment)		
In-ground		$ 18-43
Deck equipment		1.30
Paint pool, preparation & 3 coats (epoxy)		2.53
Rubber base paint		2.34
Pool Cover		.17
Swimming Pool Heaters (Cost per unit)		
(not including wiring, external piping, base or pad)		
Gas		
155 MBH		2425
190 MBH		3275
500 MBH		7950
Electric		
15 KW 7200 gallon pool		1900
24 KW 9600 gallon pool		2500
54 KW 24,000 gallon pool		3750

Wood and coal stoves

Wood Only		
Free Standing (minimum)		$ 1075
Fireplace Insert (minimum)		1235
Coal Only		
Free Standing		1395
Fireplace Insert		1528
Wood and Coal		
Free Standing		2870
Fireplace Insert		2940

Sidewalks (Cost per S.F.)

Concrete, 3000 psi with wire mesh	4" thick	$ 2.18
	5" thick	2.64
	6" thick	2.95
Precast concrete patio blocks (natural)	2" thick	7.50
Precast concrete patio blocks (colors)	2" thick	7.64
Flagstone, bluestone	1" thick	7.58
Flagstone, bluestone	1-1/2" thick	8.95
Slate (natural, irregular)	3/4" thick	6.88
Slate (random rectangular)	1/2" thick	8.51
Seeding		
Fine grading & seeding includes lime, fertilizer & seed	per S.Y.	1.60
Lawn Sprinkler System	per S.F.	.47

Fencing (Cost per L.F.)

Chain Link, 4' high, galvanized	$ 9.70
Gate, 4' high (each)	109.00
Cedar Picket, 3' high, 2 rail	8.35
Gate (each)	106.00
3 Rail, 4' high	9.30
Gate (each)	114.00
Cedar Stockade, 3 Rail, 6' high	9.50
Gate (each)	113.00
Board & Battens, 2 sides 6' high, pine	13.15
6' high, cedar	20.00
No. 1 Cedar, basketweave, 6' high	10.25
Gate, 6' high (each)	122.50

Carport (Cost per S.F.)

Economy	5.38
Average	8.20
Custom	12.48
Luxury	14.29

ADJUSTMENTS

Insurance Exclusions

Insurance exclusions are a matter of policy coverage and are not standard or universal. When making an appraisal for insurance purposes, it is recommended that some time be taken in studying the insurance policy to determine the specific items to be excluded. Most homeowners insurance policies have, as part of their provisions, statements that read "the policy permits the insured to exclude from value, in determining whether the amount of the insurance equals or exceeds 80% of its replacement cost, such items as building excavation, basement footings, underground piping and wiring, piers and other supports which are below the undersurface of the lowest floor or where there is no basement below the ground." Loss to any of these items, however, is covered.

Costs for Excavation, Spread and Strip Footings and Underground Piping

This chart shows excluded items expressed as a percentage of total building cost.

Class	Number of Stories			
	1	1-1/2	2	3
Luxury	3.4%	2.7%	2.4%	1.9%
Custom	3.5%	2.7%	2.4%	1.9%
Average	5.0%	4.3%	3.8%	3.3%
Economy	5.0%	4.4%	4.1%	3.5%

Architect/Designer Fees

Architect/designer fees as presented in the following chart are typical ranges for 4 classes of residences. Factors affecting these ranges include economic conditions, size and scope of project, site selection, and standardization. Where superior quality and detail is required or where closer supervision is required, fees may run higher than listed. Lower quality or simplicity may dictate lower fees.

The editors have included architect/designers fees from the "mean" column in calculating costs.

Listed below are average costs expressed as a range and mean of total building cost for 4 classes of residences.

Class		Range	Mean
Luxury	— Architecturally designed and supervised	5%–15%	10%
Custom	— Architecturally modified designer plans	$1500–$2000	$1800
Average	— Designer plans	$800–$1200	$1000
Economy	— Stock plans	$25–$400	$250

Depreciation as generally defined is "the loss of value due to any cause." Specifically, depreciation can be broken down into three categories: **Physical Depreciation, Functional Obsolescence,** and **Economic Obsolescence.**

Physical Depreciation is the loss of value from the wearing out of the building's components. Such causes may be decay, dry rot, cracks, or structural defects.

Functional Obsolescence is the loss of value from features which render the residence less useful or desirable. Such features may be higher than normal ceilings, design and/or style changes, or outdated mechanical systems.

Economic Obsolescence is the loss of value from external features which render the residence less useful or desirable. These features may be changes in neighborhood socio-economic grouping, zoning changes, legislation, etc.

Depreciation as it applies to the residential portion of this manual deals with the observed physical condition of the residence being appraised. It is in essence *"the cost to cure."*

The ultimate method to arrive at *"the cost to cure"* is to analyze each component of the residence.

For example:

> Component, Roof Covering − Cost = $1800
> Average Life = 15 years
> Actual Life = 5 years
> Depreciation = 5 ÷ 15 = .33 or 33%
>
> $1800 × 33% = $594 = Amount of Depreciation

The following table, however, can be used as a guide in estimating the % depreciation using the actual age and general condition of the residence.

Depreciation Table — Residential

Age in Years	Good	Average	Poor
2	2%	3%	10%
5	4	6	20
10	7	10	25
15	10	15	30
20	15	20	35
25	18	25	40
30	24	30	45
35	28	35	50
40	32	40	55
45	36	45	60
50	40	50	65

ADJUSTMENTS

Commercial/Industrial/ Institutional Section

Table of Contents

Introduction to the Commercial/ Industrial/Institutional Section

General

The Commercial/Industrial/Institutional section of this manual contains base building costs per square foot of floor area for 70 model buildings. Each model has a table of square foot costs for combinations of exterior wall and framing systems. This table is supplemented by a list of common additives and their unit costs. A breakdown of the component costs used to develop the base cost for the model is on the opposite page. The total cost derived from the base cost and additives must be modified using the appropriate location factor from the Reference Section.

This section may be used directly to estimate the construction cost of most types of buildings knowing only the floor area, exterior wall construction and framing systems. To adjust the base cost for components which are different than the model, use the tables from the Assemblies Section.

Building Identification & Model Selection

The building models in this section represent structures by use. Occupancy, however, does not necessarily identify the building, i.e., a restaurant could be a converted warehouse. In all instances, the building should be described and identified by its own physical characteristics. The model selection should also be guided by comparing specifications with the model. In the case of converted use, data from one model may be used to supplement data from another.

Adjustments

The base cost tables represent the base cost per square foot of floor area for buildings without a basement and without unusual special features. Basement costs and other common additives are listed below the base cost table. Cost adjustments can also be made to the model by using the tables from the Assemblies Section. This table is for example only.

Dimensions

All base cost tables are developed so that measurement can be readily made during the inspection process. Areas are calculated from exterior dimensions and story heights are measured from the top surface of one floor to the top surface of the floor above. Roof areas are measured by horizontal area covered and costs related to inclines are converted with appropriate factors. The precision of measurement is a matter of the user's choice and discretion. For ease in calculation, consideration should be given to measuring in tenths of a foot, i.e., 9 ft. 6 in. = 9.5 ft., 9 ft. 4 in. = 9.3 ft.

Floor Area

The expression "Floor Area" as used in this section includes the sum of floor areas at grade level and above. Basement costs are calculated separately. The user must exercise his own judgment, where the lowest level floor is slightly below grade, whether to consider it at grade level or make the basement adjustment.

How to Use the Commercial/ Industrial/Institutional Section

The following is a detailed explanation of a sample entry in the Commercial/Industrial/Institutional Square Foot Cost Section. Each bold number below corresponds to the item being described on the following page with the appropriate component or cost of the sample entry following in parenthesis.

Prices listed are costs that include overhead and profit of the installing contractor and additional mark-ups for General Conditions and Architects' Fees.

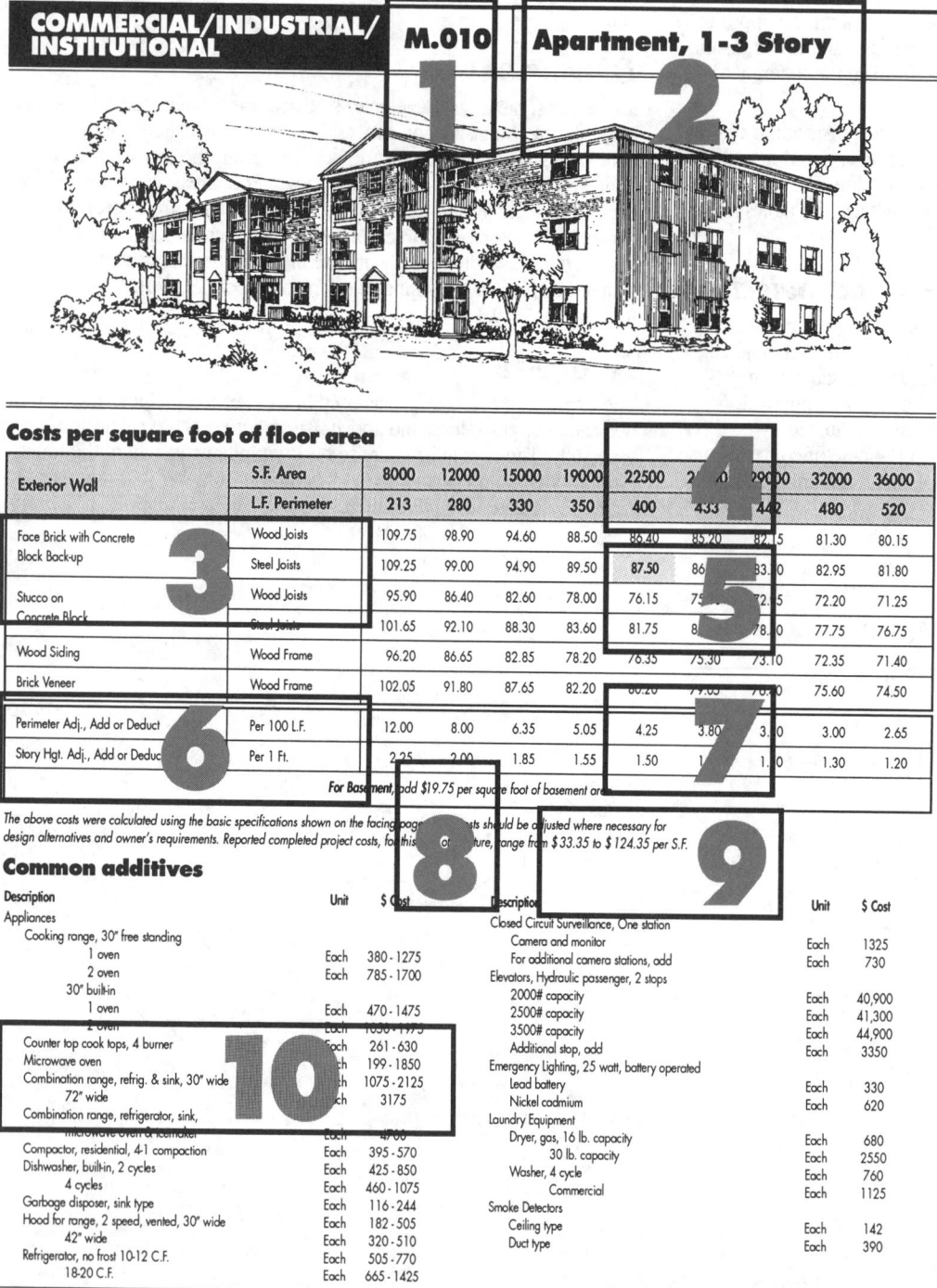

COMMERCIAL/INDUSTRIAL/ INSTITUTIONAL — **1** M.010 — **2** Apartment, 1-3 Story

Costs per square foot of floor area

Exterior Wall	S.F. Area	8000	12000	15000	19000	22500	26000	29000	32000	36000
	L.F. Perimeter	213	280	330	350	400	433	442	480	520
Face Brick with Concrete Block Back-up	Wood Joists	109.75	98.90	94.60	88.50	86.40	85.20	82.15	81.30	80.15
	Steel Joists	109.25	99.00	94.90	89.50	87.50	86.30	83.30	82.95	81.80
Stucco on Concrete Block	Wood Joists	95.90	86.40	82.60	78.00	76.15	75.00	72.15	72.20	71.25
	Steel Joists	101.65	92.10	88.30	83.60	81.75	80.50	78.30	77.75	76.75
Wood Siding	Wood Frame	96.20	86.65	82.85	78.20	76.35	75.30	73.10	72.35	71.40
Brick Veneer	Wood Frame	102.05	91.80	87.65	82.20	80.20	79.05	76.10	75.60	74.50
Perimeter Adj., Add or Deduct	Per 100 L.F.	12.00	8.00	6.35	5.05	4.25	3.80	3.40	3.00	2.65
Story Hgt. Adj., Add or Deduct	Per 1 Ft.	2.25	2.00	1.85	1.55	1.50	1.40	1.30	1.30	1.20

For Basement, add $19.75 per square foot of basement area

The above costs were calculated using the basic specifications shown on the facing page. These costs should be adjusted where necessary for design alternatives and owner's requirements. Reported completed project costs, for this type of structure, range from $33.35 to $124.35 per S.F.

Common additives

Description	Unit	$ Cost		Description	Unit	$ Cost
Appliances				Closed Circuit Surveillance, One station		
Cooking range, 30" free standing				Camera and monitor	Each	1325
1 oven	Each	380 - 1275		For additional camera stations, add	Each	730
2 oven	Each	785 - 1700		Elevators, Hydraulic passenger, 2 stops		
30" built-in				2000# capacity	Each	40,900
1 oven	Each	470 - 1475		2500# capacity	Each	41,300
2 oven	Each	1030 - 1975		3500# capacity	Each	44,900
Counter top cook tops, 4 burner	Each	261 - 630		Additional stop, add	Each	3350
Microwave oven	Each	199 - 1850		Emergency Lighting, 25 watt, battery operated		
Combination range, refrig. & sink, 30" wide	Each	1075 - 2125		Lead battery	Each	330
72" wide	Each	3175		Nickel cadmium	Each	620
Combination range, refrigerator, sink,				Laundry Equipment		
microwave oven & icemaker	Each	4700		Dryer, gas, 16 lb. capacity	Each	680
Compactor, residential, 4-1 compaction	Each	395 - 570		30 lb. capacity	Each	2550
Dishwasher, built-in, 2 cycles	Each	425 - 850		Washer, 4 cycle	Each	760
4 cycles	Each	460 - 1075		Commercial	Each	1125
Garbage disposer, sink type	Each	116 - 244		Smoke Detectors		
Hood for range, 2 speed, vented, 30" wide	Each	182 - 505		Ceiling type	Each	142
42" wide	Each	320 - 510		Duct type	Each	390
Refrigerator, no frost 10-12 C.F.	Each	505 - 770				
18-20 C.F.	Each	665 - 1425				

Important: See the Reference Section for Location Factors

1 Model Number (M.010)

"M" distinguishes this section of the book and stands for model. The number designation is a sequential number.

2 Type of Building (Apartment, 1-3 Story)

There are 43 different types of commercial/industrial/institutional buildings highlighted in this section.

3 Exterior Wall Construction and Building Framing Options (Face Brick with Concrete Block Back-up and Open Web Steel Bar Joists)

Three or more commonly used exterior walls, and in most cases, two typical building framing systems are presented for each type of building. The model selected should be based on the actual characteristics of the building being estimated.

4 Total Square Foot of Floor Area and Base Perimeter Used to Compute Base Costs (22,500 Square Feet and 400 Linear Feet)

Square foot of floor area is the total gross area of all floors at grade, and above, and does not include a basement. The perimeter in linear feet used for the base cost is for a generally rectangular economical building shape.

5 Cost per Square Foot of Floor Area ($87.50)

The highlighted cost is for a building of the selected exterior wall and framing system and floor area. Costs for buildings with floor areas other than those calculated may be interpolated between the costs shown.

6 Building Perimeter and Story Height Adjustments

Square foot costs for a building with a perimeter or floor to floor story height significantly different from the model used to calculate the base cost may be adjusted, add or deduct, to reflect the actual building geometry.

7 Cost per Square Foot of Floor Area for the Perimeter and/or Height Adjustment ($4.25 for Perimeter Difference and $1.50 for Story Height Difference)

Add (or deduct) $4.25 to the base square foot cost for each 100 feet of perimeter difference between the model and the actual building. Add (or deduct) $1.50 to the base square foot cost for each 1 foot of story height difference between the model and the actual building.

8 Optional Cost per Square Foot of Basement Floor Area ($19.75)

The cost of an unfinished basement for the building being estimated is $19.75 times the gross floor area of the basement.

9 Range of Cost per Square Foot of Floor Area for Similar Buildings ($33.35 to $124.35)

Many different buildings of the same type have been built using similar materials and systems. Means historical cost data of actual construction projects indicates a range of $33.35 to $124.35 for this type of building.

10 Common Additives

Common components and/or systems used in this type of building are listed. These costs should be added to the total building cost. Additional selections may be found in the Assemblies Section.

How to Use the Commercial/Industrial/Institutional Section (Continued)

The following is a detailed explanation of a specification and costs for a model building in the Commercial/Industrial/Institutional Square Foot Cost Section. Each bold number below corresponds to the item being described on the following page with the appropriate component of the sample entry following in parenthesis.

Prices listed are costs that include overhead and profit of the installing contractor.

Model costs calculated for a 3 story building with 10' story height and 22,500 square feet of floor area **1**

2 **Apartment, 1-3 Story**

				Unit	Unit Cost	Cost Per S.F.	% Of Sub-Total
1.0 Foundations							
	.1	Footings & Foundations	Poured concrete; strip and spread footings and 4' foundation wall	S.F. Ground	6.39	2.13	
	.4	Piles & Coissons	N/A	—	—	—	3.5%
	.9	Excavation & Backfill	Site preparation for slab and trench for foundation wall and footing	S.F. Ground	1.04	.35	
2.0 Substructure							
	.1	Slab on Grade	4" reinforced concrete with vapor barrier and granular base	S.F. Slab	3.06	1.02	1.4%
	.2	Special Substructures	N/A	—	—	—	
3.0 Superstructure							
	.1	Columns & Beams	Gypsum board fireproofing on columns; steel columns in 3.5 and 3.7	S.F. Floor	.82	.82	
	.4	Structural Walls	N/A	—	—	—	
	.5	Elevated Floors	Open web steel joists, slab form, concrete, interior steel columns	S.F. Floor	9.44	6.29	13.8%
	.7	Roof	Open web steel joists with rib metal deck, interior steel columns	S.F. Roof	4.39	1.46	
	.9	Stairs	Concrete filled metal pan	Flight	5025	1.12	
4.0 Exterior Closure							
	.1	Walls	Face brick with concrete block backup	88% of wall	S.F. Wall	15.04	7.06	
	.5	Exterior Wall Finishes	N/A	—	—	—	
	.6	Doors	Aluminum and glass	Each	1078	.19	12.0%
	.7	Windows & Glazed Walls	Aluminum horizontal sliding	12% of wall	Each	284	1.21	
5.0 Roofing							
	.1	Roof Coverings	Built-up tar and gravel with flashing	S.F. Roof	2.58	.86	
	.7	Insulation	Perlite/EPS composite	Roof	1.25		.8%
	.8	Openings & Specialties	N/A	—	—	
6.0 Interior Construction							
	.1	Partitions	Gypsum board and sound deadening board on metal studs	10 S.F. of Floor/L.F. Partition	S.F. Partition	3.23	2.69	
	.4	Interior Doors	15% solid core wood, 85% hollow core wood	80 S.F. Floor/Door	Each	368	4.62	
	.5	Wall Finishes	70% paint, 25% vinyl wall covering, 5% ceramic tile		S.F. Surface	.96	1.70	24.4%
	.6	Floor Finishes	60% carpet, 30% vinyl composition tile, 10% ceramic tile		S.F. Floor		4.02	
	.7	Ceiling Finishes	Painted gypsum board on resilient channels		S.F. Ceiling		2.73	
	.9	Interior Surface/Exterior Wall	Painted gypsum board on furring	80% of wall	S.F. Wall		1.22	
7.0 Conveying							
	.1	Elevators	One hydraulic passenger elevator	Each	60,750	2.70	
	.2	Special Conveyors	N/A	—	—	—	
8.0 Mechanical							
	.1	Plumbing	Kitchen, bathroom and service fixtures, supply and drainage	1 Fixture/200 S.F. Floor	Each	1774	8.87	
	.2	Fire Protection	Wet pipe sprinkler system	S.F. Floor	1.65	1.65	
	.3	Heating	Oil fired hot water, baseboard radiation	S.F. Floor	4.58	4.58	29.8%
	.4	Cooling	Chilled water, air cooled condenser system	S.F. Floor	5.86	5.86	
	.5	Special Systems	N/A	—	—	—	
9.0 Electrical							
	.1	Service & Distribution	400 ampere service, panel board and feeders	S.F. Floor	.78	.78	
	.2	Lighting & Power	Incandescent fixtures, receptacles, switches, A.C. and misc. power	S.F. Floor	4.12	4.12	7.8%
	.4	Special Electrical	Alarm systems and emergency lighting	S.F. Floor	.61	.61	
11.0 Special Construction							
	.1	Specialties	Kitchen cabinets	S.F. Floor	1.19	1.19	1.7%
12.0 Site Work							
	.1	Earthwork	N/A				
	.3	Utilities	N/A				0.0%
	.5	Roads & Parking	N/A				
	.7	Site Improvements	N/A				
					Sub-Total	70.45	100%
		GENERAL CONDITIONS (Overhead & Profit)			15%	10.57	
		ARCHITECT FEES			8%	6.48	
					Total Building Cost	87.50	

6 **8** **3** **4** **5** **7** **9** **10** **11** **12**

1 *Building Description*
(Model costs are calculated for a three-story apartment building with 10' story height and 22,500 square feet of floor area)

The model highlighted is described in terms of building type, number of stories, typical story height and square footage.

2 *Type of Building*
(Apartment, 1-3 Story)

3 *Division 6.0 Interior Construction*
(6.4 Interior Doors)

System costs are presented in divisions according to the 12-division UniFormat classifications. Each of the component systems are listed.

4 *Specification Highlights*
(15% solid core wood; 85% hollow core wood)

All systems in each subdivision are described with the material and proportions used.

5 *Quality Criteria*
(80 S.F. Floor/Door)

The criteria used in determining quantities for the calculations are shown.

6 *Unit (Each)*

The unit of measure shown in this column is the unit of measure of the particular system shown that corresponds to the unit cost.

7 *Unit Cost ($368)*

The cost per unit of measure of each system subdivision.

8 *Cost per Square Foot ($4.62)*

The cost per square foot for each system is the unit cost of the system times the total number of units divided by the total square feet of building area.

9 *% of Sub-Total (24.4%)*

The percent of sub-total is the total cost per square foot of all systems in the division divided by the sub-total cost per square foot of the building.

10 *Sub-Total ($70.45)*

The sub-total is the total of all the system costs per square foot.

11 *General Conditions*
(Overhead & Profit) (15%)
(Architects' Fees) (8%)

General Conditions to cover the overhead and profit of the General Contractor are added as a percentage of the sub-total. An Architect's Fee, also as a percentage of the sub-total, is also added. These values vary with the building type.

12 *Total Building Cost ($87.50)*

The total building cost per square foot of building area is the sum of the square foot costs of all the systems plus the General Contractor's overhead and profit and the Architect's fee. The total building cost is the amount which appears shaded in the Cost per Square Foot of Floor Area table shown previously.

Examples

Example 1

This example illustrates the use of the base cost tables. The base cost is adjusted for different exterior wall systems, different story height and a partial basement.

CII APPRAISAL FIELD DATA FORM

COMMERCIAL
INDUSTRIAL
INSTITUTIONAL

| Use sheets A & D for |
| App. Form 2 |
| Use sheets A,B & C for |
| App. Form 3 |

CII APPRAISAL

5. DATE: **Jan. 2, 1998**

1. SUBJECT PROPERTY: **Westernberg Holdings, Inc.**

6. APPRAISER: **SJS**

2. BUILDING: **#623**

3. ADDRESS: **55 Pine wood Rd., Bondsville, N.J. 07410**

4. BUILDING USE: **Apartments**

7. YEAR BUILT: **1972**

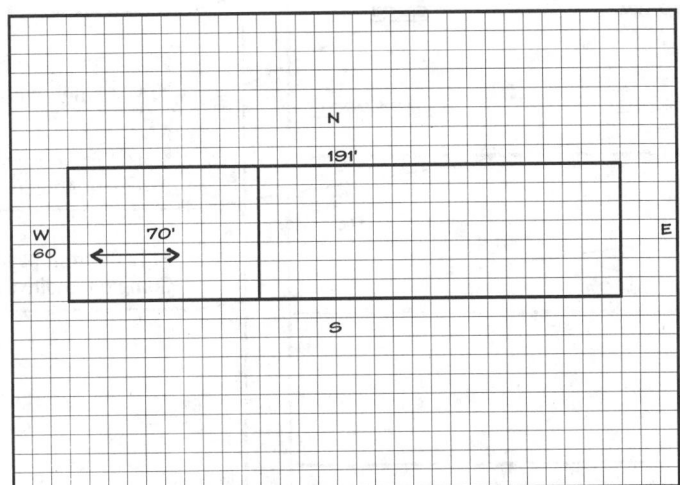

North wall - Face brick w/concrete block back up

8. EXTERIOR WALL CONSTRUCTION: **East, West & South wall - Decorative concrete block**

9. FRAME: **Steel**

10. GROUND FLOOR AREA: **60 x 191   11,460** S.F.

11. GROSS FLOOR AREA (EXCL. BASEMENT): **80, 220 (7 Floors)** S.F.

12. NUMBER OF STORIES: **7**

13. STORY HEIGHT: **11' - 4"**

14. PERIMETER: **502** L.F.

15. BASEMENT AREA: **4200'** S.F.

16. GENERAL COMMENTS:

Example 1 (continued)

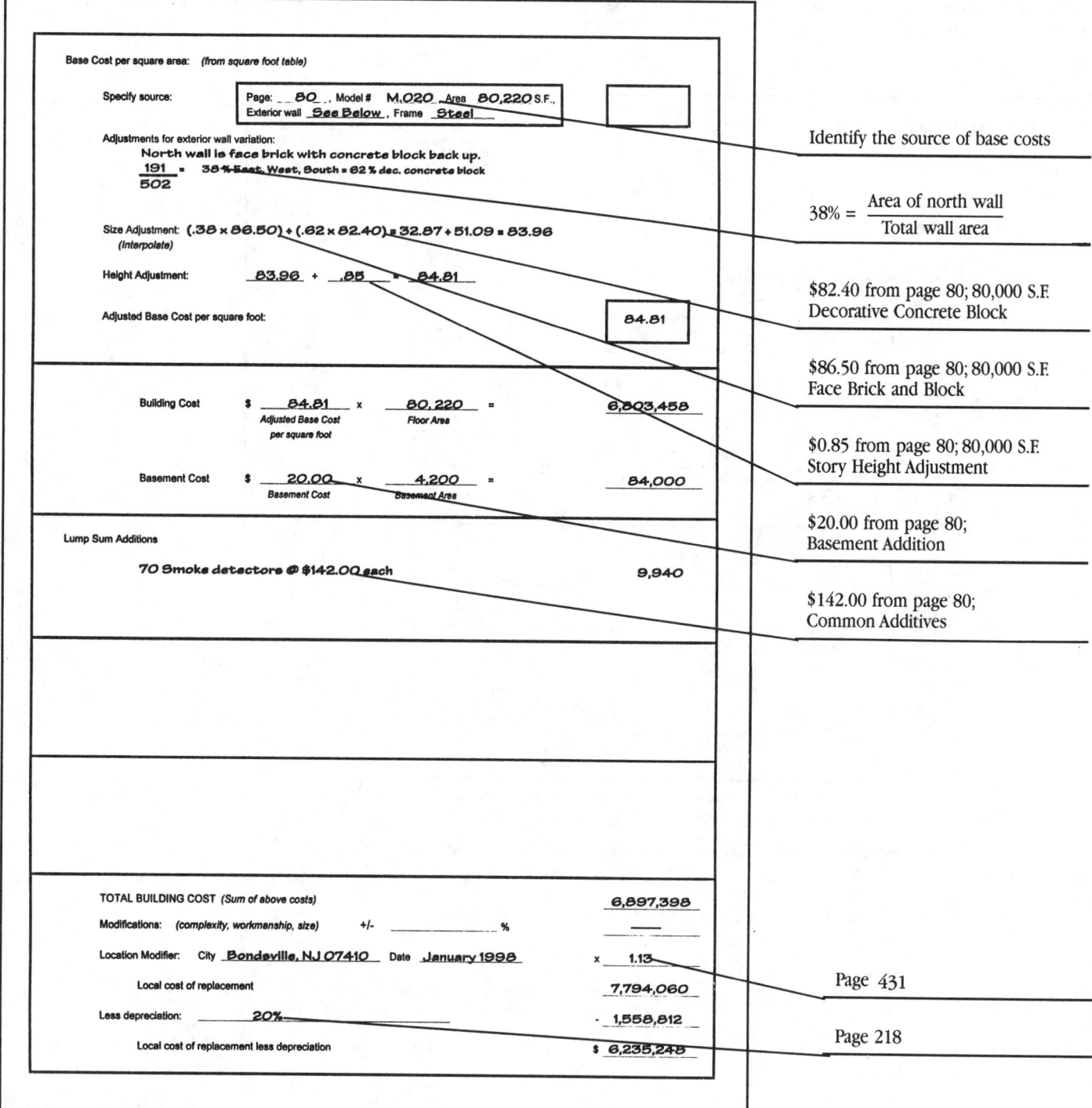

Base Cost per square area: *(from square foot table)*

Specify source:

Page: __80__, Model # __M.020__ Area __80,220__ S.F., Exterior wall __See Below__, Frame __Steel__

Adjustments for exterior wall variation:

North wall is face brick with concrete block back up.

$\frac{191}{502}$ = 38 % East, West, South = 62 % dec. concrete block

Size Adjustment: (.38 × 86.50) + (.62 × 82.40) = 32.87 + 51.09 = 83.96
(Interpolate)

Height Adjustment: __83.96__ + __.85__ = __84.81__

Adjusted Base Cost per square foot: __84.81__

Building Cost $ __84.81__ x __80,220__ = __6,803,458__
                  Adjusted Base Cost   Floor Area
                  per square foot

Basement Cost $ __20.00__ x __4,200__ = __84,000__
                  Basement Cost   Basement Area

Lump Sum Additions

70 Smoke detectors @ $142.00 each   9,940

TOTAL BUILDING COST *(Sum of above costs)*   6,897,398

Modifications: *(complexity, workmanship, size)*   +/- _____ %   ——

Location Modifier: City __Bondsville, NJ 07410__ Date __January 1998__   x __1.13__

Local cost of replacement   7,794,060

Less depreciation: __20%__   - 1,558,812

Local cost of replacement less depreciation $ 6,235,248

Identify the source of base costs

$38\% = \dfrac{\text{Area of north wall}}{\text{Total wall area}}$

$82.40 from page 80; 80,000 S.F. Decorative Concrete Block

$86.50 from page 80; 80,000 S.F. Face Brick and Block

$0.85 from page 80; 80,000 S.F. Story Height Adjustment

$20.00 from page 80; Basement Addition

$142.00 from page 80; Common Additives

Page 431

Page 218

EXAMPLES

Example 2
This example shows how to modify a model building

Model #M.020, 4–7 Story Apartment, page 80, matches the example quite closely. The model specifies a 6-story building with a 10′ story height.

The following adjustments must be made:
- add stone ashlar wall
- adjust model to a 5-story building
- change partitions
- change floor finish
- change plumbing

CII APPRAISAL

5. DATE: **Jan. 2, 1998**	
1. SUBJECT PROPERTY: **Westernberg Holdings, Inc.**	6. APPRAISER: **SJS**
2. BUILDING: **#623**	
3. ADDRESS: **55 Pine wood Rd., Bondsville, N.J. 07410**	
4. BUILDING USE: **Apartments**	7. YEAR BUILT: **1972**

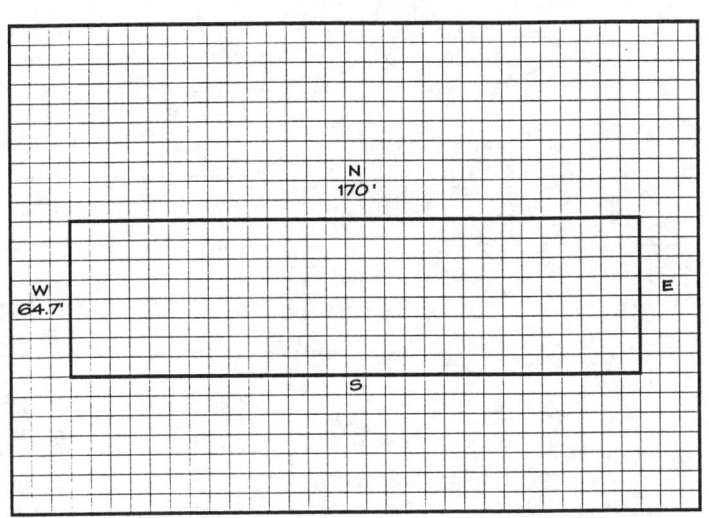

Two exterior wall types

North wall - 4" stone Ashlar w/8" block back up (split Face)

8. EXTERIOR WALL CONSTRUCTION: **East, West & South walls - 12" Decorative concrete block**

9. FRAME: **Steel**

10. GROUND FLOOR AREA: **11,000** S.F. 11. GROSS FLOOR AREA (EXCL. BASEMENT): **55,000** S.F.

12. NUMBER OF STORIES: **5** 13. STORY HEIGHT: **11 Ft.**

14. PERIMETER: **469.4** L.F. 15. BASEMENT AREA: **None** S.F.

16. GENERAL COMMENTS:

Example 2 (continued)

Square foot area (excluding basement) from item 11 __55,000__ S.F.

Perimeter from item 14 __469.4__ Lin. Ft.

Item 17—Model square foot costs (from model sub-total) $ __73.10__

FIELD DESCRIPTION & CALCULATION SECTION

NO.	SYSTEM/COMPONENT	DESCRIPTION	UNIT	UNIT COST	NEW SF COST	MODEL SF COST	+/- CHANGE
1.0	**FOUNDATION**						
.1	Footings and Foundations L.F. Length	Bay Size Same as model	S.F. Gnd.	6.54	1.31	1.09	+.20
.4	Piles & Caissons						
.9	Excavation & Backfill Depth	Area 11,000 S.F.	S.F. Gnd.	1.00	.20	.17	+.03
2.0	**SUBSTRUCTURE**						
.1	Slab on grade Material	Thickness Same as model	S.F. Slab	3.06	.61	.51	+.10
.2	Special Substructures Type						
3.0	**SUPERSTRUCTURE**						
.1	Columns, Beams, & Joists Size Type	Length Fireproof only	S.F. Floor L.F.	1.54	1.54	1.54	—
	Size Type	Length	L.F.				
.4	Structure Walls		S.F.				
.5	Elevated Floors Same as model		S.F. Floor	11.89	9.51	9.91	-.40
.7	Roof		S.F. Roof	4.89	.98	.81	+.17
.9	Stairs 13 Flights		Flight	4185	.99	1.12	-.13
4.0	**EXTERIOR CLOSURE**						
.1	Walls North wall - 4" stone ashlar Material on 8" conc. block Thickness 36 % of Wall		S.F. Wall	15.56	6.28	6.47	-.19
	East, West & South walls - 12" Material split ribbed block Thickness 64 % of Wall		S.F. Wall				
	Material Thickness % of Wall		S.F. Wall				
.5	Exterior Wall Finishes Type		S.F. Surf.				
.6	Doors Type Same as model	Number 4	Each	3110	.23	.21	+.02
	Type	Number	Each				
.7	Windows & Glazed Walls Type % of Wall		S.F. Wind.				
	Type 14 % of Wall	Each	S.F. Wind.	284	1.25	1.33	-.08
5.0	**ROOFING**						
.1	Roof Coverings Material Same as model		S.F. Roof	2.52	.50	.42	+.08
.7	Insulation Material	Thickness	S.F. Insul.	1.25	.25	.21	+.04
.8	Openings & Specialities		S.F. Opng.				
6.0	**INTERIOR CONSTRUCTION**						
.1	Partitions 9' High Material Gypsum board on metal stud	Density 8 S.F. Floor/L.F. Part.	S.F. Part.	3.60	4.05	3.65	+.40
	Material	Density	S.F. Part.				
.4	Interior Doors Type	Number but	Each	441	5.51	5.52	-.01
.5	Wall Finishes Material Same as model	% of Wall 9'-0" High	S.F. Wall	1.04	2.13	2.09	+.04
	Material		S.F. Wall				
.6	Floor Finishes Material Hardwood 60 % of Floor Carpet 30 % of Floor		S.F. Floor	3.81	5.77	3.81	+1.96
	Material Tile 10 % of Floor		S.F. Floor				
.7	Ceiling Finishes Material Same as model		S.F. Ceil.	2.69	2.69	2.69	—

Annotations (right side):

From page 81, Model #M.020 Sub-total

$$\frac{\$.85/\text{S.F. ground area}}{5 \text{ stories}}$$

$$\frac{\$.84/\text{S.F. ground area}}{6 \text{ stories}}$$

(Page 81, Line 3.1)

Area of Elevated Floors

$$\frac{\$11.89/\text{S.F.} \times 4 \times 11,000 \text{ S.F.}}{55,000}$$

Building Floor Area

$$\frac{\$4.89/\text{S.F. roof}}{5 \text{ stories}}$$

Windows are 14% of wall area, 15 S.F./window

$$\frac{284}{15} \times .14 \times 469.4 \times 11 \times 5 \div 55,000 = 1.25$$

From Table 6.1-510 page 325 and 6.1-580 page 327

$$\frac{\$3.60/\text{S.F. partition} \times 9 \text{ ft. high}}{8 \text{ S.F. floor/L.F. partition}}$$

From Table 6.6-100 pages 336 & 337

Oak Strip, Fin. (max. price)		@ $6.80/S.F.
Carpet (26 oz.)		@ $3.32/S.F.
Ceramic Tile		@ $6.90/S.F.

$$.60 \times \$6.80 = \$4.08$$
$$.30 \times \$3.32 = 1.00$$
$$.10 \times \$6.90 = \underline{.69}$$
$$\$5.77$$

Convert cost per S.F. wall to cost per S.F. of floor
Windows are 14% of wall area

Annotations (bottom left):

North wall is 36% of wall area, e.g., $\dfrac{170 \times 11}{469.4 \times 11}$

From Table 4.1-242-2350 page 289
Stone ashlar @ $21.35/S.F. wall
East, west and south walls are 64% of wall area
From Table 4.1-212-1530 page 284
Split ribbed block @ $12.30/S.F. wall

New wall cost

$$.36 \times \$21.35/\text{S.F.} + .64 \times \$12.30/\text{S.F.} = \$15.56/\text{S.F. wall}$$

$$\$6.28 \text{ S.F. floor} = \frac{\$15.56/\text{S.F. wall} \times .86 \times 469.4 \text{ L.F.} \times 11 \text{ ft. high} \times 5 \text{ stories}}{55,000 \text{ S.F. of floor}}$$

EXAMPLES

Example 2 (Continued)

NO.	SYSTEM/COMPONENET					Unit	Unit Cost	New S.F. Cost	Model S.F. Cost	+/- Change
6.0	**INTERIOR CONSTRUCTION**					S.F. Surf.				
.9	Interior Surface/Exterior Wall	Material	86 % of Wall		Board on furring	S.F. Surf.	2.86	1.15	1.14	+.01
		Material	% of Wall			S.F. Surf.				
7.0	**CONVEYING**									
.1	Elevators 2	Type Same as model	Capacity		Stops	Each	122,400	4.45	4.08	+.37
.2	Special Conveyors	Type				Each				
8.0	**MECHANICAL**					S.F. Floor				
.1	Plumbing Kitchen, bath + service fixtures	1 Fixture /200				Each	1745	8.73	8.12	+.61
.2	Fire Protection	Same as model				S.F.	1.62	1.62	1.62	—
.3	Heating	Type Same as model				S.F.	4.01	4.01	4.01	—
.4	Cooling	Type Same as model				S.F.	5.62	5.62	5.62	—
.5	Special Systems					Each				
9.0	**ELECTRICAL**									
.1	Service Distribution	Same as model				S.F.	.87	.87	.87	—
.2	Lighting & Power					S.F.	4.10	4.10	4.10	—
.4	Special Electrical					S.F.	.35	.35	.35	—
11.0	**SPECIAL CONSTRUCTION**									
.1	Specialties					S.F.	1.43	1.43	1.43	—
12.0	**SITE WORK**									
.1	Earthwork									
.3	Utilities									
.5	Roads & Parking									
.7	Site Improvement									
								Total Change	**3.22**	

ITEM			
18	Adjusted S.F. cost $ 76.32 item 17 +/- changes		
19	Building area - from item 11 55,000 S.F. x adjusted S.F. cost	$	4,197,600
20	Basement area - from item 15 _____ S.F. x S.F. cost $	$	—
21	Base building sub-total - item 20 + item 19	$	4,197,600
22	Miscellaneous addition (quality, etc.)	$	—
23	Sub-total - item 22 + 21	$	4,197,600
24	General conditions - 15 % of item 23	$	629,640
25	Sub-total - item 24 + item 23	$	4,827,240
26	Architects fees 5.5 % of item 25	$	265,498
27	Sub-total - item 26 + item 27	$	5,092,738
28	Location modifier zip code 07410	x	1.13
29	Local replacement cost - item 28 x item 27	$	5,754,794
30	Depreciation 20 % of item 29	$	1,150,959
31	Depreciated local replacement cost - item 29 less item 30	$	4,603,835
32	Exclusions	$	—
33	Net depreciated replacement cost - item 31 less item 32	$	4,603,835

469.4 L.F. × 11 ft. high × 5 stories

Unit Cost   86%

$$\frac{\$2.86/\text{S.F. wall} \times \text{total S.F. wall} \times .86}{55,000}$$

From page 431

Use table on page 218

Costs per square foot of floor area

Exterior Wall	S.F. Area	8000	12000	15000	19000	22500	25000	29000	32000	36000
	L.F. Perimeter	213	280	330	350	400	433	442	480	520
Face Brick with Concrete Block Back-up	Wood Joists	109.75	98.90	94.60	88.50	86.40	85.20	82.15	81.30	80.15
	Steel Joists	109.25	99.00	94.90	89.50	87.50	86.35	83.70	82.95	81.80
Stucco on Concrete Block	Wood Joists	95.90	86.40	82.60	78.00	76.15	75.10	72.95	72.20	71.25
	Steel Joists	101.65	92.10	88.30	83.60	81.75	80.70	78.50	77.75	76.75
Wood Siding	Wood Frame	96.20	86.65	82.85	78.20	76.35	75.30	73.10	72.35	71.40
Brick Veneer	Wood Frame	102.05	91.80	87.65	82.20	80.20	79.05	76.40	75.60	74.50
Perimeter Adj., Add or Deduct	Per 100 L.F.	12.00	8.00	6.35	5.05	4.25	3.80	3.30	3.00	2.65
Story Hgt. Adj., Add or Deduct	Per 1 Ft.	2.25	2.00	1.85	1.55	1.50	1.45	1.30	1.30	1.20

For Basement, add $19.75 per square foot of basement area

The above costs were calculated using the basic specifications shown on the facing page. These costs should be adjusted where necessary for design alternatives and owner's requirements. Reported completed project costs, for this type of structure, range from $33.35 to $124.35 per S.F.

Common additives

Description	Unit	$ Cost
Appliances		
Cooking range, 30" free standing		
1 oven	Each	380 - 1275
2 oven	Each	785 - 1700
30" built-in		
1 oven	Each	470 - 1475
2 oven	Each	1050 - 1975
Counter top cook tops, 4 burner	Each	261 - 630
Microwave oven	Each	199 - 1850
Combination range, refrig. & sink, 30" wide	Each	1075 - 2125
72" wide	Each	3175
Combination range, refrigerator, sink,		
microwave oven & icemaker	Each	4700
Compactor, residential, 4-1 compaction	Each	395 - 570
Dishwasher, built-in, 2 cycles	Each	425 - 850
4 cycles	Each	460 - 1075
Garbage disposer, sink type	Each	116 - 244
Hood for range, 2 speed, vented, 30" wide	Each	182 - 505
42" wide	Each	320 - 510
Refrigerator, no frost 10-12 C.F.	Each	505 - 770
18-20 C.F.	Each	665 - 1425

Description	Unit	$ Cost
Closed Circuit Surveillance, One station		
Camera and monitor	Each	1325
For additional camera stations, add	Each	730
Elevators, Hydraulic passenger, 2 stops		
2000# capacity	Each	40,900
2500# capacity	Each	41,300
3500# capacity	Each	44,900
Additional stop, add	Each	3350
Emergency Lighting, 25 watt, battery operated		
Lead battery	Each	330
Nickel cadmium	Each	620
Laundry Equipment		
Dryer, gas, 16 lb. capacity	Each	680
30 lb. capacity	Each	2550
Washer, 4 cycle	Each	760
Commercial	Each	1125
Smoke Detectors		
Ceiling type	Each	142
Duct type	Each	390

Important: See the Reference Section for Location Factors

Model costs calculated for a 3 story building with 10' story height and 22,500 square feet of floor area

Apartment, 1-3 Story

				Unit	Unit Cost	Cost Per S.F.	% Of Sub-Total
1.0 Foundations							
.1	Footings & Foundations	Poured concrete; strip and spread footings and 4' foundation wall		S.F. Ground	6.39	2.13	3.5%
.4	Piles & Caissons	N/A		—	—	—	
.9	Excavation & Backfill	Site preparation for slab and trench for foundation wall and footing		S.F. Ground	1.04	.35	
2.0 Substructure							
.1	Slab on Grade	4" reinforced concrete with vapor barrier and granular base		S.F. Slab	3.06	1.02	1.4%
.2	Special Substructures	N/A		—	—	—	
3.0 Superstructure							
.1	Columns & Beams	Gypsum board fireproofing on columns; steel columns in 3.5 and 3.7		S.F. Floor	.82	.82	
.4	Structural Walls	N/A		—	—	—	
.5	Elevated Floors	Open web steel joists, slab form, concrete, interior steel columns		S.F. Floor	9.44	6.29	13.8%
.7	Roof	Open web steel joists with rib metal deck, interior steel columns		S.F. Roof	4.39	1.46	
.9	Stairs	Concrete filled metal pan		Flight	5025	1.12	
4.0 Exterior Closure							
.1	Walls	Face brick with concrete block backup	88% of wall	S.F. Wall	15.04	7.06	
.5	Exterior Wall Finishes	N/A		—	—	—	12.0%
.6	Doors	Aluminum and glass		Each	1078	.19	
.7	Windows & Glazed Walls	Aluminum horizontal sliding	12% of wall	Each	284	1.21	
5.0 Roofing							
.1	Roof Coverings	Built-up tar and gravel with flashing		S.F. Roof	2.58	.86	
.7	Insulation	Perlite/EPS composite		S.F. Roof	1.25	.42	1.8%
.8	Openings & Specialties	N/A		—	—	—	
6.0 Interior Construction							
.1	Partitions	Gypsum board and sound deadening board on metal studs	10 S.F. of Floor/L.F. Partition	S.F. Partition	3.23	2.87	
.4	Interior Doors	15% solid core wood, 85% hollow core wood	80 S.F. Floor/Door	Each	368	4.62	
.5	Wall Finishes	70% paint, 25% vinyl wall covering, 5% ceramic tile		S.F. Surface	.96	1.70	24.4%
.6	Floor Finishes	60% carpet, 30% vinyl composition tile, 10% ceramic tile		S.F. Floor	4.02	4.02	
.7	Ceiling Finishes	Painted gypsum board on resilient channels		S.F. Ceiling	2.73	2.73	
.9	Interior Surface/Exterior Wall	Painted gypsum board on furring	80% of wall	S.F. Wall	2.86	1.22	
7.0 Conveying							
.1	Elevators	One hydraulic passenger elevator		Each	60,750	2.70	3.8%
.2	Special Conveyors	N/A		—	—	—	
8.0 Mechanical							
.1	Plumbing	Kitchen, bathroom and service fixtures, supply and drainage	1 Fixture/200 S.F. Floor	Each	1774	8.87	
.2	Fire Protection	Wet pipe sprinkler system		S.F. Floor	1.65	1.65	
.3	Heating	Oil fired hot water, baseboard radiation		S.F. Floor	4.58	4.58	29.8%
.4	Cooling	Chilled water, air cooled condenser system		S.F. Floor	5.86	5.86	
.5	Special Systems	N/A		—	—	—	
9.0 Electrical							
.1	Service & Distribution	400 ampere service, panel board and feeders		S.F. Floor	.78	.78	
.2	Lighting & Power	Incandescent fixtures, receptacles, switches, A.C. and misc. power		S.F. Floor	4.12	4.12	7.8%
.4	Special Electrical	Alarm systems and emergency lighting		S.F. Floor	.61	.61	
11.0 Special Construction							
.1	Specialties	Kitchen cabinets		S.F. Floor	1.19	1.19	1.7%
12.0 Site Work							
.1	Earthwork	N/A		—	—	—	
.3	Utilities	N/A		—	—	—	
.5	Roads & Parking	N/A		—	—	—	0.0%
.7	Site Improvements	N/A		—	—	—	
				Sub-Total		70.45	100%
	GENERAL CONDITIONS (Overhead & Profit)			15%		10.57	
	ARCHITECT FEES			8%		6.48	
				Total Building Cost		87.50	

Costs per square foot of floor area

Exterior Wall	S.F. Area	40000	45000	50000	55000	60000	70000	80000	90000	100000
	L.F. Perimeter	366	400	433	466	500	566	505	550	594
Face Brick with Concrete Block Back-up	Steel Frame	92.00	91.30	90.75	90.30	**89.95**	89.35	86.50	86.05	85.65
	R/Conc. Frame	95.95	95.25	94.60	94.05	93.70	93.05	89.75	89.25	88.80
Decorative Concrete Block	Steel Frame	86.90	86.35	85.85	85.45	85.15	84.65	82.40	82.00	81.70
	R/Conc. Frame	88.30	87.70	87.20	86.80	86.50	86.00	83.70	83.30	82.95
Precast Concrete Panels	Steel Frame	89.20	88.60	88.05	87.60	87.25	86.70	83.95	83.55	83.20
	R/Conc. Frame	90.40	89.75	89.25	88.75	88.45	87.90	85.15	84.70	84.35
Perimeter Adj., Add or Deduct	Per 100 L.F.	3.80	3.40	3.05	2.75	2.50	2.15	1.90	1.70	1.50
Story Hgt. Adj., Add or Deduct	Per 1 Ft.	1.30	1.25	1.25	1.15	1.15	1.10	.85	.85	.85

For Basement, add $20.00 per square foot of basement area

The above costs were calculated using the basic specifications shown on the facing page. These costs should be adjusted where necessary for design alternatives and owner's requirements. Reported completed project costs, for this type of structure, range from $37.85 to $102.95 per S.F.

Common additives

Description	Unit	$ Cost
Appliances		
Cooking range, 30" free standing		
1 oven	Each	380 - 1275
2 oven	Each	785 - 1700
30" built-in		
1 oven	Each	470 - 1475
2 oven	Each	1050 - 1975
Counter top cook tops, 4 burner	Each	261 - 630
Microwave oven	Each	199 - 1850
Combination range, refrig. & sink, 30" wide	Each	1075 - 2125
72" wide	Each	3175
Combination range, refrigerator, sink,		
microwave oven & icemaker	Each	4700
Compactor, residential, 4-1 compaction	Each	395 - 570
Dishwasher, built-in, 2 cycles	Each	425 - 850
4 cycles	Each	460 - 1075
Garbage disposer, sink type	Each	116 - 244
Hood for range, 2 speed, vented, 30" wide	Each	182 - 505
42" wide	Each	320 - 510
Refrigerator, no frost 10-12 C.F.	Each	505 - 770
18-20 C.F.	Each	665 - 1425

Description	Unit	$ Cost
Closed Circuit Surveillance, One station		
Camera and monitor	Each	1325
For additional camera stations, add	Each	730
Elevators, Electric passenger, 5 stops		
2000# capacity	Each	91,300
3500# capacity	Each	97,300
5000# capacity	Each	100,800
Additional stop, add	Each	5225
Emergency Lighting, 25 watt, battery operated		
Lead battery	Each	330
Nickel cadmium	Each	620
Laundry Equipment		
Dryer, gas, 16 lb. capacity	Each	680
30 lb. capacity	Each	2550
Washer, 4 cycle	Each	760
Commercial	Each	1125
Smoke Detectors		
Ceiling type	Each	142
Duct type	Each	390

Important: See the Reference Section for Location Factors

Model costs calculated for a 6 story building with 10'-4" story height and 60,000 square feet of floor area

BUILDING TYPES

			Unit	Unit Cost	Cost Per S.F.	% Of Sub-Total
1.0 Foundations						
.1	Footings & Foundations	Poured concrete; strip and spread footings and 4' foundation wall	S.F. Ground	6.54	1.09	
.4	Piles & Caissons	N/A	—	—	—	1.7%
.9	Excavation & Backfill	Site preparation for slab and trench for foundation wall and footing	S.F. Ground	1.00	.17	
2.0 Substructure						
.1	Slab on Grade	4" reinforced concrete with vapor barrier and granular base	S.F. Slab	3.06	.51	0.7%
.2	Special Substructures	N/A	—	—	—	
3.0 Superstructure						
.1	Columns & Beams	Gypsum board fireproofing on columns, steel columns in 3.5 and 3.7	S.F. Floor	1.54	1.54	
.4	Structural Walls	N/A	—	—	—	
.5	Elevated Floors	Open web steel joists, slab form, concrete, steel columns	S.F. Floor	11.89	9.91	18.3%
.7	Roof	Open web steel joists with rib metal deck, steel columns	S.F. Roof	4.89	.81	
.9	Stairs	Concrete filled metal pan	Flight	4185	1.12	
4.0 Exterior Closure						
.1	Walls	Face brick with concrete block backup 86% of wall	S.F. Wall	14.56	6.47	
.5	Exterior Wall Finishes	N/A	—	—	—	
.6	Doors	Aluminum and glass	Each	3110	.21	11.0%
.7	Windows & Glazed Walls	Aluminum horizontal sliding 14% of wall	Each	284	1.33	
5.0 Roofing						
.1	Roof Coverings	Built-up tar and gravel with flashing	S.F. Roof	2.52	.42	
.7	Insulation	Perlite/EPS composite	S.F. Roof	1.25	.21	0.9%
.8	Openings & Specialties	N/A	—	—	—	
6.0 Interior Construction						
.1	Partitions	Gypsum board and sound deadening board on metal studs 8 S.F. Floor/L.F. Partitions	S.F. Partition	3.65	3.65	
.4	Interior Doors	15% solid core wood, 85% hollow core wood 80 S.F Floor/Door	Each	441	5.52	
.5	Wall Finishes	70% paint, 25% vinyl wall covering, 5% ceramic tile	S.F. Surface	1.04	2.09	
.6	Floor Finishes	60% carpet, 30% vinyl composition tile, 10% ceramic tile	S.F. Floor	4.02	4.02	26.1%
.7	Ceiling Finishes	Painted gypsum board on resilient channels	S.F. Ceiling	2.69	2.69	
.9	Interior Surface/Exterior Wall	Painted gypsum board on furring 80% of wall	S.F. Wall	2.86	1.14	
7.0 Conveying						
.1	Elevators	Two geared passenger elevators	Each	122,400	4.08	5.6%
.2	Special Conveyors	N/A	—	—	—	
8.0 Mechanical						
.1	Plumbing	Kitchen, bathroom and service fixtures, supply and drainage 1 Fixture/215 S.F. Floor	Each	1745	8.12	
.2	Fire Protection	Standpipe and wet pipe sprinkler system	S.F. Floor	1.62	1.62	
.3	Heating	Oil fired hot water, baseboard radiation	S.F. Floor	4.01	4.01	26.4%
.4	Cooling	Chilled water, air cooled condenser system	S.F. Floor	5.62	5.62	
.5	Special Systems	N/A	—	—	—	
9.0 Electrical						
.1	Service & Distribution	1000 ampere service, panel board and feeders	S.F. Floor	.87	.87	
.2	Lighting & Power	Incandescent fixtures, receptacles, switches, A.C. and misc. power	S.F. Floor	4.10	4.10	7.3%
.4	Special Electrical	Alarm systems, emergency lighting, and intercom	S.F. Floor	.35	.35	
11.0 Special Construction						
.1	Specialties	Kitchen cabinets	S.F. Floor	1.43	1.43	2.0%
12.0 Site Work						
.1	Earthwork	N/A	—	—	—	
.3	Utilities	N/A	—	—	—	
.5	Roads & Parking	N/A	—	—	—	0.0%
.7	Site Improvements	N/A	—	—	—	
			Sub-Total		73.10	**100%**
	GENERAL CONDITIONS (Overhead & Profit)			15%	10.97	
	ARCHITECT FEES			7%	5.88	
			Total Building Cost		89.95	

81

Costs per square foot of floor area

Exterior Wall	S.F. Area	95000	112000	129000	145000	162000	181000	210000	220000	240000
	L.F. Perimeter	345	386	406	442	480	522	510	526	560
Ribbed Precast Concrete Panel	Steel Frame	100.40	98.70	96.75	95.85	95.05	94.35	92.05	91.75	91.30
	R/Conc. Frame	94.95	93.25	91.45	90.60	89.80	89.15	87.00	86.75	86.30
Face Brick with Concrete Block Back-up	Steel Frame	94.65	92.90	90.95	90.05	89.25	88.50	86.15	85.90	85.40
	R/Conc. Frame	96.45	94.70	92.80	91.85	91.05	90.35	88.00	87.75	87.25
Stucco on Concrete Block	Steel Frame	88.30	86.80	85.45	84.70	84.00	83.45	81.90	81.65	81.30
	R/Conc. Frame	90.05	88.65	87.25	86.50	85.85	85.30	83.75	83.55	83.15
Perimeter Adj., Add or Deduct	Per 100 L.F.	4.30	3.65	3.15	2.80	2.50	2.25	1.90	1.85	1.65
Story Hgt. Adj., Add or Deduct	Per 1 Ft.	1.45	1.35	1.25	1.20	1.15	1.15	.95	.95	.90
For Basement, add $20.15 per square foot of basement area										

The above costs were calculated using the basic specifications shown on the facing page. These costs should be adjusted where necessary for design alternatives and owner's requirements. Reported completed project costs, for this type of structure, range from $51.20 to $118.05 per S.F.

Common additives

Description	Unit	$ Cost
Appliances		
Cooking range, 30" free standing		
1 oven	Each	380 - 1275
2 oven	Each	785 - 1700
30" built-in		
1 oven	Each	470 - 1475
2 oven	Each	1050 - 1975
Counter top cook tops, 4 burner	Each	261 - 630
Microwave oven	Each	199 - 1850
Combination range, refrig. & sink, 30" wide	Each	1075 - 2125
72" wide	Each	3175
Combination range, refrigerator, sink,		
microwave oven & icemaker	Each	4700
Compactor, residential, 4-1 compaction	Each	395 - 570
Dishwasher, built-in, 2 cycles	Each	425 - 850
4 cycles	Each	460 - 1075
Garbage disposer, sink type	Each	116 - 244
Hood for range, 2 speed, vented, 30" wide	Each	182 - 505
42" wide	Each	320 - 510
Refrigerator, no frost 10-12 C.F.	Each	505 - 770
18-20 C.F.	Each	665 - 1425

Description	Unit	$ Cost
Closed Circuit Surveillance, One station		
Camera and monitor	Each	1325
For additional camera stations, add	Each	730
Elevators, Electric passenger, 10 stops		
3000# capacity	Each	198,500
4000# capacity	Each	200,500
5000# capacity	Each	204,000
Additional stop, add	Each	5225
Emergency Lighting, 25 watt, battery operated		
Lead battery	Each	330
Nickel cadmium	Each	620
Laundry Equipment		
Dryer, gas, 16 lb. capacity	Each	680
30 lb. capacity	Each	2550
Washer, 4 cycle	Each	760
Commercial	Each	1125
Smoke Detectors		
Ceiling type	Each	142
Duct type	Each	390

Important: See the Reference Section for Location Factors

Model costs calculated for a 15 story building with 10'-6" story height and 162,000 square feet of floor area

				Unit	Unit Cost	Cost Per S.F.	% Of Sub-Total
1.0 Foundations							
.1	Footings & Foundations	Poured concrete; strip and spread footings and 4' foundation wall		S.F. Ground	10.65	.71	
.4	Piles & Caissons	N/A		–	–	–	1.0%
.9	Excavation & Backfill	Site preparation for slab and trench for foundation wall and footing		S.F. Ground	1.00	.07	
2.0 Substructure							
.1	Slab on Grade	4" reinforced concrete with vapor barrier and granular base		S.F. Slab	3.06	.20	0.3%
.2	Special Substructures	N/A		–	–	–	
3.0 Superstructure							
.1	Columns & Beams	Gypsum board fireproofing on columns; steel columns in 3.5 and 3.7		S.F. Floor	2.62	2.62	
.4	Structural Walls	N/A		–	–	–	
.5	Elevated Floors	Open web steel joists, slab form, concrete, interior steel columns		S.F. Floor	9.15	8.54	16.1%
.7	Roof	Open web steel joists with rib metal deck, interior steel columns		S.F. Roof	3.73	.25	
.9	Stairs	Concrete filled metal pan		Flight	4185	1.11	
4.0 Exterior Closure							
.1	Walls	Ribbed precast concrete panel	87% of wall	S.F. Wall	15.79	6.41	
.5	Exterior Wall Finishes	N/A		–	–	–	10.9%
.6	Doors	Aluminum and glass		Each	1359	1.03	
.7	Windows & Glazed Walls	Aluminum horizontal sliding	13% of wall	Each	284	1.09	
5.0 Roofing							
.1	Roof Coverings	Built-up tar and gravel with flashing		S.F. Roof	2.40	.16	
.7	Insulation	Perlite/EPS composite		S.F. Roof	1.25	.08	0.3%
.8	Openings & Specialties	N/A		–	–	–	
6.0 Interior Construction							
.1	Partitions	Gypsum board on concrete block and metal studs	10 S.F. of Floor/L.F. Partition	S.F. Partition	7.64	7.64	
.4	Interior Doors	15% solid core wood, 85% hollow core wood	80 S.F. Floor/Door	Each	441	5.52	
.5	Wall Finishes	70% paint, 25% vinyl wall covering, 5% ceramic tile		S.F. Surface	1.03	2.06	29.4%
.6	Floor Finishes	60% carpet, 30% vinyl composition tile, 10% ceramic tile		S.F. Floor	4.02	4.02	
.7	Ceiling Finishes	Painted gypsum board on resilient channels		S.F. Ceiling	2.69	2.69	
.9	Interior Surface/Exterior Wall	Painted gypsum board on furring	80% of wall	S.F. Wall	2.86	1.02	
7.0 Conveying							
.1	Elevators	Four geared passenger elevators		Each	215,055	5.31	6.8%
.2	Special Conveyors	N/A		–	–	–	
8.0 Mechanical							
.1	Plumbing	Kitchen, bathroom and service fixtures, supply and drainage	1 Fixture/210 S.F. Floor	Each	1743	8.30	
.2	Fire Protection	Standpipe and wet pipe sprinkler systems		S.F. Floor	1.95	1.95	
.3	Heating	Oil fired hot water, baseboard radiation		S.F. Floor	4.01	4.01	25.5%
.4	Cooling	Chilled water, air cooled condenser system		S.F. Floor	5.62	5.62	
.5	Special Systems	N/A		–	–	–	
9.0 Electrical							
.1	Service & Distribution	1200 ampere service, panel board and feeders		S.F. Floor	.44	.44	
.2	Lighting & Power	Incandescent fixtures, receptacles, switches, A.C. and misc. power		S.F. Floor	4.41	4.41	8.0%
.4	Special Electrical	Alarm systems, emergency lighting, antenna, intercom and security television		S.F. Floor	1.38	1.38	
11.0 Special Construction							
.1	Specialties	Kitchen cabinets		S.F. Floor	1.33	1.33	1.7%
12.0 Site Work							
.1	Earthwork	N/A		–	–	–	
.3	Utilities	N/A		–	–	–	
.5	Roads & Parking	N/A		–	–	–	0.0%
.7	Site Improvements	N/A		–	–	–	
				Sub-Total		77.97	**100%**
	GENERAL CONDITIONS (Overhead & Profit)				15%	11.70	
	ARCHITECT FEES				6%	5.38	
				Total Building Cost		**95.05**	

Costs per square foot of floor area

Exterior Wall	S.F. Area	12000	15000	18000	21000	24000	27000	30000	33000	36000
	L.F. Perimeter	440	500	540	590	640	665	700	732	770
Face Brick with Concrete Block Back-up	Steel Frame	113.45	110.15	106.95	105.10	103.65	101.80	100.55	99.45	98.70
	Bearing Wall	109.90	106.90	104.00	102.35	101.05	99.35	98.25	97.30	96.60
Precast Concrete	Steel Frame	107.45	104.70	102.10	100.55	99.35	97.85	96.85	95.95	95.35
Decorative Concrete Block	Bearing Wall	96.40	94.25	92.20	90.95	90.05	88.85	88.05	87.40	86.85
Concrete Block	Steel Frame	103.25	101.55	99.90	98.95	98.20	97.30	96.70	96.15	95.75
	Bearing Wall	94.00	92.05	90.20	89.10	88.30	87.20	86.50	85.90	85.45
Perimeter Adj., Add or Deduct	Per 100 L.F.	7.35	5.85	4.90	4.20	3.70	3.25	2.90	2.65	2.45
Story Hgt. Adj., Add or Deduct	Per 1 Ft.	1.15	1.05	.95	.90	.85	.80	.70	.70	.65

For Basement, add $16.00 per square foot of basement area

The above costs were calculated using the basic specifications shown on the facing page. These costs should be adjusted where necessary for design alternatives and owner's requirements. Reported completed project costs, for this type of structure, range from $49.15 to $130.25 per S.F.

Common additives

Description	Unit	$ Cost
Closed Circuit Surveillance, One station		
Camera and monitor	Each	1325
For additional camera stations, add	Each	730
Emergency Lighting, 25 watt, battery operated		
Lead battery	Each	330
Nickel cadmium	Each	620
Seating		
Auditorium chair, all veneer	Each	127
Veneer back, padded seat	Each	152
Upholstered, spring seat	Each	178
Classroom, movable chair & desk	Set	65 - 120
Lecture hall, pedestal type	Each	127 - 370
Smoke Detectors		
Ceiling type	Each	142
Duct type	Each	390
Sound System		
Amplifier, 250 watts	Each	1525
Speaker, ceiling or wall	Each	131
Trumpet	Each	246

BUILDING TYPES

Important: See the Reference Section for Location Factors

Model costs calculated for a 1 story building with 24' story height and 24,000 square feet of floor area

				Unit	Unit Cost	Cost Per S.F.	% Of Sub-Total
1.0 Foundations							
.1	Footings & Foundations	Poured concrete; strip and spread footings and 4' foundation wall		S.F. Ground	2.20	2.20	
.4	Piles & Caissons	N/A		–	–	–	4.0%
.9	Excavation & Backfill	Site preparation for slab and trench for foundation wall and footing		S.F. Ground	1.04	1.04	
2.0 Substructure							
.1	Slab on Grade	6" reinforced concrete with vapor barrier and granular base		S.F. Slab	3.66	3.66	4.5%
.2	Special Substructures	N/A		–	–	–	
3.0 Superstructure							
.1	Columns & Beams	Steel columns		S.F. Floor	.38	.38	
.4	Structural Walls	N/A		–	–	–	
.5	Elevated Floors	Open web steel joists, slab form, concrete	(balcony)	S.F. Floor	10.72	1.35	13.6%
.7	Roof	Metal deck on steel truss		S.F. Roof	8.54	8.54	
.9	Stairs	Concrete filled metal pan		Flight	5900	.74	
4.0 Exterior Closure							
.1	Walls	Precast concrete panel	80% of wall (adjusted for end walls)	S.F. Wall	16.19	8.29	
.5	Exterior Wall Finishes	N/A		–	–	–	
.6	Doors	Double aluminum and glass and hollow metal		Each	2312	1.16	15.8%
.7	Windows & Glazed Walls	Glass curtain wall	20% of wall	S.F. Window	25	3.32	
5.0 Roofing							
.1	Roof Coverings	Built-up tar and gravel with flashing		S.F. Roof	2.05	2.05	
.7	Insulation	Perlite/EPS composite		S.F. Roof	1.25	1.25	4.4%
.8	Openings & Specialties	Gravel stop and roof hatches		S.F. Roof	.25	.25	
6.0 Interior Construction							
.1	Partitions	Concrete Block and toilet partitions	40 S.F. Floor/L.F. Partition	S.F. Partition	5.15	2.06	
.4	Interior Doors	Single leaf hollow metal	400 S.F. Floor/Door	Each	526	1.32	
.5	Wall Finishes	70% paint, 30% epoxy coating		S.F. Surface	2.17	1.74	
.6	Floor Finishes	70% hardwood, 30% carpet		S.F. Floor	7.93	7.93	20.3%
.7	Ceiling Finishes	Fiberglass board, suspended		S.F. Floor	2.37	2.37	
.9	Interior Surface/Exterior Wall	Paint	80% of wall	S.F. Wall	1.95	1.00	
7.0 Conveying							
.1	Elevators	One hydraulic passenger elevator		Each	54,240	2.26	2.8%
.2	Special Conveyors	N/A		–	–	–	
8.0 Mechanical							
.1	Plumbing	Toilet and service fixtures, supply and drainage	1 Fixture/800 S.F. Floor	Each	2904	3.63	
.2	Fire Protection	Wet pipe sprinkler system		S.F. Floor	1.60	1.60	
.3	Heating	Included in 8.4		–	–	–	20.0%
.4	Cooling	Single zone rooftop unit, gas heating, electric cooling		S.F. Floor	10.83	10.83	
.5	Special Systems	N/A		–	–	–	
9.0 Electrical							
.1	Service & Distribution	800 ampere service, panel board and feeders		S.F. Floor	1.35	1.35	
.2	Lighting & Power	Fluorescent fixtures, receptacles, switches, A.C. and misc. power		S.F. Floor	7.21	7.21	14.6%
.4	Special Electrical	Alarm systems and emergency lighting, and public address system		S.F. Floor	3.23	3.23	
11.0 Special Construction							
.1	Specialties	N/A		–	–	–	0.0%
12.0 Site Work							
.1	Earthwork	N/A		–	–	–	
.3	Utilities	N/A		–	–	–	
.5	Roads & Parking	N/A		–	–	–	0.0%
.7	Site Improvements	N/A		–	–	–	
			Sub-Total			80.76	100%
	GENERAL CONDITIONS (Overhead & Profit)				15%	12.11	
	ARCHITECT FEES				7%	6.48	
			Total Building Cost			99.35	

BUILDING TYPES

Costs per square foot of floor area

Exterior Wall	S.F. Area	2000	2700	3400	4100	4800	5500	6200	6900	7600
	L.F. Perimeter	180	208	236	256	280	303	317	337	357
Face Brick with Concrete Block Back-up	Steel Frame	144.70	135.90	130.75	126.25	123.60	121.45	119.05	117.60	116.40
	R/Conc. Frame	155.30	146.50	141.35	136.80	134.15	132.00	129.60	128.15	126.95
Precast Concrete Panel	Steel Frame	136.85	129.20	124.75	120.80	118.50	116.65	114.60	113.35	112.30
	R/Conc. Frame	147.45	139.80	135.30	131.40	129.05	127.25	125.15	123.90	122.85
Limestone with Concrete Block Back-up	Steel Frame	153.45	143.40	137.50	132.35	129.25	126.85	124.00	122.35	120.95
	R/Conc. Frame	164.00	153.95	148.05	142.90	139.80	137.40	134.55	132.90	131.50
Perimeter Adj., Add or Deduct	Per 100 L.F.	27.00	20.00	15.85	13.20	11.30	9.85	8.70	7.80	7.10
Story Hgt. Adj., Add or Deduct	Per 1 Ft.	2.60	2.20	2.00	1.80	1.70	1.60	1.45	1.40	1.35

For Basement, add $18.50 per square foot of basement area

The above costs were calculated using the basic specifications shown on the facing page. These costs should be adjusted where necessary for design alternatives and owner's requirements. Reported completed project costs, for this type of structure, range from $77.10 to $189.80 per S.F.

Common additives

Description	Unit	$ Cost
Bulletproof Teller Window, 44" x 60"	Each	4000
60" x 48"	Each	4975
Closed Circuit Surveillance, One station		
Camera and monitor	Each	1325
For additional camera stations, add	Each	730
Counters, Complete	Station	4200
Door & Frame, 3' x 6'-8", bullet resistant steel		
with vision panel	Each	3225 - 5425
Drive-up Window, Drawer & micr., not incl. glass	Each	6125 - 9375
Emergency Lighting, 25 watt, battery operated		
Lead battery	Each	330
Nickel cadmium	Each	620
Night Depository	Each	7675 - 12,600
Package Receiver, painted	Each	1375
stainless steel	Each	2150
Partitions, Bullet resistant to 8' high	L.F.	227 - 370
Pneumatic Tube Systems, 2 station	Each	23,600
With TV viewer	Each	43,800

Description	Unit	$ Cost
Service Windows, Pass thru, steel		
24" x 36"	Each	2150
48" x 48"	Each	2875
72" x 40"	Each	3525
For stainless steel frames, add	Each	
Smoke Detectors		
Ceiling type	Each	142
Duct type	Each	390
Twenty-four Hour Teller		
Automatic deposit cash & memo	Each	42,000
Vault Front, Door & frame		
1 hour test, 32"x 78"	Opening	3425
2 hour test, 32" door	Opening	3525
40" door	Opening	4050
4 hour test, 32" door	Opening	3675
40" door	Opening	4375
Time lock, two movement, add	Each	1575

BUILDING TYPES

Model costs calculated for a 1 story building with 14' story height and 4,100 square feet of floor area

				Unit	Unit Cost	Cost Per S.F.	% Of Sub-Total
1.0 Foundations							
.1	Footings & Foundations	Poured concrete; strip and spread footings and 4' foundation wall		S.F. Ground	5.87	5.87	
.4	Piles & Caissons	N/A		–	–	–	7.4%
.9	Excavation & Backfill	Site preparation for slab and trench for foundation wall and footing		S.F. Ground	2.05	2.05	
2.0 Substructure							
.1	Slab on Grade	4" reinforced concrete with vapor barrier and granular base		S.F. Slab	3.06	3.06	2.9%
.2	Special Substructures	N/A		–	–	–	
3.0 Superstructure							
.1	Columns & Beams	Concrete columns		L.F. Columns	67	3.29	
.4	Structural Walls	N/A		–	–	–	
.5	Elevated Floors	N/A		–	–	–	11.8%
.7	Roof	Cast-in-place concrete slab		S.F. Roof	9.37	9.37	
.9	Stairs	N/A		–	–	–	
4.0 Exterior Closure							
.1	Walls	Face brick with concrete block backup	80% of wall	S.F. Wall	19.45	13.60	
.5	Exterior Wall Finishes	N/A		–	–	–	17.4%
.6	Doors	Double aluminum and glass and hollow metal		Each	1859	.90	
.7	Windows & Glazed Walls	Horizontal aluminum sliding	20% of wall	Each	359	4.18	
5.0 Roofing							
.1	Roof Coverings	Built-up tar and gravel with flashing		S.F. Roof	2.74	2.74	
.7	Insulation	Perlite/EPS composite		S.F. Roof	1.25	1.25	4.0%
.8	Openings & Specialties	Gravel stop		L.F. Perimeter	5.60	.35	
6.0 Interior Construction							
.1	Partitions	Gypsum board on metal studs	20 S.F. of Floor/L.F. Partition	S.F. Partition	2.98	1.49	
.4	Interior Doors	Single leaf hollow metal	200 S.F. Floor/Door	Each	526	2.63	
.5	Wall Finishes	50% vinyl wall covering, 50% paint		S.F. Surface	.92	.92	13.0%
.6	Floor Finishes	50% carpet, 50% vinyl composition tile		S.F. Floor	3.85	3.85	
.7	Ceiling Finishes	Mineral fiber tile on concealed zee bars		S.F. Ceiling	3.07	3.07	
.9	Interior Surface/Exterior Wall	Painted gypsum board on furring	80% of wall	S.F. Wall	2.85	1.99	
7.0 Conveying							
.1	Elevators	N/A		–	–	–	0.0%
.2	Special Conveyors	N/A		–	–	–	
8.0 Mechanical							
.1	Plumbing	Toilet and service fixtures, supply and drainage	1 Fixture/580 S.F. Floor	Each	3294	5.68	
.2	Fire Protection	Wet pipe sprinkler system		S.F. Floor	2.09	2.09	
.3	Heating	Included in 8.4		–	–	–	16.2%
.4	Cooling	Single zone rooftop unit, gas heating, electric cooling		S.F. Floor	9.51	9.51	
.5	Special Systems	N/A		–	–	–	
9.0 Electrical							
.1	Service & Distribution	200 ampere service, panel board and feeders		S.F. Floor	1.49	1.49	
.2	Lighting & Power	Fluorescent fixtures, receptacles, switches, A.C. and misc. power		S.F. Floor	6.38	6.38	8.9%
.4	Special Electrical	Alarm systems, emergency lighting, and security television		S.F. Floor	1.72	1.72	
11.0 Special Construction							
.1	Specialties	Vault door, automatic teller, drive up window, closed circuit T.V., night depository		S.F. Floor	19.70	19.70	18.4%
12.0 Site Work							
.1	Earthwork	N/A		–	–	–	
.3	Utilities	N/A		–	–	–	
.5	Roads & Parking	N/A		–	–	–	0.0%
.7	Site Improvements	N/A		–	–	–	
				Sub-Total		107.18	100%

GENERAL CONDITIONS (Overhead & Profit)		15%	16.08
ARCHITECT FEES		11%	13.54

Total Building Cost	**136.80**

BUILDING TYPES

Costs per square foot of floor area

Exterior Wall	S.F. Area	12000	14000	16000	18000	20000	22000	24000	26000	28000
	L.F. Perimeter	460	491	520	540	566	593	620	645	670
Concrete Block	Steel Roof Deck	63.30	61.60	60.25	59.05	58.25	57.55	56.95	56.50	56.00
Decorative Concrete Block	Steel Roof Deck	65.90	64.00	62.50	61.15	60.20	59.40	58.75	58.20	57.65
Face Brick on Concrete Block	Steel Roof Deck	70.15	67.85	66.10	64.40	63.30	62.35	61.60	60.90	60.30
Jumbo Brick on Concrete Block	Steel Roof Deck	68.15	66.10	64.40	62.90	61.85	60.95	60.25	59.65	59.05
Stucco on Concrete Block	Steel Roof Deck	64.30	62.55	61.15	59.90	59.00	58.25	57.65	57.15	56.65
Precast Concrete Panel	Steel Roof Deck	64.20	62.40	61.05	59.80	58.90	58.20	57.55	57.05	56.60
Perimeter Adj., Add or Deduct	Per 100 L.F.	2.75	2.35	2.05	1.85	1.65	1.50	1.40	1.25	1.20
Story Hgt. Adj., Add or Deduct	Per 1 Ft.	.60	.55	.50	.50	.45	.45	.40	.35	.35

For Basement, add $18.15 per square foot of basement area

The above costs were calculated using the basic specifications shown on the facing page. These costs should be adjusted where necessary for design alternatives and owner's requirements. Reported completed project costs, for this type of structure, range from $34.50 to $84.50 per S.F.

Common additives

Description	Unit	$ Cost
Bowling Alleys, Incl. alley, pinsetter,		
scorer, counter & misc. supplies, average	Lane	45,000
For automatic scorer, add (maximum)	Lane	7475
Emergency Lighting, 25 watt, battery operated		
Lead battery	Each	330
Nickel cadmium	Each	620
Lockers, Steel, Single tier, 60" or 72"	Opening	135 - 216
2 tier, 60" or 72" total	Opening	83 - 105
5 tier, box lockers	Opening	44 - 57
Locker bench, lam., maple top only	L.F.	18.65
Pedestals, steel pipe	Each	43
Seating		
Auditorium chair, all veneer	Each	127
Veneer back, padded seat	Each	152
Upholstered, spring seat	Each	178
Sound System		
Amplifier, 250 watts	Each	1525
Speaker, ceiling or wall	Each	131
Trumpet	Each	246

Model costs calculated for a 1 story building with 14' story height and 20,000 square feet of floor area

			Unit	Unit Cost	Cost Per S.F.	% Of Sub-Total
1.0 Foundations						
.1	Footings & Foundations	Poured concrete; strip and spread footings and 4' foundation wall	S.F. Ground	2.01	2.01	
.4	Piles & Caissons	N/A	—	—	—	6.9%
.9	Excavation & Backfill	Site preparation for slab and trench for foundation wall and footing	S.F. Ground	1.24	1.24	
2.0 Substructure						
.1	Slab on Grade	4" reinforced concrete with vapor barrier and granular base	S.F. Slab	3.06	3.06	6.5%
.2	Special Substructures	N/A	—	—	—	
3.0 Superstructure						
.1	Columns & Beams	Columns included in 3.7	—	—	—	
.4	Structural Walls	N/A	—	—	—	
.5	Elevated Floors	N/A	—	—	—	11.2%
.7	Roof	Metal deck on open web steel joists with columns and beams	S.F. Roof	5.28	5.28	
.9	Stairs	N/A	—	—	—	
4.0 Exterior Closure						
.1	Walls	Concrete block 90% of wall	S.F. Wall	7.91	2.82	
.5	Exterior Wall Finishes	N/A	—	—	—	
.6	Doors	Double aluminum and glass and hollow metal	Each	1339	.34	10.1%
.7	Windows & Glazed Walls	Horizontal pivoted 10% of wall	Each	971	1.60	
5.0 Roofing						
.1	Roof Coverings	Built-up tar and gravel with flashing	S.F. Roof	2.09	2.09	
.7	Insulation	Perlite/EPS composite	S.F. Roof	1.25	1.25	7.2%
.8	Openings & Specialties	Gravel stop and hatches	S.F. Roof	.09	.09	
6.0 Interior Construction						
.1	Partitions	Concrete block 50 S.F. Floor/L.F. Partition	S.F. Partition	5.70	1.14	
.4	Interior Doors	Hollow metal single leaf 1000 S.F. Floor/Door	Each	526	.53	
.5	Wall Finishes	Paint and block filler	S.F. Surface	1.95	.78	
.6	Floor Finishes	Vinyl tile 25% of floor	S.F. Floor	2.00	.50	9.0%
.7	Ceiling Finishes	Suspended fiberglass board 25% of area	S.F. Ceiling	2.37	.59	
.9	Interior Surface/Exterior Wall	Paint and block filler 90% of wall	S.F. Wall	1.95	.70	
7.0 Conveying						
.1	Elevators	N/A	—	—	—	0.0%
.2	Special Conveyors	N/A	—	—	—	
8.0 Mechanical						
.1	Plumbing	Toilet and service fixtures, supply and drainage 1 Fixture/2200 S.F. Floor	Each	2772	1.26	
.2	Fire Protection	Sprinklers, light hazard	S.F. Floor	1.41	1.41	
.3	Heating	Included in 8.4	—	—	—	31.4%
.4	Cooling	Single zone rooftop unit, gas heating, electric cooling	S.F. Floor	12.25	12.25	
.5	Special Systems	N/A	—	—	—	
9.0 Electrical						
.1	Service & Distribution	800 ampere service, panel board and feeders	S.F. Floor	1.61	1.61	
.2	Lighting & Power	Fluorescent fixtures, receptacles, switches, A.C. and misc. power	S.F. Floor	6.51	6.51	17.7%
.4	Special Electrical	Alarm systems and emergency lighting	S.F. Floor	.27	.27	
11.0 Special Construction						
.1	Specialties	N/A	—	—	—	0.0%
12.0 Site Work						
.1	Earthwork	N/A	—	—	—	
.3	Utilities	N/A	—	—	—	
.5	Roads & Parking	N/A	—	—	—	0.0%
.7	Site Improvements	N/A	—	—	—	
			Sub-Total		47.33	100%
	GENERAL CONDITIONS (Overhead & Profit)			15%	7.10	
	ARCHITECT FEES			7%	3.82	
			Total Building Cost		58.25	

BUILDING TYPES

BUILDING TYPES

Costs per square foot of floor area

Exterior Wall	S.F. Area	6000	8000	10000	12000	14000	16000	18000	20000	22000
	L.F. Perimeter	320	386	453	520	540	597	610	660	710
Face Brick with Concrete Block Back-up	Bearing Walls	87.10	83.55	81.50	80.15	77.65	76.80	75.10	74.45	74.00
	Steel Frame	87.80	84.30	82.25	80.95	78.45	77.65	75.95	75.35	74.90
Decorative Concrete Block	Bearing Walls	82.50	79.45	77.65	76.45	74.35	73.60	72.20	71.65	71.25
	Steel Frame	83.20	80.15	78.40	77.20	75.15	74.45	73.05	72.50	72.15
Precast Concrete Panels	Bearing Walls	81.75	78.75	77.00	75.85	73.80	73.10	71.70	71.20	70.80
	Steel Frame	82.45	79.50	77.75	76.60	74.65	73.95	72.60	72.05	71.70
Perimeter Adj., Add or Deduct	Per 100 L.F.	7.60	5.70	4.60	3.80	3.25	2.85	2.55	2.30	2.10
Story Hgt. Adj., Add or Deduct	Per 1 Ft.	1.25	1.15	1.10	1.00	.95	.90	.80	.80	.80
Basement—Not Applicable										

The above costs were calculated using the basic specifications shown on the facing page. These costs should be adjusted where necessary for design alternatives and owner's requirements. Reported completed project costs, for this type of structure, range from $41.50 to $100.00 per S.F.

Common additives

Description	Unit	$ Cost
Directory Boards, Plastic, glass covered		
30" x 20"	Each	510
36" x 48"	Each	930
Aluminum, 24" x 18"	Each	450
36" x 24"	Each	540
48" x 32"	Each	640
48" x 60"	Each	1450
Emergency Lighting, 25 watt, battery operated		
Lead battery	Each	330
Nickel cadmium	Each	620
Benches, Hardwood	L.F.	65 - 101
Ticket Printer	Each	5325
Turnstiles, One way		
4 arm, 46" dia., manual	Each	400
Electric	Each	1500
High security, 3 arm		
65" dia., manual	Each	3150
Electric	Each	4175
Gate with horizontal bars		
65" dia., 7' high transit type	Each	3625

Important: See the Reference Section for Location Factors

Model costs calculated for a 1 story building with 14' story height and 12,000 square feet of floor area

				Unit	Unit Cost	Cost Per S.F.	% Of Sub-Total
1.0 Foundations							
.1	Footings & Foundations	Poured concrete; strip and spread footings and 4' foundation wall		S.F. Ground	3.55	3.55	
.4	Piles & Caissons	N/A		—	—	—	7.2%
.9	Excavation & Backfill	Site preparation for slab and trench for foundation wall and footing		S.F. Ground	1.11	1.11	
2.0 Substructure							
.1	Slab on Grade	6" reinforced concrete with vapor barrier and granular base		S.F. Slab	3.66	3.66	5.6%
.2	Special Substructures	N/A		—	—	—	
3.0 Superstructure							
.1	Columns & Beams	Columns included in 3.7		—	—	—	
.4	Structural Walls	N/A		—	—	—	
.5	Elevated Floors	N/A		—	—	—	5.3%
.7	Roof	Metal deck on open web steel joists with columns and beams		S.F. Roof	3.44	3.44	
.9	Stairs	N/A		—	—	—	
4.0 Exterior Closure							
.1	Walls	Face brick with concrete block backup	70% of wall	S.F. Wall	19.45	8.26	
.5	Exterior Wall Finishes	N/A		—	—	—	19.1%
.6	Doors	Double aluminum and glass		Each	3525	1.18	
.7	Windows & Glazed Walls	Store front	30% of wall	S.F. Window	16.30	2.97	
5.0 Roofing							
.1	Roof Coverings	Built-up tar and gravel with flashing		S.F. Roof	2.37	2.37	
.7	Insulation	Perlite/EPS composite		S.F. Roof	1.25	1.25	5.9%
.8	Openings & Specialties	Gravel stop		L.F. Perimeter	5.60	.24	
6.0 Interior Construction							
.1	Partitions	Lightweight concrete block	15 S.F. Floor/L.F. Partition	S.F. Partition	5.68	3.79	
.4	Interior Doors	Hollow metal	150 S.F. Floor/Door	Each	526	3.51	
.5	Wall Finishes	Glazed coating		S.F. Surface	1.14	1.52	27.2%
.6	Floor Finishes	Quarry tile and vinyl composition tile		S.F. Floor	5.32	5.32	
.7	Ceiling Finishes	Mineral fiber tile on concealed zee bars		S.F. Ceiling	3.07	3.07	
.9	Interior Surface/Exterior Wall	Glazed coating	70% of wall	S.F. Wall	1.15	.49	
7.0 Conveying							
.1	Elevators	N/A		—	—	—	0.0%
.2	Special Conveyors	N/A		—	—	—	
8.0 Mechanical							
.1	Plumbing	Toilet and service fixtures, supply and drainage	1 Fixture/850 S.F. Floor	Each	2320	2.73	
.2	Fire Protection	Wet pipe sprinkler system		S.F. Floor	1.60	1.60	
.3	Heating	Included in 8.4		—	—	—	20.6%
.4	Cooling	Single zone rooftop unit, gas heating, electric cooling		S.F. Floor	9.17	9.17	
.5	Special Systems	N/A		—	—	—	
9.0 Electrical							
.1	Service & Distribution	400 ampere service, panel board and feeders		S.F. Floor	1.03	1.03	
.2	Lighting & Power	Fluorescent fixtures, receptacles, switches, A.C. and misc. power		S.F. Floor	4.39	4.39	9.1%
.4	Special Electrical	Alarm systems and emergency lighting		S.F. Floor	.48	.48	
11.0 Special Construction							
.1	Specialties	N/A		—	—	—	0.0%
12.0 Site Work							
.1	Earthwork	N/A		—	—	—	
.3	Utilities	N/A		—	—	—	
.5	Roads & Parking	N/A		—	—	—	0.0%
.7	Site Improvements	N/A		—	—	—	
				Sub-Total		65.13	**100%**
	GENERAL CONDITIONS (Overhead & Profit)				15%	9.77	
	ARCHITECT FEES				7%	5.25	
				Total Building Cost		80.15	

Costs per square foot of floor area

Exterior Wall	S.F. Area	600	800	1000	1200	1600	2000	2400	3000	4000
	L.F. Perimeter	100	114	128	139	164	189	214	250	314
Brick with Concrete Block Back-up	Steel Frame	189.00	181.30	176.70	172.85	168.65	166.05	164.40	162.50	161.00
	Bearing Walls	187.95	180.30	175.70	171.80	167.60	165.05	163.35	161.50	160.00
Concrete Block	Steel Frame	165.55	161.25	158.70	156.55	154.20	152.75	151.85	150.80	149.95
	Bearing Walls	164.50	160.25	157.65	155.50	153.15	151.75	150.80	149.75	148.90
Galvanized Steel Siding	Steel Frame	163.90	160.20	158.00	156.15	154.10	152.90	152.10	151.20	150.50
Metal Sandwich Panel	Steel Frame	164.10	159.50	156.75	154.45	151.95	150.40	149.40	148.30	147.40
Perimeter Adj., Add or Deduct	Per 100 L.F.	52.95	39.70	31.75	26.50	19.85	15.90	13.25	10.60	7.95
Story Hgt. Adj., Add or Deduct	Per 1 Ft.	3.25	2.80	2.50	2.25	2.00	1.85	1.75	1.65	1.50
Basement—Not Applicable										

The above costs were calculated using the basic specifications shown on the facing page. These costs should be adjusted where necessary for design alternatives and owner's requirements. Reported completed project costs, for this type of structure, range from $65.75 to $195.00 per S.F.

Common additives

Description	Unit	$ Cost
Air Compressors, Electric		
1-1/2 H.P., standard controls	Each	3325
Dual controls	Each	3550
5 H.P., 115/230 volt, standard control	Each	4525
Dual controls	Each	4750
Emergency Lighting, 25 watt, battery operated		
Lead battery	Each	330
Nickel cadmium	Each	620
Fence, Chain link, 6' high		
9 ga. wire, galvanized	L.F.	13.65
6 ga. wire	L.F.	17.80
Gate	Each	247
Product Dispenser with vapor recovery		
for 6 nozzles	Each	16,500
Laundry Equipment		
Dryer, gas, 16 lb. capacity	Each	680
30 lb. capacity	Each	2550
Washer, 4 cycle	Each	760
Commercial	Each	1125

Description	Unit	$ Cost
Lockers, Steel, single tier, 60" or 72"	Opening	135 - 216
2 tier, 60" or 72" total	Opening	83 - 105
5 tier, box lockers	Opening	44 - 57
Locker bench, lam. maple top only	L.F.	18.65
Pedestals, steel pipe	Each	43
Paving, Bituminous		
Wearing course plus base course	Sq. Yard	9.75
Safe, Office type, 4 hour rating		
30" x 18" x 18"	Each	3475
62" x 33" x 20"	Each	7575
Sidewalks, Concrete 4" thick	S.F.	2.55
Yard Lighting,		
20' aluminum pole with 400 watt high pressure sodium fixture	Each	2085

Important: See the Reference Section for Location Factors

Model costs calculated for a 1 story building with 12' story height and 800 square feet of floor area

			Unit	Unit Cost	Cost Per S.F.	% Of Sub-Total
1.0 Foundations						
.1	Footings & Foundations	Poured concrete; strip and spread footings and 4' foundation wall	S.F. Ground	7.60	7.60	
.4	Piles & Caissons	N/A	—	—	—	6.2%
.9	Excavation & Backfill	Site preparation for slab and trench for foundation wall and footing	S.F. Ground	1.60	1.60	
2.0 Substructure						
.1	Slab on Grade	5" reinforced concrete with vapor barrier and granular base	S.F. Slab	3.30	3.30	2.2%
.2	Special Substructures	N/A	—	—	—	
3.0 Superstructure						
.1	Columns & Beams	Steel columns included in 3.7	—	—	—	
.4	Structural Walls	N/A	—	—	—	
.5	Elevated Floors	N/A	—	—	—	2.2%
.7	Roof	Metal deck, open web steel joists, beams, columns	S.F. Roof	3.28	3.28	
.9	Stairs	N/A	—	—	—	
4.0 Exterior Closure						
.1	Walls	Face brick with concrete block backup 70% of wall	S.F. Wall	19.45	23.28	
.5	Exterior Wall Finishes	N/A	—	—	—	
.6	Doors	Steel overhead hollow metal	Each	1694	8.48	24.1%
.7	Windows & Glazed Walls	Horizontal pivoted steel 5% of wall	Each	388	3.69	
5.0 Roofing						
.1	Roof Coverings	Built-up tar and gravel with flashing	S.F. Roof	3.76	3.76	
.7	Insulation	Perlite/EPS composite	S.F. Roof	1.25	1.25	3.4%
.8	Openings & Specialties	N/A	—	—	—	
6.0 Interior Construction						
.1	Partitions	Concrete block 20 S.F. Floor/S.F. Partition	S.F. Partition	5.84	2.92	
.4	Interior Doors	Hollow metal 600 S.F. Floor/Door	Each	526	.88	
.5	Wall Finishes	Paint	S.F. Surface	.78	.78	
.6	Floor Finishes	N/A	—	—	—	3.1%
.7	Ceiling Finishes	N/A	—	—	—	
.9	Interior Surface/Exterior Wall	N/A	—	—	—	
7.0 Conveying						
.1	Elevators	N/A	—	—	—	0.0%
.2	Special Conveyors	N/A	—	—	—	
8.0 Mechanical						
.1	Plumbing	Toilet and service fixtures, supply and drainage 1 Fixture/160 S.F. Floor	Each	5142	32.14	
.2	Fire Protection	N/A	—	—	—	
.3	Heating	Unit heaters	S.F. Floor	12.25	12.25	30.2%
.4	Cooling	N/A	—	—	—	
.5	Special Systems	N/A	—	—	—	
9.0 Electrical						
.1	Service & Distribution	400 ampere service, panel board and feeders	S.F. Floor	13.83	13.83	
.2	Lighting & Power	Fluorescent fixtures, receptacles, switches and misc. power	S.F. Floor	27	27.01	28.6%
.4	Special Electrical	Emergency power	S.F. Floor	1.31	1.31	
11.0 Special Construction						
.1	Specialties	N/A	—	—	—	0.0%
12.0 Site Work						
.1	Earthwork	N/A	—	—	—	
.3	Utilities	N/A	—	—	—	
.5	Roads & Parking	N/A	—	—	—	0.0%
.7	Site Improvements	N/A	—	—	—	
			Sub-Total		147.36	**100%**
	GENERAL CONDITIONS (Overhead & Profit)			15%	22.10	
	ARCHITECT FEES			7%	11.84	
			Total Building Cost		**181.30**	

BUILDING TYPES

Costs per square foot of floor area

Exterior Wall	S.F. Area	2000	7000	12000	17000	22000	27000	32000	37000	42000
	L.F. Perimeter	180	340	470	540	640	740	762	793	860
Decorative Concrete Brick	Wood Arch	141.80	107.90	100.40	94.70	92.60	91.30	88.60	86.75	86.05
	Steel Truss	138.55	104.65	97.15	91.45	89.35	88.05	85.35	83.50	82.80
Stone with Concrete Block Back-up	Wood Arch	162.35	119.00	109.35	101.95	99.25	97.55	94.05	91.65	90.75
	Steel Truss	155.35	111.95	102.30	94.95	92.20	90.55	87.00	84.65	83.75
Face Brick with Concrete Block Back-up	Wood Arch	157.40	116.30	107.15	100.20	97.60	96.05	92.70	90.45	89.60
	Steel Truss	150.35	109.30	100.15	93.20	90.60	89.00	85.70	83.45	82.60
Perimeter Adj., Add or Deduct	Per 100 L.F.	36.75	10.50	6.10	4.30	3.30	2.70	2.25	2.00	1.75
Story Hgt. Adj., Add or Deduct	Per 1 Ft.	2.20	1.15	.95	.75	.70	.65	.55	.55	.50
For Basement, add $19.50 per square foot of basement area										

The above costs were calculated using the basic specifications shown on the facing page. These costs should be adjusted where necessary for design alternatives and owner's requirements. Reported completed project costs, for this type of structure, range from $45.70 to $170.00 per S.F.

Common additives

Description	Unit	$ Cost
Altar, Wood, custom design, plain	Each	1625
Deluxe	Each	9200
Granite or marble, average	Each	6450
Deluxe	Each	14,700
Ark, Prefabricated, plain	Each	3100
Deluxe	Each	107,000
Baptistry, Fiberglass, incl. plumbing	Each	2400 - 4750
Bells & Carillons, 48 bells	Each	440,000
24 bells	Each	165,000
Confessional, Prefabricated wood		
Single, plain	Each	2300
Deluxe	Each	7975
Double, plain	Each	7600
Deluxe	Each	18,100
Emergency Lighting, 25 watt, battery operated		
Lead battery	Each	330
Nickel cadmium	Each	620
Lecterns, Wood, plain	Each	495
Deluxe	Each	1550

Description	Unit	$ Cost
Pews/Benches, Hardwood	L.F.	65 - 101
Pulpits, Prefabricated, hardwood	Each	1275 - 7450
Railing, Hardwood	L.F.	151
Steeples, translucent fiberglass		
30" square, 15' high	Each	2950
25' high	Each	4350
Painted fiberglass, 24" square, 14' high	Each	3000
28' high	Each	4150
Aluminum		
20' high, 3'- 6" base	Each	4275
35' high, 8'- 0" base	Each	16,200
60' high, 14'- 0" base	Each	36,400

Important: See the Reference Section for Location Factors

Model costs calculated for a 1 story building with 24' story height and 17,000 square feet of floor area

BUILDING TYPES

			Unit	Unit Cost	Cost Per S.F.	% Of Sub-Total
1.0 Foundations						
.1	Footings & Foundations	Poured concrete; strip and spread footings and 4' foundation wall	S.F. Ground	4.54	4.54	
.4	Piles & Caissons	N/A	–	–	–	7.1%
.9	Excavation & Backfill	Site preparation for slab and trench for foundation wall and footing	S.F. Ground	1.04	1.04	
2.0 Substructure						
.1	Slab on Grade	4" reinforced concrete with vapor barrier and granular base	S.F. Slab	3.06	3.06	3.9%
.2	Special Substructures	N/A	–	–	–	
3.0 Superstructure						
.1	Columns & Beams	N/A	–	–	–	
.4	Structural Walls	N/A	–	–	–	
.5	Elevated Floors	N/A	–	–	–	17.2%
.7	Roof	Wood deck on laminated wood arches	S.F. Roof	12.65	13.50	
.9	Stairs	N/A	–	–	–	
4.0 Exterior Closure						
.1	Walls	Face brick with concrete block backup 80% of wall (adjusted for end walls)	S.F. Wall	23	14.09	
.5	Exterior Wall Finishes	N/A	–	–	–	22.0%
.6	Doors	Double hollow metal swinging, single hollow metal	Each	931	.33	
.7	Windows & Glazed Walls	Aluminum, top hinged, in-swinging and curtain wall panels 20% of wall	S.F. Window	18.56	2.83	
5.0 Roofing						
.1	Roof Coverings	Asphalt shingles with flashing	S.F. Roof	1.53	1.53	
.7	Insulation	Polystyrene	S.F. Ground	1.12	1.34	3.9%
.8	Openings & Specialties	Gutters and downspouts	S.F. Ground	.23	.23	
6.0 Interior Construction						
.1	Partitions	Plaster on metal studs 40 S.F. Floor/L.F. Partitions	S.F. Partition	6.63	3.98	
.4	Interior Doors	Hollow metal 400 S.F. Floor/Door	Each	526	1.32	
.5	Wall Finishes	Paint	S.F. Surface	.73	.88	16.6%
.6	Floor Finishes	Carpet	S.F. Floor	4.99	4.99	
.7	Ceiling Finishes	N/A	–	–	–	
.9	Interior Surface/Exterior Wall	Painted plaster 80% of wall	S.F. Wall	3.12	1.90	
7.0 Conveying						
.1	Elevators	N/A	–	–	–	0.0%
.2	Special Conveyors	N/A	–	–	–	
8.0 Mechanical						
.1	Plumbing	Kitchen, toilet and service fixtures, supply and drainage 1 Fixture/2430 S.F. Floor	Each	2502	1.03	
.2	Fire Protection	Wet pipe sprinkler system	S.F. Floor	1.60	1.60	
.3	Heating	Oil fired hot water, wall fin radiation	S.F. Floor	5.76	5.76	19.8%
.4	Cooling	Split systems with air cooled condensing units	S.F. Floor	7.14	7.14	
.5	Special Systems	N/A	–	–	–	
9.0 Electrical						
.1	Service & Distribution	200 ampere service, panel board and feeders	S.F. Floor	.32	.32	
.2	Lighting & Power	Fluorescent fixtures, receptacles, switches, A.C. and misc. power	S.F. Floor	5.24	5.24	9.5%
.4	Special Electrical	Alarm systems, sound system and emergency lighting	S.F. Floor	1.86	1.86	
11.0 Special Construction						
.1	Specialties	N/A	–	–	–	0.0%
12.0 Site Work						
.1	Earthwork	N/A	–	–	–	
.3	Utilities	N/A	–	–	–	
.5	Roads & Parking	N/A	–	–	–	0.0%
.7	Site Improvements	N/A	–	–	–	

	Sub-Total	78.51	100%
GENERAL CONDITIONS (Overhead & Profit)	15%	11.78	
ARCHITECT FEES	11%	9.91	
Total Building Cost		**100.20**	

Costs per square foot of floor area

Exterior Wall	S.F. Area	2000	4000	6000	8000	12000	15000	18000	20000	22000
	L.F. Perimeter	180	280	340	386	460	535	560	600	640
Stone Ashlar Veneer On Concrete Block	Wood Truss	140.50	123.85	115.25	110.10	104.25	102.55	100.10	99.40	98.85
	Steel Joists	138.40	121.75	113.15	108.00	102.20	100.45	98.00	97.35	96.75
Stucco on Concrete Block	Wood Truss	128.00	114.15	107.35	103.40	98.95	97.60	95.80	95.25	94.85
	Steel Joists	126.15	112.30	105.55	101.55	97.15	95.75	93.95	93.45	93.00
Brick Veneer	Wood Frame	135.10	119.60	111.70	107.05	101.80	100.20	98.05	97.40	96.90
Wood Shingles	Wood Frame	128.55	114.55	107.65	103.55	99.05	97.65	95.80	95.25	94.80
Perimeter Adj., Add or Deduct	Per 100 L.F.	23.20	11.60	7.70	5.80	3.90	3.10	2.60	2.30	2.10
Story Hgt. Adj., Add or Deduct	Per 1 Ft.	2.55	2.00	1.60	1.35	1.10	1.00	.90	.85	.85
For Basement, add $16.75 per square foot of basement area										

The above costs were calculated using the basic specifications shown on the facing page. These costs should be adjusted where necessary for design alternatives and owner's requirements. Reported completed project costs, for this type of structure, range from $46.75 to $153.00 per S.F.

Common additives

Description	Unit	$ Cost
Bar, Front bar	L.F.	262
Back bar	L.F.	209
Booth, Upholstered, custom, straight	L.F.	129 - 237
"L" or "U" shaped	L.F.	133 - 225
Fireplaces, Brick not incl. chimney		
or foundation, 30" x 24" opening	Each	1900
Chimney, standard brick		
Single flue 16" x 20"	V.L.F.	55
20" x 20"	V.L.F.	63
2 flue, 20" x 32"	V.L.F.	91
Lockers, Steel, single tier, 60" or 72"	Opening	135 - 216
2 tier, 60" or 72" total	Opening	83 - 105
5 tier, box lockers	Opening	44 - 57
Locker bench, lam. maple top only	L.F.	18.65
Pedestals, steel pipe	Each	43
Refrigerators, Prefabricated, walk-in		
7'-6" High, 6' x 6'	S.F.	117
10' x 10'	S.F.	91
12' x 14'	S.F.	81
12' x 20'	S.F.	72

Description	Unit	$ Cost
Sauna, Prefabricated, complete, 6' x 4'	Each	3725
6' x 9'	Each	5850
8' x 8'	Each	6225
10' x 12'	Each	8900
Smoke Detectors		
Ceiling type	Each	142
Duct type	Each	390
Sound System		
Amplifier, 250 watts	Each	1525
Speaker, ceiling or wall	Each	131
Trumpet	Each	246
Steam Bath, Complete, to 140 C.F.	Each	980
To 300 C.F.	Each	1175
To 800 C.F.	Each	3050
To 2500 C.F.	Each	3400
Swimming Pool Complete, gunite	S.F.	45 - 55
Tennis Court, Complete with fence		
Bituminous	Each	21,400 - 26,200
Clay	Each	23,300 - 29,200

Model costs calculated for a 1 story building with 12' story height and 6,000 square feet of floor area

			Unit	Unit Cost	Cost Per S.F.	% Of Sub-Total
1.0 Foundations						
.1	Footings & Foundations	Poured concrete; strip and spread footings and 4' foundation wall	S.F. Ground	5.02	5.02	
.4	Piles & Caissons	N/A	—	—	—	6.8%
.9	Excavation & Backfill	Site preparation for slab and trench for foundation wall and footing	S.F. Ground	1.11	1.11	
2.0 Substructure						
.1	Slab on Grade	4" reinforced concrete with vapor barrier and granular base	S.F. Slab	3.06	3.06	3.4%
.2	Special Substructures		—	—	—	
3.0 Superstructure						
.1	Columns & Beams	N/A	—	—	—	
.4	Structural Walls	Load bearing partition walls, see item 6.1	—	—	—	
.5	Elevated Floors	N/A	—	—	—	4.8%
.7	Roof	Wood truss with plywood sheathing	S.F. Ground	4.33	4.33	
.9	Stairs	N/A	—	—	—	
4.0 Exterior Closure						
.1	Walls	Stone ashlar veneer on concrete block 65% of wall	S.F. Wall	21	9.44	
.5	Exterior Wall Finishes	N/A	—	—	—	17.2%
.6	Doors	Double aluminum and glass, hollow metal	Each	1570	1.57	
.7	Windows & Glazed Walls	Aluminum horizontal sliding 35% of wall	Each	284	4.51	
5.0 Roofing						
.1	Roof Coverings	Asphalt shingles	S.F. Roof	1.00	1.23	
.7	Insulation	N/A	—	—	—	1.8%
.8	Openings & Specialties	Gutters and downspouts	S.F. Roof	.40	.40	
6.0 Interior Construction						
.1	Partitions	Gypsum board on metal studs, load bearing 14 S.F. Floor/L.F. Partition	S.F. Partition	2.97	2.12	
.4	Interior Doors	Single leaf wood 140 S.F. Floor/Door	Each	408	2.91	
.5	Wall Finishes	40% vinyl wall covering, 40% paint, 20% ceramic tile	S.F. Surface	1.76	2.52	
.6	Floor Finishes	50% carpet, 30% hardwood tile, 20% ceramic tile	S.F. Floor	5.60	5.60	19.1%
.7	Ceiling Finishes	Gypsum plaster on wood furring	S.F. Ceiling	2.87	2.87	
.9	Interior Surface/Exterior Wall	Painted gypsum board on furring 65% of wall	S.F. Wall	2.85	1.26	
7.0 Conveying						
.1	Elevators	N/A	—	—	—	0.0%
.2	Special Conveyors	N/A	—	—	—	
8.0 Mechanical						
.1	Plumbing	Kitchen, toilet and service fixtures, supply and drainage 1 Fixture/125 S.F. Floor	Each	1912	15.30	
.2	Fire Protection	Wet pipe sprinkler system	S.F. Floor	2.09	2.09	
.3	Heating	Included in 8.4	—	—	—	41.4%
.4	Cooling	Multizone rooftop unit, gas heating, electric cooling	S.F. Floor	20	20.00	
.5	Special Systems	N/A	—	—	—	
9.0 Electrical						
.1	Service & Distribution	200 ampere service, panel board and feeders	S.F. Floor	1.24	1.24	
.2	Lighting & Power	Fluorescent fixtures, receptacles, switches, A.C. and misc. power	S.F. Floor	3.26	3.26	5.5%
.4	Special Electrical	Alarm systems and emergency lighting	S.F. Floor	.43	.43	
11.0 Special Construction						
.1	Specialties	N/A	—	—	—	0.0%
12.0 Site Work						
.1	Earthwork	N/A	—	—	—	
.3	Utilities	N/A	—	—	—	0.0%
.5	Roads & Parking	N/A	—	—	—	
.7	Site Improvements	N/A	—	—	—	
			Sub-Total		90.27	**100%**
	GENERAL CONDITIONS (Overhead & Profit)			15%	13.54	
	ARCHITECT FEES			11%	11.44	
			Total Building Cost		115.25	

Costs per square foot of floor area

Exterior Wall	S.F. Area	4000	8000	12000	17000	22000	27000	32000	37000	42000
	L.F. Perimeter	280	386	520	585	640	740	840	940	940
Stone Ashlar on Concrete Block	Steel Joists	112.25	96.40	92.35	86.70	83.35	82.15	81.35	80.70	79.00
	Wood Joists	109.85	94.50	90.65	85.15	81.90	80.75	79.95	79.35	77.70
Face Brick on Concrete Block	Steel Joists	110.95	95.50	91.55	86.05	82.80	81.65	80.85	80.25	78.60
	Wood Joists	108.95	93.90	90.05	84.75	81.60	80.45	79.70	79.10	77.50
Decorative Concrete Block	Steel Joists	106.20	92.20	88.60	83.70	80.80	79.75	79.05	78.50	77.05
	Wood Joists	103.75	90.20	86.70	81.95	79.15	78.15	77.45	76.95	75.55
Perimeter Adj., Add or Deduct	Per 100 L.F.	13.85	6.90	4.65	3.25	2.55	2.05	1.70	1.50	1.30
Story Hgt. Adj., Add or Deduct	Per 1 Ft.	2.30	1.55	1.40	1.10	.95	.90	.85	.80	.70

For Basement, add $16.95 per square foot of basement area

The above costs were calculated using the basic specifications shown on the facing page. These costs should be adjusted where necessary for design alternatives and owner's requirements. Reported completed project costs, for this type of structure, range from $47.00 to $122.30 per S.F.

Common additives

Description	Unit	$ Cost
Bar, Front bar	L.F.	262
Back bar	L.F.	209
Booth, Upholstered, custom, straight	L.F.	129 - 237
"L" or "U" shaped	L.F.	133 - 225
Emergency Lighting, 25 watt, battery operated		
Lead battery	Each	330
Nickel cadmium	Each	620
Flagpoles, Complete		
Aluminum, 20' high	Each	1100
40' High	Each	2475
70' High	Each	7100
Fiberglass, 23' High	Each	1475
39'-5" High	Each	2525
59' High	Each	7025
Kitchen Equipment		
Broiler	Each	3425
Coffee urn, twin 6 gallon	Each	5725
Cooler, 6 ft. long, reach-in	Each	2725
Dishwasher, 10-12 racks per hr.	Each	2600
Food warmer, counter 1.2 kw	Each	745

Description	Unit	$ Cost
Kitchen Equipment, cont.		
Freezer, 44 C.F., reach-in	Each	6625
Ice cube maker, 50 lb. per day	Each	1750
Lockers, Steel, single tier, 60" or 72"	Opening	135 - 216
2 tier, 60" or 72" total	Opening	83 - 105
5 tier, box lockers	Opening	44 - 57
Locker bench, lam. maple top only	L.F.	18.65
Pedestals, steel pipe	Each	43
Refrigerators, Prefabricated, walk-in		
7'-6" High, 6' x 6'	S.F.	117
10' x 10'	S.F.	91
12' x 14'	S.F.	81
12' x 20'	S.F.	72
Smoke Detectors		
Ceiling type	Each	142
Duct type	Each	390
Sound System		
Amplifier, 250 watts	Each	1525
Speaker, ceiling or wall	Each	131
Trumpet	Each	246

Important: See the Reference Section for Location Factors

Model costs calculated for a 1 story building with 12' story height and 22,000 square feet of floor area

				Unit	Unit Cost	Cost Per S.F.	% Of Sub-Total
1.0 Foundations							
.1	Footings & Foundations	Poured concrete; strip and spread footings and 4' foundation wall		S.F. Ground	3.17	3.17	
.4	Piles & Caissons	N/A		—	—	—	6.2%
.9	Excavation & Backfill	Site preparation for slab and trench for foundation wall and footing		S.F. Ground	1.04	1.04	
2.0 Substructure							
.1	Slab on Grade	4" reinforced concrete with vapor barrier and granular base		S.F. Slab	3.06	3.06	4.5%
.2	Special Substructures	N/A		—	—	—	
3.0 Superstructure							
.1	Columns & Beams	N/A		—	—	—	
.4	Structural Walls	Load bearing partition walls, see item 6.1		—	—	—	
.5	Elevated Floors	N/A		—	—	—	5.7%
.7	Roof	Metal deck on open web steel joists		S.F. Roof	3.88	3.88	
.9	Stairs	N/A		—	—	—	
4.0 Exterior Closure							
.1	Walls	Stone, ashlar veneer on concrete block	65% of wall	S.F. Wall	21	4.84	
.5	Exterior Wall Finishes	N/A		—	—	—	13.2%
.6	Doors	Double aluminum and glass doors		Each	1570	.43	
.7	Windows & Glazed Walls	Window wall	35% of wall	S.F. Window	29	3.65	
5.0 Roofing							
.1	Roof Coverings	Built-up tar and gravel with flashing		S.F. Roof	2.10	2.10	
.7	Insulation	Perlite/EPS composite		S.F. Roof	1.25	1.25	5.2%
.8	Openings & Specialties	Gravel stop		L.F. Perimeter	5.60	.16	
6.0 Interior Construction							
.1	Partitions	Lightweight concrete block	14 S.F. Floor/L.F. Partition	S.F. Partition	5.68	4.06	
.4	Interior Doors	Single leaf wood	140 S.F. Floor/Door	Each	408	2.91	
.5	Wall Finishes	65% paint, 25% vinyl wall covering, 10% ceramic tile		S.F. Surface	1.18	1.69	26.8%
.6	Floor Finishes	60% carpet, 35% hardwood, 15% ceramic tile		S.F. Floor	5.68	5.68	
.7	Ceiling Finishes	Mineral fiber tile on concealed zee bars		S.F. Ceiling	3.07	3.07	
.9	Interior Surface/Exterior Wall	Painted gypsum board on furring	65% of wall	S.F. Wall	3.12	.71	
7.0 Conveying							
.1	Elevators	N/A		—	—	—	0.0%
.2	Special Conveyors	N/A		—	—	—	
8.0 Mechanical							
.1	Plumbing	Kitchen, toilet and service fixtures, supply and drainage	1 Fixture/1050 S.F. Floor	Each	2268	2.16	
.2	Fire Protection	Wet pipe sprinkler system		S.F. Floor	1.60	1.60	
.3	Heating	Included in 8.4		—	—	—	32.1%
.4	Cooling	Multizone rooftop unit, gas heating, electric cooling		S.F. Floor	18.00	18.00	
.5	Special Systems	N/A		—	—	—	
9.0 Electrical							
.1	Service & Distribution	400 ampere service, panel board and feeders		S.F. Floor	.69	.69	
.2	Lighting & Power	Fluorescent fixtures, receptacles, switches, A.C. and misc. power		S.F. Floor	3.33	3.33	6.3%
.4	Special Electrical	Alarm systems and emergency lighting		S.F. Floor	.24	.24	
11.0 Special Construction							
.1	Specialties	N/A		—	—	—	0.0%
12.0 Site Work							
.1	Earthwork	N/A		—	—	—	
.3	Utilities	N/A		—	—	—	0.0%
.5	Roads & Parking	N/A		—	—	—	
.7	Site Improvements	N/A		—	—	—	
				Sub-Total		67.72	100%
	GENERAL CONDITIONS (Overhead & Profit)				15%	10.16	
	ARCHITECT FEES				7%	5.47	
				Total Building Cost		83.35	

BUILDING TYPES

Costs per square foot of floor area

Exterior Wall	S.F. Area	30000	45000	60000	75000	90000	105000	120000	135000	150000
	L.F. Perimeter	550	650	800	950	1100	1250	1210	1330	1450
Face Brick with Concrete Block Back-up	Steel Frame	95.45	89.95	87.90	86.70	85.85	85.25	83.60	83.10	82.80
	Bearing Walls	96.10	89.85	87.55	86.20	85.25	84.55	82.55	82.05	81.70
Decorative Concrete Block	Steel Frame	92.10	87.35	85.45	84.35	83.60	83.05	81.75	81.30	81.05
	Bearing Walls	92.75	87.20	85.10	83.85	83.00	82.40	80.70	80.25	79.95
Stucco on Concrete Block	Steel Frame	90.45	85.85	84.05	83.00	82.25	81.70	80.50	80.05	79.80
	Bearing Walls	91.90	86.50	84.50	83.30	82.45	81.85	80.20	79.75	79.50
Perimeter Adj., Add or Deduct	Per 100 L.F.	2.65	1.75	1.30	1.05	.85	.75	.60	.60	.55
Story Hgt. Adj., Add or Deduct	Per 1 Ft.	1.05	.80	.75	.70	.70	.65	.55	.55	.55

For Basement, add $20.20 per square foot of basement area

The above costs were calculated using the basic specifications shown on the facing page. These costs should be adjusted where necessary for design alternatives and owner's requirements. Reported completed project costs, for this type of structure, range from $65.90 to $163.05 per S.F.

Common additives

Description	Unit	$ Cost
Carrels Hardwood	Each	655 - 855
Clock System		
20 Room	Each	11,800
50 Room	Each	28,700
Elevators, Hydraulic passenger, 2 stops		
1500# capacity	Each	40,000
2500# capacity	Each	41,300
3500# capacity	Each	44,900
Additional stop, add	Each	3350
Emergency Lighting, 25 watt, battery operated		
Lead battery	Each	330
Nickel cadmium	Each	620
Flagpoles, Complete		
Aluminum, 20' high	Each	1100
40' High	Each	2475
70' High	Each	7100
Fiberglass, 23' High		1475
39'-5" High	Each	2525
59' High	Each	7025

Description	Unit	$ Cost
Lockers, Steel, single tier, 60" or 72"	Opening	135 - 216
2 tier, 60" or 72" total	Opening	83 - 105
5 tier, box lockers	Opening	44 - 57
Locker bench, lam. maple top only	L.F.	18.65
Pedestals, steel pipe	Each	43
Seating		
Auditorium chair, all veneer	Each	127
Veneer back, padded seat	Each	152
Upholstered, spring seat	Each	178
Classroom, movable chair & desk	Set	65 - 120
Lecture hall, pedestal type	Each	127 - 370
Smoke Detectors		
Ceiling type	Each	142
Duct type	Each	390
Sound System		
Amplifier, 250 watts	Each	1525
Speaker, ceiling or wall	Each	131
Trumpet	Each	246
TV Antenna, Master system, 12 outlet	Outlet	229
30 outlet	Outlet	147
100 outlet	Outlet	141

Important: See the Reference Section for Location Factors

Model costs calculated for a 2 story building with 12' story height and 90,000 square feet of floor area

				Unit	Unit Cost	Cost Per S.F.	% Of Sub-Total
1.0 Foundations							
.1	Footings & Foundations	Poured concrete; strip and spread footings and 4' foundation wall		S.F. Ground	2.44	1.22	
.4	Piles & Caissons	N/A		—	—	—	2.5%
.9	Excavation & Backfill	Site preparation for slab and trench for foundation wall and footing		S.F. Ground	1.00	.50	
2.0 Substructure							
.1	Slab on Grade	4" reinforced concrete with vapor barrier and granular base		S.F. Slab	3.06	1.53	2.3%
.2	Special Substructures	N/A		—	—	—	
3.0 Superstructure					—	—	
.1	Columns & Beams	Interior columns included in 3.7		—	—	—	
.4	Structural Walls	Concrete block bearing walls		S.F. Wall	5.68	1.67	
.5	Elevated Floors	Open web steel joists, slab form, concrete		S.F. Floor	9.66	4.83	14.3%
.7	Roof	Metal deck on open web steel joists, columns		S.F. Roof	5.04	2.52	
.9	Stairs	Concrete filled metal pan		Flight	5900	.66	
4.0 Exterior Closure							
.1	Walls	Decorative concrete block	65% of wall	S.F. Wall	9.91	1.89	
.5	Exterior Wall Finishes	N/A		—	—	—	7.0%
.6	Doors	Double glass and aluminum with transom		Each	2725	.18	
.7	Windows & Glazed Walls	Window wall	35% of wall	S.F. Wall	25	2.62	
5.0 Roofing							
.1	Roof Coverings	Built-up tar and gravel with flashing		S.F. Roof	2.00	1.00	
.7	Insulation	Perlite/EPS composite		S.F. Roof	1.25	.63	2.5%
.8	Openings & Specialties	Gravel stop		L.F. Perimeter	5.60	.07	
6.0 Interior Construction							
.1	Partitions	Concrete block	20 S.F. Floor/L.F. Partition	S.F. Partition	5.14	2.57	
.4	Interior Doors	Single leaf hollow metal	200 S.F. Floor/Door	Each	526	2.63	
.5	Wall Finishes	95% paint, 5% ceramic tile		S.F. Surface	2.12	2.12	
.6	Floor Finishes	70% vinyl composition tile, 25% carpet, 5% ceramic tile		S.F. Floor	3.23	3.23	20.7%
.7	Ceiling Finishes	Mineral fiber tile on concealed zee bars		S.F. Ceiling	3.07	3.07	
.9	Interior Surface/Exterior Wall	Paint and block filler	65% of wall	S.F. Wall	1.95	.37	
7.0 Conveying							
.1	Elevators	Two hydraulic passenger elevators		Each	49,050	1.09	1.6%
.2	Special Conveyors	N/A		—	—	—	
8.0 Mechanical							
.1	Plumbing	Toilet and service fixtures, supply and drainage	1 Fixture/455 S.F. Floor	Each	2493	5.48	
.2	Fire Protection	Sprinklers, light hazard		S.F. Floor	1.25	1.25	
.3	Heating	Included in 8.4		—	—	—	30.5%
.4	Cooling	Multizone unit, gas heating, electric cooling		S.F. Floor	13.80	13.80	
.5	Special Systems	N/A		—	—	—	
9.0 Electrical							
.1	Service & Distribution	1600 ampere service, panel board and feeders		S.F. Floor	1.23	1.23	
.2	Lighting & Power	Fluorescent fixtures, receptacles, switches, A.C. and misc. power		S.F. Floor	7.50	7.50	16.0%
.4	Special Electrical	Alarm systems, communications systems and emergency lighting		S.F. Floor	2.04	2.04	
11.0 Special Construction							
.1	Specialties	Chalkboards, counters, cabinets		S.F. Floor	1.77	1.77	2.6%
12.0 Site Work							
.1	Earthwork	N/A		—	—	—	
.3	Utilities	N/A		—	—	—	
.5	Roads & Parking	N/A		—	—	—	0.0%
.7	Site Improvements	N/A		—	—	—	
				Sub-Total		67.47	100%
	GENERAL CONDITIONS (Overhead & Profit)				15%	10.12	
	ARCHITECT FEES				7%	5.41	
				Total Building Cost		83.00	

BUILDING TYPES

Costs per square foot of floor area

Exterior Wall	S.F. Area	20000	30000	40000	50000	60000	70000	80000	90000	100000
	L.F. Perimeter	341	454	476	550	600	628	684	721	772
Face Brick with Concrete Block Back-up	R/Conc. Frame	93.25	89.75	85.70	84.30	82.95	81.65	81.05	80.35	80.00
	Steel Frame	97.55	94.05	90.00	88.60	87.25	85.95	85.40	84.65	84.25
Decorative Concrete Block	R/Conc. Frame	87.80	84.95	81.95	80.90	79.90	78.95	78.50	77.95	77.70
	Steel Frame	91.80	88.95	85.95	84.85	83.85	82.95	82.50	81.95	81.65
Precast Concrete Panels	R/Conc. Frame	90.35	87.20	83.70	82.45	81.30	80.15	79.65	79.05	78.70
	Steel Frame	94.60	91.45	87.90	86.70	85.50	84.40	83.90	83.25	82.95
Perimeter Adj., Add or Deduct	Per 100 L.F.	5.25	3.50	2.60	2.10	1.75	1.50	1.30	1.15	1.05
Story Hgt. Adj., Add or Deduct	Per 1 Ft.	1.35	1.20	.90	.85	.80	.70	.70	.60	.60

For Basement, add $19.85 per square foot of basement area

The above costs were calculated using the basic specifications shown on the facing page. These costs should be adjusted where necessary for design alternatives and owner's requirements. Reported completed project costs, for this type of structure, range from $45.05 to $125.90 per S.F.

Common additives

Description	Unit	$ Cost
Carrels Hardwood	Each	655 - 855
Closed Circuit Surveillance, One station		
Camera and monitor	Each	1325
For additional camera stations, add	Each	730
Elevators, Hydraulic passenger, 2 stops		
2000# capacity	Each	40,900
2500# capacity	Each	41,300
3500# capacity	Each	44,900
Additional stop, add	Each	3350
Emergency Lighting, 25 watt, battery operated		
Lead battery	Each	330
Nickel cadmium	Each	620
Furniture	Student	1900 - 3600
Intercom System, 25 station capacity		
Master station	Each	1825
Intercom outlets	Each	114
Handset	Each	300

Description	Unit	$ Cost
Kitchen Equipment		
Broiler	Each	3425
Coffee urn, twin 6 gallon	Each	5725
Cooler, 6 ft. long	Each	2725
Dishwasher, 10-12 racks per hr.	Each	2600
Food warmer	Each	745
Freezer, 44 C.F., reach-in	Each	6625
Ice cube maker, 50 lb. per day	Each	1750
Range with 1 oven	Each	2325
Laundry Equipment		
Dryer, gas, 16 lb. capacity	Each	680
30 lb. capacity	Each	2550
Washer, 4 cycle	Each	760
Commercial	Each	1125
Smoke Detectors		
Ceiling type	Each	142
Duct type	Each	390
TV Antenna, Master system, 12 outlet	Outlet	229
30 outlet	Outlet	147
100 outlet	Outlet	141

Important: See the Reference Section for Location Factors

Model costs calculated for a 3 story building with 12' story height and 40,000 square feet of floor area

				Unit	Unit Cost	Cost Per S.F.	% Of Sub-Total
1.0 Foundations							
.1	Footings & Foundations	Poured concrete; strip and spread footings and 4' foundation wall		S.F. Ground	4.05	1.35	
.4	Piles & Caissons	N/A		—	—	—	2.4%
.9	Excavation & Backfill	Site preparation for slab and trench for foundation wall and footing		S.F. Ground	1.04	.35	
2.0 Substructure							
.1	Slab on Grade	4" reinforced concrete with vapor barrier and granular base		S.F. Slab	3.06	1.02	1.5%
.2	Special Substructures	N/A		—	—	—	
3.0 Superstructure							
.1	Columns & Beams	Concrete columns		L.F. Column	58	2.54	
.4	Structural Walls	N/A		—	—	—	
.5	Elevated Floors	Concrete flat plate		S.F. Floor	8.48	5.65	17.1%
.7	Roof	Concrete flat plate		S.F. Roof	8.16	2.72	
.9	Stairs	Concrete		Flight	3075	1.00	
4.0 Exterior Closure							
.1	Walls	Face brick with concrete block backup	80% of wall	S.F. Wall	19.46	6.67	
.5	Exterior Wall Finishes	N/A		—	—	—	
.6	Doors	Double glass & aluminum doors		Each	3500	.53	12.8%
.7	Windows & Glazed Walls	Aluminum horizontal sliding	20% of wall	Each	461	1.72	
5.0 Roofing							
.1	Roof Coverings	Built-up tar and gravel with flashing		S.F. Roof	2.22	.74	
.7	Insulation	Perlite/EPS composite		S.F. Roof	1.25	.42	1.8%
.8	Openings & Specialties	Gravel stop		L.F. Perimeter	5.60	.07	
6.0 Interior Construction							
.1	Partitions	Gypsum board on metal studs, concrete block	9 S.F. Floor/L.F. Partition	S.F. Partition	4.06	4.51	
.4	Interior Doors	Single leaf wood	90 S.F. Floor/Door	Each	408	4.53	
.5	Wall Finishes	95% paint, 5% ceramic tile		S.F. Surface	.78	1.73	
.6	Floor Finishes	80% carpet, 10% vinyl composition tile, 10% ceramic tile		S.F. Floor	5.37	5.37	25.0%
.7	Ceiling Finishes	90% paint, 10% suspended fiberglass board		S.F. Ceiling	.59	.59	
.9	Interior Surface/Exterior Wall	Paint	80% of wall	S.F. Wall	1.95	.67	
7.0 Conveying							
.1	Elevators	One hydraulic passenger elevator		Each	63,200	1.58	2.3%
.2	Special Conveyors	N/A		—	—	—	
8.0 Mechanical							
.1	Plumbing	Toilet and service fixtures, supply and drainage	1 Fixture/455 S.F. Floor	Each	1992	4.38	
.2	Fire Protection	Wet pipe sprinkler system		S.F. Floor	1.40	1.40	
.3	Heating	Included in 8.4		—	—	—	20.3%
.4	Cooling	Rooftop multizone unit system		S.F. Floor	8.40	8.40	
.5	Special Systems	N/A		—	—	—	
9.0 Electrical							
.1	Service & Distribution	600 ampere service, panel board and feeders		S.F. Floor	.69	.69	
.2	Lighting & Power	Fluorescent fixtures, receptacles, switches, A.C. and misc. power		S.F. Floor	5.73	5.73	11.3%
.4	Special Electrical	Alarm systems, communications systems and emergency lighting		S.F. Floor	1.42	1.42	
11.0 Special Construction							
.1	Specialties	Closet shelving, mirrors		S.F. Floor	3.86	3.86	5.5%
12.0 Site Work							
.1	Earthwork	N/A		—	—	—	
.3	Utilities	N/A		—	—	—	
.5	Roads & Parking	N/A		—	—	—	0.0%
.7	Site Improvements	N/A		—	—	—	
				Sub-Total		69.64	100%
	GENERAL CONDITIONS (Overhead & Profit)			15%		10.45	
	ARCHITECT FEES			7%		5.61	
				Total Building Cost		85.70	

BUILDING TYPES

Costs per square foot of floor area

Exterior Wall	S.F. Area	50000	70000	90000	100000	110000	130000	150000	170000	190000
	L.F. Perimeter	372	461	550	533	566	633	650	703	756
Face Brick with Concrete Block Back-up	R/Conc. Frame	91.95	89.40	87.95	86.15	85.65	85.00	83.70	83.15	82.80
	Steel Frame	98.35	95.75	94.30	92.55	92.05	91.35	90.10	89.55	89.15
Decorative Concrete Block	R/Conc. Frame	87.45	85.40	84.20	82.90	82.55	82.00	81.05	80.65	80.35
	Steel Frame	93.85	91.80	90.65	89.35	89.00	88.45	87.50	87.10	86.80
Precast Concrete Panels With Exposed Aggregate	R/Conc. Frame	91.80	89.25	87.80	86.05	85.60	84.85	83.60	83.10	82.70
	Steel Frame	98.15	95.60	94.15	92.45	91.95	91.25	90.00	89.45	89.05
Perimeter Adj., Add or Deduct	Per 100 L.F.	3.95	2.80	2.25	2.00	1.80	1.50	1.35	1.20	1.05
Story Hgt. Adj., Add or Deduct	Per 1 Ft.	1.15	1.00	.90	.85	.80	.75	.70	.65	.60
For Basement, add $19.95 per square foot of basement area										

The above costs were calculated using the basic specifications shown on the facing page. These costs should be adjusted where necessary for design alternatives and owner's requirements. Reported completed project costs, for this type of structure, range from $63.10 to $138.20 per S.F.

Common additives

Description	Unit	$ Cost
Carrels Hardwood	Each	655 - 855
Closed Circuit Surveillance, One station		
Camera and monitor	Each	1325
For additional camera stations, add	Each	730
Elevators, Electric passenger, 5 stops		
2000# capacity	Each	91,300
2500# capacity	Each	94,300
3500# capacity	Each	95,300
Additional stop, add	Each	5225
Emergency Lighting, 25 watt, battery operated		
Lead battery	Each	330
Nickel cadmium	Each	620
Furniture	Student	1900 - 3600
Intercom System, 25 station capacity		
Master station	Each	1825
Intercom outlets	Each	114
Handset	Each	300

Description	Unit	$ Cost
Kitchen Equipment		
Broiler	Each	3425
Coffee urn, twin, 6 gallon	Each	5725
Cooler, 6 ft. long	Each	2725
Dishwasher, 10-12 racks per hr.	Each	2600
Food warmer	Each	745
Freezer, 44 C.F., reach-in	Each	6625
Ice cube maker, 50 lb. per day	Each	1750
Range with 1 oven	Each	2325
Laundry Equipment		
Dryer, gas, 16 lb. capacity	Each	680
30 lb. capacity	Each	2550
Washer, 4 cycle	Each	760
Commercial	Each	1125
Smoke Detectors		
Ceiling type	Each	142
Duct type	Each	390
TV Antenna, Master system, 12 outlet	Outlet	229
30 outlet	Outlet	147
100 outlet	Outlet	141

Important: See the Reference Section for Location Factors

Model costs calculated for a 6 story building with 12' story height and 110,000 square feet of floor area

				Unit	Unit Cost	Cost Per S.F.	% Of Sub-Total
1.0 Foundations							
.1	Footings & Foundations	Poured concrete; strip and spread footings and 4' foundation wall		S.F. Ground	6.24	1.04	
.4	Piles & Caissons	N/A		—	—	—	1.7%
.9	Excavation & Backfill	Site preparation for slab and trench for foundation wall and footing		S.F. Ground	1.00	.17	
2.0 Substructure							
.1	Slab on Grade	4" reinforced concrete with vapor barrier and granular base		S.F. Slab	3.06	.51	0.7%
.2	Special Substructures	N/A		—	—	—	
3.0 Superstructure							
.1	Columns & Beams	Steel columns with fireproofing		L.F. Column	124	1.81	
.4	Structural Walls	N/A		—	—	—	
.5	Elevated Floors	Concrete slab with metal deck and beams		S.F. Floor	12.29	10.24	20.7%
.7	Roof	Concrete slab with metal deck and beams		S.F. Roof	11.55	1.92	
.9	Stairs	Concrete filled metal pan		Flight	5900	.97	
4.0 Exterior Closure							
.1	Walls	Decorative concrete block	80% of wall	S.F. Wall	11.00	3.26	
.5	Exterior Wall Finishes	N/A		—	—	—	6.7%
.6	Doors	Double glass & aluminum doors		Each	2290	.17	
.7	Windows & Glazed Walls	Aluminum horizontal sliding	20% of wall	Each	284	1.40	
5.0 Roofing							
.1	Roof Coverings	Built-up tar and gravel with flashing		S.F. Roof	2.16	.36	
.7	Insulation	Perlite/EPS composite		S.F. Roof	1.25	.21	0.8%
.8	Openings & Specialties	Gravel stop		L.F. Perimeter	5.60	.03	
6.0 Interior Construction							
.1	Partitions	Concrete block	9 S.F. Floor/L.F. Partition	S.F. Partition	4.79	5.32	
.4	Interior Doors	Single leaf wood	90 S.F. Floor/Door	Each	408	4.53	
.5	Wall Finishes	95% paint, 5% ceramic tile		S.F. Surface	1.29	2.87	
.6	Floor Finishes	80% carpet, 10% vinyl composition tile, 10% ceramic tile		S.F. Floor	5.37	5.37	26.6%
.7	Ceiling Finishes	Mineral fiber tile on concealed zee bars, paint		S.F. Ceiling	.59	.59	
.9	Interior Surface/Exterior Wall	Paint and block filler	80% of wall	S.F. Wall	1.95	.58	
7.0 Conveying							
.1	Elevators	Four geared passenger elevator		Each	125,400	4.56	6.3%
.2	Special Conveyors	N/A		—	—	—	
8.0 Mechanical							
.1	Plumbing	Toilet and service fixtures, supply and drainage	1 Fixture/390 S.F. Floor	Each	1950	5.00	
.2	Fire Protection	Sprinklers, light hazard		S.F. Floor	1.35	1.35	
.3	Heating	Oil fired hot water, wall fin radiation		S.F. Floor	2.59	2.59	20.1%
.4	Cooling	Chilled water, air cooled condenser system		S.F. Floor	5.62	5.62	
.5	Special Systems	N/A		—	—	—	
9.0 Electrical							
.1	Service & Distribution	1000 ampere service, panel board and feeders		S.F. Floor	.48	.48	
.2	Lighting & Power	Fluorescent fixtures, receptacles, switches, A.C. and misc. power		S.F. Floor	5.63	5.63	11.0%
.4	Special Electrical	Alarm systems, communications systems and emergency lighting		S.F. Floor	1.83	1.83	
11.0 Special Construction							
.1	Specialties	Closet shelving, mirrors		S.F. Floor	3.91	3.91	5.4%
12.0 Site Work							
.1	Earthwork	N/A		—	—	—	
.3	Utilities	N/A		—	—	—	0.0%
.5	Roads & Parking	N/A		—	—	—	
.7	Site Improvements	N/A		—	—	—	
				Sub-Total		72.32	**100%**
	GENERAL CONDITIONS (Overhead & Profit)				15%	10.85	
	ARCHITECT FEES				7%	5.83	
				Total Building Cost		**89.00**	

BUILDING TYPES

Costs per square foot of floor area

Exterior Wall	S.F. Area	12000	20000	28000	37000	45000	57000	68000	80000	92000
	L.F. Perimeter	470	600	698	793	900	1060	1127	1200	1320
Face Brick with Concrete Brick Back-up	Steel Frame	155.30	128.30	115.90	108.05	104.10	100.25	97.05	94.60	93.10
	Bearing Walls	152.65	125.60	113.25	105.40	101.45	97.60	94.40	91.90	90.45
Decorative Concrete Block	Steel Frame	151.55	125.40	113.50	106.00	102.20	98.45	95.45	93.15	91.75
	Bearing Walls	148.65	122.50	110.65	103.10	99.30	95.60	92.60	90.25	88.85
Stucco on Concrete Block	Steel Frame	149.95	124.20	112.55	105.15	101.40	97.75	94.80	92.55	91.20
	Bearing Walls	147.05	121.30	109.65	102.25	98.50	94.85	91.90	89.65	88.30
Perimeter Adj., Add or Deduct	Per 100 L.F.	5.90	3.55	2.50	1.95	1.55	1.20	1.05	.90	.75
Story Hgt. Adj., Add or Deduct	Per 1 Ft.	1.10	.85	.70	.60	.55	.50	.50	.40	.40

For Basement, add $18.40 per square foot of basement area

The above costs were calculated using the basic specifications shown on the facing page. These costs should be adjusted where necessary for design alternatives and owner's requirements. Reported completed project costs, for this type of structure, range from $91.10 to $171.50 per S.F.

Common additives

Description	Unit	$ Cost
Cabinets, Base, door units, metal	L.F.	154
Drawer units	L.F.	267
Tall storage cabinets, open	L.F.	276
With doors	L.F.	310
Wall, metal 12-1/2" deep, open	L.F.	104
With doors	L.F.	180
Carrels Hardwood	Each	655 - 855
Countertops, not incl. base cabinets, acid proof	S.F.	23 - 33
Stainless steel	S.F.	73
Fume Hood, Not incl. ductwork	L.F.	755 - 1750
Ductwork	Hood	1675 - 6175
Glassware Washer, Distilled water rinse	Each	5275 - 19,600
Seating		
Auditorium chair, all veneer	Each	127
Veneer back, padded seat	Each	152
Upholstered, spring seat	Each	178
Classroom, movable chair & desk	Set	65 - 120
Lecture hall, pedestal type	Each	127 - 370

Description	Unit	$ Cost
Safety Equipment, Eye wash, hand held	Each	315
Deluge shower	Each	370
Sink, One piece plastic		
Flask wash, freestanding	Each	1625
Tables, acid resist. top, drawers	L.F.	134
Titration Unit, Four 2000 ml reservoirs	Each	7550
Alternate Pricing Method: As % of lab furniture		
Plumbing, final connections, simple		10%
Moderately complex		15%
Complex		20%

Important: See the Reference Section for Location Factors

Model costs calculated for a 1 story building with 12' story height and 45,000 square feet of floor area

			Unit	Unit Cost	Cost Per S.F.	% Of Sub-Total
1.0 Foundations						
.1	Footings & Foundations	Poured concrete; strip and spread footings and 4' foundation wall	S.F. Ground	5.36	5.36	
.4	Piles & Caissons	N/A	—	—	—	8.0%
.9	Excavation & Backfill	Site preparation for slab and trench for foundation wall and footing	S.F. Ground	1.04	1.04	
2.0 Substructure						
.1	Slab on Grade	4" reinforced concrete with vapor barrier and granular base	S.F. Slab	3.06	3.06	3.8%
.2	Special Substructures	N/A	—	—	—	
3.0 Superstructure						
.1	Columns & Beams	N/A	—	—	—	
.4	Structural Walls	Grouted concrete block wall	S.F. Wall	2.39	2.39	
.5	Elevated Floors	Metal deck and concrete trench cover (5680 S.F.)	S.F. Cover	3.23	.41	6.9%
.7	Roof	Metal deck on open web steel joists	S.F. Roof	2.74	2.74	
.9	Stairs	N/A	—	—	—	
4.0 Exterior Closure						
.1	Walls	Face brick with concrete block backup 75% of wall	S.F. Wall	19.44	3.50	
.5	Exterior Wall Finishes	N/A	—	—	—	7.7%
.6	Doors	Glass and metal doors and entrances with transom	Each	2522	1.12	
.7	Windows & Glazed Walls	Window wall 25% of wall	S.F. Window	25	1.53	
5.0 Roofing						
.1	Roof Coverings	Built-up tar and gravel with flashing	S.F. Roof	1.93	1.93	
.7	Insulation	Perlite/EPS composite	S.F. Roof	1.25	1.25	4.4%
.8	Openings & Specialties	Gravel stop and skylight	S.F. Roof	.31	.31	
6.0 Interior Construction						
.1	Partitions	Concrete block 10 S.F. Floor/L.F. Partition	S.F. Partition	5.41	5.41	
.4	Interior Doors	Single leaf-kalamein fire doors 820 S.F. Floor/Door	Each	630	.77	
.5	Wall Finishes	60% paint, 40% epoxy coating	S.F. Surface	1.75	3.49	
.6	Floor Finishes	60% epoxy, 20% carpet, 20% vinyl composition tile	S.F. Floor	4.17	4.17	21.5%
.7	Ceiling Finishes	Mineral fiber tile on concealed zee runners	S.F. Ceiling	3.07	3.07	
.9	Interior Surface/Exterior Wall	Paint and block filler 75% of wall	S.F. Wall	1.95	.35	
7.0 Conveying						
.1	Elevators	N/A	—	—	—	0.0%
.2	Special Conveyors	N/A	—	—	—	
8.0 Mechanical						
.1	Plumbing	Toilet and service fixtures, supply and drainage 1 Fixture/260 S.F. Floor	Each	3621	13.93	
.2	Fire Protection	Sprinklers, light hazard	S.F. Floor	1.41	1.41	
.3	Heating	Included in 8.4	—	—	—	36.3%
.4	Cooling	Multizone unit, gas heating, electric cooling	S.F. Floor	13.80	13.80	
.5	Special Systems	N/A	—	—	—	
9.0 Electrical						
.1	Service & Distribution	600 ampere service, panel board and feeders	S.F. Floor	.70	.70	
.2	Lighting & Power	Fluorescent fixtures, receptacles, switches, A.C. and misc. power	S.F. Floor	6.48	6.48	9.6%
.4	Special Electrical	Alarm systems and emergency lighting	S.F. Floor	.53	.53	
11.0 Special Construction						
.1	Specialties	Cabinets, fume hoods, lockers	S.F. Floor	1.44	1.44	1.8%
12.0 Site Work						
.1	Earthwork	N/A	—	—	—	
.3	Utilities	N/A	—	—	—	
.5	Roads & Parking	N/A	—	—	—	0.0%
.7	Site Improvements	N/A	—	—	—	
			Sub-Total		80.19	**100%**

GENERAL CONDITIONS (Overhead & Profit)		15%	12.03
ARCHITECT FEES		10%	9.23
Total Building Cost			**101.45**

BUILDING TYPES

Costs per square foot of floor area

Exterior Wall	S.F. Area	15000	20000	25000	30000	35000	40000	45000	50000	55000
	L.F. Perimeter	354	425	457	513	568	583	629	644	683
Brick Face with Concrete Block Back-up	Steel Frame	98.70	95.75	92.80	91.40	90.45	88.85	88.20	87.15	86.70
	R/Conc. Frame	95.25	92.30	89.30	87.95	86.95	85.40	84.75	83.70	83.25
Precast Concrete Panel	Steel Frame	96.65	93.95	91.25	90.00	89.10	87.70	87.05	86.10	85.70
	R/Conc. Frame	93.50	90.75	87.95	86.70	85.75	84.35	83.70	82.75	82.35
Limestone Face Concrete Block Back-up	Steel Frame	102.25	98.95	95.55	94.00	92.90	91.05	90.30	89.10	88.60
	R/Conc. Frame	98.80	95.50	92.05	90.55	89.40	87.60	86.85	85.65	85.15
Perimeter Adj., Add or Deduct	Per 100 L.F.	5.30	4.00	3.15	2.60	2.30	2.00	1.75	1.55	1.45
Story Hgt. Adj., Add or Deduct	Per 1 Ft.	1.35	1.20	1.05	1.00	.90	.85	.80	.70	.70

For Basement, add $21.05 per square foot of basement area

The above costs were calculated using the basic specifications shown on the facing page. These costs should be adjusted where necessary for design alternatives and owner's requirements. Reported completed project costs, for this type of structure, range from $73.55 to $150.90 per S.F.

Common additives

Description	Unit	$ Cost
Carrels Hardwood	Each	655 - 855
Elevators, Hydraulic passenger, 2 stops		
2000# capacity	Each	40,900
2500# capacity	Each	41,300
3500# capacity	Each	44,900
Emergency Lighting, 25 watt, battery operated		
Lead battery	Each	330
Nickel cadmium	Each	620
Escalators, Metal		
32" wide, 10' story height	Each	85,700
20' story height	Each	96,800
48" wide, 10' Story height	Each	90,200
20' story height	Each	101,300
Glass		
32" wide, 10' story height	Each	85,700
20' story height	Each	96,800
48" wide, 10' story height	Each	90,200
20' story height	Each	101,300

Description	Unit	$ Cost
Lockers, Steel, Single tier, 60" or 72"	Opening	135 - 216
2 tier, 60" or 72" total	Opening	83 - 105
5 tier, box lockers	Opening	44 - 57
Locker bench, lam. maple top only	L.F.	18.65
Pedestals, steel pipe	Each	43
Sound System		
Amplifier, 250 watts	Each	1525
Speaker, ceiling or wall	Each	131
Trumpet	Each	246

BUILDING TYPES

108

Model costs calculated for a 2 story building with 12' story height and 25,000 square feet of floor area

				Unit	Unit Cost	Cost Per S.F.	% Of Sub-Total
1.0	**Foundations**						
.1	Footings & Foundations	Poured concrete; strip and spread footings and 4' foundation wall		S.F. Ground	4.40	2.20	
.4	Piles & Caissons	N/A		—	—	—	3.7%
.9	Excavation & Backfill	Site preparation for slab and trench for foundation wall and footing		S.F. Ground	1.04	.52	
2.0	**Substructure**						
.1	Slab on Grade	4" reinforced concrete with vapor barrier and granular base		S.F. Slab	3.06	1.53	2.1%
.2	Special Substructures	N/A		—	—	—	
3.0	**Superstructure**						
.1	Columns & Beams	Concrete columns		L.F. Column	110	2.64	
.4	Structural Walls	N/A		—	—	—	
.5	Elevated Floors	Concrete flat plate		S.F. Floor	9.76	4.88	17.6%
.7	Roof	Concrete flat plate		S.F. Roof	9.00	4.50	
.9	Stairs	Concrete		Flight	4675	.75	
4.0	**Exterior Closure**						
.1	Walls	Face brick with concrete block backup	75% of wall	S.F. Wall	19.45	6.40	
.5	Exterior Wall Finishes	N/A		—	—	—	13.1%
.6	Doors	Double aluminum and glass		Each	2170	.35	
.7	Windows & Glazed Walls	Window wall	25% of wall	S.F. Window	25	2.79	
5.0	**Roofing**						
.1	Roof Coverings	Built-up tar and gravel with flashing		S.F. Roof	2.22	1.11	
.7	Insulation	Perlite/EPS composite		S.F. Roof	1.25	.63	2.6%
.8	Openings & Specialties	Gravel stop, hatches, gutters and downspouts		S.F. Roof	.30	.15	
6.0	**Interior Construction**						
.1	Partitions	Gypsum board on metal studs	14 S.F. Floor/L.F. Partition	S.F. Partition	3.89	2.78	
.4	Interior Doors	Single leaf hollow metal	140 S.F. Floor/Door	Each	526	3.76	
.5	Wall Finishes	50% paint, 50% vinyl wall covering		S.F. Surface	.92	1.31	
.6	Floor Finishes	50% carpet, 50% vinyl composition tile		S.F. Floor	4.14	4.14	21.0%
.7	Ceiling Finishes	Suspended fiberglass board		S.F. Ceiling	2.37	2.37	
.9	Interior Surface/Exterior Wall	Paint	75% of wall	S.F. Wall	1.95	.86	
7.0	**Conveying**						
.1	Elevators	One hydraulic passenger elevator		Each	48,750	1.95	2.7%
.2	Special Conveyors	N/A		—	—	—	
8.0	**Mechanical**						
.1	Plumbing	Toilet and service fixtures, supply and drainage	1 Fixture/1040 S.F. Floor	Each	1830	1.76	
.2	Fire Protection	Wet pipe sprinkler system		S.F. Floor	1.45	1.45	
.3	Heating	Included in 8.4		—	—	—	23.5%
.4	Cooling	Multizone unit, gas heating, electric cooling		S.F. Floor	13.80	13.80	
.5	Special Systems	N/A		—	—	—	
9.0	**Electrical**						
.1	Service & Distribution	600 ampere service, panel board and feeders		S.F. Floor	1.26	1.26	
.2	Lighting & Power	Fluorescent fixtures, receptacles, switches, A.C. and misc. power		S.F. Floor	7.51	7.51	13.7%
.4	Special Electrical	Alarm systems, communications systems and emergency lighting		S.F. Floor	1.19	1.19	
11.0	**Special Construction**						
.1	Specialties	N/A		—	—	—	0.0%
12.0	**Site Work**						
.1	Earthwork	N/A		—	—	—	
.3	Utilities	N/A		—	—	—	0.0%
.5	Roads & Parking	N/A		—	—	—	
.7	Site Improvements	N/A		—	—	—	
				Sub-Total		72.59	100%
	GENERAL CONDITIONS (Overhead & Profit)				15%	10.89	
	ARCHITECT FEES				7%	5.82	
				Total Building Cost		89.30	

BUILDING TYPES

Costs per square foot of floor area

Exterior Wall	S.F. Area	4000	6000	8000	10000	12000	14000	16000	18000	20000
	L.F. Perimeter	260	340	420	453	460	510	560	610	600
Face Brick with Concrete Block Back-up	Bearing Walls	93.45	88.60	86.15	82.40	78.75	77.70	76.95	76.30	74.35
	Steel Frame	90.50	86.25	84.10	80.85	77.75	76.80	76.15	75.60	73.90
Decorative Concrete Block	Bearing Walls	82.10	78.30	76.40	73.50	70.70	69.85	69.25	68.80	67.30
	Steel Frame	83.05	79.75	78.05	75.65	73.30	72.60	72.10	71.70	70.45
Tilt Up Concrete Wall Panels	Bearing Walls	82.60	79.15	77.40	74.85	72.40	71.65	71.10	70.70	69.35
	Steel Frame	79.65	76.80	75.35	73.30	71.35	70.75	70.30	69.95	68.90
Perimeter Adj., Add or Deduct	Per 100 L.F.	12.45	8.30	6.25	5.00	4.10	3.60	3.05	2.80	2.50
Story Hgt. Adj., Add or Deduct	Per 1 Ft.	1.60	1.40	1.30	1.10	.95	.90	.85	.85	.75
For Basement, add $18.15 per square foot of basement area										

The above costs were calculated using the basic specifications shown on the facing page. These costs should be adjusted where necessary for design alternatives and owner's requirements. Reported completed project costs, for this type of structure, range from $45.00 to $143.50 per S.F.

Common additives

Description	Unit	$ Cost
Bar, Front bar	L.F.	262
Back bar	L.F.	209
Booth, Upholstered, custom straight	L.F.	129 - 237
"L" or "U" shaped	L.F.	133 - 225
Bowling Alleys, incl. alley, pinsetter		
Scorer, counter & misc. supplies, average	Lane	45,000
For automatic scorer, add	Lane	7475
Emergency Lighting, 25 watt, battery operated		
Lead battery	Each	330
Nickel cadmium	Each	620
Kitchen Equipment		
Broiler	Each	3425
Coffee urn, twin 6 gallon	Each	5725
Cooler, 6 ft. long	Each	2725
Dishwasher, 10-12 racks per hr.	Each	2600
Food warmer	Each	745
Freezer, 44 C.F., reach-in	Each	6625
Ice cube maker, 50 lb. per day	Each	1750
Range with 1 oven	Each	2325

Description	Unit	$ Cost
Movie Equipment		
Projector, 35mm	Each	10,900 - 14,500
Screen, wall or ceiling hung	S.F.	7.20 - 11.25
Partitions, Folding leaf, wood		
Acoustic type	S.F.	49 - 80
Seating		
Auditorium chair, all veneer	Each	127
Veneer back, padded seat	Each	152
Upholstered, spring seat	Each	178
Classroom, movable chair & desk	Set	65 - 120
Lecture hall, pedestal type	Each	127 - 370
Sound System		
Amplifier, 250 watts	Each	1525
Speaker, ceiling or wall	Each	131
Trumpet	Each	246
Stage Curtains, Medium weight	S.F.	13.20 - 410
Curtain Track, Light duty	L.F.	51
Swimming Pools, Complete, gunite	S.F.	45 - 55

Important: See the Reference Section for Location Factors

Model costs calculated for a 1 story building with 12' story height and 10,000 square feet of floor area

					Unit	Unit Cost	Cost Per S.F.	% Of Sub-Total

1.0 Foundations

.1	Footings & Foundations	Poured concrete; strip and spread footings and 4' foundation wall			S.F. Ground	5.84	5.84	
.4	Piles & Caissons	N/A			–	–	–	10.6%
.9	Excavation & Backfill	Site preparation for slab and trench for foundation wall and footing			S.F. Ground	1.11	1.11	

2.0 Substructure

.1	Slab on Grade	4" reinforced concrete with vapor barrier and granular base			S.F. Slab	3.06	3.06	4.7%
.2	Special Substructures	N/A			–	–	–	

3.0 Superstructure

.1	Columns & Beams	N/A			–	–	–	
.4	Structural Walls	Concrete block			S.F. Wall	5.68	.93	
.5	Elevated Floors	N/A			–	–	–	7.3%
.7	Roof	Metal deck on open web steel joists			S.F. Roof	3.87	3.87	
.9	Stairs	N/A			–	–	–	

4.0 Exterior Closure

.1	Walls	Face brick with concrete block backup	80% of wall		S.F. Wall	19.45	8.46	
.5	Exterior Wall Finishes	N/A			–	–	–	15.9%
.6	Doors	Double aluminum and glass and hollow metal			Each	1426	.57	
.7	Windows & Glazed Walls	Aluminum sliding	20% of wall		Each	415	1.41	

5.0 Roofing

.1	Roof Coverings	Built-up tar and gravel with flashing			S.F. Roof	2.41	2.41	
.7	Insulation	Perlite/EPS composite			S.F. Roof	1.25	1.25	6.5%
.8	Openings & Specialties	Gravel stop and hatches			S.F. Roof	.63	.63	

6.0 Interior Construction

.1	Partitions	Gypsum board on metal studs, toilet partitions	14 S.F. Floor/L.F. Partition		S.F. Partition	5.56	3.97	
.4	Interior Doors	Single leaf hollow metal	140 S.F. Floor/Door		Each	526	3.76	
.5	Wall Finishes	Paint			S.F. Surface	.52	.74	
.6	Floor Finishes	50% carpet, 50% vinyl tile			S.F. Floor	4.14	4.14	25.2%
.7	Ceiling Finishes	Mineral fiber tile on concealed zee bars			S.F. Ceiling	3.07	3.07	
.9	Interior Surface/Exterior Wall	Paint	80% of wall		S.F. Wall	1.95	.85	

7.0 Conveying

.1	Elevators	N/A			–	–	–	0.0%
.2	Special Conveyors	N/A			–	–	–	

8.0 Mechanical

.1	Plumbing	Kitchen, toilet and service fixtures, supply and drainage	1 Fixture/910 S.F. Floor		Each	3594	3.95	
.2	Fire Protection	Wet pipe sprinkler system			S.F. Floor	1.60	1.60	
.3	Heating	Included in 8.4			–	–	–	21.2%
.4	Cooling	Single zone rooftop unit, gas heating, electric cooling			S.F. Floor	8.42	8.42	
.5	Special Systems	N/A			–	–	–	

9.0 Electrical

.1	Service & Distribution	200 ampere service, panel board and feeders			S.F. Floor	.61	.61	
.2	Lighting & Power	Incandescent fixtures, receptacles, switches, A.C. and misc. power			S.F. Floor	2.88	2.88	5.9%
.4	Special Electrical	Alarm systems and emergency lighting			S.F. Floor	.42	.42	

11.0 Special Construction

.1	Specialties	Built-in coat racks, fume hoods, freezer, kitchen equipment			S.F. Floor	1.77	1.77	2.7%

12.0 Site Work

.1	Earthwork	N/A			–	–	–	
.3	Utilities	N/A			–	–	–	
.5	Roads & Parking	N/A			–	–	–	0.0%
.7	Site Improvements	N/A			–	–	–	

						Sub-Total	65.72	100%

GENERAL CONDITIONS (Overhead & Profit)		15%	9.86
ARCHITECT FEES		9%	6.82

Total Building Cost	**82.40**

BUILDING TYPES

Costs per square foot of floor area

Exterior Wall	S.F. Area	16000	23000	30000	37000	44000	51000	58000	65000	72000
	L.F. Perimeter	597	763	821	968	954	1066	1090	1132	1220
Limestone with Concrete Block Back-up	R/Conc. Frame	120.25	116.45	112.15	111.00	107.85	107.15	105.65	104.65	104.30
	Steel Frame	120.35	116.60	112.25	111.10	107.95	107.30	105.80	104.75	104.40
Face Brick with Concrete Block Back-up	R/Conc. Frame	114.80	111.60	108.15	107.15	104.70	104.10	102.95	102.10	101.80
	Steel Frame	114.90	111.75	108.25	107.30	104.80	104.25	103.05	102.25	101.95
Stone with Concrete Block Back-up	R/Conc. Frame	115.65	112.40	108.75	107.75	105.15	104.55	103.35	102.45	102.15
	Steel Frame	115.85	112.55	108.95	107.90	105.35	104.75	103.50	102.65	102.35
Perimeter Adj., Add or Deduct	Per 100 L.F.	4.00	2.75	2.15	1.70	1.45	1.25	1.10	1.00	.90
Story Hgt. Adj., Add or Deduct	Per 1 Ft.	1.40	1.25	1.00	.95	.80	.80	.70	.65	.60

For Basement, add $17.95 per square foot of basement area

The above costs were calculated using the basic specifications shown on the facing page. These costs should be adjusted where necessary for design alternatives and owner's requirements. Reported completed project costs, for this type of structure, range from $79.30 to $147.75 per S.F.

Common additives

Description	Unit	$ Cost
Benches, Hardwood	L.F.	65 - 101
Clock System		
20 room	Each	11,800
50 room	Each	28,700
Closed Circuit Surveillance, One station		
Camera and monitor	Each	1325
For additional camera stations, add	Each	730
Directory Boards, Plastic, glass covered		
30" x 20"	Each	510
36" x 48"	Each	930
Aluminum, 24" x 18"	Each	450
36" x 24"	Each	540
48" x 32"	Each	640
48" x 60"	Each	1450
Emergency Lighting, 25 watt, battery operated		
Lead battery	Each	330
Nickel cadmium	Each	620

Description	Unit	$ Cost
Flagpoles, Complete		
Aluminum, 20' high	Each	1100
40' high	Each	2475
70' high	Each	7100
Fiberglass, 23' high	Each	1475
39'-5" high	Each	2525
59' high	Each	7025
Intercom System, 25 station capacity		
Master station	Each	1825
Intercom outlets	Each	114
Handset	Each	300
Safe, Office type, 4 hour rating		
30" x 18" x 18"	Each	3475
62" x 33" x 20"	Each	7575
Smoke Detectors		
Ceiling type	Each	142
Duct type	Each	390

Important: See the Reference Section for Location Factors

Model costs calculated for a 1 story building with 14' story height and 30,000 square feet of floor area

				Unit	Unit Cost	Cost Per S.F.	% Of Sub-Total
1.0 Foundations							
.1	Footings & Foundations	Poured concrete; strip and spread footings and 4' foundation wall		S.F. Ground	2.35	2.35	
.4	Piles & Caissons	N/A		—	—	—	3.7%
.9	Excavation & Backfill	Site preparation for slab and trench for foundation wall and footing		S.F. Ground	1.04	1.04	
2.0 Substructure							
.1	Slab on Grade	4" reinforced concrete with vapor barrier and granular base		S.F. Slab	3.06	3.06	3.4%
.2	Special Substructures	N/A		—	—	—	
3.0 Superstructure							
.1	Columns & Beams	Concrete		L.F. Column	42	1.08	
.4	Structural Walls	N/A		—	—	—	
.5	Elevated Floors	N/A		—	—	—	13.1%
.7	Roof	Cast-in-place concrete waffle slab		S.F. Roof	10.89	10.89	
.9	Stairs	N/A		—	—	—	
4.0 Exterior Closure							
.1	Walls	Limestone panels with concrete block backup	75% of wall	S.F. Wall	30	8.84	
.5	Exterior Wall Finishes	N/A		—	—	—	12.1%
.6	Doors	Double wood		Each	1277	.29	
.7	Windows & Glazed Walls	Aluminum with insulated glass	25% of wall	Each	461	1.92	
5.0 Roofing							
.1	Roof Coverings	Built-up tar and gravel with flashing		S.F. Roof	2.16	2.16	
.7	Insulation	Perlite/EPS composite		S.F. Roof	1.25	1.25	3.8%
.8	Openings & Specialties	Gravel stop, hatches		S.F. Roof	.05	.05	
6.0 Interior Construction							
.1	Partitions	Plaster on metal studs, toilet partitions	10 S.F. Floor/L.F. Partition	S.F. Partition	7.92	9.51	
.4	Interior Doors	Single leaf wood	100 S.F. Floor/Door	Each	408	4.08	
.5	Wall Finishes	70% paint, 20% wood paneling, 10% vinyl wall covering		S.F. Surface	1.75	4.20	36.8%
.6	Floor Finishes	60% hardwood, 20% carpet, 20% terrazzo		S.F. Floor	8.62	8.62	
.7	Ceiling Finishes	Gypsum plaster on metal lath, suspended		S.F. Ceiling	6.24	6.24	
.9	Interior Surface/Exterior Wall	Painted plaster	75% of wall	S.F. Wall	3.12	.90	
7.0 Conveying							
.1	Elevators	N/A		—	—	—	0.0%
.2	Special Conveyors	N/A		—	—	—	
8.0 Mechanical							
.1	Plumbing	Toilet and service fixtures, supply and drainage	1 Fixture/1110 S.F. Floor	Each	2941	2.65	
.2	Fire Protection	Wet pipe sprinkler system		S.F. Floor	1.41	1.41	
.3	Heating	Included in 8.4		—	—	—	19.6%
.4	Cooling	Multizone unit, gas heating, electric cooling		S.F. Floor	13.80	13.80	
.5	Special Systems	N/A		—	—	—	
9.0 Electrical							
.1	Service & Distribution	400 ampere service, panel board and feeders		S.F. Floor	.58	.58	
.2	Lighting & Power	Fluorescent fixtures, receptacles, switches, A.C. and misc. power		S.F. Floor	5.82	5.82	7.5%
.4	Special Electrical	Alarm systems and emergency lighting		S.F. Floor	.39	.39	
11.0 Special Construction							
.1	Specialties	N/A		—	—	—	0.0%
12.0 Site Work							
.1	Earthwork	N/A		—	—	—	
.3	Utilities	N/A		—	—	—	0.0%
.5	Roads & Parking	N/A		—	—	—	
.7	Site Improvements	N/A		—	—	—	
				Sub-Total		91.13	**100%**
	GENERAL CONDITIONS (Overhead & Profit)			15%		13.67	
	ARCHITECT FEES			7%		7.35	
				Total Building Cost		**112.15**	

Costs per square foot of floor area

Exterior Wall	S.F. Area	30000	40000	45000	50000	60000	70000	80000	90000	100000
	L.F. Perimeter	410	493	535	533	600	666	733	800	795
Limestone with Concrete Block Back-up	R/Conc. Frame	123.70	119.75	118.45	115.90	114.10	112.80	111.80	111.05	109.25
	Steel Frame	124.55	120.55	119.25	116.70	114.90	113.60	112.60	111.85	110.05
Face Brick with Concrete Block Back-up	R/Conc. Frame	118.60	115.10	113.95	111.90	110.35	109.20	108.35	107.70	106.25
	Steel Frame	119.40	115.90	114.75	112.70	111.15	110.00	109.15	108.55	107.05
Stone with Concrete Block Back-up	R/Conc. Frame	119.45	115.90	114.70	112.55	110.95	109.80	108.95	108.25	106.75
	Steel Frame	120.25	116.70	115.50	113.40	111.75	110.60	109.75	109.10	107.55
Perimeter Adj., Add or Deduct	Per 100 L.F.	5.65	4.20	3.75	3.35	2.80	2.40	2.15	1.90	1.70
Story Hgt. Adj., Add or Deduct	Per 1 Ft.	1.80	1.60	1.55	1.40	1.30	1.25	1.20	1.15	1.05

For Basement, add $18.30 per square foot of basement area

The above costs were calculated using the basic specifications shown on the facing page. These costs should be adjusted where necessary for design alternatives and owner's requirements. Reported completed project costs, for this type of structure, range from $79.75 to $148.65 per S.F.

Common additives

Description	Unit	$ Cost
Benches, Hardwood	L.F.	65 - 101
Clock System		
20 room	Each	11,800
50 room	Each	28,700
Closed Circuit Surveillance, One station		
Camera and monitor	Each	1325
For additional camera stations, add	Each	730
Directory Boards, Plastic, glass covered		
30" x 20"	Each	510
36" x 48"	Each	930
Aluminum, 24" x 18"	Each	450
36" x 24"	Each	540
48" x 32"	Each	640
48" x 60"	Each	1450
Elevators, Hydraulic passenger, 2 stops		
1500# capacity	Each	40,000
2500# capacity	Each	41,300
3500# capacity	Each	44,900
Additional stop, add	Each	3350

Description	Unit	$ Cost
Emergency Lighting, 25 watt, battery operated		
Lead battery	Each	330
Nickel cadmium	Each	620
Flagpoles, Complete		
Aluminum, 20' high	Each	1100
40' high	Each	2475
70' high	Each	7100
Fiberglass, 23' high	Each	1475
39'-5" high	Each	2525
59' high	Each	7025
Intercom System, 25 station capacity		
Master station	Each	1825
Intercom outlets	Each	114
Handset	Each	300
Safe, Office type, 4 hour rating		
30" x 18" x 18"	Each	3475
62" x 33" x 20"	Each	7575
Smoke Detectors		
Ceiling type	Each	142
Duct type	Each	390

Important: See the Reference Section for Location Factors

Model costs calculated for a 3 story building with 12' story height and 60,000 square feet of floor area

				Unit	Unit Cost	Cost Per S.F.	% Of Sub-Total
1.0 Foundations							
.1	Footings & Foundations	Poured concrete; strip and spread footings and 4' foundation wall		S.F. Ground	3.90	1.30	
.4	Piles & Caissons	N/A		—	—	—	1.8%
.9	Excavation & Backfill	Site preparation for slab and trench for foundation wall and footing		S.F. Ground	1.04	.35	
2.0 Substructure							
.1	Slab on Grade	4" reinforced concrete with vapor barrier and granular base		S.F. Slab	3.06	1.02	1.1%
.2	Special Substructures	N/A		—	—	—	
3.0 Superstructure							
.1	Columns & Beams	Steel columns		L.F. Column	44	1.35	
.4	Structural Walls	N/A		—	—	—	
.5	Elevated Floors	Concrete slab with metal deck and beams		S.F. Floor	13.64	9.09	16.7%
.7	Roof	Concrete slab with metal deck and beams		S.F. Roof	10.61	3.54	
.9	Stairs	Concrete filled metal pan		Flight	6800	1.13	
4.0 Exterior Closure							
.1	Walls	Face brick with concrete block backup	75% of wall	S.F. Wall	20	5.62	
.5	Exterior Wall Finishes	N/A		—	—	—	10.5%
.6	Doors	Double aluminum and glass and hollow metal		Each	2137	.22	
.7	Windows & Glazed Walls	Horizontal pivoted steel	25% of wall	Each	364	3.65	
5.0 Roofing							
.1	Roof Coverings	Built-up tar and gravel with flashing		S.F. Roof	2.13	.71	
.7	Insulation	Perlite/EPS composite		S.F. Roof	1.25	.42	1.3%
.8	Openings & Specialties	Gravel stop		L.F. Perimeter	5.60	.06	
6.0 Interior Construction							
.1	Partitions	Plaster on metal studs, toilet partitions	10 S.F. Floor/L.F. Partition	S.F. Partition	7.79	7.79	
.4	Interior Doors	Single leaf wood	100 S.F. Floor/Door	Each	408	4.08	
.5	Wall Finishes	70% paint, 20% wood paneling, 10% vinyl wall covering		S.F. Surface	1.75	3.50	34.4%
.6	Floor Finishes	60% hardwood, 20% terrazzo, 20% carpet		S.F. Floor	8.62	8.62	
.7	Ceiling Finishes	Gypsum plaster on metal lath, suspended		S.F. Ceiling	6.24	6.24	
.9	Interior Surface/Exterior Wall	Painted plaster	75% of wall	S.F. Wall	3.12	.84	
7.0 Conveying							
.1	Elevators	Five hydraulic passenger elevator		Each	72,480	6.04	6.7%
.2	Special Conveyors	N/A		—	—	—	
8.0 Mechanical							
.1	Plumbing	Toilet and service fixtures, supply and drainage	1 Fixture/665 S.F. Floor	Each	1655	2.49	
.2	Fire Protection	Wet pipe sprinkler system		S.F. Floor	1.40	1.40	
.3	Heating	Included in 8.4		—	—	—	19.7%
.4	Cooling	Multizone unit, gas heating, electric cooling		S.F. Floor	13.80	13.80	
.5	Special Systems	N/A		—	—	—	
9.0 Electrical							
.1	Service & Distribution	800 ampere service, panel board and feeders		S.F. Floor	.67	.67	
.2	Lighting & Power	Fluorescent fixtures, receptacles, switches, A.C. and misc. power		S.F. Floor	5.90	5.90	7.8%
.4	Special Electrical	Alarm systems and emergency lighting		S.F. Floor	.49	.49	
11.0 Special Construction							
.1	Specialties	N/A		—	—	—	0.0%
12.0 Site Work							
.1	Earthwork	N/A		—	—	—	
.3	Utilities	N/A		—	—	—	
.5	Roads & Parking	N/A		—	—	—	0.0%
.7	Site Improvements	N/A		—	—	—	
				Sub-Total		90.32	**100%**
	GENERAL CONDITIONS (Overhead & Profit)				15%	13.55	
	ARCHITECT FEES				7%	7.28	
				Total Building Cost		111.15	

BUILDING TYPES

115

Costs per square foot of floor area

Exterior Wall	S.F. Area	12000	18000	24000	30000	36000	42000	48000	54000	60000
	L.F. Perimeter	460	580	713	730	826	880	965	1006	1045
Concrete Block	Steel Frame	67.05	63.45	61.85	59.50	58.70	57.80	57.30	56.65	56.15
	Bearing Walls	66.15	62.55	60.95	58.60	57.80	56.90	56.40	55.75	55.25
Precast Concrete Panels	Steel Frame	70.60	66.45	64.60	61.75	60.85	59.75	59.15	58.40	57.75
Insulated Metal Panels	Steel Frame	68.10	64.35	62.70	60.20	59.35	58.40	57.90	57.20	56.65
Face Brick on Common Brick	Steel Frame	78.00	72.65	70.35	66.45	65.25	63.75	63.05	61.95	61.10
Tilt-up Concrete Panel	Steel Frame	67.60	63.95	62.30	59.85	59.05	58.10	57.60	56.90	56.40
Perimeter Adj., Add or Deduct	Per 100 L.F.	3.10	2.05	1.55	1.20	1.00	.85	.75	.70	.65
Story Hgt. Adj., Add or Deduct	Per 1 Ft.	.45	.40	.35	.30	.25	.25	.25	.20	.20

For Basement, add $18.05 per square foot of basement area

The above costs were calculated using the basic specifications shown on the facing page. These costs should be adjusted where necessary for design alternatives and owner's requirements. Reported completed project costs, for this type of structure, range from $25.50 to $98.60 per S.F.

Common additives

Description	Unit	$ Cost
Clock System		
20 room	Each	11,800
50 room	Each	28,700
Dock Bumpers, Rubber blocks		
4-1/2" thick, 10" high, 14" long	Each	55
24" long	Each	77
36" long	Each	96
12" high, 14" long	Each	79
24" long	Each	89
36" long	Each	104
6" thick, 10" high, 14" long	Each	77
24" long	Each	98
36" long	Each	124
20" high, 11" long	Each	128
Dock Boards, Heavy		
60" x 60" Aluminum, 5,000# cap.	Each	1125
9000# cap.	Each	1400
15,000# cap.	Each	1550

Description	Unit	$ Cost
Dock Levelers, Hinged 10 ton cap.		
6' x 8'	Each	4050
7' x 8'	Each	4275
Partitions, Woven wire, 10 ga., 1-1/2" mesh		
4' wide x 7' high	Each	117
8' high	Each	124
10' High	Each	151
Platform Lifter, Portable, 6'x 6'		
3000# cap.	Each	6000
4000# cap.	Each	8425
Fixed, 6' x 8', 5000# cap.	Each	8675

Important: See the Reference Section for Location Factors

Model costs calculated for a 1 story building with 20' story height and 30,000 square feet of floor area

				Unit	Unit Cost	Cost Per S.F.	% Of Sub-Total
1.0 Foundations							
.1	Footings & Foundations	Poured concrete; strip and spread footings and 4' foundation wall		S.F. Ground	2.17	2.17	
.4	Piles & Caissons	N/A		–	–	–	6.6%
.9	Excavation & Backfill	Site preparation for slab and trench for foundation wall and footing		S.F. Ground	1.04	1.04	
2.0 Substructure							
.1	Slab on Grade	4" reinforced concrete with vapor barrier and granular base		S.F. Slab	3.06	3.06	6.3%
.2	Special Substructures	N/A		–	–	–	
3.0 Superstructure							
.1	Columns & Beams	Steel columns included in 3.7		–	–	–	
.4	Structural Walls	N/A		–	–	–	
.5	Elevated Floors	N/A		–	–	–	10.2%
.7	Roof	Metal deck, open web steel joists, beams and columns		S.F. Roof	4.93	4.93	
.9	Stairs	N/A		–	–	–	
4.0 Exterior Closure							
.1	Walls	Concrete block	75% of wall	S.F. Wall	4.79	1.75	
.5	Exterior Wall Finishes	N/A		–	–	–	9.5%
.6	Doors	Double aluminum and glass, hollow metal, steel overhead		Each	1275	.64	
.7	Windows & Glazed Walls	Industrial horizontal pivoted steel	25% of wall	Each	580	2.21	
5.0 Roofing							
.1	Roof Coverings	Built-up tar and gravel with flashing		S.F. Roof	2.01	2.01	
.7	Insulation	Perlite/EPS composite		S.F. Roof	1.25	1.25	7.4%
.8	Openings & Specialties	Gravel stop, hatches, gutters and downspouts		S.F. Roof	.33	.33	
6.0 Interior Construction							
.1	Partitions	Concrete block, toilet partitions	60 S.F. Floor/L.F. Partition	S.F. Partition	8.75	1.75	
.4	Interior Doors	Single leaf hollow metal and fire doors	600 S.F. Floor/Door	Each	526	.88	
.5	Wall Finishes	Paint		S.F. Surface	1.10	.44	
.6	Floor Finishes	Vinyl composition tile	10% of floor	S.F. Floor	2.70	.27	9.0%
.7	Ceiling Finishes	Fiberglass board on exposed grid system	10% of area	S.F. Ceiling	3.07	.31	
.9	Interior Surface/Exterior Wall	Paint and block filler	75% of wall	S.F. Wall	1.95	.71	
7.0 Conveying							
.1	Elevators	N/A		–	–	–	0.0%
.2	Special Conveyors	N/A		–	–	–	
8.0 Mechanical							
.1	Plumbing	Toilet and service fixtures, supply and drainage	1 Fixture/1000 S.F. Floor	Each	2850	2.85	
.2	Fire Protection	Sprinklers, ordinary hazard		S.F. Floor	2.04	2.04	
.3	Heating	Oil fired hot water, unit heaters		S.F. Floor	5.53	5.53	37.1%
.4	Cooling	Chilled water, air cooled condenser system		S.F. Floor	7.46	7.46	
.5	Special Systems	N/A		–	–	–	
9.0 Electrical							
.1	Service & Distribution	600 ampere service, panel board and feeders		S.F. Floor	.75	.75	
.2	Lighting & Power	High intensity discharge fixtures, receptacles, switches, A.C. and misc. power		S.F. Floor	5.69	5.69	13.9%
.4	Special Electrical	Alarm systems and emergency lighting		S.F. Floor	.28	.28	
11.0 Special Construction							
.1	Specialties	N/A		–	–	–	0.0%
12.0 Site Work							
.1	Earthwork	N/A		–	–	–	
.3	Utilities	N/A		–	–	–	
.5	Roads & Parking	N/A		–	–	–	0.0%
.7	Site Improvements	N/A		–	–	–	
				Sub-Total		48.35	100%
	GENERAL CONDITIONS (Overhead & Profit)				15%	7.25	
	ARCHITECT FEES				7%	3.90	
				Total Building Cost		59.50	

BUILDING TYPES

Costs per square foot of floor area

Exterior Wall	S.F. Area	20000	30000	40000	50000	60000	70000	80000	90000	100000
	L.F. Perimeter	362	410	493	576	600	628	660	700	744
Face Brick Common Brick Back-up	Steel Frame	82.75	75.55	73.10	71.55	69.35	67.80	66.70	65.95	65.40
	Concrete Frame	81.75	74.55	72.10	70.60	68.35	66.80	65.70	64.95	64.45
Face Brick Concrete Block Back-up	Steel Frame	81.85	74.90	72.55	71.10	68.90	67.45	66.35	65.65	65.15
	Concrete Frame	80.75	73.80	71.40	69.95	67.80	66.30	65.25	64.50	64.00
Stucco on Concrete Block	Steel Frame	75.20	69.85	68.00	66.80	65.25	64.10	63.30	62.80	62.40
	Concrete Frame	74.00	68.70	66.80	65.65	64.05	62.95	62.15	61.60	61.25
Perimeter Adj., Add or Deduct	Per 100 L.F.	6.50	4.30	3.25	2.60	2.15	1.85	1.60	1.45	1.30
Story Hgt. Adj., Add or Deduct	Per 1 Ft.	1.80	1.35	1.20	1.15	1.00	.90	.85	.75	.75
For Basement, add $19.10 per square foot of basement area										

The above costs were calculated using the basic specifications shown on the facing page. These costs should be adjusted where necessary for design alternatives and owner's requirements. Reported completed project costs, for this type of structure, range from $25.50 to $98.60 per S.F.

Common additives

Description	Unit	$ Cost
Clock System		
20 room	Each	11,800
50 room	Each	28,700
Dock Bumpers, Rubber blocks		
4-1/2" thick, 10" high, 14" long	Each	55
24" long	Each	77
36" long	Each	96
12" high, 14" long	Each	79
24" long	Each	89
36" long	Each	104
6" thick, 10" high, 14" long	Each	77
24" long	Each	98
36" long	Each	124
20" high, 11" long	Each	128
Dock Boards, Heavy		
60" x 60" Aluminum, 5,000# cap.	Each	1125
9000# cap.	Each	1400
15,000# cap.	Each	1550

Description	Unit	$ Cost
Dock Levelers, Hinged 10 ton cap.		
6' x 8'	Each	4050
7' x 8'	Each	4275
Elevator, Hydraulic freight, 2 stops		
3500# capacity	Each	54,700
4000# capacity	Each	57,600
Additional stop, add	Each	3325
Partitions, Woven wire, 10 ga., 1-1/2" mesh		
4' Wide x 7" high	Each	117
8' High	Each	124
10' High	Each	151
Platform Lifter, Portable, 6' x 6'		
3000# cap.	Each	6000
4000# cap.	Each	8425
Fixed, 6' x 8', 5000# cap.	Each	8675

Important: See the Reference Section for Location Factors

Model costs calculated for a 3 story building with 12' story height and 90,000 square feet of floor area

			Unit	Unit Cost	Cost Per S.F.	% Of Sub-Total
1.0 Foundations						
.1	Footings & Foundations	Poured concrete; strip and spread footings and 4' foundation wall	S.F. Ground	4.98	1.66	
.4	Piles & Caissons	N/A	—	—	—	3.8%
.9	Excavation & Backfill	Site preparation for slab and trench for foundation wall and footing	S.F. Ground	1.04	.35	
2.0 Substructure						
.1	Slab on Grade	4" reinforced concrete with vapor barrier and granular base	S.F. Slab	5.16	1.72	3.2%
.2	Special Substructures	N/A	—	—	—	
3.0 Superstructure						
.1	Columns & Beams	Concrete columns	L.F. Column	98	2.47	
.4	Structural Walls	N/A	—	—	—	
.5	Elevated Floors	Concrete flat slab	S.F. Floor	10.82	7.21	24.9%
.7	Roof	Concrete flat slab	S.F. Roof	9.54	3.18	
.9	Stairs	Concrete	Flight	4675	.42	
4.0 Exterior Closure						
.1	Walls	Face brick with common brick backup 70% of wall	S.F. Wall	21	4.17	
.5	Exterior Wall Finishes	N/A	—	—	—	14.6%
.6	Doors	Double aluminum & glass, hollow metal, overhead doors	Each	1195	.22	
.7	Windows & Glazed Walls	Industrial, horizontal pivoted steel 30% of wall	Each	649	3.41	
5.0 Roofing						
.1	Roof Coverings	Built-up tar and gravel with flashing	S.F. Roof	1.92	.64	
.7	Insulation	Perlite/EPS composite	S.F. Roof	1.25	.42	2.2%
.8	Openings & Specialties	Gravel stop, hatches, gutters and downspouts	S.F. Roof	.30	.10	
6.0 Interior Construction						
.1	Partitions	Gypsum board on metal studs, toilet partition 50 S.F. Floor/L.F. Partition	S.F. Partition	4.95	.99	
.4	Interior Doors	Single leaf fire doors 500 S.F. Floor/Door	S.F. Door	630	1.26	
.5	Wall Finishes	Paint	S.F. Surface	.52	.21	8.3%
.6	Floor Finishes	90% metallic hardener, 10% vinyl composition tile	S.F. Floor	1.71	1.71	
.7	Ceiling Finishes	Fiberglass board on exposed grid systems 10% of area	S.F. Ceiling	2.37	.24	
.9	Interior Surface/Exterior Wall	N/A	—	—	—	
7.0 Conveying						
.1	Elevators	Two hydraulic freight elevators	Each	80,550	1.79	3.4%
.2	Special Conveyors	N/A	—	—	—	
8.0 Mechanical						
.1	Plumbing	Toilet and service fixtures, supply and drainage 1 Fixture/1345 S.F. Floor	Each	2784	2.07	
.2	Fire Protection	Wet pipe sprinkler system	S.F. Floor	1.83	1.83	
.3	Heating	Oil fired hot water, unit heaters	S.F. Floor	2.85	2.85	26.6%
.4	Cooling	Chilled water, air cooled condenser system	S.F. Floor	7.46	7.46	
.5	Special Systems	N/A	—	—	—	
9.0 Electrical						
.1	Service & Distribution	800 ampere service, panel board and feeders	S.F. Floor	.44	.44	
.2	Lighting & Power	High intensity discharge fixtures, switches, A.C. and misc. power	S.F. Floor	6.03	6.03	13.0%
.4	Special Electrical	Alarm systems and emergency lighting	S.F. Floor	.44	.44	
11.0 Special Construction						
.1	Specialties	N/A	—	—	—	0.0%
12.0 Site Work						
.1	Earthwork	N/A	—	—	—	
.3	Utilities	N/A	—	—	—	0.0%
.5	Roads & Parking	N/A	—	—	—	
.7	Site Improvements	N/A	—	—	—	
			Sub-Total		53.29	100%

GENERAL CONDITIONS (Overhead & Profit)		15%	7.99
ARCHITECT FEES		6%	3.67
Total Building Cost			**64.95**

Costs per square foot of floor area

Exterior Wall	S.F. Area	4000	4500	5000	5500	6000	6500	7000	7500	8000
	L.F. Perimeter	260	280	300	320	320	336	353	370	386
Face Brick Concrete Block Back-up	Steel Joists	92.00	90.25	88.80	87.70	85.15	84.10	83.40	82.70	82.05
	Precast Conc.	92.95	91.25	89.80	88.65	86.10	85.10	84.40	83.70	83.05
Decorative Concrete Block	Steel Joists	83.90	82.50	81.35	80.45	78.50	77.65	77.10	76.55	76.05
	Precast Conc.	85.05	83.60	82.50	81.55	79.65	78.80	78.25	77.70	77.20
Limestone with Concrete Block Back-up	Steel Joists	97.75	95.75	94.15	92.85	89.85	88.70	87.85	87.10	86.35
	Precast Conc.	98.90	96.90	95.30	94.00	91.00	89.85	89.00	88.20	87.45
Perimeter Adj., Add or Deduct	Per 100 L.F.	11.80	10.45	9.40	8.60	7.85	7.30	6.70	6.30	5.90
Story Hgt. Adj., Add or Deduct	Per 1 Ft.	1.55	1.45	1.45	1.35	1.25	1.25	1.20	1.15	1.15
For Basement, add $21.00 per square foot of basement area										

The above costs were calculated using the basic specifications shown on the facing page. These costs should be adjusted where necessary for design alternatives and owner's requirements. Reported completed project costs, for this type of structure, range from $ 41.75 to $ 123.45 per S.F.

Common additives

Description	Unit	$ Cost
Appliances		
Cooking range, 30" free standing		
1 oven	Each	380 - 1275
2 oven	Each	785 - 1700
30" built-in		
1 oven	Each	470 - 1475
2 oven	Each	1050 - 1975
Counter top cook tops, 4 burner	Each	261 - 630
Microwave oven	Each	199 - 1850
Combination range, refrig. & sink, 30" wide	Each	1075 - 2125
60" wide	Each	2775
72" wide	Each	3175
Combination range refrigerator, sink		
microwave oven & icemaker	Each	4700
Compactor, residential, 4-1 compaction	Each	395 - 570
Dishwasher, built-in, 2 cycles	Each	425 - 850
4 cycles	Each	460 - 1075
Garbage disposer, sink type	Each	116 - 244
Hood for range, 2 speed, vented, 30" wide	Each	182 - 505
42" wide	Each	320 - 510

Description	Unit	$ Cost
Appliances, cont.		
Refrigerator, no frost 10-12 C.F.	Each	505 - 770
14-16 C.F.	Each	580 - 850
18-20 C.F.	Each	665 - 1425
Lockers, Steel, single tier, 60" or 72"	Opening	135 - 216
2 tier, 60" or 72" total	Opening	83 - 105
5 tier, box lockers	Opening	44 - 57
Locker bench, lam. maple top only	L.F.	18.65
Pedestals, steel pipe	Each	43
Sound System		
Amplifier, 250 watts	Each	1525
Speaker, ceiling or wall	Each	131
Trumpet	Each	246

Important: See the Reference Section for Location Factors

Model costs calculated for a 1 story building with 14' story height and 6,000 square feet of floor area

				Unit	Unit Cost	Cost Per S.F.	% Of Sub-Total
1.0 Foundations							
.1	Footings & Foundations	Poured concrete; strip and spread footings and 4' foundation wall		S.F. Ground	4.37	4.37	
.4	Piles & Caissons	N/A		–	–	–	8.2%
.9	Excavation & Backfill	Site preparation for slab and trench for foundation wall and footing		S.F. Ground	1.24	1.24	
2.0 Substructure							
.1	Slab on Grade	4" reinforced concrete with vapor barrier and granular base		S.F. Slab	3.66	3.66	5.3%
.2	Special Substructures	N/A		–	–	–	
3.0 Superstructure							
.1	Columns & Beams	N/A		–	–	–	
.4	Structural Walls	N/A		–	–	–	
.5	Elevated Floors	N/A		–	–	–	3.8%
.7	Roof	Metal deck, open web steel joists, beams		S.F. Roof	2.58	2.58	
.9	Stairs	N/A		–	–	–	
4.0 Exterior Closure							
.1	Walls	Face brick with concrete block backup	75% of wall	S.F. Wall	19.45	10.89	
.5	Exterior Wall Finishes	N/A		–	–	–	21.1%
.6	Doors	Single aluminum and glass, overhead, hollow metal	15% of wall	S.F. Door	20	2.24	
.7	Windows & Glazed Walls	Aluminum insulated glass	10% of wall	Each	580	1.35	
5.0 Roofing							
.1	Roof Coverings	Built-up tar and gravel with flashing		S.F. Roof	2.57	2.57	
.7	Insulation	Perlite/EPS composite		S.F. Roof	1.25	1.25	6.7%
.8	Openings & Specialties	Gravel stop		S.F. Roof	.76	.76	
6.0 Interior Construction							
.1	Partitions	Concrete block, toilet partitions	17 S.F. Floor/L.F. Partition	S.F. Partition	5.81	3.42	
.4	Interior Doors	Single leaf hollow metal	500 S.F. Floor/Door	Each	526	1.05	
.5	Wall Finishes	Paint		S.F. Surface	1.09	1.28	
.6	Floor Finishes	50% vinyl tile, 50% paint		S.F. Floor	1.88	1.88	14.3%
.7	Ceiling Finishes	Fiberglass board on exposed grid, suspended	50% of area	S.F. Ceiling	3.07	1.54	
.9	Interior Surface/Exterior Wall	Acrylic glazed coating	75% of wall	S.F. Wall	1.15	.64	
7.0 Conveying							
.1	Elevators	N/A		–	–	–	0.0%
.2	Special Conveyors	N/A		–	–	–	
8.0 Mechanical							
.1	Plumbing	Kitchen, toilet and service fixtures, supply and drainage	1 Fixture/375 S.F. Floor	Each	2385	6.36	
.2	Fire Protection	Wet pipe sprinkler system		S.F. Floor	2.09	2.09	
.3	Heating	Included in 8.4		–	–	–	33.1%
.4	Cooling	Rooftop multizone unit system		S.F. Floor	14.25	14.25	
.5	Special Systems	N/A		–	–	–	
9.0 Electrical							
.1	Service & Distribution	200 ampere service, panel board and feeders		S.F. Floor	.89	.89	
.2	Lighting & Power	Fluorescent fixtures, receptacles, switches, A.C. and misc. power		S.F. Floor	3.96	3.96	7.5%
.4	Special Electrical	Alarm systems		S.F. Floor	.28	.28	
11.0 Special Construction							
.1	Specialties	N/A		–	–	–	0.0%
12.0 Site Work							
.1	Earthwork	N/A		–	–	–	
.3	Utilities	N/A		–	–	–	
.5	Roads & Parking	N/A		–	–	–	0.0%
.7	Site Improvements	N/A		–	–	–	
				Sub-Total		68.55	100%
	GENERAL CONDITIONS (Overhead & Profit)				15%	10.28	
	ARCHITECT FEES				8%	6.32	
				Total Building Cost		85.15	

Costs per square foot of floor area

Exterior Wall	S.F. Area	6000	7000	8000	9000	10000	11000	12000	13000	14000
	L.F. Perimeter	220	240	260	280	286	303	320	336	353
Face Brick with Concrete Block Back-up	Steel Joists	95.20	92.35	90.25	88.70	86.25	85.10	84.05	83.20	82.50
	Precast Conc.	99.45	96.70	94.60	93.05	90.60	89.45	88.45	87.60	86.90
Decorative Concrete Block	Steel Joists	87.10	84.75	83.00	81.65	79.65	78.70	77.85	77.10	76.55
	Precast Conc.	92.65	90.35	88.60	87.25	85.30	84.35	83.50	82.80	82.20
Limestone with Concrete Block Back-up	Steel Joists	101.70	98.45	96.05	94.20	91.30	89.95	88.80	87.75	86.95
	Precast Conc.	105.95	102.75	100.35	98.50	95.65	94.30	93.15	92.15	91.35
Perimeter Adj., Add or Deduct	Per 100 L.F.	13.30	11.45	10.05	8.90	8.05	7.25	6.70	6.15	5.70
Perimeter Adj., Add or Deduct	Per 100 L.F.	1.70	1.65	1.55	1.45	1.35	1.30	1.25	1.25	1.20
For Basement, add $20.15 per square foot of basement area										

The above costs were calculated using the basic specifications shown on the facing page. These costs should be adjusted where necessary for design alternatives and owner's requirements. Reported completed project costs, for this type of structure, range from $41.75 to $123.45 per S.F.

Common additives

Description	Unit	$ Cost
Appliances		
Cooking range, 30" free standing		
1 oven	Each	380 - 1275
2 oven	Each	785 - 1700
30" built-in		
1 oven	Each	470 - 1475
2 oven	Each	1050 - 1975
Counter top cook tops, 4 burner	Each	261 - 630
Microwave oven	Each	199 - 1850
Combination range, refrig. & sink, 30" wide	Each	1075 - 2125
60" wide	Each	2775
72" wide	Each	3175
Combination range, refrigerator, sink,		
microwave oven & icemaker	Each	4700
Compactor, residential, 4-1 compaction	Each	395 - 570
Dishwasher, built-in, 2 cycles	Each	425 - 850
4 cycles	Each	460 - 1075
Garbage disposer, sink type	Each	116 - 244
Hood for range, 2 speed, vented, 30" wide	Each	182 - 505
42" wide	Each	320 - 510

Description	Unit	$ Cost
Appliances, cont.		
Refrigerator, no frost 10-12 C.F.	Each	505 - 770
14-16 C.F.	Each	580 - 850
18-20 C.F.	Each	665 - 1425
Elevators, Hydraulic passenger, 2 stops		
1500# capacity	Each	40,000
2500# capacity	Each	41,300
3500# capacity	Each	44,900
Lockers, Steel, single tier, 60" or 72"	Opening	135 - 216
2 tier, 60" or 72" total	Opening	83 - 105
5 tier, box lockers	Opening	44 - 57
Locker bench, lam. maple top only	L.F.	18.65
Pedestals, steel pipe	Each	43
Sound System		
Amplifier, 250 watts	Each	1525
Speaker, ceiling or wall	Each	131
Trumpet	Each	246

Important: See the Reference Section for Location Factors

BUILDING TYPES

Model costs calculated for a 2 story building with 14' story height and 10,000 square feet of floor area

				Unit	Unit Cost	Cost Per S.F.	% Of Sub-Total
1.0 Foundations							
.1	Footings & Foundations	Poured concrete; strip and spread footings and 4' foundation wall		S.F. Ground	4.72	2.36	
.4	Piles & Caissons	N/A		—	—	—	4.6%
.9	Excavation & Backfill	Site preparation for slab and trench for foundation wall and footing		S.F. Ground	1.24	.62	
2.0 Substructure							
.1	Slab on Grade	4" reinforced concrete with vapor barrier and granular base		S.F. Slab	3.06	1.53	2.4%
.2	Special Substructures	N/A		—	—	—	
3.0 Superstructure							
.1	Columns & Beams	N/A		—	—	—	
.4	Structural Walls	Included in 6.1		—	—	—	
.5	Elevated Floors	Open web steel joists, slab form, concrete		S.F. Floor	8.17	4.09	10.6%
.7	Roof	Metal deck on open web steel joists		S.F. Roof	2.74	1.37	
.9	Stairs	Concrete filled metal pan		Flight	6800	1.36	
4.0 Exterior Closure							
.1	Walls	Decorative concrete block	75% of wall	S.F. Wall	12.35	7.42	
.5	Exterior Wall Finishes	N/A		—	—	—	16.9%
.6	Doors	Single aluminum and glass, steel overhead, hollow metal	15% of wall	S.F. Door	14.82	1.78	
.7	Windows & Glazed Walls	Aluminum insulated glass	10% of wall	Each	461	1.61	
5.0 Roofing							
.1	Roof Coverings	Tar and gravel with flashing		S.F. Roof	2.62	1.31	
.7	Insulation	Perlite/EPS composite		S.F. Roof	1.25	.63	3.3%
.8	Openings & Specialties	Gravel stop		L.F. Perimeter	5.60	.16	
6.0 Interior Construction							
.1	Partitions	Concrete block, toilet partitions	10 S.F. Floor/L.F. Partition	S.F. Partition	5.75	3.38	
.4	Interior Doors	Single leaf hollow metal	500 S.F. Floor/Door	Each	630	1.26	
.5	Wall Finishes	Paint		S.F. Surface	.52	.61	
.6	Floor Finishes	50% vinyl tile, 50% paint		S.F. Floor	1.88	1.88	14.0%
.7	Ceiling Finishes	Fiberglass board on exposed grid, suspended	50% of area	S.F. Ceiling	2.37	1.19	
.9	Interior Surface/Exterior Wall	Acrylic glazed coating	75% of wall	S.F. Wall	1.15	.69	
7.0 Conveying							
.1	Elevators	One hydraulic passenger elevator		Each	48,800	4.88	7.6%
.2	Special Conveyors	N/A		—	—	—	
8.0 Mechanical							
.1	Plumbing	Kitchen toilet and service fixtures, supply and drainage	1 Fixture/400 S.F. Floor	Each	1904	4.76	
.2	Fire Protection	Wet pipe sprinkler system		S.F. Floor	1.77	1.77	
.3	Heating	Included in 8.4		—	—	—	32.4%
.4	Cooling	Rooftop multizone unit system		S.F. Floor	14.25	14.25	
.5	Special Systems	N/A		—	—	—	
9.0 Electrical							
.1	Service & Distribution	100 ampere service, panel board and feeders		S.F. Floor	.61	.61	
.2	Lighting & Power	Fluorescent fixtures, receptacles, switches, A.C. and misc. power		S.F. Floor	4.10	4.10	8.2%
.4	Special Electrical	Alarm systems and emergency lighting		S.F. Floor	.52	.52	
11.0 Special Construction							
.1	Specialties	N/A		—	—	—	0.0%
12.0 Site Work							
.1	Earthwork	N/A		—	—	—	
.3	Utilities	N/A		—	—	—	0.0%
.5	Roads & Parking	N/A		—	—	—	
.7	Site Improvements	N/A		—	—	—	
			Sub-Total			64.14	100%
	GENERAL CONDITIONS (Overhead & Profit)				15%	9.62	
	ARCHITECT FEES				8%	5.89	
			Total Building Cost			79.65	

BUILDING TYPES

Costs per square foot of floor area

Exterior Wall	S.F. Area	4000	5000	6000	8000	10000	12000	14000	16000	18000
	L.F. Perimeter	180	205	230	260	300	340	353	386	420
Cedar Beveled Siding	Wood Frame	96.10	91.10	87.75	82.65	79.95	78.20	76.25	75.20	74.45
Aluminum Siding	Wood Frame	93.85	89.00	85.75	80.90	78.30	76.60	74.75	73.75	73.00
Board and Batten	Wood Frame	93.65	89.45	86.65	82.25	79.95	78.45	76.70	75.80	75.15
Face Brick on Block	Wood Joists	107.15	101.15	97.15	90.65	87.30	85.15	82.45	81.15	80.15
Stucco on Block	Wood Joists	95.70	90.70	87.35	82.25	79.55	77.80	75.80	74.80	74.00
Decorative Block	Wood Joists	99.35	94.05	90.50	85.00	82.15	80.25	78.05	76.95	76.15
Perimeter Adj., Add or Deduct	Per 100 L.F.	8.70	6.95	5.80	4.35	3.50	2.90	2.50	2.20	1.95
Story Hgt. Adj., Add or Deduct	Per 1 Ft.	1.10	1.00	.95	.80	.75	.70	.65	.60	.55
For Basement, add $13.15 per square foot of basement area										

The above costs were calculated using the basic specifications shown on the facing page. These costs should be adjusted where necessary for design alternatives and owner's requirements. Reported completed project costs, for this type of structure, range from $59.95 to $115.60 per S.F.

Common additives

Description	Unit	$ Cost
Appliances		
Cooking range, 30" free standing		
1 oven	Each	380 - 1275
2 oven	Each	785 - 1700
30" built-in		
1 oven	Each	470 - 1475
2 oven	Each	1050 - 1975
Counter top cook tops, 4 burner	Each	261 - 630
Microwave oven	Each	199 - 1850
Combination range, refrig. & sink, 30" wide	Each	1075 - 2125
60" wide	Each	2775
72" wide	Each	3175
Combination range, refrigerator, sink,		
microwave oven & icemaker	Each	4700
Compactor, residential, 4-1 compaction	Each	395 - 570
Dishwasher, built-in, 2 cycles	Each	425 - 850
4 cycles	Each	460 - 1075
Garbage disposer, sink type	Each	116 - 244
Hood for range, 2 speed, vented, 30" wide	Each	182 - 505
42" wide	Each	320 - 510

Description	Unit	$ Cost
Appliances, cont.		
Refrigerator, no frost 10-12 C.F.	Each	505 - 770
14-16 C.F.	Each	580 - 850
18-20 C.F.	Each	665 - 1425
Elevators, Hydraulic passenger, 2 stops		
1500# capacity	Each	40,000
2500# capacity	Each	41,300
3500# capacity	Each	44,900
Laundry Equipment		
Dryer, gas, 16 lb. capacity	Each	680
30 lb. capacity	Each	2550
Washer, 4 cycle	Each	760
Commercial	Each	1125
Sound System		
Amplifier, 250 watts	Each	1525
Speaker, ceiling or wall	Each	131
Trumpet	Each	246

Important: See the Reference Section for Location Factors

Model costs calculated for a 2 story building with 10' story height and 10,000 square feet of floor area

Fraternity/Sorority House

			Unit	Unit Cost	Cost Per S.F.	% Of Sub-Total
1.0 Foundations						
.1	Footings & Foundations	Poured concrete; strip and spread footings and 4' foundation wall	S.F. Ground	4.22	2.11	
.4	Piles & Caissons	N/A	—	—	—	4.3%
.9	Excavation & Backfill	Site preparation for slab and trench for foundation wall and footing	S.F. Ground	1.24	.62	
2.0 Substructure						
.1	Slab on Grade	4" reinforced concrete with vapor barrier and granular base	S.F. Slab	3.06	1.53	2.4%
.2	Special Substructures	N/A	—	—	—	
3.0 Superstructure						
.1	Columns & Beams	N/A	—	—	—	
.4	Structural Walls	Included in 6.1	—	—	—	
.5	Elevated Floors	Plywood on wood joists	S.F. Floor	2.86	1.43	5.2%
.7	Roof	Plywood on wood rafters (pitched)	S.F. Roof	2.42	1.35	
.9	Stairs	Wood	Flight	1311	.52	
4.0 Exterior Closure						
.1	Walls	Cedar bevel siding on wood studs, insulated 80% of wall	S.F. Wall	7.40	3.55	
.5	Exterior Wall Finishes	N/A	—	—	—	9.7%
.6	Doors	Solid core wood	Each	1383	.96	
.7	Windows & Glazed Walls	Double hung wood 20% of wall	Each	344	1.65	
5.0 Roofing						
.1	Roof Coverings	Asphalt shingles with flashing (pitched)	S.F. Ground	1.32	.66	
.7	Insulation	Fiberglass sheets	S.F. Ground	.98	.49	2.1%
.8	Openings & Specialties	Gutters and downspouts	S.F. Ground	.42	.21	
6.0 Interior Construction						
.1	Partitions	Gypsum board on wood studs 25 S.F. Floor/L.F. Partition	S.F. Partition	3.09	.99	
.4	Interior Doors	Single leaf wood 200 S.F. Floor/Door	Each	408	2.04	
.5	Wall Finishes	Paint	S.F. Surface	.52	.33	
.6	Floor Finishes	50% hardwood, 50% carpet	S.F. Floor	6.37	6.37	20.9%
.7	Ceiling Finishes	Gypsum board on wood furring	S.F. Ceiling	2.87	2.87	
.9	Interior Surface/Exterior Wall	Painted gypsum board 80% of wall	S.F. Wall	1.20	.72	
7.0 Conveying						
.1	Elevators	One hydraulic passenger elevator	Each	48,800	4.88	7.6%
.2	Special Conveyors	N/A	—	—	—	
8.0 Mechanical						
.1	Plumbing	Kitchen toilet and service fixtures, supply and drainage 1 Fixture/150 S.F. Floor	Each	774	5.16	
.2	Fire Protection	Wet pipe sprinkler system	S.F. Floor	1.77	1.77	
.3	Heating	Oil fired hot water, baseboard radiation	S.F. Floor	4.18	4.18	25.3%
.4	Cooling	Split system with air cooled condensing unit	S.F. Floor	5.04	5.04	
.5	Special Systems	N/A	—	—	—	
9.0 Electrical						
.1	Service & Distribution	600 ampere service, panel board and feeders	S.F. Floor	2.74	2.74	
.2	Lighting & Power	Fluorescent fixtures, receptacles, switches, A.C. and misc. power	S.F. Floor	5.96	5.96	22.5%
.4	Special Electrical	Alarm, communication system and generator set	S.F. Floor	5.67	5.67	
11.0 Special Construction						
.1	Specialties	N/A	—	—	—	0.0%
12.0 Site Work						
.1	Earthwork	N/A	—	—	—	
.3	Utilities	N/A	—	—	—	
.5	Roads & Parking	N/A	—	—	—	0.0%
.7	Site Improvements	N/A	—	—	—	
		Sub-Total			63.80	100%
	GENERAL CONDITIONS (Overhead & Profit)			15%	9.57	
	ARCHITECT FEES			9%	6.58	
		Total Building Cost			79.95	

BUILDING TYPES

125

Costs per square foot of floor area

Exterior Wall	S.F. Area	4000	5000	6000	7000	8000	9000	10000	11000	12000
	L.F. Perimeter	260	300	340	353	386	420	425	435	460
Vertical Redwood Siding	Wood Frame	82.85	79.95	78.05	75.75	74.60	73.75	72.40	71.40	70.85
Brick Veneer	Wood Frame	88.40	85.10	82.90	80.15	78.85	77.85	76.15	74.95	74.25
Aluminum Siding	Wood Frame	81.20	78.45	76.65	74.55	73.50	72.70	71.45	70.55	70.05
Brick on Block	Wood Truss	92.40	88.90	86.50	83.45	82.05	81.00	79.10	77.75	77.00
Limestone on Block	Wood Truss	99.85	95.75	93.00	89.20	87.50	86.30	83.90	82.20	81.30
Stucco on Block	Wood Truss	83.45	80.65	78.75	76.55	75.45	74.60	73.25	72.30	71.75
Perimeter Adj., Add or Deduct	Per 100 L.F.	5.90	4.70	3.95	3.40	2.95	2.65	2.35	2.15	1.95
Story Hgt. Adj., Add or Deduct	Per 1 Ft.	.90	.80	.75	.70	.65	.60	.60	.55	.50

For Basement, add $17.25 per square foot of basement area

The above costs were calculated using the basic specifications shown on the facing page. These costs should be adjusted where necessary for design alternatives and owner's requirements. Reported completed project costs, for this type of structure, range from $58.50 to $165.15 per S.F.

Common additives

Description	Unit	$ Cost
Autopsy Table, Standard	Each	6975
Deluxe	Each	8900
Directory Boards, Plastic, glass covered		
30" x 20"	Each	510
36" x 48"	Each	930
Aluminum, 24" x 18"	Each	450
36" x 24"	Each	540
48" x 32"	Each	640
48" x 60"	Each	1450
Emergency Lighting, 25 watt, battery operated		
Lead battery	Each	330
Nickel cadmium	Each	620
Mortuary Refrigerator, End operated		
Two capacity	Each	10,900
Six capacity	Each	20,700

Description	Unit	$ Cost
Planters, Precast concrete		
48" diam., 24" high	Each	505
7" diam., 36" high	Each	1975
Fiberglass, 36" diam., 24" high	Each	380
60" diam., 24" high	Each	970
Smoke Detectors		
Ceiling type	Each	142
Duct type	Each	390

Important: See the Reference Section for Location Factors

Model costs calculated for a 1 story building with 10' story height and 10,000 square feet of floor area

				Unit	Unit Cost	Cost Per S.F.	% Of Sub-Total
1.0 Foundations							
.1	Footings & Foundations	Poured concrete; strip and spread footings and 4' foundation wall		S.F. Ground	2.97	2.97	
.4	Piles & Caissons	N/A		—	—	—	7.1%
.9	Excavation & Backfill	Site preparation for slab and trench for foundation wall and footing		S.F. Ground	1.11	1.11	
2.0 Substructure							
.1	Slab on Grade	4" reinforced concrete with vapor barrier and granular base		S.F. Slab	3.06	3.06	5.3%
.2	Special Substructures	N/A		—	—	—	
3.0 Superstructure							
.1	Columns & Beams	N/A		—	—	—	
.4	Structural Walls	N/A		—	—	—	
.5	Elevated Floors	N/A		—	—	—	5.9%
.7	Roof	Plywood on wood rafters (pitched)		S.F. Ground	3.42	3.42	
.9	Stairs	N/A		—	—	—	
4.0 Exterior Closure							
.1	Walls	1" x 4" vertical T & G redwood siding on wood studs	90% of wall	S.F. Wall	8.18	3.13	
.5	Exterior Wall Finishes	N/A		—	—	—	8.6%
.6	Doors	Wood swinging double doors, single leaf hollow metal		Each	1335	.80	
.7	Windows & Glazed Walls	Double hung wood	10% of wall	Each	297	1.05	
5.0 Roofing							
.1	Roof Coverings	Asphalt shingles with flashing		S.F. Roof	1.56	1.56	
.7	Insulation	Fiberglass sheets		S.F. Ground	.87	.87	4.7%
.8	Openings & Specialties	Gutters and downspouts		S.F. Ground	.30	.30	
6.0 Interior Construction							
.1	Partitions	Gypsum board on wood studs with sound deadening board	15 S.F. Floor/L.F. Partition	S.F. Partition	4.84	2.58	
.4	Interior Doors	Single leaf wood	150 S.F. Floor/Door	Each	408	2.72	
.5	Wall Finishes	50% wallpaper, 25% wood paneling, 25% paint		S.F. Surface	1.91	2.04	
.6	Floor Finishes	70% carpet, 30% terrazzo		S.F. Floor	8.87	8.87	32.6%
.7	Ceiling Finishes	Mineral fiberboard on wood furring		S.F. Ceiling	2.15	2.15	
.9	Interior Surface/Exterior Wall	Painted gypsum board on wood furring	90% of wall	S.F. Wall	1.50	.45	
7.0 Conveying							
.1	Elevators	N/A		—	—	—	0.0%
.2	Special Conveyors	N/A		—	—	—	
8.0 Mechanical							
.1	Plumbing	Toilet and service fixtures, supply and drainage	1 Fixture/770 S.F. Floor	Each	2987	3.88	
.2	Fire Protection	Wet pipe sprinkler system		S.F. Floor	1.60	1.60	
.3	Heating	Included in 8.4		—	—	—	28.9%
.4	Cooling	Multizone rooftop unit, gas heating, electric cooling		S.F. Floor	11.18	11.18	
.5	Special Systems	N/A		—	—	—	
9.0 Electrical							
.1	Service & Distribution	200 ampere service, panel board and feeders		S.F. Floor	.53	.53	
.2	Lighting & Power	Fluorescent fixtures, receptacles, switches, A.C. and misc. power		S.F. Floor	3.26	3.26	6.9%
.4	Special Electrical	Alarm systems and emergency lighting		S.F. Floor	.22	.22	
11.0 Special Construction							
.1	Specialties	N/A		—	—	—	0.0%
12.0 Site Work							
.1	Earthwork	N/A		—	—	—	
.3	Utilities	N/A		—	—	—	
.5	Roads & Parking	N/A		—	—	—	0.0%
.7	Site Improvements	N/A		—	—	—	
				Sub-Total		57.75	100%
	GENERAL CONDITIONS (Overhead & Profit)				15%	8.66	
	ARCHITECT FEES				9%	5.99	
				Total Building Cost		72.40	

BUILDING TYPES

Costs per square foot of floor area

Exterior Wall	S.F. Area	12000	14000	16000	19000	21000	23000	26000	28000	30000
	L.F. Perimeter	440	474	510	556	583	607	648	670	695
Metal Panel Curtain Walls	Steel Frame	63.15	61.55	60.35	58.85	58.05	57.35	56.55	55.95	55.60
Tilt-up Concrete Wall	Steel Frame	61.65	60.15	59.05	57.65	56.90	56.25	55.50	55.00	54.65
Face Brick with Concrete Block Back-up	Bearing Walls	65.50	63.50	62.05	60.20	59.20	58.35	57.35	56.65	56.15
	Steel Frame	67.80	65.80	64.35	62.50	61.50	60.65	59.65	58.95	58.45
Stucco on Concrete Block	Bearing Walls	60.00	58.40	57.25	55.80	55.00	54.30	53.55	53.00	52.65
	Steel Frame	62.60	61.00	59.85	58.40	57.60	56.90	56.10	55.60	55.20
Perimeter Adj., Add or Deduct	Per 100 L.F.	3.50	3.00	2.60	2.25	2.00	1.85	1.60	1.50	1.40
Story Hgt. Adj., Add or Deduct	Per 1 Ft.	.85	.75	.70	.65	.60	.60	.55	.55	.50

For Basement, add $19.80 per square foot of basement area

The above costs were calculated using the basic specifications shown on the facing page. These costs should be adjusted where necessary for design alternatives and owner's requirements. Reported completed project costs, for this type of structure, range from $31.45 to $90.50 per S.F.

Common additives

Description	Unit	$ Cost
Emergency Lighting, 25 watt, battery operated		
Lead battery	Each	330
Nickel cadmium	Each	620
Smoke Detectors		
Ceiling type	Each	142
Duct type	Each	390
Sound System		
Amplifier, 250 watts	Each	1525
Speaker, ceiling or wall	Each	131
Trumpet	Each	246

Important: See the Reference Section for Location Factors

Model costs calculated for a 1 story building with 14' story height and 21,000 square feet of floor area

				Unit	Unit Cost	Cost Per S.F.	% Of Sub-Total
1.0 Foundations							
.1	Footings & Foundations	Poured concrete; strip and spread footings and 4' foundation wall		S.F. Ground	2.11	2.11	
.4	Piles & Caissons	N/A		–	–	–	6.8%
.9	Excavation & Backfill	Site preparation for slab and trench for foundation wall and footing		S.F. Ground	1.11	1.11	
2.0 Substructure							
.1	Slab on Grade	4" reinforced concrete with vapor barrier and granular base		S.F. Slab	3.66	3.66	7.8%
.2	Special Substructures	N/A		–	–	–	
3.0 Superstructure							
.1	Columns & Beams	Steel columns included in 3.7		–	–	–	
.4	Structural Walls	Metal siding support		S.F. Wall	3.13	.85	
.5	Elevated Floors	N/A		–	–	–	13.2%
.7	Roof	Metal deck, open web steel joists, beams, columns		S.F. Floor	5.36	5.36	
.9	Stairs	N/A		–	–	–	
4.0 Exterior Closure							
.1	Walls	Metal panel	70% of wall	S.F. Wall	8.75	2.38	
.5	Exterior Wall Finishes	N/A		–	–	–	17.4%
.6	Doors	Double aluminum and glass, hollow metal, steel overhead		Each	2409	2.06	
.7	Windows & Glazed Walls	Window wall	30% of wall	S.F. Window	32	3.77	
5.0 Roofing							
.1	Roof Coverings	Built-up tar and gravel with flashing		S.F. Roof	1.98	1.98	
.7	Insulation	Perlite/EPS composite		S.F. Roof	1.25	1.25	7.2%
.8	Openings & Specialties	Gravel stop and skylight		L.F. Perimeter	5.60	.16	
6.0 Interior Construction							
.1	Partitions	Gypsum board on metal studs	28 S.F. Floor/L.F. Partition	S.F. Partition	2.96	1.27	
.4	Interior Doors	Hollow metal	280 S.F. Floor/Door	Each	526	1.88	
.5	Wall Finishes	Paint		S.F. Surface	.53	.45	14.1%
.6	Floor Finishes	50% vinyl tile, 50% paint		S.F. Floor	1.88	1.88	
.7	Ceiling Finishes	Fiberglass board on exposed grid, suspended	50% of area	S.F. Ceiling	2.37	1.19	
.9	Interior Surface/Exterior Wall	N/A		–	–	–	
7.0 Conveying							
.1	Elevators	N/A		–	–	–	0.0%
.2	Special Conveyors	N/A		–	–	–	
8.0 Mechanical							
.1	Plumbing	Toilet and service fixtures, supply and drainage	1 Fixture/1500 S.F. Floor	Each	3015	2.01	
.2	Fire Protection	Wet pipe sprinkler system		S.F. Floor	2.13	2.13	
.3	Heating	Gas fired hot water, unit heaters (service area)		S.F. Floor	2.52	2.52	22.0%
.4	Cooling	Single zone rooftop unit, gas heating, electric (office and showroom)		S.F. Floor	3.52	3.52	
.5	Special Systems	Underfloor garage exhaust system		S.F. Floor	.22	.22	
9.0 Electrical							
.1	Service & Distribution	200 ampere service, panel board and feeders		S.F. Floor	.27	.27	
.2	Lighting & Power	Fluorescent fixtures, receptacles, switches, A.C. and misc. power		S.F. Floor	3.99	3.99	9.5%
.4	Special Electrical	Alarm systems and emergency lighting		S.F. Floor	.20	.20	
11.0 Special Construction							
.1	Specialties	Hoists, compressor, fuel pump		S.F. Floor	.95	.95	2.0%
12.0 Site Work							
.1	Earthwork	N/A		–	–	–	
.3	Utilities	N/A		–	–	–	0.0%
.5	Roads & Parking	N/A		–	–	–	
.7	Site Improvements	N/A		–	–	–	
				Sub-Total		47.17	**100%**
	GENERAL CONDITIONS (Overhead & Profit)				15%	7.08	
	ARCHITECT FEES				7%	3.80	
				Total Building Cost		58.05	

Costs per square foot of floor area

Exterior Wall	S.F. Area	85000	115000	145000	175000	205000	235000	265000	295000	325000
	L.F. Perimeter	529	638	723	823	923	951	1037	1057	1132
Face Brick with Concrete Block Back-up	Steel Frame	32.15	31.60	31.10	30.85	30.65	30.30	30.20	29.95	29.85
	R/Conc. Frame	27.05	26.50	26.00	25.75	25.60	25.20	25.10	24.85	24.75
Precast Concrete	Steel Frame	32.80	32.15	31.65	31.40	31.20	30.80	30.75	30.50	30.35
	R/Conc. Frame	27.25	26.65	26.15	25.90	25.75	25.35	25.25	24.95	24.90
Reinforced Concrete	Steel Frame	31.80	31.30	30.90	30.70	30.55	30.25	30.20	30.00	29.90
	R/Conc. Frame	26.10	25.60	25.20	25.00	24.90	24.55	24.50	24.30	24.20
Perimeter Adj., Add or Deduct	Per 100 L.F.	.85	.65	.50	.40	.35	.30	.30	.25	.25
Story Hgt. Adj., Add or Deduct	Per 1 Ft.	.30	.25	.25	.20	.20	.20	.20	.15	.15
Basement—Not Applicable										

The above costs were calculated using the basic specifications shown on the facing page. These costs should be adjusted where necessary for design alternatives and owner's requirements. Reported completed project costs, for this type of structure, range from $17.85 to $73.95 per S.F.

Common additives

Description	Unit	$ Cost
Automatic Gates, 8' arm, 1 way	Each	3450
2 way	Each	3550
Booth for attendant, average	Each	4400 - 7325
Elevators, Electric passenger, 5 stops		
2000# capacity	Each	91,300
3500# capacity	Each	97,300
5000# capacity	Each	100,800
Fee Indicator, 1" display	Each	1750
Ticket Printer & Dispenser, Standard	Each	5325
Rate computing	Each	6950
Card control station, single period	Each	860
4 period	Each	950
Key station on pedestal	Each	540
Coin station, multiple coins	Each	3975
Painting, Parking stalls	Stall	6.85
Parking Barriers		
Timber with saddles, 4" x 4"	L.F.	5.35
Precast concrete, 6" x 10" x 6'	Each	34
Traffic Signs, directional, 12" x 18", high densit	Each	40

Model costs calculated for a 5 story building with 10' story height and 145,000 square feet of floor area

			Unit	Unit Cost	Cost Per S.F.	% Of Sub-Total
1.0 Foundations						
.1	Footings & Foundations	Poured concrete; strip and spread footings and 4' foundation wall	S.F. Ground	8.15	1.63	8.6%
.4	Piles & Caissons	N/A	—	—	—	
.9	Excavation & Backfill	Site preparation for slab and trench for foundation wall and footing	S.F. Ground	1.04	.21	
2.0 Substructure						
.1	Slab on Grade	6" reinforced concrete with vapor barrier and granular base	S.F. Slab	3.80	.76	3.6%
.2	Special Substructures	N/A	—			
3.0 Superstructure						
.1	Columns & Beams	Precast concrete columns	S.F. Floor	4.90	4.90	
.4	Structural Walls	Concrete block elevator shaft	S.F. Wall	9.46	.49	
.5	Elevated Floors	Double tee precast concrete slab	S.F. Floor	7.97	6.38	56.0%
.7	Roof	N/A	—	—	—	
.9	Stairs	Concrete	Flight	2510	.17	
4.0 Exterior Closure						
.1	Walls	Face brick with concrete block backup _40% of story height_	S.F. Wall	19.45	1.94	
.5	Exterior Wall Finishes	N/A	—	—	—	9.1%
.6	Doors	N/A	—	—	—	
.7	Windows & Glazed Walls	N/A	—	—	—	
5.0 Roofing						
.1	Roof Coverings	N/A	—	—	—	
.7	Insulation	N/A	—	—	—	0.0%
.8	Openings & Specialties	N/A	—	—	—	
6.0 Interior Construction						
.1	Partitions	N/A	—	—	—	
.4	Interior Doors	N/A	—	—	—	
.5	Wall Finishes	N/A	—	—	—	
.6	Floor Finishes	N/A	—	—	—	0.0%
.7	Ceiling Finishes	N/A	—	—	—	
.9	Interior Surface/Exterior Wall	N/A	—	—	—	
7.0 Conveying						
.1	Elevators	Two hydraulic passenger elevators	Each	81,925	1.13	5.3%
.2	Special Conveyors	N/A	—	—	—	
8.0 Mechanical						
.1	Plumbing	Toilet and service fixtures, supply and drainage _1 Fixture/18,125 S.F. Floor_	Each	13,412	.74	
.2	Fire Protection	Standpipes and hose systems	S.F. Floor	.04	.04	
.3	Heating	N/A	—	—	—	3.7%
.4	Cooling	N/A	—	—	—	
.5	Special Systems	N/A	—	—	—	
9.0 Electrical						
.1	Service & Distribution	400 ampere service, panel board and feeders	S.F. Floor	.14	.14	
.2	Lighting & Power	Fluorescent fixtures, receptacles, switches and misc. power	S.F. Floor	1.77	1.77	9.5%
.4	Special Electrical	Alarm systems and emergency lighting	S.F. Floor	.12	.12	
11.0 Special Construction						
.1	Specialties	Ticket dispensers, booths automatic gates	S.F. Floor	.90	.90	4.2%
12.0 Site Work						
.1	Earthwork	N/A	—	—	—	
.3	Utilities	N/A	—	—	—	
.5	Roads & Parking	N/A	—	—	—	0.0%
.7	Site Improvements	N/A	—	—	—	
		Sub-Total			21.32	100%
	GENERAL CONDITIONS (Overhead & Profit)			15%	3.20	
	ARCHITECT FEES			6%	1.48	
		Total Building Cost			26.00	

BUILDING TYPES

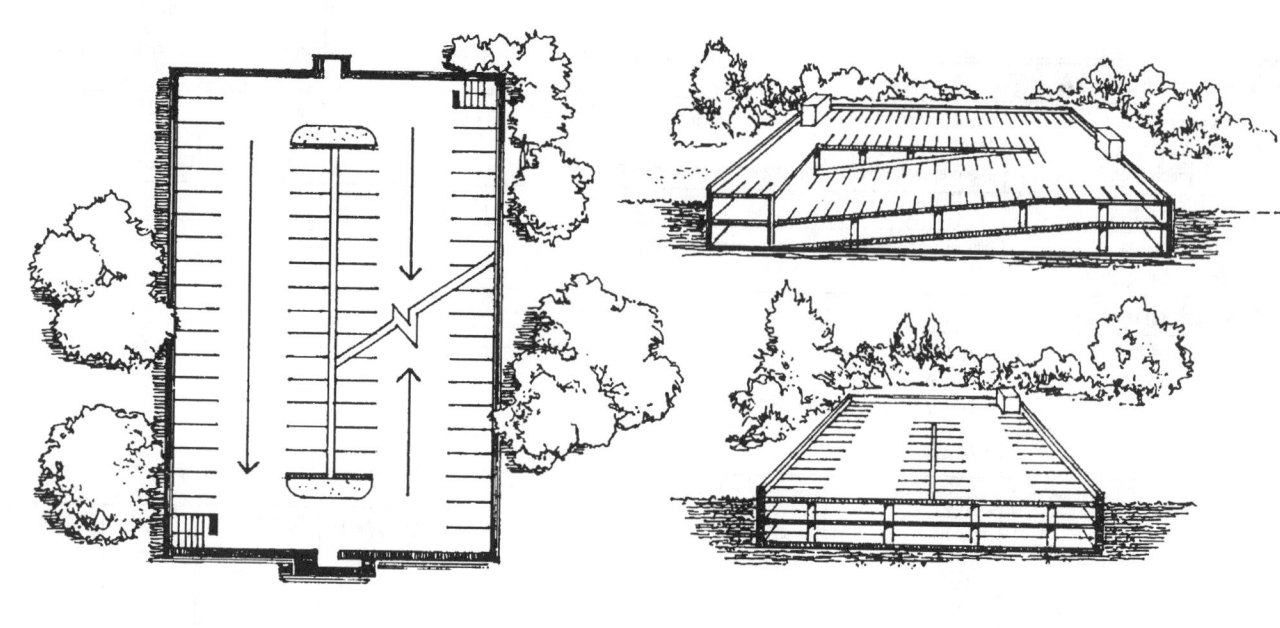

BUILDING TYPES

Costs per square foot of floor area

Exterior Wall	S.F. Area	20000	30000	40000	50000	75000	100000	125000	150000	175000
	L.F. Perimeter	400	500	600	650	775	900	1000	1100	1185
Reinforced Concrete	R/Conc. Frame	46.75	44.35	43.10	41.80	40.00	39.15	38.45	38.00	37.65
Perimeter Adj., Add or Deduct	Per 100 L.F.	3.05	2.05	1.55	1.20	.85	.60	.50	.40	.35
Story Hgt. Adj., Add or Deduct	Per 1 Ft.	1.15	.95	.90	.75	.65	.50	.45	.45	.40
Basement—Not Applicable										

The above costs were calculated using the basic specifications shown on the facing page. These costs should be adjusted where necessary for design alternatives and owner's requirements. Reported completed project costs, for this type of structure, range from $27.90 to $66.20 per S.F.

Common additives

Description	Unit	$ Cost
Automatic Gates, 8' arm, 1 way	Each	3450
2 way	Each	3550
Booth for attendant	Each	4400-7325
Elevators, Hydraulic passenger, 2 stops		
1500# capacity	Each	40,000
2500# capacity	Each	41,300
3500# capacity	Each	44,900
Fee Indicator, 1" display	Each	1750
Ticket Printer & Dispenser, Standard	Each	5325
Rate computing	Each	6950
Card control station, single period	Each	860
4 period	Each	950
Key station on pedestal	Each	540
Coin station, multiple coins	Each	3975
Painting, Parking stalls	Stall	6.85
Parking Barriers		
Timber with saddles, 4" x 4"	L.F.	5.35
Precast concrete, 6" x 10" x 6'	Each	34
Traffic Signs, directional, 12" x 18"	Each	40

Model costs calculated for a 2 story building with 10' story height and 100,000 square feet of floor area

			Unit	Unit Cost	Cost Per S.F.	% Of Sub-Total
1.0 Foundations						
.1	Footings & Foundations	Poured concrete; strip and spread footings and waterproofing	S.F. Ground	4.72	2.36	18.3%
.4	Piles & Caissons	N/A	–	–	–	
.9	Excavation & Backfill	Excavation 24' deep	S.F. Ground	6.85	3.42	
2.0 Substructure						
.1	Slab on Grade	5" reinforced concrete with vapor barrier and granular base	S.F. Slab	3.30	1.65	5.2%
.2	Special Substructures	N/A	–	–	–	
3.0 Superstructure						
.1	Columns & Beams	Concrete columns	L.F. Column	117	.59	
.4	Structural Walls	N/A	–	–	–	
.5	Elevated Floors	Cast-in-place concrete beam and slab	S.F. Floor	18.83	7.60	50.6%
.7	Roof	Cast-in-place concrete beam and slab	S.F. Roof	15.15	7.57	
.9	Stairs	Concrete	Flight	3850	.19	
4.0 Exterior Closure						
.1	Walls	Cast-in place concrete	S.F. Wall	13.17	2.37	
.5	Exterior Wall Finishes	N/A	–	–	–	7.9%
.6	Doors	Steel overhead, hollow metal	Each	2784	.11	
.7	Windows & Glazed Walls	N/A	–	–	–	
5.0 Roofing						
.1	Roof Coverings	Neoprene membrane	S.F. Roof	3.18	1.59	
.7	Insulation	N/A	–	–	–	5.0%
.8	Openings & Specialties	N/A	–	–	–	
6.0 Interior Construction						
.1	Partitions	N/A	–	–	–	
.4	Interior Doors	N/A	–	–	–	
.5	Wall Finishes	N/A	–	–	–	
.6	Floor Finishes	N/A	–	–	–	0.0%
.7	Ceiling Finishes	N/A	–	–	–	
.9	Interior Surface/Exterior Wall	N/A	–	–	–	
7.0 Conveying						
.1	Elevators	Two hydraulic passenger elevators	Each	49,000	.98	3.1%
.2	Special Conveyors	N/A	–	–	–	
8.0 Mechanical						
.1	Plumbing	Drainage in parking areas, toilets, & service fixtures *1 Fixture/5000 S.F. Floor*	Each	.75	.75	
.2	Fire Protection	Dry standpipe system, class 1	S.F. Floor	.08	.08	
.3	Heating	N/A	–	–	–	3.1%
.4	Cooling	Exhaust fans	S.F. Floor	.09	.09	
.5	Special Systems	N/A	–	–	–	
9.0 Electrical						
.1	Service & Distribution	200 ampere service, panel board and feeders	S.F. Floor	.08	.08	
.2	Lighting & Power	Fluorescent fixtures, receptacles, switches and misc. power	S.F. Floor	1.50	1.50	6.0%
.4	Special Electrical	Alarm systems and emergency lighting	S.F. Floor	.32	.32	
11.0 Special Construction						
.1	Specialties	Ticket dispensers, booths, automatic gates	S.F. Floor	.25	.25	0.8%
12.0 Site Work						
.1	Earthwork	N/A	–	–	–	
.3	Utilities	N/A	–	–	–	0.0%
.5	Roads & Parking	N/A	–	–	–	
.7	Site Improvements	N/A	–	–	–	
			Sub-Total		31.50	**100%**
	GENERAL CONDITIONS (Overhead & Profit)			15%	4.73	
	ARCHITECT FEES			8%	2.92	
			Total Building Cost		39.15	

BUILDING TYPES

133

Costs per square foot of floor area

Exterior Wall	S.F. Area	2000	4000	6000	8000	10000	12000	14000	16000	18000
	L.F. Perimeter	180	260	340	420	500	580	586	600	610
Concrete Block	Wood Joists	89.70	77.45	73.45	71.40	70.15	69.35	67.25	65.85	64.70
	Steel Joists	89.65	77.85	73.95	71.95	70.75	69.95	67.90	66.50	65.40
Poured Concrete	Wood Joists	95.35	81.75	77.25	74.95	73.60	72.65	70.15	68.40	67.05
	Steel Joists	96.05	82.45	77.90	75.65	74.25	73.35	70.85	69.10	67.75
Insulated Metal Panels	Wood Frame	84.85	74.15	70.60	68.80	67.75	67.00	65.25	64.05	63.10
	Steel Frame	87.30	76.60	73.05	71.30	70.20	69.50	67.70	66.50	65.55
Perimeter Adj., Add or Deduct	Per 100 L.F.	14.15	7.05	4.70	3.55	2.85	2.35	2.05	1.80	1.55
Story Hgt. Adj., Add or Deduct	Per 1 Ft.	1.10	.80	.70	.65	.60	.60	.55	.45	.40

For Basement, add $18.85 per square foot of basement area

The above costs were calculated using the basic specifications shown on the facing page. These costs should be adjusted where necessary for design alternatives and owner's requirements. Reported completed project costs, for this type of structure, range from $39.80 to $119.55 per S.F.

Common additives

Description	Unit	$ Cost
Air Compressors		
Electric 1-1/2 H.P., standard controls	Each	3325
Dual controls	Each	3550
5 H.P. 115/230 Volt, standard controls	Each	4525
Dual controls	Each	4750
Product Dispenser		
with vapor recovery for 6 nozzles	Each	16,500
Hoists, Single post		
8000# cap., swivel arm	Each	6200
Two post, adjustable frames, 11,000# cap.	Each	8725
24,000# cap.	Each	12,900
7500# Frame support	Each	7400
Four post, roll on ramp	Each	6850
Lockers, Steel, single tier, 60" or 72"	Opening	135 - 216
2 tier, 60" or 72" total	Opening	83 - 105
5 tier, box lockers	Opening	44 - 57
Locker bench, lam. maple top only	L.F.	18.65
Pedestals, steel pipe	Each	43
Lube Equipment		
3 reel type, with pumps, no piping	Each	8500
Spray Painting Booth, 26' long, complete	Each	15,600

Important: See the Reference Section for Location Factors

Model costs calculated for a 1 story building with 14' story height and 4,000 square feet of floor area

				Unit	Unit Cost	Cost Per S.F.	% Of Sub-Total
1.0 Foundations							
.1	Footings & Foundations	Poured concrete; strip and spread footings and 4' foundation wall		S.F. Ground	4.34	4.34	
.4	Piles & Caissons	N/A		—	—	—	8.7%
.9	Excavation & Backfill	Site preparation for slab and trench for foundation wall and footing		S.F. Ground	1.11	1.11	
2.0 Substructure							
.1	Slab on Grade	4" reinforced concrete with vapor barrier and granular base		S.F. Slab	3.06	3.06	4.9%
.2	Special Substructures	N/A		—	—	—	
3.0 Superstructure							
.1	Columns & Beams	N/A		—	—	—	
.4	Structural Walls	N/A		—	—	—	
.5	Elevated Floors	N/A		—	—	—	4.9%
.7	Roof	Metal deck on open web steel joists		S.F. Roof	3.05	3.05	
.9	Stairs	N/A		—	—	—	
4.0 Exterior Closure							
.1	Walls	Concrete block	80% of wall	L.F. Wall	7.93	5.77	
.5	Exterior Wall Finishes	N/A		—	—	—	
.6	Doors	Steel overhead and hollow metal	15% of wall	S.F. Door	13.04	1.78	13.4%
.7	Windows & Glazed Walls	Hopper type commercial steel	5% of wall	Each	284	.86	
5.0 Roofing							
.1	Roof Coverings	Built-up tar and gravel		S.F. Roof	2.56	2.56	
.7	Insulation	Perlite/EPS composite		S.F. Roof	1.25	1.25	6.7%
.8	Openings & Specialties	Gravel stop and skylight		S.F. Roof	.42	.42	
6.0 Interior Construction							
.1	Partitions	Concrete block, toilet partitions	50 S.F. Floor/L.F. Partition	S.F. Partition	5.85	1.17	
.4	Interior Doors	Single leaf hollow metal	3000 S.F. Floor/Door	Each	526	.18	
.5	Wall Finishes	Paint		S.F. Surface	.90	.36	
.6	Floor Finishes	90% metallic floor hardener, 10% vinyl composition tile		S.F. Floor	.66	.66	6.5%
.7	Ceiling Finishes	Gypsum board on wood joists in office and washrooms	10% of area	S.F. Ceiling	2.87	.29	
.9	Interior Surface/Exterior Wall	Paint	80% of wall	S.F. Wall	1.95	1.42	
7.0 Conveying							
.1	Elevators	N/A		—	—	—	0.0%
.2	Special Conveyors	N/A		—	—	—	
8.0 Mechanical							
.1	Plumbing	Toilet and service fixtures, supply and drainage	1 Fixture/500 S.F. Floor	Each	2730	5.46	
.2	Fire Protection	Sprinklers, ordinary hazard		S.F. Floor	2.27	2.27	
.3	Heating	Oil fired hot water, unit heaters		S.F. Floor	5.03	5.03	32.1%
.4	Cooling	Split systems with air cooled condensing units		S.F. Floor	6.53	6.53	
.5	Special Systems	Garage exhaust system		S.F. Floor	.82	.82	
9.0 Electrical							
.1	Service & Distribution	100 ampere service, panel board and feeders		S.F. Floor	.48	.48	
.2	Lighting & Power	Fluorescent fixtures, receptacles, switches, A.C. and misc. power		S.F. Floor	4.05	4.05	8.0%
.4	Special Electrical	Alarm systems and emergency lighting		S.F. Floor	.47	.47	
11.0 Special Construction							
.1	Specialties	Hoists		S.F. Floor	9.30	9.30	14.8%
12.0 Site Work							
.1	Earthwork	N/A		—	—	—	
.3	Utilities	N/A		—	—	—	
.5	Roads & Parking	N/A		—	—	—	0.0%
.7	Site Improvements	N/A		—	—	—	
				Sub-Total		62.69	100%
	GENERAL CONDITIONS (Overhead & Profit)				15%	9.40	
	ARCHITECT FEES				8%	5.76	
				Total Building Cost		77.85	

Costs per square foot of floor area

Exterior Wall	S.F. Area	600	800	1000	1200	1400	1600	1800	2000	2200
	L.F. Perimeter	100	120	126	140	153	160	170	180	190
Face Brick with Concrete Block Back-up	Wood Truss	121.85	112.45	102.20	97.60	94.05	90.20	87.70	85.70	84.10
	Steel Joists	117.25	107.90	97.65	93.05	89.50	85.65	83.15	81.15	79.50
Enameled Sandwich Panel Tile on Concrete Block	Steel Frame	107.10	98.80	90.10	86.15	83.05	79.80	77.65	75.90	74.55
	Steel Joists	127.50	117.20	105.65	100.55	96.55	92.20	89.40	87.15	85.35
Aluminum Siding Wood Siding	Wood Frame	104.65	97.00	89.20	85.60	82.75	79.90	77.95	76.40	75.15
	Wood Frame	105.55	97.80	89.90	86.20	83.35	80.40	78.45	76.90	75.60
Perimeter Adj., Add or Deduct	Per 100 L.F.	55.00	41.30	33.00	27.50	23.60	20.60	18.30	16.55	15.00
Story Hgt. Adj., Add or Deduct	Per 1 Ft.	4.20	3.80	3.20	2.95	2.75	2.55	2.35	2.25	2.15
Basement—Not Applicable										

The above costs were calculated using the basic specifications shown on the facing page. These costs should be adjusted where necessary for design alternatives and owner's requirements. Reported completed project costs, for this type of structure, range from $25.80 to $119.65 per S.F.

Common additives

Description	Unit	$ Cost
Air Compressors		
Electric 1-1/2 H.P., standard controls	Each	3325
Dual controls	Each	3550
5 H.P. 115/230 volt, standard controls	Each	4525
Dual controls	Each	4750
Product Dispenser		
with vapor recovery for 6 nozzles	Each	16,500
Hoists, Single post		
8000# cap. swivel arm	Each	6200
Two post, adjustable frames, 11,000# cap.	Each	8725
24,000# cap.	Each	12,900
7500# cap.	Each	7400
Four post, roll on ramp	Each	6850
Lockers, Steel, single tier, 60" or 72"	Opening	135 - 216
2 tier, 60" or 72" total	Opening	83 - 105
5 tier, box lockers	Each	44 - 57
Locker bench, lam. maple top only	L.F.	18.65
Pedestals, steel pipe	Each	43
Lube Equipment		
3 reel type, with pumps, no piping	Each	8500

Important: See the Reference Section for Location Factors

Model costs calculated for a 1 story building with 10' story height and 1,400 square feet of floor area

				Unit	Unit Cost	Cost Per S.F.	% Of Sub-Total
1.0 Foundations							
.1	Footings & Foundations	Poured concrete; strip and spread footings and 4' foundation wall		S.F. Ground	6.92	6.92	
.4	Piles & Caissons	N/A		—	—	—	11.3%
.9	Excavation & Backfill	Site preparation for slab and trench for foundation wall and footing		S.F. Ground	1.60	1.60	
2.0 Substructure							
.1	Slab on Grade	4" reinforced concrete with vapor barrier and granular base		S.F. Slab	3.06	3.06	4.0%
.2	Special Substructures	N/A		—			
3.0 Superstructure							
.1	Columns & Beams	N/A		—	—	—	
.4	Structural Walls	N/A		—	—	—	
.5	Elevated Floors	N/A		—	—	—	5.7%
.7	Roof	Plywood on wood trusses		S.F. Ground	4.33	4.33	
.9	Stairs	N/A		—	—	—	
4.0 Exterior Closure							
.1	Walls	Face brick with concrete block backup	60% of wall	S.F. Wall	19.44	12.75	
.5	Exterior Wall Finishes	N/A		—	—	—	
.6	Doors	Steel overhead, aluminum & glass and hollow metal	20% of wall	S.F. Door	24	5.42	31.4%
.7	Windows & Glazed Walls	Store front and metal top hinged outswinging	20% of wall	S.F. Window	25	5.57	
5.0 Roofing							
.1	Roof Coverings	Asphalt shingles with flashing		S.F. Ground	1.16	1.16	
.7	Insulation	Perlite/EPS composite		S.F. Ground	.87	.87	2.7%
.8	Openings & Specialties	N/A		—	—	—	
6.0 Interior Construction							
.1	Partitions	Concrete block, toilet partitions	25 S.F. Floor/L.F. Partition	S.F. Partition	9.66	3.09	
.4	Interior Doors	Single leaf hollow metal	700 S.F. Floor/Door	Each	526	.75	
.5	Wall Finishes	Paint		S.F. Surface	1.09	.70	
.6	Floor Finishes	Vinyl composition tile	35% of floor area	S.F. Floor	2.66	.93	10.2%
.7	Ceiling Finishes	Painted gypsum board on wood joists in sales area & washrooms	35% of floor area	S.F. Ceiling	2.87	1.00	
.9	Interior Surface/Exterior Wall	Paint	60% of wall	S.F. Wall	1.95	1.28	
7.0 Conveying							
.1	Elevators	N/A		—	—	—	0.0%
.2	Special Conveyors	N/A		—	—	—	
8.0 Mechanical							
.1	Plumbing	Toilet and service fixtures, supply and drainage	1 Fixture/235 S.F. Floor	Each	1760	7.49	
.2	Fire Protection	N/A		—	—	—	
.3	Heating	Oil fired hot water, wall fin radiation		S.F. Floor	5.03	5.03	24.9%
.4	Cooling	Split systems with air cooled condensing units		S.F. Floor	6.32	6.32	
.5	Special Systems	N/A		—	—	—	
9.0 Electrical							
.1	Service & Distribution	100 ampere service, panel board and feeders		S.F. Floor	2.15	2.15	
.2	Lighting & Power	Fluorescent fixtures, receptacles, switches, A.C. and misc. power		S.F. Floor	4.43	4.43	9.8%
.4	Special Electrical	Alarm systems and emergency lighting		S.F. Floor	.87	.87	
11.0 Special Construction							
.1	Specialties	N/A		—	—	—	0.0%
12.0 Site Work							
.1	Earthwork	N/A		—	—	—	
.3	Utilities	N/A		—	—	—	
.5	Roads & Parking	N/A		—	—	—	0.0%
.7	Site Improvements	N/A		—	—	—	
				Sub-Total		75.72	**100%**
	GENERAL CONDITIONS (Overhead & Profit)				15%	11.36	
	ARCHITECT FEES				8%	6.97	
				Total Building Cost		**94.05**	

BUILDING TYPES

Costs per square foot of floor area

Exterior Wall	S.F. Area	12000	16000	20000	25000	30000	35000	40000	45000	50000
	L.F. Perimeter	440	520	600	700	708	780	841	910	979
Reinforced Concrete Block	Lam. Wood Arches	90.20	87.20	85.45	84.05	81.70	80.85	80.10	79.60	79.20
	Rigid Steel Frame	87.35	84.35	82.60	81.20	78.90	78.05	77.30	76.75	76.35
Face Brick with Concrete Block Back-up	Lam. Wood Arches	104.90	100.25	97.50	95.25	91.20	89.85	88.55	87.75	87.05
	Rigid Steel Frame	102.10	97.45	94.65	92.45	88.35	87.00	85.70	84.90	84.25
Metal Sandwich Panels	Lam. Wood Arches	87.30	84.65	83.10	81.80	79.85	79.10	78.45	78.00	77.65
	Rigid Steel Frame	84.45	81.80	80.25	79.00	77.00	76.30	75.60	75.15	74.80
Perimeter Adj., Add or Deduct	Per 100 L.F.	3.75	2.85	2.25	1.80	1.50	1.30	1.10	1.00	.90
Story Hgt. Adj., Add or Deduct	Per 1 Ft.	.50	.45	.45	.40	.35	.30	.30	.30	.30
Basement—Not Applicable										

The above costs were calculated using the basic specifications shown on the facing page. These costs should be adjusted where necessary for design alternatives and owner's requirements. Reported completed project costs, for this type of structure, range from $43.55 to $130.70 per S.F.

Common additives

Description	Unit	$ Cost
Bleachers, Telescoping, manual		
To 15 tier	Seat	76 - 105
16-20 tier	Seat	155 - 190
21-30 tier	Seat	165 - 199
For power operation, add	Seat	30 - 47
Gym Divider Curtain, Mesh top		
Manual roll-up	S.F.	8.50
Gym Mats		
2" naugahyde covered	S.F.	4.19
2" nylon	S.F.	7.90
1-1/2" wall pads	S.F.	7.70
1" wrestling mats	S.F.	4.76
Scoreboard		
Basketball, one side	Each	3050 - 21,100
Basketball Backstop		
Wall mtd., 6' extended, fixed	Each	1125 - 1900
Swing up, wall mtd.	Each	1700 - 2500

Description	Unit	$ Cost
Lockers, Steel, single tier, 60" or 72"	Opening	135 - 216
2 tier, 60" or 72" total	Opening	83 - 105
5 tier, box lockers	Opening	44 - 57
Locker bench, lam. maple top only	L.F.	18.65
Pedestals, steel pipe	Each	43
Sound System		
Amplifier, 250 watts	Each	1525
Speaker, ceiling or wall	Each	131
Trumpet	Each	246
Emergency Lighting, 25 watt, battery operated		
Lead battery	Each	330
Nickel cadmium	Each	620

Important: See the Reference Section for Location Factors

Model costs calculated for a 1 story building with 25' story height and 20,000 square feet of floor area

				Unit	Unit Cost	Cost Per S.F.	% Of Sub-Total
1.0 Foundations							
.1	Footings & Foundations	Poured concrete; strip and spread footings and 4' foundation wall		S.F. Ground	2.46	2.46	5.0%
.4	Piles & Caissons	N/A		—	—	—	
.9	Excavation & Backfill	Site preparation for slab and trench for foundation wall and footing		S.F. Ground	1.04	1.04	
2.0 Substructure							
.1	Slab on Grade	4" reinforced concrete with vapor barrier and granular base		S.F. Slab	3.06	3.06	4.4%
.2	Special Substructures	N/A		—	—	—	
3.0 Superstructure							
.1	Columns & Beams	N/A		—	—	—	
.4	Structural Walls	N/A		—	—	—	
.5	Elevated Floors	N/A		—	—	—	18.0%
.7	Roof	Wood deck on laminated wood arches		S.F. Ground	12.49	12.49	
.9	Stairs	N/A		—	—	—	
4.0 Exterior Closure							
.1	Walls	Reinforced concrete block (end walls included)	90% of wall	S.F. Wall	7.96	5.37	
.5	Exterior Wall Finishes	N/A		—	—	—	11.2%
.6	Doors	Aluminum and glass, hollow metal, steel overhead		Each	907	.27	
.7	Windows & Glazed Walls	Metal horizontal pivoted	10% of wall	Each	282	2.12	
5.0 Roofing							
.1	Roof Coverings	EPDM, 60 mils, fully adhered		S.F. Ground	1.69	1.69	
.7	Insulation	Polyisocyanurate		—	1.12	1.15	4.1%
.8	Openings & Specialties	N/A		—	—	—	
6.0 Interior Construction							
.1	Partitions	Concrete block, toilet partitions	50 S.F. Floor/L.F. Partition	S.F. Partition	6.35	1.27	
.4	Interior Doors	Single leaf hollow metal	500 S.F. Floor/Door	Each	526	1.05	
.5	Wall Finishes	50% paint, 50% ceramic tile		S.F. Surface	3.13	1.25	21.6%
.6	Floor Finishes	90% hardwood, 10% ceramic tile		S.F. Floor	9.62	9.62	
.7	Ceiling Finishes	Mineral fiber tile on concealed zee bars	15% of area	S.F. Ceiling	3.07	.46	
.9	Interior Surface/Exterior Wall	Paint	90% of wall	S.F. Wall	1.95	1.32	
7.0 Conveying							
.1	Elevators	N/A		—	—	—	0.0%
.2	Special Conveyors	N/A		—	—	—	
8.0 Mechanical							
.1	Plumbing	Toilet and service fixtures, supply and drainage	1 Fixture/515 S.F. Floor	Each	2950	5.73	
.2	Fire Protection	Wet pipe sprinkler system		S.F. Floor	1.60	1.60	
.3	Heating	Included in 8.4		—	—	—	23.7%
.4	Cooling	Single zone rooftop unit, gas heating, electric cooling		S.F. Floor	9.17	9.17	
.5	Special Systems	N/A		—	—	—	
9.0 Electrical							
.1	Service & Distribution	400 ampere service, panel board and feeders		S.F. Floor	.61	.61	
.2	Lighting & Power	Fluorescent fixtures, receptacles, switches, A.C. and misc. power		S.F. Floor	5.33	5.33	10.5%
.4	Special Electrical	Alarm systems, sound system and emergency lighting		S.F. Floor	1.33	1.33	
11.0 Special Construction							
.1	Specialties	Bleachers, sauna, weight room		S.F. Floor	1.05	1.05	1.5%
12.0 Site Work							
.1	Earthwork	N/A		—	—	—	
.3	Utilities	N/A		—	—	—	0.0%
.5	Roads & Parking	N/A		—	—	—	
.7	Site Improvements	N/A		—	—	—	
				Sub-Total		69.44	100%
	GENERAL CONDITIONS (Overhead & Profit)				15%	10.42	
	ARCHITECT FEES				7%	5.59	
				Total Building Cost		85.45	

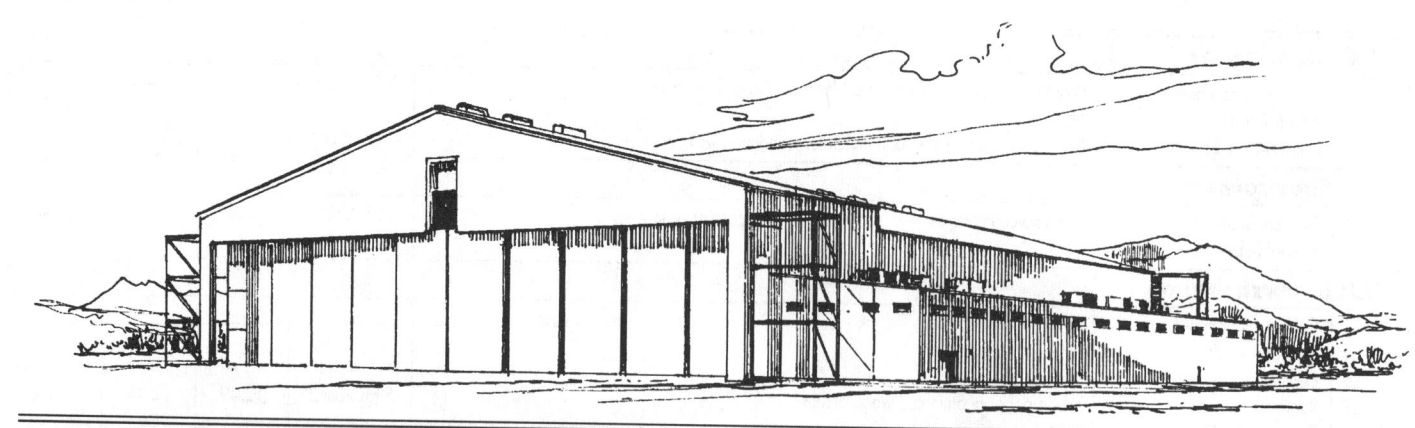

Costs per square foot of floor area

Exterior Wall	S.F. Area	5000	10000	15000	20000	30000	40000	50000	75000	100000
	L.F. Perimeter	300	410	500	580	710	830	930	1150	1300
Concrete Block Reinforced	Steel Frame	72.60	64.50	61.25	59.40	57.10	55.90	54.95	53.55	52.55
	Bearing Walls	75.85	67.60	64.35	62.45	60.15	58.90	57.95	56.55	55.55
Precast Concrete	Steel Frame	77.70	68.00	64.05	61.85	59.15	57.65	56.55	54.85	53.65
	Bearing Walls	80.65	70.95	67.05	64.80	62.10	60.60	59.50	57.80	56.60
Galv. Steel Siding	Steel Frame	72.10	64.15	60.95	59.15	56.90	55.70	54.80	53.45	52.45
Metal Sandwich Panel	Steel Frame	82.25	73.20	69.55	67.50	64.95	63.55	62.55	60.95	59.85
Perimeter Adj., Add or Deduct	Per 100 L.F.	8.50	4.25	2.85	2.10	1.40	1.05	.85	.60	.45
Story Hgt. Adj., Add or Deduct	Per 1 Ft.	.85	.55	.45	.40	.35	.30	.25	.20	.20
Basement—Not Applicable										

The above costs were calculated using the basic specifications shown on the facing page. These costs should be adjusted where necessary for design alternatives and owner's requirements. Reported completed project costs, for this type of structure, range from $25.00 to $113.70 per S.F.

Common additives

Description	Unit	$ Cost
Closed Circuit Surveillance, One station		
Camera and monitor	Each	1325
For additional camera stations, add	Each	730
Emergency Lighting, 25 watt, battery operated		
Lead battery	Each	330
Nickel cadmium	Each	620
Lockers, Steel, single tier, 60" or 72"	Opening	135 - 216
2 tier, 60" or 72" total	Opening	83 - 105
5 tier, box lockers	Opening	44 - 57
Locker bench, lam. maple top only	L.F.	18.65
Pedestals, steel pipe	Each	43
Safe, Office type, 4 hour rating		
30" x 18" x 18"	Each	3475
62" x 33" x 20"	Each	7575
Sound System		
Amplifier, 250 watts	Each	1525
Speaker, ceiling or wall	Each	131
Trumpet	Each	246

BUILDING TYPES

Important: See the Reference Section for Location Factors

Model costs calculated for a 1 story building with 24' story height and 20,000 square feet of floor area

				Unit	Unit Cost	Cost Per S.F.	% Of Sub-Total
1.0 Foundations							
.1	Footings & Foundations	Poured concrete; strip and spread footings and 4' foundation wall		S.F. Ground	2.45	2.45	
.4	Piles & Caissons	N/A			—	—	7.4%
.9	Excavation & Backfill	Site preparation for slab and trench for foundation wall and footing		S.F. Ground	1.11	1.11	
2.0 Substructure							
.1	Slab on Grade	6" reinforced concrete with vapor barrier and granular base		S.F. Slab	4.46	4.46	9.3%
.2	Special Substructures	N/A		—	—	—	
3.0 Superstructure							
.1	Columns & Beams	Steel columns included in 3.7		—	—	—	
.4	Structural Walls	Metal siding support		S.F. Floor	.28	.28	
.5	Elevated Floors	N/A		—	—	—	11.2%
.7	Roof	Metal deck, open web steel joists, beams, columns		S.F. Roof	5.12	5.12	
.9	Stairs	N/A		—	—	—	
4.0 Exterior Closure							
.1	Walls	Concrete block	50% of wall	S.F. Wall	5.17	1.80	
.5	Exterior Wall Finishes	N/A		—	—	—	27.7%
.6	Doors	Steel overhead and sliding	30% of wall	S.F. Door	28	5.87	
.7	Windows & Glazed Walls	Industrial horizontal pivoted steel	20% of wall	Each	971	5.63	
5.0 Roofing							
.1	Roof Coverings	Elastomeric membrane		S.F. Roof	1.86	1.86	
.7	Insulation	Fiberboard		S.F. Roof	.94	.94	6.3%
.8	Openings & Specialties	Gravel stop and hatches		S.F. Roof	.21	.21	
6.0 Interior Construction							
.1	Partitions	Concrete block, toilet partitions	200 S.F. Floor/L.F. Partition	S.F. Partition	5.83	.47	
.4	Interior Doors	Single leaf hollow metal	5000 S.F. Floor/Door	Each	526	.11	
.5	Wall Finishes	Paint		S.F. Surface	.80	.08	
.6	Floor Finishes	N/A		—	—	—	1.4%
.7	Ceiling Finishes	N/A		—	—	—	
.9	Interior Surface/Exterior Wall	N/A		—	—	—	
7.0 Conveying							
.1	Elevators	N/A		—	—	—	0.0%
.2	Special Conveyors	N/A		—	—	—	
8.0 Mechanical							
.1	Plumbing	Toilet and service fixtures, supply and drainage	1 Fixture/1000 S.F. Floor	Each	2090	2.09	
.2	Fire Protection	Sprinklers, extra hazard		S.F. Floor	3.28	3.28	
.3	Heating	Unit heaters		S.F. Floor	4.73	4.73	21.3%
.4	Cooling	Exhaust fan		S.F. Floor	.16	.16	
.5	Special Systems	N/A		—	—	—	
9.0 Electrical							
.1	Service & Distribution	200 ampere service, panel board and feeders		S.F. Floor	.53	.53	
.2	Lighting & Power	High intensity discharge fixtures, receptacles, switches and misc. power		S.F. Floor	6.52	6.52	15.4%
.4	Special Electrical	Alarm systems and emergency lighting		S.F. Floor	.36	.36	
11.0 Special Construction							
.1	Specialties	N/A		—	—	—	0.0%
12.0 Site Work							
.1	Earthwork	N/A		—	—	—	
.3	Utilities	N/A		—	—	—	0.0%
.5	Roads & Parking	N/A		—	—	—	
.7	Site Improvements	N/A		—	—	—	
				Sub-Total		48.06	100%
	GENERAL CONDITIONS (Overhead & Profit)			15%		7.21	
	ARCHITECT FEES			7%		3.88	
				Total Building Cost		59.15	

Costs per square foot of floor area

Exterior Wall	S.F. Area	25000	40000	55000	70000	85000	100000	115000	130000	145000
	L.F. Perimeter	388	520	566	666	766	866	878	962	1045
Face Brick with Structural Facing Tile	Steel Frame	133.80	126.85	121.60	119.65	118.35	117.45	115.80	115.20	114.75
	R/Conc. Frame	138.95	132.05	126.75	124.80	123.50	122.60	120.95	120.35	119.95
Face Brick with Concrete Block Back-up	Steel Frame	128.85	122.75	118.30	116.60	115.45	114.70	113.35	112.85	112.50
	R/Conc. Frame	134.00	127.90	123.45	121.75	120.65	119.85	118.50	118.00	117.65
Precast Concrete Panels	Steel Frame	129.15	123.00	118.50	116.80	115.65	114.90	113.50	113.00	112.60
	R/Conc. Frame	134.30	128.15	123.65	121.95	120.80	120.05	118.65	118.15	117.75
Perimeter Adj., Add or Deduct	Per 100 L.F.	5.40	3.40	2.45	1.90	1.60	1.35	1.15	1.05	.95
Story Hgt. Adj., Add or Deduct	Per 1 Ft.	1.60	1.35	1.05	.95	.95	.90	.75	.80	.75

For Basement, add $19.65 per square foot of basement area

The above costs were calculated using the basic specifications shown on the facing page. These costs should be adjusted where necessary for design alternatives and owner's requirements. Reported completed project costs, for this type of structure, range from $97.35 to $215.20 per S.F.

Common additives

Description	Unit	$ Cost
Cabinet, Base, door units, metal	L.F.	154
Drawer units	L.F.	267
Tall storage cabinets, 7' high, open	L.F.	276
With doors	L.F.	310
Wall, metal 12-1/2" deep, open	L.F.	104
With doors	L.F.	180
Closed Circuit TV (Patient monitoring)		
One station camera & monitor	Each	1325
For additional camera, add	Each	730
For automatic iris for low light, add	Each	1900
Doctors In-Out Register, 200 names	Each	12,700
Comb. control & recall, 200 names	Each	15,900
Recording register	Each	6025
Transformers	Each	270
Pocket pages	Each	920
Hubbard Tank, with accessories		
Stainless steel, 125 GPM 45 psi	Each	19,600
For electric hoist, add	Each	2150
Mortuary Refrigerator, End operated		
2 capacity	Each	10,900
6 capacity	Each	20,700

Description	Unit	$ Cost
Nurses Call Station		
Single bedside call station	Each	200
Ceiling speaker station	Each	97
Emergency call station	Each	133
Pillow speaker	Each	210
Double bedside call station	Each	360
Duty station	Each	221
Standard call button	Each	108
Master control station for 20 stations	Each	3825
Sound System		
Amplifier, 250 watts	Each	1525
Speaker, ceiling or wall	Each	131
Trumpet	Each	246
Station, Dietary with ice	Each	11,800
Sterilizers		
Single door, steam	Each	120,000
Double door, steam	Each	154,000
Portable, countertop, steam	Each	3025 - 4725
Gas	Each	29,700
Automatic washer/sterilizer	Each	40,900

Model costs calculated for a 3 story building with 12' story height and 55,000 square feet of floor area

			Unit	Unit Cost	Cost Per S.F.	% Of Sub-Total
1.0 Foundations						
.1	Footings & Foundations	Poured concrete; strip and spread footings and 4' foundation wall	S.F. Ground	4.68	1.56	
.4	Piles & Caissons	N/A	—	—	—	1.9%
.9	Excavation & Backfill	Site preparation for slab and trench for foundation wall and footing	S.F. Ground	1.04	.35	
2.0 Substructure						
.1	Slab on Grade	4" reinforced concrete with vapor barrier and granular base	S.F. Slab	3.06	1.02	1.0%
.2	Special Substructures	N/A	—	—	—	
3.0 Superstructure						
.1	Columns & Beams	Concrete columns	L.F. Column	72	1.31	
.4	Structural Walls	N/A	—	—	—	
.5	Elevated Floors	Cast-in-place concrete slab	S.F. Floor	11.99	7.99	13.8%
.7	Roof	Cast-in-place concrete slab	S.F. Roof	11.41	3.80	
.9	Stairs	Concrete filled metal pan	Flight	3775	.89	
4.0 Exterior Closure						
.1	Walls	Face brick and structural facing tile 85% of wall	S.F. Wall	27	8.74	
.5	Exterior Wall Finishes	N/A	—	—	—	9.9%
.6	Doors	Double aluminum and glass and sliding doors	Each	1893	.21	
.7	Windows & Glazed Walls	Aluminum sliding 15% of wall	Each	461	1.11	
5.0 Roofing						
.1	Roof Coverings	Built-up tar and gravel with flashing	S.F. Roof	2.16	.72	
.7	Insulation	Perlite/EPS composite	S.F. Roof	1.26	.42	1.2%
.8	Openings & Specialties	Gravel stop and hatches	S.F. Roof	.21	.07	
6.0 Interior Construction						
.1	Partitions	Concrete block, gypsum board on metal studs 9 S.F. Floor/L.F. Partition	S.F. Partition	3.61	4.01	
.4	Interior Doors	Single leaf hollow metal 90 S.F. Floor/Door	Each	526	5.84	
.5	Wall Finishes	40% vinyl wall covering, 35% ceramic tile, 25% epoxy coating	S.F. Surface	2.46	5.47	24.5%
.6	Floor Finishes	60% vinyl tile, 20% ceramic, 20% terrazzo	S.F. Floor	6.15	6.15	
.7	Ceiling Finishes	Plaster on suspended metal lath	S.F. Ceiling	3.07	3.07	
.9	Interior Surface/Exterior Wall	Glazed coating 85% of wall	S.F. Wall	1.15	.28	
7.0 Conveying						
.1	Elevators	Two hydraulic hospital elevators	Each	68,750	2.50	2.5%
.2	Special Conveyors	N/A	—	—	—	
8.0 Mechanical						
.1	Plumbing	Kitchen, toilet and service fixtures, supply and drainage 1 Fixture/265 S.F. Floor	Each	4144	15.64	
.2	Fire Protection	Wet pipe sprinkler system	S.F. Floor	1.40	1.40	
.3	Heating	Oil fired hot water, wall fin radiation	S.F. Floor	2.85	2.85	28.6%
.4	Cooling	Chilled water, fan coil units	S.F. Floor	8.93	8.93	
.5	Special Systems	N/A	—	—	—	
9.0 Electrical						
.1	Service & Distribution	1200 ampere service, panel board and feeders	S.F. Floor	1.02	1.02	
.2	Lighting & Power	Fluorescent fixtures, receptacles, switches, A.C. and misc. power	S.F. Floor	7.29	7.29	10.6%
.4	Special Electrical	Alarm systems, communications systems and emergency lighting	S.F. Floor	2.38	2.38	
11.0 Special Construction						
.1	Specialties	Conductive flooring, oxygen piping, curtain partitions	S.F. Floor	6.09	6.09	6.0%
12.0 Site Work						
.1	Earthwork	N/A	—	—	—	
.3	Utilities	N/A	—	—	—	
.5	Roads & Parking	N/A	—	—	—	0.0%
.7	Site Improvements	N/A	—	—	—	
			Sub-Total		101.11	**100%**
	GENERAL CONDITIONS (Overhead & Profit)		15%		15.17	
	ARCHITECT FEES		9%		10.47	
			Total Building Cost		**126.75**	

BUILDING TYPES

BUILDING TYPES

Costs per square foot of floor area

Exterior Wall	S.F. Area	100000	125000	150000	175000	200000	225000	250000	275000	300000
	L.F. Perimeter	594	705	816	783	866	950	1033	1116	1200
Face Brick with Structural Facing Tile	Steel Frame	116.55	114.45	113.05	109.90	109.05	108.40	107.85	107.45	107.10
	R/Conc. Frame	118.00	115.90	114.45	111.35	110.50	109.80	109.30	108.85	108.50
Face Brick with Concrete Block Back-up	Steel Frame	113.45	111.50	110.15	107.55	106.80	106.15	105.70	105.30	105.00
	R/Conc. Frame	114.90	112.95	111.60	109.00	108.20	107.60	107.10	106.70	106.45
Precast Concrete Panels With Exposed Aggregate	Steel Frame	112.65	110.80	109.50	107.10	106.35	105.80	105.30	104.95	104.65
	R/Conc. Frame	113.50	111.60	110.30	107.95	107.20	106.60	106.15	105.75	105.50
Perimeter Adj., Add or Deduct	Per 100 L.F.	2.60	2.05	1.70	1.50	1.30	1.15	1.00	.95	.85
Story Hgt. Adj., Add or Deduct	Per 1 Ft.	1.25	1.15	1.10	.95	.90	.90	.85	.80	.85

For Basement, add $20.95 per square foot of basement area

The above costs were calculated using the basic specifications shown on the facing page. These costs should be adjusted where necessary for design alternatives and owner's requirements. Reported completed project costs, for this type of structure, range from $97.75 to $219.65 per S.F.

Common additives

Description	Unit	$ Cost
Cabinets, Base, door units, metal	L.F.	154
Drawer units	L.F.	267
Tall storage cabinets, 7' high, open	L.F.	276
With doors	L.F.	310
Wall, metal 12-1/2" deep, open	L.F.	104
With doors	L.F.	180
Closed Circuit TV (Patient monitoring)		
One station camera & monitor	Each	1325
For additional camera add	Each	730
For automatic iris for low light add	Each	1900
Doctors In-Out Register, 200 names	Each	12,700
Comb. control & recall, 200 names	Each	15,900
Recording register	Each	6025
Transformers	Each	270
Pocket pages	Each	920
Hubbard Tank, with accessories		
Stainless steel, 125 GPM 45 psi	Each	19,600
For electric hoist, add	Each	2150
Mortuary Refrigerator, End operated		
2 capacity	Each	10,900
6 capacity	Each	20,700

Description	Unit	$ Cost
Nurses Call Station		
Single bedside call station	Each	200
Ceiling speaker station	Each	97
Emergency call station	Each	133
Pillow speaker	Each	210
Double bedside call station	Each	360
Duty station	Each	221
Standard call button	Each	108
Master control station for 20 stations	Each	3825
Sound System		
Amplifier, 250 watts	Each	1525
Speaker, ceiling or wall	Each	131
Trumpet	Each	246
Station, Dietary with ice	Each	11,800
Sterilizers		
Single door, steam	Each	120,000
Double door, steam	Each	154,000
Portable, counter top, steam	Each	3025 - 4725
Gas	Each	29,700
Automatic washer/sterilizer	Each	40,900

Important: See the Reference Section for Location Factors

Model costs calculated for a 6 story building with 12' story height and 200,000 square feet of floor area

			Unit	Unit Cost	Cost Per S.F.	% Of Sub-Total
1.0 Foundations						
.1	Footings & Foundations	Poured concrete; strip and spread footings and 4' foundation wall	S.F. Ground	4.20	.70	
.4	Piles & Caissons	N/A	—	—	—	1.0%
.9	Excavation & Backfill	Site preparation for slab and trench for foundation wall and footing	S.F. Ground	1.04	.17	
2.0 Substructure						
.1	Slab on Grade	4" reinforced concrete with vapor barrier and granular base	S.F. Slab	3.06	.51	0.6%
.2	Special Substructures	N/A	—	—	—	
3.0 Superstructure						
.1	Columns & Beams	Fireproofed steel columns	L.F. Column	56	.98	
.4	Structural Walls	N/A	—	—	—	
.5	Elevated Floors	Concrete slab with metal deck and beams	S.F. Floor	10.28	8.57	12.5%
.7	Roof	Metal deck, open web steel joists, beams, interior columns	S.F. Roof	4.70	.78	
.9	Stairs	Concrete filled metal pan	Flight	4210	.55	
4.0 Exterior Closure						
.1	Walls	Face brick and structural facing tile 70% of wall	S.F. Wall	27	6.06	
.5	Exterior Wall Finishes	N/A	—	—	—	10.1%
.6	Doors	Double aluminum and glass and sliding doors	Each	3700	.52	
.7	Windows & Glazed Walls	Aluminum sliding 30% of wall	Each	359	2.24	
5.0 Roofing						
.1	Roof Coverings	Built-up tar and gravel with flashing	S.F. Roof	2.16	.36	
.7	Insulation	Perlite/EPS composite	S.F. Roof	1.25	.21	0.7%
.8	Openings & Specialties	Gravel stop and hatches	S.F. Roof	.12	.02	
6.0 Interior Construction						
.1	Partitions	Gypsum board on metal studs with sound deadening board 9 S.F. Floor/L.F. Partition	S.F. Partition	4.83	5.37	
.4	Interior Doors	Single leaf hollow metal 90 S.F. Floor/Door	Each	526	5.84	
.5	Wall Finishes	40% vinyl wall covering, 35% ceramic tile, 25% epoxy coating	S.F. Surface	2.46	5.47	
.6	Floor Finishes	60% vinyl tile, 20% ceramic, 20% terrazzo	S.F. Floor	6.15	6.15	30.1%
.7	Ceiling Finishes	Plaster on suspended metal lath	S.F. Ceiling	3.07	3.07	
.9	Interior Surface/Exterior Wall	Glazed coating 70% of wall	S.F. Wall	1.15	.25	
7.0 Conveying						
.1	Elevators	Six geared hospital elevators	Each	134,333	4.03	4.6%
.2	Special Conveyors	N/A	—	—	—	
8.0 Mechanical						
.1	Plumbing	Kitchen, toilet and service fixtures, supply and drainage 1 Fixture/275 S.F. Floor	Each	4182	15.21	
.2	Fire Protection	Standpipe and wet pipe sprinkler systems	S.F. Floor	1.23	1.23	
.3	Heating	Oil fired hot water, wall fin radiation	S.F. Floor	2.85	2.85	24.5%
.4	Cooling	Chilled water, fan coil units	S.F. Floor	2.03	2.03	
.5	Special Systems	N/A	—	—	—	
9.0 Electrical						
.1	Service & Distribution	3600 ampere service, panel board and feeders	S.F. Floor	1.00	1.00	
.2	Lighting & Power	Fluorescent fixtures, receptacles, switches, A.C. and misc. power	S.F. Floor	6.70	6.70	11.5%
.4	Special Electrical	Alarm systems, communications systems and emergency lighting	S.F. Floor	2.31	2.31	
11.0 Special Construction						
.1	Specialties	Conductive flooring, oxygen piping, curtain partitions	S.F. Floor	3.82	3.82	4.4%
12.0 Site Work						
.1	Earthwork	N/A	—	—	—	
.3	Utilities	N/A	—	—	—	0.0%
.5	Roads & Parking	N/A	—	—	—	
.7	Site Improvements	N/A	—	—	—	
			Sub-Total		87.00	**100%**
	GENERAL CONDITIONS (Overhead & Profit)			15%	13.05	
	ARCHITECT FEES			9%	9.00	
			Total Building Cost		109.05	

145

Costs per square foot of floor area

Exterior Wall	S.F. Area	35000	55000	75000	95000	115000	135000	155000	175000	195000
	L.F. Perimeter	314	401	497	555	639	722	716	783	850
Face Brick with Concrete Block Back-up	Steel Frame	101.85	95.15	92.30	89.80	88.65	87.85	86.10	85.60	85.15
	R/Conc. Frame	105.00	98.35	95.45	93.00	91.85	91.00	89.30	88.80	88.30
Glass and Metal Curtain Walls	Steel Frame	96.85	91.10	88.60	86.55	85.55	84.85	83.55	83.10	82.70
	R/Conc. Frame	100.15	94.40	91.90	89.90	88.85	88.20	86.85	86.40	86.05
Precast Concrete Panels	Steel Frame	105.15	97.85	94.70	91.95	90.70	89.80	87.80	87.25	86.75
	R/Conc. Frame	109.05	101.65	98.45	95.65	94.35	93.50	91.45	90.85	90.35
Perimeter Adj., Add or Deduct	Per 100 L.F.	5.60	3.50	2.60	2.10	1.70	1.45	1.25	1.10	1.00
Story Hgt. Adj., Add or Deduct	Per 1 Ft.	1.60	1.25	1.15	1.05	1.00	.95	.85	.80	.80
For Basement, add $19.75 per square foot of basement area										

The above costs were calculated using the basic specifications shown on the facing page. These costs should be adjusted where necessary for design alternatives and owner's requirements. Reported completed project costs, for this type of structure, range from $64.55 to $124.40 per S.F.

Common additives

Description	Unit	$ Cost
Bar, Front bar	L.F.	262
Back bar	L.F.	209
Booth, Upholstered, custom, straight	L.F.	129 - 237
"L" or "U" shaped	L.F.	133 - 225
Closed Circuit Surveillance, One station		
Camera and monitor	Each	1325
For additional camera stations, add	Each	730
Directory Boards, Plastic, glass covered		
30" x 20"	Each	510
36" x 48"	Each	930
Aluminum, 24" x 18"	Each	450
48" x 32"	Each	640
48" x 60"	Each	1450
Elevators, Electric passenger, 5 stops		
3500# capacity	Each	97,300
5000# capacity	Each	100,800
Additional stop, add	Each	5225
Emergency Lighting, 25 watt, battery operated		
Lead battery	Each	330
Nickel cadmium	Each	620

Description	Unit	$ Cost
Laundry Equipment		
Folders, blankets & sheets, king size	Each	51,000
Ironers, 110" single roll	Each	27,700
Combination washer extractor 50#	Each	9925
125#	Each	24,900
Sauna, Prefabricated, complete		
6' x 4'	Each	3725
6' x 6'	Each	4400
6' x 9'	Each	5850
8' x 8'	Each	6225
10' x 12'	Each	8900
Smoke Detectors		
Ceiling type	Each	142
Duct type	Each	390
Sound System		
Amplifier, 250 watts	Each	1525
Speaker, ceiling or wall	Each	131
Trumpet	Each	246
TV Antenna, Master system, 12 outlet	Outlet	229
30 outlet	Outlet	147
100 outlet	Outlet	141

Important: See the Reference Section for Location Factors

Model costs calculated for a 6 story building with 10' story height and 135,000 square feet of floor area

				Unit	Unit Cost	Cost Per S.F.	% Of Sub-Total
1.0 Foundations							
.1	Footings & Foundations	Poured concrete; strip and spread footings and 4' foundation wall		S.F. Ground	6.60	1.10	
.4	Piles & Caissons	N/A		–	–	–	1.8%
.9	Excavation & Backfill	Site preparation for slab and trench for foundation wall and footing		S.F. Ground	1.04	.17	
2.0 Substructure							
.1	Slab on Grade	4" reinforced concrete with vapor barrier and granular base		S.F. Slab	3.06	.51	0.7%
.2	Special Substructures	N/A		–	–	–	
3.0 Superstructure							
.1	Columns & Beams	Included in 3.5 and 3.7		–	–	–	
.4	Structural Walls	N/A		–	–	–	
.5	Elevated Floors	Concrete slab with metal deck, beams, columns		S.F. Floor	10.31	8.59	13.4%
.7	Roof	Metal deck, open web steel joists, beams, columns		S.F. Roof	4.70	.78	
.9	Stairs	Concrete		Flight	3075	.32	
4.0 Exterior Closure							
.1	Walls	Face brick with concrete block backup	80% of wall	S.F. Wall	19.44	4.99	
.5	Exterior Wall Finishes	N/A		–	–	–	
.6	Doors	Glass amd metal doors and entrances		Each	2032	.15	9.9%
.7	Windows & Glazed Walls	Window wall	20% of wall	S.F. Window	31	2.01	
5.0 Roofing							
.1	Roof Coverings	Built-up tar and gravel with flashing		S.F. Roof	2.16	.36	
.7	Insulation	Perlite/EPS composite		S.F. Roof	1.25	.21	0.8%
.8	Openings & Specialties	Gravel stop and hatches		S.F. Roof	.24	.04	
6.0 Interior Construction							
.1	Partitions	Gypsum board and sound deadening board, steel studs	9 S.F. Floor/L.F. Partition	S.F. Partition	4.83	4.29	
.4	Interior Doors	Single leaf hollow metal	90 S.F. Floor/Door	Each	526	5.84	
.5	Wall Finishes	50% paint, 45% vinyl cover, 5% ceramic tile		S.F. Surface	1.11	1.98	
.6	Floor Finishes	80% carpet, 10% vinyl composition tile, 10% ceramic tile		S.F. Floor	4.28	4.28	25.4%
.7	Ceiling Finishes	Mineral fiber tile applied with adhesive		S.F. Ceiling	1.20	1.20	
.9	Interior Surface/Exterior Wall	Painted gypsum board on furring	80% of wall	S.F. Wall	2.85	.73	
7.0 Conveying							
.1	Elevators	Four geared passenger elevators		Each	122,512	3.63	5.0%
.2	Special Conveyors	N/A		–	–	–	
8.0 Mechanical							
.1	Plumbing	Kitchen, toilet and service fixtures, supply and drainage	1 Fixture/155 S.F. Floor	Each	1684	10.87	
.2	Fire Protection	Standpipes and hose systems and sprinklers, light hazard		S.F. Floor	1.42	1.42	
.3	Heating	Oil fired hot water, wall fin radiation		S.F. Floor	2.85	2.85	32.4%
.4	Cooling	Chilled water, fan coil units		S.F. Floor	8.12	8.12	
.5	Special Systems	N/A		–	–	–	
9.0 Electrical							
.1	Service & Distribution	2000 ampere service, panel board and feeders		S.F. Floor	.78	.78	
.2	Lighting & Power	Fluorescent fixtures, receptacles, switches, A.C. and misc. power		S.F. Floor	5.06	5.06	10.6%
.4	Special Electrical	Alarm systems, communications systems and emergency lighting		S.F. Floor	1.79	1.79	
11.0 Special Construction							
.1	Specialties	N/A		–	–	–	0.0%
12.0 Site Work							
.1	Earthwork	N/A		–	–	–	
.3	Utilities	N/A		–	–	–	
.5	Roads & Parking	N/A		–	–	–	0.0%
.7	Site Improvements	N/A		–	–	–	
				Sub-Total		72.07	**100%**
	GENERAL CONDITIONS (Overhead & Profit)				15%	10.81	
	ARCHITECT FEES				6%	4.97	
				Total Building Cost		87.85	

Costs per square foot of floor area

Exterior Wall	S.F. Area	140000	243000	346000	450000	552000	655000	760000	860000	965000
	L.F. Perimeter	403	587	672	800	936	1073	1213	1195	1312
Face Brick with Concrete Block Back-up	Steel Frame	86.55	82.40	79.45	78.25	77.65	77.15	76.80	75.75	75.50
	R/Conc. Frame	88.50	84.35	81.35	80.20	79.55	79.05	78.75	77.65	77.45
Face Brick Veneer On Steel Studs	Steel Frame	85.20	81.25	78.50	77.40	76.85	76.35	76.05	75.10	74.85
	R/Conc. Frame	87.45	83.55	80.80	79.70	79.10	78.60	78.30	77.35	77.10
Glass and Metal Curtain Walls	Steel Frame	93.85	88.65	85.65	84.30	83.55	83.00	82.60	81.75	81.50
	R/Conc. Frame	96.10	90.90	87.90	86.55	85.80	85.25	84.85	83.95	83.75
Perimeter Adj., Add or Deduct	Per 100 L.F.	3.25	1.85	1.30	1.00	.80	.70	.60	.55	.50
Story Hgt. Adj., Add or Deduct	Per 1 Ft.	1.25	1.05	.85	.75	.75	.75	.70	.60	.60

For Basement, add $20.25 per square foot of basement area

The above costs were calculated using the basic specifications shown on the facing page. These costs should be adjusted where necessary for design alternatives and owner's requirements. Reported completed project costs, for this type of structure, range from $71.95 to $125.80 per S.F.

Common additives

Description	Unit	$ Cost		Description	Unit	$ Cost
Bar, Front bar	L.F.	262		Laundry Equipment		
Back bar	L.F.	209		Folders, blankets & sheets, king size	Each	51,000
Booth, Upholstered, custom, straight	L.F.	129 - 237		Ironers, 110" single roll	Each	27,700
"L" or "U" shaped	L.F.	133 - 225		Combination washer & extractor 50#	Each	9925
Closed Circuit Surveillance, One station				125#	Each	24,900
Camera and monitor	Each	1325		Sauna, Prefabricated, complete		
For additional camera stations, add	Each	730		6' x 4'	Each	3725
Directory Boards, Plastic, glass covered				6' x 6'	Each	4400
30" x 20"	Each	510		6' x 9'	Each	5850
36" x 48"	Each	930		8' x 8'	Each	6225
Aluminum, 24" x 18"	Each	450		10' x 12'	Each	8900
48" x 32"	Each	640		Smoke Detectors		
48" x 60"	Each	1450		Ceiling type	Each	142
Elevators, Electric passenger, 10 stops				Duct type	Each	390
3500# capacity	Each	198,500		Sound System	Each	
5000# capacity	Each	204,000		Amplifier, 250 watts	Each	1525
Additional stop, add	Each	5225		Speaker, ceiling or wall	Each	131
Emergency Lighting, 25 watt, battery operated				Trumpet	Each	246
Lead battery	Each	330		TV Antenna, Master system, 12 outlet	Outlet	229
Nickel cadmium	Each	620		30 outlet	Outlet	147
				100 outlet	Outlet	141

Model costs calculated for a 15 story building with 10' story height and 450,000 square feet of floor area

			Unit	Unit Cost	Cost Per S.F.	% Of Sub-Total
1.0 Foundations						
.1	Footings & Foundations	Poured concrete; strip and spread footings and 4' foundation wall	S.F. Ground	12.00	.80	
.4	Piles & Caissons	N/A	–	–	–	1.3%
.9	Excavation & Backfill	Site preparation for slab and trench for foundation wall and footing	S.F. Ground	1.04	.07	
2.0 Substructure						
.1	Slab on Grade	4" reinforced concrete with vapor barrier and granular base	S.F. Slab	3.06	.20	0.3%
.2	Special Substructures	N/A	–	–	–	
3.0 Superstructure						
.1	Columns & Beams	Steel columns included in 3.5 and 3.7	–	–	–	
.4	Structural Walls	N/A	–	–	–	
.5	Elevated Floors	Open web steel joists, slab form, concrete, columns	S.F. Floor	10.15	9.47	15.5%
.7	Roof	Metal deck, open web steel joists, beams, columns	S.F. Roof	3.73	.25	
.9	Stairs	Concrete filled metal pan	Flight	5025	1.02	
4.0 Exterior Closure						
.1	Walls	N/A	–	–	–	
.5	Exterior Wall Finishes	N/A	–	–	–	5.7%
.6	Doors	Glass and metal doors and entrances	Each	1824	.13	
.7	Windows & Glazed Walls	Glass and metal curtain walls 100% of wall	S.F. Wall	14.20	3.79	
5.0 Roofing						
.1	Roof Coverings	Built-up tar and gravel with flashing	S.F. Roof	2.10	.14	
.7	Insulation	Perlite/EPS composite	S.F. Roof	1.25	.08	0.3%
.8	Openings & Specialties	Gravel stop and hatches	S.F. Roof	.15	.01	
6.0 Interior Construction						
.1	Partitions	Gypsum board and sound deadening board, steel studs 9 S.F. Floor/L.F. Partition	S.F. Partition	4.83	4.29	
.4	Interior Doors	Single leaf hollow metal 90 S.F. Floor/Door	Each	526	5.84	
.5	Wall Finishes	50% paint, 45% vinyl cover, 5% ceramic tile	S.F. Surface	1.11	1.98	30.2%
.6	Floor Finishes	80% carpet, 10% vinyl composition tile, 10% ceramic tile	S.F. Floor	4.28	4.28	
.7	Ceiling Finishes	Mineral fiber tile applied with adhesive to suspended gypsum board	S.F. Ceiling	3.89	3.89	
.9	Interior Surface/Exterior Wall	Painted gypsum board on furring 80% of wall	S.F. Wall	2.85	.61	
7.0 Conveying						
.1	Elevators	One geared freight, six geared passenger elevators	Each	216,750	2.89	4.2%
.2	Special Conveyors	N/A	–	–	–	
8.0 Mechanical						
.1	Plumbing	Kitchen, toilet and service fixtures, supply and drainage 1 Fixture/165 S.F. Floor	Each	1673	10.14	
.2	Fire Protection	Standpipes and hose systems and sprinklers, light hazard	S.F. Floor	2.22	2.22	
.3	Heating	Oil fired hot water, wall fin radiation	S.F. Floor	1.42	1.42	31.6%
.4	Cooling	Chilled water, fan coil units	S.F. Floor	8.12	8.12	
.5	Special Systems	N/A	–	–	–	
9.0 Electrical						
.1	Service & Distribution	2400 ampere service, panel board and feeders	S.F. Floor	.84	.84	
.2	Lighting & Power	Fluorescent fixtures, receptacles, switches, A.C. and misc. power	S.F. Floor	4.99	4.99	10.9%
.4	Special Electrical	Alarm systems, communications systems and emergency lighting	S.F. Floor	1.69	1.69	
11.0 Special Construction						
.1	Specialties	N/A	–	–	–	0.0%
12.0 Site Work						
.1	Earthwork	N/A	–	–	–	
.3	Utilities	N/A	–	–	–	0.0%
.5	Roads & Parking	N/A	–	–	–	
.7	Site Improvements	N/A	–	–	–	
			Sub-Total		69.16	**100%**
	GENERAL CONDITIONS (Overhead & Profit)			15%	10.37	
	ARCHITECT FEES			6%	4.77	
			Total Building Cost		84.30	

BUILDING TYPES

Costs per square foot of floor area

Exterior Wall	S.F. Area	7000	13000	20000	26000	32000	45000	58000	72000	85000
	L.F. Perimeter	196	273	366	408	475	620	609	713	766
Face Brick with Concrete Block Back-up	Steel Frame	197.75	180.20	173.25	168.15	166.10	163.55	158.25	157.00	155.50
	R/Conc. Frame	194.00	176.50	169.50	164.45	162.35	159.80	154.55	153.25	151.75
Stucco on Concrete Block	Steel Frame	187.80	172.80	166.80	162.65	160.95	158.75	154.65	153.60	152.40
	R/Conc. Frame	183.90	168.90	162.90	158.75	157.00	154.85	150.70	149.70	148.50
Reinforced Concrete	Steel Frame	191.25	175.35	169.05	164.60	162.75	160.45	155.95	154.80	153.50
	R/Conc. Frame	187.30	171.45	165.10	160.70	158.85	156.50	152.00	150.90	149.60
Perimeter Adj., Add or Deduct	Per 100 L.F.	20.65	11.10	7.25	5.55	4.50	3.20	2.50	2.00	1.70
Story Hgt. Adj., Add or Deduct	Per 1 Ft.	3.10	2.30	2.05	1.75	1.65	1.50	1.20	1.10	1.00
For Basement, add $26.70 per square foot of basement area										

The above costs were calculated using the basic specifications shown on the facing page. These costs should be adjusted where necessary for design alternatives and owner's requirements. Reported completed project costs, for this type of structure, range from $101.50 to $232.75 per S.F.

Common additives

Description	Unit	$ Cost
Clock System		
20 room	Each	11,800
50 room	Each	28,700
Closed Circuit Surveillance, One station		
Camera and monitor	Each	1325
For additional camera stations, add	Each	730
Elevators, Hydraulic passenger, 2 stops		
1500# capacity	Each	40,000
2500# capacity	Each	41,300
3500# capacity	Each	44,900
Additional stop, add	Each	3350
Emergency Generators, Complete system, gas		
15 KW	Each	12,700
70 KW	Each	25,300
85 KW	Each	28,000
115 KW	Each	48,400
170 KW	Each	71,500
Diesel, 50 KW	Each	24,100
100 KW	Each	34,800
150 KW	Each	42,600
350 KW	Each	64,500

Description	Unit	$ Cost
Emergency Lighting, 25 watt, battery operated		
Nickel cadmium	Each	620
Flagpoles, Complete		
Aluminum, 20' high	Each	1100
40' High	Each	2475
70' High	Each	7100
Fiberglass, 23' High	Each	1475
39'-5" High	Each	2525
59' High	Each	7025
Laundry Equipment	Each	
Folders, blankets, & sheets	Each	51,000
Ironers, 110" single roll	Each	27,700
Combination washer & extractor, 50#	Each	9925
125#	Each	24,900
Safe, Office type, 4 hour rating		
30" x 18" x 18"	Each	3475
62" x 33" x 20"	Each	7575
Sound System		
Amplifier, 250 watts	Each	1525
Speaker, ceiling or wall	Each	131
Trumpet	Each	246

Important: See the Reference Section for Location Factors

Model costs calculated for a 3 story building with 12' story height and 32,000 square feet of floor area

				Unit	Unit Cost	Cost Per S.F.	% Of Sub-Total
1.0 Foundations							
.1	Footings & Foundations	Poured concrete; strip and spread footings and 4' foundation wall		S.F. Ground	3.21	1.07	
.4	Piles & Caissons	N/A		–	–	–	1.1%
.9	Excavation & Backfill	Site preparation for slab and trench for foundation wall and footing		S.F. Ground	1.11	.37	
2.0 Substructure							
.1	Slab on Grade	4" reinforced concrete with vapor barrier and granular base		S.F. Slab	3.06	1.02	0.8%
.2	Special Substructures	N/A		–	–	–	
3.0 Superstructure							
.1	Columns & Beams	Steel columns		L.F. Column	42	1.42	
.4	Structural Walls	N/A		–	–	–	
.5	Elevated Floors	Concrete slab with metal deck and beams		S.F. Floor	13.85	9.23	11.6%
.7	Roof	Concrete slab with metal deck and beams		S.F. Roof	12.40	4.13	
.9	Stairs	Concrete filled metal pan		Flight	5900	.92	
4.0 Exterior Closure							
.1	Walls	Face brick with reinforced concrete block backup	85% of wall	S.F. Wall	19.44	8.83	
.5	Exterior Wall Finishes	N/A		–	–	–	11.3%
.6	Doors	Metal		Each	3525	.11	
.7	Windows & Glazed Walls	Bullet resisting and metal horizontal pivoted	15% of wall	S.F. Window	79	6.34	
5.0 Roofing							
.1	Roof Coverings	Built-up tar and gravel with flashing		S.F. Roof	2.25	.75	
.7	Insulation	Perlite/EPS composite		S.F. Roof	1.25	.42	0.9%
.8	Openings & Specialties	Gravel stop and hatches		S.F. Roof	.30	.10	
6.0 Interior Construction							
.1	Partitions	Concrete block	45 S.F. of Floor/L.F. Partition	S.F. Partition	5.66	1.51	
.4	Interior Doors	Hollow metal fire doors	930 S.F. Floor/Door	Each	526	.57	
.5	Wall Finishes	Paint		S.F. Surface	1.09	.58	5.1%
.6	Floor Finishes	70% vinyl composition tile, 20% carpet, 10% ceramic tile		S.F. Floor	2.46	2.46	
.7	Ceiling Finishes	Mineral tile on zee runners	30% of area	S.F. Ceiling	3.07	.92	
.9	Interior Surface/Exterior Wall	Paint	85% of wall	S.F. Wall	1.95	.89	
7.0 Conveying							
.1	Elevators	One hydraulic passenger elevator		Each	93,760	2.93	2.2%
.2	Special Conveyors	N/A		–	–	–	
8.0 Mechanical							
.1	Plumbing	Toilet and service fixtures, supply and drainage	1 Fixture/78 S.F. Floor	Each	1944	24.93	
.2	Fire Protection	Standpipes and hose systems		S.F. Floor	.36	.36	
.3	Heating	Included in 8.4		–	–	–	26.5%
.4	Cooling	Rooftop multizone unit systems		S.F. Floor	10.50	10.50	
.5	Special Systems	N/A		–	–	–	
9.0 Electrical							
.1	Service & Distribution	600 ampere service, panel board and feeders		S.F. Floor	.85	.85	
.2	Lighting & Power	Fluorescent fixtures, receptacles, switches, A.C. and misc. power		S.F. Floor	5.96	5.96	6.0%
.4	Special Electrical	Alarm systems and emergency lighting		S.F. Floor	1.26	1.26	
11.0 Special Construction							
.1	Specialties	Prefabricated cells, visitor cubicles		S.F. Floor	46	46.56	34.5%
12.0 Site Work							
.1	Earthwork	N/A		–	–	–	
.3	Utilities	N/A		–	–	–	0.0%
.5	Roads & Parking	N/A		–	–	–	
.7	Site Improvements	N/A		–	–	–	
				Sub-Total		134.99	**100%**
	GENERAL CONDITIONS (Overhead & Profit)				15%	20.25	
	ARCHITECT FEES				7%	10.86	
				Total Building Cost		**166.10**	

Costs per square foot of floor area

Exterior Wall	S.F. Area	1000	2000	3000	4000	5000	10000	15000	20000	25000
	L.F. Perimeter	126	179	219	253	283	400	490	568	632
Decorative Concrete Block	Steel Frame	110.85	99.85	94.90	91.90	89.90	84.90	82.70	81.40	80.45
	Bearing Walls	116.15	104.55	99.30	96.15	94.10	88.80	86.45	85.10	84.10
Face Brick with Concrete Block Back-up	Steel Frame	131.90	115.90	108.65	104.35	101.45	94.15	90.90	89.05	87.70
	Bearing Walls	132.15	115.90	108.55	104.20	101.30	93.85	90.60	88.70	87.30
Metal Sandwich Panel Precast Concrete Panel	Steel Frame	114.40	103.55	98.65	95.70	93.75	88.80	86.65	85.40	84.45
	Bearing Walls	111.75	101.40	96.75	94.00	92.10	87.40	85.30	84.10	83.20
Perimeter Adj., Add or Deduct	Per 100 L.F.	30.20	15.10	10.05	7.55	6.05	3.00	2.00	1.55	1.15
Story Hgt. Adj., Add or Deduct	Per 1 Ft.	2.05	1.45	1.20	1.05	.90	.65	.55	.50	.40

For Basement, add $17.90 per square foot of basement area

The above costs were calculated using the basic specifications shown on the facing page. These costs should be adjusted where necessary for design alternatives and owner's requirements. Reported completed project costs, for this type of structure, range from $58.25 to $141.95 per S.F.

Common additives

Description	Unit	$ Cost
Closed Circuit Surveillance, One station		
Camera and monitor	Each	1325
For additional camera stations, add	Each	730
Emergency Lighting, 25 watt, battery operated		
Lead battery	Each	330
Nickel cadmium	Each	620
Laundry Equipment		
Dryers, coin operated 30 lb.	Each	2550
Double stacked	Each	5575
50 lb.	Each	2925
Dry cleaner 20 lb.	Each	34,100
30 lb.	Each	46,500
Washers, coin operated	Each	1125
Washer/extractor 20 lb.	Each	3975
30 lb.	Each	9350
50 lb.	Each	9925
75 lb.	Each	18,600
Smoke Detectors		
Ceiling type	Each	142
Duct type	Each	390

BUILDING TYPES

Important: See the Reference Section for Location Factors

Model costs calculated for a 1 story building with 12' story height and 3,000 square feet of floor area

				Unit	Unit Cost	Cost Per S.F.	% Of Sub-Total
1.0 Foundations							
.1	Footings & Foundations	Poured concrete; strip and spread footings and 4' foundation wall		S.F. Ground	4.92	4.92	
.4	Piles & Caissons	N/A		—	—	—	8.0%
.9	Excavation & Backfill	Site preparation for slab and trench for foundation wall and footing		S.F. Ground	1.24	1.24	
2.0 Substructure							
.1	Slab on Grade	5" reinforced concrete with vapor barrier and granular base		S.F. Slab	3.30	3.30	4.3%
.2	Special Substructures	N/A		—	—	—	
3.0 Superstructure							
.1	Columns & Beams	Fireproofing; steel columns included in 3.7		—	—	—	
.4	Structural Walls	N/A		—	—	—	
.5	Elevated Floors	N/A		—	—	—	4.0%
.7	Roof	Metal deck, open web steel joists, beams, columns		S.F. Roof	3.05	3.05	
.9	Stairs	N/A		—	—	—	
4.0 Exterior Closure							
.1	Walls	Decorative concrete block	90% of wall	S.F. Wall	8.85	6.98	
.5	Exterior Wall Finishes	N/A		—	—	—	13.3%
.6	Doors	Double aluminum and glass		Each	2725	.91	
.7	Windows & Glazed Walls	Store front	10% of wall	S.F. Window	26	2.33	
5.0 Roofing							
.1	Roof Coverings	Built-up tar and gravel with flashing		S.F. Roof	3.19	3.19	
.7	Insulation	Perlite/EPS composite		S.F. Roof	1.25	1.25	5.9%
.8	Openings & Specialties	Gravel stop, gutters and downspouts		S.F. Roof	.11	.11	
6.0 Interior Construction							
.1	Partitions	Gypsum board on metal studs	60 S.F. Floor/L.F. Partition	S.F. Partition	3.00	.50	
.4	Interior Doors	Single leaf wood	750 S.F. Floor/Door	Each	408	.54	
.5	Wall Finishes	Paint		S.F. Surface	.51	.17	
.6	Floor Finishes	Vinyl composition tile		S.F. Floor	1.37	1.37	8.6%
.7	Ceiling Finishes	Fiberglass board on exposed grid system		S.F. Ceiling	1.80	1.80	
.9	Interior Surface/Exterior Wall	Painted gypsum board on furring	90% of wall	S.F. Wall	2.86	2.25	
7.0 Conveying							
.1	Elevators	N/A		—	—	—	0.0%
.2	Special Conveyors	N/A		—	—	—	
8.0 Mechanical							
.1	Plumbing	Toilet and service fixtures, supply and drainage	1 Fixture/600 S.F. Floor	Each	10,350	17.25	
.2	Fire Protection	Sprinkler, ordinary hazard		S.F. Floor	2.27	2.27	
.3	Heating	Included in 8.4		—	—	—	34.7%
.4	Cooling	Rooftop single zone unit systems		S.F. Floor	7.34	7.34	
.5	Special Systems	N/A		—	—	—	
9.0 Electrical							
.1	Service & Distribution	200 ampere service, panel board and feeders		S.F. Floor	2.24	2.24	
.2	Lighting & Power	Fluorescent fixtures, receptacles, switches, A.C. and misc. power		S.F. Floor	13.74	13.74	21.2%
.4	Special Electrical	Alarm systems and emergency lighting		S.F. Floor	.35	.35	
11.0 Special Construction							
.1	Specialties	N/A		—	—	—	0.0%
12.0 Site Work							
.1	Earthwork	N/A		—	—	—	
.3	Utilities	N/A		—	—	—	
.5	Roads & Parking	N/A		—	—	—	0.0%
.7	Site Improvements	N/A		—	—	—	
				Sub-Total		77.10	100%
	GENERAL CONDITIONS (Overhead & Profit)				15%	11.57	
	ARCHITECT FEES				7%	6.23	
				Total Building Cost		94.90	

BUILDING TYPES

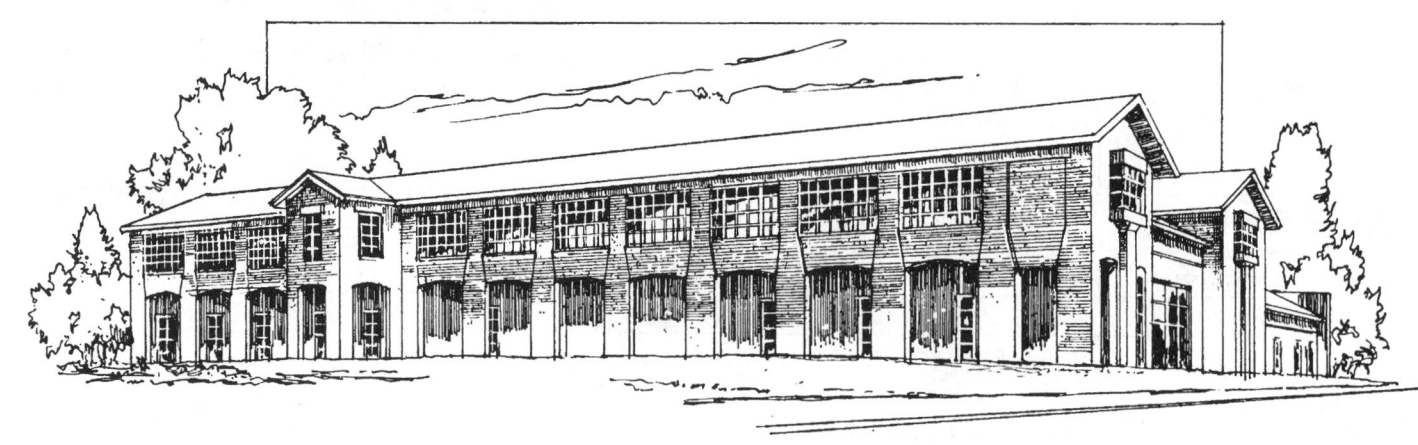

Costs per square foot of floor area

Exterior Wall	S.F. Area	7000	10000	13000	16000	19000	22000	25000	28000	31000
	L.F. Perimeter	240	300	336	386	411	435	472	510	524
Face Brick with Concrete Block Back-up	R/Conc. Frame	105.15	99.50	94.75	92.60	89.90	87.90	86.85	86.10	84.70
	Steel Frame	103.80	98.15	93.40	91.25	88.55	86.55	85.50	84.70	83.30
Limestone with Concrete Block	R/Conc. Frame	116.45	109.40	103.25	100.55	97.05	94.40	93.05	92.05	90.25
	Steel Frame	115.10	108.00	101.90	99.15	95.65	93.05	91.70	90.70	88.85
Precast Concrete Panels	R/Conc. Frame	99.75	94.75	90.65	88.75	86.50	84.75	83.85	83.20	82.00
	Steel Frame	98.35	93.40	89.30	87.40	85.10	83.40	82.50	81.80	80.65
Perimeter Adj., Add or Deduct	Per 100 L.F.	13.25	9.25	7.10	5.80	4.90	4.20	3.70	3.30	3.00
Story Hgt. Adj., Add or Deduct	Per 1 Ft.	2.00	1.75	1.50	1.40	1.30	1.15	1.10	1.05	1.00
For Basement, add $26.45 per square foot of basement area										

The above costs were calculated using the basic specifications shown on the facing page. These costs should be adjusted where necessary for design alternatives and owner's requirements. Reported completed project costs, for this type of structure, range from $55.85 to $143.25 per S.F.

Common additives

Description	Unit	$ Cost
Carrels Hardwood	Each	655 - 855
Closed Circuit Surveillance, One station		
Camera and monitor	Each	1325
For additional camera stations, add	Each	730
Elevators, Hydraulic passenger, 2 stops		
1500# capacity	Each	40,000
2500# capacity	Each	41,300
3500# capacity	Each	44,900
Emergency Lighting, 25 watt, battery operated		
Lead battery	Each	330
Nickel cadmium	Each	620
Flagpoles, Complete		
Aluminum, 20' high	Each	1100
40' high	Each	2475
70' high	Each	7100
Fiberglass, 23' high	Each	1475
39'-5" high	Each	2525
59' high	Each	7025

Description	Unit	$ Cost
Library Furnishings		
Bookshelf, 90" high, 10" shelf double face	L.F.	123
single face	L.F.	96
Charging desk, built-in with counter		
Plastic laminated top	L.F.	465
Reading table, laminated		
top 60" x 36"	Each	625

Model costs calculated for a 2 story building with 14' story height and 22,000 square feet of floor area

				Unit	Unit Cost	Cost Per S.F.	% Of Sub-Total
1.0 Foundations							
.1	Footings & Foundations	Poured concrete; strip and spread footings and 4' foundation wall		S.F. Ground	4.08	2.04	
.4	Piles & Caissons	N/A		–	–	–	3.7%
.9	Excavation & Backfill	Site preparation for slab and trench for foundation wall and footing		S.F. Ground	1.11	.56	
2.0 Substructure							
.1	Slab on Grade	4" reinforced concrete with vapor barrier and granular base		S.F. Slab	3.06	1.53	2.2%
.2	Special Substructures	N/A		–	–	–	
3.0 Superstructure							
.1	Columns & Beams	Concrete columns		L.F. Column	55	1.71	
.4	Structural Walls	N/A		–	–	–	
.5	Elevated Floors	Concrete waffle slab		S.F. Floor	11.43	5.72	18.7%
.7	Roof	Concrete waffle slab		S.F. Roof	10.59	5.30	
.9	Stairs	Concrete filled metal pan		Flight	5475	.50	
4.0 Exterior Closure							
.1	Walls	Face brick with concrete block backup	90% of wall	S.F. Wall	19.61	9.77	
.5	Exterior Wall Finishes	N/A		–	–	–	
.6	Doors	Double aluminum and glass, single leaf hollow metal		Each	3525	.32	16.8%
.7	Windows & Glazed Walls	Window wall	10% of wall	S.F. Wall	32	1.80	
5.0 Roofing							
.1	Roof Coverings	Built-up tar and gravel with flashing		S.F. Roof	2.16	1.08	
.7	Insulation	Perlite/EPS composite		S.F. Roof	1.25	.63	2.6%
.8	Openings & Specialties	Gravel stop and hatches		L.F. Roof	.26	.13	
6.0 Interior Construction							
.1	Partitions	Gypsum board on metal studs	30 S.F. Floor/L.F. Partition	S.F. Partition	3.90	1.56	
.4	Interior Doors	Single leaf wood	300 S.F. Floor/Door	Each	408	1.36	
.5	Wall Finishes	Paint		S.F. Surface	.53	.42	
.6	Floor Finishes	50% carpet, 50% vinyl tile		S.F. Floor	3.72	3.72	16.3%
.7	Ceiling Finishes	Mineral fiber on concealed zee bars		S.F. Ceiling	3.07	3.07	
.9	Interior Surface/Exterior Wall	Painted gypsum board on furring	90% of wall	S.F. Wall	2.85	1.42	
7.0 Conveying							
.1	Elevators	One hydraulic passenger elevator		Each	50,820	2.31	3.3%
.2	Special Conveyors	N/A		–	–	–	
8.0 Mechanical							
.1	Plumbing	Toilet and service fixtures, supply and drainage	1 Fixture/1835 S.F. Floor	Each	2715	1.48	
.2	Fire Protection	Wet pipe sprinkler system		S.F. Floor	1.45	1.45	
.3	Heating	Included in 8.4		–	–	–	25.3%
.4	Cooling	Multizone unit, gas heating, electric cooling		S.F. Floor	15.00	15.00	
.5	Special Systems	N/A		–	–	–	
9.0 Electrical							
.1	Service & Distribution	400 ampere service, panel board and feeders		S.F. Floor	.66	.66	
.2	Lighting & Power	Fluorescent fixtures, receptacles, switches, A.C. and misc. power		S.F. Floor	6.70	6.70	11.1%
.4	Special Electrical	Alarm systems and emergency lighting		S.F. Floor	.53	.53	
11.0 Special Construction							
.1	Specialties	N/A		–	–	–	0.0%
12.0 Site Work							
.1	Earthwork	N/A		–	–	–	
.3	Utilities	N/A		–	–	–	
.5	Roads & Parking	N/A		–	–	–	0.0%
.7	Site Improvements	N/A		–	–	–	

					Sub-Total	70.77	**100%**
	GENERAL CONDITIONS (Overhead & Profit)				15%	10.62	
	ARCHITECT FEES				8%	6.51	

					Total Building Cost	87.90	

Costs per square foot of floor area

Exterior Wall	S.F. Area	4000	5500	7000	8500	10000	11500	13000	14500	16000
	L.F. Perimeter	280	320	380	440	453	503	510	522	560
Face Brick with Concrete Block Back-up	Steel Joists	100.30	94.80	92.75	91.35	88.70	87.90	86.00	84.70	84.20
	Wood Joists	104.25	98.80	96.70	95.35	92.65	91.85	90.00	88.65	88.15
Stucco on Concrete Block	Steel Joists	92.80	88.60	86.90	85.85	83.85	83.20	81.85	80.85	80.45
	Wood Joists	96.80	92.55	90.90	89.80	87.80	87.15	85.80	84.80	84.40
Brick Veneer	Wood Frame	101.70	96.65	94.70	93.45	91.00	90.25	88.55	87.35	86.85
Wood Siding	Wood Frame	95.85	91.80	90.15	89.10	87.20	86.60	85.25	84.30	83.95
Perimeter Adj., Add or Deduct	Per 100 L.F.	9.45	6.90	5.40	4.45	3.75	3.25	2.90	2.60	2.35
Story Hgt. Adj., Add or Deduct	Per 1 Ft.	2.00	1.65	1.55	1.45	1.30	1.25	1.10	1.00	1.00
For Basement, add $16.95 per square foot of basement area										

The above costs were calculated using the basic specifications shown on the facing page. These costs should be adjusted where necessary for design alternatives and owner's requirements. Reported completed project costs, for this type of structure, range from $47.15 to $121.15 per S.F.

Common additives

Description	Unit	$ Cost
Cabinets, Hospital, base		
Laminated plastic	L.F.	255
Stainless steel	L.F.	300
Counter top, laminated plastic	L.F.	49
Stainless steel	L.F.	107
For drop-in sink, add	Each	460
Nurses station, door type		
Laminated plastic	L.F.	285
Enameled steel	L.F.	256
Stainless steel	L.F.	320
Wall cabinets, laminated plastic	L.F.	185
Enameled steel	L.F.	202
Stainless steel	L.F.	289

Description	Unit	$ Cost
Directory Boards, Plastic, glass covered		
30" x 20"	Each	510
36" x 48"	Each	930
Aluminum, 24" x 18"	Each	450
36" x 24"	Each	540
48" x 32"	Each	640
48" x 60"	Each	1450
Heat Therapy Unit		
Humidified, 26" x 78" x 28"	Each	2525
Smoke Detectors		
Ceiling type	Each	142
Duct type	Each	390
Tables, Examining, vinyl top		
with base cabinets	Each	1200 - 3250
Utensil Washer, Sanitizer	Each	8450
X-Ray, Mobile	Each	10,100 - 57,000

Model costs calculated for a 1 story building with 10′ story height and 7,000 square feet of floor area

				Unit	Unit Cost	Cost Per S.F.	% Of Sub-Total
1.0 Foundations							
.1	Footings & Foundations	Poured concrete; strip and spread footings and 4′ foundation wall		S.F. Ground	3.44	3.44	
.4	Piles & Caissons	N/A		—	—	—	5.9%
.9	Excavation & Backfill	Site preparation for slab and trench for foundation wall and footing		S.F. Ground	1.11	1.11	
2.0 Substructure							
.1	Slab on Grade	4″ reinforced concrete with vapor barrier and granular base		S.F. Slab	3.06	3.06	4.0%
.2	Special Substructures	N/A		—	—	—	
3.0 Superstructure							
.1	Columns & Beams	N/A		—	—	—	
.4	Structural Walls	N/A		—	—	—	
.5	Elevated Floors	N/A		—	—	—	5.6%
.7	Roof	Plywood on wood trusses		S.F. Ground	4.33	4.33	
.9	Stairs	N/A		—	—	—	
4.0 Exterior Closure							
.1	Walls	Face brick with concrete block backup	70% of wall	S.F. Wall	19.61	7.45	
.5	Exterior Wall Finishes	N/A		—	—	—	16.7%
.6	Doors	Aluminum and glass doors and entrance with transoms		Each	931	1.60	
.7	Windows & Glazed Walls	Wood double hung	30% of wall	Each	397	3.80	
5.0 Roofing							
.1	Roof Coverings	Asphalt shingles with flashing (Pitched)		S.F. Ground	1.24	1.24	
.7	Insulation	Fiber glass sheeets		S.F. Ground	.87	.87	3.2%
.8	Openings & Specialties	Gutters and downspouts		S.F. Ground	.38	.38	
6.0 Interior Construction							
.1	Partitions	Gypsum bd. & sound deadening bd. on wood studs w/insul.	6 S.F. Floor/L.F. Partition	S.F. Partition	4.83	6.44	
.4	Interior Doors	Single leaf wood	60 S.F. Floor/Door	Each	408	6.80	
.5	Wall Finishes	50% paint, 50% vinyl wall covering		S.F. Surface	.92	2.45	30.1%
.6	Floor Finishes	50% carpet, 50% vinyl composition tile		S.F. Floor	4.27	4.27	
.7	Ceiling Finishes	Mineral fiber tile on concealed zee bars		S.F. Ceiling	2.15	2.15	
.9	Interior Surface/Exterior Wall	Painted gypsum board on furring	70% of wall	S.F. Wall	2.86	1.09	
7.0 Conveying							
.1	Elevators	N/A		—	—	—	0.0%
.2	Special Conveyors	N/A		—	—	—	
8.0 Mechanical							
.1	Plumbing	Toilet and service fixtures, supply and drainage	1 Fixture/195 S.F. Floor	Each	1677	8.60	
.2	Fire Protection	Wet pipe sprinkler system		S.F. Floor	2.09	2.09	
.3	Heating	Included in 8.4		—	—	—	26.3%
.4	Cooling	Multizone unit, gas heating, electric cooling		S.F. Floor	9.68	9.68	
.5	Special Systems	N/A		—	—	—	
9.0 Electrical							
.1	Service & Distribution	200 ampere service, panel board and feeders		S.F. Floor	.90	.90	
.2	Lighting & Power	Fluorescent fixtures, receptacles, switches, A.C. and misc. power		S.F. Floor	4.47	4.47	8.2%
.4	Special Electrical	Alarm systems and emergency lighting		S.F. Floor	.92	.92	
11.0 Special Construction							
.1	Specialties	N/A		—	—	—	0.0%
12.0 Site Work							
.1	Earthwork	N/A		—	—	—	
.3	Utilities	N/A		—	—	—	
.5	Roads & Parking	N/A		—	—	—	0.0%
.7	Site Improvements	N/A		—	—	—	
				Sub-Total		77.14	100%
	GENERAL CONDITIONS (Overhead & Profit)				15%	11.57	
	ARCHITECT FEES				9%	7.99	
				Total Building Cost		96.70	

BUILDING TYPES

Costs per square foot of floor area

Exterior Wall	S.F. Area	4000	5500	7000	8500	10000	11500	13000	14500	16000
	L.F. Perimeter	180	210	240	270	286	311	336	361	386
Face Brick with Concrete Block Back-up	Steel Joists	115.70	110.05	106.85	104.80	102.45	101.15	100.20	99.45	98.85
	Wood Joists	117.90	112.25	109.05	106.95	104.60	103.30	102.40	101.65	101.05
Stucco on Concrete Block	Steel Joists	112.25	107.10	104.20	102.30	100.20	99.05	98.20	97.50	96.95
	Wood Joists	110.30	105.80	103.25	101.60	99.80	98.80	98.05	97.45	97.00
Brick Veneer	Wood Frame	114.70	109.50	106.60	104.70	102.55	101.40	100.55	99.90	99.35
Wood Siding	Wood Frame	109.70	105.30	102.80	101.15	99.40	98.40	97.70	97.10	96.65
Perimeter Adj., Add or Deduct	Per 100 L.F.	16.40	11.90	9.35	7.70	6.50	5.70	5.05	4.50	4.10
Story Hgt. Adj., Add or Deduct	Per 1 Ft.	2.40	2.05	1.85	1.70	1.50	1.45	1.40	1.35	1.30

For Basement, add $18.90 per square foot of basement area

The above costs were calculated using the basic specifications shown on the facing page. These costs should be adjusted where necessary for design alternatives and owner's requirements. Reported completed project costs, for this type of structure, range from $47.15 to $121.15 per S.F.

Common additives

Description	Unit	$ Cost
Cabinets, Hospital, base		
Laminated plastic	L.F.	255
Stainless steel	L.F.	300
Counter top, laminated plastic	L.F.	49
Stainless steel	L.F.	107
For drop-in sink, add	Each	460
Nurses station, door type		
Laminated plastic	L.F.	285
Enameled steel	L.F.	256
Stainless steel	L.F.	320
Wall cabinets, laminated plastic	L.F.	185
Enameled steel	L.F.	202
Stainless steel	L.F.	289
Elevators, Hydraulic passenger, 2 stops		
1500# capacity	Each	40,000
2500# capacity	Each	41,300
3500# capacity	Each	44,900

Description	Unit	$ Cost
Directory Boards, Plastic, glass covered		
30" x 20"	Each	510
36" x 48"	Each	930
Aluminum, 24" x 18"	Each	450
36" x 24"	Each	540
48" x 32"	Each	640
48" x 60"	Each	1450
Emergency Lighting, 25 watt, battery operated		
Lead battery	Each	330
Nickel cadmium	Each	620
Heat Therapy Unit		
Humidified, 26" x 78" x 28"	Each	2525
Smoke Detectors		
Ceiling type	Each	142
Duct type	Each	390
Tables, Examining, vinyl top		
with base cabinets	Each	1200 - 3250
Utensil Washer, Sanitizer	Each	8450
X-Ray, Mobile	Each	10,100 - 57,000

BUILDING TYPES

Model costs calculated for a 2 story building with 10' story height and 7,000 square feet of floor area

Medical Office, 2 Story

				Unit	Unit Cost	Cost Per S.F.	% Of Sub-Total
1.0 Foundations							
.1	Footings & Foundations	Poured concrete; strip and spread footings and 4' foundation wall		S.F. Ground	5.26	2.63	
.4	Piles & Caissons	N/A		—	—	—	3.8%
.9	Excavation & Backfill	Site preparation for slab and trench for foundation wall and footing		S.F. Ground	1.04	.52	
2.0 Substructure							
.1	Slab on Grade	4" reinforced concrete with vapor barrier and granular base		S.F. Slab	3.06	1.53	1.8%
.2	Special Substructures	N/A		—	—	—	
3.0 Superstructure							
.1	Columns & Beams	Included in 3.5 and 3.7		—	—	—	
.4	Structural Walls	N/A		—	—	—	
.5	Elevated Floors	Open web steel joists, slab form, concrete, columns		S.F. Floor	7.75	3.88	8.2%
.7	Roof	Metal deck, open web steel joists, beams, columns		S.F. Roof	2.95	1.48	
.9	Stairs	Concrete filled metal pan		Flight	5025	1.44	
4.0 Exterior Closure							
.1	Walls	Concrete block, insulated	70% of wall	S.F. Floor	4.55	4.55	
.5	Exterior Wall Finishes	Stucco on concrete block	70% of wall	S.F. Wall	5.75	2.76	14.7%
.6	Doors	Aluminum and glass doors with transoms		Each	3525	1.01	
.7	Windows & Glazed Walls	Outward projecting metal	30% of wall	Each	284	3.89	
5.0 Roofing							
.1	Roof Coverings	Built-up tar and gravel with flashing		S.F. Roof	2.60	1.30	
.7	Insulation	Perlite/EPS composite		S.F. Roof	1.25	.63	2.6%
.8	Openings & Specialties	Gravel stop and hatches		S.F. Roof	.52	.26	
6.0 Interior Construction							
.1	Partitions	Gypsum bd. & sound deadening bd. on wood studs w/insul.	6 S.F. Floor/L.F. Partition	S.F. Partition	4.83	6.44	
.4	Interior Doors	Single leaf wood	60 S.F. Floor/Door	Each	408	6.80	
.5	Wall Finishes	50% paint, 50% vinyl wall coating, 5% ceramic tile		S.F. Surface	.92	2.45	28.2%
.6	Floor Finishes	50% carpet, 50% vinyl asbestos tile		S.F. Floor	4.27	4.27	
.7	Ceiling Finishes	Mineral fiber tile on concealed zee bars		S.F. Ceiling	2.15	2.15	
.9	Interior Surface/Exterior Wall	Painted gypsum board on furring	70% of wall	S.F. Wall	2.86	1.37	
7.0 Conveying							
.1	Elevators	One hydraulic hospital elevator		Each	59,010	8.43	10.1%
.2	Special Conveyors	N/A		—	—	—	
8.0 Mechanical							
.1	Plumbing	Toilet and service fixtures, supply and drainage	1 Fixture/160 S.F. Floor	Each	1214	7.59	
.2	Fire Protection	Wet pipe sprinkler system		S.F. Floor	1.77	1.77	
.3	Heating	Included in 8.4		—	—	—	23.0%
.4	Cooling	Multizone unit, gas heating, electric cooling		S.F. Floor	9.68	9.68	
.5	Special Systems	N/A		—	—	—	
9.0 Electrical							
.1	Service & Distribution	200 ampere service, panel board and feeders		S.F. Floor	.90	.90	
.2	Lighting & Power	Fluorescent fixtures, receptacles, switches, A.C. and misc. power		S.F. Floor	4.47	4.47	7.6%
.4	Special Electrical	Alarm systems and emergency lighting		S.F. Floor	.92	.92	
11.0 Special Construction							
.1	Specialties	N/A		—	—	—	0.0%
12.0 Site Work							
.1	Earthwork	N/A		—	—	—	
.3	Utilities	N/A		—	—	—	0.0%
.5	Roads & Parking	N/A		—	—	—	
.7	Site Improvements	N/A		—	—	—	
				Sub-Total		83.12	**100%**
	GENERAL CONDITIONS (Overhead & Profit)				15%	12.47	
	ARCHITECT FEES				9%	8.61	
				Total Building Cost		**104.20**	

Costs per square foot of floor area

Exterior Wall	S.F. Area	2000	3000	4000	6000	8000	10000	12000	14000	16000
	L.F. Perimeter	240	260	280	380	480	560	580	660	740
Brick Veneer	Wood Frame	104.55	91.80	85.45	82.35	80.85	79.25	76.55	76.05	75.65
Aluminum Siding	Wood Frame	94.10	84.25	79.35	76.85	75.65	74.40	72.35	71.95	71.65
Wood Siding	Wood Frame	94.40	84.50	79.55	77.05	75.80	74.55	72.50	72.10	71.75
Wood Shingles	Wood Frame	96.80	86.20	80.95	78.30	77.00	75.65	73.45	73.00	72.70
Precast Concrete Block	Wood Truss	93.25	83.65	78.85	76.45	75.20	74.00	72.00	71.65	71.30
Brick on Concrete Block	Wood Truss	108.85	94.90	87.95	84.65	83.00	81.25	78.30	77.75	77.35
Perimeter Adj., Add or Deduct	Per 100 L.F.	16.50	11.00	8.25	5.50	4.10	3.30	2.75	2.35	2.05
Story Hgt. Adj., Add or Deduct	Per 1 Ft.	2.85	2.05	1.65	1.50	1.45	1.35	1.15	1.10	1.10

For Basement, add $13.05 per square foot of basement area

The above costs were calculated using the basic specifications shown on the facing page. These costs should be adjusted where necessary for design alternatives and owner's requirements. Reported completed project costs, for this type of structure, range from $38.65 to $93.35 per S.F.

Common additives

Description	Unit	$ Cost
Closed Circuit Surveillance, One station		
Camera and monitor	Each	1325
For additional camera stations, add	Each	730
Emergency Lighting, 25 watt, battery operated		
Lead battery	Each	330
Nickel cadmium	Each	620
Laundry Equipment		
Dryer, gas, 16 lb. capacity	Each	680
30 lb. capacity	Each	2550
Washer, 4 cycle	Each	760
Commercial	Each	1125
Sauna, Prefabricated, complete		
6' x 4'	Each	3725
6' x 6'	Each	4400
6' x 9'	Each	5850
8' x 8'	Each	6225
8' x 10'	Each	6850
10' x 12'	Each	8900
Smoke Detectors		
Ceiling type	Each	142
Duct type	Each	390

Description	Unit	$ Cost
Swimming Pools, Complete, gunite	S.F.	45 - 55
TV Antenna, Master system, 12 outlet	Outlet	229
30 outlet	Outlet	147
100 outlet	Outlet	141

Important: See the Reference Section for Location Factors

Model costs calculated for a 1 story building with 9' story height and 8,000 square feet of floor area

				Unit	Unit Cost	Cost Per S.F.	% Of Sub-Total
1.0 Foundations							
.1	Footings & Foundations	Poured concrete; strip and spread footings and 4' foundation wall		S.F. Ground	5.21	5.21	9.6%
.4	Piles & Caissons	N/A		–	–	–	
.9	Excavation & Backfill	Site preparation for slab and trench for foundation wall and footing		S.F. Ground	1.11	1.11	
2.0 Substructure							
.1	Slab on Grade	4" reinforced concrete with vapor barrier and granular base		S.F. Slab	3.06	3.06	4.7%
.2	Special Substructures	N/A		–	–	–	
3.0 Superstructure							
.1	Columns & Beams	N/A		–	–	–	
.4	Structural Walls	N/A		–	–	–	
.5	Elevated Floors	N/A		–	–	–	6.6%
.7	Roof	Plywood on wood trusses		S.F. Ground	4.33	4.33	
.9	Stairs	N/A		–	–	–	
4.0 Exterior Closure							
.1	Walls	Face brick on wood studs with sheathing, insulation and paper	80% of wall	S.F. Wall	15.39	6.65	
.5	Exterior Wall Finishes	N/A		–	–	–	18.2%
.6	Doors	Wood solid core		Each	931	2.56	
.7	Windows & Glazed Walls	Wood double hung	20% of wall	Each	353	2.72	
5.0 Roofing							
.1	Roof Coverings	Asphalt shingles with flashing (pitched)		S.F. Ground	1.05	1.05	
.7	Insulation	Fiberglass sheets		S.F. Ground	.87	.87	3.6%
.8	Openings & Specialties	Gutters and downspouts		S.F. Ground	.43	.43	
6.0 Interior Construction							
.1	Partitions	Gypsum bd. and sound deadening bd. on wood studs	9 S.F. Floor/L.F. Partition	S.F. Partition	4.83	4.29	
.4	Interior Doors	Single leaf hollow metal	300 S.F. Floor/Door	Each	362	1.21	
.5	Wall Finishes	90% paint, 10% ceramic tile		S.F. Surface	.98	1.75	25.9%
.6	Floor Finishes	85% carpet, 15% vinyl composition tile		S.F. Floor	5.80	5.80	
.7	Ceiling Finishes	Painted gypsum board on furring		S.F. Ceiling	2.87	2.87	
.9	Interior Surface/Exterior Wall	Painted gypsum board on furring	80% of wall	S.F. Wall	2.86	1.08	
7.0 Conveying							
.1	Elevators	N/A		–	–	–	0.0%
.2	Special Conveyors	N/A		–	–	–	
8.0 Mechanical							
.1	Plumbing	Toilet and service fixtures, supply and drainage	1 Fixture/90 S.F. Floor	Each	954	10.60	
.2	Fire Protection	Wet pipe sprinkler system		S.F. Floor	2.09	2.09	
.3	Heating	Included in 8.4		–	–	–	23.9%
.4	Cooling	Through the wall electric heating and cooling units		S.F. Floor	3.09	3.09	
.5	Special Systems	N/A		–	–	–	
9.0 Electrical							
.1	Service & Distribution	100 ampere service, panel board and feeders		S.F. Floor	.53	.53	
.2	Lighting & Power	Fluorescent fixtures, receptacles, switches and misc. power		S.F. Floor	4.14	4.14	7.5%
.4	Special Electrical	Alarm systems		S.F. Floor	.27	.27	
11.0 Special Construction							
.1	Specialties	N/A		–	–	–	0.0%
12.0 Site Work							
.1	Earthwork	N/A		–	–	–	
.3	Utilities	N/A		–	–	–	0.0%
.5	Roads & Parking	N/A		–	–	–	
.7	Site Improvements	N/A		–	–	–	
				Sub-Total		65.71	**100%**
	GENERAL CONDITIONS (Overhead & Profit)				15%	9.86	
	ARCHITECT FEES				7%	5.28	
				Total Building Cost		80.85	

Costs per square foot of floor area

Exterior Wall	S.F. Area	25000	37000	49000	61000	73000	81000	88000	96000	104000
	L.F. Perimeter	433	593	606	720	835	911	978	1054	1074
Decorative Concrete Block	Wood Joists	76.30	74.35	71.40	70.70	70.20	70.00	69.85	69.65	69.15
	Precast Conc.	81.75	79.80	76.85	76.15	75.65	75.45	75.30	75.10	74.60
Stucco on Concrete Block	Wood Joists	75.20	73.35	70.50	69.80	69.35	69.10	69.00	68.80	68.30
	Precast Conc.	81.20	79.30	76.45	75.75	75.30	75.10	74.95	74.75	74.30
Wood Siding	Wood Frame	74.80	72.95	70.20	69.50	69.05	68.85	68.75	68.55	68.10
Brick Veneer	Wood Frame	77.90	75.80	72.40	71.60	71.10	70.85	70.70	70.50	69.90
Perimeter Adj., Add or Deduct	Per 100 L.F.	2.65	1.80	1.35	1.10	.95	.80	.75	.70	.65
Story Hgt. Adj., Add or Deduct	Per 1 Ft.	.95	.90	.70	.65	.65	.65	.60	.65	.60

For Basement, add $17.05 per square foot of basement area

The above costs were calculated using the basic specifications shown on the facing page. These costs should be adjusted where necessary for design alternatives and owner's requirements. Reported completed project costs, for this type of structure, range from $39.15 to $94.05 per S.F.

Common additives

Description	Unit	$ Cost
Closed Circuit Surveillance, One station		
Camera and monitor	Each	1325
For additional camera station, add	Each	730
Elevators, Hydraulic passenger, 2 stops		
1500# capacity	Each	40,000
2500# capacity	Each	41,300
3500# capacity	Each	44,900
Additional stop, add	Each	3350
Emergency Lighting, 25 watt, battery operated		
Lead battery	Each	330
Nickel cadmium	Each	620
Laundry Equipment		
Dryer, gas, 16 lb. capacity	Each	680
30 lb. capacity	Each	2550
Washer, 4 cycle	Each	760
Commercial	Each	1125

Description	Unit	$ Cost
Sauna, Prefabricated, complete		
6' x 4'	Each	3725
6' x 6'	Each	4400
6' x 9'	Each	5850
8' x 8'	Each	6225
8' x 10'	Each	6850
10' x 12'	Each	8900
Smoke Detectors		
Ceiling type	Each	142
Duct type	Each	390
Swimming Pools, Complete, gunite	S.F.	45 - 55
TV Antenna, Master system, 12 outlet	Outlet	229
30 outlet	Outlet	147
100 outlet	Outlet	141

Important: See the Reference Section for Location Factors

Model costs calculated for a 3 story building with 9' story height and 73,000 square feet of floor area

Motel, 2-3 Story

				Unit	Unit Cost	Cost Per S.F.	% Of Sub-Total
1.0 Foundations							
.1	Footings & Foundations	Poured concrete; strip and spread footings and 4' foundation wall		S.F. Ground	3.57	1.19	
.4	Piles & Caissons	N/A		—	—	—	2.5%
.9	Excavation & Backfill	Site preparation for slab and trench for foundation wall and footing		S.F. Ground	1.04	.35	
2.0 Substructure							
.1	Slab on Grade	4" reinforced concrete with vapor barrier and granular base		S.F. Slab	3.06	1.02	1.6%
.2	Special Substructures	N/A		—		—	
3.0 Superstructure							
.1	Columns & Beams	N/A		—	—	—	
.4	Structural Walls	Reinforced concrete block		S.F. Wall	9.46	1.56	
.5	Elevated Floors	Precast concrete plank		S.F. Floor	5.97	3.98	13.6%
.7	Roof	Precast concrete plank		S.F. Roof	6.25	2.08	
.9	Stairs	Concrete filled metal pan		Flight	5025	.83	
4.0 Exterior Closure							
.1	Walls	Face brick with concrete block backup	85% of wall	S.F. Wall	10.97	2.88	
.5	Exterior Wall Finishes	N/A		—	—	—	10.3%
.6	Doors	Aluminum and glass doors and entrance with transom		Each	1085	2.42	
.7	Windows & Glazed Walls	Aluminum sliding	15% of wall	Each	359	1.11	
5.0 Roofing							
.1	Roof Coverings	Built-up tar and gravel with flashing		S.F. Roof	2.10	.70	
.7	Insulation	Perlite/EPS composite		S.F. Roof	1.25	.42	2.0%
.8	Openings & Specialties	Gravel stop, hatches, gutters and downspouts		S.F. Roof	.45	.15	
6.0 Interior Construction							
.1	Partitions	Gypsum board and sound deadening board on wood studs	7 S.F. Floor/L.F. Partition	S.F. Partition	4.83	5.52	
.4	Interior Doors	Wood hollow core	70 S.F. Floor/Door	Each	362	5.17	
.5	Wall Finishes	90% paint, 10% ceramic tile		S.F. Surface	.98	2.25	36.0%
.6	Floor Finishes	85% carpet, 15% vinyl composition tile		S.F. Floor	5.80	5.80	
.7	Ceiling Finishes	Painted gypsum board on furring		S.F. Ceiling	2.87	2.87	
.9	Interior Surface/Exterior Wall	Painted gypsum board on furring	85% of wall	S.F. Wall	2.86	.75	
7.0 Conveying							
.1	Elevators	Two hydraulic passenger elevators		Each	60,955	1.67	2.7%
.2	Special Conveyors	N/A		—	—	—	
8.0 Mechanical							
.1	Plumbing	Toilet and service fixtures, supply and drainage	1 Fixture/180 S.F. Floor	Each	1764	9.80	
.2	Fire Protection	Sprinklers, light hazard		S.F. Floor	1.17	1.17	
.3	Heating	Included in 8.4		—	—	—	23.0%
.4	Cooling	Through the wall electric heating and cooling units		S.F. Floor	3.23	3.23	
.5	Special Systems	N/A		—	—	—	
9.0 Electrical							
.1	Service & Distribution	400 ampere service, panel board and feeders		S.F. Floor	.30	.30	
.2	Lighting & Power	Fluorescent fixtures, receptacles, switches and misc. power		S.F. Floor	4.52	4.52	8.3%
.4	Special Electrical	Alarm systems and emergency lighting		S.F. Floor	.33	.33	
11.0 Special Construction							
.1	Specialties	N/A		—	—	—	0.0%
12.0 Site Work							
.1	Earthwork	N/A		—	—	—	
.3	Utilities	N/A		—	—	—	0.0%
.5	Roads & Parking	N/A		—	—	—	
.7	Site Improvements	N/A		—	—	—	
				Sub-Total		62.07	100%
	GENERAL CONDITIONS (Overhead & Profit)				15%	9.31	
	ARCHITECT FEES				6%	4.27	
				Total Building Cost		75.65	

BUILDING TYPES

163

Costs per square foot of floor area

Exterior Wall	S.F. Area	9000	10000	12000	13000	14000	15000	16000	18000	20000
	L.F. Perimeter	385	410	460	460	480	500	510	547	583
Decorative Concrete Block	Steel Joists	90.15	87.70	84.05	81.75	80.45	79.35	78.05	76.35	74.95
Painted Concrete Block	Steel Joists	85.50	83.25	79.90	77.90	76.70	75.70	74.55	73.00	71.80
Face Brick on Conc. Block	Steel Joists	97.65	94.90	90.75	87.95	86.45	85.15	83.60	81.65	80.05
Precast Concrete Panels	Steel Joists	93.70	91.10	87.25	84.70	83.30	82.10	80.70	78.85	77.35
Tilt-up Panels	Steel Joists	88.80	86.40	82.85	80.60	79.35	78.25	77.00	75.35	74.00
Metal Sandwich Panels	Steel Joists	81.80	79.70	76.60	74.85	73.75	72.85	71.85	70.40	69.30
Perimeter Adj., Add or Deduct	Per 100 L.F.	5.15	4.65	3.85	3.60	3.30	3.10	2.90	2.55	2.35
Story Hgt. Adj., Add or Deduct	Per 1 Ft.	.75	.70	.65	.60	.60	.60	.60	.50	.50
Basement—Not Applicable										

The above costs were calculated using the basic specifications shown on the facing page. These costs should be adjusted where necessary for design alternatives and owner's requirements. Reported completed project costs, for this type of structure, range from $46.10 to $120.40 per S.F.

Common additives

Description	Unit	$ Cost
Emergency Lighting, 25 watt, battery operated		
Lead battery	Each	330
Nickel cadmium	Each	620
Seating		
Auditorium chair, all veneer	Each	127
Veneer back, padded seat	Each	152
Upholstered, spring seat	Each	178
Classroom, movable chair & desk	Set	65 - 120
Lecture hall, pedestal type	Each	127 - 370
Smoke Detectors		
Ceiling type	Each	142
Duct type	Each	390
Sound System		
Amplifier, 250 watts	Each	1525
Speaker, ceiling or wall	Each	131
Trumpet	Each	246

BUILDING TYPES

Important: See the Reference Section for Location Factors

Model costs calculated for a 1 story building with 20' story height and 12,000 square feet of floor area

			Unit	Unit Cost	Cost Per S.F.	% Of Sub-Total
1.0 Foundations						
.1	Footings & Foundations	Poured concrete; strip and spread footings and 4' foundation wall	S.F. Ground	2.43	2.43	
.4	Piles & Caissons	N/A	—	—	—	7.0%
.9	Excavation & Backfill	Site preparation for slab and trench for foundation wall and footing	S.F. Ground	2.36	2.36	
2.0 Substructure						
.1	Slab on Grade	4" reinforced concrete with vapor barrier and granular base	S.F. Slab	3.06	3.06	4.5%
.2	Special Substructures	N/A	—	—	—	
3.0 Superstructure						
.1	Columns & Beams	N/A	—	—	—	
.4	Structural Walls	N/A	—	—	—	
.5	Elevated Floors	Open web steel joists, slab form, concrete	mezzanine 2250 S.F.	S.F. Mezz	6.33	1.19	11.6%
.7	Roof	Metal deck on open web steel joists	S.F. Roof	5.73	5.73	
.9	Stairs	Concrete filled metal pan	Flight	5900	.98	
4.0 Exterior Closure						
.1	Walls	Decorative concrete block	100% of wall	S.F. Wall	12.35	9.47	
.5	Exterior Wall Finishes	N/A	—	—	—	14.9%
.6	Doors	Sliding mallfront aluminum and glass and hollow metal	Each	1415	.70	
.7	Windows & Glazed Walls	N/A	—	—	—	
5.0 Roofing						
.1	Roof Coverings	Built-up tar and gravel with flashing	S.F. Roof	2.14	2.14	
.7	Insulation	Perlite/EPS composite	S.F. Roof	1.25	1.25	5.7%
.8	Openings & Specialties	Gravel stop, hatches, gutters and downspouts	S.F. Roof	.49	.49	
6.0 Interior Construction						
.1	Partitions	Concrete block, toilet partitions	40 S.F. Floor/L.F. Partition	S.F. Partition	6.15	2.46	
.4	Interior Doors	Single leaf hollow metal	705 S.F. Floor/Door	Each	526	.75	
.5	Wall Finishes	Paint		S.F. Surface	1.09	.87	16.8%
.6	Floor Finishes	Carpet	50% of area	S.F. Floor	5.62	2.81	
.7	Ceiling Finishes	Mineral fiber tile on concealed zee runners suspended		S.F. Ceiling	3.07	3.07	
.9	Interior Surface/Exterior Wall	Paint	100% of wall	S.F. Wall	1.95	1.50	
7.0 Conveying						
.1	Elevators	N/A	—	—	—	0.0%
.2	Special Conveyors	N/A	—	—	—	
8.0 Mechanical						
.1	Plumbing	Toilet and service fixtures, supply and drainage	1 Fixture/500 S.F. Floor	Each	1440	2.88	
.2	Fire Protection	Wet pipe sprinkler system		S.F. Floor	1.60	1.60	
.3	Heating	Included in 8.4		—	—	—	18.8%
.4	Cooling	Single zone rooftop unit, gas heating, electric cooling		S.F. Floor	8.42	8.42	
.5	Special Systems	N/A		—	—	—	
9.0 Electrical						
.1	Service & Distribution	400 ampere service, panel board and feeders	S.F. Floor	.62	.62	
.2	Lighting & Power	Fluorescent fixtures, receptacles, switches, A.C. and misc. power	S.F. Floor	2.78	2.78	5.6%
.4	Special Electrical	Alarm systems, sound system and emergency lighting	S.F. Floor	.45	.45	
11.0 Special Construction						
.1	Specialties	Projection equipment, screen, seating	S.F. Floor	10.31	10.31	15.1%
12.0 Site Work						
.1	Earthwork	N/A	—	—	—	
.3	Utilities	N/A	—	—	—	0.0%
.5	Roads & Parking	N/A	—	—	—	
.7	Site Improvements	N/A	—	—	—	
		Sub-Total			68.32	100%
	GENERAL CONDITIONS (Overhead & Profit)		15%		10.25	
	ARCHITECT FEES		7%		5.48	
		Total Building Cost			84.05	

BUILDING TYPES

165

Costs per square foot of floor area

Exterior Wall	S.F. Area	10000	15000	20000	25000	30000	35000	40000	45000	50000
	L.F. Perimeter	286	370	453	457	513	568	624	680	735
Redwood Siding	Wood Frame	87.65	84.45	82.80	80.50	79.70	79.10	78.65	78.30	78.05
Brick Veneer	Wood Frame	93.40	89.40	87.35	84.15	83.15	82.35	81.80	81.35	81.00
Face Brick with Concrete Block Back-up	Wood Joists	96.00	91.65	89.45	85.80	84.70	83.85	83.25	82.75	82.35
	Steel Joists	103.40	99.05	96.85	93.20	92.10	91.25	90.65	90.15	89.75
Stucco on Concrete Block	Wood Joists	89.95	86.45	84.65	81.95	81.10	80.40	79.95	79.55	79.25
	Steel Joists	97.35	93.85	92.00	89.35	88.50	87.80	87.30	86.95	86.65
Perimeter Adj., Add or Deduct	Per 100 L.F.	4.20	2.80	2.10	1.65	1.40	1.20	1.05	.95	.85
Story Hgt. Adj., Add or Deduct	Per 1 Ft.	.80	.70	.65	.50	.50	.45	.45	.45	.45

For Basement, add $18.15 per square foot of basement area

The above costs were calculated using the basic specifications shown on the facing page. These costs should be adjusted where necessary for design alternatives and owner's requirements. Reported completed project costs, for this type of structure, range from $49.85 to $123.80 per S.F.

Common additives

Description	Unit	$ Cost
Beds, Manual	Each	865 - 1500
Doctors In-Out Register, 200 names	Each	12,700
Elevators, Hydraulic passenger, 2 stops		
1500# capacity	Each	40,000
2500# capacity	Each	41,300
3500# capacity	Each	44,900
Emergency Lighting, 25 watt, battery operated		
Lead battery	Each	330
Nickel cadmium	Each	620
Intercom System, 25 station capacity		
Master station	Each	1825
Intercom outlets	Each	114
Handset	Each	300
Kitchen Equipment		
Broiler	Each	3425
Coffee urn, twin 6 gallon	Each	5725
Cooler, 6 ft. long	Each	2725
Dishwasher, 10-12 racks per hr.	Each	2600
Food warmer	Each	745
Freezer, 44 C.F., reach-in	Each	6625

Description	Unit	$ Cost
Kitchen Equipment, cont.		
Ice cube maker, 50 lb. per day	Each	1750
Range with 1 oven	Each	2325
Laundry Equipment		
Dryer, gas, 16 lb. capacity	Each	680
30 lb. capacity	Each	2550
Washer, 4 cycle	Each	760
Commercial	Each	1125
Nurses Call System		
Single bedside call station	Each	200
Pillow speaker	Each	210
Refrigerator, Prefabricated, walk-in		
7'-6" high, 6' x 6'	S.F.	117
10' x 10'	S.F.	91
12' x 14'	S.F.	81
12' x 20'	S.F.	72
TV Antenna, Master system, 12 outlet	Outlet	229
30 outlet	Outlet	147
100 outlet	Outlet	141
Whirlpool Bath, Mobile, 18" x 24" x 60"	Each	3475
X-Ray, Mobile	Each	10,100 - 57,000

Important: See the Reference Section for Location Factors

Model costs calculated for a 2 story building with 10' story height and 25,000 square feet of floor area

			Unit	Unit Cost	Cost Per S.F.	% Of Sub-Total
1.0 Foundations						
.1	Footings & Foundations	Poured concrete; strip and spread footings and 4' foundation wall	S.F. Ground	2.76	1.38	
.4	Piles & Caissons	N/A	—	—	—	3.1%
.9	Excavation & Backfill	Site preparation for slab and trench for foundation wall and footing	S.F. Ground	1.11	.56	
2.0 Substructure						
.1	Slab on Grade	4" reinforced concrete with vapor barrier and granular base	S.F. Slab	3.06	1.53	2.4%
.2	Special Substructures	N/A	—	—	—	
3.0 Superstructure						
.1	Columns & Beams	N/A	—	—	—	
.4	Structural Walls	N/A	—	—	—	
.5	Elevated Floors	Plywood on wood joists	S.F. Floor	3.50	1.75	5.5%
.7	Roof	Plywood on wood joists	S.F. Roof	2.73	1.37	
.9	Stairs	Wood	Flight	1403	.34	
4.0 Exterior Closure						
.1	Walls	Vertical T & G redwood siding 85% of wall	S.F. Wall	6.11	1.90	
.5	Exterior Wall Finishes	N/A	—	—	—	5.6%
.6	Doors	Double aluminum & glass doors, single leaf hollow metal	Each	1449	.29	
.7	Windows & Glazed Walls	Wood double hung 15% of wall	Each	359	1.31	
5.0 Roofing						
.1	Roof Coverings	Built-up tar and gravel with flashing	S.F. Roof	2.10	1.05	
.7	Insulation	Perlite/EPS composite	S.F. Roof	1.25	.63	3.1%
.8	Openings & Specialties	Gravel stop, hatches, gutters and downspouts	S.F. Roof	.50	.25	
6.0 Interior Construction						
.1	Partitions	Gypsum board on wood studs 8 S.F. Floor/L.F. Partition	S.F. Partition	3.09	3.09	
.4	Interior Doors	Single leaf wood 80 S.F. Floor/Door	Each	408	5.10	
.5	Wall Finishes	50% vinyl wall coverings, 45% paint, 5% ceramic tile	S.F. Surface	1.15	2.31	
.6	Floor Finishes	50% carpet, 45% vinyl tile, 5% ceramic tile	S.F. Floor	4.49	4.49	29.7%
.7	Ceiling Finishes	Painted gypsum board on wood furring	S.F. Ceiling	2.87	2.87	
.9	Interior Surface/Exterior Wall	Painted gypsum board on wood furring 85% of wall	S.F. Wall	2.86	.89	
7.0 Conveying						
.1	Elevators	One hydraulic hospital elevator	Each	59,000	2.36	3.7%
.2	Special Conveyors	N/A	—	—	—	
8.0 Mechanical						
.1	Plumbing	Kitchen, toilet and service fixtures, supply and drainage 1 Fixture/230 S.F. Floor	Each	2293	9.97	
.2	Fire Protection	Sprinkler, light hazard	S.F. Floor	2.90	2.90	
.3	Heating	Oil fired hot water, wall fin radiation	S.F. Floor	4.37	4.37	35.1%
.4	Cooling	Split systems with air cooled condensing units	S.F. Floor	4.88	4.88	
.5	Special Systems	N/A	—	—	—	
9.0 Electrical						
.1	Service & Distribution	600 ampere service, panel board and feeders	S.F. Floor	.90	.90	
.2	Lighting & Power	Incandescent fixtures, receptacles, switches, A.C. and misc. power	S.F. Floor	5.58	5.58	11.8%
.4	Special Electrical	Alarm systems and emergency lighting	S.F. Floor	.97	.97	
11.0 Special Construction						
.1	Specialties	N/A	—	—	—	0.0%
12.0 Site Work						
.1	Earthwork	N/A	—	—	—	
.3	Utilities	N/A	—	—	—	
.5	Roads & Parking	N/A	—	—	—	0.0%
.7	Site Improvements	N/A	—	—	—	
			Sub-Total		63.04	**100%**
	GENERAL CONDITIONS (Overhead & Profit)			15%	9.46	
	ARCHITECT FEES			11%	8.00	
			Total Building Cost		**80.50**	

Costs per square foot of floor area

Exterior Wall	S.F. Area	10000	22000	34000	46000	58000	63000	68000	73000	78000
	L.F. Perimeter	246	393	443	543	562	590	603	624	645
Face Brick with Concrete Block Back-up	Wood Joists	91.45	78.30	71.35	69.15	66.45	65.90	65.25	64.85	64.40
	Steel Joists	92.95	79.75	72.80	70.65	67.90	67.40	66.75	66.30	65.90
Glass and Metal Curtain Wall	Steel Frame	90.25	78.35	72.30	70.35	68.00	67.55	66.95	66.60	66.25
	R/Conc. Frame	92.45	80.55	74.45	72.50	70.15	69.70	69.15	68.80	68.40
Wood Siding	Wood Frame	79.00	68.30	63.15	61.40	59.45	59.05	58.55	58.25	57.95
Brick Veneer	Wood Frame	84.50	72.30	66.05	64.05	61.60	61.15	60.55	60.20	59.80
Perimeter Adj., Add or Deduct	Per 100 L.F.	10.75	4.90	3.15	2.35	1.85	1.70	1.60	1.50	1.40
Story Hgt. Adj., Add or Deduct	Per 1 Ft.	2.00	1.45	1.05	.95	.75	.75	.70	.70	.70

For Basement, add $19.85 per square foot of basement area

The above costs were calculated using the basic specifications shown on the facing page. These costs should be adjusted where necessary for design alternatives and owner's requirements. Reported completed project costs, for this type of structure, range from $39.10 to $110.75 per S.F.

Common additives

Description	Unit	$ Cost
Clock System		
20 room	Each	11,800
50 room	Each	28,700
Closed Circuit Surveillance, One station		
Camera and monitor	Each	1325
For additional camera stations, add	Each	730
Directory Boards, Plastic, glass covered		
30" x 20"	Each	510
36" x 48"	Each	930
Aluminum, 24" x 18"	Each	450
36" x 24"	Each	540
48" x 32"	Each	640
48" x 60"	Each	1450
Elevators, Hydraulic passenger, 2 stops		
1500# capacity	Each	40,000
2500# capacity	Each	41,300
3500# capacity	Each	44,900
Additional stop, add	Each	3350
Emergency Lighting, 25 watt, battery operated		
Lead battery	Each	330
Nickel cadmium	Each	620

Description	Unit	$ Cost
Smoke Detectors		
Ceiling type	Each	142
Duct type	Each	390
Sound System		
Amplifier, 250 watts	Each	1525
Speaker, ceiling or wall	Each	131
Trumpet	Each	246
TV Antenna, Master system, 12 outlet	Outlet	229
30 outlet	Outlet	147
100 outlet	Outlet	141

Important: See the Reference Section for Location Factors

Model costs calculated for a 3 story building with 12' story height and 58,000 square feet of floor area

				Unit	Unit Cost	Cost Per S.F.	% Of Sub-Total
1.0 Foundations							
.1	Footings & Foundations	Poured concrete; strip and spread footings and 4' foundation wall		S.F. Ground	3.24	1.08	
.4	Piles & Caissons	N/A		—	—	—	2.6%
.9	Excavation & Backfill	Site preparation for slab and trench for foundation wall and footing		S.F. Ground	1.04	.35	
2.0 Substructure							
.1	Slab on Grade	4" reinforced concrete with vapor barrier and granular base		S.F. Slab	3.06	1.02	1.8%
.2	Special Substructures	N/A		—	—	—	
3.0 Superstructure							
.1	Columns & Beams	Fireproofing; interior columns included in 3.5 and 3.7		L.F. Columns	21	.59	
.4	Structural Walls	N/A		—	—	—	
.5	Elevated Floors	Open web steel joists, slab form, concrete, columns		S.F. Floor	7.67	5.11	13.6%
.7	Roof	Metal deck, open web steel joists, columns		S.F. Roof	3.50	1.17	
.9	Stairs	Concrete filled metal pan		Flight	5025	.61	
4.0 Exterior Closure							
.1	Walls	Face brick with concrete block backup	80% of wall	S.F. Wall	19.46	5.43	
.5	Exterior Wall Finishes	N/A		—	—	—	
.6	Doors	Aluminum and glass, hollow metal		Each	2252	.23	12.8%
.7	Windows & Glazed Walls	Steel outward projecting	20% of wall	Each	461	1.40	
5.0 Roofing							
.1	Roof Coverings	Built-up tar and gravel with flashing		S.F. Roof	2.10	.70	
.7	Insulation	Perlite/EPS composite		S.F. Roof	1.25	.42	2.0%
.8	Openings & Specialties	N/A		—	—	—	
6.0 Interior Construction							
.1	Partitions	Gypsum board on metal studs, toilet partitions	20 S.F. Floor/L.F. Partition	S.F. Partition	2.97	1.49	
.4	Interior Doors	Single leaf hollow metal	200 S.F. Floor/Door	Each	526	2.63	
.5	Wall Finishes	60% vinyl wall covering, 40% paint		S.F. Surface	1.00	.80	24.7%
.6	Floor Finishes	60% carpet, 30% vinyl composition tile, 10% ceramic tile		S.F. Floor	4.85	4.85	
.7	Ceiling Finishes	Mineral fiber tile on concealed zee bars		S.F. Ceiling	3.07	3.07	
.9	Interior Surface/Exterior Wall	Painted gypsum board on furring	80% of wall	S.F. Wall	2.86	.80	
7.0 Conveying							
.1	Elevators	Two hydraulic passenger elevators		Each	63,510	2.19	4.0%
.2	Special Conveyors	N/A		—	—	—	
8.0 Mechanical							
.1	Plumbing	Toilet and service fixtures, supply and drainage	1 Fixture/1320 S.F. Floor	Each	2046	1.55	
.2	Fire Protection	Standpipes and hose systems		S.F. Floor	.20	.20	
.3	Heating	Included in 8.4		—	—	—	23.8%
.4	Cooling	Multizone unit gas heating, electric cooling		S.F. Floor	11.40	11.40	
.5	Special Systems	N/A		—	—	—	
9.0 Electrical							
.1	Service & Distribution	1000 ampere service, panel board and feeders		S.F. Floor	.92	.92	
.2	Lighting & Power	Fluorescent fixtures, receptacles, switches, A.C. and misc. power		S.F. Floor	6.98	6.98	14.7%
.4	Special Electrical	Alarm systems and emergency lighting		S.F. Floor	.20	.20	
11.0 Special Construction							
.1	Specialties	N/A		—	—	—	0.0%
12.0 Site Work							
.1	Earthwork	N/A		—	—	—	
.3	Utilities	N/A		—	—	—	
.5	Roads & Parking	N/A		—	—	—	0.0%
.7	Site Improvements	N/A		—	—	—	
				Sub-Total		55.19	100%
	GENERAL CONDITIONS (Overhead & Profit)			15%		8.28	
	ARCHITECT FEES			7%		4.43	
				Total Building Cost		67.90	

BUILDING TYPES

Costs per square foot of floor area

Exterior Wall	S.F. Area	50000	60000	70000	80000	90000	100000	110000	120000	130000
	L.F. Perimeter	328	370	378	410	441	450	475	500	510
Precast Concrete Panel	Steel Frame	81.70	80.15	77.90	76.90	76.15	74.95	74.35	73.80	73.15
	R/Conc. Frame	82.10	80.55	78.25	77.25	76.50	75.25	74.70	74.15	73.45
Face Brick with Concrete Block Back-up	Steel Frame	83.20	81.60	79.15	78.10	77.25	75.95	75.30	74.75	74.00
	R/Conc. Frame	83.50	81.85	79.40	78.35	77.50	76.20	75.60	75.00	74.25
Limestone Panel Concrete Block Back-up	Steel Frame	90.15	88.10	84.85	83.50	82.45	80.70	79.90	79.15	78.15
	R/Conc. Frame	90.40	88.40	85.10	83.80	82.70	81.00	80.15	79.45	78.40
Perimeter Adj., Add or Deduct	Per 100 L.F.	5.10	4.25	3.65	3.20	2.85	2.55	2.30	2.15	1.95
Story Hgt. Adj., Add or Deduct	Per 1 Ft.	1.35	1.25	1.10	1.05	1.00	.90	.85	.85	.80

For Basement, add $21.50 per square foot of basement area

The above costs were calculated using the basic specifications shown on the facing page. These costs should be adjusted where necessary for design alternatives and owner's requirements. Reported completed project costs, for this type of structure, range from $44.50 to $130.80 per S.F.

Common additives

Description	Unit	$ Cost
Clock System		
20 room	Each	11,800
50 room	Each	28,700
Closed Circuit Surveillance, One station		
Camera and monitor	Each	1325
For additional camera stations, add	Each	730
Directory Boards, Plastic, glass covered		
30" x 20"	Each	510
36" x 48"	Each	930
Aluminum, 24" x 18"	Each	450
36" x 24"	Each	540
48" x 32"	Each	640
48" x 60"	Each	1450
Elevators, Electric passenger, 5 stops		
2000# capacity	Each	91,300
3500# capacity	Each	97,300
5000# capacity	Each	100,800
Additional stop, add	Each	5225
Emergency Lighting, 25 watt, battery operated		
Lead battery	Each	330
Nickel cadmium	Each	620

Description	Unit	$ Cost
Intercom System, 25 station capacity		
Master station	Each	1825
Intercom outlets	Each	114
Handset	Each	300
Smoke Detectors		
Ceiling type	Each	142
Duct type	Each	390
Sound System		
Amplifier, 250 watts	Each	1525
Speaker, ceiling or wall	Each	131
Trumpet	Each	246
TV Antenna, Master system, 12 oulet	Outlet	229
30 outlet	Outlet	147
100 outlet	Outlet	141

Important: See the Reference Section for Location Factors

Model costs calculated for an 8 story building with 12' story height and 100,000 square feet of floor area.

			Unit	Unit Cost	Cost Per S.F.	% Of Sub-Total
1.0 Foundations						
.1	Footings & Foundations	Poured concrete; strip and spread footings and 4' foundation wall	S.F. Ground	7.68	.96	
.4	Piles & Caissons	N/A	—	—	—	1.8%
.9	Excavation & Backfill	Site preparation for slab and trench for foundation wall and footing	S.F. Ground	1.11	.14	
2.0 Substructure						
.1	Slab on Grade	4" reinforced concrete with vapor barrier and granular base	S.F. Slab	3.06	.38	0.6%
.2	Special Substructures	N/A	—		—	
3.0 Superstructure						
.1	Columns & Beams	Steel columns w/ fireproofing	L.F. Columns	71	1.94	
.4	Structural Walls	N/A				
.5	Elevated Floors	Concrete slab with metal deck and beams	S.F. Floor	10.12	8.85	19.7%
.7	Roof	Metal deck, open web steel joists, interior columns	S.F. Roof	3.73	.47	
.9	Stairs	Concrete filled metal pan	Flight	5025	.85	
4.0 Exterior Closure						
.1	Walls	Precast concrete panels 80% of wall	S.F. Wall	17.19	5.94	
.5	Exterior Wall Finishes	N/A	—	—	—	13.2%
.6	Doors	Double aluminum and glass doors and entrance with transoms	Each	2512	.13	
.7	Windows & Glazed Walls	Vertical pivoted steel 20% of wall	Each	359	2.07	
5.0 Roofing						
.1	Roof Coverings	Built-up tar and gravel with flashing	S.F. Roof	2.24	.28	
.7	Insulation	Perlite/EPS composite	S.F. Roof	1.25	.16	0.7%
.8	Openings & Specialties	N/A	—	—	—	
6.0 Interior Construction						
.1	Partitions	Gypsum board on metal studs, toilet partitions 30 S.F. Floor/L.F. Partition	S.F. Partition	2.97	1.46	
.4	Interior Doors	Single leaf hollow metal 400 S.F. Floor/Door	Each	526	1.32	
.5	Wall Finishes	60% vinyl wall covering, 40% paint	S.F. Surface	1.00	.67	
.6	Floor Finishes	60% carpet, 30% vinyl composition tile, 10% ceramic tile	S.F. Floor	4.85	4.85	20.1%
.7	Ceiling Finishes	Mineral fiber tile on concealed zee bars	S.F. Ceiling	3.07	3.07	
.9	Interior Surface/Exterior Wall	Painted gypsum board on furring 80% of wall	S.F. Wall	2.86	.99	
7.0 Conveying						
.1	Elevators	Four geared passenger elevators	Each	145,000	5.80	9.4%
.2	Special Conveyors	N/A	—	—	—	
8.0 Mechanical						
.1	Plumbing	Toilet and service fixtures, supply and drainage 1 Fixture/1370 S.F. Floor	Each	1507	1.10	
.2	Fire Protection	Standpipes and hose systems	S.F. Floor	.14	.14	
.3	Heating	Included in 8.4	—	—	—	20.7%
.4	Cooling	Multizone unit gas heating, electric cooling	S.F. Floor	11.40	11.40	
.5	Special Systems	N/A				
9.0 Electrical						
.1	Service & Distribution	1600 ampere service, panel board and feeders	S.F. Floor	.80	.80	
.2	Lighting & Power	Fluorescent fixtures, receptacles, switches, A.C. and misc. power	S.F. Floor	7.00	7.00	13.8%
.4	Special Electrical	Alarm systems and emergency lighting	S.F. Floor	.70	.70	
11.0 Special Construction						
.1	Specialties	N/A	—	—	—	0.0%
12.0 Site Work						
.1	Earthwork	N/A	—	—	—	
.3	Utilities	N/A	—	—	—	0.0%
.5	Roads & Parking	N/A	—	—	—	
.7	Site Improvements	N/A	—	—	—	
			Sub-Total		61.47	100%
	GENERAL CONDITIONS (Overhead & Profit)			15%	9.22	
	ARCHITECT FEES			6%	4.26	
			Total Building Cost		74.95	

Costs per square foot of floor area

Exterior Wall	S.F. Area	90000	100000	110000	120000	130000	140000	150000	160000	170000
	L.F. Perimeter	320	342	364	360	376	393	400	413	426
Double Glazed Heat Absorbing Tinted Plate Glass Panels	Steel Frame	96.15	94.80	93.65	91.60	90.60	89.85	88.85	88.10	87.50
	R/Conc. Frame	99.10	97.15	95.55	93.05	91.70	90.65	89.35	88.45	87.60
Face Brick with Concret Block Back-up	Steel Frame	99.05	97.20	95.65	93.45	92.20	91.20	90.05	89.15	88.45
	R/Conc. Frame	95.20	93.35	91.85	89.65	88.40	87.45	86.30	85.45	84.75
Precast Concrete Panel With Exposed Aggregate	Steel Frame	97.45	95.65	94.15	92.10	90.85	89.95	88.85	88.00	87.30
	R/Conc. Frame	93.90	92.15	90.70	88.65	87.45	86.55	85.45	84.65	83.95
Perimeter Adj., Add or Deduct	Per 100 L.F.	6.00	5.40	4.90	4.50	4.20	3.85	3.60	3.40	3.20
Story Hgt. Adj., Add or Deduct	Per 1 Ft.	1.85	1.80	1.70	1.55	1.50	1.45	1.40	1.35	1.30
	For Basement, add $21.00 per square foot of basement area									

The above costs were calculated using the basic specifications shown on the facing page. These costs should be adjusted where necessary for design alternatives and owner's requirements. Reported completed project costs, for this type of structure, range from $56.60 to $137.95 per S.F.

Common additives

Description	Unit	$ Cost
Clock System		
20 room	Each	11,800
50 room	Each	28,700
Directory Boards, Plastic, glass covered		
30" x 20"	Each	510
36" x 48"	Each	930
Aluminum, 24" x 18"	Each	450
36" x 24"	Each	540
48" x 32"	Each	640
48" x 60"	Each	1450
Elevators, Electric passenger, 10 stops		
3000# capacity	Each	198,500
4000# capacity	Each	200,500
5000# capacity	Each	204,000
Additional stop, add	Each	5225
Emergency Lighting, 25 watt, battery operated		
Lead battery	Each	330
Nickel cadmium	Each	620

Description	Unit	$ Cost
Escalators, Metal		
32" wide, 10' story height	Each	85,700
20' story height	Each	96,800
48" wide, 10' story height	Each	90,200
20' story height	Each	101,300
Glass		
32" wide, 10' story height	Each	85,700
20' story height	Each	96,800
48" wide, 10' story height	Each	90,200
20' story height	Each	101,300
Smoke Detectors		
Ceiling type	Each	142
Duct type	Each	390
Sound System		
Amplifier, 250 watts	Each	1525
Speaker, ceiling or wall	Each	131
Trumpet	Each	246
TV Antenna, Master system, 12 oulet	Outlet	229
30 oulet	Outlet	147
100 outlet	Outlet	141

Model costs calculated for a 15 story building with 10' story height and 140,000 square feet of floor area

			Unit	Unit Cost	Cost Per S.F.	% Of Sub-Total
1.0 Foundations						
.1	Footings & Foundations	Poured concrete; strip and spread footings and 4' foundation wall	S.F. Ground	18.75	1.25	
.4	Piles & Caissons	N/A	—	—	—	1.8%
.9	Excavation & Backfill	Site preparation for slab and trench for foundation wall and footing	S.F. Ground	1.11	.07	
2.0 Substructure						
.1	Slab on Grade	4" reinforced concrete with vapor barrier and granular base	S.F. Slab	3.06	.20	0.3%
.2	Special Substructures	N/A	—	—	—	
3.0 Superstructure						
.1	Columns & Beams	Steel columns with fireproofing	L.F. Column	109	2.86	
.4	Structural Walls	N/A	—	—	—	
.5	Elevated Floors	Concrete slab, metal deck, beams	S.F. Floor	13.10	12.23	22.9%
.7	Roof	Metal deck, open web steel joists, beams, columns	S.F. Roof	4.39	.29	
.9	Stairs	Concrete filled metal pan	Flight	5900	1.48	
4.0 Exterior Closure						
.1	Walls	N/A	—	—	—	
.5	Exterior Wall Finishes	N/A	—	—	—	16.0%
.6	Doors	Double aluminum & glass doors	Each	3896	.78	
.7	Windows & Glazed Walls	Double glazed heat absorbing, tinted plate glass wall panels 100% of wall	S.F. Wall	26	10.99	
5.0 Roofing						
.1	Roof Coverings	Built-up tar and gravel with flashing	S.F. Roof	4.05	.27	
.7	Insulation	Perlite/EPS composite	S.F. Roof	1.25	.08	0.5%
.8	Openings & Specialties	N/A	—	—	—	
6.0 Interior Construction						
.1	Partitions	Gypsum board on metal studs, toilet partitions 30 S.F. Floor/L.F. Partition	S.F. Partition	2.97	1.42	
.4	Interior Doors	Single leaf hollow metal 400 S.F. Floor/Door	Each	526	1.32	
.5	Wall Finishes	60% vinyl wall covering, 40% paint	S.F. Surface	.99	.53	16.5%
.6	Floor Finishes	60% carpet, 30% vinyl composition tile, 10% ceramic tile	S.F. Floor	4.85	4.85	
.7	Ceiling Finishes	Mineral fiber tile on concealed zee bars	S.F. Ceiling	3.07	3.07	
.9	Interior Surface/Exterior Wall	Painted drywall on metal furring 80% of wall	S.F. Wall	2.86	.96	
7.0 Conveying						
.1	Elevators	Four geared passenger elevators	Each	214,900	6.14	8.3%
.2	Special Conveyors	N/A	—	—	—	
8.0 Mechanical						
.1	Plumbing	Toilet and service fixtures, supply and drainage 1 Fixture/1345 S.F. Floor	Each	1990	1.48	
.2	Fire Protection	Standpipes and hose systems and sprinklers, light hazard	S.F. Floor	2.81	2.81	
.3	Heating	Oil fired hot water	S.F. Floor	2.85	2.85	21.8%
.4	Cooling	Chilled water, fan coil units	S.F. Floor	8.97	8.97	
.5	Special Systems	N/A	—	—	—	
9.0 Electrical						
.1	Service & Distribution	2400 ampere service, panel board and feeders	S.F. Floor	1.14	1.14	
.2	Lighting & Power	Fluorescent fixtures, receptacles, switches, A.C. and misc. power	S.F. Floor	6.98	6.98	11.9%
.4	Special Electrical	Alarm systems and emergency lighting	S.F. Floor	.68	.68	
11.0 Special Construction						
.1	Specialties	N/A	—	—	—	0.0%
12.0 Site Work						
.1	Earthwork	N/A	—	—	—	
.3	Utilities	N/A	—	—	—	0.0%
.5	Roads & Parking	N/A	—	—	—	
.7	Site Improvements	N/A	—	—	—	
			Sub-Total		73.70	100%
	GENERAL CONDITIONS (Overhead & Profit)			15%	11.06	
	ARCHITECT FEES			6%	5.09	
			Total Building Cost		89.85	

Costs per square foot of floor area

Exterior Wall	S.F. Area	7000	9000	11000	13000	15000	17000	19000	21000	23000
	L.F. Perimeter	240	280	303	325	354	372	397	422	447
Limestone with Concrete Block Back-up	Bearing Walls	132.80	124.15	116.85	111.65	108.40	105.15	103.05	101.25	99.90
	R/Conc. Frame	139.55	131.30	124.50	119.65	116.60	113.60	111.65	110.00	108.70
Face Brick with Concrete Block Back-up	Bearing Walls	118.15	110.80	104.90	100.80	98.10	95.55	93.85	92.45	91.35
	R/Conc. Frame	130.20	122.85	117.00	112.85	110.20	107.65	105.95	104.50	103.40
Decorative Concrete Block	Bearing Walls	112.30	105.45	100.20	96.50	94.10	91.85	90.25	89.00	88.00
	R/Conc. Frame	124.35	117.55	112.30	108.60	106.15	103.90	102.35	101.10	100.10
Perimeter Adj., Add or Deduct	Per 100 L.F.	17.25	13.45	11.00	9.25	8.05	7.10	6.35	5.80	5.25
Story Hgt. Adj., Add or Deduct	Per 1 Ft.	3.05	2.75	2.45	2.20	2.10	1.95	1.85	1.80	1.75

For Basement, add $15.70 per square foot of basement area

The above costs were calculated using the basic specifications shown on the facing page. These costs should be adjusted where necessary for design alternatives and owner's requirements. Reported completed project costs, for this type of structure, range from $65.45 to $168.10 per S.F.

Common additives

Description	Unit	$ Cost
Cells Prefabricated, 5'-6' wide, 7'-8' high, 7'-8' deep	Each	8200
Elevators, Hydraulic passenger, 2 stops		
1500# capacity	Each	40,000
2500# capacity	Each	41,300
3500# capacity	Each	44,900
Emergency Lighting, 25 watt, battery operated		
Lead battery	Each	330
Nickel cadmium	Each	620
Flagpoles, Complete		
Aluminum, 20' high	Each	1100
40' high	Each	2475
70' high	Each	7100
Fiberglass, 23' high	Each	1475
39'-5" high	Each	2525
59' high	Each	7025

Description	Unit	$ Cost
Lockers, Steel, Single tier, 60" to 72"	Opening	135 - 216
2 tier, 60" or 72" total	Opening	83 - 105
5 tier, box lockers	Opening	44 - 57
Locker bench, lam. maple top only	L.F.	18.65
Pedestals, steel pipe	Each	43
Safe, Office type, 4 hour rating		
30" x 18" x 18"	Each	3475
62" x 33" x 20"	Each	7575
Shooting Range, Incl. bullet traps, target provisions, and controls, not incl. structural shell	Each	21,800
Smoke Detectors		
Ceiling type	Each	142
Duct type	Each	390
Sound System		
Amplifier, 250 watts	Each	1525
Speaker, ceiling or wall	Each	131
Trumpet	Each	246

BUILDING TYPES

Important: See the Reference Section for Location Factors

Model costs calculated for a 2 story building with 12' story height and 11,000 square feet of floor area

				Unit	Unit Cost	Cost Per S.F.	% Of Sub-Total
1.0 Foundations							
.1	Footings & Foundations	Poured concrete; strip and spread footings and 4' foundation wall		S.F. Ground	4.98	2.49	
.4	Piles & Caissons	N/A		—	—	—	3.3%
.9	Excavation & Backfill	Site preparation for slab and trench for foundation wall and footing		S.F. Ground	1.11	.56	
2.0 Substructure							
.1	Slab on Grade	4" reinforced concrete with vapor barrier and granular base		S.F. Slab	3.06	1.53	1.6%
.2	Special Substructures	N/A		—	—	—	
3.0 Superstructure							
.1	Columns & Beams	N/A		—	—	—	
.4	Structural Walls	N/A		—	—	—	
.5	Elevated Floors	Open web steel joists, slab form, concrete		S.F. Floor	6.63	3.32	6.1%
.7	Roof	Metal deck on open web steel joists		S.F. Roof	2.68	1.34	
.9	Stairs	Concrete filled metal pan		Flight	5900	1.07	
4.0 Exterior Closure							
.1	Walls	Limestone with concrete block backup	80% of wall	S.F. Wall	30	16.26	
.5	Exterior Wall Finishes	N/A		—	—	—	25.0%
.6	Doors	Hollow metal		Each	1415	1.03	
.7	Windows & Glazed Walls	Metal horizontal sliding	20% of wall	Each	685	6.04	
5.0 Roofing							
.1	Roof Coverings	Built-up tar and gravel with flashing		S.F. Roof	2.60	1.30	
.7	Insulation	Perlite/EPS composite		S.F. Roof	1.25	.63	2.2%
.8	Openings & Specialties	Gravel stop		L.F. Perimeter	5.60	.15	
6.0 Interior Construction							
.1	Partitions	Concrete block, toilet partitions	20 S.F. Floor/L.F. Partition	S.F. Partition	5.14	3.00	
.4	Interior Doors	Single leaf kalamein fire door	200 S.F. Floor/Door	Each	526	2.63	
.5	Wall Finishes	90% paint, 10% ceramic tile		S.F. Surface	1.34	1.34	
.6	Floor Finishes	70% vinyl asbestos tile, 20% carpet, 10% ceramic tile		S.F. Floor	3.55	3.55	15.7%
.7	Ceiling Finishes	Mineral fiber tile on concealed zee bars		S.F. Ceiling	3.07	3.07	
.9	Interior Surface/Exterior Wall	Paint	80% of wall	S.F. Wall	1.95	1.03	
7.0 Conveying							
.1	Elevators	One hydraulic passenger elevator		Each	48,840	4.44	4.8%
.2	Special Conveyors	N/A		—	—	—	
8.0 Mechanical							
.1	Plumbing	Toilet and service fixtures, supply and drainage	1 Fixture/580 S.F. Floor	Each	2395	4.13	
.2	Fire Protection	Wet pipe sprinkler system		S.F. Floor	1.77	1.77	
.3	Heating	Oil fired hot water, wall fin radiation		S.F. Floor	6.34	6.34	20.5%
.4	Cooling	Split systems with air cooled condensing units		S.F. Floor	6.83	6.83	
.5	Special Systems	N/A		—	—	—	
9.0 Electrical							
.1	Service & Distribution	400 ampere service, panel board and feeders		S.F. Floor	1.13	1.13	
.2	Lighting & Power	Fluorescent fixtures, receptacles, switches, A.C. and misc. power		S.F. Floor	6.58	6.58	9.0%
.4	Special Electrical	Alarm systems and emergency lighting		S.F. Floor	.68	.68	
11.0 Special Construction							
.1	Specialties	Lockers, detention rooms, cells		S.F. Floor	10.97	10.97	11.8%
12.0 Site Work							
.1	Earthwork	N/A		—	—	—	
.3	Utilities	N/A		—	—	—	
.5	Roads & Parking	N/A		—	—	—	0.0%
.7	Site Improvements	N/A		—	—	—	
				Sub-Total		93.21	**100%**
	GENERAL CONDITIONS (Overhead & Profit)			15%		13.98	
	ARCHITECT FEES			9%		9.66	
				Total Building Cost		**116.85**	

BUILDING TYPES

Costs per square foot of floor area

Exterior Wall	S.F. Area	5000	7000	9000	11000	13000	15000	17000	19000	21000
	L.F. Perimeter	300	380	420	486	468	513	540	580	620
Face Brick with Concrete Block Back-up	Steel Frame	86.80	82.45	77.80	76.05	71.55	70.40	68.95	68.20	67.55
	Bearing Walls	85.65	81.30	76.65	74.90	70.40	69.25	67.80	67.05	66.40
Limestone with Concrete Block Back-up	Steel Frame	96.30	91.10	85.20	83.05	77.30	75.85	74.00	73.05	72.25
	Bearing Walls	94.70	89.45	83.55	81.40	75.65	74.20	72.35	71.40	70.60
Decorative Concrete Block	Steel Frame	80.80	77.05	73.15	71.65	67.95	67.00	65.75	65.15	64.60
	Bearing Walls	79.65	75.90	71.95	70.50	66.85	65.85	64.60	64.00	63.45
Perimeter Adj., Add or Deduct	Per 100 L.F.	10.10	7.20	5.60	4.60	3.85	3.35	2.95	2.65	2.40
Story Hgt. Adj., Add or Deduct	Per 1 Ft.	1.55	1.40	1.25	1.20	.95	.90	.85	.80	.80

For Basement, add $16.45 per square foot of basement area

The above costs were calculated using the basic specifications shown on the facing page. These costs should be adjusted where necessary for design alternatives and owner's requirements. Reported completed project costs, for this type of structure, range from $53.40 to $137.85 per S.F.

Common additives

Description	Unit	$ Cost
Closed Circuit Surveillance, One station		
Camera and monitor	Each	1325
For additional camera stations, add	Each	730
Emergency Lighting, 25 watt, battery operated		
Lead battery	Each	330
Nickel cadmium	Each	620
Flagpoles, Complete		
Aluminum, 20' high	Each	1100
40' high	Each	2475
70' high	Each	7100
Fiberglass, 23' high	Each	1475
39'-5" high	Each	2525
59' high	Each	7025

Description	Unit	$ Cost
Mail Boxes, Horizontal, key lock, 15" x 6" x 5"	Each	47
Double 15" x 12" x 5"	Each	80
Quadruple 15" x 12" x 10"	Each	144
Vertical, 6" x 5" x 15", aluminum	Each	38
Bronze	Each	63
Steel, enameled	Each	38
Scales, Dial type, 5 ton cap.		
8' x 6' platform	Each	7050
9' x 7' platform	Each	10,900
Smoke Detectors		
Ceiling type	Each	142
Duct type	Each	390

BUILDING TYPES

Model costs calculated for a 1 story building with 14' story height and 13,000 square feet of floor area

Post Office

			Unit	Unit Cost	Cost Per S.F.	% Of Sub-Total
1.0 Foundations						
.1	Footings & Foundations	Poured concrete; strip and spread footings and 4' foundation wall	S.F. Ground	3.28	3.28	
.4	Piles & Caissons	N/A	–	–	–	7.7%
.9	Excavation & Backfill	Site preparation for slab and trench for foundation wall and footing	S.F. Ground	1.11	1.11	
2.0 Substructure						
.1	Slab on Grade	4" reinforced concrete with vapor barrier and granular base	S.F. Slab	3.06	3.06	5.4%
.2	Special Substructures	N/A	–	–	–	
3.0 Superstructure						
.1	Columns & Beams	Fireproofing, steel columns included in 3.7	L.F. Column	21	.07	
.4	Structural Walls	N/A	–	–	–	
.5	Elevated Floors	N/A	–	–	–	8.8%
.7	Roof	Metal deck, open web steel joists, columns	S.F. Roof	4.93	4.93	
.9	Stairs	N/A	–	–	–	
4.0 Exterior Closure						
.1	Walls	Face brick with concrete block backup 80% of wall	S.F. Wall	19.44	7.84	
.5	Exterior Wall Finishes	N/A	–	–	–	
.6	Doors	Double aluminum & glass, single aluminum, hollow metal, steel overhead	Each	1622	.75	18.6%
.7	Windows & Glazed Walls	Double strength window glass 20% of wall	Each	461	2.02	
5.0 Roofing						
.1	Roof Coverings	Built-up tar and gravel with flashing	S.F. Roof	2.24	2.24	
.7	Insulation	Perlite/EPS composite	S.F. Roof	1.25	1.25	6.5%
.8	Openings & Specialties	Gravel stop	L.F. Perimeter	5.60	.20	
6.0 Interior Construction						
.1	Partitions	Concrete block, toilet partitions 15 S.F. Floor/L.F. Partition	S.F. Partition	5.14	4.48	
.4	Interior Doors	Single leaf hollow metal 150 S.F. Floor/Door	Each	526	3.51	
.5	Wall Finishes	Paint	S.F. Surface	.91	1.46	
.6	Floor Finishes	50% vinyl tile, 50% paint	S.F. Floor	1.88	1.88	22.6%
.7	Ceiling Finishes	Mineral fiber tile on concealed zee bars 25% of area	S.F. Ceiling	3.07	.77	
.9	Interior Surface/Exterior Wall	Paint 80% of wall	S.F. Wall	1.95	.79	
7.0 Conveying						
.1	Elevators	N/A	–	–	–	0.0%
.2	Special Conveyors	N/A	–	–	–	
8.0 Mechanical						
.1	Plumbing	Toilet and service fixtures, supply and drainage 1 Fixture/1180 S.F. Floor	Each	1923	1.63	
.2	Fire Protection	Wet pipe sprinkler system	S.F. Floor	1.60	1.60	
.3	Heating	Included in 8.4	–	–	–	17.8%
.4	Cooling	Single zone, gas heating, electric cooling	S.F. Floor	7.03	7.03	
.5	Special Systems	N/A	–	–	–	
9.0 Electrical						
.1	Service & Distribution	400 ampere service, panel board and feeders	S.F. Floor	.95	.95	
.2	Lighting & Power	Fluorescent fixtures, receptacles, switches, A.C. and misc. power	S.F. Floor	5.34	5.34	11.7%
.4	Special Electrical	Alarm systems and emergency lighting	S.F. Floor	.40	.40	
11.0 Special Construction						
.1	Specialties	Cabinets, lockers, shelving	S.F. Floor	.50	.50	0.9%
12.0 Site Work						
.1	Earthwork	N/A	–	–	–	
.3	Utilities	N/A	–	–	–	
.5	Roads & Parking	N/A	–	–	–	0.0%
.7	Site Improvements	N/A	–	–	–	
			Sub-Total		57.09	100%
	GENERAL CONDITIONS (Overhead & Profit)			15%	8.56	
	ARCHITECT FEES			9%	5.90	
			Total Building Cost		71.55	

Costs per square foot of floor area

Exterior Wall	S.F. Area	5000	10000	15000	21000	25000	30000	40000	50000	60000
	L.F. Perimeter	287	400	500	600	700	700	834	900	1000
Face Brick with Concrete Block Back-up	Steel Frame	136.45	115.00	107.15	101.80	100.80	96.40	93.75	91.00	89.65
	Bearing Walls	132.15	111.90	104.40	99.30	98.40	94.05	91.50	88.80	87.50
Concrete Block	Steel Frame	118.90	102.50	96.60	92.70	91.85	88.95	87.05	85.20	84.25
Brick Veneer	Steel Frame	129.85	109.30	101.85	96.80	95.80	91.80	89.30	86.80	85.55
Galvanized Steel Siding	Steel Frame	111.35	97.10	92.05	88.75	87.95	85.60	84.00	82.50	81.75
Metal Sandwich Panel	Steel Frame	111.20	96.50	91.25	87.85	87.05	84.60	82.95	81.40	80.60
Perimeter Adj., Add or Deduct	Per 100 L.F.	16.95	8.50	5.65	4.05	3.40	2.85	2.10	1.70	1.40
Story Hgt. Adj., Add or Deduct	Per 1 Ft.	3.50	2.40	2.05	1.75	1.70	1.45	1.25	1.10	1.00
For Basement, add $15.85 per square foot of basement area										

The above costs were calculated using the basic specifications shown on the facing page. These costs should be adjusted where necessary for design alternatives and owner's requirements. Reported completed project costs, for this type of structure, range from $50.40 to $127.15 per S.F.

Common additives

Description	Unit	$ Cost
Bar, Front Bar	L.F.	262
Back Bar	L.F.	209
Booth, Upholstered, custom straight	L.F.	129 - 237
"L" or "U" shaped	L.F.	133 - 225
Bleachers, Telescoping, manual		
To 15 tier	Seat	76 - 105
21-30 tier	Seat	165 - 199
Courts		
Ceiling	Court	5225
Floor	Court	9275
Walls	Court	18,400
Emergency Lighting, 25 watt, battery operated		
Lead battery	Each	330
Nickel cadmium	Each	620
Kitchen Equipment		
Broiler	Each	3425
Cooler, 6 ft. long, reach-in	Each	2725
Dishwasher, 10-12 racks per hr.	Each	2600
Food warmer, counter 1.2 KW	Each	745
Freezer, reach-in, 44 C.F.	Each	6625
Ice cube maker, 50 lb. per day	Each	1750

Description	Unit	$ Cost
Lockers, Steel, single tier, 60" or 72"	Opening	135 - 216
2 tier, 60" or 72" total	Opening	83 - 105
5 tier, box lockers	Opening	44 - 57
Locker bench, lam. maple top only	L.F.	18.65
Pedestals, steel pipe	Each	43
Sauna, Prefabricated, complete		
6' x 4'	Each	3725
6' x 9'	Each	5850
8' x 8'	Each	6225
8' x 10'	Each	6850
10' x 12'	Each	8900
Sound System		
Amplifier, 250 watts	Each	1525
Speaker, ceiling or wall	Each	131
Trumpet	Each	246
Steam Bath, Complete, to 140 C.F.	Each	980
To 300 C.F.	Each	1175
To 800 C.F.	Each	3050
To 2500 C.F.	Each	3400

Important: See the Reference Section for Location Factors

Model costs calculated for a 2 story building with 12' story height and 30,000 square feet of floor area

			Unit	Unit Cost	Cost Per S.F.	% Of Sub-Total
1.0 Foundations						
.1	Footings & Foundations	Poured concrete; strip and spread footings and 4' foundation wall	S.F. Ground	3.94	1.97	
.4	Piles & Caissons	N/A	–	–	–	3.5%
.9	Excavation & Backfill	Site preparation for slab and trench for foundation wall and footing	S.F. Ground	1.11	.72	
2.0 Substructure						
.1	Slab on Grade	6" reinforced concrete with vapor barrier and granular base	S.F. Slab	6.12	3.98	5.1%
.2	Special Substructures	N/A	–	–	–	
3.0 Superstructure						
.1	Columns & Beams	Steel columns included in 3.5 and 3.7	–	–	–	
.4	Structural Walls	N/A	–	–	–	
.5	Elevated Floors	Open web steel joists, slab form, concrete, columns _50% of area_	S.F. Floor	13.16	4.61	9.5%
.7	Roof	Metal deck on open web steel joists, columns	S.F. Roof	4.54	2.27	
.9	Stairs	Concrete filled metal pan	Flight	5025	.50	
4.0 Exterior Closure						
.1	Walls	Face brick with concrete block backup _95% of wall_	S.F. Wall	19.61	10.43	
.5	Exterior Wall Finishes	N/A	–	–	–	16.0%
.6	Doors	Aluminum and glass and hollow metal	Each	1844	.25	
.7	Windows & Glazed Walls	Storefront _5% of wall_	S.F. Window	62	1.74	
5.0 Roofing						
.1	Roof Coverings	Built-up tar and gravel with flashing	S.F. Roof	2.74	1.37	
.7	Insulation	Perlite/EPS composite	S.F. Roof	1.25	.81	3.0%
.8	Openings & Specialties	Gravel stop and hatches	S.F. Roof	.34	.17	
6.0 Interior Construction						
.1	Partitions	Concrete block, gypsum board on metal studs _25 S.F. Floor/L.F. Partition_	S.F. Partition	5.14	2.17	
.4	Interior Doors	Single leaf hollow metal _810 S.F. Floor/Door_	Each	526	.65	
.5	Wall Finishes	Paint	S.F. Surface	.86	.69	
.6	Floor Finishes	80% carpet, 20% ceramic tile _50% of floor area_	S.F. Floor	5.18	2.59	12.3%
.7	Ceiling Finishes	Mineral fiber tile on concealed zee bars _60% of area_	S.F. Ceiling	3.07	1.84	
.9	Interior Surface/Exterior Wall	Painted gypsum board on furring _95% of wall_	S.F. Wall	2.86	1.60	
7.0 Conveying						
.1	Elevators	N/A	–	–	–	0.0%
.2	Special Conveyors	N/A	–	–	–	
8.0 Mechanical						
.1	Plumbing	Kitchen, bathroom and service fixtures, supply and drainage _1 Fixture/1000 S.F. Floor_	Each	3190	3.19	
.2	Fire Protection	Sprinklers, light hazard	S.F. Floor	.14	.14	
.3	Heating	Included in 8.4	–	–	–	30.5%
.4	Cooling	Multizone unit, gas heating, electric cooling	S.F. Floor	20	20.35	
.5	Special Systems	N/A	–	–	–	
9.0 Electrical						
.1	Service & Distribution	400 ampere service, panel board and feeders	S.F. Floor	.51	.51	
.2	Lighting & Power	Fluorescent and high intensity disch. fixtures, receptacles, switches, A.C. and misc. power	S.F. Floor	3.99	3.99	6.1%
.4	Special Electrical	Alarm systems and emergency lighting	S.F. Floor	.25	.25	
11.0 Special Construction						
.1	Specialties	Courts, sauna baths	S.F. Floor	10.83	10.83	14.0%
12.0 Site Work						
.1	Earthwork	N/A	–	–	–	
.3	Utilities	N/A	–	–	–	0.0%
.5	Roads & Parking	N/A	–	–	–	
.7	Site Improvements	N/A	–	–	–	
			Sub-Total		77.62	**100%**
	GENERAL CONDITIONS (Overhead & Profit)			15%	11.64	
	ARCHITECT FEES			8%	7.14	
			Total Building Cost		**96.40**	

BUILDING TYPES

Costs per square foot of floor area

Exterior Wall	S.F. Area	5000	6000	7000	8000	9000	10000	11000	12000	13000
	L.F. Perimeter	286	320	353	386	397	425	454	460	486
Face Brick with Concrete Block Back-up	Steel Joists	100.25	97.85	96.10	94.80	92.60	91.70	90.95	89.40	88.90
	Wood Joists	99.00	96.55	94.70	93.30	91.05	90.10	89.30	87.75	87.15
Stucco on Concrete Block	Steel Joists	95.70	93.60	92.05	90.90	89.05	88.30	87.65	86.35	85.90
	Wood Joists	92.00	90.00	88.50	87.40	85.65	84.90	84.25	83.05	82.55
Limestone with Concrete Block Back-up	Steel Joists	110.95	107.85	105.55	103.80	100.85	99.65	98.65	96.60	95.85
	Wood Joists	107.30	104.25	102.00	100.30	97.45	96.25	95.30	93.25	92.55
Perimeter Adj., Add or Deduct	Per 100 L.F.	9.25	7.70	6.65	5.80	5.15	4.60	4.20	3.90	3.55
Story Hgt. Adj., Add or Deduct	Per 1 Ft.	1.60	1.50	1.45	1.40	1.25	1.20	1.15	1.10	1.05

For Basement, add $16.70 per square foot of basement area

The above costs were calculated using the basic specifications shown on the facing page. These costs should be adjusted where necessary for design alternatives and owner's requirements. Reported completed project costs, for this type of structure, range from $43.95 to $100.25 per S.F.

Common additives

Description	Unit	$ Cost
Carrels Hardwood	Each	655 - 855
Emergency Lighting, 25 watt, battery operated		
Lead battery	Each	330
Nickel cadmium	Each	620
Flagpoles, Complete		
Aluminum, 20' high	Each	1100
40' high	Each	2475
70' high	Each	7100
Fiberglass, 23' high	Each	1475
39'-5" high	Each	2525
59' high	Each	7025
Gym Floor, Incl. sleepers and finish, maple	S.F.	9.20
Intercom System, 25 Station capacity		
Master station	Each	1825
Intercom outlets	Each	114
Handset	Each	300

Description	Unit	$ Cost
Lockers, Steel, single tier, 60" to 72"	Opening	135 - 216
2 tier, 60" to 72" total	Opening	83 - 105
5 tier, box lockers	Opening	44 - 57
Locker bench, lam. maple top only	L.F.	18.65
Pedestals, steel pipe	Each	43
Seating		
Auditorium chair, all veneer	Each	127
Veneer back, padded seat	Each	152
Upholstered, spring seat	Each	178
Classroom, movable chair & desk	Set	65 - 120
Lecture hall, pedestal type	Each	127 - 370
Smoke Detectors		
Ceiling type	Each	142
Duct type	Each	390
Sound System		
Amplifier, 250 watts	Each	1525
Speaker, ceiling or wall	Each	131
Trumpet	Each	246
Swimming Pools, Complete, gunite	S.F.	45 - 55

BUILDING TYPES

Important: See the Reference Section for Location Factors

Model costs calculated for a 1 story building with 12' story height and 10,000 square feet of floor area

			Unit	Unit Cost	Cost Per S.F.	% Of Sub-Total
1.0 Foundations						
.1	Footings & Foundations	Poured concrete; strip and spread footings and 4' foundation wall	S.F. Ground	3.35	3.35	6.1%
.4	Piles & Caissons	N/A	—	—	—	
.9	Excavation & Backfill	Site preparation for slab and trench for foundation wall and footing	S.F. Ground	1.11	1.11	
2.0 Substructure						
.1	Slab on Grade	4" reinforced concrete with vapor barrier and granular base	S.F. Slab	3.06	3.06	4.2%
.2	Special Substructures	N/A	—			
3.0 Superstructure						
.1	Columns & Beams	Interior columns included in 3.7	—	—	—	
.4	Structural Walls	N/A	—	—	—	
.5	Elevated Floors	N/A	—	—	—	5.1%
.7	Roof	Metal deck, open web steel joists, beams, interior columns	S.F. Roof	3.73	3.73	
.9	Stairs	N/A	—	—	—	
4.0 Exterior Closure						
.1	Walls	Face brick with concrete block backup 85% of wall	S.F. Wall	19.45	8.43	
.5	Exterior Wall Finishes	N/A	—	—	—	16.8%
.6	Doors	Double aluminum and glass	Each	3962	1.59	
.7	Windows & Glazed Walls	Window wall 15% of wall	S.F. Window	29	2.29	
5.0 Roofing						
.1	Roof Coverings	Built-up tar and gravel with flashing	S.F. Roof	2.21	2.21	
.7	Insulation	Perlite/EPS composite	S.F. Roof	1.25	1.25	5.1%
.8	Openings & Specialties	Gravel stop	L.F. Perimeter	5.60	.24	
6.0 Interior Construction						
.1	Partitions	Concrete block, toilet partitions 8 S.F. Floor/S.F. Partition	S.F. Partition	5.14	7.38	
.4	Interior Doors	Single leaf hollow metal 700 S.F. Floor/Door	Each	526	.75	
.5	Wall Finishes	Paint	S.F. Surface	1.09	2.73	25.9%
.6	Floor Finishes	50% vinyl tile, 50% carpet	S.F. Floor	4.14	4.14	
.7	Ceiling Finishes	Mineral fiber tile on concealed zee bars	S.F. Ceiling	3.07	3.07	
.9	Interior Surface/Exterior Wall	Paint 85% of wall	S.F. Wall	1.95	.85	
7.0 Conveying						
.1	Elevators	N/A	—	—	—	0.0%
.2	Special Conveyors	N/A	—	—	—	
8.0 Mechanical						
.1	Plumbing	Toilet and service fixtures, supply and drainage 1 Fixture/455 S.F. Floor	Each	1829	4.02	
.2	Fire Protection	Sprinkler, light hazard	S.F. Floor	1.60	1.60	
.3	Heating	Oil fired hot water, wall fin radiation	S.F. Floor	6.34	6.34	27.8%
.4	Cooling	Split systems with air cooled condensing units	S.F. Floor	8.39	8.39	
.5	Special Systems	N/A	—	—	—	
9.0 Electrical						
.1	Service & Distribution	200 ampere service, panel board and feeders	S.F. Floor	.61	.61	
.2	Lighting & Power	Fluorescent fixtures, receptacles, switches, A.C. and misc. power	S.F. Floor	5.65	5.65	9.0%
.4	Special Electrical	Alarm systems and emergency lighting	S.F. Floor	.35	.35	
11.0 Special Construction						
.1	Specialties	N/A	—	—	—	0.0%
12.0 Site Work						
.1	Earthwork	N/A	—	—	—	
.3	Utilities	N/A	—	—	—	0.0%
.5	Roads & Parking	N/A	—	—	—	
.7	Site Improvements	N/A	—	—	—	
			Sub-Total		73.14	**100%**
	GENERAL CONDITIONS (Overhead & Profit)			15%	10.97	
	ARCHITECT FEES			9%	7.59	
			Total Building Cost		**91.70**	

BUILDING TYPES

Costs per square foot of floor area

Exterior Wall	S.F. Area	2000	2800	3500	4200	5000	5800	6500	7200	8000
	L.F. Perimeter	180	212	240	268	300	314	336	344	368
Wood Siding	Wood Frame	128.15	118.95	114.30	111.25	108.80	106.05	104.65	102.95	101.90
Brick Veneer	Wood Frame	135.45	125.05	119.80	116.35	113.60	110.35	108.75	106.75	105.55
Face Brick with Concrete Block Back-up	Wood Joists	139.40	128.40	122.90	119.20	116.30	112.80	111.10	108.95	107.70
	Steel Joists	141.50	129.45	123.40	119.40	116.20	112.30	110.40	107.95	106.55
Stucco on Concrete Block	Wood Joists	130.50	120.90	116.05	112.90	110.35	107.45	105.95	104.20	103.10
	Steel Joists	126.10	116.50	111.65	108.45	105.90	103.00	101.55	99.75	98.70
Perimeter Adj., Add or Deduct	Per 100 L.F.	15.40	10.95	8.75	7.35	6.15	5.30	4.75	4.25	3.85
Story Hgt. Adj., Add or Deduct	Per 1 Ft.	1.65	1.40	1.25	1.15	1.10	1.00	.95	.90	.85

For Basement, add $19.40 per square foot of basement area

The above costs were calculated using the basic specifications shown on the facing page. These costs should be adjusted where necessary for design alternatives and owner's requirements. Reported completed project costs, for this type of structure, range from $67.85 to $158.50 per S.F.

Common additives

Description	Unit	$ Cost
Bar, Front Bar	L.F.	262
Back bar	L.F.	209
Booth, Upholstered, custom straight	L.F.	129 - 237
"L" or "U" shaped	L.F.	133 - 225
Cupola, Stock unit, redwood		
30" square, 37" high, aluminum roof	Each	465
Copper roof	Each	235
Fiberglass, 5'-0" base, 63" high	Each	2800 - 3400
6'-0" base, 63" high	Each	4225 - 4625
Decorative Wood Beams, Non load bearing		
Rough sawn, 4" x 6"	L.F.	7.55
4" x 8"	L.F.	9.00
4" x 10"	L.F.	10.45
4" x 12"	L.F.	12.00
8" x 8"	L.F.	14.80
Emergency Lighting, 25 watt, battery operated		
Lead battery	Each	330
Nickel cadmium	Each	620

Description	Unit	$ Cost
Fireplace, Brick, not incl. chimney or foundation		
30" x 29" opening	Each	1900
Chimney, standard brick		
Single flue, 16" x 20"	V.L.F.	55
20" x 20"	V.L.F.	63
2 Flue, 20" x 24"	V.L.F.	76
20" x 32"	V.L.F.	91
Kitchen Equipment		
Broiler	Each	3425
Coffee urn, twin 6 gallon	Each	5725
Cooler, 6 ft. long	Each	2725
Dishwasher, 10-12 racks per hr.	Each	2600
Food warmer, counter, 1.2 KW	Each	745
Freezer, 44 C.F., reach-in	Each	6625
Ice cube maker, 50 lb. per day	Each	1750
Range with 1 oven	Each	2325
Refrigerators, Prefabricated, walk-in		
7'-6" high, 6' x 6'	S.F.	117
10' x 10'	S.F.	91
12' x 14'	S.F.	81
12' x 20'	S.F.	72

Model costs calculated for a 1 story building with 12' story height and 5,000 square feet of floor area

				Unit	Unit Cost	Cost Per S.F.	% Of Sub-Total
1.0 Foundations							
.1	Footings & Foundations	Poured concrete; strip and spread footings and 4' foundation wall		S.F. Ground	3.97	3.97	
.4	Piles & Caissons	N/A		–	–	–	5.9%
.9	Excavation & Backfill	Site preparation for slab and trench for foundation wall and footing		S.F. Ground	1.24	1.24	
2.0 Substructure							
.1	Slab on Grade	4" reinforced concrete with vapor barrier and granular base		S.F. Slab	3.06	3.06	3.5%
.2	Special Substructures	N/A		–	–	–	
3.0 Superstructure							
.1	Columns & Beams	Wood columns		S.F. Ground	.34	.34	
.4	Structural Walls	N/A		–	–	–	
.5	Elevated Floors	N/A		–	–	–	2.9%
.7	Roof	Plywood on wood rafters (pitched)		S.F. Roof	2.02	2.26	
.9	Stairs	N/A		–	–	–	
4.0 Exterior Closure							
.1	Walls	Cedar siding on wood studs with insulation	70% of wall	S.F. Wall	7.36	3.71	
.5	Exterior Wall Finishes	N/A		–	–	–	14.1%
.6	Doors	Aluminum and glass doors and entrance with transom		Each	3183	3.19	
.7	Windows & Glazed Walls	Storefront windows	30% of wall	S.F. Window	25	5.59	
5.0 Roofing							
.1	Roof Coverings	Cedar shingles with flashing (pitched)		S.F. Roof	3.25	3.63	
.7	Insulation	Fiberglass sheets		S.F. Roof	.87	.97	5.7%
.8	Openings & Specialties	Gutters and downspouts and skylight		S.F. Roof	.48	.48	
6.0 Interior Construction							
.1	Partitions	Gypsum board on wood studs, toilet partition	25 S.F. Floor/L.F. Partition	S.F. Partition	735	1.83	
.4	Interior Doors	Hollow core wood	250 S.F. Floor/Door	Each	362	1.45	
.5	Wall Finishes	75% paint, 25% ceramic tile		S.F. Surface	1.67	1.34	17.8%
.6	Floor Finishes	65% carpet, 35% quarry tile		S.F. Floor	6.65	6.65	
.7	Ceiling Finishes	Mineral fiber tile on concealed zee bars		S.F. Ceiling	3.07	3.07	
.9	Interior Surface/Exterior Wall	Painted gypsum board on furring	70% of wall	S.F. Wall	2.85	1.44	
7.0 Conveying							
.1	Elevators	N/A		–	–	–	0.0%
.2	Special Conveyors	N/A		–	–	–	
8.0 Mechanical							
.1	Plumbing	Kitchen, bathroom and service fixtures, supply and drainage	1 Fixture/355 S.F. Floor	Each	2861	8.06	
.2	Fire Protection	Sprinklers, light hazard		S.F. Floor	1.60	1.60	
.3	Heating	Included in 8.4		–	–	–	39.5%
.4	Cooling	Multizone unit, gas heating, electric cooling		S.F. Floor	25	25.15	
.5	Special Systems	N/A		–	–	–	
9.0 Electrical							
.1	Service & Distribution	400 ampere service, panel board and feeders		S.F. Floor	2.59	2.59	
.2	Lighting & Power	Fluorescent fixtures, receptacles, switches, A.C. and misc. power		S.F. Floor	6.08	6.08	10.6%
.4	Special Electrical	Alarm systems and emergency lighting		S.F. Floor	.71	.71	
11.0 Special Construction							
.1	Specialties	N/A		–	–	–	0.0%
12.0 Site Work							
.1	Earthwork	N/A		–	–	–	
.3	Utilities	N/A		–	–	–	
.5	Roads & Parking	N/A		–	–	–	0.0%
.7	Site Improvements	N/A		–	–	–	
				Sub-Total		88.41	**100%**
	GENERAL CONDITIONS (Overhead & Profit)				15%	13.26	
	ARCHITECT FEES				7%	7.13	
				Total Building Cost		108.80	

Costs per square foot of floor area

Exterior Wall	S.F. Area	2000	2800	3500	4000	5000	5800	6500	7200	8000
	L.F. Perimeter	180	212	240	260	300	314	336	344	368
Face Brick with Concrete Block Back-up	Bearing Walls	116.85	108.85	104.85	102.85	100.05	97.20	95.95	94.10	93.20
	Steel Frame	120.10	112.05	108.10	106.10	103.25	100.45	99.20	97.35	96.40
Concrete Block With Stucco	Bearing Walls	108.85	102.10	98.75	97.05	94.70	92.40	91.35	89.85	89.10
	Steel Frame	111.55	104.90	101.55	99.90	97.55	95.30	94.30	92.80	92.05
Wood Siding	Wood Frame	109.40	102.80	99.55	97.90	95.60	93.40	92.35	90.90	90.15
Brick Veneer	Steel Frame	116.65	109.15	105.40	103.60	100.95	98.35	97.20	95.50	94.65
Perimeter Adj., Add or Deduct	Per 100 L.F.	20.25	14.50	11.55	10.10	8.10	6.95	6.20	5.65	5.05
Story Hgt. Adj., Add or Deduct	Per 1 Ft.	2.65	2.25	2.00	1.95	1.80	1.60	1.50	1.45	1.35
Basement—Not Applicable										

The above costs were calculated using the basic specifications shown on the facing page. These costs should be adjusted where necessary for design alternatives and owner's requirements. Reported completed project costs, for this type of structure, range from $67.75 to $131.90 per S.F.

Common additives

Description	Unit	$ Cost		Description	Unit	$ Cost
Bar, Front Bar	L.F.	262		Refrigerators, Prefabricated, walk-in		
Back bar	L.F.	209		7'-6" High, 6' x 6'	S.F.	117
Booth, Upholstered, custom straight	L.F.	129 - 237		10' x 10'	S.F.	91
"L" or "U" shaped	L.F.	133 - 225		12' x 14'	S.F.	81
Drive-up Window	Each	6125 - 9375		12' x 20'	S.F.	72
Emergency Lighting, 25 watt, battery operated				Serving		
Lead battery	Each	330		Counter top (Stainless steel)	L.F.	107
Nickel cadmium	Each	620		Base cabinets	L.F.	255 - 300
Kitchen Equipment				Sound System		
Broiler	Each	3425		Amplifier, 250 watts	Each	1525
Coffee urn, twin 6 gallon	Each	5725		Speaker, ceiling or wall	Each	131
Cooler, 6 ft. long	Each	2725		Trumpet	Each	246
Dishwasher, 10-12 racks per hr.	Each	2600		Storage		
Food warmer, counter, 1.2 KW	Each	745		Shelving	S.F.	11.90
Freezer, 44 C.F., reach-in	Each	6625		Washing		
Ice cube maker, 50 lb. per day	Each	1750		Stainless steel counter	L.F.	107
Range with 1 oven	Each	2325				

Model costs calculated for a 1 story building with 10' story height and 4,000 square feet of floor area

				Unit	Unit Cost	Cost Per S.F.	% Of Sub-Total
1.0 Foundations							
.1	Footings & Foundations	Poured concrete; strip and spread footings and 4' foundation wall		S.F. Ground	4.43	4.43	6.8%
.4	Piles & Caissons	N/A		–	–	–	
.9	Excavation & Backfill	Site preparation for slab and trench for foundation wall and footing		S.F. Ground	1.24	1.24	
2.0 Substructure							
.1	Slab on Grade	4" reinforced concrete with vapor barrier and granular base		S.F. Slab	3.06	3.06	3.7%
.2	Special Substructures	N/A		–	–	–	
3.0 Superstructure							
.1	Columns & Beams	N/A		–	–	–	
.4	Structural Walls	N/A		–	–	–	
.5	Elevated Floors	N/A		–	–	–	3.7%
.7	Roof	Metal deck on open web steel joists		S.F. Roof	3.05	3.05	
.9	Stairs	N/A		–	–	–	
4.0 Exterior Closure							
.1	Walls	Face brick with concrete block backup	70% of wall	S.F. Wall	16.42	8.85	
.5	Exterior Wall Finishes	N/A		–	–	–	23.6%
.6	Doors	Aluminum and glass		Each	2459	4.92	
.7	Windows & Glazed Walls	Window wall	30% of wall	S.F Window	29	5.79	
5.0 Roofing							
.1	Roof Coverings	Built-up tar and gravel with flashing		S.F. Roof	2.79	2.79	
.7	Insulation	Perlite/EPS composite		S.F. Roof	1.25	1.25	5.2%
.8	Openings & Specialties	Gravel stop and hatches		S.F. Roof	.28	.28	
6.0 Interior Construction							
.1	Partitions	Gypsum board on metal studs	25 S.F. Floor/L.F. Partition	S.F. Partition	3.89	1.40	
.4	Interior Doors	Hollow core wood	1000 S.F. Floor/Door	Each	362	.36	
.5	Wall Finishes	Paint		S.F. Surface	.51	.37	
.6	Floor Finishes	Quarry tile		S.F. Floor	8.60	8.60	17.7%
.7	Ceiling Finishes	Mineral fiber tile on concealed zee bars		S.F. Floor	3.07	3.07	
.9	Interior Surface/Exterior Wall	Paint	70% of wall	S.F. Wall	1.95	.89	
7.0 Conveying							
.1	Elevators	N/A		–	–	–	0.0%
.2	Special Conveyors	N/A		–	–	–	
8.0 Mechanical							
.1	Plumbing	Kitchen, bathroom and service fixtures, supply and drainage	1 Fixture/400 S.F. Floor	Each	2640	6.60	
.2	Fire Protection	Sprinklers, light hazard		S.F. Floor	2.09	2.09	
.3	Heating	Included in 8.4		–	–	–	22.1%
.4	Cooling	Multizone unit, gas heating, electric cooling		S.F. Floor	9.51	9.51	
.5	Special Systems	N/A		–	–	–	
9.0 Electrical							
.1	Service & Distribution	400 ampere service, panel board and feeders		S.F. Floor	2.70	2.70	
.2	Lighting & Power	Fluorescent fixtures, receptacles, switches, A.C. and misc. power		S.F. Floor	6.63	6.63	12.0%
.4	Special Electrical	Alarm systems and emergency lighting		S.F. Floor	.60	.60	
11.0 Special Construction							
.1	Specialties	Walk-in refrigerator		Each	17,304	4.33	5.2%
12.0 Site Work							
.1	Earthwork	N/A		–	–	–	
.3	Utilities	N/A		–	–	–	
.5	Roads & Parking	N/A		–	–	–	0.0%
.7	Site Improvements	N/A		–	–	–	
				Sub-Total		82.81	**100%**
	GENERAL CONDITIONS (Overhead & Profit)				15%	12.42	
	ARCHITECT FEES				8%	7.62	
				Total Building Cost		102.85	

Costs per square foot of floor area

Exterior Wall	S.F. Area	10000	15000	20000	25000	30000	35000	40000	45000	50000
	L.F. Perimeter	450	500	600	700	740	822	890	920	966
Face Brick with Concrete Block Back-up	Steel Frame	119.05	108.40	104.95	102.90	100.05	98.85	97.75	96.20	95.25
	Lam. Wood Truss	113.15	102.90	99.55	97.60	94.85	93.70	92.65	91.20	90.25
Concrete Block	Steel Frame	98.85	92.50	90.30	89.00	87.40	86.65	86.00	85.15	84.55
	Lam. Wood Truss	99.55	93.20	91.00	89.70	88.05	87.35	86.70	85.80	85.25
Galvanized Steel Siding	Steel Frame	90.70	85.75	83.95	82.90	81.65	81.05	80.55	79.90	79.50
Metal Sandwich Panel	Steel Joists	91.60	86.45	84.55	83.45	82.15	81.55	81.00	80.30	79.90
Perimeter Adj., Add or Deduct	Per 100 L.F.	7.65	5.10	3.85	3.05	2.50	2.20	1.90	1.70	1.50
Story Hgt. Adj., Add or Deduct	Per 1 Ft.	1.20	.90	.80	.75	.65	.65	.60	.55	.50
Basement—Not Applicable										

The above costs were calculated using the basic specifications shown on the facing page. These costs should be adjusted where necessary for design alternatives and owner's requirements. Reported completed project costs, for this type of structure, range from $40.55 to $118.75 per S.F.

Common additives

Description	Unit	$ Cost
Bar, Front Bar	L.F.	262
Back bar	L.F.	209
Booth, Upholstered, custom straight	L.F.	129 - 237
"L" or "U" shaped	L.F.	133 - 225
Bleachers, Telescoping, manual		
To 15 tier	Seat	76 - 105
16-20 tier	Seat	155 - 190
21-30 tier	Seat	165 - 199
For power operation, add	Seat	30 - 47
Emergency Lighting, 25 watt, battery operated		
Lead battery	Each	330
Nickel cadmium	Each	620
Lockers, Steel, single tier, 60" or 72"	Opening	135 - 216
2 tier, 60" or 72" total	Opening	83 - 105
5 tier, box lockers	Opening	44 - 57
Locker bench, lam. maple top only	L.F.	18.65
Pedestals, steel pipe	Each	43

Description	Unit	$ Cost
Rink		
Dasher boards & top guard	Each	149,000
Mats, rubber	S.F.	15.65
Score Board	Each	13,300 - 34,600

Rink, Hockey/Indoor Soccer

Model costs calculated for a 1 story building with 24' story height and 30,000 square feet of floor area

BUILDING TYPES

				Unit	Unit Cost	Cost Per S.F.	% Of Sub-Total
1.0 Foundations							
.1	Footings & Foundations	Poured concrete; strip and spread footings and 4' foundation wall		S.F. Ground	2.50	2.50	
.4	Piles & Caissons	N/A		–	–	–	5.0%
.9	Excavation & Backfill	Site preparation for slab and trench for foundation wall and footing		S.F. Ground	1.04	1.04	
2.0 Substructure							
.1	Slab on Grade	6" reinforced concrete with vapor barrier and granular base		S.F. Slab	3.66	3.66	5.2%
.2	Special Substructures	N/A		–	–	–	
3.0 Superstructure							
.1	Columns & Beams	Wide flange beams and columns		S.F. Ground	8.52	8.52	
.4	Structural Walls	N/A		–	–	–	
.5	Elevated Floors	N/A		–	–	–	18.9%
.7	Roof	Metal deck on steel joist		S.F. Roof	4.88	4.88	
.9	Stairs	N/A		–	–	–	
4.0 Exterior Closure							
.1	Walls	Concrete block	95% of wall	S.F. Wall	7.50	4.22	
.5	Exterior Wall Finishes	N/A		–	–	–	
.6	Doors	Aluminum and glass, hollow metal, overhead		Each	1832	.48	7.8%
.7	Windows & Glazed Walls	Store front	5% of wall	S.F. Window	29	.87	
5.0 Roofing							
.1	Roof Coverings	Elastomeric neoprene membrane with flashing		S.F. Roof	1.70	1.70	
.7	Insulation	Perlite/EPS composite		S.F. Roof	1.25	1.25	4.5%
.8	Openings & Specialties	Hatches		S.F. Roof	.27	.27	
6.0 Interior Construction							
.1	Partitions	Concrete block	140 S.F. Floor/L.F. Partition	S.F. Partition	5.13	.44	
.4	Interior Doors	Hollow metal	2500 S.F. Floor/Door	Each	526	.21	
.5	Wall Finishes	Paint		S.F. Surface	1.11	.19	
.6	Floor Finishes	80% rubber mat, 20% paint	50% of floor area	S.F. Floor	4.60	2.30	6.4%
.7	Ceiling Finishes	Mineral fiber tile on concealed zee bar	10% of area	S.F. Ceiling	3.07	.31	
.9	Interior Surface/Exterior Wall	Paint	95% of wall	S.F. Wall	1.95	1.10	
7.0 Conveying							
.1	Elevators	N/A		–	–	–	0.0%
.2	Special Conveyors	N/A		–	–	–	
8.0 Mechanical							
.1	Plumbing	Toilet and service fixtures, supply and drainage	1 Fixture/1070 S.F. Floor	Each	3755	3.51	
.2	Fire Protection	Standpipes and hose systems		–	–	–	
.3	Heating	Oil fired hot water, unit heaters	10% of area	S.F. Floor	.50	.50	21.1%
.4	Cooling	Single zone, electric cooling	90% of area	S.F. Floor	11.03	11.03	
.5	Special Systems	N/A		–	–	–	
9.0 Electrical							
.1	Service & Distribution	400 ampere service, panel board and feeders		S.F. Floor	.58	.58	
.2	Lighting & Power	Fluorescent and high intensity disch. fixtures, receptacles, switches, A.C. and misc. power		S.F. Floor	4.82	4.82	8.8%
.4	Special Electrical	Alarm systems, emergency lighting and public address		S.F. Floor	.83	.83	
11.0 Special Construction							
.1	Specialties	Dasher boards and rink (including ice making system)		S.F. Floor	15.80	15.80	22.3%
12.0 Site Work							
.1	Earthwork	N/A		–	–	–	
.3	Utilities	N/A		–	–	–	
.5	Roads & Parking	N/A		–	–	–	0.0%
.7	Site Improvements	N/A		–	–	–	
				Sub-Total		71.01	**100%**
	GENERAL CONDITIONS (Overhead & Profit)				15%	10.65	
	ARCHITECT FEES				7%	5.74	
				Total Building Cost		**87.40**	

187

Costs per square foot of floor area

Exterior Wall	S.F. Area	25000	30000	35000	40000	45000	50000	55000	60000	65000
	L.F. Perimeter	700	740	823	906	922	994	1033	1100	1166
Face Brick with Concrete Block Back-up	Steel Frame	81.95	79.70	78.80	78.10	76.70	76.30	75.55	75.20	74.90
	Bearing Walls	79.60	77.35	76.40	75.70	74.35	73.90	73.20	72.80	72.50
Stucco on Concrete Block	Steel Frame	79.20	77.30	76.45	75.85	74.70	74.30	73.70	73.40	73.15
	Bearing Walls	76.80	74.90	74.10	73.45	72.30	71.90	71.30	71.00	70.75
Decorative Concrete Block	Steel Frame	79.90	77.90	77.10	76.45	75.25	74.85	74.20	73.85	73.60
	Bearing Walls	77.50	75.50	74.70	74.05	72.85	72.45	71.80	71.45	71.20
Perimeter Adj., Add or Deduct	Per 100 L.F.	2.25	1.85	1.60	1.40	1.25	1.10	1.00	.95	.85
Story Hgt. Adj., Add or Deduct	Per 1 Ft.	.75	.65	.65	.60	.55	.55	.50	.50	.50

For Basement, add $14.35 per square foot of basement area

The above costs were calculated using the basic specifications shown on the facing page. These costs should be adjusted where necessary for design alternatives and owner's requirements. Reported completed project costs, for this type of structure, range from $45.45 to $121.15 per S.F.

Common additives

Description	Unit	$ Cost
Bleachers, Telescoping, manual		
To 15 tier	Seat	76 - 105
16-20 tier	Seat	155 - 190
21-30 tier	Seat	165 - 199
For power operation, add	Seat	30 - 47
Carrels Hardwood	Each	655 - 855
Clock System		
20 room	Each	11,800
50 room	Each	28,700
Emergency Lighting, 25 watt, battery operated		
Lead battery	Each	330
Nickel cadmium	Each	620
Flagpoles, Complete		
Aluminum, 20' high	Each	1100
40' high	Each	2475
Fiberglass, 23' high	Each	1475
39'-5" high	Each	2525
Kitchen Equipment		
Broiler	Each	3425
Cooler, 6 ft. long, reach-in	Each	2725

Description	Unit	$ Cost
Kitchen Equipment, cont.		
Dishwasher, 10-12 racks per hr.	Each	2600
Food warmer, counter, 1.2 KW	Each	745
Freezer, 44 C.F., reach-in	Each	6625
Ice cube maker, 50 lb. per day	Each	1750
Range with 1 oven	Each	2325
Lockers, Steel, single tier, 60" to 72"	Opening	135 - 216
2 tier, 60" to 72" total	Opening	83 - 105
5 tier, box lockers	Opening	44 - 57
Locker bench, lam. maple top only	L.F.	18.65
Pedestals, steel pipe	Each	43
Seating		
Auditorium chair, all veneer	Each	127
Veneer back, padded seat	Each	152
Upholstered, spring seat	Each	178
Classroom, movable chair & desk	Set	65 - 120
Lecture hall, pedestal type	Each	127 - 370
Sound System		
Amplifier, 250 watts	Each	1525
Speaker, ceiling or wall	Each	131
Trumpet	Each	246

Important: See the Reference Section for Location Factors

Model costs calculated for a 1 story building with 12' story height and 45,000 square feet of floor area

BUILDING TYPES

			Unit	Unit Cost	Cost Per S.F.	% Of Sub-Total
1.0 Foundations						
.1	Footings & Foundations	Poured concrete; strip and spread footings and 4' foundation wall	S.F. Ground	3.36	3.36	
.4	Piles & Caissons	N/A	–	–	–	7.3%
.9	Excavation & Backfill	Site preparation for slab and trench for foundation wall and footing	S.F. Ground	1.04	1.04	
2.0 Substructure						
.1	Slab on Grade	4" reinforced concrete with vapor barrier and granular base	S.F. Slab	3.06	3.06	5.1%
.2	Special Substructures	N/A	–	–	–	
3.0 Superstructure						
.1	Columns & Beams	N/A	–	–	–	
.4	Structural Walls	N/A	–	–	–	
.5	Elevated Floors	N/A	–	–	–	4.0%
.7	Roof	Metal deck on open web steel joists	S.F. Roof	2.44	2.44	
.9	Stairs	N/A	–	–	–	
4.0 Exterior Closure						
.1	Walls	Face brick with concrete block backup	70% of wall	S.F. Wall	19.46	3.35	
.5	Exterior Wall Finishes	N/A		–	–	–	8.7%
.6	Doors	Metal and glass	5% of wall	Each	2259	.40	
.7	Windows & Glazed Walls	Steel outward projecting	25% of wall	Each	461	1.48	
5.0 Roofing						
.1	Roof Coverings	Built-up tar and gravel with flashing	S.F. Roof	1.94	1.94	
.7	Insulation	Perlite/EPS composite	S.F. Roof	1.25	1.25	5.5%
.8	Openings & Specialties	Gravel stop	L.F. Perimeter	5.60	.11	
6.0 Interior Construction						
.1	Partitions	Concrete block, toilet partitions	20 S.F. Floor/L.F. Partition	S.F. Partition	5.14	3.21	
.4	Interior Doors	Single leaf kalamein fire doors	700 S.F. Floor/Door	Each	526	.75	
.5	Wall Finishes	75% paint, 15% glazed coating, 10% ceramic tile		S.F. Surface	1.42	1.42	22.6%
.6	Floor Finishes	65% vinyl composition tile, 25% carpet, 10% terrazzo		S.F. Floor	4.72	4.72	
.7	Ceiling Finishes	Mineral fiber tile on concealed zee bars		S.F. Ceiling	3.07	3.07	
.9	Interior Surface/Exterior Wall	Painted gypsum board on furring	70% of wall	S.F. Wall	2.86	.49	
7.0 Conveying						
.1	Elevators	N/A	–	–	–	0.0%
.2	Special Conveyors	N/A	–	–	–	
8.0 Mechanical						
.1	Plumbing	Kitchen, bathroom and service fixtures, supply and drainage	1 Fixture/625 S.F. Floor	Each	2262	3.62	
.2	Fire Protection	Sprinklers, light hazard		S.F. Floor	1.41	1.41	
.3	Heating	Oil fired hot water, wall fin radiation		S.F. Floor	5.76	5.76	32.1%
.4	Cooling	Split systems with air cooled condensing units		S.F. Floor	8.62	8.62	
.5	Special Systems	N/A		–	–	–	
9.0 Electrical						
.1	Service & Distribution	600 ampere service, panel board and feeders	S.F. Floor	.61	.61	
.2	Lighting & Power	Fluorescent fixtures, receptacles, switches, A.C. and misc. power	S.F. Floor	6.28	6.28	13.3%
.4	Special Electrical	Alarm systems, communications systems and emergency lighting	S.F. Floor	1.16	1.16	
11.0 Special Construction						
.1	Specialties	Chalkboards	S.F. Floor	.86	.86	1.4%
12.0 Site Work						
.1	Earthwork	N/A	–	–	–	
.3	Utilities	N/A	–	–	–	0.0%
.5	Roads & Parking	N/A	–	–	–	
.7	Site Improvements	N/A	–	–	–	
			Sub-Total		60.41	**100%**
	GENERAL CONDITIONS (Overhead & Profit)			15%	9.06	
	ARCHITECT FEES			7%	4.88	
			Total Building Cost		**74.35**	

Costs per square foot of floor area

Exterior Wall	S.F. Area	50000	70000	90000	110000	130000	150000	170000	190000	210000
	L.F. Perimeter	816	1083	1100	1300	1290	1450	1433	1566	1700
Face Brick with Concrete Block Back-up	Steel Frame	83.15	81.20	77.55	76.75	74.75	74.30	73.00	72.75	72.50
	R/Conc. Frame	85.45	83.75	80.25	79.55	77.55	77.15	75.90	75.65	75.40
Decorative Concrete Block	Steel Frame	79.80	78.20	75.20	74.60	72.90	72.50	71.45	71.20	71.05
	R/Conc. Frame	82.90	81.30	78.30	77.70	76.00	75.65	74.55	74.30	74.15
Limestone with Concrete Block Back-up	Steel Frame	86.45	84.50	80.20	79.40	76.95	76.45	74.90	74.60	74.35
	R/Conc. Frame	90.15	88.15	83.85	83.05	80.60	80.15	78.55	78.25	78.00
Perimeter Adj., Add or Deduct	Per 100 L.F.	1.80	1.25	1.00	.85	.70	.60	.55	.45	.40
Story Hgt. Adj., Add or Deduct	Per 1 Ft.	1.10	1.00	.80	.75	.65	.65	.55	.55	.55

For Basement, add $18.85 per square foot of basement area

The above costs were calculated using the basic specifications shown on the facing page. These costs should be adjusted where necessary for design alternatives and owner's requirements. Reported completed project costs, for this type of structure, range from $55.05 to $127.15 per S.F.

Common additives

Description	Unit	$ Cost
Bleachers, Telescoping, manual		
To 15 tier	Seat	76 - 105
16-20 tier	Seat	155 - 190
21-30 tier	Seat	165 - 199
For power operation, add	Seat	30 - 47
Carrels Hardwood	Each	655 - 855
Clock System		
20 room	Each	11,800
50 room	Each	28,700
Elevators, Hydraulic passenger, 2 stops		
1500# capacity	Each	40,000
2500# capacity	Each	41,300
Emergency Lighting, 25 watt, battery operated		
Lead battery	Each	330
Nickel cadmium	Each	620
Flagpoles, Complete		
Aluminum, 20' high	Each	1100
40' high	Each	2475
Fiberglass, 23' high	Each	1475
39'-5" high	Each	2525

Description	Unit	$ Cost
Kitchen Equipment		
Broiler	Each	3425
Cooler, 6 ft. long, reach-in	Each	2725
Dishwasher, 10-12 racks per hr.	Each	2600
Food warmer, counter, 1.2 KW	Each	745
Freezer, 44 C.F., reach-in	Each	6625
Lockers, Steel, single tier, 60" or 72"	Opening	135 - 216
2 tier, 60" or 72" total	Opening	83 - 105
5 tier, box lockers	Opening	44 - 57
Locker bench, lam. maple top only	L.F.	18.65
Pedestals, steel pipe	Each	43
Seating		
Auditorium chair, all veneer	Each	127
Veneer back, padded seat	Each	152
Upholstered, spring seat	Each	178
Classroom, movable chair & desk	Set	65 - 120
Lecture hall, pedestal type	Each	127 - 370
Sound System		
Amplifier, 250 watts	Each	1525
Speaker, ceiling or wall	Each	131
Trumpet	Each	246

BUILDING TYPES

Model costs calculated for a 2 story building with 12' story height and 130,000 square feet of floor area

			Unit	Unit Cost	Cost Per S.F.	% Of Sub-Total
1.0 Foundations						
.1	Footings & Foundations	Poured concrete; strip and spread footings and 4' foundation wall	S.F. Ground	2.20	1.10	
.4	Piles & Caissons	N/A	—	—	—	2.6%
.9	Excavation & Backfill	Site preparation for slab and trench for foundation wall and footing	S.F. Ground	1.04	.52	
2.0 Substructure						
.1	Slab on Grade	4" reinforced concrete with vapor barrier and granular base	S.F. Slab	3.06	1.53	2.4%
.2	Special Substructures	N/A	—	—	—	
3.0 Superstructure						
.1	Columns & Beams	Concrete columns	L.F. Column	73	1.80	
.4	Structural Walls	N/A	—	—	—	
.5	Elevated Floors	Concrete slab without drop panel	S.F. Floor	9.37	4.68	17.8%
.7	Roof	Concrete slab without drop panel	S.F. Roof	9.00	4.50	
.9	Stairs	Concrete filled metal pan	Flight	5025	.23	
4.0 Exterior Closure						
.1	Walls	Face brick with concrete block backup 75% of wall	S.F. Wall	19.43	3.47	
.5	Exterior Wall Finishes	N/A	—	—	—	9.7%
.6	Doors	Metal and glass	Each	1235	.27	
.7	Windows & Glazed Walls	Window wall 25% of wall	S.F. Window	40	2.40	
5.0 Roofing						
.1	Roof Coverings	Built-up tar and gravel with flashing	S.F. Roof	1.84	.92	
.7	Insulation	Perlite/EPS composite	S.F. Roof	1.25	.63	2.6%
.8	Openings & Specialties	Gravel stop and hatches	S.F. Roof	.18	.09	
6.0 Interior Construction						
.1	Partitions	Concrete block, toilet partitions 25 S.F. Floor/L.F. Partition	S.F. Partition	5.14	2.52	
.4	Interior Doors	Single leaf kalamein fire doors 700 S.F. Floor/Door	Each	526	.75	
.5	Wall Finishes	75% paint, 15% glazed coating, 10% ceramic tile	S.F. Surface	1.42	1.14	19.9%
.6	Floor Finishes	70% vinyl composition tile, 20% carpet, 10% terrazzo	S.F. Floor	4.57	4.57	
.7	Ceiling Finishes	Mineral fiber tile on concealed zee bars	S.F. Ceiling	3.07	3.07	
.9	Interior Surface/Exterior Wall	Painted gypsum board on furring 75% of wall	S.F. Wall	2.86	.51	
7.0 Conveying						
.1	Elevators	One hydraulic passenger elevator	Each	49,400	.38	0.6%
.2	Special Conveyors	N/A	—	—	—	
8.0 Mechanical						
.1	Plumbing	Kitchen, bathroom and service fixtures, supply and drainage 1 Fixture/860 S.F. Floor	Each	2528	2.94	
.2	Fire Protection	Sprinklers, light hazard	S.F. Floor	1.25	1.25	
.3	Heating	Oil fired hot water, wall fin radiation	S.F. Floor	2.98	2.98	28.8%
.4	Cooling	Chilled water, cooling tower systems	S.F. Floor	10.98	10.98	
.5	Special Systems	N/A	—	—	—	
9.0 Electrical						
.1	Service & Distribution	1200 ampere service, panel board and feeders	S.F. Floor	.43	.43	
.2	Lighting & Power	Fluorescent fixtures, receptacles, switches, A.C. and misc. power	S.F. Floor	6.22	6.22	14.4%
.4	Special Electrical	Alarm systems, communications systems and emergency lighting	S.F. Floor	2.41	2.41	
11.0 Special Construction						
.1	Specialties	Chalkboards, laboratory counters, built-in athletic equipment	S.F. Floor	.73	.73	1.2%
12.0 Site Work						
.1	Earthwork	N/A	—	—	—	
.3	Utilities	N/A	—	—	—	0.0%
.5	Roads & Parking	N/A	—	—	—	
.7	Site Improvements	N/A	—	—	—	
			Sub-Total		63.02	100%
	GENERAL CONDITIONS (Overhead & Profit)			15%	9.45	
	ARCHITECT FEES			7%	5.08	
			Total Building Cost		77.55	

Costs per square foot of floor area

Exterior Wall	S.F. Area	50000	65000	80000	95000	110000	125000	140000	155000	170000
	L.F. Perimeter	816	1016	1000	1150	1890	2116	2287	2438	2552
Face Brick with Concrete Block Back-up	Steel Frame	81.90	81.05	78.30	77.85	82.00	81.70	81.15	80.60	80.00
	Bearing Walls	79.25	78.40	75.65	75.20	79.35	79.05	78.50	77.95	77.35
Concrete Block Stucco Face	Steel Frame	78.35	77.65	75.55	75.20	78.25	78.00	77.60	77.20	76.70
	Bearing Walls	75.70	75.05	72.90	72.55	75.65	75.35	74.95	74.55	74.10
Decorative Concrete Block	Steel Frame	79.25	78.50	76.25	75.85	79.20	78.95	78.50	78.05	77.55
	Bearing Walls	76.20	75.45	73.20	72.80	76.15	75.90	75.45	75.00	74.50
Perimeter Adj., Add or Deduct	Per 100 L.F.	1.65	1.30	1.05	.85	.75	.65	.60	.55	.45
Story Hgt. Adj., Add or Deduct	Per 1 Ft.	1.00	.95	.75	.75	1.00	1.00	1.00	.95	.90
For Basement, add $19.40 per square foot of basement area										

The above costs were calculated using the basic specifications shown on the facing page. These costs should be adjusted where necessary for design alternatives and owner's requirements. Reported completed project costs, for this type of structure, range from $52.25 to $119.40 per S.F.

Common additives

Description	Unit	$ Cost		Description	Unit	$ Cost
Bleachers, Telescoping, manual				Kitchen Equipment		
To 15 tier	Seat	76 - 105		Broiler	Each	3425
16-20 tier	Seat	155 - 190		Cooler, 6 ft. long, reach-in	Each	2725
21-30 tier	Seat	165 - 199		Dishwasher, 10-12 racks per hr.	Each	2600
For power operation, add	Seat	30 - 47		Food warmer, counter, 1.2 KW	Each	745
Carrels Hardwood	Each	655 - 855		Freezer, 44 C.F., reach-in	Each	6625
Clock System				Lockers, Steel, single tier, 60" to 72"	Opening	135 - 216
20 room	Each	11,800		2 tier, 60" to 72" total	Opening	83 - 105
50 room	Each	28,700		5 tier, box lockers	Opening	44 - 57
Elevators, Hydraulic passenger, 2 stops				Locker bench, lam. maple top only	L.F.	18.65
1500# capacity	Each	40,000		Pedestals, steel pipe	Each	43
2500# capacity	Each	41,300		Seating		
Emergency Lighting, 25 watt, battery operated				Auditorium chair, all veneer	Each	127
Lead battery	Each	330		Veneer back, padded seat	Each	152
Nickel cadmium	Each	620		Upholstered, spring seat	Each	178
Flagpoles, Complete				Classroom, movable chair & desk	Set	65 - 120
Aluminum, 20' high	Each	1100		Lecture hall, pedestal type	Each	127 - 370
40' high	Each	2475		Sound System		
Fiberglass, 23' high	Each	1475		Amplifier, 250 watts	Each	1525
39'-5" high	Each	2525		Speaker, ceiling or wall	Each	131
				Trumpet	Each	246

Important: See the Reference Section for Location Factors

Model costs calculated for a 2 story building with 12' story height and 110,000 square feet of floor area

BUILDING TYPES

				Unit	Unit Cost	Cost Per S.F.	% Of Sub-Total
1.0 Foundations							
.1	Footings & Foundations	Poured concrete; strip and spread footings and 4' foundation wall		S.F. Ground	3.04	1.52	
.4	Piles & Caissons	N/A		—	—	—	3.1%
.9	Excavation & Backfill	Site preparation for slab and trench for foundation wall and footing		S.F. Ground	1.04	.52	
2.0 Substructure							
.1	Slab on Grade	4" reinforced concrete with vapor barrier and granular base		S.F. Slab	3.06	1.53	2.3%
.2	Special Substructures	N/A		—	—	—	
3.0 Superstructure							
.1	Columns & Beams	Fireproofing, steel columns included in 3.5 and 3.7		L.F. Column	21	.30	
.4	Structural Walls	N/A		—	—	—	
.5	Elevated Floors	Open web steel joists, slab form, concrete, columns		S.F. Floor	13.61	6.80	15.1%
.7	Roof	Metal deck, open web steel joists, columns		S.F. Roof	5.36	2.68	
.9	Stairs	Concrete filled metal pan		Flight	5025	.27	
4.0 Exterior Closure							
.1	Walls	Face brick with concrete block backup	75% of wall	S.F. Wall	19.47	6.02	
.5	Exterior Wall Finishes	N/A		—	—	—	
.6	Doors	Double aluminum & glass		Each	1315	.35	14.2%
.7	Windows & Glazed Walls	Window wall	25% of wall	S.F. Window	29	3.06	
5.0 Roofing							
.1	Roof Coverings	Built-up tar and gravel with flashing		S.F. Roof	2.20	1.10	
.7	Insulation	Perlite/EPS composite		S.F. Roof	1.25	.63	2.8%
.8	Openings & Specialties	Gravel stop and hatches		S.F. Roof	.24	.12	
6.0 Interior Construction							
.1	Partitions	Concrete block, toilet partitions	20 S.F. Floor/L.F. Partition	S.F. Partition	5.14	3.00	
.4	Interior Doors	Single leaf kalamein fire doors	750 S.F. Floor/Door	Each	526	.70	
.5	Wall Finishes	50% paint, 40% glazed coatings, 10% ceramic tile		S.F. Surface	1.58	1.58	
.6	Floor Finishes	50% vinyl composition tile, 30% carpet, 20% terrrazzo		S.F. Floor	6.18	6.18	23.1%
.7	Ceiling Finishes	Mineral fiberboard on concealed zee bars		S.F. Ceiling	3.07	3.07	
.9	Interior Surface/Exterior Wall	Painted gypsum board on furring	75% of wall	S.F. Wall	2.86	.88	
7.0 Conveying							
.1	Elevators	One hydraulic passenger elevator		Each	48,400	.44	0.7%
.2	Special Conveyors	N/A		—	—	—	
8.0 Mechanical							
.1	Plumbing	Kitchen, toilet and service fixtures, supply and drainage	1 Fixture/1170 S.F. Floor	Each	2784	2.38	
.2	Fire Protection	Sprinklers, light hazard	10% of area	S.F. Floor	.25	.25	
.3	Heating	Included in 8.4		—	—	—	24.5%
.4	Cooling	Multizone unit, gas heating, electric cooling		S.F. Floor	13.80	13.80	
.5	Special Systems	N/A		—	—	—	
9.0 Electrical							
.1	Service & Distribution	1000 ampere service, panel board and feeders		S.F. Floor	.48	.48	
.2	Lighting & Power	Fluorescent fixtures, receptacles, switches, A.C. and misc. power		S.F. Floor	6.31	6.31	13.1%
.4	Special Electrical	Alarm systems, communications systems and emergency lighting		S.F. Floor	1.96	1.96	
11.0 Special Construction							
.1	Specialties	Chalkboards, laboratory counters		S.F. Floor	.71	.71	1.1%
12.0 Site Work							
.1	Earthwork	N/A		—	—	—	
.3	Utilities	N/A		—	—	—	
.5	Roads & Parking	N/A		—	—	—	0.0%
.7	Site Improvements	N/A		—	—	—	
				Sub-Total		66.64	100%

GENERAL CONDITIONS (Overhead & Profit)		15%	10.00
ARCHITECT FEES		7%	5.36
Total Building Cost			**82.00**

BUILDING TYPES

Costs per square foot of floor area

Exterior Wall	S.F. Area	20000	30000	40000	50000	60000	70000	80000	90000	100000
	L.F. Perimeter	400	500	685	700	800	900	1000	1100	1200
Face Brick with Concrete Block Back-up	Steel Frame	85.20	81.95	82.10	79.40	78.75	78.25	77.90	77.65	77.45
	Bearing Walls	83.45	80.20	80.30	77.60	76.95	76.50	76.10	75.85	75.65
Decorative Concrete Block	Steel Frame	80.40	77.95	78.00	76.05	75.55	75.20	74.90	74.70	74.55
	Bearing Walls	78.65	76.20	76.20	74.25	73.75	73.40	73.10	72.95	72.80
Steel Siding on Steel Studs	Steel Frame	77.75	75.75	75.70	74.15	73.75	73.45	73.25	73.05	72.95
Metal Sandwich Panel	Steel Frame	77.65	75.70	75.60	74.10	73.70	73.40	73.20	73.05	72.90
Perimeter Adj., Add or Deduct	Per 100 L.F.	4.05	2.70	2.00	1.60	1.35	1.15	1.05	.90	.80
Story Hgt. Adj., Add or Deduct	Per 1 Ft.	1.15	.95	1.00	.80	.75	.75	.70	.70	.70
For Basement, add $18.95 per square foot of basement area										

The above costs were calculated using the basic specifications shown on the facing page. These costs should be adjusted where necessary for design alternatives and owner's requirements. Reported completed project costs, for this type of structure, range from $46.50 to $132.75 per S.F.

Common additives

Description	Unit	$ Cost
Carrels Hardwood	Each	655 - 855
Clock System		
20 room	Each	11,800
50 room	Each	28,700
Directory Boards, Plastic, glass covered		
30" x 20"	Each	510
36" x 48"	Each	930
Aluminum, 24" x 18"	Each	450
36" x 24"	Each	540
48" x 32"	Each	640
48" x 60"	Each	1450
Elevators, Hydraulic passenger, 2 stops		
1500# capacity	Each	40,000
2500# capacity	Each	41,300
3500# capacity	Each	44,900
Emergency Lighting, 25 watt, battery operated		
Lead battery	Each	330
Nickel cadmium	Each	620

Description	Unit	$ Cost
Flagpoles, Complete		
Aluminum, 20' high	Each	1100
40' high	Each	2475
Fiberglass, 23' high	Each	1475
39'-5" high	Each	2525
Seating		
Auditorium chair, all veneer	Each	127
Veneer back, padded seat	Each	152
Upholstered, spring seat	Each	178
Classroom, movable chair & desk	Set	65 - 120
Lecture hall, pedestal type	Each	127 - 370
Shops & Workroom:		
Benches, metal	Each	380
Parts bins 6'-3" high, 3' wide, 12" deep, 72 bins	Each	550
Shelving, metal 1' x 3'	S.F.	9.15
Wide span 6' wide x 24" deep	S.F.	11.90
Sound System		
Amplifier, 250 watts	Each	1525
Speaker, ceiling or wall	Each	131
Trumpet	Each	246

Model costs calculated for a 2 story building with 12' story height and 40,000 square feet of floor area

				Unit	Unit Cost	Cost Per S.F.	% Of Sub-Total
1.0 Foundations							
.1	Footings & Foundations	Poured concrete; strip and spread footings and 4' foundation wall		S.F. Ground	3.12	1.56	
.4	Piles & Caissons	N/A		—	—	—	3.2%
.9	Excavation & Backfill	Site preparation for slab and trench for foundation wall and footing		S.F. Ground	1.04	.52	
2.0 Substructure							
.1	Slab on Grade	5" reinforced concrete with vapor barrier and granular base		S.F. Slab	6.12	3.06	4.7%
.2	Special Substructures	N/A		—	—	—	
3.0 Superstructure							
.1	Columns & Beams	Interior steel columns included in 3.5 and 3.7		—	—	—	
.4	Structural Walls	N/A		—	—	—	
.5	Elevated Floors	Open web steel joists, slab form, concrete, beams, columns		S.F. Floor	9.58	4.79	10.6%
.7	Roof	Metal deck, open web steel joists, beams, columns		S.F. Roof	3.26	1.63	
.9	Stairs	Concrete filled metal pan		Flight	5025	.50	
4.0 Exterior Closure							
.1	Walls	Face brick with concrete block backup	85% of wall	S.F. Wall	19.44	6.79	
.5	Exterior Wall Finishes	N/A		—	—	—	13.7%
.6	Doors	Metal and glass doors		Each	1501	.30	
.7	Windows & Glazed Walls	Steel outward projecting	15% of wall	S.F. Wall	29	1.83	
5.0 Roofing							
.1	Roof Coverings	Built-up tar and gravel with flashing		S.F. Roof	2.20	1.10	
.7	Insulation	Perlite/EPS composite		S.F. Roof	1.25	.63	2.9%
.8	Openings & Specialties	Gravel stop and hatches		S.F. Roof	.38	.19	
6.0 Interior Construction							
.1	Partitions	Concrete block, toilet partitions	20 S.F. Floor/L.F. Partition	S.F. Partition	5.14	3.17	
.4	Interior Doors	Single leaf kalamein fire doors	600 S.F. Floor/Door	Each	526	.88	
.5	Wall Finishes	50% paint, 40% glazed coating, 10% ceramic tile		S.F. Surface	1.58	1.58	20.0%
.6	Floor Finishes	70% vinyl composition tile, 20% carpet, 10% terrazzo		S.F. Floor	4.57	4.57	
.7	Ceiling Finishes	Mineral fiber tile on concealed zee bars	60% of area	S.F. Ceiling	3.07	1.84	
.9	Interior Surface/Exterior Wall	Painted gypsum board on furring	85% of wall	S.F. Wall	2.86	1.00	
7.0 Conveying							
.1	Elevators	One hydraulic passenger elevator		Each	48,800	1.22	1.9%
.2	Special Conveyors	N/A		—	—	—	
8.0 Mechanical							
.1	Plumbing	Toilet and service fixtures, supply and drainage	1 Fixture/700 S.F. Floor	Each	2303	3.29	
.2	Fire Protection	Sprinklers, light hazard		S.F. Floor	.82	.82	
.3	Heating	Oil fired hot water, wall fin radiation and unit heaters		S.F. Floor	10.79	5.47	29.1%
.4	Cooling	Chilled water, cooling tower systems		S.F. Floor	9.42	9.42	
.5	Special Systems	N/A		—	—	—	
9.0 Electrical							
.1	Service & Distribution	600 ampere service, panel board and feeders		S.F. Floor	.69	.69	
.2	Lighting & Power	Fluorescent fixtures, receptacles, switches, A.C. and misc. power		S.F. Floor	6.31	6.31	12.9%
.4	Special Electrical	Alarm systems, communications systems and emergency lighting		S.F. Floor	1.43	1.43	
11.0 Special Construction							
.1	Specialties	Chalkboards		S.F. Floor	.66	.66	1.0%
12.0 Site Work							
.1	Earthwork	N/A		—	—	—	
.3	Utilities	N/A		—	—	—	
.5	Roads & Parking	N/A		—	—	—	0.0%
.7	Site Improvements	N/A		—	—	—	
				Sub-Total		65.25	100%
	GENERAL CONDITIONS (Overhead & Profit)				15%	9.79	
	ARCHITECT FEES				7%	5.26	
				Total Building Cost		80.30	

BUILDING TYPES

Costs per square foot of floor area

Exterior Wall	S.F. Area	1000	2000	3000	4000	6000	8000	10000	12000	15000
	L.F. Perimeter	126	179	219	253	310	358	400	438	490
Wood Siding	Wood Frame	74.40	66.70	63.25	61.20	58.75	57.30	56.30	55.55	54.75
Face Brick Veneer	Wood Frame	90.95	78.30	72.60	69.15	65.15	62.75	61.10	59.90	58.55
Stucco on Concrete Block	Steel Frame	84.15	74.00	69.45	66.70	63.55	61.60	60.30	59.30	58.25
	Bearing Walls	82.80	72.65	68.10	65.35	62.20	60.25	58.90	57.95	56.90
Metal Sandwich Panel	Steel Frame	86.55	75.80	71.00	68.10	64.70	62.70	61.30	60.25	59.15
Precast Concrete	Steel Frame	97.20	83.15	76.80	73.00	68.55	65.85	64.05	62.70	61.20
Perimeter Adj., Add or Deduct	Per 100 L.F.	21.05	10.55	7.00	5.25	3.50	2.65	2.10	1.75	1.40
Story Hgt. Adj., Add or Deduct	Per 1 Ft.	1.55	1.15	.90	.80	.65	.55	.50	.45	.40
For Basement, add $14.25 per square foot of basement area										

The above costs were calculated using the basic specifications shown on the facing page. These costs should be adjusted where necessary for design alternatives and owner's requirements. Reported completed project costs, for this type of structure, range from $38.25 to $105.50 per S.F.

Common additives

Description	Unit	$ Cost
Check Out Counter		
Single belt	Each	2175
Double belt	Each	3675
Emergency Lighting, 25 watt, battery operated		
Lead battery	Each	330
Nickel cadmium	Each	620
Refrigerators, Prefabricated, walk-in		
7'-6" high, 6' x 6'	S.F.	117
10' x 10'	S.F.	91
12' x 14'	S.F.	81
12' x 20'	S.F.	72
Refrigerated Food Cases		
Dairy, multi deck, 12' long	Each	6975
Delicatessen case, single deck, 12' long	Each	4925
Multi deck, 18 S.F. shelf display	Each	7425
Freezer, self-contained chest type, 30 C.F.	Each	2950
Glass door upright, 78 C.F.	Each	6525

Description	Unit	$ Cost
Refrigerated Food Cases, cont.		
Frozen food, chest type, 12' long	Each	4775
Glass door reach-in, 5 door	Each	9025
Island case 12' long, single deck	Each	5400
Multi deck	Each	11,100
Meat cases, 12' long, single deck	Each	3925
Multi deck	Each	6700
Produce, 12' long single deck	Each	5175
Multi deck	Each	5775
Safe, Office type, 4 hour rating		
30" x 18" x 18"	Each	3475
62" x 33" x 20"	Each	7575
Smoke Detectors		
Ceiling type	Each	142
Duct type	Each	390
Sound System		
Amplifier, 250 watts	Each	1525
Speaker, ceiling or wall	Each	131
Trumpet	Each	246

Important: See the Reference Section for Location Factors

Model costs calculated for a 1 story building with 12' story height and 4,000 square feet of floor area

				Unit	Unit Cost	Cost Per S.F.	% Of Sub-Total
1.0 Foundations							
.1	Footings & Foundations	Poured concrete; strip and spread footings and 4' foundation wall		S.F. Ground	3.34	3.34	
.4	Piles & Caissons	N/A		—	—	—	9.2%
.9	Excavation & Backfill	Site preparation for slab and trench for foundation wall and footing		S.F. Ground	1.24	1.24	
2.0 Substructure							
.1	Slab on Grade	4" reinforced concrete with vapor barrier and granular base		S.F. Slab	3.06	3.06	6.2%
.2	Special Substructures	N/A		—	—	—	
3.0 Superstructure							
.1	Columns & Beams	N/A		—	—	—	
.4	Structural Walls	N/A		—	—	—	
.5	Elevated Floors	N/A		—	—	—	8.7%
.7	Roof	Wood truss with plywood sheathing		S.F. Ground	4.33	4.33	
.9	Stairs	N/A		—	—	—	
4.0 Exterior Closure							
.1	Walls	Wood siding on wood studs, insulated	80% of wall	S.F. Wall	6.29	3.82	
.5	Exterior Wall Finishes	N/A		—	—	—	16.4%
.6	Doors	Double aluminum and glass, solid core wood		Each	1929	1.45	
.7	Windows & Glazed Walls	Storefront	20% of wall	S.F. Window	29	2.91	
5.0 Roofing							
.1	Roof Coverings	Asphalt shingles		S.F. Roof	1.65	1.65	
.7	Insulation	Fiberglass sheets		S.F. Roof	.87	.87	5.3%
.8	Openings & Specialties	Gutters and downspouts		S.F. Roof	.10	.10	
6.0 Interior Construction							
.1	Partitions	Gypsum board on wood studs	60 S.F. Floor/L.F. Partition	S.F. Partition	3.12	.52	
.4	Interior Doors	Single leaf wood, hollow metal	1300 S.F. Floor/Door	Each	624	.48	
.5	Wall Finishes	Paint		S.F. Surface	.72	.24	12.5%
.6	Floor Finishes	Asphalt tile		S.F. Floor	1.90	1.90	
.7	Ceiling Finishes	Mineral fiber tile on wood furring		S.F. Ceiling	2.15	2.15	
.9	Interior Surface/Exterior Wall	Painted gypsum board on furring	80% of wall	S.F. Wall	1.50	.91	
7.0 Conveying							
.1	Elevators	N/A		—	—	—	0.0%
.2	Special Conveyors	N/A		—	—	—	
8.0 Mechanical							
.1	Plumbing	Toilet and service fixtures, supply and drainage	1 Fixture/1000 S.F. Floor	Each	1920	1.92	
.2	Fire Protection	Sprinkler, ordinary hazard		S.F. Floor	2.09	2.09	
.3	Heating	Included in 8.4		—	—	—	19.8%
.4	Cooling	Single zone rooftop unit, gas heating, electric cooling		S.F. Floor	5.89	5.89	
.5	Special Systems	N/A		—	—	—	
9.0 Electrical							
.1	Service & Distribution	200 ampere service, panel board and feeders		S.F. Floor	1.67	1.67	
.2	Lighting & Power	Fluorescent fixtures, receptacles, switches, A.C. and misc. power		S.F. Floor	6.36	6.36	21.9%
.4	Special Electrical	Alarm systems and emergency lighting		S.F. Floor	2.84	2.84	
11.0 Special Construction							
.1	Specialties	N/A		—	—	—	0.0%
12.0 Site Work							
.1	Earthwork	N/A		—	—	—	
.3	Utilities	N/A		—	—	—	0.0%
.5	Roads & Parking	N/A		—	—	—	
.7	Site Improvements	N/A		—	—	—	
				Sub-Total		49.74	**100%**
	GENERAL CONDITIONS (Overhead & Profit)			15%		7.46	
	ARCHITECT FEES			7%		4.00	
				Total Building Cost		**61.20**	

BUILDING TYPES

Costs per square foot of floor area

Exterior Wall	S.F. Area	50000	65000	80000	95000	110000	125000	140000	155000	170000
	L.F. Perimeter	920	1065	1167	1303	1333	1433	1533	1633	1733
Face Brick with Concrete Block Back-up	R/Conc. Frame	62.55	61.05	59.85	59.15	58.20	57.70	57.40	57.10	56.85
	Steel Frame	51.60	50.10	48.85	48.20	47.20	46.75	46.40	46.10	45.85
Decorative Concrete Block	R/Conc. Frame	59.65	58.45	57.55	57.00	56.25	55.90	55.65	55.45	55.25
	Steel Joists	49.25	47.95	46.95	46.40	45.60	45.25	44.95	44.70	44.50
Precast Concrete Panels	R/Conc. Frame	59.95	58.75	57.85	57.30	56.55	56.20	55.95	55.75	55.55
	Steel Joists	48.60	47.40	46.50	45.95	45.25	44.85	44.60	44.40	44.20
Perimeter Adj., Add or Deduct	Per 100 L.F.	1.05	.80	.65	.55	.45	.45	.40	.35	.30
Story Hgt. Adj., Add or Deduct	Per 1 Ft.	.55	.50	.45	.40	.35	.35	.35	.35	.30

For Basement, add $14.00 per square foot of basement area

The above costs were calculated using the basic specifications shown on the facing page. These costs should be adjusted where necessary for design alternatives and owner's requirements. Reported completed project costs, for this type of structure, range from $33.05 to $73.00 per S.F.

Common additives

Description	Unit	$ Cost
Closed Circuit Surveillance, One station		
Camera and monitor	Each	1325
For additional camera stations, add	Each	730
Directory Boards, Plastic, glass covered		
30" x 20"	Each	510
36" x 48"	Each	930
Aluminum, 24" x 18"	Each	450
36" x 24"	Each	540
48" x 32"	Each	640
48" x 60"	Each	1450
Emergency Lighting, 25 watt, battery operated		
Lead battery	Each	330
Nickel cadmium	Each	620
Safe, Office type, 4 hour rating		
30" x 18" x 18"	Each	3475
62" x 33" x 20"	Each	7575
Sound System		
Amplifier, 250 watts	Each	1525
Speaker, ceiling or wall	Each	131
Trumpet	Each	246

BUILDING TYPES

Important: See the Reference Section for Location Factors

Model costs calculated for a 1 story building with 14' story height and 110,000 square feet of floor area

Store, Department, 1 Story

				Unit	Unit Cost	Cost Per S.F.	% Of Sub-Total
1.0 Foundations							
.1	Footings & Foundations	Poured concrete; strip and spread footings and 4' foundation wall		S.F. Ground	1.04	1.04	
.4	Piles & Caissons	N/A		—	—	—	4.3%
.9	Excavation & Backfill	Site preparation for slab and trench for foundation wall and footing		S.F. Ground	1.00	1.00	
2.0 Substructure							
.1	Slab on Grade	4" reinforced concrete with vapor barrier and granular base		S.F. Slab	3.06	3.06	6.4%
.2	Special Substructures	N/A		—	—	—	
3.0 Superstructure							
.1	Columns & Beams	Concrete columns		L.F. Column	42	.55	
.4	Structural Walls	N/A		—	—	—	
.5	Elevated Floors	N/A		—	—	—	28.8%
.7	Roof	Precast concrete beam and plank		S.F. Roof	13.22	13.22	
.9	Stairs	N/A		—	—	—	
4.0 Exterior Closure							
.1	Walls	Face brick with concrete block backup	90% of wall	S.F. Wall	19.58	2.99	
.5	Exterior Wall Finishes	N/A		—	—	—	
.6	Doors	Sliding electric operated entrance, hollow metal		Each	4378	.24	8.1%
.7	Windows & Glazed Walls	Storefront	10% of wall	S.F. Window	38	.65	
5.0 Roofing							
.1	Roof Coverings	Built-up tar and gravel with flashing		S.F. Roof	1.74	1.74	
.7	Insulation	Perlite/EPS composite		S.F. Roof	1.25	1.25	6.5%
.8	Openings & Specialties	Gravel stop and hatches		S.F. Roof	.10	.10	
6.0 Interior Construction							
.1	Partitions	Gypsum board on metal studs	60 S.F. Floor/L.F. Partition	S.F. Partition	3.90	.65	
.4	Interior Doors	Single leaf hollow metal	600 S.F. Floor/Door	Each	526	.88	
.5	Wall Finishes	Paint		S.F. Surface	.51	.17	15.7%
.6	Floor Finishes	50% vinyl tile, 50% carpet		S.F. Floor	4.14	4.14	
.7	Ceiling Finishes	Fiberglass board on exposed grid system, suspended		S.F. Ceiling	1.20	1.20	
.9	Interior Surface/Exterior Wall	Painted gypsum board on furring	90% of wall	S.F. Wall	2.86	.44	
7.0 Conveying							
.1	Elevators	N/A		—	—	—	0.0%
.2	Special Conveyors	N/A		—	—	—	
8.0 Mechanical							
.1	Plumbing	Toilet and service fixtures, supply and drainage	1 Fixture/4075 S.F. Floor	Each	2730	.67	
.2	Fire Protection	Sprinklers, light hazard		S.F. Floor	1.41	1.41	
.3	Heating	Included in 8.4		—	—	—	17.9%
.4	Cooling	Single zone unit gas heating, electric cooling		S.F. Floor	6.46	6.46	
.5	Special Systems	N/A		—	—	—	
9.0 Electrical							
.1	Service & Distribution	1200 ampere service, panel board and feeders		S.F. Floor	.51	.51	
.2	Lighting & Power	Fluorescent fixtures, receptacles, switches, A.C. and misc. power		S.F. Floor	5.07	5.07	12.3%
.4	Special Electrical	Alarm systems and emergency lighting		S.F. Floor	.30	.30	
11.0 Special Construction							
.1	Specialties	N/A		—	—	—	0.0%
12.0 Site Work							
.1	Earthwork	N/A		—	—	—	
.3	Utilities	N/A		—	—	—	
.5	Roads & Parking	N/A		—	—	—	0.0%
.7	Site Improvements	N/A		—	—	—	
				Sub-Total		47.74	100%
	GENERAL CONDITIONS (Overhead & Profit)				15%	7.16	
	ARCHITECT FEES				6%	3.30	
				Total Building Cost		58.20	

Costs per square foot of floor area

Exterior Wall	S.F. Area	50000	65000	80000	95000	110000	125000	140000	155000	170000
	L.F. Perimeter	533	593	670	715	778	840	871	923	976
Face Brick with Concrete Block Back-up	Steel Frame	74.90	71.75	70.10	68.50	67.55	66.85	65.95	65.45	65.00
	R/Conc. Frame	77.05	73.90	72.25	70.70	69.75	69.00	68.10	67.60	67.20
Face Brick on Steel Studs	Steel Frame	72.45	69.65	68.20	66.80	65.95	65.30	64.50	64.10	63.70
	R/Conc. Frame	75.65	72.85	71.35	69.95	69.15	68.50	67.70	67.25	66.90
Precast Concrete Panels Exposed Aggregate	Steel Frame	72.60	69.80	68.30	66.90	66.05	65.40	64.65	64.20	63.80
	R/Conc. Frame	75.45	72.65	71.15	69.75	68.90	68.25	67.50	67.05	66.65
Perimeter Adj., Add or Deduct	Per 100 L.F.	2.80	2.20	1.75	1.50	1.30	1.15	1.00	.95	.85
Story Hgt. Adj., Add or Deduct	Per 1 Ft.	.85	.75	.70	.60	.55	.55	.50	.50	.45
For Basement, add $23.80 per square foot of basement area										

The above costs were calculated using the basic specifications shown on the facing page. These costs should be adjusted where necessary for design alternatives and owner's requirements. Reported completed project costs, for this type of structure, range from $34.20 to $91.55 per S.F.

Common additives

Description	Unit	$ Cost	Description	Unit	$ Cost
Closed Circuit Surveillance, One station			Escalators, Metal		
Camera and monitor	Each	1325	32" wide, 10' story height	Each	85,700
For additional camera stations, add	Each	730	20' story height	Each	96,800
Directory Boards, Plastic, glass covered			48" wide, 10' story height	Each	90,200
30" x 20"	Each	510	20' story height	Each	101,300
36" x 48"	Each	930	Glass		
Aluminum, 24" x 18"	Each	450	32" wide, 10' story height	Each	85,700
36" x 24"	Each	540	20' story height	Each	96,800
48" x 32"	Each	640	48" wide, 10' story height	Each	90,200
48" x 60"	Each	1450	20' story height	Each	101,300
Elevators, Hydraulic passenger, 2 stops			Safe, Office type, 4 hour rating		
1500# capacity	Each	40,000	30" x 18" x 18"	Each	3475
2500# capacity	Each	41,300	62" x 33" x 20"	Each	7575
3500# capacity	Each	44,900	Sound System		
Additional stop, add	Each	3350	Amplifier, 250 watts	Each	1525
Emergency Lighting, 25 watt, battery operated			Speaker, ceiling or wall	Each	131
Lead battery	Each	330	Trumpet	Each	246
Nickel cadmium	Each	620			

Important: See the Reference Section for Location Factors

BUILDING TYPES

Model costs calculated for a 3 story building with 16' story height and 95,000 square feet of floor area

			Unit	Unit Cost	Cost Per S.F.	% Of Sub-Total
1.0 Foundations						
.1	Footings & Foundations	Poured concrete; strip and spread footings and 4' foundation wall	S.F. Ground	3.27	1.09	
.4	Piles & Caissons	N/A	–	–	–	2.6%
.9	Excavation & Backfill	Site preparation for slab and trench for foundation wall and footing	S.F. Ground	1.04	.35	
2.0 Substructure						
.1	Slab on Grade	4" reinforced concrete with vapor barrier and granular base	S.F. Slab	3.06	1.02	1.8%
.2	Special Substructures	N/A	–	–	–	
3.0 Superstructure						
.1	Columns & Beams	Steel columns with sprayed fiber fireproofing	L.F. Column	101	1.20	
.4	Structural Walls	N/A	–	–	–	
.5	Elevated Floors	Concrete slab with metal deck and beams	S.F. Floor	13.64	9.09	22.8%
.7	Roof	Metal deck, open web steel joists, beams, columns	S.F. Roof	5.36	1.79	
.9	Stairs	Concrete filled metal pan	Flight	6800	.72	
4.0 Exterior Closure						
.1	Walls	Face brick with concrete block backup 90% of wall	S.F. Wall	19.59	6.37	
.5	Exterior Wall Finishes	N/A	–	–	–	15.0%
.6	Doors	Revolving and sliding panel, mall-front	Each	6858	1.01	
.7	Windows & Glazed Walls	Storefront 10% of wall	S.F. Window	28	1.03	
5.0 Roofing						
.1	Roof Coverings	Built-up tar and gravel with flashing	S.F. Roof	2.01	.67	
.7	Insulation	Perlite/EPS composite	S.F. Roof	1.25	.42	2.1%
.8	Openings & Specialties	Gravel stop and hatches	S.F. Roof	.24	.08	
6.0 Interior Construction						
.1	Partitions	Gypsum board on metal studs 60 S.F. Floor/L.F. Partition	S.F. Partition	3.90	.65	
.4	Interior Doors	Single leaf hollow metal 600 S.F. Floor/Door	Each	526	.88	
.5	Wall Finishes	70% paint, 20% vinyl wall covering, 10% ceramic tile	S.F. Surface	1.14	.38	19.9%
.6	Floor Finishes	50% carpet, 40% vinyl tile, 10% terrazzo	S.F. Floor	5.59	5.59	
.7	Ceiling Finishes	Mineral fiber tile on concealed zee bars	S.F. Ceiling	3.07	3.07	
.9	Interior Surface/Exterior Wall	Paint 90% of wall	S.F. Wall	1.95	.63	
7.0 Conveying						
.1	Elevators	One hydraulic passenger, one hydraulic freight	Each	135,850	1.43	9.4%
.2	Special Conveyors	Four escalators	Each	–	3.83	
8.0 Mechanical						
.1	Plumbing	Toilet and service fixtures, supply and drainage 1 Fixture/2570 S.F. Floor	Each	1567	.61	
.2	Fire Protection	Sprinklers, light hazard	S.F. Floor	1.17	1.17	
.3	Heating	Included in 8.4	–	–	–	14.5%
.4	Cooling	Multizone rooftop unit, gas heating, electric cooling	S.F. Floor	6.46	6.46	
.5	Special Systems	N/A	–	–	–	
9.0 Electrical						
.1	Service & Distribution	1200 ampere service, panel board and feeders	S.F. Floor	.77	.77	
.2	Lighting & Power	Fluorescent fixtures, receptacles, switches, A.C. and misc. power	S.F. Floor	5.37	5.37	11.9%
.4	Special Electrical	Alarm systems and emergency lighting	S.F. Floor	.52	.52	
11.0 Special Construction						
.1	Specialties	N/A	–	–	–	0.0%
12.0 Site Work						
.1	Earthwork	N/A	–	–	–	
.3	Utilities	N/A	–	–	–	0.0%
.5	Roads & Parking	N/A	–	–	–	
.7	Site Improvements	N/A	–	–	–	
			Sub-Total		56.20	**100%**
	GENERAL CONDITIONS (Overhead & Profit)			15%	8.43	
	ARCHITECT FEES			6%	3.87	
			Total Building Cost		**68.50**	

Costs per square foot of floor area

Exterior Wall	S.F. Area	4000	6000	8000	10000	12000	15000	18000	20000	22000
	L.F. Perimeter	260	340	360	410	440	490	540	565	594
Split Face Concrete Block	Steel Joists	78.20	71.50	65.45	62.95	60.65	58.45	57.05	56.15	55.50
Stucco on Concrete Block	Steel Joists	74.80	68.55	63.10	60.75	58.70	56.70	55.45	54.65	54.05
Painted Concrete Block	Steel Joists	73.25	66.90	61.40	59.05	56.95	55.00	53.70	52.85	52.25
Face Brick on Concrete Block	Steel Joists	86.90	79.10	71.45	68.40	65.50	62.80	61.05	59.90	59.05
Painted Reinforced Concrete	Steel Joists	82.15	74.95	68.20	65.40	62.85	60.45	58.85	57.80	57.10
Tilt-up Concrete Panels	Steel Joists	74.05	67.90	62.60	60.30	58.25	56.35	55.10	54.30	53.70
Perimeter Adj., Add or Deduct	Per 100 L.F.	8.90	5.95	4.45	3.55	3.00	2.40	1.95	1.80	1.65
Story Hgt. Adj., Add or Deduct	Per 1 Ft.	1.15	1.00	.80	.75	.70	.60	.55	.50	.50

For Basement, add $19.65 per square foot of basement area

The above costs were calculated using the basic specifications shown on the facing page. These costs should be adjusted where necessary for design alternatives and owner's requirements. Reported completed project costs, for this type of structure, range from $33.30 to $87.75 per S.F.

Common additives

Description	Unit	$ Cost
Emergency Lighting, 25 watt, battery operated		
Lead battery	Each	330
Nickel cadmium	Each	620
Safe, Office type, 4 hour rating		
30" x 18" x 18"	Each	3475
62" x 33" x 20"	Each	7575
Smoke Detectors		
Ceiling type	Each	142
Duct type	Each	390
Sound System		
Amplifier, 250 watts	Each	1525
Speaker, ceiling or wall	Each	131
Trumpet	Each	246

Important: See the Reference Section for Location Factors

BUILDING TYPES

Model costs calculated for a 1 story building with 14' story height and 8,000 square feet of floor area

				Unit	Unit Cost	Cost Per S.F.	% Of Sub-Total
1.0 Foundations							
.1	Footings & Foundations	Poured concrete; strip and spread footings and 4' foundation wall		S.F. Ground	2.94	2.94	
.4	Piles & Caissons	N/A		–	–	–	7.7%
.9	Excavation & Backfill	Site preparation for slab and trench for foundation wall and footing		S.F. Ground	1.11	1.11	
2.0 Substructure							
.1	Slab on Grade	4" reinforced concrete with vapor barrier and granular base		S.F. Slab	3.06	3.06	5.8%
.2	Special Substructures	N/A		–	–	–	
3.0 Superstructure							
.1	Columns & Beams	Interior columns included in 3.7		–	–	–	
.4	Structural Walls	N/A		–	–	–	
.5	Elevated Floors	N/A		–	–	–	6.8%
.7	Roof	Metal deck, open web steel joists, beams, interior columns		S.F. Roof	3.56	3.56	
.9	Stairs	N/A		–	–	–	
4.0 Exterior Closure							
.1	Walls	Decorative concrete block	90% of wall	S.F. Wall	10.99	6.23	
.5	Exterior Wall Finishes	N/A		–	–	–	16.0%
.6	Doors	Sliding entrance door and hollow metal service doors		Each	1581	.39	
.7	Windows & Glazed Walls	Storefront windows	10% of wall	S.F. Window	28	1.79	
5.0 Roofing							
.1	Roof Coverings	Built-up tar and gravel with flashing		S.F. Roof	2.25	2.25	
.7	Insulation	Perlite/EPS composite		S.F. Roof	1.25	1.25	7.2%
.8	Openings & Specialties	Gravel stop and hatches		S.F. Roof	.31	.31	
6.0 Interior Construction							
.1	Partitions	Gypsum board on metal studs	60 S.F. Floor/L.F. Partition	S.F. Partition	3.90	.65	
.4	Interior Doors	Single leaf hollow metal	600 S.F. Floor/Door	Each	526	.88	
.5	Wall Finishes	Paint		S.F. Surface	.51	.17	16.2%
.6	Floor Finishes	Vinyl tile		S.F. Floor	2.66	2.66	
.7	Ceiling Finishes	Mineral fiber tile on concealed zee bars		S.F. Ceiling	3.07	3.07	
.9	Interior Surface/Exterior Wall	Paint	90% of wall	S.F. Wall	1.95	1.11	
7.0 Conveying							
.1	Elevators	N/A		–	–	–	0.0%
.2	Special Conveyors	N/A		–	–	–	
8.0 Mechanical							
.1	Plumbing	Toilet and service fixtures, supply and drainage	1 Fixture/890 S.F. Floor	Each	3889	4.37	
.2	Fire Protection	Wet pipe sprinkler system		S.F. Floor	2.13	2.13	
.3	Heating	Included in 8.4		–	–	–	24.5%
.4	Cooling	Single zone unit, gas heating, electric cooling		S.F. Floor	6.46	6.46	
.5	Special Systems	N/A		–	–	–	
9.0 Electrical							
.1	Service & Distribution	400 ampere service, panel board and feeders		S.F. Floor	1.55	1.55	
.2	Lighting & Power	Fluorescent fixtures, receptacles, switches, A.C. and misc. power		S.F. Floor	6.29	6.29	15.8%
.4	Special Electrical	Alarm systems and emergency lighting		S.F. Floor	.48	.48	
11.0 Special Construction							
.1	Specialties	N/A		–	–	–	0.0%
12.0 Site Work							
.1	Earthwork	N/A		–	–	–	
.3	Utilities	N/A		–	–	–	
.5	Roads & Parking	N/A		–	–	–	0.0%
.7	Site Improvements	N/A		–	–	–	
				Sub-Total		52.71	**100%**
	GENERAL CONDITIONS (Overhead & Profit)				15%	7.91	
	ARCHITECT FEES				8%	4.83	
				Total Building Cost		65.45	

Costs per square foot of floor area

Exterior Wall	S.F. Area	6000	8000	12000	16000	20000	25000	32000	38000	45000
	L.F. Perimeter	**310**	**360**	**460**	**520**	**600**	**656**	**773**	**806**	**900**
Face Brick with Concrete Block Back-up	Steel Frame	87.45	81.25	75.00	70.30	68.10	65.20	63.50	61.30	60.35
	Bearing Walls	86.00	79.75	73.55	68.80	66.65	63.75	62.00	59.85	58.90
Stucco on Concrete Block	Steel Frame	78.10	73.10	68.10	64.40	62.70	60.45	59.10	57.50	56.75
	Bearing Walls	78.70	73.70	68.65	65.00	63.25	61.05	59.70	58.05	57.30
Precast Concrete Panels	Steel Frame	83.80	78.05	72.30	68.00	66.00	63.35	61.75	59.80	58.95
Metal Sandwich Panels	Steel Frame	72.85	68.50	64.15	61.10	59.65	57.80	56.65	55.30	54.70
Perimeter Adj., Add or Deduct	Per 100 L.F.	10.80	8.10	5.45	4.05	3.25	2.60	2.00	1.75	1.40
Story Hgt. Adj., Add or Deduct	Per 1 Ft.	1.55	1.35	1.15	1.00	.85	.80	.75	.65	.60

For Basement, add $14.20 per square foot of basement area

The above costs were calculated using the basic specifications shown on the facing page. These costs should be adjusted where necessary for design alternatives and owner's requirements. Reported completed project costs, for this type of structure, range from $39.00 to $85.05 per S.F.

Common additives

Description	Unit	$ Cost
Check Out Counter		
Single belt	Each	2175
Double belt	Each	3675
Scanner, registers, guns & memory 2 lanes	Each	13,200
10 lanes	Each	125,500
Power take away	Each	4575
Emergency Lighting, 25 watt, battery operated		
Lead battery	Each	330
Nickel cadmium	Each	620
Refrigerators, Prefabricated, walk-in		
7'-6" High, 6' x 6'	S.F.	117
10' x 10'	S.F.	91
12' x 14'	S.F.	81
12' x 20'	S.F.	72
Refrigerated Food Cases		
Dairy, multi deck, 12' long	Each	6975
Delicatessen case, single deck, 12' long	Each	4925
Multi deck, 18 S.F. shelf display	Each	7425
Freezer, self-contained chest type, 30 C.F.	Each	2950
Glass door upright, 78 C.F.	Each	6525

Description	Unit	$ Cost
Refrigerated Food Cases, cont.		
Frozen food, chest type, 12' long	Each	4775
Glass door reach in, 5 door	Each	9025
Island case 12' long, single deck	Each	5400
Multi deck	Each	11,100
Meat case 12' long, single deck	Each	3925
Multi deck	Each	6700
Produce, 12' long single deck	Each	5175
Multi deck	Each	5775
Safe, Office type, 4 hour rating		
30" x 18" x 18"	Each	3475
62" x 33" x 20"	Each	7575
Smoke Detectors		
Ceiling type	Each	142
Duct type	Each	390
Sound System		
Amplifier, 250 watts	Each	1525
Speaker, ceiling or wall	Each	131
Trumpet	Each	246

Model costs calculated for a 1 story building with 18' story height and 20,000 square feet of floor area

				Unit	Unit Cost	Cost Per S.F.	% Of Sub-Total
1.0 Foundations							
.1	Footings & Foundations	Poured concrete; strip and spread footings and 4' foundation wall		S.F. Ground	2.25	2.25	
.4	Piles & Caissons	N/A		–	–	–	6.1%
.9	Excavation & Backfill	Site preparation for slab and trench for foundation wall and footing		S.F. Ground	1.04	1.04	
2.0 Substructure							
.1	Slab on Grade	4" reinforced concrete with vapor barrier and granular base		S.F. Slab	3.06	3.06	5.6%
.2	Special Substructures	N/A		–	–	–	
3.0 Superstructure							
.1	Columns & Beams	Interior columns included in 3.7		–	–	–	
.4	Structural Walls	N/A		–	–	–	
.5	Elevated Floors	N/A		–	–	–	7.2%
.7	Roof	Metal deck, open web steel joists, beams, interior columns		S.F. Roof	3.92	3.92	
.9	Stairs	N/A		–	–	–	
4.0 Exterior Closure							
.1	Walls	Face brick with concrete block backup	85% of wall	S.F. Wall	19.46	8.93	
.5	Exterior Wall Finishes	N/A		–	–	–	25.8%
.6	Doors	Sliding entrance doors with electrical operator, hollow metal		Each	3072	1.69	
.7	Windows & Glazed Walls	Storefront windows	15% of wall	S.F. Window	41	3.35	
5.0 Roofing							
.1	Roof Coverings	Built-up tar and gravel with flashing		S.F. Roof	2.02	2.02	
.7	Insulation	Perlite/EPS composite		S.F. Roof	1.25	1.25	6.6%
.8	Openings & Specialties	Gravel stop and hatches		S.F. Roof	.32	.32	
6.0 Interior Construction							
.1	Partitions	50% concrete block, 50% gypsum board on metal studs	50 S.F. Floor/L.F. Partition	S.F. Partition	4.54	1.09	
.4	Interior Doors	Single leaf hollow metal	2000 S.F. Floor/Door	Each	526	.26	
.5	Wall Finishes	Paint		S.F. Surface	.71	.34	15.4%
.6	Floor Finishes	Vinyl composition tile		S.F. Floor	2.66	2.66	
.7	Ceiling Finishes	Mineral fiber tile on concealed zee bars		S.F. Ceiling	3.07	3.07	
.9	Interior Surface/Exterior Wall	Paint	85% of wall	S.F. Wall	1.95	.90	
7.0 Conveying							
.1	Elevators	N/A		–	–	–	0.0%
.2	Special Conveyors	N/A		–	–	–	
8.0 Mechanical							
.1	Plumbing	Toilet and service fixtures, supply and drainage	1 Fixture/1820 S.F. Floor	Each	3185	1.75	
.2	Fire Protection	Sprinklers, light hazard		S.F. Floor	1.60	1.60	
.3	Heating	Included in 8.4		–	–	–	17.8%
.4	Cooling	Single zone rooftop unit, gas heating, electric cooling		S.F. Floor	6.27	6.27	
.5	Special Systems	N/A		–	–	–	
9.0 Electrical							
.1	Service & Distribution	400 ampere service, panel board and feeders		S.F. Floor	.85	.85	
.2	Lighting & Power	Fluorescent fixtures, receptacles, switches, A.C. and misc. power		S.F. Floor	7.19	7.19	15.5%
.4	Special Electrical	Alarm systems and emergency lighting		S.F. Floor	.35	.35	
11.0 Special Construction							
.1	Specialties	N/A		–	–	–	0.0%
12.0 Site Work							
.1	Earthwork	N/A		–	–	–	
.3	Utilities	N/A		–	–	–	0.0%
.5	Roads & Parking	N/A		–	–	–	
.7	Site Improvements	N/A		–	–	–	
				Sub-Total		54.16	**100%**
	GENERAL CONDITIONS (Overhead & Profit)			15%		8.12	
	ARCHITECT FEES			7%		4.37	
				Total Building Cost		66.65	

BUILDING TYPES

Costs per square foot of floor area

Exterior Wall	S.F. Area	16000	18000	20000	22000	24000	26000	28000	30000	32000
	L.F. Perimeter	560	560	600	640	680	673	706	740	737
Face Brick with Concrete Block Back-up	Wood Truss	115.50	111.50	110.15	109.10	108.15	105.70	104.90	104.25	102.60
	Precast Conc.	132.80	128.55	127.10	125.95	125.00	122.35	121.50	120.85	119.10
Metal Sandwich Panel	Wood Truss	118.25	115.50	114.50	113.75	113.05	111.40	110.85	110.40	109.25
Wood Siding	Wood Frame	121.30	118.35	117.30	116.45	115.75	113.90	113.30	112.85	111.60
Painted Concrete Block	Wood Frame	120.55	117.65	116.60	115.80	115.10	113.35	112.75	112.30	111.10
	Precast Conc.	122.60	119.45	118.35	117.50	116.75	114.80	114.15	113.65	112.35
Perimeter Adj., Add or Deduct	Per 100 L.F.	5.90	5.25	4.75	4.25	3.95	3.60	3.35	3.15	2.95
Story Hgt. Adj., Add or Deduct	Per 1 Ft.	1.00	.90	.85	.80	.80	.75	.70	.70	.65

For Basement, add $21.35 per square foot of basement area

The above costs were calculated using the basic specifications shown on the facing page. These costs should be adjusted where necessary for design alternatives and owner's requirements. Reported completed project costs, for this type of structure, range from $61.50 to $180.75 per S.F.

Common additives

Description	Unit	$ Cost
Bleachers, Telescoping, manual		
To 15 tier	Seat	76 - 105
16-20 tier	Seat	155 - 190
21-30 tier	Seat	165 - 199
For power operation, add	Seat	30 - 47
Emergency Lighting, 25 watt, battery operated		
Lead battery	Each	330
Nickel cadmium	Each	620
Lockers, Steel, single tier, 60" or 72"	Opening	135 - 216
2 tier, 60" or 72" total	Opening	83 - 105
5 tier, box lockers	Opening	44 - 57
Locker bench, lam. maple top only	L.F.	18.65
Pedestal, steel pipe	Each	43
Pool Equipment		
Diving stand, 3 meter	Each	7125
1 meter	Each	4875
Diving board, 16' aluminum	Each	2300
Fiberglass	Each	1775
Lifeguard chair, fixed	Each	1275
Portable	Each	2950
Lights, underwater, 12 volt, 300 watt	Each	1100

Description	Unit	$ Cost
Sauna, Prefabricated, complete		
6' x 4'	Each	3725
6' x 6'	Each	4400
6' x 9'	Each	5850
8' x 8'	Each	6225
8' x 10'	Each	6850
10' x 12'	Each	8900
Sound System		
Amplifier, 250 watts	Each	1525
Speaker, ceiling or wall	Each	131
Trumpet	Each	246
Steam Bath, Complete, to 140 C.F.	Each	980
To 300 C.F.	Each	1175
To 800 C.F.	Each	3050
To 2500 C.F.	Each	3400

Important: See the Reference Section for Location Factors

BUILDING TYPES

Model costs calculated for a 1 story building with 24' story height and 20,000 square feet of floor area

Swimming Pool, Enclosed

				Unit	Unit Cost	Cost Per S.F.	% Of Sub-Total
1.0 Foundations							
.1	Footings & Foundations	Poured concrete; strip and spread footings and 4' foundation wall		S.F. Ground	3.98	3.98	
.4	Piles & Caissons	N/A		–	–	–	12.2%
.9	Excavation & Backfill	Site preparation for slab and trench for foundation wall and footing		S.F. Ground	6.85	6.85	
2.0 Substructure							
.1	Slab on Grade	4" reinforced concrete with vapor barrier and granular base		S.F. Slab	3.06	3.06	6.2%
.2	Special Substructures	Pool walls		S.F. Ground	2.42	2.42	
3.0 Superstructure							
.1	Columns & Beams	Wood columns included in 3.7		–	–	–	
.4	Structural Walls	N/A		–	–	–	
.5	Elevated Floors	N/A		–	–	–	13.4%
.7	Roof	Wood deck on laminated wood truss		S.F. Ground	11.91	11.91	
.9	Stairs	N/A		–	–	–	
4.0 Exterior Closure							
.1	Walls	Face brick with concrete block backup	80% of wall	S.F. Wall	19.60	11.29	
.5	Exterior Wall Finishes	N/A		–	–	–	17.8%
.6	Doors	Double aluminum and glass, hollow metal		Each	1449	.37	
.7	Windows & Glazed Walls	Outward projecting steel	20% of wall	S.F. Window	28	4.11	
5.0 Roofing							
.1	Roof Coverings	Asphalt shingles with flashing		S.F. Ground	1.03	1.03	
.7	Insulation	Perlite/EPS composite		S.F. Ground	–	–	1.3%
.8	Openings & Specialties	Gutters and downspouts		S.F. Ground	.16	.16	
6.0 Interior Construction							
.1	Partitions	Concrete block, toilet partitions	100 S.F. Floor/L.F. Partition	S.F. Partition	5.14	.66	
.4	Interior Doors	Single leaf hollow metal	1000 S.F. Floor/Door	Each	362	.36	
.5	Wall Finishes	Acrylic glazed coating		S.F. Surface	1.50	.30	17.9%
.6	Floor Finishes	70% terrazzo, 30% ceramic tile		S.F. Floor	13.17	13.17	
.7	Ceiling Finishes	Mineral fiber tile on concealed zee bars	20% of area	S.F. Ceiling	1.20	.24	
.9	Interior Surface/Exterior Wall	Paint	80% of wall	S.F. Wall	1.95	1.12	
7.0 Conveying							
.1	Elevators	N/A		–	–	–	0.0%
.2	Special Conveyors	N/A		–	–	–	
8.0 Mechanical							
.1	Plumbing	Toilet and service fixtures, supply and drainage	1 Fixture/210 S.F. Floor	Each	2032	9.68	
.2	Fire Protection	N/A		–	–	–	
.3	Heating	Terminal unit heaters	10% of area	–	.50	.50	23.9%
.4	Cooling	Single zone unit gas heating, electric cooling		S.F. Floor	11.03	11.03	
.5	Special Systems	N/A		–	–	–	
9.0 Electrical							
.1	Service & Distribution	200 ampere service, panel board and feeders		S.F. Floor	.64	.64	
.2	Lighting & Power	Fluorescent fixtures, receptacles, switches, A.C. and misc. power		S.F. Floor	5.50	5.50	7.3%
.4	Special Electrical	Alarm systems and emergency lighting		S.F. Floor	.30	.30	
11.0 Special Construction							
.1	Specialties	N/A		–	–	–	0.0%
12.0 Site Work							
.1	Earthwork	N/A		–	–	–	
.3	Utilities	N/A		–	–	–	0.0%
.5	Roads & Parking	N/A		–	–	–	
.7	Site Improvements	N/A		–	–	–	
				Sub-Total		88.68	**100%**
	GENERAL CONDITIONS (Overhead & Profit)			15%		13.30	
	ARCHITECT FEES			8%		8.17	
			Total Building Cost			**110.15**	

Costs per square foot of floor area

Exterior Wall	S.F. Area	2000	3000	4000	5000	6000	7000	8000	9000	10000
	L.F. Perimeter	180	220	260	286	320	353	368	397	425
Face Brick with Concrete Block Back-up	Steel Frame	112.75	101.00	95.05	90.05	87.40	85.45	82.80	81.55	80.50
	Bearing Walls	111.30	99.50	93.60	88.60	85.95	84.00	81.35	80.10	79.05
Limestone with Concrete Block Back-up	Steel Frame	125.25	111.15	104.05	98.00	94.80	92.45	89.20	87.65	86.40
	Bearing Walls	123.75	109.70	102.60	96.50	93.35	91.00	87.70	86.20	84.95
Decorative Concrete Block	Steel Frame	104.95	94.60	89.40	85.10	82.80	81.05	78.80	77.70	76.80
	Bearing Walls	103.45	93.15	87.95	83.60	81.30	79.60	77.35	76.25	75.35
Perimeter Adj., Add or Deduct	Per 100 L.F.	26.40	17.60	13.15	10.55	8.80	7.50	6.60	5.85	5.25
Story Hgt. Adj., Add or Deduct	Per 1 Ft.	3.15	2.55	2.25	2.00	1.85	1.75	1.60	1.50	1.45

For Basement, add $20.40 per square foot of basement area

The above costs were calculated using the basic specifications shown on the facing page. These costs should be adjusted where necessary for design alternatives and owner's requirements. Reported completed project costs, for this type of structure, range from $53.85 to $190.15 per S.F.

Common additives

Description	Unit	$ Cost
Emergency Lighting, 25 watt, battery operated		
Lead battery	Each	330
Nickel cadmium	Each	620
Smoke Detectors		
Ceiling type	Each	142
Duct type	Each	390
Emergency Generators, complete system, gas		
15 kw	Each	12,700
85 kw	Each	28,000
170 kw	Each	71,500
Diesel, 50 kw	Each	24,100
150 kw	Each	42,600
350 kw	Each	64,500

BUILDING TYPES

Model costs calculated for a 1 story building with 12' story height and 5,000 square feet of floor area

				Unit	Unit Cost	Cost Per S.F.	% Of Sub-Total
1.0 Foundations							
.1	Footings & Foundations	Poured concrete; strip and spread footings and 4' foundation wall		S.F. Ground	4.23	4.23	
.4	Piles & Caissons	N/A		–	–	–	7.8%
.9	Excavation & Backfill	Site preparation for slab and trench for foundation wall and footing		S.F. Ground	1.24	1.24	
2.0 Substructure							
.1	Slab on Grade	4" reinforced concrete with vapor barrier and granular base		S.F. Slab	3.06	3.06	4.3%
.2	Special Substructures	N/A		–			
3.0 Superstructure							
.1	Columns & Beams	Steel columns included in 3.7		–	–	–	
.4	Structural Walls	N/A		–	–	–	
.5	Elevated Floors	N/A		–	–	–	6.2%
.7	Roof	Metal deck, open web steel joists, beams, columns		S.F. Roof	4.39	4.39	
.9	Stairs	N/A		–	–	–	
4.0 Exterior Closure							
.1	Walls	Face brick with concrete block backup	80% of wall	S.F. Wall	19.45	10.68	
.5	Exterior Wall Finishes	N/A		–	–	–	
.6	Doors	Single aluminum glass with transom		Each	2246	.90	26.0%
.7	Windows & Glazed Walls	Outward projecting steel	20% of wall	Each	988	6.78	
5.0 Roofing							
.1	Roof Coverings	Built-up tar and gravel with flashing		S.F. Roof	2.44	2.44	
.7	Insulation	Perlite/EPS composite		S.F. Roof	1.25	1.25	5.7%
.8	Openings & Specialties	Gravel stop and hatches		S.F. Roof	.32	.32	
6.0 Interior Construction							
.1	Partitions	Double layer gypsum board on metal studs, toilet partitions	15 S.F. Floor/L.F. Partition	S.F. Partition	5.31	3.54	
.4	Interior Doors	Single leaf hollow metal	150 S.F. Floor/Door	Each	526	3.51	
.5	Wall Finishes	Paint		S.F. Surface	.52	.69	23.7%
.6	Floor Finishes	90% carpet, 10% terrazzo		S.F. Floor	5.55	5.55	
.7	Ceiling Finishes	Fiberglass board on exposed grid system, suspended		S.F. Ceiling	2.37	2.37	
.9	Interior Surface/Exterior Wall	Paint	80% of wall	S.F. Floor	1.95	1.07	
7.0 Conveying							
.1	Elevators	N/A		–	–	–	0.0%
.2	Special Conveyors	N/A		–	–	–	
8.0 Mechanical							
.1	Plumbing	Kitchen, toilet and service fixtures, supply and drainage	1 Fixture/715 S.F. Floor	Each	2809	3.93	
.2	Fire Protection	Wet pipe sprinkler system		S.F. Floor	2.27	2.27	
.3	Heating	Included in 8.4		–	–	–	18.2%
.4	Cooling	Single zone unit, gas heating, electric cooling		S.F. Floor	6.59	6.59	
.5	Special Systems	N/A		–	–	–	
9.0 Electrical							
.1	Service & Distribution	200 ampere service, panel board and feeders		S.F. Floor	1.07	1.07	
.2	Lighting & Power	Fluorescent fixtures, receptacles, switches, A.C. and misc. power		S.F. Floor	3.63	3.63	8.1%
.4	Special Electrical	Alarm systems and emergency lighting		S.F. Floor	1.04	1.04	
11.0 Special Construction							
.1	Specialties	N/A		–	–	–	0.0%
12.0 Site Work							
.1	Earthwork	N/A		–	–	–	
.3	Utilities	N/A		–	–	–	
.5	Roads & Parking	N/A		–	–	–	0.0%
.7	Site Improvements	N/A		–	–	–	
				Sub-Total		70.55	**100%**
	GENERAL CONDITIONS (Overhead & Profit)				15%	10.58	
	ARCHITECT FEES				11%	8.92	
				Total Building Cost		90.05	

Costs per square foot of floor area

Exterior Wall	S.F. Area	5000	6500	8000	9500	11000	14000	17500	21000	24000
	L.F. Perimeter	300	360	386	396	435	510	550	620	680
Face Brick with Concrete Block Back-up	Steel Joists	89.95	86.50	82.50	79.05	77.65	75.70	73.05	71.95	71.25
	Wood Joists	89.05	85.55	81.40	77.85	76.45	74.45	71.75	70.60	69.85
Stone with Concrete Block Back-up	Steel Joists	91.15	87.65	83.50	79.90	78.45	76.45	73.70	72.55	71.80
	Wood Joists	90.25	86.65	82.40	78.70	77.25	75.20	72.40	71.20	70.45
Brick Veneer	Wood Frame	86.25	83.05	79.40	76.20	74.95	73.15	70.75	69.70	69.05
Wood Siding	Wood Frame	82.25	79.35	76.20	73.45	72.30	70.70	68.65	67.75	67.20
Perimeter Adj., Add or Deduct	Per 100 L.F.	8.85	6.85	5.50	4.65	4.05	3.15	2.55	2.10	1.85
Story Hgt. Adj., Add or Deduct	Per 1 Ft.	1.65	1.50	1.30	1.10	1.10	1.00	.85	.80	.75

For Basement, add $17.05 per square foot of basement area

The above costs were calculated using the basic specifications shown on the facing page. These costs should be adjusted where necessary for design alternatives and owner's requirements. Reported completed project costs, for this type of structure, range from $47.95 to $135.30 per S.F.

Common additives

Description	Unit	$ Cost
Directory Boards, Plastic, glass covered		
30" x 20"	Each	510
36" x 48"	Each	930
Aluminum, 24" x 18"	Each	450
36" x 24"	Each	540
48" x 32"	Each	640
48" x 60"	Each	1450
Emergency Lighting, 25 watt, battery operated		
Lead battery	Each	330
Nickel cadmium	Each	620
Flagpoles, Complete		
Aluminum, 20' high	Each	1100
40' high	Each	2475
70' high	Each	7100
Fiberglass, 23' high	Each	1475
39'-5" high	Each	2525
59' high	Each	7025
Safe, Office type, 4 hour rating		
30" x 18" x 18"	Each	3475
62" x 33" x 20"	Each	7575

Description	Unit	$ Cost
Smoke Detectors		
Ceiling type	Each	142
Duct type	Each	390
Vault Front, Door & frame		
1 Hour test, 32" x 78"	Opening	3425
2 Hour test, 32" door	Opening	3525
40" door	Opening	4050
4 Hour test, 32" door	Opening	3675
40" door	Opening	4375
Time lock movement; two movement	Each	1575

BUILDING TYPES

Model costs calculated for a 1 story building with 12' story height and 11,000 square feet of floor area

Town Hall, 1 Story

				Unit	Unit Cost	Cost Per S.F.	% Of Sub-Total
1.0 Foundations							
.1	Footings & Foundations	Poured concrete; strip and spread footings and 4' foundation wall		S.F. Ground	2.79	2.79	
.4	Piles & Caissons	N/A		—	—	—	6.3%
.9	Excavation & Backfill	Site preparation for slab and trench for foundation wall and footing		S.F. Ground	1.11	1.11	
2.0 Substructure							
.1	Slab on Grade	4" reinforced concrete with vapor barrier and granular base		S.F. Slab	3.06	3.06	4.9%
.2	Special Substructures	N/A		—	—	—	
3.0 Superstructure							
.1	Columns & Beams	Steel columns included in 3.7		—	—	—	
.4	Structural Walls	N/A		—	—	—	
.5	Elevated Floors	N/A		—	—	—	6.1%
.7	Roof	Metal deck, open web steel joists, beams, interior columns		S.F. Roof	3.75	3.75	
.9	Stairs	N/A		—	—	—	
4.0 Exterior Closure							
.1	Walls	Face brick with concrete block backup	70% of wall	S.F. Wall	19.45	6.46	
.5	Exterior Wall Finishes	N/A		—	—	—	16.0%
.6	Doors	Metal and glass with transom		Each	1626	.59	
.7	Windows & Glazed Walls	Metal outward projecting	30% of wall	Each	461	2.85	
5.0 Roofing							
.1	Roof Coverings	Built-up tar and gravel with flashing		S.F. Roof	2.30	2.30	
.7	Insulation	Perlite/EPS composite		S.F. Roof	1.25	1.25	6.1%
.8	Openings & Specialties	Gravel stop and hatches		S.F. Roof	.22	.22	
6.0 Interior Construction							
.1	Partitions	Gypsum board on metal studs, toilet partition	20 S.F. Floor/L.F. Partition	S.F. Partition	3.89	2.35	
.4	Interior Doors	Wood solid core	200 S.F. Floor/Door	Each	408	2.04	
.5	Wall Finishes	90% paint, 10% ceramic tile		S.F. Surface	.99	.99	26.3%
.6	Floor Finishes	70% carpet, 15% terrazzo, 15% vinyl composition tile		S.F. Floor	6.89	6.89	
.7	Ceiling Finishes	Mineral fiber tile on concealed zee bars		S.F. Ceiling	3.07	3.07	
.9	Interior Surface/Exterior Wall	Painted gypsum board on furring	70% of wall	S.F. Wall	2.86	.95	
7.0 Conveying							
.1	Elevators	N/A		—	—	—	0.0%
.2	Special Conveyors	N/A		—	—	—	
8.0 Mechanical							
.1	Plumbing	Kitchen, toilet and service fixtures, supply and drainage	1 Fixture/500 S.F. Floor	Each	2415	4.83	
.2	Fire Protection	Wet pipe sprinkler system		S.F. Floor	1.60	1.60	
.3	Heating	Included in 8.4		—	—	—	21.7%
.4	Cooling	Multizone unit, gas heating, electric cooling		S.F. Floor	7.03	7.03	
.5	Special Systems	N/A		—	—	—	
9.0 Electrical							
.1	Service & Distribution	400 ampere service, panel board and feeders		S.F. Floor	1.13	1.13	
.2	Lighting & Power	Fluorescent fixtures, receptacles, switches, A.C. and misc. power		S.F. Floor	6.14	6.14	12.6%
.4	Special Electrical	Alarm systems and emergency lighting		S.F. Floor	.56	.56	
11.0 Special Construction							
.1	Specialties	N/A		—	—	—	0.0%
12.0 Site Work							
.1	Earthwork	N/A		—	—	—	
.3	Utilities	N/A		—	—	—	
.5	Roads & Parking	N/A		—	—	—	0.0%
.7	Site Improvements	N/A		—	—	—	

	Sub-Total	61.96	100%
GENERAL CONDITIONS (Overhead & Profit)	15%	9.29	
ARCHITECT FEES	9%	6.40	
Total Building Cost		77.65	

BUILDING TYPES

211

Costs per square foot of floor area

Exterior Wall	S.F. Area	8000	10000	12000	15000	18000	24000	28000	35000	40000
	L.F. Perimeter	206	233	260	300	320	360	393	451	493
Face Brick with Concrete Block Back-up	Steel Frame	106.35	101.25	97.95	94.50	91.00	86.70	85.05	83.15	82.15
	R/Conc. Frame	107.45	102.35	99.00	95.60	92.10	87.75	86.15	84.20	83.25
Stone with Concrete Block Back-up	Steel Frame	101.75	97.30	94.35	91.35	88.25	84.45	83.00	81.30	80.45
	R/Conc. Frame	109.00	103.75	100.30	96.80	93.15	88.65	87.00	85.00	84.00
Limestone with Concrete Block Back-up	Steel Frame	115.55	109.60	105.65	101.65	97.35	92.05	90.10	87.75	86.55
	R/Conc. Frame	116.65	110.65	106.75	102.70	98.45	93.10	91.15	88.80	87.65
Perimeter Adj., Add or Deduct	Per 100 L.F.	13.65	10.95	9.05	7.25	6.05	4.55	3.90	3.10	2.75
Story Hgt. Adj., Add or Deduct	Per 1 Ft.	2.10	1.90	1.75	1.65	1.45	1.20	1.15	1.05	1.00

For Basement, add $16.55 per square foot of basement area

The above costs were calculated using the basic specifications shown on the facing page. These costs should be adjusted where necessary for design alternatives and owner's requirements. Reported completed project costs, for this type of structure, range from $47.95 to $135.30 per S.F.

Common additives

Description	Unit	$ Cost
Directory Boards, Plastic, glass covered		
30" x 20"	Each	510
36" x 48"	Each	930
Aluminum, 24" x 18"	Each	450
36" x 24"	Each	540
48" x 32"	Each	640
48" x 60"	Each	1450
Elevators, Hydraulic passenger, 2 stops		
1500# capacity	Each	40,000
2500# capacity	Each	41,300
3500# capacity	Each	44,900
Additional stop, add	Each	3350
Emergency Lighting, 25 watt, battery operated		
Lead battery	Each	330
Nickel cadmium	Each	620

Description	Unit	$ Cost
Flagpoles, Complete		
Aluminum, 20' high	Each	1100
40' high	Each	2475
70' high	Each	7100
Fiberglass, 23' high	Each	1475
39'-5" high	Each	2525
59' high	Each	7025
Safe, Office type, 4 hour rating		
30" x 18" x 18"	Each	3475
62" x 33" x 20"	Each	7575
Smoke Detectors		
Ceiling type	Each	142
Duct type	Each	390
Vault Front, Door & frame		
1 Hour test, 32" x 78"	Opening	3425
2 Hour test, 32" door	Opening	3525
40" door	Opening	4050
4 Hour test, 32" door	Opening	3675
40" door	Opening	4375
Time lock movement; two movement	Each	1575

Important: See the Reference Section for Location Factor

BUILDING TYPES

Model costs calculated for a 3 story building with 12' story height and 18,000 square feet of floor area

BUILDING TYPES

				Unit	Unit Cost	Cost Per S.F.	% Of Sub-Total
1.0 Foundations							
.1	Footings & Foundations	Poured concrete; strip and spread footings and 4' foundation wall		S.F. Ground	5.10	1.70	
.4	Piles & Caissons	N/A		—	—	—	2.9%
.9	Excavation & Backfill	Site preparation for slab and trench for foundation wall and footing		S.F. Ground	1.04	.35	
2.0 Substructure							
.1	Slab on Grade	4" reinforced concrete with vapor barrier and granular base		S.F. Slab	3.06	1.02	1.4%
.2	Special Substructures	N/A		—	—	—	
3.0 Superstructure							
.1	Columns & Beams	Wide flange steel columns with gypsum board fireproofing		L.F. Column	48	1.45	
.4	Structural Walls	N/A		—	—	—	
.5	Elevated Floors	Open web steel joists, slab form, concrete		S.F. Floor	11.62	7.75	17.9%
.7	Roof	Metal deck, open web steel joists, beams, interior columns		S.F. Roof	4.39	1.46	
.9	Stairs	Concrete filled metal pan		Flight	5900	1.97	
4.0 Exterior Closure							
.1	Walls	Stone with concrete block backup	70% of wall	S.F. Wall	21	9.56	
.5	Exterior Wall Finishes	N/A		—	—	—	
.6	Doors	Metal and glass with transoms		Each	1499	.42	16.2%
.7	Windows & Glazed Walls	Metal outward projecting	10% of wall	Each	634	1.46	
5.0 Roofing							
.1	Roof Coverings	Built-up tar and gravel with flashing		S.F. Roof	2.58	.86	
.7	Insulation	Perlite/EPS composite		S.F. Roof	1.25	.42	1.8%
.8	Openings & Specialties	N/A		—	—	—	
6.0 Interior Construction							
.1	Partitions	Gypsum board on metal studs, toilet partitions	20 S.F. Floor/L.F. Partition	S.F. Partition	3.89	2.19	
.4	Interior Doors	Wood solid core	200 S.F. Floor/Door	Each	408	2.04	
.5	Wall Finishes	90% paint, 10% ceramic tile		S.F. Surface	.99	.99	
.6	Floor Finishes	70% carpet, 15% terrazzo, 15% vinyl composition tile		S.F. Floor	6.89	6.89	23.4%
.7	Ceiling Finishes	Mineral fiber tile on concealed zee bars		S.F. Ceiling	3.07	3.07	
.9	Interior Surface/Exterior Wall	Painted gypsum board on furring	70% of wall	S.F. Wall	2.86	1.28	
7.0 Conveying							
.1	Elevators	Two hydraulic elevators		Each	63,360	7.04	10.0%
.2	Special Conveyors	N/A		—	—	—	
8.0 Mechanical							
.1	Plumbing	Toilet and service fixtures, supply and drainage	1 Fixture/1385 S.F. Floor	Each	2160	1.56	
.2	Fire Protection	Sprinklers, light hazard		S.F. Floor	1.54	1.54	
.3	Heating	Included in 8.4		—	—	—	14.5%
.4	Cooling	Multizone unit, gas heating, electric cooling		S.F. Floor	7.03	7.03	
.5	Special Systems	N/A		—	—	—	
9.0 Electrical							
.1	Service & Distribution	400 ampere service, panel board and feeders		S.F. Floor	.95	.95	
.2	Lighting & Power	Fluorescent fixtures, receptacles, switches, A.C. and misc. power		S.F. Floor	6.83	6.83	11.9%
.4	Special Electrical	Alarm systems and emergency lighting		S.F. Floor	.59	.59	
11.0 Special Construction							
.1	Specialties	N/A		—	—	—	0.0%
12.0 Site Work							
.1	Earthwork	N/A		—	—	—	
.3	Utilities	N/A		—	—	—	
.5	Roads & Parking	N/A		—	—	—	0.0%
.7	Site Improvements	N/A		—	—	—	
				Sub-Total		70.42	**100%**
	GENERAL CONDITIONS (Overhead & Profit)				15%	10.56	
	ARCHITECT FEES				9%	7.27	
				Total Building Cost		88.25	

Costs per square foot of floor area

Exterior Wall	S.F. Area	10000	15000	20000	25000	30000	35000	40000	50000	60000
	L.F. Perimeter	410	500	600	700	700	766	833	966	1000
Brick with Concrete Block Back-up	Steel Frame	71.55	64.20	60.90	58.90	55.15	53.85	52.90	51.55	49.40
	Bearing Walls	70.95	63.40	60.00	57.95	54.15	52.80	51.80	50.45	48.25
Concrete Block	Steel Frame	58.65	53.65	51.35	50.00	47.70	46.85	46.20	45.35	44.05
	Bearing Walls	57.75	52.70	50.35	48.95	46.65	45.75	45.15	44.25	42.95
Galvanized Steel Siding	Steel Frame	58.80	54.15	52.00	50.70	48.65	47.85	47.25	46.50	45.30
Metal Sandwich Panels	Steel Frame	60.10	54.85	52.45	50.95	48.55	47.65	46.95	46.05	44.65
Perimeter Adj., Add or Deduct	Per 100 L.F.	7.30	4.85	3.65	2.90	2.45	2.10	1.80	1.45	1.20
Story Hgt. Adj., Add or Deduct	Per 1 Ft.	1.05	.85	.75	.70	.60	.55	.50	.45	.40

For Basement, add $15.95 per square foot of basement area

The above costs were calculated using the basic specifications shown on the facing page. These costs should be adjusted where necessary for design alternatives and owner's requirements. Reported completed project costs, for this type of structure, range from $20.35 to $81.50 per S.F.

Common additives

Description	Unit	$ Cost
Dock Leveler, 10 ton cap.		
6' x 8'	Each	4050
7' x 8'	Each	4275
Emergency Lighting, 25 watt, battery operated		
Lead battery	Each	330
Nickel cadmium	Each	620
Fence, Chain link, 6' high		
9 ga. wire	L.F.	13.65
6 ga. wire	L.F.	17.80
Gate	Each	247
Flagpoles, Complete		
Aluminum, 20' high	Each	1100
40' high	Each	2475
70' high	Each	7100
Fiberglass, 23' high	Each	1475
39'-5" high	Each	2525
59' high	Each	7025
Paving, Bituminous		
Wearing course plus base course	S.Y.	9.75
Sidewalks, Concrete 4" thick	S.F.	2.55

Description	Unit	$ Cost
Sound System		
Amplifier, 250 watts	Each	1525
Speaker, ceiling or wall	Each	131
Trumpet	Each	246
Yard Lighting, 20' aluminum pole with 400 watt high pressure sodium fixture.	Each	2085

Important: See the Reference Section for Location Factors

BUILDING TYPES

Model costs calculated for a 1 story building with 24' story height and 30,000 square feet of floor area

			Unit	Unit Cost	Cost Per S.F.	% Of Sub-Total
1.0 Foundations						
.1	Footings & Foundations	Poured concrete; strip and spread footings and 4' foundation wall	S.F. Ground	2.22	2.22	
.4	Piles & Caissons	N/A	–	–	–	8.6%
.9	Excavation & Backfill	Site preparation for slab and trench for foundation wall and footing	S.F. Ground	1.04	1.04	
2.0 Substructure						
.1	Slab on Grade	5" reinforced concrete with vapor barrier and granular base	S.F. Slab	6.12	6.12	16.1%
.2	Special Substructures	N/A	–	–	–	
3.0 Superstructure						
.1	Columns & Beams	Steel columns included in 3.5 and 3.7	–	–	–	
.4	Structural Walls	N/A		–	–	
.5	Elevated Floors	Open web steel joists, slab form, concrete beams, columns, (above office) 10% of area	S.F. Floor	8.92	1.16	13.1%
.7	Roof	Metal deck, open web steel joists, beams, columns	S.F. Roof	3.50	3.50	
.9	Stairs	Steel gate with rails	Flight	4750	.32	
4.0 Exterior Closure						
.1	Walls	Concrete block 95% of wall	S.F. Wall	7.91	4.21	
.5	Exterior Wall Finishes	N/A	–	–	–	12.6%
.6	Doors	Steel overhead, hollow metal 5% of wall	Each	1587	.58	
.7	Windows & Glazed Walls	N/A	–	–	–	
5.0 Roofing						
.1	Roof Coverings	Built-up tar and gravel with flashing	S.F. Roof	1.91	1.91	
.7	Insulation	Perlite/EPS composite	S.F. Roof	1.25	1.25	9.2%
.8	Openings & Specialties	Gravel stop, hatches and skylight	S.F. Roof	.31	.31	
6.0 Interior Construction						
.1	Partitions	Concrete block (office and washrooms) 100 S.F. Floor/L.F. Partition	S.F. Partition	5.13	.41	
.4	Interior Doors	Single leaf hollow metal 5000 S.F. Floor/Door	Each	526	.11	
.5	Wall Finishes	Paint	S.F. Surface	.94	.15	
.6	Floor Finishes	90% hardener, 10% vinyl composition tile	S.F. Floor	.99	.99	7.9%
.7	Ceiling Finishes	Suspended mineral tile on zee channels in office area 10% of area	S.F. Ceiling	3.07	.31	
.9	Interior Surface/Exterior Wall	Paint 95% of wall	S.F. Wall	1.95	1.04	
7.0 Conveying						
.1	Elevators	N/A	–	–	–	0.0%
.2	Special Conveyors	N/A	–	–	–	
8.0 Mechanical						
.1	Plumbing	Toilet and service fixtures, supply and drainage 1 Fixture/2500 S.F. Floor	Each	3500	1.40	
.2	Fire Protection	Sprinklers, ordinary hazard	S.F. Floor	1.79	1.79	
.3	Heating	Oil fired hot water, unit heaters 90% of area	S.F. Floor	3.24	3.24	18.9%
.4	Cooling	Single zone unit gas, heating, electric cooling 10% of area	S.F. Floor	.70	.70	
.5	Special Systems	N/A	–	–	–	
9.0 Electrical						
.1	Service & Distribution	200 ampere service, panel board and feeders	S.F. Floor	.30	.30	
.2	Lighting & Power	Fluorescent fixtures, receptacles, switches, A.C. and misc. power	S.F. Floor	2.94	2.94	9.3%
.4	Special Electrical	Alarm systems	S.F. Floor	.28	.28	
11.0 Special Construction						
.1	Specialties	Dock boards, dock levelers	S.F. Floor	1.64	1.64	4.3%
12.0 Site Work						
.1	Earthwork	N/A	–	–	–	
.3	Utilities	N/A	–	–	–	
.5	Roads & Parking	N/A	–	–	–	0.0%
.7	Site Improvements	N/A	–	–	–	
			Sub-Total		37.92	**100%**
	GENERAL CONDITIONS (Overhead & Profit)		15%		5.69	
	ARCHITECT FEES		7%		3.04	
			Total Building Cost		**46.65**	

BUILDING TYPES

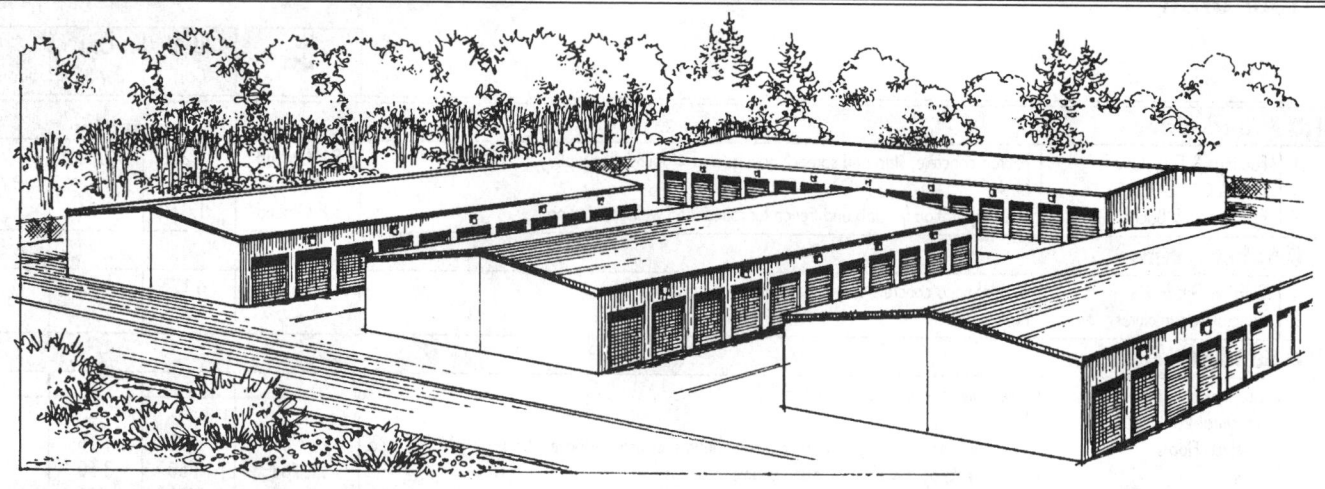

Costs per square foot of floor area

Exterior Wall	S.F. Area	2000	3000	5000	8000	12000	20000	30000	50000	100000
	L.F. Perimeter	179	220	284	350	438	566	693	894	1266
Concrete Block	Steel Frame	176.25	131.65	95.05	73.55	61.60	51.40	46.05	41.50	37.70
	R/Conc. Frame	90.20	82.70	75.75	70.85	68.10	65.30	63.60	61.95	60.35
Metal Sandwich Panel	Steel Frame	82.95	75.10	67.75	62.60	59.75	56.75	54.95	53.20	51.55
Tilt-up Concrete Panel	R/Conc. Frame	90.50	82.95	75.90	70.95	68.15	65.30	63.55	61.90	60.30
Precast Concrete Panel	Steel Frame	82.95	75.05	67.60	62.35	59.35	56.30	54.45	52.70	50.95
	Bearing Wall	95.40	86.85	78.75	73.05	69.80	66.45	64.35	62.40	60.45
Perimeter Adj., Add or Deduct	Per 100 L.F.	12.75	8.50	5.10	3.15	2.10	1.30	.85	.50	.25
Story Hgt. Adj., Add or Deduct	Per 1 Ft.	.75	.60	.45	.35	.30	.25	.20	.15	.10
Basement—Not Applicable										

The above costs were calculated using the basic specifications shown on the facing page. These costs should be adjusted where necessary for design alternatives and owner's requirements. Reported completed project costs, for this type of structure, range from $16.95 to $71.90 per S.F.

Common additives

Description	Unit	$ Cost
Dock Leveler, 10 ton cap.		
6' x 8'	Each	4050
7' x 8'	Each	4275
Emergency Lighting, 25 watt, battery operated		
Lead battery	Each	330
Nickel cadmium	Each	620
Fence, Chain link, 6' high		
9 ga. wire	L.F.	13.65
6 ga. wire	L.F.	17.80
Gate	Each	247
Flagpoles, Complete		
Aluminum, 20' high	Each	1100
40' high	Each	2475
70' high	Each	7100
Fiberglass, 23' high	Each	1475
39'-5" high	Each	2525
59' high	Each	7025
Paving, Bituminous		
Wearing course plus base course	S.Y.	9.75
Sidewalks, Concrete 4" thick	S.F.	2.55

Description	Unit	$ Cost
Sound System		
Amplifier, 250 watts	Each	1525
Speaker, ceiling or wall	Each	131
Trumpet	Each	246
Yard Lighting,	Each	2085
20' aluminum pole with 400 watt high pressure sodium fixture		

Important: See the Reference Section for Location Factor

Model costs calculated for a 1 story building with 12' story height and 20,000 square feet of floor area

			Unit	Unit Cost	Cost Per S.F.	% Of Sub-Total
1.0 Foundations						
.1	Footings & Foundations	Poured concrete; strip and spread footings and 4' foundation wall	S.F. Ground	.65	.65	
.4	Piles & Caissons	N/A	–	–	–	4.2%
.9	Excavation & Backfill	Site preparation for slab and trench for foundation wall and footing	S.F. Ground	1.11	1.11	
2.0 Substructure						
.1	Slab on Grade	4" reinforced concrete with vapor barrier and granular base	S.F. Slab	5.16	5.16	12.4%
.2	Special Substructures	N/A	–	–	–	
3.0 Superstructure						
.1	Columns & Beams	Column fireproofing, steel columns included in 3.7	S.F. Floor	2.46	2.46	
.4	Structural Walls	N/A	–	–	–	
.5	Elevated Floors	N/A	–	–	–	18.7%
.7	Roof	Metal deck, open web steel joists, beams, columns	S.F. Roof	5.36	5.36	
.9	Stairs	N/A	–	–	–	
4.0 Exterior Closure						
.1	Walls	Concrete block 70% of wall	S.F. Wall	9.63	2.29	
.5	Exterior Wall Finishes	N/A	–	–	–	
.6	Doors	Steel overhead, hollow metal 25% of wall	Each	801	6.49	21.6%
.7	Windows & Glazed Walls	Aluminum projecting 5% of wall	Each	592	.24	
5.0 Roofing						
.1	Roof Coverings	Built-up tar and gravel with flashing	S.F. Roof	1.93	1.93	
.7	Insulation	Perlite/EPS composite	S.F. Roof	.78	.78	6.5%
.8	Openings & Specialties	N/A	–	–	–	
6.0 Interior Construction						
.1	Partitions	Concrete block, gypsum board on metal studs 10.65 S.F. Floor/L.F. Partition	S.F. Partition	7.95	5.67	
.4	Interior Doors	Single leaf hollow metal 1835 S.F. Floor/Door	Each	66	.22	
.5	Wall Finishes	Paint	S.F. Floor	.40	.40	15.7%
.6	Floor Finishes	Carpet, asphalt tile, ceramic tile	S.F. Floor	.25	.25	
.7	Ceiling Finishes	N/A	–	–	–	
.9	Interior Surface/Exterior Wall	N/A	–	–	–	
7.0 Conveying						
.1	Elevators	N/A	–	–	–	0.0%
.2	Special Conveyors	N/A	–	–	–	
8.0 Mechanical						
.1	Plumbing	Toilet and service fixtures, supply and drainage 1 Fixture/5000 S.F. Floor	Each	5100	1.02	
.2	Fire Protection	Wet pipe sprinkler system	S.F. Floor	2.13	2.13	
.3	Heating	Oil fired hot water, unit heaters	S.F. Floor	11,157	.56	8.8%
.4	Cooling	N/A	–	–	–	
.5	Special Systems	N/A	–	–	–	
9.0 Electrical						
.1	Service & Distribution	200 ampere service, panel board and feeders	S.F. Floor	.96	.96	
.2	Lighting & Power	Fluorescent fixtures, receptacles, switches and misc. power	S.F. Floor	3.23	3.23	10.7%
.4	Special Electrical	Alarm systems	S.F. Floor	.26	.26	
11.0 Special Construction						
.1	Specialties	Bathroom access, appliances, cabinets	S.F. Floor	.60	.60	1.4%
12.0 Site Work						
.1	Earthwork	N/A	–	–	–	
.3	Utilities	N/A	–	–	–	
.5	Roads & Parking	N/A	–	–	–	0.0%
.7	Site Improvements	N/A	–	–	–	
		Sub-Total			41.77	100%
	GENERAL CONDITIONS (Overhead & Profit)			15%	6.27	
	ARCHITECT FEES			7%	3.36	
		Total Building Cost			51.40	

BUILDING TYPES

Deterioration occurs within the structure itself and is determined by the observation of both materials and equipment.

Curable deterioration can be remedied either by maintenance, repair or replacement, within prudent economic limits.

Incurable deterioration that has progressed to the point of actually affecting the structural integrity of the structure, making repair or replacement not economically feasible.

Actual Versus Observed Age

The observed age of a structure refers to the age that the structure appears to be. Periodic maintenance, remodeling and renovation all tend to reduce the amount of deterioration that has taken place, thereby decreasing the observed age. Actual age on the other hand relates solely to the year that the structure was built.

The Depreciation Table shown here relates to the observed age.

Obsolescence arises from conditions either occurring within the structure (functional) or caused by factors outside the limits of the structure (economic).

Functional obsolescence is any inadequacy caused by outmoded design, dated construction materials or oversized or undersized areas, all of which cause excessive operation costs.

Incurable is so costly as not to economically justify the capital expenditure required to correct the deficiency.

Economic obsolescence is caused by factors outside the limits of the structure. The prime causes of economic obsolescence are:

- zoning and environmental laws
- government legislation
- negative neighborhood influences
- business climate
- proximity to transportation facilities

Depreciation Table
Commercial/Industrial/Institutional

	Building Material		
Observed Age (Years)	Frame	Masonry On Wood	Masonry On Masonry or Steel
1	1%	0%	0%
2	2	1	0
3	3	2	1
4	4	3	2
5	6	5	3
10	20	15	8
15	25	20	15
20	30	25	20
25	35	30	25
30	40	35	30
35	45	40	35
40	50	45	40
45	55	50	45
50	60	55	50
55	65	60	55
60	70	65	60

DEPRECIATION

Assemblies Section

Table of Contents

Table of Contents

Introduction to the Assemblies Section

This section of the manual contains the technical description and installed cost of all components used in the commercial/industrial/institutional section. In most cases, there is an illustration of the component to aid in identification.

All Costs Include:

Materials purchased in lot sizes typical of normal construction projects with discounts appropriate to established contractors.

Installation work performed by union labor working under standard conditions at a normal pace.

Installing Contractor's Overhead & Profit

Standard installing contractor's overhead & profit are included in the assemblies costs.

These assemblies costs contain no provisions for General Contractor's overhead and profit or Architect's fees, nor do they include premiums for labor or materials or savings which may be realized under certain economic situations.

The assemblies tables are arranged by physical size (dimensions) of the component wherever possible.

How to Use the Assemblies Cost Tables

The following is a detailed explanation of a sample Assemblies Cost Table. Most Assembly Tables are separated into two parts: 1) an illustration of the system to be estimated with descriptive notes; and 2) the costs for similar systems with dimensional and/or size variations. Next to each bold number below is the item being described with the appropriate component of the sample entry following in parenthesis. In most cases, if the work is to be subcontracted, the general contractor will need to add an additional markup (R.S. Means suggests using 10%) to the "Total" figures.

1 System/Line Numbers (A2.1-200-2240)

Each Assemblies Cost Line has been assigned a unique identification number based on the UniFormat classification system.

UniFormat Division

2.1 200 2240

Means Subdivision
Means Major Classification
Means Individual Line Number

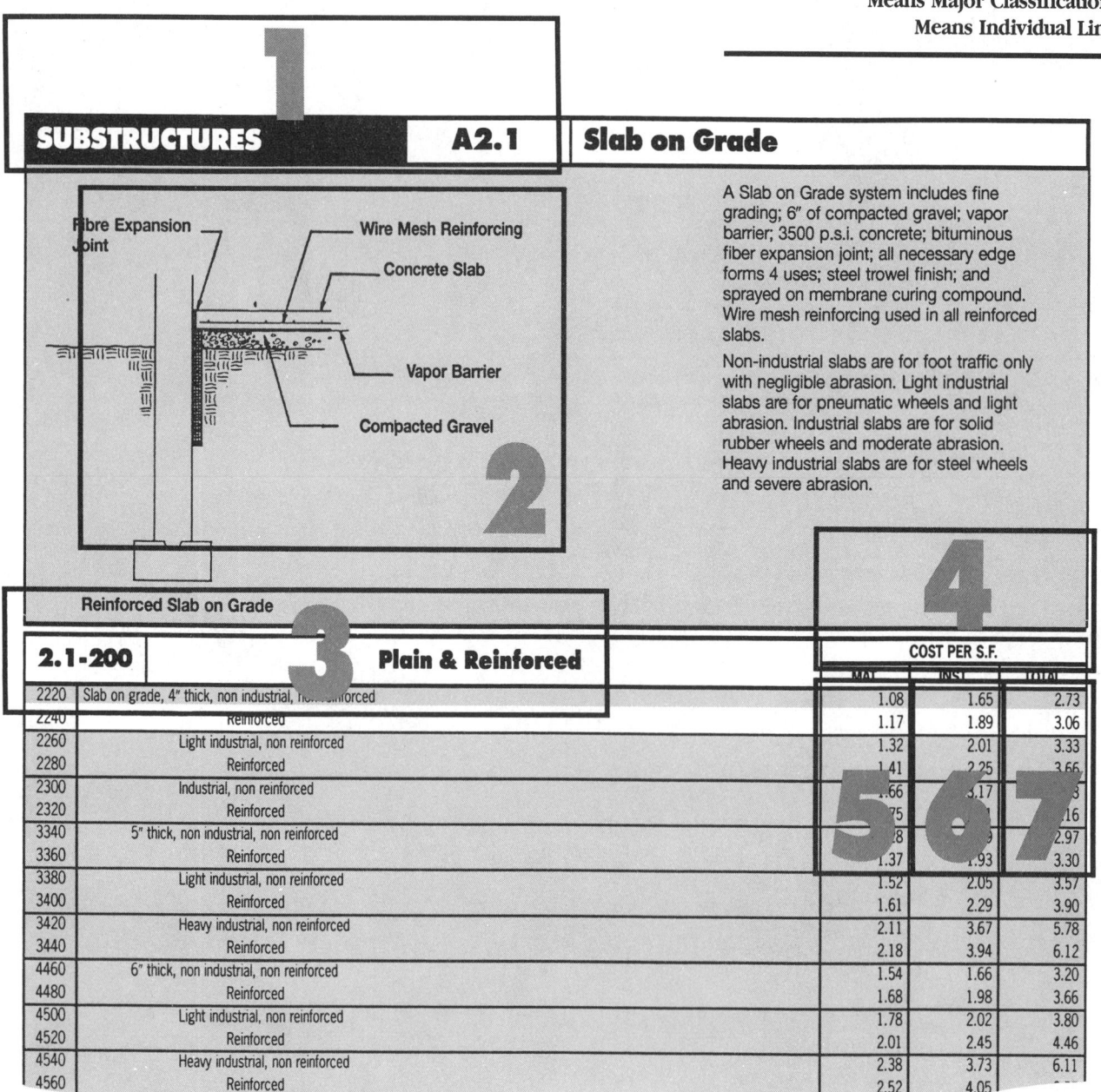

| SUBSTRUCTURES | A2.1 | Slab on Grade |

A Slab on Grade system includes fine grading; 6" of compacted gravel; vapor barrier; 3500 p.s.i. concrete; bituminous fiber expansion joint; all necessary edge forms 4 uses; steel trowel finish; and sprayed on membrane curing compound. Wire mesh reinforcing used in all reinforced slabs.

Non-industrial slabs are for foot traffic only with negligible abrasion. Light industrial slabs are for pneumatic wheels and light abrasion. Industrial slabs are for solid rubber wheels and moderate abrasion. Heavy industrial slabs are for steel wheels and severe abrasion.

Reinforced Slab on Grade

2.1-200	Plain & Reinforced	COST PER S.F.		
		MAT.	INST.	TOTAL
2220	Slab on grade, 4" thick, non industrial, non reinforced	1.08	1.65	2.73
2240	Reinforced	1.17	1.89	3.06
2260	Light industrial, non reinforced	1.32	2.01	3.33
2280	Reinforced	1.41	2.25	3.66
2300	Industrial, non reinforced	1.66	3.17	
2320	Reinforced	.75		.16
3340	5" thick, non industrial, non reinforced	8	9	2.97
3360	Reinforced	1.37	1.93	3.30
3380	Light industrial, non reinforced	1.52	2.05	3.57
3400	Reinforced	1.61	2.29	3.90
3420	Heavy industrial, non reinforced	2.11	3.67	5.78
3440	Reinforced	2.18	3.94	6.12
4460	6" thick, non industrial, non reinforced	1.54	1.66	3.20
4480	Reinforced	1.68	1.98	3.66
4500	Light industrial, non reinforced	1.78	2.02	3.80
4520	Reinforced	2.01	2.45	4.46
4540	Heavy industrial, non reinforced	2.38	3.73	6.11
4560	Reinforced	2.52	4.05	

Illustration

At the top of most assembly pages is an illustration, a brief description, and the design criteria used to develop the cost.

System Description

The components of a typical system are listed in the description to show what has been included in the development of the total system price. The rest of the table contains prices for other similar systems with dimensional and/or size variations.

Unit of Measure for Each System (S.F.)

Costs shown in the three right hand columns have been adjusted by the component quantity and unit of measure for the entire system. In this example, "Cost per S.F." is the unit of measure for this system or "assembly."

Materials
(1.17)

This column contains the Materials Cost of each component. These cost figures are bare costs plus 10% for profit.

Installation
(1.89)

Installation includes labor and equipment plus the installing contractor's overhead and profit. Equipment costs are the bare rental costs plus 10% for profit. The labor overhead and profit is defined on the inside back cover of this book.

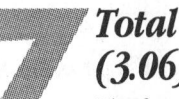

Total
(3.06)

The figure in this column is the sum of the material and installation costs.

Material Cost	+	Installation Cost	=	Total
$1.17	+	$1.89	=	$3.06

Example

Assemblies costs can be used to calculate the cost for each component of a building and, accumulated with appropriate markups, produce a cost for the complete structure. Components from a model building with similar characteristics in conjunction with Assemblies components may be used to calculate costs for a complete structure.

Example:

Outline Specifications

General	Building size—60' x 100', 8 suspended floors, 12' floor to floor, 4' high parapet above roof, full basement 11'-8" floor to floor, bay size 25' x 30', ceiling heights - 9' in office area and 8' in core area.
1.0 Foundations	Concrete, spread and strip footings and concrete walls waterproofed.
2.0 Substructures	4" concrete slab on grade.
3.0 Superstructures	Columns, wide flange; 3 hr. fire rated; Floors, composite steel frame & deck with concrete slab; Roof, steel beams, open web joists and deck.
4.0 Exterior Closure	Walls; North, East & West, brick & lightweight concrete block with 2" cavity insulation, 25% window; South, 8" lightweight concrete block insulated, 10% window. Doors, aluminum & glass. Windows, aluminum, 3'-0" x 5'-4", insulating glass.
5.0 Roofing	Tar & gravel, 2" rigid insulation, R 12.5.
6.0 Interior Construction	Core—6" lightweight concrete block, full height with plaster on exposed faces to ceiling height.
	Corridors—1st & 2nd floor; 3-5/8" steel studs with F.R. gypsum board, full height.
	Exterior Wall—plaster, ceiling height.
	Doors—hollow metal.
	Wall Finishes—lobby, mahogany paneling on furring, remainder plaster & gypsum board, paint.
	Floor Finishes—1st floor lobby, corridors & toilet rooms, terrazzo; remainder, concrete, tenant developed. 2nd thru 8th, toilet rooms, ceramic tile; office & corridor, carpet.
	Ceiling Finishes—24" x 48" fiberglass board on Tee grid.
7.0 Conveying Systems	2-2500 lb. capacity, 200 F.P.M. geared elevators, 9 stops.
8.0 Mechanical	Fixtures, see sketch.
	Roof Drains—2-4" C.I. pipe.
	Fire Protection—4" standpipe, 9 hose cabinets.
	Heating—fin tube radiation, forced hot water.
	Air Conditioning—chilled water with water cooled condenser.
9.0 Electrical	Lighting, 1st thru 8th, 15 fluorescent fixtures/1000 S.F., 3 Watts/S.F.
	Basement, 10 fluorescent fixtures/1000 S.F., 2 Watts/S.F.
	Receptacles, 1st thru 8th, 16.5/1000 S.F., 2 Watts/S.F.
	Basement, 10 receptacles/1000 S.F., 1.2 Watts/S.F.
	Air Conditioning, 4 Watts/S.F.
	Miscellaneous Connections, 1.2 Watts/S.F.
	Elevator Power, 2-10 H.P., 230 volt motors.
	Wall Switches, 2/1000 S.F.
	Service, panel board & feeder, 2000 Amp.
	Fire Detection System, pull stations, signals, smoke and heat detectors.
	Emergency Lighting Generator, 30 KW.
10.0 General Conditions & Profit	See Summary Sheet.
11.0 Special Construction	Toilet Accessories & Directory Boards.
12.0 Site Work	NA.

Front Elevation

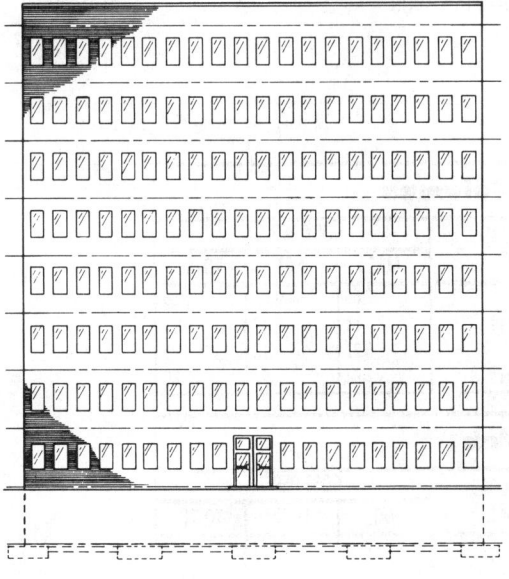

Basement Plan

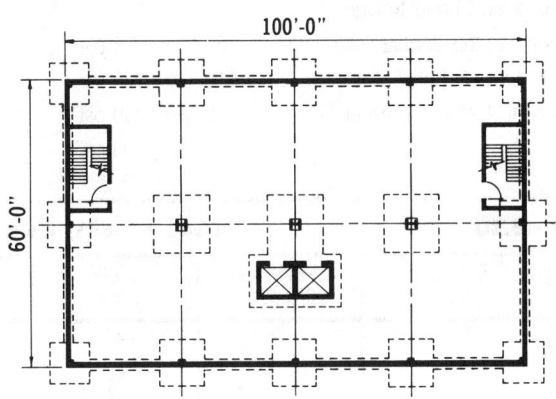

Typical Floor Plan

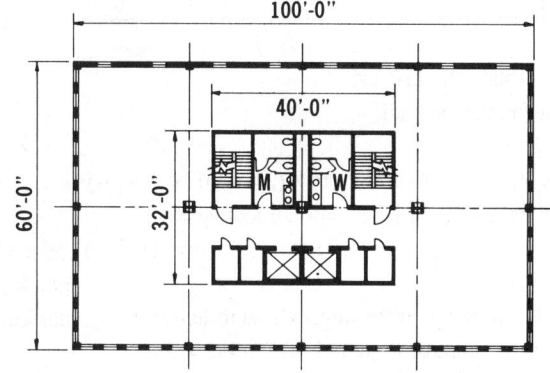

Ground Floor Plan

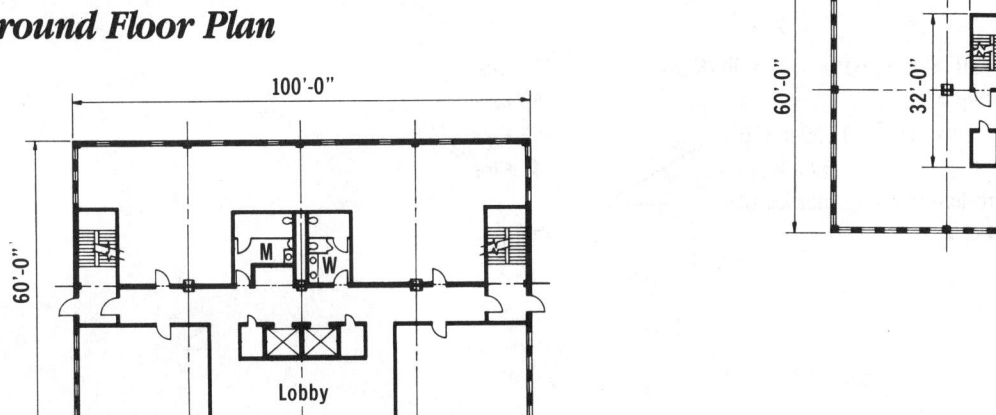

Example

If spread footing & column sizes are unknown, develop approximate loads as follows. Enter tables with these loads to determine costs.

Superimposed Load Ranges

Apartments & Residential Structures	65	to	75 psf
Assembly Areas & Retail Stores	110	to	125 psf
Commercial & Manufacturing	150	to	250 psf
Offices	75	to	100 psf

Approximate loads/S.F. for roof & floors.

Roof. Assume 40 psf superimposed load.

Steel joists, beams & deck.

Table 3.7-420—Line 3900

3.7-420		Steel Joists, Beams, & Deck on Columns						
	BAY SIZE (FT.)	SUPERIMPOSED LOAD (P.S.F.)	DEPTH (IN.)	TOTAL LOAD (P.S.F.)	COLUMN ADD	COST PER S.F.		
						MAT.	INST.	TOTAL
3500	25x30	20	22	40		2.66	1	3.66
3600					columns	.54	.17	.71
3900		40	25	60		3.21	1.18	4.39
4000					columns	.64	.21	.85

3.5-540		Composite Beams, Deck & Slab						
	BAY SIZE (FT.)	SUPERIMPOSED LOAD (P.S.F.)	SLAB THICKNESS (IN.)	TOTAL DEPTH (FT. - IN.)	TOTAL LOAD (P.S.F.)	COST PER S.F.		
						MAT.	INST.	TOTAL
3400	25x30	40	5-1/2	1 - 11-1/2	83	6	3.64	9.64
3600		75	5-1/2	1 - 11-1/2	119	6.45	3.69	10.14
3900		125	5-1/2	1 - 11-1/2	170	7.45	4.17	11.62
4000		200	6-1/4	2 - 6-1/4	252	9.35	4.74	14.09

Floors—Total load, 119 psf.

Interior foundation load.

Roof

$[(25' \times 30' \times 60 \text{ psf}) + 8 \text{ floors} \times (25' \times 30' \times 119 \text{ psf})] \times 1/1000 \text{ lb./Kip} =$	759 Kips
Approximate Footing Loads, Interior footing =	759 Kips
Exterior footing (1/2 bay) 759 k × .6 =	455 Kips
Corner footing (1/4 bay) 759 k × .45 =	342 Kips
[Factors to convert Interior load to Exterior & Corner loads]	
Approximate average Column load 759 k/2 =	379 Kips

Means Forms

PRELIMINARY ESTIMATE

PROJECT *Office Building*	TOTAL SITE AREA
BUILDING TYPE	OWNER
LOCATION	ARCHITECT
DATE OF CONSTRUCTION	ESTIMATED CONSTRUCTION PERIOD

BRIEF DESCRIPTION *Building size: 60' x 100' - 8 structural floors - 12' floor to floor*

4' high parapet above roof, full basement 11' - 8" floor to floor, bay size

25' x 30', ceiling heights - 9' in office area + 8' in core area

| TYPE OF PLAN | TYPE OF CONSTRUCTION |
| QUALITY | BUILDING CAPACITY |

Floor			**Wall Area**				
Below Grade Levels			Foundation Walls	L.F.		Ht.	S.F.
Area		S.F.	Frost Walls	L.F.		Ht.	S.F.
Area		S.F.	Exterior Closure		Total		S.F.
Total Area		S.F.	Comment				
Ground Floor			Fenestration		%		S.F.
Area		S.F.			%		S.F.
Area		S.F.	Exterior Wall		%		S.F.
Total Area		S.F.			%		S.F.
Supported Levels			**Site Work**				
Area			Parking		S.F. (For		Cars)
Area		S.F.	Access Roads		L.F. (X		Ft. Wide)
Area		S.F.	Sidewalk		L.F. (X		Ft. Wide)
Area		S.F.	Landscaping		S.F. (		% Unbuilt Site)
Area		S.F.	**Building Codes**				
Total Area		S.F.	City		Country		
Miscellaneous			National		Other		
Area		S.F.	**Loading**				
Area		S.F.	Roof	psf	Ground Floor		psf
Area		S.F.	Supported Floors	psf	Corridor		psf
Area		S.F.	Balcony	psf	Partition, allow		psf
Total Area		S.F.	Miscellaneous				psf
Net Finished Area		S.F.	Live Load Reduction				
Net Floor Area		S.F.	Wind				
Gross Floor Area	**54,000**	S.F.	Earthquake		Zone		
Roof			Comment				
Total Area		S.F.	Soil Type				
Comments			Bearing Capacity				K.S.F.
			Frost Depth				Ft.
Volume			**Frame**				
Depth of Floor System			Type		Bay Spacing		
Minimum		In.	Foundation				
Maximum		In.	Special				
Foundation Wall Height		Ft.	Substructure				
Floor to Floor Height		Ft.	Comment				
Floor to Ceiling Height		Ft.	Superstructure, Vertical				
Subgrade Volume		C.F.	Fireproofing		☐ Columns		Hrs.
Above Grade Volume		C.F.	☐ Girders	Hrs.	☐ Beams		Hrs.
Total Building Volume	**648,000**	C.F.	☐ Floor	Hrs.	☐ None		

ASSEMBLY NUMBER		QTY.	UNIT	TOTAL COST		COST PER S.F.
				UNIT	TOTAL	
1.0	**Foundations**					
1.1-120-7900	Corner Footings 8'-6" SQ. x 27"	4	Ea.	1140	4,560	
-8010	Exterior 9'-6" SQ. x 30"	8		1515	12,120	
-8300	Interior 12" SQ.	3		2775	8,325	
1.1-140-2700	Strip 2' Wide x 1' Thick					
	320 L.F. [(4 x 8.5) + 8 x 9.5)] =	210	L.F.	23.00	4,830	
1.1-210-7262	Foundation Wall 12' High, 1' Thick			140.50	29,505	
1.1-292-2800	Foundation Waterproofing			11.72	2,461	
1.9-100-3440	Building Excavation + Backfill	6,000	S.F.	4.24	25,440	
-3500	(Interpolated ; 12' Between					
-4620	8' and 16' ; 6,000 Between					
-4680	4,000 and 10,000 S.F.					
	Total				87,241	1.62
2.0	**Slab on Grade**					
2.1-200-2240	4", non-industrial, reinf.	6,000	S.F.	3.06	18,360	
	Total				18,360	.34
3.0	**Superstructure**					
3.1-130-5800	Columns: Load 379K → Use 400K					
	Exterior 12 x 96' + Interior 3 x 108'	1,476	V.L.F.	68.65	101,372	
3.1-190-3650	Column Fireproofing - Interior 4 sides					
-3700	Exterior 1/2 (Interpolated for 12")	900	V.L.F	21.52	19,368	
3.1-540-3600	Floors: Composite steel + lt. wt. conc.	48,000	S.F.	10.14	486,720	
3.7-420-3900	Roof: Open web joists, beams + decks	6,000	S.F.	4.39	26,340	
3.9-100-0760	Stairs: Steel w/conc. fill	18	Flt.	5900	106,200	
	Total				740,000	13.70
4.0	**Exterior Closure**					
4.1-273-1200	4" Brick + 6" Block - Insulated					
	75% NE + W Walls 220' x 100' x .75	16,500	S.F.	18.10	298,650	
4.1-211-3410	8" Block - Insulated 90%					
	100' x 100' x .9	9,000	S.F.	6.49	58,410	
4.6-100-6950	Double Aluminum + Glass Door	1	Pr.	3525	3,525	
-6300	Single	2	Ea.	1575	3,150	
4.7-110-8800	Aluminum windows - Insul. glass					
	(5,500 S.F. + 1,000 S.F.)/3 x 5.33	406	Ea.	622	252,532	
4.1-140-6776	Precast concrete coping	320	L.F.	25.25	8,080	
	Total				624,347	11.56
5.0	**Roofing**					
5.7-101-4020	Insulation: Perlite/Polyisocyanurate	6,000	S.F.	1.25	7,500	
5.1-103-1400	Roof: 4 Ply T & G composite	6,000		1.55	9,300	
5.1-620-0300	Flashing: Aluminum - fabric backed	640		2.09	1,338	
5.8-100-0500	Roof hatch	1	Ea.	757	757	
	Total				18,895	.35
	Page Total					

Means Forms

ASSEMBLY NUMBER		QTY.	UNIT	TOTAL COST UNIT	TOTAL COST TOTAL	COST PER S.F.
6.0	**Interior Construction**					
6.1-210-5500	Core partitions: 6" Lt. Wt. concrete					
	block 288 L.F. x 11.5' x 8 Floors	26,449	S.F.	5.14	135,948	
6.1-680-0920	Plaster: [(196 L.F. x 8') + (144 L.F. x 9')] 8					
	+ Ext. Wall [(320 x 9 x 8 Floors) - 6500]	39,452	S.F.	1.92	75,748	
6.1-510-5400	Corridor partitions 1st + 2nd floors					
	steel studs + F.R. gypsum board	2,530	S.F.	2.87	7,261	
6.4-100-2600	Interior doors	66	Ea.	526	34,716	
	Wall finishes - Lobby: Mahogany w/furring					
6.1-580-0652	Furring: 150 L.F. x 9' + Ht.	1,350	S.F.	.98	1,323	
6.5-100-1662	Mahogany paneling	1,350	S.F.	3.77	5,090	
	Paint, plaster + gypsum board					
6.5-100-0008	39,452 + [(220' x 9' x 2) - 1350] + 72	42,134	S.F.	.73	30,758	
	Floor Finishes					
6.6-100-1100	1st floor lobby + terrazzo	2175	S.F.	7.67	16,682	
	Remainder: Concrete - Tenant finished					
6.6-100-006	2nd - 8th: Carpet - 4725 S.F. x 7	33,075	S.F.	3.43	113,447	
6.6-100-1720	Ceramic Tile - 300 S.F. x 7	2,100	S.F.	6.90	14,490	
6.7-810-2780 3260	Suspended Ceiling - 5200 S.F. x 8	41,600	S.F.	2.03	84,448	
6.1-870-0700	Toilet partitions	31	Ea.	660	20,460	
-0760	Handicapped addition	16	Ea.	268	4,288	
	Total				544,659	10.09
7.0	**Conveying**					
	2 Elevators - 2500# 200'/min. Geared					
7.1-2001600	5 Floors $94,300					
-1800	15 Floors $177,000					
	Diff. $82,700/10 = 8,270/Flr.					
	$94,300 + (4 x 8,270) = $127,380/Ea.	2	Ea.	127,380	254,760	
	Total				254,760	4.72
8.0	**Mechanical System**					
8.1-433-2120	Plumbing - Lavatories	31	Ea.	870	26,970	
-434-4340	-Service Sink	8	Ea.	1490	11,920	
-450-2000	-Urinals	8	Ea.	955	7,640	
-470-2080	-Water Closets	31	Ea.	1095	33,945	
	Water Control, Waste Vent Piping	45%	OF TOTAL ABOVE		36,200	
-310-4200	- Roof Drains, 4" C.I.	2	Ea.	815	1,630	
-310-4240	- Pipe, 9 Flr. x 12' x 12' Ea.	216	L.F.	21.65	4,676	
	Fire Protection					
8.2-310-0560	Wet Stand Pipe: 4" x 10' 1st floor	12/10	Ea.	3850	4,620	
-310-0580	Additional Floors: (12'/10 x 8)	9.6	Ea.	1175	11,280	
-390-8400	Cabinet Assembly	9	Ea.	933	8,397	
	Heating - Fin Tube Radiation					
				Page Total		

ASSEMBLY NUMBER		QTY.	UNIT	TOTAL COST		COST PER S.F.
				UNIT	TOTAL	
8.3-161-2000	10,000 S.F. @ $5.76/S.F. ⌉ Interpolate					
-2040	10,000 S.F. @ $2.59/S.F. ⌡	48,000	S.F.	4.42	212,160	
	Cooling - Chilled Water, Air Cool, Cond.					
8.4-120-4000	40,000 S.F. @8.73/S.F. ⌉ Interpolate	48,000	S.F.	8.89	426,720	
-4040	60,000 S.F. @8.97/S.F. ⌡					
	Total				786,158	14.56
9.0	**Electrical**					
9.2-213-0280	Office Lighting, 15/1000 S.F. - 3 Watts/S.F.	48,000	S.F.	4.80	230,400	
-0240	Basement Lighting, 10/1000 S.F. - 2 Watts/S.F.	6,000	S.F.	3.19	19,140	
-522-0640	Office Receptacles 16.5/1000 S.F. - 2 Watts/S.F.	48,000	S.F.	2.71	130,080	
-0560	Basement Receptacles 10/1000 S.F. - 1.2 Watts/S.	6,000	S.F.	2.05	12,300	
-610-0280	Central A.C. - 4 Watts/S.F.	48,000	S.F.	.35	16,800	
-582-0320	Misc. Connections - 1.2 Watts/S.F.	48,000	S.F.	.19	9,120	
-710-0680	Elevator Motor Power - 10 H.P.	2	Ea.	1845	3,690	
-542-0280	Wall Switches - 2/1000 S.F.	54,000	S.F.	.26	14,040	
9.1-210-0560	2000 Amp Service	1	Ea.	22,125	22,125	
-410-0400	Switchgear	1	Ea.	35,700	35,700	
-310-0560	Feeder	50	L.F.	309	15,450	
9.1-100-0400	Fire Detection System - 50 Detectors	1	Ea.	20,500	20,500	
-310-0320	Emergency Generator - 30 kW	30	kW	505	15,150	
	Total				544,495	10.08
10.0	**General Conditions**					
11.0	**Special Construction**					
11.0-100-1100	Towel Dispenser	16	Ea.	66.00	1,056	
-1200	Grab Bar	32	Ea.	52.90	1,693	
-1300	Mirror	31	Ea.	116.15	3,601	
-1400	Toilet Tissue Dispenser	31	Ea.	21.65	671	
-2100	Directory Board	8	Ea.	181.50	1,452	
	Total				8,473	.16
12.0	**Site Work**					
13.0	**Miscellaneous**					
		Page Total				

 Means Forms

PRELIMINARY
ESTIMATE

PROJECT	*Office Building*	TOTAL AREA	**54,000 S.F.**	SHEET NO.
LOCATION		TOTAL VOLUME	**648,000 C.F.**	ESTIMATE NO.
ARCHITECT		COST PER S.F.		DATE
OWNER		COST PER C.F.		NO. OF STORIES
QUANTITIES BY		EXTENSIONS BY		CHECKED BY

NO.	DESCRIPTION	SUBTOTAL COST	COST/S.F.	%
1.0	Foundation	87,241	1.62	
2.0	Substructure	18,360	.34	
3.0	Superstructure	740,000	13.70	
4.0	Exterior Closure	624,347	11.56	
5.0	Roofing	18,895	.35	
6.0	Interior Construction	544,659	10.09	
7.0	Conveying	254,760	4.72	
8.0	Mechanical System	786,158	14.56	
9.0	Electrical	544,495	10.08	
10.0	General Conditions			
11.0	Special Construction	8,473	.16	
12.0	Site Work			

	Building Subtotal	3,627,388		$	3,627,388
Sales Tax N/A	% x Subtotal $	N/A	/2	$	
General Conditions (%) 5	% x Subtotal $	3,627,388 =	181,369		
		General Conditions	$	181,369
		Subtotal "A"	$	3,808,757
Overhead 7	% x Subtotal "A" $ 3,808,757		$	266,613
		Subtotal "B"	$	4,075,370
Profit 3	% x Subtotal "B" $ 4,075,370		$	122,261
		Subtotal "C"	$	4,197,632
Location Factor N/A	% x Subtotal "C" $ N/A			
		Adjusted Building Cost	$	4,197,632
Architect's Fee 6.5	% x Adjusted Building Cost	4,197,632 = $		272,846
Contingency N/A	% x Adjusted Building Cost	N/A = $		
		Total Cost		4,470,478

Square Foot Cost $ *4,470,478* / *54,000* S.F. *82.79* $/S.F.

Cubic Foot Cost $ *4,470,478* / *648,000* C.F. *6.90* $/S.F.

Division 1
Foundations

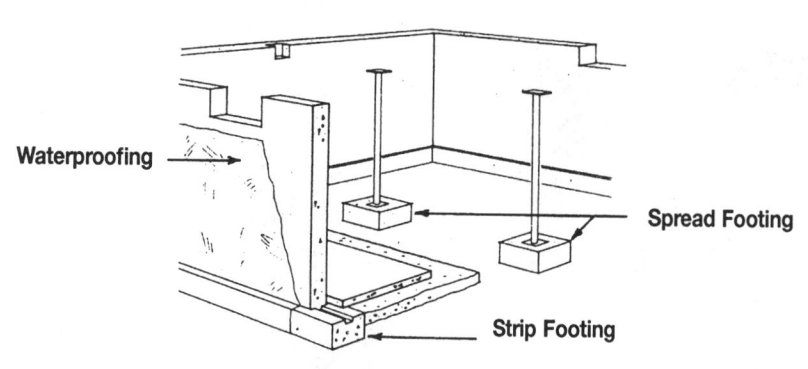

Waterproofing →

Spread Footing ←

Strip Footing ←

1.1-120	Spread Footings	COST EACH		
		MAT.	INST.	TOTAL
7090	Spread footings, 3000 psi concrete, chute delivered			
7100	Load 25K, soil capacity 3 KSF, 3'-0" sq. x 12" deep	36	73	109
7150	Load 50K, soil capacity 3 KSF, 4'-6" sq. x 12" deep	74.50	127	201.50
7200	Load 50K, soil capacity 6 KSF, 3'-0" sq. x 12" deep	36	73	109
7250	Load 75K, soil capacity 3 KSF, 5'-6" sq. x 13" deep	117	179	296
7300	Load 75K, soil capacity 6 KSF, 4'-0" sq. x 12" deep	61	109	170
7350	Load 100K, soil capacity 3 KSF, 6'-0" sq. x 14" deep	148	215	363
7410	Load 100K, soil capacity 6 KSF, 4'-6" sq. x 15" deep	91.50	148	239.50
7450	Load 125K, soil capacity 3 KSF, 7'-0" sq. x 17" deep	233	305	538
7500	Load 125K, soil capacity 6 KSF, 5'-0" sq. x 16" deep	118	179	297
7550	Load 150K, soil capacity 3 KSF 7'-6" sq. x 18" deep	280	360	640
7610	Load 150K, soil capacity 6 KSF, 5'-6" sq. x 18" deep	156	225	381
7650	Load 200K, soil capacity 3 KSF, 8'-6" sq. x 20" deep	395	480	875
7700	Load 200K, soil capacity 6 KSF, 6'-0" sq. x 20" deep	203	277	480
7750	Load 300K, soil capacity 3 KSF, 10'-6" sq. x 25" deep	725	780	1,505
7810	Load 300K, soil capacity 6 KSF, 7'-6" sq. x 25" deep	380	470	850
7850	Load 400K, soil capacity 3 KSF, 12'-6" sq. x 28" deep	1,150	1,150	2,300
7900	Load 400K, soil capacity 6 KSF, 8'-6" sq. x 27" deep	530	610	1,140
8010	Load 500K, soil capacity 6 KSF, 9'-6" sq. x 30" deep	725	790	1,515
8100	Load 600K, soil capacity 6 KSF, 10'-6" sq. x 33" deep	975	1,025	2,000
8200	Load 700K, soil capacity 6 KSF, 11'-6" sq. x 36" deep	1,250	1,250	2,500
8300	Load 800K, soil capacity 6 KSF, 12'-0" sq. x 37" deep	1,400	1,375	2,775
8400	Load 900K, soil capacity 6 KSF, 13'-0" sq. x 39" deep	1,725	1,650	3,375
8500	Load 1000K, soil capacity 6 KSF, 13'-6" sq. x 41" deep	1,950	1,825	3,775

1.1-140	Strip Footings	COST PER L.F.		
		MAT.	INST.	TOTAL
2100	Strip footing, load 2.6KLF, soil capacity 3KSF, 16"wide x 8"deep plain	4.30	8.65	12.95
2300	Load 3.9 KLF, soil capacity, 3 KSF, 24"wide x 8"deep, plain	5.40	9.55	14.95
2500	Load 5.1KLF, soil capacity 3 KSF, 24"wide x 12"deep, reinf.	8.55	14.45	23
2700	Load 11.1KLF, soil capacity 6 KSF, 24"wide x 12"deep, reinf.	8.55	14.45	23
2900	Load 6.8 KLF, soil capacity 3 KSF, 32"wide x 12"deep, reinf.	10.45	15.85	26.30
3100	Load 14.8 KLF, soil capacity 6 KSF, 32"wide x 12"deep, reinf.	10.45	15.85	26.30
3300	Load 9.3 KLF, soil capacity 3 KSF, 40"wide x 12"deep, reinf.	12.35	17.20	29.55
3500	Load 18.4 KLF, soil capacity 6 KSF, 40"wide x 12"deep, reinf.	12.45	17.35	29.80
4500	Load 10KLF, soil capacity 3 KSF, 48"wide x 16"deep, reinf.	17.55	21.50	39.05
4700	Load 22KLF, soil capacity 6 KSF, 48"wide, 16"deep, reinf.	17.95	22	39.95
5700	Load 15KLF, soil capacity 3 KSF, 72"wide x 20"deep, reinf.	30.50	30.50	61
5900	Load 33KLF, soil capacity 6 KSF, 72"wide x 20"deep, reinf.	32.50	33	65.50

Important: See the Reference Section for critical supporting data - Location Factors & Historical Cost Indexes

1.1-210 — Walls, Cast in Place

WALL HEIGHT (FT.)	PLACING METHOD	CONCRETE (C.Y./L.F.)	REINFORCING (LBS./L.F.)	WALL THICKNESS (IN.)	COST PER L.F. MAT.	COST PER L.F. INST.	COST PER L.F. TOTAL
1500	4'	direct chute	.074	3.3	6	9.35	28	37.35
1520			.099	4.8	8	11.35	29	40.35
1540			.123	6.0	10	13.20	29.50	42.70
1561			.148	7.2	12	15.15	30	45.15
1580			.173	8.1	14	17	30.50	47.50
1600			.197	9.44	16	18.90	31.50	50.40
3000	6'	direct chute	.111	4.95	6	14	42	56
3020			.149	7.20	8	17.05	43.50	60.55
3040			.184	9.00	10	19.80	44	63.80
3061			.222	10.8	12	22.50	45	67.50
5000	8'	direct chute	.148	6.6	6	18.65	56	74.65
5020			.199	9.6	8	23	57.50	80.50
5040			.250	12	10	26.50	59.50	86
5061			.296	14.39	12	30.50	61	91.50
6020	10'	direct chute	.248	12	8	28.50	72.50	101
6040			.307	14.99	10	33	74	107
6061			.370	17.99	12	38	76	114
7220	12'	pumped	.298	14.39	8	34	90	124
7240			.369	17.99	10	39.50	92	131.50
7262			.444	21.59	12	45.50	95	140.50
9220	16'	pumped	.397	19.19	8	45.50	120	165.50
9240			.492	23.99	10	53	123	176
9260			.593	28.79	12	60.50	127	187.50

1.1-292 — Foundation Dampproofing

	COST PER L.F. MAT.	COST PER L.F. INST.	COST PER L.F. TOTAL
1000 Foundation dampproofing, bituminous, 1 coat, 4' high	.28	2.74	3.02
1400 8' high	.56	5.50	6.06
1800 12' high	.84	8.50	9.34
2000 2 coats, 4' high	.44	3.38	3.82
2400 8' high	.88	6.75	7.63
2800 12' high	1.32	10.40	11.72
3000 Asphalt with fibers, 1/16" thick, 4' high	.68	3.38	4.06
3400 8' high	1.36	6.75	8.11
3800 12' high	2.04	10.40	12.44
4000 1/8" thick, 4' high	1.16	4.02	5.18
4400 8' high	2.32	8.05	10.37
4800 12' high	3.48	12.30	15.78
5000 Asphalt coated board and mastic, 1/4" thick, 4' high	2.60	3.66	6.26
5400 8' high	5.20	7.30	12.50
5800 12' high	7.80	11.25	19.05
6000 1/2" thick, 4' high	3.90	5.10	9
6400 8' high	7.80	10.20	18
6800 12' high	11.70	15.55	27.25
7000 Cementitious coating, on walls, 1/8" thick coating, 4' high	6.30	4.32	10.62
7400 8' high	12.65	8.65	21.30
7800 12' high	18.95	12.95	31.90
8000 Cementitious/metallic slurry, 2 coat, 1/4" thick, 2' high	37.50	122	159.50
8400 4' high	75	244	319
8800 6' high	113	365	478

FOUNDATIONS

1

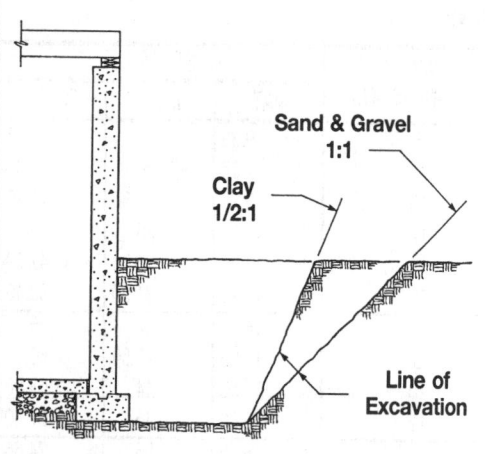

Sand & Gravel
1:1

Clay
1/2:1

Line of
Excavation

In general, the following items are accounted for in the table below.

Costs:
1) Excavation for building or other structure to depth and extent indicated.
2) Backfill compacted in place
3) Haul of excavated waste
4) Replacement of unsuitable backfill material with bank run gravel.

1.9-100	Building Excavation & Backfill	COST PER S.F.		
		MAT.	INST.	TOTAL
2220	Excav & fill, 1000 S.F. 4' sand, gravel, or common earth, on site storage		1.60	1.60
2240	Off site storage		3.37	3.37
2260	Clay excavation, bank run gravel borrow for backfill	.91	3.03	3.94
2280	8' deep, sand, gravel, or common earth, on site storage		3.73	3.73
2300	Off site storage		8.60	8.60
2320	Clay excavation, bank run gravel borrow for backfill	2.19	6.90	9.09
2340	16' deep, sand, gravel, or common earth, on site storage		10.25	10.25
2350	Off site storage		21	21
2360	Clay excavation, bank run gravel borrow for backfill	6.05	17.70	23.75
3380	4000 S.F., 4' deep, sand, gravel, or common earth, on site storage		1.24	1.24
3400	Off site storage		2.05	2.05
3420	Clay excavation, bank run gravel borrow for backfill	.42	1.91	2.33
3440	8' deep, sand, gravel, or common earth, on site storage		2.70	2.70
3460	Off site storage		4.86	4.86
3480	Clay excavation, bank run gravel borrow for backfill	1	4.18	5.18
3500	16' deep, sand, gravel, or common earth, on site storage		6.50	6.50
3520	Off site storage		13.30	13.30
3540	Clay, excavation, bank run gravel borrow for backfill	2.67	9.90	12.57
4560	10,000 S.F., 4' deep, sand gravel, or common earth, on site storage		1.11	1.11
4580	Off site storage		1.60	1.60
4600	Clay excavation, bank run gravel borrow for backfill	.26	1.53	1.79
4620	8' deep, sand, gravel, or common earth, on site storage		2.36	2.36
4640	Off site storage		3.65	3.65
4660	Clay excavation, bank run gravel borrow for backfill	.60	3.26	3.86
4680	16' deep, sand, gravel, or common earth, on site storage		5.40	5.40
4700	Off site storage		9.35	9.35
4720	Clay excavation, bank run gravel borrow for backfill	1.60	7.45	9.05
5740	30,000 S.F., 4' deep, sand, gravel, or common earth, on site storage		1.04	1.04
5760	Off site storage		1.32	1.32
5780	Clay excavation, bank run gravel borrow for backfill	.15	1.27	1.42
5860	16' deep, sand, gravel, or common earth, on site storage		4.68	4.68
5880	Off site storage		6.85	6.85
5900	Clay excavation, bank run gravel borrow for backfill	.89	5.80	6.69
6910	100,000 S.F., 4' deep, sand, gravel, or common earth, on site storage		1	1
6940	8' deep, sand, gravel, or common earth, on site storage		2.03	2.03
6970	16' deep, sand, gravel, or common earth, on site storage		4.30	4.30
6980	Off site storage		5.45	5.45
6990	Clay excavation, bank run gravel borrow for backfill	.48	4.86	5.34

For information about Means Estimating Seminars, see yellow pages 11 and 12 in back of book

Important: See the Reference Section for critical supporting data - Location Factors & Historical Cost Indexes

Division 2
Substructures

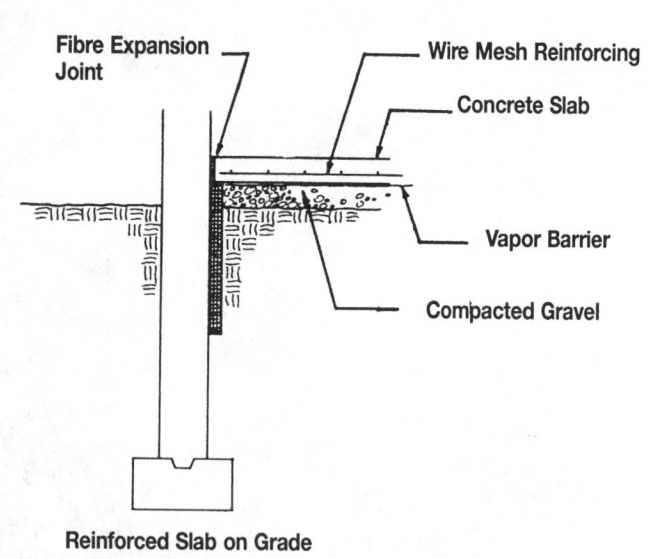

Fibre Expansion Joint

Wire Mesh Reinforcing

Concrete Slab

Vapor Barrier

Compacted Gravel

Reinforced Slab on Grade

A Slab on Grade system includes fine grading; 6" of compacted gravel; vapor barrier; 3500 p.s.i. concrete; bituminous fiber expansion joint; all necessary edge forms 4 uses; steel trowel finish; and sprayed on membrane curing compound. Wire mesh reinforcing used in all reinforced slabs.

Non-industrial slabs are for foot traffic only with negligible abrasion. Light industrial slabs are for pneumatic wheels and light abrasion. Industrial slabs are for solid rubber wheels and moderate abrasion. Heavy industrial slabs are for steel wheels and severe abrasion.

2.1-200	Plain & Reinforced	COST PER S.F.		
		MAT.	INST.	TOTAL
2220	Slab on grade, 4" thick, non industrial, non reinforced	1.08	1.65	2.73
2240	Reinforced	1.17	1.89	3.06
2260	Light industrial, non reinforced	1.32	2.01	3.33
2280	Reinforced	1.41	2.25	3.66
2300	Industrial, non reinforced	1.66	3.17	4.83
2320	Reinforced	1.75	3.41	5.16
3340	5" thick, non industrial, non reinforced	1.28	1.69	2.97
3360	Reinforced	1.37	1.93	3.30
3380	Light industrial, non reinforced	1.52	2.05	3.57
3400	Reinforced	1.61	2.29	3.90
3420	Heavy industrial, non reinforced	2.11	3.67	5.78
3440	Reinforced	2.18	3.94	6.12
4460	6" thick, non industrial, non reinforced	1.54	1.66	3.20
4480	Reinforced	1.68	1.98	3.66
4500	Light industrial, non reinforced	1.78	2.02	3.80
4520	Reinforced	2.01	2.45	4.46
4540	Heavy industrial, non reinforced	2.38	3.73	6.11
4560	Reinforced	2.52	4.05	6.57
5580	7" thick, non industrial, non reinforced	1.75	1.72	3.47
5600	Reinforced	1.95	2.06	4.01
5620	Light industrial, non reinforced	2	2.08	4.08
5640	Reinforced	2.20	2.42	4.62
5660	Heavy industrial, non reinforced	2.59	3.67	6.26
5680	Reinforced	2.73	3.99	6.72
6700	8" thick, non industrial, non reinforced	1.95	1.75	3.70
6720	Reinforced	2.12	2.03	4.15
6740	Light industrial, non reinforced	2.20	2.11	4.31
6760	Reinforced	2.37	2.39	4.76
6780	Heavy industrial, non reinforced	2.81	3.73	6.54
6800	Reinforced	3.03	4.04	7.07

For information about Means Estimating Seminars, see yellow pages 11 and 12 in back of book

Important: See the Reference Section for critical supporting data - Location Factors & Historical Cost Indexes

Division 3
Superstructures

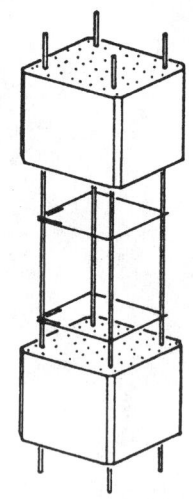

General: It is desirable for purposes of consistency and simplicity to maintain constant column sizes throughout building height. To do this, concrete strength may be varied (higher strength concrete at lower stories and lower strength concrete at upper stories), as well as varying the amount of reinforcing.

The table provides probably minimum column sizes with related costs and weight per lineal foot of story height.

3.1-114 — C.I.P. Column, Square Tied

	LOAD (KIPS)	STORY HEIGHT (FT.)	COLUMN SIZE (IN.)	COLUMN WEIGHT (P.L.F.)	CONCRETE STRENGTH (PSI)	COST PER V.L.F. MAT.	INST.	TOTAL
0640	100	10	10	96	4000	7	28	35
0680		12	10	97	4000	6.85	28	34.85
0720		14	12	142	4000	8.85	34	42.85
0840	200	10	12	140	4000	8.95	34	42.95
0860		12	12	142	4000	8.90	34	42.90
0900		14	14	196	4000	11.15	39	50.15
0920	300	10	14	192	4000	11.40	39.50	50.90
0960		12	14	194	4000	11.30	39.50	50.80
0980		14	16	253	4000	12.75	43	55.75
1020	400	10	16	248	4000	13.90	45	58.90
1060		12	16	251	4000	13.75	45	58.75
1080		14	16	253	4000	13.65	44.50	58.15
1200	500	10	18	315	4000	18.30	54	72.30
1250		12	20	394	4000	18.60	56	74.60
1300		14	20	397	4000	18.50	56	74.50
1350	600	10	20	388	4000	21.50	62	83.50
1400		12	20	394	4000	21.50	61.50	83
1600		14	20	397	4000	21.50	61	82.50
3400	900	10	24	560	4000	31	80.50	111.50
3800		12	24	567	4000	30.50	79.50	110
4000		14	24	571	4000	30	79	109
7300	300	10	14	192	6000	11.30	39.50	50.80
7500		12	14	194	6000	11.20	39	50.20
7600		14	14	196	6000	11.15	39	50.15
8000	500	10	16	248	6000	13.90	45	58.90
8050		12	16	251	6000	13.75	45	58.75
8100		14	16	253	6000	13.65	44.50	58.15
8200	600	10	18	315	6000	16.70	51.50	68.20
8300		12	18	319	6000	16.55	51.50	68.05
8400		14	18	321	6000	16.40	51	67.40
8800	800	10	20	388	6000	18.80	56.50	75.30
8900		12	20	394	6000	18.60	56	74.60
9000		14	20	397	6000	18.50	56	74.50

Important: See the Reference Section for critical supporting data - Location Factors & Historical Cost Indexes

3.1-114   C.I.P. Column, Square Tied

	LOAD (KIPS)	STORY HEIGHT (FT.)	COLUMN SIZE (IN.)	COLUMN WEIGHT (P.L.F.)	CONCRETE STRENGTH (PSI)	COST PER V.L.F.		
						MAT.	INST.	TOTAL
9100	900	10	20	388	6000	27	71	98
9300		12	20	394	6000	26.50	70	96.50
9600		14	20	397	6000	26	69.50	95.50

3.1-114   C.I.P. Column, Square Tied-Minimum Reinforcing

	LOAD (KIPS)	STORY HEIGHT (FT.)	COLUMN SIZE (IN.)	COLUMN WEIGHT (P.L.F.)	CONCRETE STRENGTH (PSI)	COST PER V.L.F.		
						MAT.	INST.	TOTAL
9912	150	10-14	12	135	4000	8.70	33.50	42.20
9918	300	10-14	16	240	4000	12.65	43	55.65
9924	500	10-14	20	375	4000	18.30	55.50	73.80
9930	700	10-14	24	540	4000	26	72	98
9936	1000	10-14	28	740	4000	32.50	85	117.50
9942	1400	10-14	32	965	4000	43	99	142
9948	1800	10-14	36	1220	4000	52	114	166

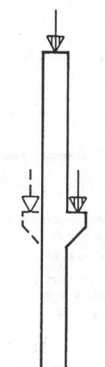

Concentric Load

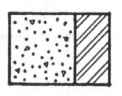

Eccentric Load

General: Data presented here is for plant produced members transported 50 miles to 100 miles to the site and erected.

Design and pricing assumptions:
Normal wt. concrete, f'c = 5 KSI

Main reinforcement, fy = 60 KSI
Ties, fy = 40 KSI

Minimum design eccentricity, 0.1t.

Concrete encased structural steel haunches are assumed where practical; otherwise galvanized rebar haunches are assumed.

Base plates are integral with columns.

Foundation anchor bolts, nuts and washers are included in price.

3.1-120 | Tied, Concentric Loaded Precast Concrete Columns

	LOAD (KIPS)	STORY HEIGHT (FT.)	COLUMN SIZE (IN.)	COLUMN WEIGHT (P.L.F.)	LOAD LEVELS	COST PER V.L.F.		
						MAT.	INST.	TOTAL
0560	100	10	12x12	164	2	34.50	8.15	42.65
0570		12	12x12	162	2	34	6.80	40.80
0580		14	12x12	161	2	33	6.80	39.80
0590	150	10	12x12	166	3	33.50	6.80	40.30
0600		12	12x12	169	3	31.50	6.10	37.60
0610		14	12x12	162	3	32.50	6.10	38.60
0620	200	10	12x12	168	4	34.50	7.45	41.95
0630		12	12x12	170	4	35	6.80	41.80
0640		14	14x14	220	4	37.50	6.80	44.30

3.1-120 | Tied, Eccentric Loaded Precast Concrete Columns

	LOAD (KIPS)	STORY HEIGHT (FT.)	COLUMN SIZE (IN.)	COLUMN WEIGHT (P.L.F.)	LOAD LEVELS	COST PER V.L.F.		
						MAT.	INST.	TOTAL
1130	100	10	12x12	161	2	31.50	8.15	39.65
1140		12	12x12	159	2	33.50	6.80	40.30
1150		14	12x12	159	2	32.50	6.80	39.30
1390	600	10	18x18	385	4	62	7.45	69.45
1400		12	18x18	380	4	60.50	6.80	67.30
1410		14	18x18	375	4	60	6.80	66.80
1480	800	10	20x20	490	4	79.50	7.45	86.95
1490		12	20x20	480	4	78.50	6.80	85.30
1500		14	20x20	475	4	78.50	6.80	85.30

Important: See the Reference Section for critical supporting data - Location Factors & Historical Cost Indexes

SUPERSTRUCTURES 3

(A) Wide Flange

 (B) Pipe

 (C) Pipe, Concrete Filled

 (G) Rectangular Tube, Concrete Filled

3.1-130		Steel Columns					
	LOAD (KIPS)	UNSUPPORTED HEIGHT (FT.)	WEIGHT (P.L.F.)	SIZE (IN.)	TYPE	COST PER V.L.F.		
						MAT.	INST.	TOTAL
1000	25	10	13	4	A	10.20	6.65	16.85
1020			7.58	3	B	9.25	6.65	15.90
1040			15	3-1/2	C	11.25	6.65	17.90
1120			20	4x3	G	12.50	6.65	19.15
1200		16	16	5	A	11.65	4.97	16.62
1220			10.79	4	B	12.20	4.97	17.17
1240			36	5-1/2	C	16.90	4.97	21.87
1320			64	8x6	G	25	4.97	29.97
1600	50	10	16	5	A	12.60	6.65	19.25
1620			14.62	5	B	17.85	6.65	24.50
1640			24	4-1/2	C	13.40	6.65	20.05
1720			28	6x3	G	16.60	6.65	23.25
1800		16	24	8	A	17.45	4.97	22.42
1840			36	5-1/2	C	16.90	4.97	21.87
1920			64	8x6	G	25	4.97	29.97
2000	50	20	28	8	A	19.30	4.97	24.27
2040			49	6-5/8	C	21	4.97	25.97
2120			64	8x6	G	24	4.97	28.97
2200	75	10	20	6	A	15.75	6.65	22.40
2240			36	4-1/2	C	34	6.65	40.65
2320			35	6x4	G	18.70	6.65	25.35
2400		16	31	8	A	22.50	4.97	27.47
2440			49	6-5/8	C	22	4.97	26.97
2520			64	8x6	G	25	4.97	29.97
2600		20	31	8	A	21.50	4.97	26.47
2640			81	8-5/8	C	31.50	4.97	36.47
2720			64	8x6	G	24	4.97	28.97
2800	100	10	24	8	A	18.90	6.65	25.55
2840			35	4-1/2	C	34	6.65	40.65
2920			46	8x4	G	23	6.65	29.65

SUPERSTRUCTURES

3

	LOAD (KIPS)	UNSUPPORTED HEIGHT (FT.)	WEIGHT (P.L.F.)	SIZE (IN.)	TYPE	COST PER V.L.F.		
						MAT.	INST.	TOTAL
3000	100	16	31	8	A	22.50	4.97	27.47
3040			56	6-5/8	C	33	4.97	37.97
3120			64	8x6	G	25	4.97	29.97
3200		20	40	8	A	27.50	4.97	32.47
3240			81	8-5/8	C	31.50	4.97	36.47
3320			70	8x6	G	34	4.97	38.97
3400	125	10	31	8	A	24.50	6.65	31.15
3440			81	8	C	36	6.65	42.65
3520			64	8x6	G	27	6.65	33.65
3600	125	16	40	8	A	29	4.97	33.97
3640			81	8	C	33	4.97	37.97
3720			64	8x6	G	25	4.97	29.97
3800		20	48	8	A	33	4.97	37.97
3840			81	8	C	31.50	4.97	36.47
3920			60	8x6	G	34	4.97	38.97
4000	150	10	35	8	A	27.50	6.65	34.15
4040			81	8-5/8	C	36	6.65	42.65
4120			64	8x6	G	27	6.65	33.65
4200		16	45	10	A	33	4.97	37.97
4240			81	8-5/8	C	33	4.97	37.97
4320			70	8x6	G	36	4.97	40.97
4400		20	49	10	A	34	4.97	38.97
4440			123	10-3/4	C	45	4.97	49.97
4520			86	10x6	G	33.50	4.97	38.47
4600	200	10	45	10	A	35.50	6.65	42.15
4640			81	8-5/8	C	36	6.65	42.65
4720			70	8x6	G	39	6.65	45.65
4800		16	49	10	A	35.50	4.97	40.47
4840			123	10-3/4	C	47	4.97	51.97
4920			85	10x6	G	42	4.97	46.97
5200	300	10	61	14	A	48	6.65	54.65
5240			169	12-3/4	C	62.50	6.65	69.15
5320			86	10x6	G	58	6.65	64.65
5400		16	72	12	A	52.50	4.97	57.47
5440			169	12-3/4	C	58	4.97	62.97
5600		20	79	12	A	54.50	4.97	59.47
5640			169	12-3/4	C	55	4.97	59.97
5800	400	10	79	12	A	62	6.65	68.65
5840			178	12-3/4	C	82	6.65	88.65
6000		16	87	12	A	63.50	4.97	68.47
6040			178	12-3/4	C	76	4.97	80.97
6400	500	10	99	14	A	78	6.65	84.65
6600		16	109	14	A	79.50	4.97	84.47
6800		20	120	12	A	82.50	4.97	87.47
7000	600	10	120	12	A	94.50	6.65	101.15
7200		16	132	14	A	96	4.97	100.97
7400		20	132	14	A	91	4.97	95.97
7600	700	10	136	12	A	107	6.65	113.65
7800		16	145	14	A	106	4.97	110.97
8000		20	145	14	A	100	4.97	104.97
8200	800	10	145	14	A	114	6.65	120.65
8300		16	159	14	A	116	4.97	120.97
8400		20	176	14	A	121	4.97	125.97

	LOAD (KIPS)	UNSUPPORTED HEIGHT (FT.)	WEIGHT (P.L.F.)	SIZE (IN.)	TYPE	COST PER V.L.F.		
						MAT.	INST.	TOTAL
8800	900	10	159	14	A	125	6.65	131.65
8900		16	176	14	A	128	4.97	132.97
9000		20	193	14	A	133	4.97	137.97
9100	1000	10	176	14	A	138	6.65	144.65
9200		16	193	14	A	141	4.97	145.97
9300		20	211	14	A	145	4.97	149.97

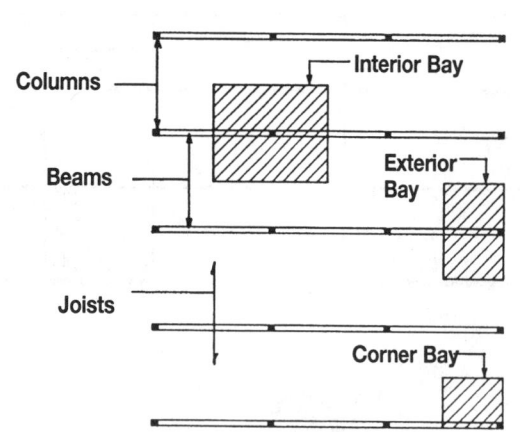

Columns

Interior Bay

Beams

Exterior Bay

Joists

Corner Bay

Description: Table below lists costs per S.F. of bay size for wood columns of various sizes and unsupported heights and the maximum allowable total load per S.F. per bay size.

3.1-140 — Wood Columns

	NOMINAL COLUMN SIZE (IN.)	BAY SIZE (FT.)	UNSUPPORTED HEIGHT (FT.)	MATERIAL (BF/M.S.F.)	TOTAL LOAD (P.S.F.)	COST PER S.F.		
						MAT.	INST.	TOTAL
1000	4 x 4	10 x 8	8	133	100	.18	.15	.33
1050			10	167	60	.23	.18	.41
1200		10 x 10	8	106	80	.15	.12	.27
1250			10	133	50	.18	.15	.33
1400		10 x 15	8	71	50	.10	.08	.18
1450			10	88	30	.12	.10	.22
1600		15 x 15	8	47	30	.06	.05	.11
1650			10	59	15	.08	.06	.14
2000	6 x 6	10 x 15	8	160	230	.33	.16	.49
2050			10	200	210	.42	.21	.63
2200		15 x 15	8	107	150	.22	.11	.33
2250			10	133	140	.28	.14	.42
2400		15 x 20	8	80	110	.17	.08	.25
2450			10	100	100	.21	.10	.31
2600		20 x 20	8	60	80	.12	.06	.18
2650			10	75	70	.16	.08	.24
2800		20 x 25	8	48	60	.10	.05	.15
2850			10	60	50	.12	.06	.18
3400	8 x 8	20 x 20	8	107	160	.24	.10	.34
3450			10	133	160	.30	.13	.43
3600		20 x 25	8	85	130	.12	.09	.21
3650			10	107	130	.15	.12	.27
3800		25 x 25	8	68	100	.14	.06	.20
3850			10	85	100	.18	.08	.26
4200	10 x 10	20 x 25	8	133	210	.28	.12	.40
4250			10	167	210	.35	.15	.50
4400		25 x 25	8	107	160	.22	.10	.32
4450			10	133	160	.28	.12	.40
4700	12 x 12	20 x 25	8	192	310	.40	.16	.56
4750			10	240	310	.50	.20	.70
4900		25 x 25	8	154	240	.32	.13	.45
4950			10	192	240	.40	.16	.56

Important: See the Reference Section for critical supporting data - Reference Numbers and City Cost Indexes

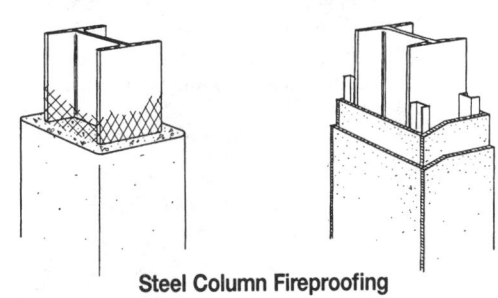

Steel Column Fireproofing

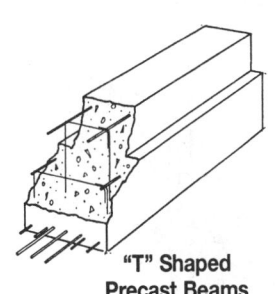

"T" Shaped Precast Beams

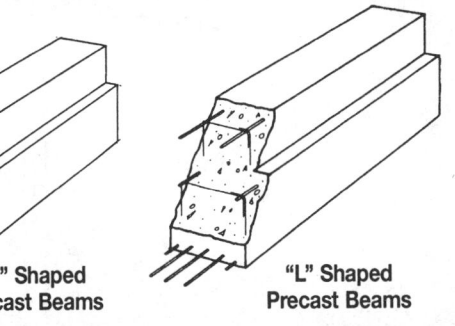

"L" Shaped Precast Beams

3.1-190 — Steel Column Fireproofing

	ENCASEMENT SYSTEM	COLUMN SIZE (IN.)	THICKNESS (IN.)	FIRE RATING (HRS.)	WEIGHT (P.L.F.)	COST PER V.L.F. MAT.	INST.	TOTAL
3000	Concrete	8	1	1	110	3.92	21.50	25.42
3300		14	1	1	258	6.40	31	37.40
3400			2	3	325	7.70	35.50	43.20
3450	Gypsum board	8	1/2	2	8	2.44	12.90	15.34
3550	1 layer	14	1/2	2	18	2.68	13.80	16.48
3600	Gypsum board	8	1	3	14	3.41	16.50	19.91
3650	1/2" fire rated	10	1	3	17	3.69	17.50	21.19
3700	2 layers	14	1	3	22	3.84	18	21.84
3750	Gypsum board	8	1-1/2	3	23	4.59	21	25.59
3800	1/2" fire rated	10	1-1/2	3	27	5.20	23	28.20
3850	3 layers	14	1-1/2	3	35	5.80	25	30.80
3900	Sprayed fiber	8	1-1/2	2	6.3	3.05	4.88	7.93
3950	Direct application		2	3	8.3	4.21	6.75	10.96
4050		10	1-1/2	2	7.9	3.69	5.90	9.59
4200		14	1-1/2	2	10.8	4.58	7.35	11.93

3.1-224 — "T" Shaped Precast Beams

	SPAN (FT.)	SUPERIMPOSED LOAD (K.L.F.)	SIZE W X D (IN.)	BEAM WEIGHT (P.L.F.)	TOTAL LOAD (K.L.F.)	COST PER L.F. MAT.	INST.	TOTAL
2300	15	2.8	12x16	260	3.06	96	14.50	110.50
2500		8.37	12x28	515	8.89	123	14.25	137.25
8900	45	3.34	12x60	1165	4.51	220	13.60	233.60
9900		6.5	24x60	1915	8.42	286	13.60	299.60

3.1-226 — "L" Shaped Precast Beams

	SPAN (FT.)	SUPERIMPOSED LOAD (K.L.F.)	SIZE W X D (IN.)	BEAM WEIGHT (P.L.F.)	TOTAL LOAD (K.L.F.)	COST PER L.F. MAT.	INST.	TOTAL
2250	15	2.58	12x16	230	2.81	72.50	14.50	87
2400		5.92	12x24	370	6.29	88.50	13.60	102.10
4000	25	2.64	12x28	435	3.08	98	8.15	106.15
4450		6.44	18x36	790	7.23	134	10.20	144.20
5300	30	2.80	12x36	565	3.37	117	8.15	125.15
6400		8.66	24x44	1245	9.90	187	14.95	201.95

SUPERSTRUCTURES

3

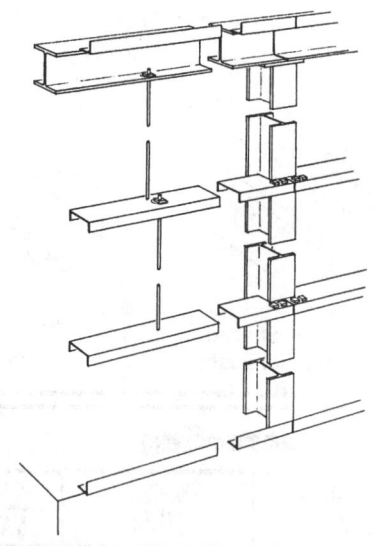

Description: The table below lists costs, $/S.F., for channel girts with sag rods and connector angles top and bottom for various column spacings, building heights and wind loads. Additive costs are shown for wind columns.

3.4-300 Metal Siding Support

	BLDG. HEIGHT (FT.)	WIND LOAD (P.S.F.)	COL. SPACING (FT.)		INTERMEDIATE COLUMNS	COST PER S.F.		
						MAT.	INST.	TOTAL
3000	18	20	20			1.31	.71	2.02
3100					wind cols.	.56	.18	.74
3200		20	25			1.41	.74	2.15
3300					wind cols.	.44	.14	.58
3400		20	30			1.53	.78	2.31
3500					wind cols.	.37	.12	.49
3600		20	35			1.67	.82	2.49
3700					wind cols.	.32	.10	.42
3800		30	20			1.43	.74	2.17
3900					wind cols.	.56	.18	.74
4000		30	25			1.53	.78	2.31
4100					wind cols.	.44	.14	.58
4200		30	30			1.67	.83	2.50
4300					wind cols.	.50	.16	.66
4600	30	20	20			1.20	.57	1.77
4700					wind cols.	.79	.25	1.04
4800		20	25			1.33	.61	1.94
4900					wind cols.	.63	.20	.83
5000		20	30			1.48	.66	2.14
5100					wind cols.	.62	.20	.82
5200		20	35			1.63	.72	2.35
5300					wind cols.	.61	.20	.81
5400		30	20			1.35	.62	1.97
5500					wind cols.	.93	.30	1.23
5600		30	25			1.48	.66	2.14
5700					wind cols.	.89	.29	1.18
5800		30	30			1.64	.72	2.36
5900					wind cols.	.83	.27	1.10

Important: See the Reference Section for critical supporting data - Reference Numbers and City Cost Indexes

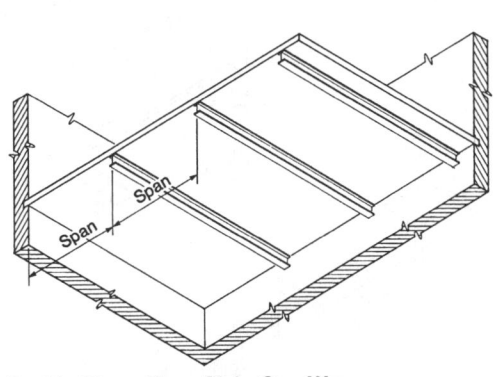

Cast in Place Floor Slab, One Way
General: Solid concrete slabs of uniform depth reinforced for flexure in one direction and for temperature and shrinkage in the other direction.

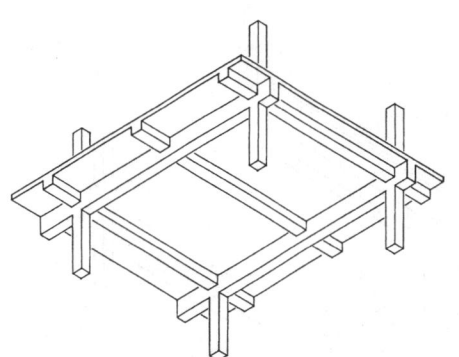

Cast in Place Beam & Slab, One Way
General: Solid concrete one way slab cast monolithically with reinforced concrete beams and girders.

3.5-110 — Cast in Place Slabs, One Way

	SLAB DESIGN & SPAN (FT.)	SUPERIMPOSED LOAD (P.S.F.)	THICKNESS (IN.)	TOTAL LOAD (P.S.F.)	COST PER S.F. MAT.	COST PER S.F. INST.	COST PER S.F. TOTAL
2500	Single 8	40	4	90	1.98	4.88	6.86
2600		75	4	125	2.05	5.30	7.35
2700		125	4-1/2	181	2.15	5.35	7.50
2800		200	5	262	2.41	5.55	7.96
3000	Single 10	40	4	90	2.12	5.25	7.37
3100		75	4	125	2.12	5.25	7.37
3200		125	5	188	2.38	5.40	7.78
3300		200	7-1/2	293	3.21	6	9.21
3500	Single 15	40	5-1/2	90	2.56	5.35	7.91
3600		75	6-1/2	156	2.86	5.60	8.46
3700		125	7-1/2	219	3.13	5.75	8.88
3800		200	8-1/2	306	3.41	5.90	9.31
4000	Single 20	40	7-1/2	115	3.12	5.65	8.77
4100		75	9	200	3.45	5.65	9.10
4200		125	10	250	3.83	5.90	9.73
4300		200	10	324	4.12	6.15	10.27

3.5-120 — Cast in Place Beam & Slab, One Way

	BAY SIZE (FT.)	SUPERIMPOSED LOAD (P.S.F.)	MINIMUM COL. SIZE (IN.)	SLAB THICKNESS (IN.)	TOTAL LOAD (P.S.F.)	COST PER S.F. MAT.	COST PER S.F. INST.	COST PER S.F. TOTAL
5000	20x25	40	12	5-1/2	121	3.02	6.75	9.77
5200		125	16	5-1/2	215	3.57	7.75	11.32
5500	25x25	40	12	6	129	3.19	6.60	9.79
5700		125	18	6	227	4.05	8.15	12.20
7500	30x35	40	16	8	158	3.96	7.45	11.41
7600		75	18	8	196	4.24	7.75	11.99
7700		125	22	8	254	4.74	8.65	13.39
7800		200	26	8	332	5.15	8.95	14.10
8000	35x35	40	16	9	169	4.43	7.75	12.18
8200		75	20	9	213	4.78	8.45	13.23
8400		125	24	9	272	5.25	8.75	14
8600		200	26	9	355	5.75	9.40	15.15

SUPERSTRUCTURES

3

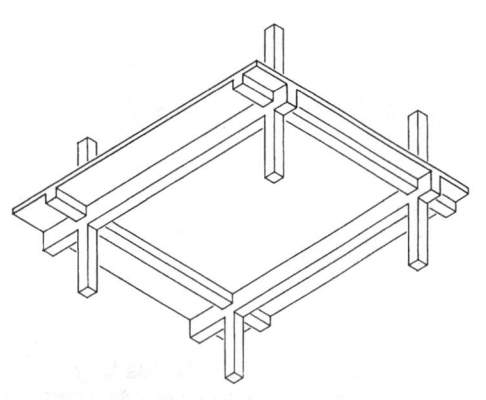

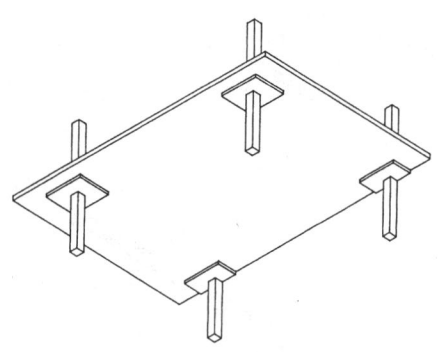

General: Solid concrete two way slab cast monolithically with reinforced concrete support beams and girders.

General: Flat Slab: Solid uniform depth concrete two way slabs with drop panels at columns and no column capitals.

3.5-130 — Cast in Place Beam & Slab, Two Way

	BAY SIZE (FT.)	SUPERIMPOSED LOAD (P.S.F.)	MINIMUM COL. SIZE (IN.)	SLAB THICKNESS (IN.)	TOTAL LOAD (P.S.F.)	COST PER S.F.		
						MAT.	INST.	TOTAL
4000	20 x 25	40	12	7	141	3.50	6.80	10.30
4300		75	14	7	181	3.99	7.45	11.44
4500		125	16	7	236	4.07	7.65	11.72
5100	25 x 25	40	12	7-1/2	149	3.65	6.90	10.55
5200		75	16	7-1/2	185	3.97	7.35	11.32
5300		125	18	7-1/2	250	4.31	8.05	12.36
7600	30 x 35	40	16	10	188	4.81	7.80	12.61
7700		75	18	10	225	5.10	8.10	13.20
8000		125	22	10	282	5.65	8.80	14.45
8500	35 x 35	40	16	10-1/2	193	5.15	8	13.15
8600		75	20	10-1/2	233	5.40	8.30	13.70
9000		125	24	10-1/2	287	6	8.95	14.95

3.5-140 — Cast in Place Flat Slab with Drop Panels

	BAY SIZE (FT.)	SUPERIMPOSED LOAD (P.S.F.)	MINIMUM COL. SIZE (IN.)	SLAB & DROP (IN.)	TOTAL LOAD (P.S.F.)	COST PER S.F.		
						MAT.	INST.	TOTAL
1960	20 x 20	40	12	7 - 3	132	3.31	5.45	8.76
1980		75	16	7 - 4	168	3.48	5.55	9.03
2000		125	18	7 - 6	221	3.85	5.75	9.60
3200	25 x 25	40	12	8-1/2 - 5-1/2	154	3.84	5.70	9.54
4000		125	20	8-1/2 - 8-1/2	243	4.31	6.10	10.41
4400		200	24	9 - 8-1/2	329	4.52	6.30	10.82
5000	25 x 30	40	14	9-1/2 - 7	168	4.15	5.95	10.10
5200		75	18	9-1/2 - 7	203	4.42	6.20	10.62
5600		125	22	9-1/2 - 8	256	4.61	6.35	10.96
6400	30 x 30	40	14	10-1/2 - 7-1/2	182	4.48	6.10	10.58
6600		75	18	10-1/2 - 7-1/2	217	4.76	6.35	11.11
6800		125	22	10-1/2 - 9	269	4.96	6.50	11.46
7400	30 x 35	40	16	11-1/2 - 9	196	4.86	6.35	11.21
7900		75	20	11-1/2 - 9	231	5.15	6.60	11.75
8000		125	24	11-1/2 - 11	284	5.35	6.80	12.15
9000	35 x 35	40	16	12 - 9	202	4.98	6.40	11.38
9400		75	20	12 - 11	240	5.35	6.70	12.05
9600		125	24	12 - 11	290	5.50	6.85	12.35

Important: See the Reference Section for critical supporting data - Location Factors & Historical Cost Indexes

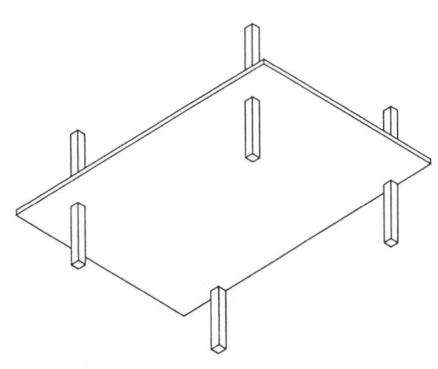

General: Flat Plates: Solid uniform depth concrete two way slab without drops or interior beams. Primary design limit is shear at columns.

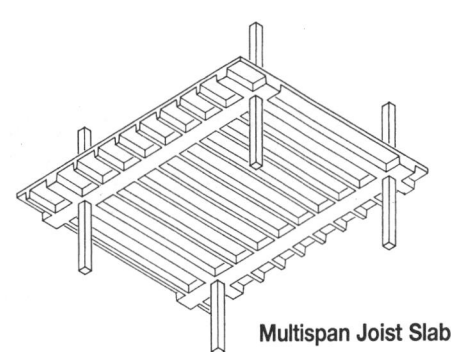

Multispan Joist Slab

General: Combination of thin concrete slab and monolithic ribs at uniform spacing to reduce dead weight and increase rigidity.

3.5-150 — Cast in Place Flat Plate

	BAY SIZE (FT.)	SUPERIMPOSED LOAD (P.S.F.)	MINIMUM COL. SIZE (IN.)	SLAB THICKNESS (IN.)	TOTAL LOAD (P.S.F.)	COST PER S.F.		
						MAT.	INST.	TOTAL
3000	15 x 20	40	14	7	127	2.92	5.25	8.17
3400		75	16	7-1/2	169	3.11	5.35	8.46
3600		125	22	8-1/2	231	3.41	5.50	8.91
3800		175	24	8-1/2	281	3.43	5.50	8.93
4200	20 x 20	40	16	7	127	2.91	5.25	8.16
4400		75	20	7-1/2	175	3.13	5.35	8.48
4600		125	24	8-1/2	231	3.41	5.45	8.86
5000		175	24	8-1/2	281	3.45	5.50	8.95
5600	20 x 25	40	18	8-1/2	146	3.39	5.45	8.84
6000		75	20	9	188	3.51	5.50	9.01
6400		125	26	9-1/2	244	3.80	5.70	9.50
6600		175	30	10	300	3.94	5.80	9.74
7000	25 x 25	40	20	9	152	3.50	5.50	9
7400		75	24	9-1/2	194	3.72	5.65	9.37
7600		125	30	10	250	3.96	5.80	9.76

3.5-160 — Cast in Place Multispan Joist Slab

	BAY SIZE (FT.)	SUPERIMPOSED LOAD (P.S.F.)	MINIMUM COL. SIZE (IN.)	RIB DEPTH (IN.)	TOTAL LOAD (P.S.F.)	COST PER S.F.		
						MAT.	INST.	TOTAL
2000	15 x 15	40	12	8	115	2.98	6.35	9.33
2100		75	12	8	150	2.99	6.40	9.39
2200		125	12	8	200	3.07	6.45	9.52
2300		200	14	8	275	3.19	6.70	9.89
2600	15 x 20	40	12	8	115	3.03	6.35	9.38
2800		75	12	8	150	3.11	6.50	9.61
3000		125	14	8	200	3.24	6.90	10.14
3300		200	16	8	275	3.42	7	10.42
3600	20 x 20	40	12	10	120	3.11	6.30	9.41
3900		75	14	10	155	3.27	6.65	9.92
4000		125	16	10	205	3.30	6.75	10.05
4100		200	18	10	280	3.49	7.05	10.54
6200	30 x 30	40	14	14	131	3.51	6.55	10.06
6400		75	18	14	166	3.64	6.75	10.39
6600		125	20	14	216	3.88	7.15	11.03
6700		200	24	16	297	4.19	7.45	11.64

SUPERSTRUCTURES

3

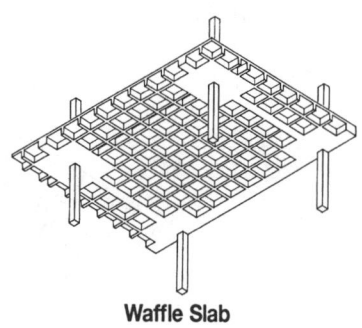

Waffle Slab

3.5-170 — Cast in Place Waffle Slab

	BAY SIZE (FT.)	SUPERIMPOSED LOAD (P.S.F.)	MINIMUM COL. SIZE (IN.)	RIB DEPTH (IN.)	TOTAL LOAD (P.S.F.)	COST PER S.F.		
						MAT.	INST.	TOTAL
3900	20 x 20	40	12	8	144	3.92	6.30	10.22
4000		75	12	8	179	4.01	6.40	10.41
4100		125	16	8	229	4.10	6.45	10.55
4200		200	18	8	304	4.36	6.70	11.06
4400	20 x 25	40	12	8	146	4.02	6.35	10.37
4500		75	14	8	181	4.13	6.45	10.58
4600		125	16	8	231	4.23	6.55	10.78
4700		200	18	8	306	4.46	6.75	11.21
4900	25 x 25	40	12	10	150	4.14	6.45	10.59
5000		75	16	10	185	4.29	6.60	10.89
5300		125	18	10	235	4.43	6.75	11.18
5500		200	20	10	310	4.58	6.85	11.43
5700	25 x 30	40	14	10	154	4.24	6.50	10.74
5800		75	16	10	189	4.38	6.65	11.03
5900		125	18	10	239	4.51	6.75	11.26
6000		200	20	12	329	5.05	7.05	12.10
6400	30 x 30	40	14	12	169	4.57	6.65	11.22
6500		75	18	12	204	4.69	6.75	11.44
6600		125	20	12	254	4.78	6.80	11.58
6700		200	24	12	329	5.25	7.20	12.45
6900	30 x 35	40	16	12	169	4.67	6.70	11.37
7000		75	18	12	204	4.67	6.70	11.37
7100		125	22	12	254	4.94	6.95	11.89
7200		200	26	14	334	5.55	7.40	12.95
7400	35 x 35	40	16	14	174	4.98	6.90	11.88
7500		75	20	14	209	5.10	7	12.10
7600		125	24	14	259	5.25	7.15	12.40
7700		200	26	16	346	5.65	7.50	13.15
8000	35 x 40	40	18	14	176	5.15	7.05	12.20
8300		75	22	14	211	5.30	7.20	12.50
8500		125	26	16	271	5.60	7.35	12.95
8750		200	30	20	372	6.25	7.80	14.05
9200	40 x 40	40	18	14	176	5.30	7.20	12.50
9400		75	24	14	211	5.55	7.40	12.95
9500		125	26	16	271	5.70	7.45	13.15
9700	40 x 45	40	20	16	186	5.55	7.30	12.85
9800		75	24	16	221	5.80	7.50	13.30
9900		125	28	16	271	5.90	7.60	13.50

Important: See the Reference Section for critical supporting data - Location Factors & Historical Cost Indexes

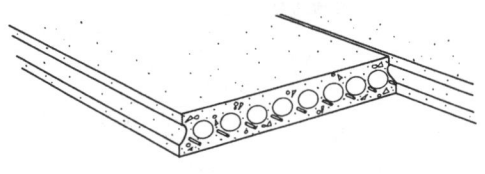

Precast Plank with No Topping

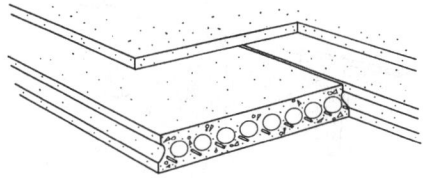

Precast Plank with 2" Concrete Topping

3.5-210   Precast Plank with No Topping

	SPAN (FT.)	SUPERIMPOSED LOAD (P.S.F.)	TOTAL DEPTH (IN.)	DEAD LOAD (P.S.F.)	TOTAL LOAD (P.S.F.)	COST PER S.F.		
						MAT.	INST.	TOTAL
0720	10	40	4	50	90	5.25	1.51	6.76
0750		75	6	50	125	5.05	1.20	6.25
0770		100	6	50	150	5.05	1.20	6.25
0800	15	40	6	50	90	5.05	1.20	6.25
0820		75	6	50	125	5.05	1.20	6.25
0850		100	6	50	150	5.05	1.20	6.25
0950	25	40	6	50	90	5.05	1.20	6.25
0970		75	8	55	130	5	.97	5.97
1000		100	8	55	155	5	.97	5.97
1200	30	40	8	55	95	5	.97	5.97
1300		75	8	55	130	5	.97	5.97
1400		100	10	70	170	5.85	.61	6.46
1500	40	40	10	70	110	5.85	.61	6.46
1600		75	12	70	145	5.25	.67	5.92
1700	45	40	12	70	110	5.25	.67	5.92

3.5-210   Precast Plank with 2" Concrete Topping

	SPAN (FT.)	SUPERIMPOSED LOAD (P.S.F.)	TOTAL DEPTH (IN.)	DEAD LOAD (P.S.F.)	TOTAL LOAD (P.S.F.)	COST PER S.F.		
						MAT.	INST.	TOTAL
2000	10	40	6	75	115	5.80	2.71	8.51
2100		75	8	75	150	5.60	2.40	8
2200		100	8	75	175	5.60	2.40	8
2500	15	40	8	75	115	5.60	2.40	8
2600		75	8	75	150	5.60	2.40	8
2700		100	8	75	175	5.60	2.40	8
3100	25	40	8	75	115	5.60	2.40	8
3200		75	8	75	150	5.60	2.40	8
3300		100	10	80	180	5.55	2.17	7.72
3400	30	40	10	80	120	5.55	2.17	7.72
3500		75	10	80	155	5.55	2.17	7.72
3600		100	10	80	180	5.55	2.17	7.72
4000	40	40	12	95	135	6.40	1.81	8.21
4500		75	14	95	170	5.80	1.87	7.67
5000	45	40	14	95	135	5.80	1.87	7.67

Most widely used for moderate span floors and moderate and long span roofs. At shorter spans, they tend to be competitive with hollow core slabs. They are also used as wall panels.

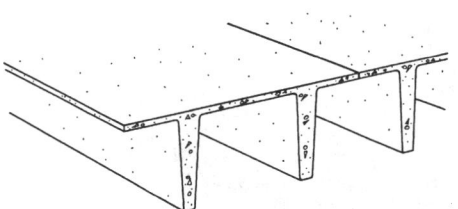

Precast Double "T" Beams with No Topping

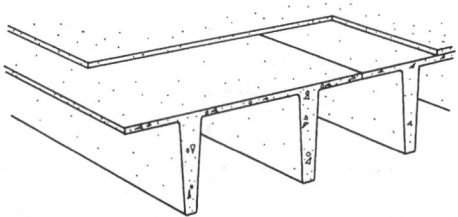

Precast Double "T" Beams with 2" Topping

3.5-230 — Precast Double "T" Beams with No Topping

	SPAN (FT.)	SUPERIMPOSED LOAD (P.S.F.)	DBL. "T" SIZE D (IN.) W (FT.)	CONCRETE "T" TYPE	TOTAL LOAD (P.S.F.)	COST PER S.F.		
						MAT.	INST.	TOTAL
4300	50	30	20x8	Lt. Wt.	66	5.15	1.03	6.18
4400		40	20x8	Lt. Wt.	76	5.50	.94	6.44
4500		50	20x8	Lt. Wt.	86	5.75	1.13	6.88
4600		75	20x8	Lt. Wt.	111	6	1.29	7.29
5600	70	30	32x10	Lt. Wt.	78	5.95	.71	6.66
5750		40	32x10	Lt. Wt.	88	6.15	.87	7.02
5900		50	32x10	Lt. Wt.	98	6.35	.99	7.34
6000		75	32x10	Lt. Wt.	123	6.70	1.20	7.90
6100		100	32x10	Lt. Wt.	148	7.35	1.62	8.97
6200	80	30	32x10	Lt. Wt.	78	6.35	.99	7.34
6300		40	32x10	Lt. Wt.	88	6.70	1.19	7.89
6400		50	32x10	Lt. Wt.	98	7	1.41	8.41

3.5-230 — Precast Double "T" Beams With 2" Topping

	SPAN (FT.)	SUPERIMPOSED LOAD (P.S.F.)	DBL. "T" SIZE D (IN.) W (FT.)	CONCRETE "T" TYPE	TOTAL LOAD (P.S.F.)	COST PER S.F.		
						MAT.	INST.	TOTAL
7100	40	30	18x8	Reg. Wt.	120	4.76	2.02	6.78
7200		40	20x8	Reg. Wt.	130	4.55	1.93	6.48
7300		50	20x8	Reg. Wt.	140	4.84	2.11	6.95
7400		75	20x8	Reg. Wt.	165	4.97	2.20	7.17
7500		100	20x8	Reg. Wt.	190	5.40	2.46	7.86
7550	50	30	24x8	Reg. Wt.	120	5.05	2.02	7.07
7600		40	24x8	Reg. Wt.	130	5.15	2.09	7.24
7750		50	24x8	Reg. Wt.	140	5.20	2.11	7.31
7800		75	24x8	Reg. Wt.	165	5.60	2.37	7.97
7900		100	32x10	Reg. Wt.	189	5.95	1.93	7.88

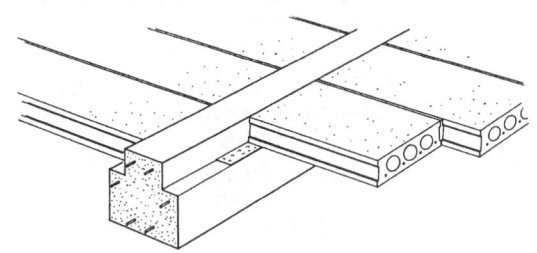

Precast Beam and Plank with No Topping

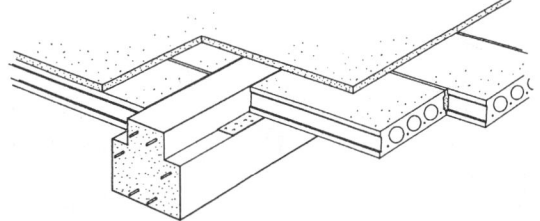

Precast Beam and Plank with 2" Topping

3.5-242 — Precast Beam & Plank with No Topping

	BAY SIZE (FT.)	SUPERIMPOSED LOAD (P.S.F.)	PLANK THICKNESS (IN.)	TOTAL DEPTH (IN.)	TOTAL LOAD (P.S.F.)	COST PER S.F.		
						MAT.	INST.	TOTAL
6200	25x25	40	8	28	118	10.70	2.33	13.03
6400		75	8	36	158	11.25	2.33	13.58
6500		100	8	36	183	11.25	2.33	13.58
7000	25x30	40	8	36	110	10.25	2.33	12.58
7200		75	10	36	159	11.35	1.97	13.32
7400		100	12	36	188	11.25	2.03	13.28
7600	30x30	40	8	36	121	10.60	2.33	12.93
8000		75	10	44	140	11.20	2.33	13.53
8250		100	12	52	206	12.35	2.03	14.38
8500	30x35	40	12	44	135	10.40	2.03	12.43
8750		75	12	52	176	11.15	2.03	13.18
9000	35x35	40	12	52	141	11.15	2.03	13.18
9250		75	12	60	181	11.85	2.03	13.88
9500	35x40	40	12	52	137	10.65	2.71	13.36

3.5-244 — Precast Beam & Plank with 2" Topping

	BAY SIZE (FT.)	SUPERIMPOSED LOAD (P.S.F.)	PLANK THICKNESS (IN.)	TOTAL DEPTH (IN.)	TOTAL LOAD (P.S.F.)	COST PER S.F.		
						MAT.	INST.	TOTAL
4300	20x20	40	6	22	135	11.25	3.75	15
4400		75	6	24	173	11.90	3.75	15.65
4500		100	6	28	200	12.30	3.75	16.05
4600	20x25	40	6	26	134	10.60	3.73	14.33
5000		75	8	30	177	11	3.50	14.50
5200		100	8	30	202	11	3.50	14.50
5400	25x25	40	6	38	143	11.85	3.71	15.56
5600		75	8	38	183	11.85	3.71	15.56
6000		100	8	46	216	12.75	3.48	16.23
6200	25x30	40	8	38	144	10.80	3.46	14.26
6400		75	10	46	200	12.40	3.10	15.50
6600		100	10	46	225	12.40	3.10	15.50
7000	30x30	40	8	46	150	11.75	3.45	15.20
7200		75	10	54	181	12.60	3.45	16.05
7600		100	10	54	231	12.60	3.45	16.05
7800	30x35	40	10	54	166	12.30	3.09	15.39
8000		75	12	54	200	11.45	3.15	14.60

SUPERSTRUCTURES

3

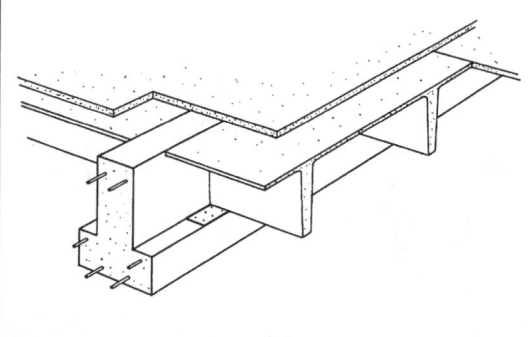

With Topping

General: Beams and double tees priced here are for plant produced prestressed members transported to the site and erected.

The 2″ structural topping is applied after the beams and double tees are in place and is reinforced with W.W.F.

Note: Deduct from prices 20% for Southern States. Add to prices 10% for Western States.

3.5-254 — Precast Double "T" & 2″ Topping on Precast Beams

	BAY SIZE (FT.)	SUPERIMPOSED LOAD (P.S.F.)	DEPTH (IN.)		TOTAL LOAD (P.S.F.)	COST PER S.F.		
						MAT.	INST.	TOTAL
3000	25x30	40	38		130	10.35	3.62	13.97
3100		75	38		168	10.35	3.62	13.97
3300		100	46		196	10.85	3.62	14.47
3600	30x30	40	46		150	11.25	3.60	14.85
3750		75	46		174	11.25	3.60	14.85
4000		100	54		203	11.85	3.60	15.45
4100	30x40	40	46		136	9.15	3.40	12.55
4300		75	54		173	9.60	3.40	13
4400		100	62		204	10.15	3.40	13.55
4600	30x50	40	54		138	8.70	3.31	12.01
4800		75	54		181	9.25	3.31	12.56
5000		100	54		219	10.15	3.10	13.25
5200	30x60	40	62		151	9.15	3.10	12.25
5400		75	62		192	9.70	3.10	12.80
5600		100	62		215	9.70	3.10	12.80
5800	35x40	40	54		139	9.80	3.40	13.20
6000		75	62		179	10.80	3.32	14.12
6250		100	62		212	11.10	3.32	14.42
6500	35x50	40	62		142	9.35	3.31	12.66
6750		75	62		186	9.90	3.31	13.21
7300		100	62		231	11.60	3.10	14.70
7600	35x60	40	54		154	10.50	3.02	13.52
7750		75	54		179	10.85	3.02	13.87
8000		100	62		224	11.50	3.02	14.52
8250	40x40	40	62		145	10.90	4.07	14.97
8400		75	62		187	11.50	4.08	15.58
8750		100	62		223	12.55	4.08	16.63
9000	40x50	40	62		151	10.25	3.97	14.22
9300		75	62		193	10.95	3.42	14.37
9800	40x60	40	62		164	13.15	3.53	16.68

Important: See the Reference Section for critical supporting data - Location Factors & Historical Cost Indexes

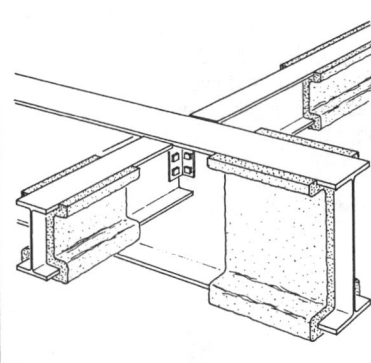

General: The following table is based upon structural wide flange (WF) beam and girder framing. Non-composite action is assumed between WF framing and decking. Deck costs not included.

The deck spans the short direction. The steel beams and girders are fireproofed with sprayed fiber fireproofing.

No columns included in price.

3.5-310 — W Shape Beams & Girders

	BAY SIZE (FT.) BEAM X GIRD	SUPERIMPOSED LOAD (P.S.F.)	STEEL FRAMING DEPTH (IN.)	FIREPROOFING (S.F./S.F.)	TOTAL LOAD (P.S.F.)	COST PER S.F.		
						MAT.	INST.	TOTAL
1350	15x20	40	12	.535	50	2.35	1.24	3.59
1400		40	16	.65	90	3.06	1.60	4.66
1450		75	18	.694	125	4.01	2.02	6.03
1500		125	24	.796	175	5.50	2.76	8.26
2000	20x20	40	12	.55	50	2.63	1.37	4
2050		40	14	.579	90	3.56	1.77	5.33
2100		75	16	.672	125	4.26	2.10	6.36
2150		125	16	.714	175	5.05	2.53	7.58
2900	20x25	40	16	.53	50	2.88	1.46	4.34
2950		40	18	.621	96	4.50	2.17	6.67
3000		75	18	.651	131	5.15	2.46	7.61
3050		125	24	.77	200	6.80	3.29	10.09
5200	25x25	40	18	.486	50	3.12	1.53	4.65
5300		40	18	.592	96	4.63	2.22	6.85
5400		75	21	.668	131	5.60	2.64	8.24
5450		125	24	.738	191	7.35	3.51	10.86
6300	25x30	40	21	.568	50	3.56	1.76	5.32
6350		40	21	.694	90	4.80	2.34	7.14
6400		75	24	.776	125	6.15	2.94	9.09
6450		125	30	.904	175	7.40	3.62	11.02
7700	30x30	40	21	.52	50	3.79	1.83	5.62
7750		40	24	.629	103	5.80	2.72	8.52
7800		75	30	.715	138	6.90	3.21	10.11
7850		125	36	.822	206	9.10	4.29	13.39
8250	30x35	40	21	.508	50	4.32	2.04	6.36
8300		40	24	.651	109	6.35	2.95	9.30
8350		75	33	.732	150	7.70	3.54	11.24
8400		125	36	.802	225	9.60	4.50	14.10
9550	35x35	40	27	.560	50	4.47	2.13	6.60
9600		40	36	.706	109	8.10	3.70	11.80
9650		75	36	.750	150	8.40	3.83	12.23
9820		125	36	.797	225	11.20	5.15	16.35

For expanded coverage of these items see *Means Assemblies Cost Data 1998*

SUPERSTRUCTURES

3

Description: Table below lists costs for light gage CEE or PUNCHED DOUBLE joists to suit the span and loading with the minimum thickness subfloor required by the joist spacing.

3.5-360		Light Gauge Steel Floor Systems						
	SPAN (FT.)	SUPERIMPOSED LOAD (P.S.F.)	FRAMING DEPTH (IN.)	FRAMING SPAC. (IN.)	TOTAL LOAD (P.S.F.)	COST PER S.F.		
						MAT.	INST.	TOTAL
1500	15	40	8	16	54	1.69	1.32	3.01
1550			8	24	54	1.79	1.36	3.15
1600		65	10	16	80	2.02	1.61	3.63
1650			10	24	80	2.02	1.58	3.60
1700		75	10	16	90	2.02	1.61	3.63
1750			10	24	90	2.02	1.58	3.60
1800		100	10	16	116	2.58	2.05	4.63
1850			10	24	116	2.02	1.58	3.60
1900		125	10	16	141	2.58	2.05	4.63
1950			10	24	141	2.02	1.58	3.60
2500	20	40	8	16	55	1.93	1.49	3.42
2550			8	24	55	1.79	1.36	3.15
2600		65	8	16	80	2.22	1.69	3.91
2650			10	24	80	2.03	1.59	3.62
2700		75	10	16	90	1.91	1.53	3.44
2750			12	24	90	2.12	1.37	3.49
2800		100	10	16	115	1.91	1.53	3.44
2850			10	24	116	2.36	1.77	4.13
2900		125	12	16	142	2.73	1.73	4.46
2950			12	24	141	2.43	1.87	4.30
3500	25	40	10	16	55	2.02	1.61	3.63
3550			10	24	55	2.02	1.58	3.60
3600		65	10	16	81	2.58	2.05	4.63
3650			12	24	81	2.33	1.49	3.82
3700		75	12	16	92	2.72	1.73	4.45
3750			12	24	91	2.43	1.87	4.30
3800		100	12	16	117	3.04	1.91	4.95
3850		125	12	16	143	3.20	2.49	5.69
4500	30	40	12	16	57	2.72	1.73	4.45
4550			12	24	56	2.32	1.49	3.81
4600		65	12	16	82	3.03	1.90	4.93
4650		75	12	16	92	3.03	1.90	4.93

Important: See the Reference Section for critical supporting data - Reference Numbers and City Cost Indexes

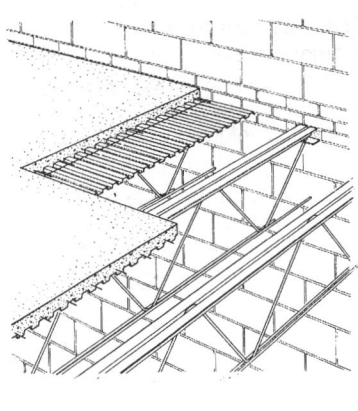

Description: The table below lists costs per S.F. for a floor system on bearing walls using open web steel joists, galvanized steel slab form and 2-1/2″ concrete slab reinforced with welded wire fabric. Costs of the bearing walls are not included.

3.5-420 — Deck & Joists on Bearing Walls

	SPAN (FT.)	SUPERIMPOSED LOAD (P.S.F.)	JOIST SPACING FT./IN.	DEPTH (IN.)	TOTAL LOAD (P.S.F.)	COST PER S.F. MAT.	COST PER S.F. INST.	COST PER S.F. TOTAL
1050	20	40	2-0	14-1/2	83	2.93	2.36	5.29
1070		65	2-0	16-1/2	109	3.20	2.49	5.69
1100		75	2-0	16-1/2	119	3.20	2.49	5.69
1120		100	2-0	18-1/2	145	3.22	2.50	5.72
1150		125	1-9	18-1/2	170	4.12	2.96	7.08
1170	25	40	2-0	18-1/2	84	3.69	2.77	6.46
1200		65	2-0	20-1/2	109	3.84	2.83	6.67
1220		75	2-0	20-1/2	119	3.59	2.65	6.24
1250		100	2-0	22-1/2	145	3.65	2.68	6.33
1270		125	1-9	22-1/2	170	4.21	2.93	7.14
1300	30	40	2-0	22-1/2	84	3.56	2.44	6
1320		65	2-0	24-1/2	110	3.88	2.55	6.43
1350		75	2-0	26-1/2	121	4.03	2.60	6.63
1370		100	2-0	26-1/2	146	4.32	2.69	7.01
1400		125	2-0	24-1/2	172	4.84	3.52	8.36
1420	35	40	2-0	26-1/2	85	4.56	3.37	7.93
1450		65	2-0	28-1/2	111	4.72	3.45	8.17
1470		75	2-0	28-1/2	121	4.72	3.45	8.17
1500		100	1-11	28-1/2	147	5.15	3.68	8.83
1520		125	1-8	28-1/2	172	5.70	3.96	9.66
1550	5/8″ gyp. fireproof.							
1560	On metal furring, add					.53	1.83	2.36

SUPERSTRUCTURES

3

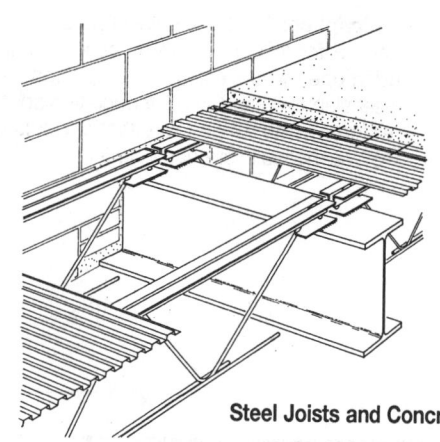

Steel Joists and Concrete Slab on Walls and Beams

The table below lists costs for a floor system on exterior bearing walls and interior columns and beams using open web steel joists, galvanized steel slab form, 2-1/2″ concrete slab reinforced with welded wire fabric.

Columns are sized to accommodate one floor plus roof loading but costs for columns are from suspended floor to slab on grade only. Costs of the bearing walls are not included.

3.5-440 — Steel Joists on Beam & Wall

	BAY SIZE (FT.)	SUPERIMPOSED LOAD (P.S.F.)	DEPTH (IN.)	TOTAL LOAD (P.S.F.)	COLUMN ADD	COST PER S.F. MAT.	COST PER S.F. INST.	COST PER S.F. TOTAL
1720	25x25	40	23	84		4.44	2.96	7.40
1730					columns	.19	.08	.27
1750	25x25	65	29	110		4.68	3.07	7.75
1760					columns	.19	.08	.27
1770	25x25	75	26	120		4.68	2.99	7.67
1780					columns	.22	.09	.31
1800	25x25	100	29	145		5.30	3.26	8.56
1810					columns	.22	.09	.31
1820	25x25	125	29	170		5.55	3.37	8.92
1830					columns	.24	.09	.33
2020	30x30	40	29	84		4.52	2.71	7.23
2030					columns	.17	.07	.24
2050	30x30	65	29	110		5.15	2.94	8.09
2060					columns	.17	.07	.24
2070	30x30	75	32	120		5.30	2.99	8.29
2080					columns	.19	.08	.27
2100	30x30	100	35	145		5.85	3.20	9.05
2110					columns	.22	.09	.31
2120	30x30	125	35	172		6.55	4.10	10.65
2130					columns	.27	.10	.37
2170	30x35	40	29	85		5.15	2.93	8.08
2180					columns	.16	.06	.22
2200	30x35	65	29	111		5.90	3.83	9.73
2210					columns	.18	.08	.26
2220	30x35	75	32	121		5.90	3.83	9.73
2230					columns	.19	.08	.27
2250	30x35	100	35	148		6.25	3.33	9.58
2260					columns	.23	.09	.32
2320	35x35	40	32	85		5.30	2.99	8.29
2330					columns	.16	.07	.23
2350	35x35	65	35	111		6.25	3.97	10.22
2360					columns	.20	.08	.28
2370	35x35	75	35	121		6.35	4.02	10.37
2380					columns	.20	.08	.28
2400	35x35	100	38	148		6.65	4.17	10.82
2410					columns	.24	.09	.33
2460	5/8 gyp. fireproof.							
2475	On metal furring, add					.53	1.83	2.36

Important: See the Reference Section for critical supporting data - Reference Numbers and City Cost Indexes

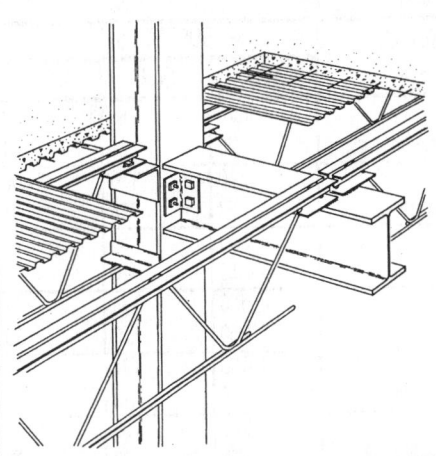

The table below lists costs per S.F. for a floor system on steel columns and beams using open web steel joists, galvanized steel slab form, 2-1/2″ concrete slab reinforced with welded wire fabric.

Columns are sized to accommodate one floor plus roof loading but costs for columns are from floor to grade only.

Steel Joists and Concrete Slab on Steel Columns and Beams

3.5-460 — Steel Joists, Beams & Slab on Columns

	BAY SIZE (FT.)	SUPERIMPOSED LOAD (P.S.F.)	DEPTH (IN.)	TOTAL LOAD (P.S.F.)	COLUMN ADD	COST PER S.F.		
						MAT.	INST.	TOTAL
2350	15x20	40	17	83		4.09	2.72	6.81
2400					column	.61	.25	.86
2450	15x20	65	19	108		4.51	2.88	7.39
2500					column	.61	.25	.86
2550	15x20	75	19	119		4.72	2.97	7.69
2600					column	.67	.27	.94
2650	15x20	100	19	144		5	3.10	8.10
2700					column	.67	.27	.94
2750	15x20	125	19	170		5.80	3.49	9.29
2800					column	.89	.36	1.25
2850	20x20	40	19	83		4.41	2.84	7.25
2900					column	.50	.20	.70
2950	20x20	65	23	109		4.87	3.04	7.91
3000					column	.67	.27	.94
3100	20x20	75	26	119		5.15	3.14	8.29
3200					column	.67	.27	.94
3400	20x20	100	23	144		5.30	3.22	8.52
3450					column	.67	.27	.94
3500	20x20	125	23	170		5.95	3.48	9.43
3600					column	.80	.32	1.12
3700	20x25	40	44	83		5.20	3.27	8.47
3800					column	.54	.22	.76
3900	20x25	65	26	110		5.70	3.45	9.15
4000					column	.54	.22	.76
4100	20x25	75	26	120		5.45	3.28	8.73
4200					column	.64	.26	.90
4300	20x25	100	26	145		5.75	3.42	9.17
4400					column	.64	.26	.90
4500	20x25	125	29	170		6.45	3.70	10.15
4600					column	.75	.30	1.05
4700	25x25	40	23	84		5.60	3.39	8.99
4800					column	.51	.21	.72
4900	25x25	65	29	110		5.90	3.54	9.44
5000					column	.51	.21	.72
5100	25x25	75	26	120		6	3.51	9.51
5200					column	.60	.24	.84

	BAY SIZE (FT.)	SUPERIMPOSED LOAD (P.S.F.)	DEPTH (IN.)	TOTAL LOAD (P.S.F.)	COLUMN ADD	COST PER S.F.		
						MAT.	INST.	TOTAL
5300	25x25	100	29	145		6.70	3.80	10.50
5400					column	.60	.24	.84
5500	25x25	125	32	170		7.10	3.96	11.06
5600					column	.67	.26	.93
5700	25x30	40	29	84		5.65	3.59	9.24
5800					column	.50	.20	.70
5900	25x30	65	29	110		5.90	3.74	9.64
6000					column	.50	.20	.70
6050	25x30	75	29	120		6.25	3.37	9.62
6100					column	.55	.22	.77
6150	25x30	100	29	145		6.80	3.56	10.36
6200					column	.55	.22	.77
6250	25x30	125	32	170		7.40	4.40	11.80
6300					column	.64	.26	.90
6350	30x30	40	29	84		5.80	3.21	9.01
6400					column	.46	.18	.64
6500	30x30	65	29	110		6.60	3.52	10.12
6600					column	.46	.18	.64
6700	30x30	75	32	120		6.75	3.56	10.31
6800					column	.53	.21	.74
6900	30x30	100	35	145		7.50	3.85	11.35
7000					column	.62	.25	.87
7100	30x30	125	35	172		8.35	4.81	13.16
7200					column	.69	.27	.96
7300	30x35	40	29	85		6.60	3.49	10.09
7400					column	.40	.16	.56
7500	30x35	65	29	111		7.45	4.44	11.89
7600					column	.51	.21	.72
7700	30x35	75	32	121		7.45	4.44	11.89
7800					column	.52	.21	.73
7900	30x35	100	35	148		7.95	3.99	11.94
8000					column	.64	.26	.90
8100	30x35	125	38	173		8.85	4.31	13.16
8200					column	.65	.26	.91
8300	35x35	40	32	85		6.75	3.57	10.32
8400					column	.46	.18	.64
8500	35x35	65	35	111		7.85	4.60	12.45
8600					column	.55	.22	.77
9300	35x35	75	38	121		8.05	4.68	12.73
9400					column	.55	.22	.77
9500	35x35	100	38	148		8.65	4.96	13.61
9600					column	.67	.27	.94
9750	35x35	125	41	173		9.20	4.47	13.67
9800					column	.69	.27	.96
9810	5/8 gyp. fireproof.							
9815	On metal furring, add					.53	1.83	2.36

Important: See the Reference Section for critical supporting data - Reference Numbers and City Cost Indexes

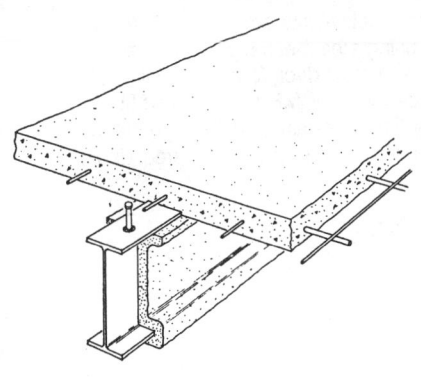

General: Composite construction of wide flange beams and concrete slabs is most efficiently used when loads are heavy and spans are moderately long. It is stiffer with less deflection than non-composite construction of similar depth and spans.

Composite Beam & Cast in Place Slab

3.5-520	Composite Beam & Cast in Place Slab						
	BAY SIZE (FT.)	SUPERIMPOSED LOAD (P.S.F.)	SLAB THICKNESS (IN.)	TOTAL DEPTH (FT. - IN.)	TOTAL LOAD (P.S.F.)	COST PER S.F.	
						MAT.	INST.	TOTAL

	BAY SIZE (FT.)	SUPERIMPOSED LOAD (P.S.F.)	SLAB THICKNESS (IN.)	TOTAL DEPTH (FT. - IN.)	TOTAL LOAD (P.S.F.)	MAT.	INST.	TOTAL
3800	20x25	40	4	1 - 8	94	4.83	6.40	11.23
3900		75	4	1 - 8	130	5.60	6.85	12.45
4000		125	4	1 - 10	181	6.55	7.35	13.90
4100		200	5	2 - 2	272	8.50	8.35	16.85
4200	25x25	40	4-1/2	1 - 8-1/2	99	5.05	6.45	11.50
4300		75	4-1/2	1 - 10-1/2	136	6.10	7	13.10
4400		125	5-1/2	2 - 0-1/2	200	7.20	7.65	14.85
4500		200	5-1/2	2 - 2-1/2	278	8.95	8.55	17.50
4600	25x30	40	4-1/2	1 - 8-1/2	100	5.65	6.75	12.40
4700		75	4-1/2	1 - 10-1/2	136	6.60	7.25	13.85
4800		125	5-1/2	2 - 2-1/2	202	8.10	8	16.10
4900		200	5-1/2	2 - 5-1/2	279	9.65	8.80	18.45
5000	30x30	40	4	1 - 8	95	5.65	6.75	12.40
5200		75	4	2 - 1	131	6.50	7.20	13.70
5400		125	4	2 - 4	183	8.05	8	16.05
5600		200	5	2 - 10	274	10.25	9.10	19.35
5800	30x35	40	4	2 - 1	95	5.95	6.90	12.85
6000		75	4	2 - 4	139	7.15	7.50	14.65
6250		125	4	2 - 4	185	9.30	8.50	17.80
6500		200	5	2 - 11	276	11.35	9.55	20.90
8000	35x35	40	4	2 - 1	96	6.30	7.05	13.35
8250		75	4	2 - 4	133	7.45	7.65	15.10
8500		125	4	2 - 10	185	8.95	8.40	17.35
8750		200	5	3 - 5	276	11.65	9.65	21.30
9000	35x40	40	4	2 - 4	97	6.95	7.40	14.35
9250		75	4	2 - 4	134	8.15	7.85	16
9500		125	4	2 - 7	186	9.90	8.80	18.70
9750		200	5	3 - 5	278	12.80	10.15	22.95

SUPERSTRUCTURES

3

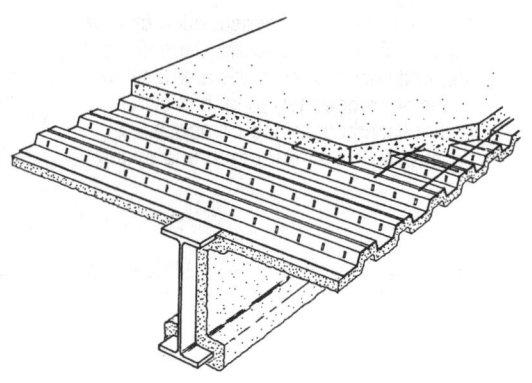

The table below lists costs per S.F. for floors using steel beams and girders, composite steel deck, concrete slab reinforced with W.W.F. and sprayed fiber fireproofing (non- asbestos) on the steel beams and girders and on the steel deck.

Wide Flange, Composite Deck & Slab

3.5-530 — W Shape, Composite Deck, & Slab

	BAY SIZE (FT.) BEAM X GIRD	SUPERIMPOSED LOAD (P.S.F.)	SLAB THICKNESS (IN.)	TOTAL DEPTH (FT.-IN.)	TOTAL LOAD (P.S.F.)	COST PER S.F. MAT.	COST PER S.F. INST.	COST PER S.F. TOTAL
0540	15x20	40	5	1-7	89	5.95	4.21	10.16
0560		75	5	1-9	125	6.45	4.47	10.92
0580		125	5	1-11	176	7.45	4.87	12.32
0600		200	5	2-2	254	9.20	5.55	14.75
0700	20x20	40	5	1-9	90	6.40	4.40	10.80
0720		75	5	1-9	126	7.15	4.75	11.90
0740		125	5	1-11	177	8.05	5.20	13.25
0760		200	5	2-5	255	9.75	5.75	15.50
0780	20x25	40	5	1-9	90	6.65	4.37	11.02
0800		75	5	1-9	127	8.10	5.05	13.15
0820		125	5	1-11	180	9.80	5.50	15.30
0840		200	5	2-5	256	10.40	6.20	16.60
0940	25x25	40	5	1-11	91	7.40	4.69	12.09
0960		75	5	2-5	178	8.45	5.15	13.60
0980		125	5	2-5	181	10.40	5.70	16.10
1000		200	5-1/2	2-8-1/2	263	11.50	6.65	18.15
1400	25x30	40	5	2-5	91	7.25	4.82	12.07
1500		75	5	2-5	128	8.85	5.35	14.20
1600		125	5	2-8	180	9.85	5.95	15.80
1700		200	5	2-11	259	12.25	6.85	19.10
2200	30x30	40	5	2-2	92	8	5.10	13.10
2300		75	5	2-5	129	9.25	5.65	14.90
2400		125	5	2-11	182	10.85	6.40	17.25
2500		200	5	3-2	263	14.60	7.95	22.55
2600	30x35	40	5	2-5	94	8.65	5.40	14.05
2700		75	5	2-11	131	10	5.95	15.95
2800		125	5	3-2	183	11.65	6.70	18.35
2900		200	5-1/2	3-5-1/2	268	14.25	7.65	21.90
3800	35x35	40	5	2-8	94	9	5.40	14.40
3900		75	5	2-11	131	10.35	6	16.35
4000		125	5	3-5	184	12.30	6.85	19.15
4100		200	5-1/2	3-5-1/2	270	15.90	8.10	24

Important: See the Reference Section for critical supporting data - Reference Numbers and City Cost Indexes

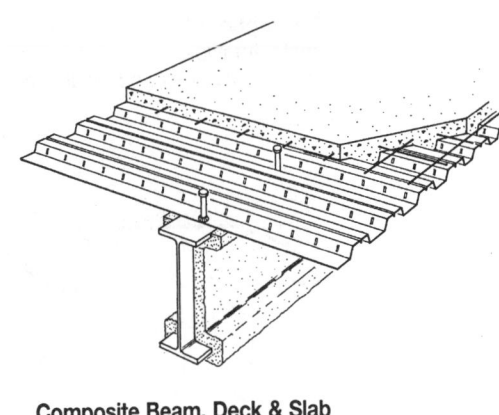

The table below lists costs per S.F. for floors using composite steel beams with welded shear studs, composite steel deck and light weight concrete slab reinforced with W.W.F. and sprayed fiber fireproofing (non-asbestos) on the steel beams.

Composite Beam, Deck & Slab

3.5-540 — Composite Beams, Deck & Slab

	BAY SIZE (FT.)	SUPERIMPOSED LOAD (P.S.F.)	SLAB THICKNESS (IN.)	TOTAL DEPTH (FT. - IN.)	TOTAL LOAD (P.S.F.)	COST PER S.F.		
						MAT.	INST.	TOTAL
2400	20x25	40	5-1/2	1 - 5-1/2	80	6	3.86	9.86
2500		75	5-1/2	1 - 9-1/2	115	6.25	3.87	10.12
2750		125	5-1/2	1 - 9-1/2	167	7.40	4.56	11.96
2900		200	6-1/4	1 - 11-1/2	251	8.35	4.92	13.27
3000	25x25	40	5-1/2	1 - 9-1/2	82	5.90	3.68	9.58
3100		75	5-1/2	1 - 11-1/2	118	6.55	3.74	10.29
3200		125	5-1/2	2 - 2-1/2	169	6.85	4.06	10.91
3300		200	6-1/4	2 - 6-1/4	252	9.30	4.72	14.02
3400	25x30	40	5-1/2	1 - 11-1/2	83	6	3.64	9.64
3600		75	5-1/2	1 - 11-1/2	119	6.45	3.69	10.14
3900		125	5-1/2	1 - 11-1/2	170	7.45	4.17	11.62
4000		200	6-1/4	2 - 6-1/4	252	9.35	4.74	14.09
4200	30x30	40	5-1/2	1 - 11-1/2	81	6.05	3.78	9.83
4400		75	5-1/2	2 - 2-1/2	116	6.35	3.93	10.28
4500		125	5-1/2	2 - 5-1/2	168	7.65	4.43	12.08
4700		200	6-1/4	2 - 9-1/4	252	9.15	5.15	14.30
4900	30x35	40	5-1/2	2 - 2-1/2	82	6.35	3.90	10.25
5100		75	5-1/2	2 - 5-1/2	117	6.90	4	10.90
5300		125	5-1/2	2 - 5-1/2	169	7.90	4.52	12.42
5500		200	6-1/4	2 - 9-1/4	254	9.40	5.15	14.55
5750	35x35	40	5-1/2	2 - 5-1/2	84	6.70	3.91	10.61
6000		75	5-1/2	2 - 5-1/2	121	7.60	4.19	11.79
7000		125	5-1/2	2 - 8-1/2	170	8.85	4.79	13.64
7200		200	5-1/2	2 - 11-1/2	254	10.40	5.30	15.70
7400	35x40	40	5-1/2	2 - 5-1/2	85	7.35	4.20	11.55
7600		75	5-1/2	2 - 5-1/2	121	7.95	4.34	12.29
8000		125	5-1/2	2 - 5-1/2	171	9.05	4.85	13.90
9000		200	5-1/2	2 - 11-1/2	255	11.15	5.55	16.70

SUPERSTRUCTURES

3

Description: Table 3.5-580 lists S.F. costs for steel deck and concrete slabs for various spans.

Description: Table 3.5-710 lists the S.F. costs for wood joists and a minimum thickness plywood subfloor.

Description: Table 3.5-720 lists the S.F. costs, total load, and member sizes, for various bay sizes and loading conditions.

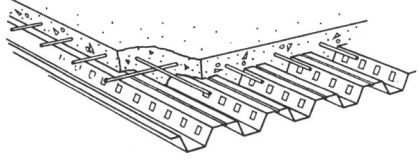

Metal Deck/Concrete Fill

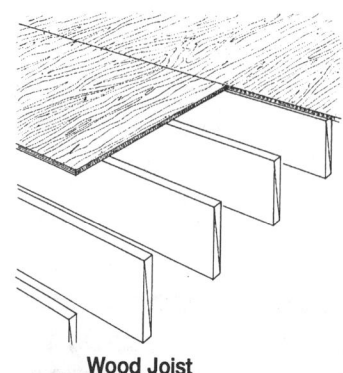

Wood Joist

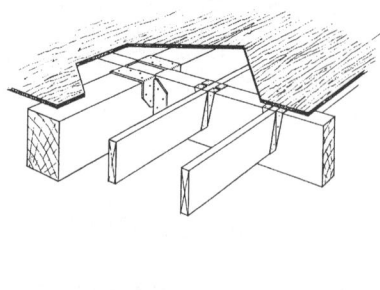

Wood Beam and Joist

3.5-580 — Metal Deck/Concrete Fill

	SUPERIMPOSED LOAD (P.S.F.)	DECK SPAN (FT.)	DECK GAGE DEPTH	SLAB THICKNESS (IN.)	TOTAL LOAD (P.S.F.)	COST PER S.F.		
						MAT.	INST.	TOTAL
0900	125	6	22 1-1/2	4	164	1.51	1.55	3.06
0920		7	20 1-1/2	4	164	1.60	1.63	3.23
0950		8	20 1-1/2	4	165	1.60	1.63	3.23
0970		9	18 1-1/2	4	165	1.88	1.63	3.51
1000		10	18 2	4	165	1.97	1.71	3.68
1020		11	18 3	5	169	2.45	1.86	4.31

3.5-710 — Wood Joist

		COST PER S.F.		
		MAT.	INST.	TOTAL
2902	Wood joist, 2" x 8", 12" O.C.	1.42	1.12	2.54
2950	16" O.C.	1.19	.96	2.15
3000	24" O.C.	1.17	.90	2.07
3300	2"x10", 12" O.C.	1.81	1.28	3.09
3350	16" O.C.	1.48	1.07	2.55
3400	24" O.C.	1.36	.96	2.32
3700	2"x12", 12" O.C.	2.20	1.30	3.50
3750	16" O.C.	1.77	1.09	2.86
3800	24" O.C.	1.55	.98	2.53
4100	2"x14", 12" O.C.	2.58	1.41	3.99
4150	16" O.C.	2.06	1.17	3.23
4200	24" O.C.	1.75	1.04	2.79
7101	Note: Subfloor cost is included in these prices			

3.5-720 — Wood Beam & Joist

	BAY SIZE (FT.)	SUPERIMPOSED LOAD (P.S.F.)	GIRDER BEAM (IN.)	JOISTS (IN.)	TOTAL LOAD (P.S.F.)	COST PER S.F.		
						MAT.	INST.	TOTAL
2000	15x15	40	8 x 12 4 x 12	2 x 6 @ 16	53	4.58	2.62	7.20
2050		75	8 x 16 4 x 16	2 x 8 @ 16	90	6.35	2.86	9.21
2100		125	12 x 16 6 x 16	2 x 8 @ 12	144	9.60	3.69	13.29
2150		200	14 x 22 12 x 16	2 x 10 @ 12	227	17.70	5.65	23.35
3000	20x20	40	10 x 14 10 x 12	2 x 8 @ 16	63	6.70	2.63	9.33
3050		75	12 x 16 8 x 16	2 x 10 @ 16	102	8.70	2.96	11.66
3100		125	14 x 22 12 x 16	2 x 10 @ 12	163	14.30	4.70	19

Important: See the Reference Section for critical supporting data - Reference Numbers and City Cost Indexes

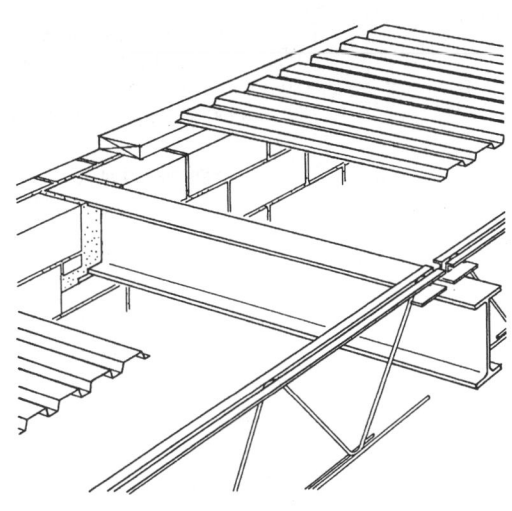

The table below lists the cost per S.F. for a roof system with steel columns, beams and deck using open web steel joists and 1-1/2″ galvanized metal deck. Perimeter of system is supported on bearing walls.

Fireproofing is not included. Costs/S.F. are based on a building 4 bays long and 4 bays wide.

Column costs are additive. Costs for the bearing walls are not included.

Steel Joists, Beams and Deck on Bearing Walls

3.7-410		**Steel Joists, Beams & Deck on Columns & Walls**						
	BAY SIZE (FT.)	SUPERIMPOSED LOAD (P.S.F.)	DEPTH (IN.)	TOTAL LOAD (P.S.F.)	COLUMN ADD	COST PER S.F.		
						MAT.	INST.	TOTAL
3000	25x25	20	18	40		2.08	.87	2.95
3100					columns	.17	.05	.22
3200		30	22	50		2.41	1.03	3.44
3300					columns	.23	.08	.31
3400		40	20	60		2.48	1.02	3.50
3500					columns	.23	.08	.31
3600	25x30	20	22	40		2.20	.84	3.04
3700					columns	.19	.06	.25
3800		30	20	50		2.42	1.11	3.53
3900					columns	.19	.06	.25
4000		40	25	60		2.59	.97	3.56
4100					columns	.23	.08	.31
4200	30x30	20	25	42		2.40	.91	3.31
4300					columns	.16	.05	.21
4400		30	22	52		2.62	.99	3.61
4500					columns	.19	.06	.25
4600		40	28	62		2.72	1.03	3.75
4700					columns	.19	.06	.25
4800	30x35	20	22	42		2.51	.95	3.46
4900					columns	.17	.05	.22
5000		30	28	52		2.64	.99	3.63
5100					columns	.17	.05	.22
5200		40	25	62		2.86	1.06	3.92
5300					columns	.19	.06	.25
5400	35x35	20	28	42		2.52	.96	3.48
5500					columns	.14	.04	.18

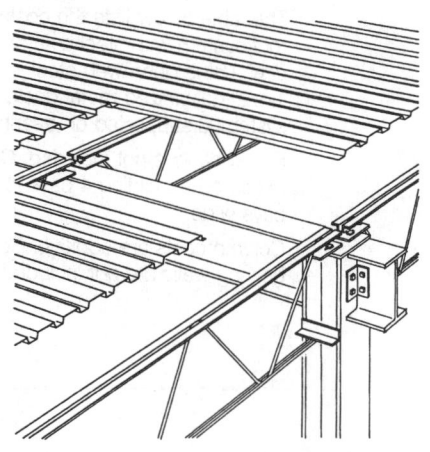

Description: The table below lists the cost per S.F. for a roof system with steel columns, beams, and deck, using open web steel joists and 1-1/2″ galvanized metal deck.

Column costs are additive. Fireproofing is not included.

Steel Joists, Beams and Deck on Columns

3.7-420 — Steel Joists, Beams, & Deck on Columns

	BAY SIZE (FT.)	SUPERIMPOSED LOAD (P.S.F.)	DEPTH (IN.)	TOTAL LOAD (P.S.F.)	COLUMN ADD	COST PER S.F. MAT.	INST.	TOTAL
1100	15x20	20	16	40		2.06	.86	2.92
1200					columns	1	.33	1.33
1500		40	18	60		2.31	.97	3.28
1600					columns	1	.33	1.33
2300	20x25	20	18	40		2.35	.96	3.31
2400					columns	.60	.20	.80
2700		40	20	60		2.64	1.09	3.73
2800					columns	.80	.26	1.06
2900	25x25	20	18	40		2.72	1.09	3.81
3000					columns	.48	.16	.64
3100		30	22	50		3.01	1.24	4.25
3200					columns	.64	.21	.85
3300		40	20	60		3.15	1.24	4.39
3400					columns	.64	.21	.85
3500	25x30	20	22	40		2.66	1	3.66
3600					columns	.54	.17	.71
3900		40	25	60		3.21	1.18	4.39
4000					columns	.64	.21	.85
4100	30x30	20	25	42		3	1.10	4.10
4200					columns	.45	.14	.59
4300		30	22	52		3.30	1.20	4.50
4400					columns	.54	.17	.71
4500		40	28	62		3.45	1.25	4.70
4600					columns	.54	.17	.71
5300	35x35	20	28	42		3.28	1.20	4.48
5400					columns	.39	.13	.52
5500		30	25	52		3.65	1.32	4.97
5600					columns	.46	.15	.61
5700		40	28	62		3.94	1.42	5.36
5800					columns	.51	.17	.68

Important: See the Reference Section for critical supporting data - Reference Numbers and City Cost Indexes

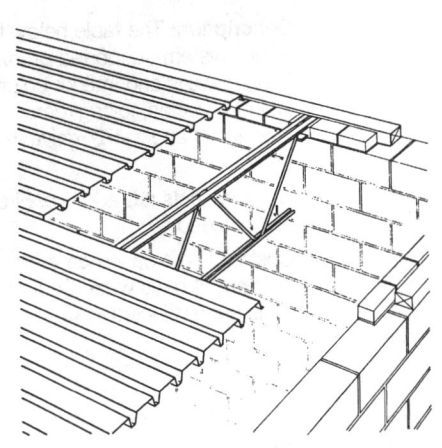

Description: The table below lists the cost per S.F. for a roof system using open web steel joists and 1-1/2" galvanized metal deck. The system is assumed supported on bearing walls or other suitable support. Costs for the supports are not included.

3.7-430 — Steel Joists & Deck on Bearing Walls

	BAY SIZE (FT.)	SUPERIMPOSED LOAD (P.S.F.)	DEPTH (IN.)	TOTAL LOAD (P.S.F.)		COST PER S.F. MAT.	INST.	TOTAL
1100	20	20	13-1/2	40		1.37	.64	2.01
1200		30	15-1/2	50		1.40	.65	2.05
1300		40	15-1/2	60		1.51	.70	2.21
1400	25	20	17-1/2	40		1.51	.70	2.21
1500		30	17-1/2	50		1.64	.75	2.39
1600		40	19-1/2	60		1.67	.77	2.44
1700	30	20	19-1/2	40		1.66	.76	2.42
1800		30	21-1/2	50		1.70	.78	2.48
1900		40	23-1/2	60		1.84	.84	2.68
2000	35	20	23-1/2	40		1.78	.71	2.49
2100		30	25-1/2	50		1.84	.74	2.58
2200		40	25-1/2	60		1.96	.78	2.74
2300	40	20	25-1/2	41		1.99	.79	2.78
2400		30	25-1/2	51		2.12	.83	2.95
2500		40	25-1/2	61		2.19	.86	3.05
2600	45	20	27-1/2	41		2.28	1.18	3.46
2700		30	31-1/2	51		2.42	1.24	3.66
2800		40	31-1/2	61		2.55	1.32	3.87
2900	50	20	29-1/2	42		2.56	1.32	3.88
3000		30	31-1/2	52		2.80	1.44	4.24
3100		40	31-1/2	62		2.98	1.54	4.52
3200	60	20	37-1/2	42		2.95	1.33	4.28
3300		30	37-1/2	52		3.27	1.47	4.74
3400		40	37-1/2	62		3.27	1.47	4.74
3500	70	20	41-1/2	42		3.26	1.47	4.73
3600		30	41-1/2	52		3.48	1.56	5.04
3700		40	41-1/2	64		4.23	1.89	6.12
3800	80	20	45-1/2	44		4.46	2.11	6.57
3900		30	45-1/2	54		4.46	2.11	6.57
4000		40	45-1/2	64		4.93	2.34	7.27
4400	100	20	57-1/2	44		3.89	1.74	5.63
4500		30	57-1/2	54		4.64	2.07	6.71
4600		40	57-1/2	65		5.10	2.28	7.38
4700	125	20	69-1/2	44		5.50	2.39	7.89
4800		30	69-1/2	56		6.45	2.77	9.22
4900		40	69-1/2	67		7.35	3.16	10.51

SUPERSTRUCTURES

3

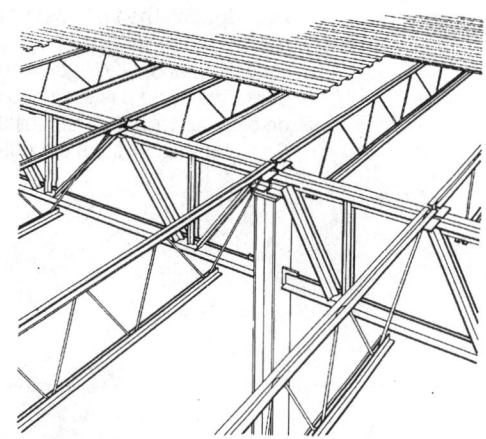

Description: The table below lists costs for a roof system supported on exterior bearing walls and interior columns. Costs include bracing, joist girders, open web steel joists and 1-1/2″ galvanized metal deck.

Column costs are additive. Fireproofing is not included.

Costs/S.F. are based on a building 4 bays long and 4 bays wide. Costs for bearing walls are not included.

3.7-440 | Steel Joists, Joist Girders & Deck on Columns & Walls

	BAY SIZE (FT.) GIRD X JOISTS	SUPERIMPOSED LOAD (P.S.F.)	DEPTH (IN.)	TOTAL LOAD (P.S.F.)	COLUMN ADD	COST PER S.F. MAT.	COST PER S.F. INST.	COST PER S.F. TOTAL
2350	30x35	20	32-1/2	40		1.97	.95	2.92
2400					columns	.17	.05	.22
2550		40	36-1/2	60		2.21	1.05	3.26
2600					columns	.19	.06	.25
3000	35x35	20	36-1/2	40		2.13	1.23	3.36
3050					columns	.14	.05	.19
3200		40	36-1/2	60		2.42	1.34	3.76
3250					columns	.18	.06	.24
3300	35x40	20	36-1/2	40		2.18	1.07	3.25
3350					columns	.14	.05	.19
3500		40	36-1/2	60		2.46	1.19	3.65
3550					columns	.16	.05	.21
3900	40x40	20	40-1/2	41		2.48	1.37	3.85
3950					columns	.14	.05	.19
4100		40	40-1/2	61		2.78	1.48	4.26
4150					columns	.14	.05	.19
5100	45x50	20	52-1/2	41		2.80	1.58	4.38
5150					columns	.10	.04	.14
5300		40	52-1/2	61		3.41	1.87	5.28
5350					columns	.14	.05	.19
5400	50x45	20	56-1/2	41		2.71	1.52	4.23
5450					columns	.10	.04	.14
5600		40	56-1/2	61		3.49	1.91	5.40
5650					columns	.14	.04	.18
5700	50x50	20	56-1/2	42		3.06	1.72	4.78
5750					columns	.10	.04	.14
5900		40	59	64		3.66	1.92	5.58
5950					columns	.14	.05	.19
6300	60x50	20	62-1/2	43		3.08	1.75	4.83
6350					columns	.16	.05	.21
6500		40	71	65		3.75	1.98	5.73
6550					columns	.21	.07	.28

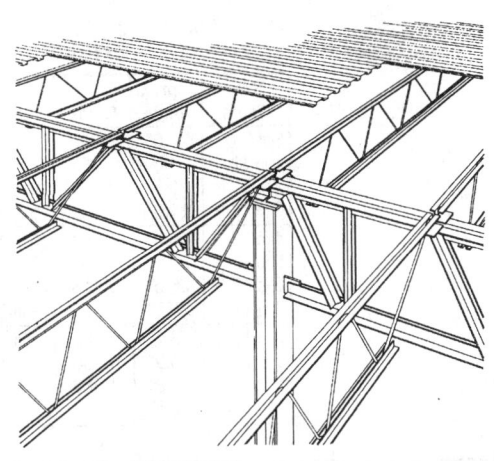

Description: The table below lists the cost per S.F. for a roof system supported on columns. Costs include joist girders, open web steel joists and 1-1/2″ galvanized metal deck. Column costs are additive.

Fireproofing is not included. Costs/S.F. are based on a building 4 bays long and 4 bays wide.

Costs for wind columns are not included.

3.7-450 — Steel Joists & Joist Girders on Columns

	BAY SIZE (FT.) GIRD X JOISTS	SUPERIMPOSED LOAD (P.S.F.)	DEPTH (IN.)	TOTAL LOAD (P.S.F.)	COLUMN ADD	COST PER S.F. MAT.	COST PER S.F. INST.	COST PER S.F. TOTAL
2000	30x30	20	17-1/2	40		2	1.15	3.15
2050					columns	.54	.17	.71
2100		30	17-1/2	50		2.19	1.21	3.40
2150					columns	.54	.17	.71
2200		40	21-1/2	60		2.24	1.25	3.49
2250					columns	.54	.17	.71
2300	30x35	20	32-1/2	40		2.11	1.08	3.19
2350					columns	.46	.14	.60
2500		40	36-1/2	60		2.39	1.20	3.59
2550					columns	.54	.17	.71
3200	35x40	20	36-1/2	40		2.71	1.69	4.40
3250					columns	.40	.13	.53
3400		40	36-1/2	60		3.06	1.85	4.91
3450					columns	.45	.14	.59
3800	40x40	20	40-1/2	41		2.74	1.68	4.42
3850					columns	.39	.13	.52
4000		40	40-1/2	61		3.08	1.85	4.93
4050					columns	.39	.13	.52
4100	40x45	20	40-1/2	41		2.95	1.94	4.89
4150					columns	.35	.12	.47
4200		30	40-1/2	51		3.21	2.06	5.27
4250					columns	.35	.12	.47
4300		40	40-1/2	61		3.58	2.23	5.81
4350					columns	.39	.13	.52
5000	45x50	20	52-1/2	41		3.12	1.94	5.06
5050					columns	.28	.09	.37
5200		40	52-1/2	61		3.83	2.30	6.13
5250					columns	.36	.12	.48
5300	50x45	20	56-1/2	41		3.03	1.90	4.93
5350					columns	.28	.09	.37
5500		40	56-1/2	61		3.92	2.34	6.26
5550					columns	.36	.12	.48
5600	50x50	20	56-1/2	42		3.38	2.08	5.46
5650					columns	.28	.09	.37
5800		40	59	64		3.22	1.90	5.12
5850					columns	.39	.13	.52

The table below lists prices per S.F. for roof rafters and sheathing by nominal size and spacing. Sheathing is 5/16" CDX for 12" and 16" spacing and 3/8" CDX for 24" spacing.

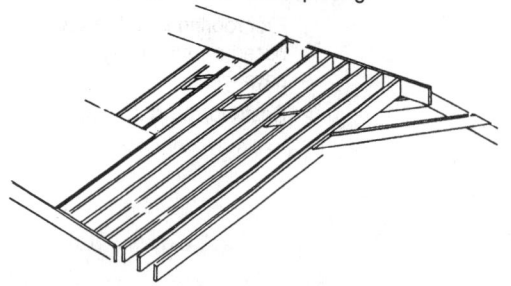

Factors for Converting Inclined to Horizontal

Roof Slope	Approx. Angle	Factor	Roof Slope	Approx. Angle	Factor
Flat	0°	1.000	12 in 12	45.0°	1.414
1 in 12	4.8°	1.003	13 in 12	47.3°	1.474
2 in 12	9.5°	1.014	14 in 12	49.4°	1.537
3 in 12	14.0°	1.031	15 in 12	51.3°	1.601
4 in 12	18.4°	1.054	16 in 12	53.1°	1.667
5 in 12	22.6°	1.083	17 in 12	54.8°	1.734
6 in 12	26.6°	1.118	18 in 12	56.3°	1.803
7 in 12	30.3°	1.158	19 in 12	57.7°	1.873
8 in 12	33.7°	1.202	20 in 12	59.0°	1.943
9 in 12	36.9°	1.250	21 in 12	60.3°	2.015
10 in 12	39.8°	1.302	22 in 12	61.4°	2.088
11 in 12	42.5°	1.357	23 in 12	62.4°	2.162

3.7-510	Wood/Flat or Pitched	COST PER S.F.		
		MAT.	INST.	TOTAL
2500	Flat rafter, 2"x4", 12" O.C.	.84	1.03	1.87
2550	16" O.C.	.73	.88	1.61
2600	24" O.C.	.62	.76	1.38
2900	2"x6", 12" O.C.	1.04	1.03	2.07
2950	16" O.C.	.88	.88	1.76
3000	24" O.C.	.72	.75	1.47
3300	2"x8", 12" O.C.	1.31	1.10	2.41
3350	16" O.C.	1.08	.94	2.02
3400	24" O.C.	.86	.80	1.66
3700	2"x10", 12" O.C.	1.70	1.26	2.96
3750	16" O.C.	1.37	1.05	2.42
3800	24" O.C.	1.05	.86	1.91
4100	2"x12", 12" O.C.	2.09	1.28	3.37
4150	16" O.C.	1.66	1.07	2.73
4200	24" O.C.	1.24	.88	2.12
4500	2"x14", 12" O.C.	2.47	1.86	4.33
4550	16" O.C.	1.95	1.51	3.46
4600	24" O.C.	1.44	1.18	2.62
4900	3"x6", 12" O.C.	2.36	1.08	3.44
4950	16" O.C.	1.87	.92	2.79
5000	24" O.C.	1.39	.78	2.17
5300	3"x8", 12" O.C.	3.02	1.21	4.23
5350	16" O.C.	2.36	1.02	3.38
5400	24" O.C.	1.71	.85	2.56
5700	3"x10", 12" O.C.	3.68	1.38	5.06
5750	16" O.C.	2.86	1.14	4
5800	24" O.C.	2.05	.93	2.98
6100	3"x12", 12" O.C.	4.34	1.66	6
6150	16" O.C.	3.36	1.36	4.72
6200	24" O.C.	2.37	1.07	3.44
7001	Wood truss, 4 in 12 slope, 24" O.C., 24' to 29' span	2.61	1.55	4.16
7100	30' to 43' span	2.78	1.55	4.33
7200	44' to 60' span	2.94	1.55	4.49

Important: See the Reference Section for critical supporting data - Location Factors & Historical Cost Indexes

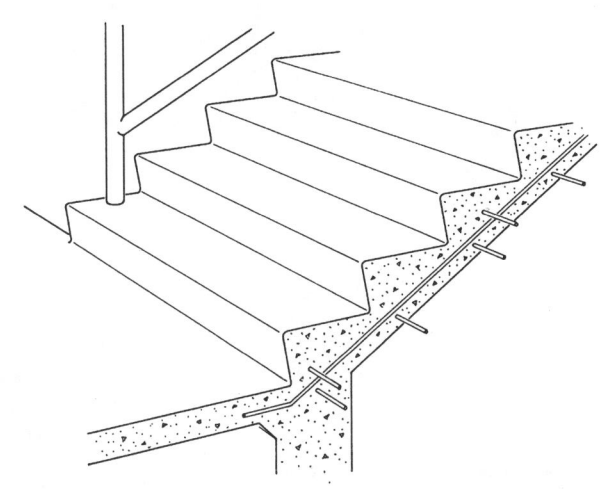

The table below lists the cost per flight for 4'-0" wide stairs. Side walls are not included. Railings are included in the prices.

3.9-100	Stairs	COST PER FLIGHT		
		MAT.	INST.	TOTAL
0470	Stairs, C.I.P. concrete, w/o landing, 12 risers, w/o nosing	655	1,425	2,080
0480	With nosing	1,000	1,625	2,625
0550	W/landing, 12 risers, w/o nosing	735	1,775	2,510
0560	With nosing	1,100	1,975	3,075
0570	16 risers, w/o nosing	910	2,200	3,110
0580	With nosing	1,375	2,475	3,850
0590	20 risers, w/o nosing	1,100	2,675	3,775
0600	With nosing	1,675	3,000	4,675
0610	24 risers, w/o nosing	1,275	3,100	4,375
0620	With nosing	1,975	3,500	5,475
0630	Steel, grate type w/nosing & rails, 12 risers, w/o landing	1,625	515	2,140
0640	With landing	2,525	800	3,325
0660	16 risers, with landing	3,075	965	4,040
0680	20 risers, with landing	3,625	1,125	4,750
0700	24 risers, with landing	4,175	1,325	5,500
0701				
0710	Cement fill metal pan & picket rail, 12 risers, w/o landing	2,100	515	2,615
0720	With landing	3,300	885	4,185
0740	16 risers, with landing	3,975	1,050	5,025
0760	20 risers, with landing	4,675	1,225	5,900
0780	24 risers, with landing	5,375	1,425	6,800
0790	Cast iron tread & pipe rail, 12 risers, w/o landing	2,125	515	2,640
0800	With landing	3,325	885	4,210
1120	Wood, prefab box type, oak treads, wood rails 3'-6" wide, 14 risers	1,125	278	1,403
1150	Prefab basement type, oak treads, wood rails 3'-0" wide, 14 risers	815	223	1,038

For information about Means Estimating Seminars, see yellow pages 11 and 12 in back of book

SUPERSTRUCTURES

3

Division 4
Exterior Closure

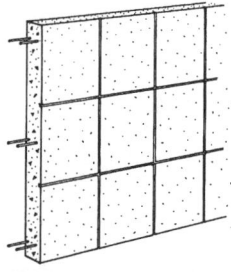

Cast in Place Concrete Wall

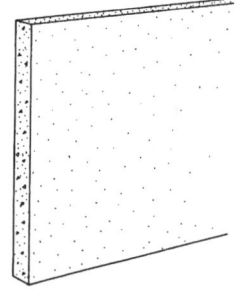

Flat Precast Concrete Wall

Table 4.1-110 describes a concrete wall system for exterior closure. There are several types of wall finishes priced from plain finish to a finish with 3/4" rustication strip.

In Table 4.1-140, precast concrete wall panels are either solid or insulated. Prices below are based on delivery within 50 miles of a plant. Small jobs can double the prices below. For large, highly repetitive jobs, deduct up to 15%.

4.1-110	Cast In Place Concrete	COST PER S.F.		
		MAT.	INST.	TOTAL
2100	Conc wall reinforced, 8' high, 6" thick, plain finish, 3000 PSI	2.77	9.60	12.37
2700	Aged wood liner, 3000 PSI	4.53	10.05	14.58
3000	Sand blast light 1 side, 3000 PSI	2.99	10.40	13.39
3700	3/4" bevel rustication strip, 3000 PSI	2.93	11.75	14.68
4000	8" thick, plain finish, 3000 PSI	3.25	9.85	13.10
4100	4000 PSI	3.33	9.85	13.18
4300	Rub concrete 1 side, 3000 PSI	3.33	11.60	14.93
4550	8" thick, aged wood liner, 3000 PSI	5	10.30	15.30
4750	Sand blast light 1 side, 3000 PSI	3.47	10.65	14.12
5300	3/4" bevel rustication strip, 3000 PSI	3.41	12	15.41
5600	10" thick, plain finish, 3000 PSI	3.72	10.10	13.82
5900	Rub concrete 1 side, 3000 PSI	3.80	11.85	15.65
6200	Aged wood liner, 3000 PSI	5.50	10.55	16.05
6500	Sand blast light 1 side, 3000 PSI	3.94	10.90	14.84
7100	3/4" bevel rustication strip, 3000 PSI	3.88	12.25	16.13
7400	12" thick, plain finish, 3000 PSI	4.26	10.40	14.66
7700	Rub concrete 1 side, 3000 PSI	4.34	12.15	16.49
8000	Aged wood liner, 3000 PSI	6	10.85	16.85
8300	Sand blast light 1 side, 3000 PSI	4.48	11.20	15.68
8900	3/4" bevel rustication strip, 3000 PSI	4.42	12.55	16.97

4.1-140			Flat Precast Concrete					
	THICKNESS (IN.)	PANEL SIZE (FT.)	FINISHES	RIGID INSULATION (IN)	TYPE	COST PER S.F.		
						MAT.	INST.	TOTAL
3000	4	5x18	smooth gray	none	low rise	5.20	5.20	10.40
3050		6x18				4.35	4.34	8.69
3100		8x20				6.35	1.62	7.97
3150		12x20				6	1.53	7.53
3200	6	5x18	smooth gray	2	low rise	6	5.50	11.50
3250		6x18				5.15	4.65	9.80
3300		8x20				7.25	2.01	9.26
3350		12x20				6.60	1.85	8.45
3400	8	5x18	smooth gray	2	low rise	9.30	2.52	11.82
3450		6x18				8.80	2.40	11.20
3500		8x20				8.10	2.21	10.31
3550		12x20				7.40	2.04	9.44
3600	4	4x8	white face	none	low rise	15.90	2.99	18.89
3650		8x8				11.90	2.24	14.14
3700		10x10				10.45	1.97	12.42
3750		20x10				9.45	1.78	11.23

Important: See the Reference Section for critical supporting data - Location Factors & Historical Cost Indexes

EXTERIOR CLOSURE 4

4.1-140 — Flat Precast Concrete

	THICKNESS (IN.)	PANEL SIZE (FT.)	FINISHES	RIGID INSULATION (IN)	TYPE	COST PER S.F.		
						MAT.	INST.	TOTAL
3800	5	4x8	white face	none	low rise	16.20	3.06	19.26
3850		8x8				12.25	2.30	14.55
3900		10x10				10.85	2.04	12.89
3950		20x20				9.95	1.87	11.82
4000	6	4x8	white face	none	low rise	16.80	3.16	19.96
4050		8x8				12.75	2.40	15.15
4100		10x10				11.25	2.12	13.37
4150		20x10				10.35	1.94	12.29
4200	6	4x8	white face	2	low rise	17.65	3.53	21.18
4250		8x8				13.60	2.77	16.37
4300		10x10				12.10	2.49	14.59
4350		20x10				10.35	1.94	12.29
4400	7	4x8	white face	none	low rise	17.20	3.24	20.44
4450		8x8				13.20	2.49	15.69
4500		10x10				11.85	2.23	14.08
4550		20x10				10.85	2.04	12.89
4600	7	4x8	white face	2	low rise	18.05	3.61	21.66
4650		8x8				14.05	2.86	16.91
4700		10x10				12.70	2.60	15.30
4750		20x10				11.70	2.41	14.11
4800	8	4x8	white face	none	low rise	17.60	3.31	20.91
4850		8x8				13.55	2.55	16.10
4900		10x10				12.20	2.30	14.50
4950		20x10				11.25	2.11	13.36
5000	8	4x8	white face	2	low rise	18.45	3.68	22.13
5050		8x8				14.40	2.92	17.32
5100		10x10				13.05	2.67	15.72
5150		20x10				12.05	2.48	14.53

4.1-140 — Fluted Window or Mullion Precast Concrete

	THICKNESS (IN.)	PANEL SIZE (FT.)	FINISHES	RIGID INSULATION (IN)	TYPE	COST PER S.F.		
						MAT.	INST.	TOTAL
5200	4	4x8	smooth gray	none	high rise	11.90	11.90	23.80
5250		8x8				8.50	8.45	16.95
5300		10x10				12.15	3.10	15.25
5350		20x10				10.65	2.72	13.37
5400	5	4x8	smooth gray	none	high rise	12.10	12.05	24.15
5450		8x8				8.80	8.75	17.55
5500		10x10				12.70	3.23	15.93
5550		20x10				11.20	2.86	14.06
5600	6	4x8	smooth gray	none	high rise	12.45	12.40	24.85
5650		8x8				9.10	9.05	18.15
5700		10x10				13.05	3.33	16.38
5750		20x10				11.60	2.95	14.55
5800	6	4x8	smooth gray	2	high rise	13.25	12.75	26
5850		8x8				9.95	9.45	19.40
5900		10x10				13.90	3.70	17.60
5950		20x10				12.45	3.32	15.77
6000	7	4x8	smooth gray	none	high rise	12.70	12.65	25.35
6050		8x8				9.35	9.35	18.70
6100		10x10				13.60	3.48	17.08
6150		20x10				11.95	3.05	15

EXTERIOR CLOSURE

4

4.1-140 | Fluted Window or Mullion Precast Concrete

	THICKNESS (IN.)	PANEL SIZE (FT.)	FINISHES	RIGID INSULATION (IN)	TYPE	COST PER S.F.		
						MAT.	INST.	TOTAL
6200	7	4x8	smooth gray	2	high rise	13.55	13.05	26.60
6250		8x8				10.20	9.70	19.90
6300		10x10				14.45	3.85	18.30
6350		20x10				12.80	3.42	16.22
6400	8	4x8	smooth gray	none	high rise	12.85	12.85	25.70
6450		8x8				9.60	9.55	19.15
6500		10x10				14.05	3.58	17.63
6550		20x10				12.55	3.20	15.75
6600	8	4x8	smooth gray	2	high rise	13.70	13.20	26.90
6650		8x8				10.45	9.90	20.35
6700		10x10				14.90	3.95	18.85
6750		20x10				13.40	3.57	16.97

4.1-140 | Precast Concrete Specialties

	TYPE	SIZE				COST PER L.F.		
						MAT.	INST.	TOTAL
6773	Coping, precast	6"wide				10.55	7.95	18.50
6774	Stock units	10"wide				11.10	8.55	19.65
6775		12"wide				10	9.20	19.20
6776		14" wide				15.30	9.95	25.25
6777	Window sills	6" wide				9.85	8.55	18.40
6778	Precast	10" wide				9.15	9.95	19.10
6779		14" wide				13.35	11.95	25.30

Important: See the Reference Section for critical supporting data - Location Factors & Historical Cost Indexes

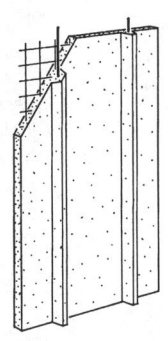

Ribbed Precast Panel

4.1-140		Ribbed Precast Concrete						
	THICKNESS (IN.)	PANEL SIZE (FT.)	FINISHES	RIGID INSULATION(IN.)	TYPE	COST PER S.F.		
						MAT.	INST.	TOTAL
6800	4	4x8	aggregate	none	high rise	12.95	11.85	24.80
6850		8x8				9.45	8.70	18.15
6900		10x10				13.45	3.13	16.58
6950		20x10				11.90	2.76	14.66
7000	5	4x8	aggregate	none	high rise	13.20	12.05	25.25
7050		8x8				9.75	8.90	18.65
7100		10x10				14	3.25	17.25
7150		20x10				12.50	2.91	15.41
7200	6	4x8	aggregate	none	high rise	13.45	12.30	25.75
7250		8x8				10	9.15	19.15
7300		10x10				14.45	3.36	17.81
7350		20x10				12.95	3.01	15.96
7400	6	4x8	aggregate	2	high rise	14.30	12.70	27
7450		8x8				10.85	9.55	20.40
7500		10x10				15.30	3.73	19.03
7550		20x10				13.80	3.38	17.18
7600	7	4x8	aggregate	none	high rise	13.75	12.55	26.30
7650		8x8				10.25	9.35	19.60
7700		10x10				14.90	3.46	18.36
7750		20x10				13.30	3.10	16.40
7800	7	4x8	aggregate	2	high rise	14.60	12.95	27.55
7850		8x8				11.10	9.75	20.85
7900		10x10				15.75	3.83	19.58
7950		20x10				14.15	3.47	17.62
8000	8	4x8	aggregate	none	high rise	14	12.80	26.80
8050		8x8				10.55	9.65	20.20
8100		10x10				15.40	3.58	18.98
8150		20x10				13.90	3.23	17.13
8200	8	4x8	aggregate	2	high rise	14.85	13.20	28.05
8250		8x8				11.40	10.05	21.45
8300		10x10				16.25	3.95	20.20
8350		20x10				14.75	3.60	18.35

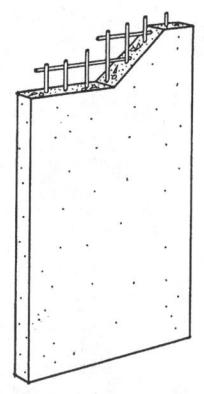

The advantage of tilt up construction is in the low cost of forms and placing of concrete and reinforcing. Tilt up has been used for several types of buildings, including warehouses, stores, offices, and schools. The panels are cast in forms on the ground, or floor slab. Most jobs use 5-1/2" thick solid reinforced concrete panels.

Tilt Up Concrete Panel

4.1-160	Tilt-Up Concrete Panel	COST PER S.F.		
		MAT.	INST.	TOTAL
3200	Tilt up conc panels, broom finish, 5-1/2" thick, 3000 PSI	2.40	2.87	5.27
3250	5000 PSI	2.47	2.80	5.27
3300	6" thick, 3000 PSI	2.63	2.97	5.60
3350	5000 PSI	2.69	2.90	5.59
3400	7-1/2" thick, 3000 PSI	3.27	3.11	6.38
3450	5000 PSI	3.36	3.04	6.40
3500	8" thick, 3000 PSI	3.50	3.21	6.71
3550	5000 PSI	3.61	3.14	6.75
3700	Steel trowel finish, 5-1/2" thick, 3000 PSI	2.41	2.97	5.38
3750	5000 PSI	2.47	2.90	5.37
3800	6" thick, 3000 PSI	2.63	3.07	5.70
3850	5000 PSI	2.69	3	5.69
3900	7-1/2" thick, 3000 PSI	3.27	3.21	6.48
3950	5000 PSI	3.36	3.14	6.50
4000	8" thick, 3000 PSI	3.50	3.31	6.81
4050	5000 PSI	3.61	3.24	6.85
4200	Exp. aggregate finish, 5-1/2" thick, 3000 PSI	2.87	2.98	5.85
4250	5000 PSI	2.92	2.91	5.83
4300	6" thick, 3000 PSI	3.09	3.08	6.17
4350	5000 PSI	3.15	3.01	6.16
4400	7-1/2" thick, 3000 PSI	3.73	3.22	6.95
4450	5000 PSI	3.82	3.15	6.97
4500	8" thick, 3000 PSI	3.95	3.32	7.27
4550	5000 PSI	4.07	3.25	7.32
4600	Exposed aggregate & vert. rustication 5-1/2" thick, 3000 PSI	4.22	5.40	9.62
4650	5000 PSI	4.27	5.35	9.62
4700	6" thick, 3000 PSI	4.44	5.50	9.94
4750	5000 PSI	4.50	5.45	9.95
4800	7-1/2" thick, 3000 PSI	5.10	5.65	10.75
4850	5000 PSI	5.15	5.55	10.70
4900	8" thick, 3000 PSI	5.30	5.75	11.05
4950	5000 PSI	5.40	5.65	11.05
5000	Vertical rib & light sandblast, 5-1/2" thick, 3000 PSI	4.51	3.70	8.21
5100	6" thick, 3000 PSI	4.73	3.80	8.53
5200	7-1/2" thick, 3000 PSI	5.35	3.94	9.29
5300	8" thick, 3000 PSI	5.60	4.04	9.64
6000	Broom finish w/2" urethane insulation, 6" thick, 3000 PSI	2.39	3.71	6.10
6100	Broom finish 2" fiberplank insulation, 6" thick, 3000 PSI	2.89	3.66	6.55
6200	Exposed aggregate w/2" urethane insulation, 6" thick, 3000 PSI	2.79	3.86	6.65
6300	Exposed aggregate 2" fiberplank insulation, 6" thick, 3000 PSI	3.29	3.81	7.10

Important: See the Reference Section for critical supporting data - Location Factors & Historical Cost Indexes

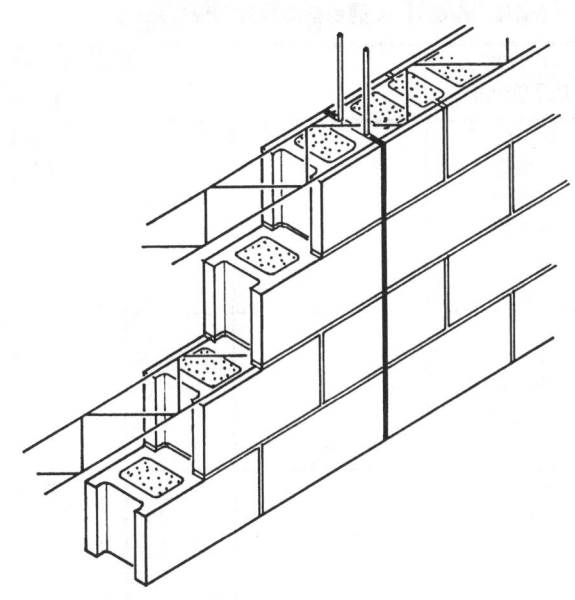

Exterior concrete block walls are defined in the following terms; structural reinforcement, weight, percent solid, size, strength and insulation. Within each of these categories, two to four variations are shown. No costs are included for brick shelf or relieving angles.

4.1-211 — Concrete Block Wall - Regular Weight

	TYPE	SIZE (IN.)	STRENGTH (P.S.I.)	CORE FILL		MAT.	INST.	TOTAL
1200	Hollow	4x8x16	2,000	none		.92	3.69	4.61
1250			4,500	none		1.37	3.69	5.06
1400		8x8x16	2,000	perlite		2	4.48	6.48
1410				styrofoam		2.41	4.23	6.64
1440				none		1.60	4.23	5.83
1450			4,500	perlite		2.83	4.48	7.31
1460				styrofoam		3.24	4.23	7.47
1490				none		2.43	4.23	6.66
2000	75% solid	4x8x16	2,000	none		1.08	3.73	4.81
2100	75% solid	6x8x16	2,000	perlite		1.44	4.10	5.54
2500	Solid	4x8x16	2,000	none		1.35	3.82	5.17
2700		8x8x16	2,000	none		2.21	4.40	6.61

COST PER S.F.

4.1-211 — Concrete Block Wall - Lightweight

	TYPE	SIZE (IN.)	WEIGHT (P.C.F.)	CORE FILL		MAT.	INST.	TOTAL
3100	Hollow	8x4x16	105	perlite		1.45	4.23	5.68
3110				styrofoam		1.86	3.98	5.84
3200		4x8x16	105	none		1.03	3.61	4.64
3250			85	none		1.27	3.53	4.80
3300		6x8x16	105	perlite		1.55	4.06	5.61
3310				styrofoam		2.09	3.86	5.95
3340				none		1.28	3.86	5.14
3400		8x8x16	105	perlite		1.96	4.37	6.33
3410				styrofoam		2.37	4.12	6.49
3440				none		1.56	4.12	5.68
3450			85	perlite		2.15	4.27	6.42
4000	75% solid	4x8x16	105	none		1.37	3.65	5.02
4050			85	none		2.13	3.57	5.70
4100		6x8x16	105	perlite		2.17	4.01	6.18
4500	Solid	4x8x16	105	none		1.39	3.77	5.16
4700		8x8x16	105	none		2.30	4.34	6.64

COST PER S.F.

EXTERIOR CLOSURE

4

4.1-211 | Reinforced Concrete Block Wall - Regular Weight

	TYPE	SIZE (IN.)	STRENGTH (P.S.I.)	VERT. REINF & GROUT SPACING		COST PER S.F. MAT.	COST PER S.F. INST.	COST PER S.F. TOTAL
5200	Hollow	4x8x16	2,000	#4 @ 48"		1.01	4.09	5.10
5300		6x8x16	2,000	#4 @ 48"		1.28	4.38	5.66
5330				#5 @ 32"		1.41	4.56	5.97
5340				#5 @ 16"		1.70	5.15	6.85
5350			4,500	#4 @ 28"		1.92	4.38	6.30
5390				#5 @ 16"		2.34	5.15	7.49
5400	Hollow	8x8x16	2,000	#4 @ 48"		1.83	4.67	6.50
5430				#5 @ 32"		1.94	4.95	6.89
5440				#5 @ 16"		2.29	5.70	7.99
5450		8x8x16	4,500	#4 @ 48"		2.62	4.73	7.35
5490				#5 @ 16"		3.12	5.70	8.82
5500		12x8x16	2,000	#4 @ 48"		2.43	5.95	8.38
5540				#5 @ 16"		3.11	6.95	10.06
6100	75% solid	6x8x16	2,000	#4 @ 48"		1.38	4.34	5.72
6140				#5 @ 16"		1.64	4.96	6.60
6150			4,500	#4 @ 48"		2.06	4.34	6.40
6190				#5 @ 16"		2.32	4.96	7.28
6200		8x8x16	2,000	#4 @ 48"		1.64	4.69	6.33
6230				#5 @ 32"		1.74	4.86	6.60
6240				#5 @ 16"		1.92	5.45	7.37
6250			4,500	#4 @ 48"		2.96	4.69	7.65
6280				#5 @ 32"		3.06	4.86	7.92
6290				#5 @ 16"		3.24	5.45	8.69
6500	Solid-double	2-4x8x16	2,000	#4 @ 48" E.W.		3.23	8.65	11.88
6530	Wythe			#5 @ 16" E.W.		3.63	9.20	12.83
6550			4,500	#4 @ 48" E.W.		3.49	8.55	12.04
6580				#5 @ 16" E.W.		3.89	9.10	12.99

4.1-211 | Reinforced Concrete Block Wall - Lightweight

	TYPE	SIZE (IN.)	WEIGHT (P.C.F.)	VERT REINF. & GROUT SPACING		COST PER S.F. MAT.	COST PER S.F. INST.	COST PER S.F. TOTAL
7100	Hollow	8x4x16	105	#4 @ 48"		1.24	4.48	5.72
7140				#5 @ 16"		1.74	5.45	7.19
7150			85	#4 @ 48"		3.02	4.38	7.40
7190				#5 @ 16"		3.52	5.35	8.87
7400		8x8x16	105	#4 @ 48"		1.75	4.62	6.37
7440				#5 @ 16"		2.25	5.60	7.85
7450		8x8x16	85	#4 @ 48"		1.94	4.52	6.46
7490				#5 @ 16"		2.44	5.50	7.94
7800		8x8x24	105	#4 @ 48"		2.26	4.96	7.22
7840				#5 @ 16"		2.76	5.90	8.66
7850			85	#4 @ 48"		3.30	4.25	7.55
7890				#5 @ 16"		3.80	5.20	9
8100	75% solid	6x8x16	105	#4 @ 48"		2.11	4.25	6.36
8130				#5 @ 32"		2.20	4.39	6.59
8150			85	#4 @ 48"		2.53	4.16	6.69
8180				#5 @ 32"		2.62	4.30	6.92
8200		8x8x16	105	#4 @ 48"		3.01	4.58	7.59
8230				#5 @ 32"		3.11	4.75	7.86
8250			85	#4 @ 48"		3.43	4.47	7.90
8280				#5 @ 32"		3.53	4.64	8.17

Important: See the Reference Section for critical supporting data - Location Factors & Historical Cost Indexes

4.1-211 Reinforced Concrete Block Wall - Lightweight

	TYPE	SIZE (IN.)	WEIGHT (P.C.F.)	VERT REINF. & GROUT SPACING		COST PER S.F.		
						MAT.	INST.	TOTAL
8500	Solid-double	2-4x8x16	105	#4 @ 48"		3.31	8.55	11.86
8530	Wythe		105	#5 @ 16"		3.71	9.10	12.81
8600		2-6x8x16	105	#4 @ 48"		4.54	9.15	13.69
8630				#5 @ 16"		4.94	9.70	14.64
8650			85	#4 @ 48"		5.85	8.75	14.60

EXTERIOR CLOSURE

4

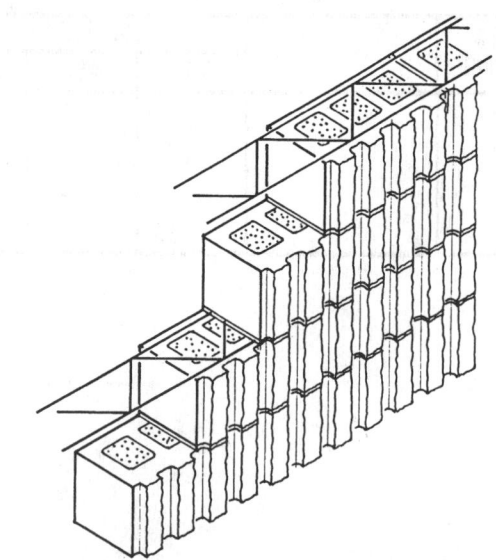

Exterior split ribbed block walls are defined in the following terms; structural reinforcement, weight, percent solid, size, number of ribs and insulation. Within each of these categories two to four variations are shown. No costs are included for brick shelf or relieving angles. Costs include control joints every 20' and horizontal reinforcing.

4.1-212 — Split Ribbed Block Wall - Regular Weight

	TYPE	SIZE (IN.)	RIBS	CORE FILL		COST PER S.F. MAT.	INST.	TOTAL
1220	Hollow	4x8x16	4	none		2.05	4.56	6.61
1250			8	none		2.39	4.56	6.95
1280			16	none		1.94	4.63	6.57
1430		8x8x16	8	perlite		4.10	5.40	9.50
1440				styrofoam		4.51	5.15	9.66
1450				none		3.70	5.15	8.85
1530		12x8x16	8	perlite		5.15	7.15	12.30
1540				styrofoam		5.45	6.65	12.10
1550				none		4.47	6.65	11.12
2120	75% solid	4x8x16	4	none		2.57	4.63	7.20
2150			8	none		3.01	4.63	7.64
2180			16	none		2.46	4.70	7.16
2520	Solid	4x8x16	4	none		3.11	4.70	7.81
2550			8	none		3.65	4.70	8.35
2580			16	none		2.94	4.77	7.71

4.1-212 — Reinforced Split Ribbed Block Wall - Regular Weight

	TYPE	SIZE (IN.)	RIBS	VERT. REINF. & GROUT SPACING		COST PER S.F. MAT.	INST.	TOTAL
5200	Hollow	4x8x16	4	#4 @ 48"		2.14	4.96	7.10
5230			8	#4 @ 48"		2.48	4.96	7.44
5260			16	#4 @ 48"		2.03	5.05	7.08
5430		8x8x16	8	#4 @ 48"		3.89	5.65	9.54
5440				#5 @ 32"		4.04	5.85	9.89
5450				#5 @ 16"		4.39	6.60	10.99
5530		12x8x16	8	#4 @ 48"		4.75	7.20	11.95
5540				#5 @ 32"		4.95	7.40	12.35
5550				#5 @ 16"		5.45	8.20	13.65
6230	75% solid	6x8x16	8	#4 @ 48"		3.83	5.25	9.08
6240				#5 @ 32"		3.92	5.40	9.32
6250				#5 @ 16"		4.09	5.85	9.94
6330		8x8x16	8	#4 @ 48"		4.80	5.65	10.45
6340				#5 @ 32"		4.90	5.80	10.70
6350				#5 @ 16"		5.10	6.40	11.50

Important: See the Reference Section for critical supporting data - Location Factors & Historical Cost Indexes

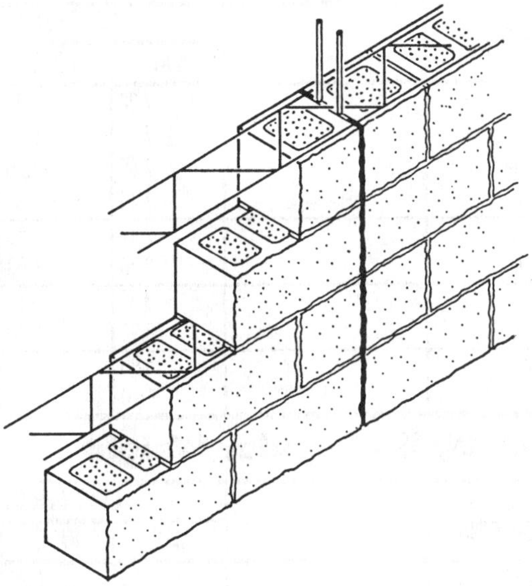

Exterior split face block walls are defined in the following terms; structural reinforcement, weight, percent solid, size, scores and insulation. Within each of these categories two to four variations are shown. No costs are included for brick shelf or relieving angles. Costs include control joints every 20′ and horizontal reinforcing.

4.1-213 Split Face Block Wall - Regular Weight

	TYPE	SIZE (IN.)	SCORES	CORE FILL		COST PER S.F. MAT.	COST PER S.F. INST.	COST PER S.F. TOTAL
1200	Hollow	8x4x16	0	perlite		5.95	5.95	11.90
1210				styrofoam		6.40	5.70	12.10
1240				none		5.55	5.70	11.25
1250			1	perlite		6.35	5.95	12.30
1260				styrofoam		6.80	5.70	12.50
1290				none		5.95	5.70	11.65
1300		12x4x16	0	perlite		7.75	6.75	14.50
1310				styrofoam		8.10	6.25	14.35
1340				none		7.10	6.25	13.35
1350			1	perlite		8.20	6.75	14.95
1360				styrofoam		8.55	6.25	14.80
1390				none		7.55	6.25	13.80
1400		4x8x16	0	none		2.05	4.50	6.55
1450			1	none		2.31	4.56	6.87
1500		6x8x16	0	perlite		2.92	5.20	8.12
1510				styrofoam		3.46	4.99	8.45
1540				none		2.65	4.99	7.64
1550			1	perlite		3.17	5.25	8.42
1560				styrofoam		3.71	5.05	8.76
1590				none		2.90	5.05	7.95
1600		8x8x16	0	perlite		3.53	5.60	9.13
1610				styrofoam		3.94	5.35	9.29
1640				none		3.13	5.35	8.48
1650		8x8x16	1	perlite		3.74	5.70	9.44
1660				styrofoam		4.15	5.45	9.60
1690				none		3.34	5.45	8.79
1700		12x8x16	0	perlite		4.52	7.30	11.82
1710				styrofoam		4.86	6.80	11.66
1740				none		3.86	6.80	10.66
1750			1	perlite		4.73	7.40	12.13
1760				styrofoam		5.05	6.90	11.95
1790				none		4.07	6.90	10.97

4.1-213 — Split Face Block Wall - Regular Weight

	TYPE	SIZE (IN.)	SCORES	CORE FILL		COST PER S.F. MAT.	COST PER S.F. INST.	COST PER S.F. TOTAL
1800	75% solid	8x4x16	0	perlite		7.20	5.95	13.15
1840				none		7	5.80	12.80
1852		8x4x16	1	perlite		7.65	5.95	13.60
1890				none		7.45	5.80	13.25
2000		4x8x16	0	none		2.57	4.56	7.13
2050			1	none		2.78	4.63	7.41
2400	Solid	8x4x16	0	none		8.55	5.90	14.45
2450			1	none		8.95	5.90	14.85
2900		12x8x16	0	none		5.90	7.05	12.95
2950			1	none		6.10	7.15	13.25

4.1-213 — Reinforced Split Face Block Wall - Regular Weight

	TYPE	SIZE (IN.)	SCORES	VERT. REINF. & GROUT SPACING		COST PER S.F. MAT.	COST PER S.F. INST.	COST PER S.F. TOTAL
5200	Hollow	8x4x16	0	#4 @ 48"		5.75	6.20	11.95
5210				#5 @ 32"		5.90	6.40	12.30
5240				#5 @ 16"		6.25	7.15	13.40
5250			1	#4 @ 48"		6.15	6.20	12.35
5260				#5 @ 32"		6.30	6.40	12.70
5290				#5 @ 16"		6.65	7.15	13.80
5700		12x8x16	0	#4 @ 48"		4.14	7.35	11.49
5710				#5 @ 32"		4.34	7.55	11.89
5740				#5 @ 16"		4.82	8.35	13.17
5750			1	#4 @ 48"		4.35	7.45	11.80
5760				#5 @ 32"		4.55	7.65	12.20
5790				#5 @ 16"		5.05	8.45	13.50
6000	75% solid	8x4x16	0	#4 @ 48"		7.10	6.20	13.30
6010				#5 @ 32"		7.20	6.35	13.55
6040				#5 @ 16"		7.40	6.95	14.35
6050			1	#4 @ 48"		7.55	6.20	13.75
6060				#5 @ 32"		7.65	6.35	14
6090				#5 @ 16"		7.85	6.95	14.80
6100		12x4x16	0	#4 @ 48"		9.15	6.85	16
6110				#5 @ 32"		9.25	7	16.25
6140				#5 @ 16"		9.50	7.50	17
6150			1	#4 @ 48"		9.50	6.85	16.35
6160				#5 @ 32"		9.60	7	16.60
6190				#5 @ 16"		9.90	7.60	17.50
6700	Solid-double Wythe	2-4x8x16	0	#4 @ 48" E.W.		6.75	10.30	17.05
6710				#5 @ 32" E.W.		6.90	10.40	17.30
6740				#5 @ 16" E.W.		7.15	10.80	17.95
6750			1	#4 @ 48" E.W.		7.15	10.45	17.60
6760				#5 @ 32" E.W.		7.30	10.50	17.80
6790				#5 @ 16" E.W.		7.55	10.95	18.50
6800		2-6x8x16	0	#4 @ 48" E.W.		8.75	11.30	20.05
6810				#5 @ 32" E.W.		8.90	11.40	20.30
6840				#5 @ 16" E.W.		9.15	11.85	21
6850			1	#4 @ 48" E.W.		9.20	11.50	20.70
6860				#5 @ 32" E.W.		9.35	11.60	20.95
6890				#5 @ 16" E.W.		9.60	12.05	21.65

EXTERIOR CLOSURE 4

Important: See the Reference Section for critical supporting data - Location Factors & Historical Cost Indexes

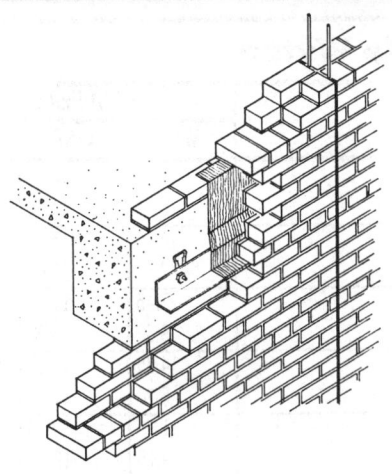

Exterior solid brick walls are defined in the following terms; structural reinforcement, type of brick, thickness, and bond. Nine different types of face bricks are presented, with single wythes shown in four different bonds. Shelf angles are included in system components. These walls do not include ties and as such are not tied to a backup wall.

4.1-231 — Solid Brick Walls - Single Wythe

	TYPE	THICKNESS (IN.)	BOND			COST PER S.F.		
						MAT.	INST.	TOTAL
1000	Common	4	running			3.01	8.35	11.36
1010			common			3.53	9.55	13.08
1050			Flemish			3.86	11.55	15.41
1100			English			4.26	12.25	16.51
1150	Standard	4	running			3.68	8.65	12.33
1160			common			4.27	9.95	14.22
1200			Flemish			4.67	11.90	16.57
1250			English			5.20	12.65	17.85
1300	Glazed	4	running			8.70	9	17.70
1310			common			10.25	10.45	20.70
1350			Flemish			11.35	12.65	24
1400			English			12.65	13.45	26.10
1600	Economy	4	running			4.01	6.65	10.66
1610			common			4.64	7.60	12.24
1650			Flemish			5.10	9	14.10
1700			English			9.05	9.55	18.60
1900	Fire	4-1/2	running			7.65	7.60	15.25
1910			common			9.05	8.65	17.70
1950			Flemish			9.95	10.45	20.40
2000			English			11.15	10.95	22.10
2050	King	3-1/2	running			3.36	6.80	10.16
2060			common			3.88	7.80	11.68
2100			Flemish			4.63	9.35	13.98
2150			English			5.10	9.75	14.85
2200	Roman	4	running			5.75	7.80	13.55
2210			common			6.80	9	15.80
2250			Flemish			7.45	10.70	18.15
2300			English			8.35	11.55	19.90
2350	Norman	4	running			4.86	6.50	11.36
2360			common			5.70	7.35	13.05
2400			Flemish			9.55	8.80	18.35
2450			English			6.95	9.35	16.30
2500	Norwegian	4	running			4.11	5.80	9.91
2510			common			4.78	6.60	11.38
2550			Flemish			5.20	7.80	13
2600			English			5.80	8.20	14

4.1-231   Solid Brick Walls - Double Wythe

	TYPE	THICKNESS (IN.)	COLLAR JOINT THICKNESS (IN.)			COST PER S.F.		
						MAT.	INST.	TOTAL
4100	Common	8	3/4			5.60	14	19.60
4150	Standard	8	1/2			6.80	14.50	21.30
4200	Glazed	8	1/2			16.80	15	31.80
4250	Engineer	8	1/2			7.40	12.80	20.20
4300	Economy	8	3/4			7.45	11	18.45
4350	Double	8	3/4			9.60	8.90	18.50
4400	Fire	8	3/4			14.75	12.80	27.55
4450	King	7	3/4			6.20	11.25	17.45
4500	Roman	8	1			11	13.20	24.20
4550	Norman	8	3/4			9.15	10.75	19.90
4600	Norwegian	8	3/4			7.65	9.55	17.20
4650	Utility	8	3/4			7	8.25	15.25
4700	Triple	8	3/4			5.55	7.45	13
4750	SCR	12	3/4			10.25	11.05	21.30
4800	Norwegian	12	3/4			9.20	9.75	18.95
4850	Jumbo	12	3/4			9.80	8.55	18.35

4.1-231   Reinforced Brick Walls

	TYPE	THICKNESS (IN.)	WYTHE	REINF. & SPACING		COST PER S.F.		
						MAT.	INST.	TOTAL
7200	Common	4"	1	#4 @ 48"vert		3.16	8.70	11.86
7220				#5 @ 32"vert		3.25	8.80	12.05
7230				#5 @ 16"vert		3.41	9.20	12.61
7700	King	9"	2	#4 @ 48" E.W.		6.45	11.80	18.25
7720				#5 @ 32" E.W.		6.60	11.90	18.50
7730				#5 @ 16" E.W.		6.85	12.25	19.10
7800	Common	10"	2	#4 @ 48" E.W.		5.85	14.55	20.40
7820				#5 @ 32" E.W.		6	14.65	20.65
7830				#5 @ 16" E.W.		6.25	15	21.25
7900	Standard	10"	2	#4 @ 48" E.W.		7.10	15.05	22.15
7920				#5 @ 32" E.W.		7.25	15.15	22.40
7930				#5 @ 16" E.W.		7.50	15.50	23
8100	Economy	10"	2	#4 @ 48" E.W.		7.75	11.55	19.30
8120				#5 @ 32" E.W.		7.90	11.65	19.55
8130				#5 @ 16" E.W.		8.15	12	20.15
8200	Double	10"	2	#4 @ 48" E.W.		9.90	9.45	19.35
8220				#5 @ 32" E.W.		10.05	9.55	19.60
8230				#5 @ 16" E.W.		10.30	9.90	20.20
8300	Fire	10"	2	#4 @ 48" E.W.		15.05	13.35	28.40
8320				#5 @ 32" E.W.		15.20	13.45	28.65
8330				#5 @ 16" E.W.		15.45	13.80	29.25
8400	Roman	10"	2	#4 @ 48" E.W.		11.30	13.75	25.05
8420				#5 @ 32" E.W.		11.45	13.85	25.30
8430				#5 @ 16" E.W.		11.70	14.20	25.90
8500	Norman	10"	2	#4 @ 48" E.W.		9.45	11.30	20.75
8520				#5 @ 32" E.W.		9.60	11.40	21
8530				#5 @ 16" E.W.		9.85	11.75	21.60
8600	Norwegian	10"	2	#4 @ 48" E.W.		7.95	10.10	18.05
8620				#5 @ 32" E.W.		8.10	10.20	18.30
8630				#5 @ 16" E.W.		8.35	10.55	18.90
8800	Triple	10"	2	#4 @ 48" E.W.		5.80	8	13.80
8820				#5 @ 32" E.W.		5.95	8.10	14.05
8830				#5 @ 16" E.W.		6.20	8.45	14.65

Important: See the Reference Section for critical supporting data - Location Factors & Historical Cost Indexes

EXTERIOR CLOSURE 4

The table below lists costs per S.F. for stone veneer walls on various backup using different stone.

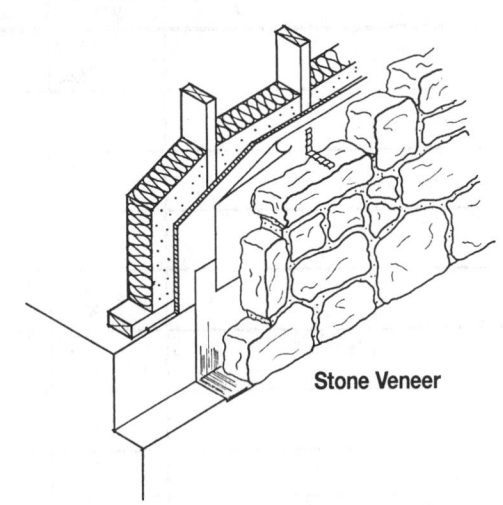

Stone Veneer

4.1-242	Stone Veneer	COST PER S.F.		
		MAT.	INST.	TOTAL
2000	Ashlar veneer,4″, 2″x4″ stud backup,16″ O.C., 8′ high, low priced stone	6.50	11.30	17.80
2100	Metal stud backup, 8′ high, 16″ O.C.	7	11.80	18.80
2150	24″ O.C.	6.85	11.55	18.40
2200	Conc. block backup, 4″ thick	6.75	13.95	20.70
2300	6″ thick	6.90	14.10	21
2350	8″ thick	7.05	14.30	21.35
2400	10″ thick	7.50	14.40	21.90
2500	12″ thick	7.55	15.30	22.85
3100	High priced stone, wood stud backup, 10′ high, 16″ O.C.	11.20	12.75	23.95
3200	Metal stud backup, 10′ high, 16″ O.C.	11.65	13.25	24.90
3250	24″ O.C.	11.50	13	24.50
3300	Conc. block backup, 10′ high, 4″ thick	11.40	15.40	26.80
3350	6″ thick	11.55	15.45	27
3400	8″ thick	11.70	15.75	27.45
3450	10″ thick	12.15	15.85	28
3500	12″ thick	12.20	16.75	28.95
4000	Indiana limestone 2″ thick,sawn finish,wood stud backup,10′ high,16″ O.C.	13.05	9.65	22.70
4100	Metal stud backup, 10′ high, 16″ O.C.	13.55	10.10	23.65
4150	24″ O.C.	13.40	9.85	23.25
4200	Conc. block backup, 4″ thick	13.30	12.30	25.60
4250	6″ thick	13.45	12.45	25.90
4300	8″ thick	13.60	12.65	26.25
4350	10″ thick	14.05	12.70	26.75
4400	12″ thick	14.10	13.65	27.75
4450	2″ thick, smooth finish, wood stud backup, 8′ high, 16″ O.C.	13.05	9.65	22.70
4550	Metal stud backup, 8′ high, 16″ O.C.	13.55	10	23.55
4600	24″ O.C.	13.40	9.75	23.15
4650	Conc. block backup, 4″ thick	13.30	12.15	25.45
4700	6″ thick	13.45	12.35	25.80
4750	8″ thick	13.60	12.50	26.10
4800	10″ thick	14.05	12.70	26.75
4850	12″ thick	14.10	13.65	27.75
5350	4″ thick, smooth finish, wood stud backup, 8′ high, 16″ O.C.	17.55	9.65	27.20
5450	Metal stud backup, 8′ high, 16″ O.C.	18.05	10.10	28.15
5500	24″ O.C.	17.90	9.85	27.75
5550	Conc. block backup, 4″ thick	17.80	12.30	30.10
5600	6″ thick	17.95	12.45	30.40
5650	8″ thick	18.10	12.65	30.75
5700	10″ thick	18.55	12.70	31.25
5750	12″ thick	18.60	13.65	32.25

EXTERIOR CLOSURE

4

4.1-242	Stone Veneer	COST PER S.F.		
		MAT.	INST.	TOTAL
6000	Granite, grey or pink, 2" thick, wood stud backup, 8' high, 16" O.C.	18.45	16.15	34.60
6100	Metal studs, 8' high, 16" O.C.	18.95	16.60	35.55
6150	24" O.C.	18.80	16.35	35.15
6200	Conc. block backup, 4" thick	18.70	18.80	37.50
6250	6" thick	18.85	18.95	37.80
6300	8" thick	19	19.15	38.15
6350	10" thick	19.45	19.20	38.65
6400	12" thick	19.50	20	39.50
6900	4" thick, wood stud backup, 8' high, 16" O.C.	30	18.25	48.25
7000	Metal studs, 8' high, 16" O.C.	30.50	18.70	49.20
7050	24" O.C.	30.50	18.45	48.95
7100	Conc. block backup, 4" thick	30	21	51
7150	6" thick	30.50	21	51.50
7200	8" thick	30.50	21	51.50
7250	10" thick	31	21.50	52.50
7300	12" thick	31	22	53

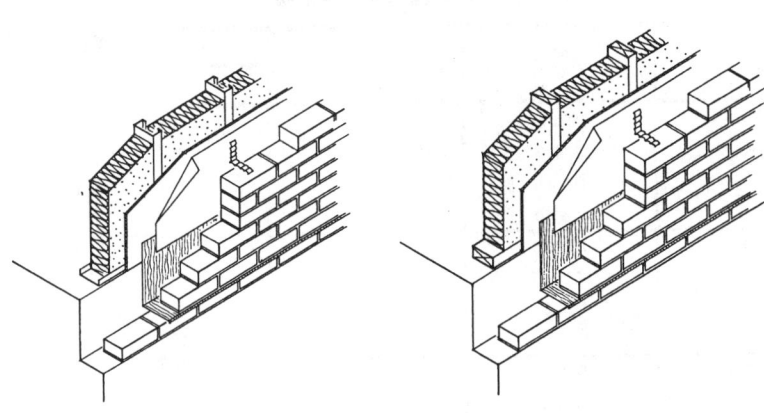

Exterior brick veneer/stud backup walls are defined in the following terms; type of brick and studs, stud spacing and bond. All systems include a brick shelf, ties to the backup and all necessary dampproofing and insulation.

4.1-252 — Brick Veneer/Wood Stud Backup

	FACE BRICK	STUD BACKUP	STUD SPACING (IN.)	BOND		COST PER S.F. MAT.	INST.	TOTAL
1100	Standard	2x4-wood	16	running		4.91	10.20	15.11
1120				common		5.50	11.50	17
1140				Flemish		5.90	13.45	19.35
1160				English		6.40	14.20	20.60
1400		2x6-wood	16	running		5.10	10.30	15.40
1420				common		5.70	11.60	17.30
1440				Flemish		6.10	13.55	19.65
1460				English		6.60	14.30	20.90
1700	Glazed	2x4-wood	16	running		9.95	10.55	20.50
1720				common		11.50	12	23.50
1740				Flemish		12.60	14.20	26.80
1760				English		13.90	15	28.90
2300	Engineer	2x4-wood	16	running		5.20	9.15	14.35
2320				common		5.85	10.20	16.05
2340				Flemish		6.30	12	18.30
2360				English		6.85	12.50	19.35
2900	Roman	2x4-wood	16	running		7	9.35	16.35
2920				common		8.05	10.55	18.60
2940				Flemish		8.70	12.25	20.95
2960				English		9.60	13.10	22.70
4100	Norwegian	2x4-wood	16	running		5.35	7.35	12.70
4120				common		6	8.15	14.15
4140				Flemish		6.45	9.35	15.80
4160				English		7.05	9.75	16.80

EXTERIOR CLOSURE

4

4.1-252 — Brick Veneer/Metal Stud Backup

	FACE BRICK	STUD BACKUP	STUD SPACING (IN.)	BOND		COST PER S.F.		
						MAT.	INST.	TOTAL
5100	Standard	25ga.x6"NLB	24	running		4.62	10.25	14.87
5120				common		5.20	11.55	16.75
5140				Flemish		5.35	12.85	18.20
5160				English		6.10	14.25	20.35
5200		20ga.x3-5/8"NLB	16	running		4.97	10.30	15.27
5220				common		5.55	11.60	17.15
5240				Flemish		5.95	13.55	19.50
5260				English		6.45	14.30	20.75
5400		16ga.x3-5/8"LB	16	running		5.35	10.85	16.20
5420				common		5.90	12.15	18.05
5440				Flemish		6.30	14.10	20.40
5460				English		6.85	14.85	21.70
5700	Glazed	25ga.x6"NLB	24	running		9.65	10.60	20.25
5720				common		11.20	12.05	23.25
5740				Flemish		12.30	14.25	26.55
5760				English		13.60	15.05	28.65
5800		20ga.x3-5/8"NLB	24	running		9.85	10.55	20.40
5820				common		11.40	12	23.40
5840				Flemish		12.50	14.20	26.70
5860				English		13.80	15	28.80
6000		16ga.x3-5/8"LB	16	running		10.35	11.20	21.55
6020				common		11.90	12.65	24.55
6040				Flemish		13	14.85	27.85
6060				English		14.30	15.65	29.95
6300	Engineer	25ga.x6"NLB	24	running		4.91	9.20	14.11
6320				common		5.55	10.25	15.80
6340				Flemish		6	12.05	18.05
6360				English		6.55	12.55	19.10
6400		20ga.x3-5/8"NLB	16	running		5.25	9.25	14.50
6420				common		5.90	10.30	16.20
6440				Flemish		6.35	12.10	18.45
6460				English		6.90	12.60	19.50
6900	Roman	25ga.x6"NLB	24	running		6.70	9.40	16.10
6920				common		7.75	10.60	18.35
6940				Flemish		8.40	12.30	20.70
6960				English		9.30	13.15	22.45
7000		20ga.x3-5/8"NLB	16	running		7.05	9.45	16.50
7020				common		8.10	10.65	18.75
7040				Flemish		8.75	12.35	21.10
7060				English		9.65	13.20	22.85
7500	Norman	25ga.x6"NLB	24	running		5.80	8.05	13.85
7520				common		6.60	8.95	15.55
7540				Flemish		10.50	10.40	20.90
7560				English		7.90	10.95	18.85
7600		20ga.x3-5/8"NLB	24	running		6	8.05	14.05
7620				common		6.85	8.90	15.75
7640				Flemish		10.70	10.35	21.05
7660				English		8.10	10.90	19
8100	Norwegian	25ga.x6"NLB	24	running		5.05	7.35	12.40
8120				common		5.70	8.15	13.85
8140				Flemish		6.15	9.40	15.55
8160				English		6.75	9.80	16.55

Important: See the Reference Section for critical supporting data - Location Factors & Historical Cost Indexes

4.1-252		Brick Veneer/Metal Stud Backup						
	FACE BRICK	STUD BACKUP	STUD SPACING (IN.)	BOND		COST PER S.F.		
						MAT.	INST.	TOTAL
8400	Norwegian	16ga.x3-5/8"LB	16	running		5.75	8	13.75
8420				common		6.45	8.75	15.20
8440				Flemish		6.85	10	16.85
8460				English		7.45	10.40	17.85

EXTERIOR CLOSURE

4

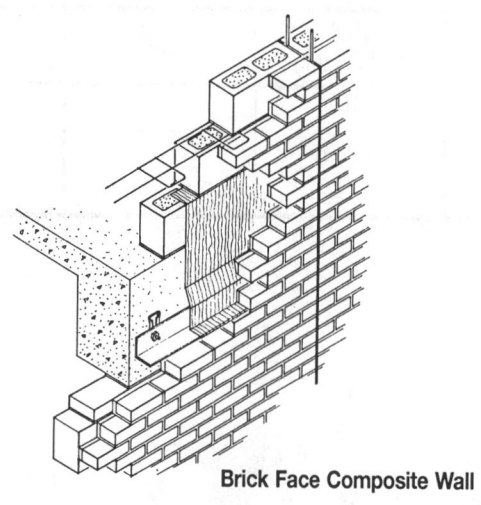

Brick Face Composite Wall

4.1-272 — Brick Face Composite Wall - Double Wythe

	FACE BRICK	BACKUP MASONRY	BACKUP THICKNESS (IN.)	BACKUP CORE FILL		COST PER S.F.		
						MAT.	INST.	TOTAL
1000	Standard	common brick	4	none		6.10	15.85	21.95
1040		SCR brick	6	none		8.45	14.15	22.60
1080		conc. block	4	none		4.84	12.75	17.59
1120			6	perlite		5.30	13.25	18.55
1160				styrofoam		5.85	13.05	18.90
1200			8	perlite		5.90	13.55	19.45
1240				styrofoam		6.30	13.30	19.60
1520		glazed block	4	none		9.65	13.65	23.30
1560			6	perlite		10.60	14.05	24.65
1600				styrofoam		11.15	13.85	25
1640			8	perlite		11.10	14.40	25.50
1680				styrofoam		11.50	14.15	25.65
1720		clay tile	4	none		8.15	12.25	20.40
1760			6	none		8.30	12.60	20.90
1800			8	none		9.25	13.05	22.30
1840		glazed tile	4	none		11.65	16.15	27.80
1880								
2000	Glazed	common brick	4	none		11.10	16.20	27.30
2040		SCR brick	6	none		13.50	14.50	28
2080		conc. block	4	none		9.85	13.10	22.95
2200			8	perlite		10.95	13.90	24.85
2240				styrofoam		11.35	13.65	25
2280		L.W. block	4	none		9.95	13.05	23
2400			8	perlite		10.90	13.80	24.70
2520		glazed block	4	none		14.65	14	28.65
2560			6	perlite		15.60	14.40	30
2600				styrofoam		16.15	14.20	30.35
2640			8	perlite		16.10	14.75	30.85
2680				styrofoam		16.50	14.50	31
2720		clay tile	4	none		13.15	12.60	25.75
2760			6	none		13.35	12.95	26.30
2800			8	none		14.25	13.40	27.65
2840		glazed tile	4	none		16.65	16.50	33.15
3000	Engineer	common brick	4	none		6.40	14.80	21.20
3040		SCR brick	6	none		8.75	13.10	21.85

4.1-272 | Brick Face Composite Wall - Double Wythe

	FACE BRICK	BACKUP MASONRY	BACKUP THICKNESS (IN.)	BACKUP CORE FILL		COST PER S.F.		
						MAT.	INST.	TOTAL
3080		conc. block	4	none		5.15	11.70	16.85
3200			8	perlite		6.20	12.50	18.70
3280		L.W. block	4	none		4.41	8.35	12.76
3322	Engineer	L.W. block	6	perlite		5.75	12.10	17.85
3520		glazed block	4	none		9.90	12.60	22.50
3560			6	perlite		10.90	13	23.90
3600				styrofoam		11.45	12.80	24.25
3640			8	perlite		11.35	13.35	24.70
3680				styrofoam		11.80	13.10	24.90
3720		clay tile	4	none		8.45	11.20	19.65
3800			8	none		9.50	12	21.50
3840		glazed tile	4	none		11.90	15.10	27
4000	Roman	common brick	4	none		8.20	15	23.20
4040		SCR brick	6	none		10.55	13.30	23.85
4080		conc. block	4	none		6.95	11.90	18.85
4120			6	perlite		7.40	12.40	19.80
4200			8	perlite		8	12.70	20.70
5000	Norman	common brick	4	none		7.25	13.70	20.95
5040		SCR brick	6	none		9.65	12	21.65
5280		L.W. block	4	none		6.15	10.50	16.65
5320			6	perlite		6.65	11.10	17.75
5720		clay tile	4	none		9.35	10.10	19.45
5760			6	none		9.50	10.45	19.95
5840		glazed tile	4	none		12.80	14	26.80
6000	Norwegian	common brick	4	none		6.50	13	19.50
6040		SCR brick	6	none		8.90	11.30	20.20
6080		conc. block	4	none		5.25	9.90	15.15
6120			6	perlite		5.75	10.35	16.10
6160				styrofoam		6.30	10.15	16.45
6200			8	perlite		6.35	10.65	17
6520		glazed block	4	none		10.05	10.75	20.80
6560			6	perlite		11.05	11.20	22.25
7000	Utility	common brick	4	none		6.20	12.30	18.50
7040		SCR brick	6	none		8.55	10.60	19.15
7080		conc. block	4	none		4.93	9.20	14.13
7120			6	perlite		5.40	9.70	15.10
7160				styrofoam		5.95	9.50	15.45
7200			8	perlite		6	10	16
7240				styrofoam		6.40	9.75	16.15
7280		L.W. block	4	none		5.05	9.15	14.20
7320			6	perlite		5.55	9.60	15.15
7520		glazed block	4	none		9.70	10.10	19.80
7560			6	perlite		10.70	10.50	21.20
7720		clay tile	4	none		8.25	8.70	16.95
7760			6	none		8.40	9.05	17.45
7840		glazed tile	4	none		11.70	12.60	24.30

4.1-272 | Brick Face Composite Wall - Triple Wythe

	FACE BRICK	MIDDLE WYTHE	INSIDE MASONRY	TOTAL THICKNESS (IN.)		COST PER S.F.		
						MAT.	INST.	TOTAL
8000	Standard	common brick	standard brick	12		8.95	22.50	31.45
8100		4" conc. brick	standard brick	12		9.25	23	32.25
8120		4" conc. brick	common brick	12		8.60	23	31.60

EXTERIOR CLOSURE

4

4.1-272		Brick Face Composite Wall - Triple Wythe							
	FACE BRICK	MIDDLE WYTHE	INSIDE MASONRY	TOTAL THICKNESS (IN.)			COST PER S.F.		
							MAT.	INST.	TOTAL
8200	Glazed	common brick	standard brick	12			14	23	37
8300		4" conc. brick	standard brick	12			14.30	23.50	37.80
8320		4" conc. brick	glazed brick	12			13.60	23.50	37.10

Important: See the Reference Section for critical supporting data - Location Factors & Historical Cost Indexes

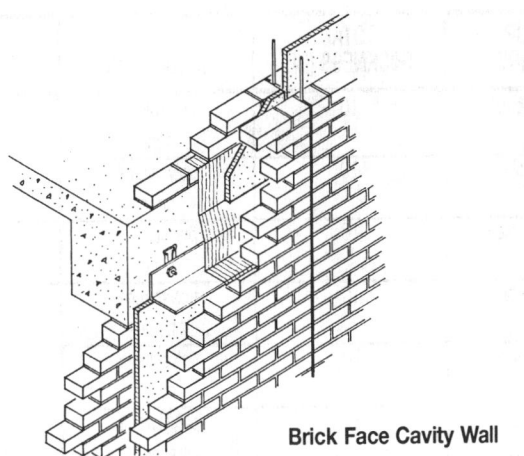

Brick Face Cavity Wall

4.1-273 — Brick Face Cavity Wall

	FACE BRICK	BACKUP MASONRY	TOTAL THICKNESS (IN.)	CAVITY INSULATION		COST PER S.F.		
						MAT.	INST.	TOTAL
1000	Standard	4"common brick	10	polystyrene		6	15.90	21.90
1020				none		5.80	15.45	21.25
1040		6"SCR brick	12	polystyrene		8.40	14.15	22.55
1060				none		8.20	13.75	21.95
1080		4"conc. block	10	polystyrene		4.77	12.80	17.57
1100				none		4.57	12.35	16.92
1120		6"conc. block	12	polystyrene		4.99	13.05	18.04
1140				none		4.79	12.65	17.44
1160		4"L.W. block	10	polystyrene		4.88	12.70	17.58
1180				none		4.68	12.30	16.98
1200		6"L.W. block	12	polystyrene		5.15	12.95	18.10
1220				none		4.95	12.55	17.50
1240		4"glazed block	10	polystyrene		9.55	13.65	23.20
1260				none		9.35	13.25	22.60
1280		6"glazed block	12	polystyrene		9.60	13.65	23.25
1300				none		9.40	13.25	22.65
1320		4"clay tile	10	polystyrene		8.10	12.30	20.40
1340				none		7.90	11.85	19.75
1360		4"glazed tile	10	polystyrene		11.55	16.20	27.75
1380				none		11.35	15.75	27.10
1500	Glazed	4" common brick	10	polystyrene		11.05	16.25	27.30
1520				none		10.85	15.80	26.65
1580		4" conc. block	10	polystyrene		9.80	13.15	22.95
1600				none		9.60	12.70	22.30
1660		4" L.W. block	10	polystyrene		9.90	13.05	22.95
1680				none		9.70	12.65	22.35
1740		4" glazed block	10	polystyrene		14.60	14	28.60
1760				none		14.40	13.60	28
1820		4" clay tile	10	polystyrene		13.10	12.65	25.75
1840				none		12.90	12.20	25.10
1860		4" glazed tile	10	polystyrene		13.25	13	26.25
1880				none		13.05	12.55	25.60
2000	Engineer	4" common brick	10	polystyrene		6.30	14.85	21.15
2020				none		6.10	14.40	20.50
2080		4" conc. block	10	polystyrene		5.05	11.75	16.80
2100				none		4.86	11.30	16.16
2162		4" L.W. block	10	polystyrene		5.15	11.65	16.80
2180				none		4.97	11.25	16.22

	FACE BRICK	BACKUP MASONRY	TOTAL THICKNESS (IN.)	CAVITY INSULATION		COST PER S.F.		
						MAT.	INST.	TOTAL
2240	Engineer	4" glazed block	10	polystyrene		9.85	12.60	22.45
2260				none		9.65	12.20	21.85
2320		4" clay tile	10	polystyrene		8.35	11.25	19.60
2340				none		8.15	10.80	18.95
2360		4" glazed tile	10	polystyrene		11.85	15.15	27
2380				none		11.65	14.70	26.35
2500	Roman	4" common brick	10	polystyrene		8.10	15.05	23.15
2520				none		7.90	14.60	22.50
2580		4" conc. block	10	polystyrene		6.85	11.95	18.80
2600				none		6.65	11.50	18.15
2660		4" L.W. block	10	polystyrene		6.95	11.85	18.80
2680				none		6.75	11.45	18.20
2740		4" glazed block	10	polystyrene		11.65	12.80	24.45
2760				none		11.45	12.40	23.85
2820		4"clay tile	10	polystyrene		10.15	11.45	21.60
2840				none		9.95	11	20.95
2860		4"glazed tile	10	polystyrene		13.65	15.35	29
2880				none		13.45	14.90	28.35
3000	Norman	4" common brick	10	polystyrene		7.20	13.70	20.90
3020				none		7	13.30	20.30
3080		4" conc. block	10	polystyrene		5.95	10.60	16.55
3100				none		5.75	10.20	15.95
3160		4" L.W. block	10	polystyrene		6.05	10.55	16.60
3180				none		5.85	10.10	15.95
3240		4" glazed block	10	polystyrene		10.75	11.50	22.25
3260				none		10.55	11.05	21.60
3320		4" clay tile	10	polystyrene		9.25	10.10	19.35
3340				none		9.05	9.70	18.75
3360		4" glazed tile	10	polystyrene		12.75	14	26.75
3380				none		12.55	13.60	26.15
3500	Norwegian	4" common brick	10	polystyrene		6.45	13	19.45
3520				none		6.25	12.60	18.85
3580		4" conc. block	10	polystyrene		5.20	9.90	15.10
3600				none		5	9.50	14.50
3660		4" L.W. block	10	polystyrene		5.30	9.85	15.15
3680				none		5.10	9.40	14.50
3740		4" glazed block	10	polystyrene		10	10.80	20.80
3760				none		9.80	10.35	20.15
3820		4" clay tile	10	polystyrene		8.50	9.40	17.90
3840				none		8.30	9	17.30
3860		4" glazed tile	10	polystyrene		12	13.30	25.30
3880				none		11.80	12.90	24.70
4000	Utility	4" common brick	10	polystyrene		6.10	12.35	18.45
4020				none		5.90	11.90	17.80
4080		4" conc. block	10	polystyrene		4.86	9.25	14.11
4100				none		4.66	8.80	13.46
4160		4" L.W. block	10	polystyrene		4.97	9.15	14.12
4180				none		4.77	8.75	13.52
4240		4" glazed block	10	polystyrene		9.65	10.10	19.75
4260				none		9.45	9.70	19.15
4320		4" clay tile	10	polystyrene		8.15	8.75	16.90
4340				none		7.95	8.30	16.25
4360		4" glazed tile	10	polystyrene		11.65	12.65	24.30
4380				none		11.45	12.20	23.65

Important: See the Reference Section for critical supporting data - Location Factors & Historical Cost Indexes

4.1-273　Brick Face Cavity Wall - Insulated Backup

	FACE BRICK	BACKUP MASONRY	TOTAL THICKNESS (IN.)	BACKUP CORE FILL		MAT.	INST.	TOTAL
5100	Standard	6" conc. block	10	perlite		5.05	12.85	17.90
5120				styrofoam		5.60	12.65	18.25
5180		6" L.W. block	10	perlite		5.20	12.75	17.95
5200				styrofoam		5.75	12.55	18.30
5260		6" glazed block	10	perlite		10.35	13.65	24
5280				styrofoam		10.85	13.45	24.30
5340		6" clay tile	10	none		8.05	12.20	20.25
5360		8" clay tile	12	none		9	12.65	21.65
5600	Glazed	6" conc. block	10	perlite		10.05	13.20	23.25
5620				styrofoam		10.60	13	23.60
5680		6" L.W. block	10	perlite		10.20	13.10	23.30
5700				styrofoam		10.75	12.90	23.65
5760		6" glazed block	10	perlite		15.35	14	29.35
5780				styrofoam		15.90	13.80	29.70
5840		6" clay tile	10	none		13.05	12.55	25.60
5860		8"clay tile	8	none		14	13	27
6100	Engineer	6" conc. block	10	perlite		5.35	11.80	17.15
6120				styrofoam		5.85	11.60	17.45
6180		6" L.W. block	10	perlite		5.50	11.70	17.20
6200				styrofoam		6.05	11.50	17.55
6260		6" glazed block	10	perlite		10.60	12.60	23.20
6280				styrofoam		11.15	12.40	23.55
6340		6" clay tile	10	none		8.35	11.15	19.50
6360		8" clay tile	12	none		9.25	11.60	20.85
6600	Roman	6" conc. block	10	perlite		7.15	12	19.15
6620				styrofoam		7.65	11.80	19.45
6680		6" L.W. block	10	perlite		7.30	11.90	19.20
6700				styrofoam		7.85	11.70	19.55
6760		6" glazed block	10	perlite		12.40	12.80	25.20
6780				styrofoam		12.95	12.60	25.55
6840		6" clay tile	10	none		10.15	11.35	21.50
6860		8" clay tile	12	none		11.05	11.80	22.85
7100	Norman	6" conc. block	10	perlite		6.20	10.65	16.85
7120				styrofoam		6.75	10.45	17.20
7180		6" L.W. block	10	perlite		6.40	10.55	16.95
7200				styrofoam		6.90	10.35	17.25
7260		6" glazed block	10	perlite		11.50	11.50	23
7280				styrofoam		12.05	11.30	23.35
7340		6" clay tile	10	none		9.20	10.05	19.25
7360		8" clay tile	12	none		10.15	10.45	20.60
7600	Norwegian	6" conc. block	10	perlite		5.45	9.95	15.40
7620				styrofoam		6	9.75	15.75
7680		6" L.W. block	10	perlite		5.65	9.85	15.50
7700				styrofoam		6.15	9.65	15.80
7760		6" glazed block	10	perlite		10.75	10.80	21.55
7780				styrofoam		11.30	10.60	21.90
7840		6" clay tile	10	none		8.45	9.35	17.80
7860		8" clay tile	12	none		9.40	9.75	19.15
8100	Utility	6" conc. block	10	perlite		5.15	9.30	14.45
8120				styrofoam		5.65	9.10	14.75
8180		6" L.W. block	10	perlite		5.30	9.20	14.50
8200				styrofoam		5.85	9	14.85
8260		6" glazed block	10	perlite		10.40	10.10	20.50
8280				styrofoam		10.95	9.90	20.85

EXTERIOR CLOSURE

4

4.1-273   Brick Face Cavity Wall - Insulated Backup

	FACE BRICK	BACKUP MASONRY	TOTAL THICKNESS (IN.)	BACKUP CORE FILL		COST PER S.F.		
						MAT.	INST.	TOTAL
8340	Utility	6" clay tile	10	none		8.15	8.65	16.80
8360		8" clay tile	12	none		9.05	9.10	18.15

Important: See the Reference Section for critical supporting data - Location Factors & Historical Cost Indexes

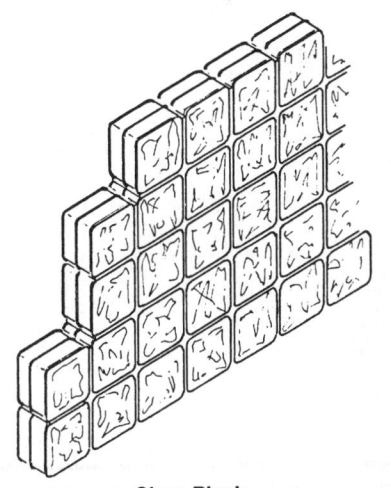

Glass Block

4.1-282	Glass Block	COST PER S.F.		
		MAT.	INST.	TOTAL
2300	Glass block 4" thick, 6"x6" plain, under 1,000 S.F.	17.70	14.85	32.55
2400	1,000 to 5,000 S.F.	16.60	12.90	29.50
2500	Over 5,000 S.F.	15.50	12.10	27.60
2600	Solar reflective, under 1,000 S.F.	25	20	45
2700	1,000 to 5,000 S.F.	23	17.40	40.40
2800	Over 5,000 S.F.	21.50	16.30	37.80
3500	8"x8" plain, under 1,000 S.F.	12.50	11.10	23.60
3600	1,000 to 5,000 S.F.	11.85	9.60	21.45
3700	Over 5,000 S.F.	11.25	8.65	19.90
3800	Solar reflective, under 1,000 S.F.	17.45	14.90	32.35
3900	1,000 to 5,000 S.F.	16.55	12.80	29.35
4000	Over 5,000 S.F.	15.70	11.45	27.15
5000	12"x12" plain, under 1,000 S.F.	14.95	10.30	25.25
5100	1,000 to 5,000 S.F.	14.15	8.65	22.80
5200	Over 5,000 S.F.	13.30	7.95	21.25
5300	Solar reflective, under 1,000 S.F.	21	13.80	34.80
5400	1,000 to 5,000 S.F.	19.80	11.45	31.25
5600	Over 5,000 S.F.	18.60	10.50	29.10
5800	3" thinline, 6"x6" plain, under 1,000 S.F.	14.85	14.85	29.70
5900	Over 5,000 S.F.	14.85	14.85	29.70
6000	Solar reflective, under 1,000 S.F.	21	20	41
6100	Over 5,000 S.F.	20	16.30	36.30
6200	8"x8" plain, under 1,000 S.F.	9.15	11.10	20.25
6300	Over 5,000 S.F.	8.65	8.65	17.30
6400	Solar reflective, under 1,000 S.F.	12.80	14.90	27.70
6500	Over 5,000 S.F.	12.10	11.45	23.55

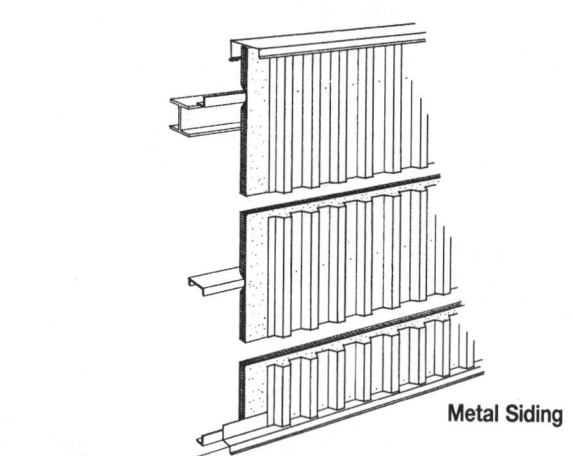

The table below lists costs for metal siding of various descriptions, not including the steel frame, or the structural steel, of a building. Costs are per S.F. including all accessories and insulation.

Metal Siding

4.1-384	Metal Siding Panel	COST PER S.F.		
		MAT.	INST.	TOTAL
1400	Metal siding aluminum panel, corrugated, .024" thick, natural	1.43	2.41	3.84
1450	Painted	1.65	2.41	4.06
1500	.032" thick, natural	1.84	2.41	4.25
1550	Painted	2.19	2.41	4.60
1600	Ribbed 4" pitch, .032" thick, natural	1.86	2.41	4.27
1650	Painted	2.15	2.41	4.56
1700	.040" thick, natural	2.20	2.41	4.61
1750	Painted	2.52	2.41	4.93
1800	.050" thick, natural	2.52	2.41	4.93
1850	Painted	2.82	2.41	5.23
1900	8" pitch panel, .032" thick, natural	1.77	2.31	4.08
1950	Painted	2.06	2.32	4.38
2000	.040" thick, natural	2.11	2.33	4.44
2050	Painted	2.41	2.35	4.76
2100	.050" thick, natural	2.43	2.34	4.77
2150	Painted	2.75	2.36	5.11
3000	Steel, corrugated or ribbed, 29 Ga. .0135" thick, galvanized	1.22	2.15	3.37
3050	Colored	1.62	2.19	3.81
3100	26 Ga. .0179" thick, galvanized	1.26	2.16	3.42
3150	Colored	1.55	2.20	3.75
3200	24 Ga. .0239" thick, galvanized	1.34	2.17	3.51
3250	Colored	1.73	2.21	3.94
3300	22 Ga. .0299" thick, galvanized	1.52	2.18	3.70
3350	Colored	2.07	2.22	4.29
3400	20 Ga. .0359" thick, galvanized	1.52	2.18	3.70
3450	Colored	2.20	2.35	4.55
4100	Sandwich panels, factory fab., 1" polystyrene, steel core, 26 Ga., galv.	2.96	3.36	6.32
4200	Colored, 1 side	3.72	3.36	7.08
4300	2 sides	4.85	3.36	8.21
4400	2" polystyrene, steel core, 26 Ga., galvanized	3.53	3.36	6.89
4500	Colored, 1 side	4.29	3.36	7.65
4600	2 sides	5.40	3.36	8.76
4700	22 Ga., baked enamel exterior	7.20	3.55	10.75
4800	Polyvinyl chloride exterior	7.60	3.55	11.15
5100	Textured aluminum, 4' x 8' x 5/16" plywood backing, single face	2.40	2.02	4.42
5200	Double face	3.61	2.02	5.63

Important: See the Reference Section for critical supporting data - Location Factors & Historical Cost Indexes

The table below lists costs per S.F. for exterior walls with wood siding. A variety of systems are presented using both wood and metal studs at 16" and 24" O.C.

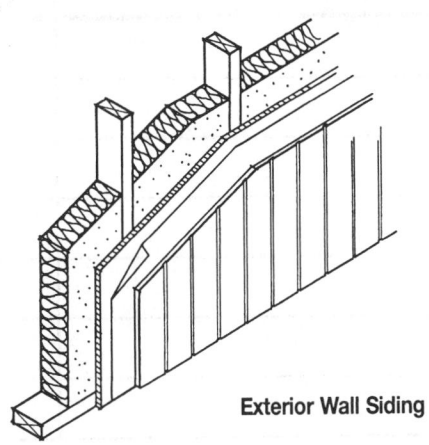

Exterior Wall Siding

4.1-412	Wood & Other Siding	COST PER S.F.		
		MAT.	INST.	TOTAL
1400	Wood siding w/2"x4"studs, 16"O.C., insul. wall, 5/8"text 1-11 fir plywood	2.48	3.41	5.89
1450	5/8" text 1-11 cedar plywood	3.02	3.30	6.32
1500	1" x 4" vert T.&G. redwood	4.38	4.64	9.02
1600	1" x 8" vert T.&G. redwood	3.97	4.20	8.17
1650	1" x 5" rabbetted cedar bev. siding	3.64	3.75	7.39
1700	1" x 6" cedar drop siding	3.69	3.78	7.47
1750	1" x 12" rough sawn cedar	2.66	3.44	6.10
1800	1" x 12" sawn cedar, 1" x 4" battens	3.66	3.70	7.36
1850	1" x 10" redwood shiplap siding	3.90	3.59	7.49
1900	18" no. 1 red cedar shingles, 5-1/2" exposed	3.65	4.49	8.14
1950	6" exposed	3.49	4.34	7.83
2000	6-1/2" exposed	3.33	4.20	7.53
2100	7" exposed	3.16	4.05	7.21
2150	7-1/2" exposed	3	3.90	6.90
3000	8" wide aluminum siding	2.59	3.01	5.60
3150	8" plain vinyl siding	1.92	3.03	4.95
3250	8" insulated vinyl siding	2.08	3.03	5.11
3300				
3400	2" x 6" studs, 16" O.C., insul. wall, w/ 5/8" text 1-11 fir plywood	2.74	3.55	6.29
3500	5/8" text 1-11 cedar plywood	3.28	3.55	6.83
3600	1" x 4" vert T.&G. redwood	4.64	4.78	9.42
3700	1" x 8" vert T.&G. redwood	4.23	4.34	.8.57
3800	1" x 5" rabbetted cedar bev siding	3.90	3.89	7.79
3900	1" x 6" cedar drop siding	3.95	3.92	7.87
4000	1" x 12" rough sawn cedar	2.92	3.58	6.50
4200	1" x 12" sawn cedar, 1" x 4" battens	3.92	3.84	7.76
4500	1" x 10" redwood shiplap siding	4.16	3.73	7.89
4550	18" no. 1 red cedar shingles, 5-1/2" exposed	3.91	4.63	8.54
4600	6" exposed	3.75	4.48	8.23
4650	6-1/2" exposed	3.59	4.34	7.93
4700	7" exposed	3.42	4.19	7.61
4750	7-1/2" exposed	3.26	4.04	7.30
4800	8" wide aluminum siding	2.85	3.15	6
4850	8" plain vinyl siding	2.18	3.17	5.35
4900	8" insulated vinyl siding	2.34	3.17	5.51
4910				
5000	2" x 6" studs, 24" O.C., insul. wall, 5/8" text 1-11, fir plywood	2.59	3.35	5.94
5050	5/8" text 1-11 cedar plywood	3.13	3.35	6.48
5100	1" x 4" vert T.&G. redwood	4.49	4.58	9.07
5150	1" x 8" vert T.&G. redwood	4.08	4.14	8.22
5200	1" x 5" rabbetted cedar bev siding	3.75	3.69	7.44
5250	1" x 6" cedar drop siding	3.80	3.72	7.52

For expanded coverage of these items see *Means Assemblies Cost Data 1998*

EXTERIOR CLOSURE

4

4.1-412	Wood & Other Siding	COST PER S.F.		
		MAT.	INST.	TOTAL
5300	1" x 12" rough sawn cedar	2.77	3.38	6.15
5400	1" x 12" sawn cedar, 1" x 4" battens	3.77	3.64	7.41
5450	1" x 10" redwood shiplap siding	4.01	3.53	7.54
5500	18" no. 1 red cedar shingles, 5-1/2" exposed	3.76	4.43	8.19
5550	6" exposed	3.60	4.28	7.88
5650	7" exposed	3.27	3.99	7.26
5700	7-1/2" exposed	3.11	3.84	6.95
5750	8" wide aluminum siding	2.70	2.95	5.65
5800	8" plain vinyl siding	2.03	2.97	5
5850	8" insulated vinyl siding	2.19	2.97	5.16
5900	3-5/8" metal studs, 16 Ga., 16" OC insul.wall, 5/8"text 1-11 fir plywood	3.01	3.89	6.90
5950	5/8" text 1-11 cedar plywood	3.55	3.89	7.44
6000	1" x 4" vert T.&G. redwood	4.91	5.10	10.01
6050	1" x 8" vert T.&G. redwood	4.50	4.68	9.18
6100	1" x 5" rabbetted cedar bev siding	4.17	4.23	8.40
6150	1" x 6" cedar drop siding	4.22	4.26	8.48
6200	1" x 12" rough sawn cedar	3.19	3.92	7.11
6250	1" x 12" sawn cedar, 1" x 4" battens	4.19	4.18	8.37
6300	1" x 10" redwood shiplap siding	4.43	4.07	8.50
6350	18" no. 1 red cedar shingles, 5-1/2" exposed	4.18	4.97	9.15
6500	6" exposed	4.02	4.82	8.84
6550	6-1/2" exposed	3.86	4.68	8.54
6600	7" exposed	3.69	4.53	8.22
6650	7-1/2" exposed	3.69	4.42	8.11
6700	8" wide aluminum siding	3.12	3.49	6.61
6750	8" plain vinyl siding	2.45	3.40	5.85
6800	8" insulated vinyl siding	2.61	3.40	6.01
7000	3-5/8" metal studs, 16 Ga. 24" OC insul wall, 5/8" text 1-11 fir plywood	2.86	3.53	6.39
7050	5/8" text 1-11 cedar plywood	3.40	3.53	6.93
7100	1" x 4" vert T.&G. redwood	4.76	4.87	9.63
7150	1" x 8" vert T.&G. redwood	4.35	4.43	8.78
7200	1" x 5" rabbetted cedar bev siding	4.02	3.98	8
7250	1" x 6" cedar drop siding	4.06	4.01	8.07
7300	1" x 12" rough sawn cedar	3.04	3.67	6.71
7350	1" x 12" sawn cedar 1" x 4" battens	4.04	3.93	7.97
7400	1" x 10" redwood shiplap siding	4.28	3.82	8.10
7450	18" no. 1 red cedar shingles, 5-1/2" exposed	4.20	4.87	9.07
7500	6" exposed	4.03	4.72	8.75
7550	6-1/2" exposed	3.87	4.57	8.44
7600	7" exposed	3.71	4.43	8.14
7650	7-1/2" exposed	3.38	4.13	7.51
7700	8" wide aluminum siding	2.97	3.24	6.21
7750	8" plain vinyl siding	2.30	3.26	5.56
7800	8" insul. vinyl siding	2.46	3.26	5.72

Important: See the Reference Section for critical supporting data - Location Factors & Historical Cost Indexes

The table below lists costs for some typical stucco walls including all the components as demonstrated in the component block below. Prices are presented for backup walls using wood studs, metal studs and CMU.

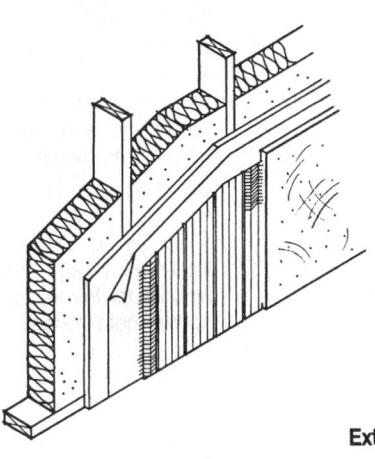

Exterior Stucco Wall

4.5-110	Stucco Wall	COST PER S.F.		
		MAT.	INST.	TOTAL
2100	Cement stucco, 7/8" th., plywood sheathing, stud wall, 2" x 4", 16" O.C.	2.19	3.83	6.02
2200	24" O.C.	2.09	3.67	5.76
2300	2" x 6", 16" O.C.	2.45	3.97	6.42
2400	24" O.C.	2.30	3.77	6.07
2500	No sheathing, metal lath on stud wall, 2" x 4", 16" O.C.	1.60	3.11	4.71
2600	24" O.C.	1.50	2.95	4.45
2700	2" x 6", 16" O.C.	1.86	3.25	5.11
2800	24" O.C.	1.71	3.05	4.76
2900	1/2" gypsum sheathing, 3-5/8" metal studs, 16" O.C.	2.16	3.59	5.75
2950	24" O.C.	2.01	3.34	5.35
3000	Cement stucco, 5/8" th., 2 coats on std. CMU block, 8"x 16", 8" thick	1.93	5.50	7.43
3100	10" Thick	2.74	5.75	8.49
3200	12" Thick	2.81	6.85	9.66
3300	Std. light Wt. block 8" x 16", 8" Thick	1.75	5.30	7.05
3400	10" Thick	2.27	5.45	7.72
3500	12" Thick	2.52	6.45	8.97
3600	3 coat stucco, self furring metal lath 3.4 Lb/SY, on 8" x 16", 8" thick	1.97	5.70	7.67
3700	10" Thick	2.78	5.90	8.68
3800	12" Thick	2.85	7	9.85
3900	Lt. Wt. block, 8" Thick	1.79	5.45	7.24
4000	10" Thick	2.31	5.65	7.96
4100	12" Thick	2.56	6.65	9.21

EXTERIOR CLOSURE

4

Exterior Door System

The table below lists exterior door systems by material, type and size. Prices between sizes listed can be interpolated with reasonable accuracy. Prices are per opening for a complete door system including frame and required hardware.

Wood doors in this table are designed with wood frames and metal doors with hollow metal frames. Depending upon quality the total material and installation cost of a wood door frame is about the same as a hollow metal frame.

4.6-100   Wood, Steel & Aluminum

	MATERIAL	TYPE	DOORS	SPECIFICATION	OPENING	COST PER OPNG.		
						MAT.	INST.	TOTAL
2350	Birch	solid core	single door	hinged	2'-6" x 6'-8"	735	181	916
2400					2'-6" x 7'-0"	740	183	923
2450					2'-8" x 7'-0"	745	183	928
2500					3'-0" x 7'-0"	745	186	931
2550			double door	hinged	2'-6" x 6'-8"	1,375	330	1,705
2600					2'-6" x 7'-0"	1,400	335	1,735
2650					2'-8" x 7'-0"	1,400	335	1,735
2700					3'-0" x 7'-0"	1,400	340	1,740
2750	Wood	combination	storm & screen	hinged	3'-0" x 6'-8"	229	44.50	273.50
2800					3'-0" x 7'-0"	252	49	301
2850		overhead	panels, H.D.	manual oper.	8'-0" x 8'-0"	415	335	750
2900					10'-0" x 10'-0"	635	370	1,005
2950					12'-0" x 12'-0"	885	445	1,330
3000					14'-0" x 14'-0"	1,525	515	2,040
3050					20'-0" x 16'-0"	3,175	1,025	4,200
3100				electric oper.	8'-0" x 8'-0"	820	500	1,320
3150					10'-0" x 10'-0"	1,050	535	1,585
3200					12'-0" x 12'-0"	1,300	610	1,910
3250					14'-0" x 14'-0"	1,925	680	2,605
3300					20'-0" x 16'-0"	3,600	1,350	4,950
3350	Steel 18 Ga.	hollow metal	1 door w/frame	no label	2'-6" x 7'-0"	805	188	993
3400					2'-8" x 7'-0"	805	188	993
3450					3'-0" x 7'-0"	805	188	993
3500					3'-6" x 7'-0"	880	198	1,078
3550		hollow metal	1 door w/frame	no label	4'-0" x 8'-0"	1,100	199	1,299
3600			2 doors w/frame	no label	5'-0" x 7'-0"	1,550	350	1,900
3650					5'-4" x 7'-0"	1,550	350	1,900
3700					6'-0" x 7'-0"	1,550	350	1,900
3750					7'-0" x 7'-0"	1,700	370	2,070
3800					8'-0" x 8'-0"	2,125	370	2,495
3850			1 door w/frame	"A" label	2'-6" x 7'-0"	970	227	1,197
3900					2'-8" x 7'-0"	975	230	1,205
3950					3'-0" x 7'-0"	975	230	1,205
4000					3'-6" x 7'-0"	1,050	235	1,285
4050					4'-0" x 8'-0"	1,050	246	1,296
4100			2 doors w/frame	"A" label	5'-0" x 7'-0"	1,875	420	2,295
4150					5'-4" x 7'-0"	1,850	425	2,275
4200					6'-0" x 7'-0"	1,850	425	2,275

	MATERIAL	TYPE	DOORS	SPECIFICATION	OPENING	COST PER OPNG.		
						MAT.	INST.	TOTAL
4250	Steel 18 Ga.				7'-0" x 7'-0"	2,000	435	2,435
4300					8'-0" x 8'-0"	2,000	435	2,435
4350	Steel 24 Ga.	overhead	sectional	manual oper.	8'-0" x 8'-0"	370	335	705
4400					10'-0" x 10'-0"	480	370	850
4450					12'-0" x 12'-0"	650	445	1,095
4500					20'-0" x 14'-0"	1,625	950	2,575
4550				electric oper.	8'-0" x 8'-0"	775	500	1,275
4600					10'-0" x 10'-0"	885	535	1,420
4650					12'-0" x 12'-0"	1,050	610	1,660
4700					20'-0" x 14'-0"	2,050	1,275	3,325
4750	Steel	overhead	rolling	manual oper.	8'-0" x 8'-0"	710	545	1,255
4800					10'-0" x 10'-0"	950	620	1,570
4850					12'-0" x 12'-0"	1,250	725	1,975
4900					14'-0" x 14'-0"	1,700	1,100	2,800
4950					20'-0" x 12'-0"	2,525	970	3,495
5000					20'-0" x 16'-0"	3,575	1,450	5,025
5050				electric oper.	8'-0" x 8'-0"	1,650	720	2,370
5100					10'-0" x 10'-0"	1,900	795	2,695
5150					12'-0" x 12'-0"	2,200	900	3,100
5200					14'-0" x 14'-0"	2,650	1,275	3,925
5250					20'-0" x 12'-0"	3,500	1,150	4,650
5300					20'-0" x 16'-0"	4,550	1,625	6,175
5350				fire rated	10'-0" x 10'-0"	1,675	790	2,465
5400			rolling grill	manual oper.	10'-0" x 10'-0"	1,600	870	2,470
5450					15'-0" x 8'-0"	1,875	1,100	2,975
5500		vertical lift	1 door w/frame	motor operator	16"-0" x 16'-0"	16,100	3,475	19,575
5550					32'-0" x 24'-0"	29,900	2,325	32,225
5600	St. Stl. & glass	revolving	stock unit	manual oper.	6'-0" x 7'-0"	26,400	5,800	32,200
5650				auto Cntrls.	6'-10" x 7'-0"	37,400	6,175	43,575
5700	Bronze	revolving	stock unit	manual oper.	6'-10" x 7'-0"	30,800	11,600	42,400
5750				auto Cntrls.	6'-10" x 7'-0"	41,800	12,000	53,800
5800	St. Stl. & glass	balanced	standard	economy	3'-0" x 7'-0"	5,700	975	6,675
5850				premium	3'-0" x 7'-0"	9,875	1,275	11,150
6000	Aluminum	combination	storm & screen	hinged	3'-0" x 6'-8"	188	47.50	235.50
6050					3'-0" x 7'-0"	207	52.50	259.50
6100		overhead	rolling grill	manual oper.	12'-0" x 12'-0"	2,575	1,525	4,100
6150				motor oper.	12'-0" x 12'-0"	3,375	1,700	5,075
6200	Alum. & Fbrgls.	overhead	heavy duty	manual oper.	12'-0" x 12'-0"	1,300	445	1,745
6250				electric oper.	12'-0" x 12'-0"	1,700	610	2,310
6300	Alum. & glass	w/o transom	narrow stile	w/panic Hrdwre.	3'-0" x 7'-0"	945	630	1,575
6350				dbl. door, Hrdwre.	6'-0" x 7'-0"	1,700	1,025	2,725
6400			wide stile	hdwre.	3'-0" x 7'-0"	1,350	620	1,970
6450				dbl. door, Hdwre.	6'-0" x 7'-0"	2,700	1,225	3,925
6500			full vision	hdwre.	3'-0" x 7'-0"	1,400	990	2,390
6550				dbl. door, Hdwre.	6'-0" x 7'-0"	2,100	1,400	3,500
6600			non-standard	hdwre.	3'-0" x 7'-0"	1,525	620	2,145
6650				dbl. door, Hdwre.	6'-0" x 7'-0"	3,025	1,225	4,250
6700			bronze fin.	hdwre.	3'-0" x 7'-0"	1,550	620	2,170
6750				dbl. door, Hrdwre.	6'-0" x 7'-0"	3,075	1,225	4,300
6800			black fin.	hdwre.	3'-0" x 7'-0"	1,600	620	2,220
6850				dbl. door, Hdwre.	6'-0" x 7'-0"	3,175	1,225	4,400
6900		w/transom	narrow stile	hdwre.	3'-0" x 10'-0"	1,350	715	2,065
6950				dbl. door, Hdwre.	6'-0" x 10'-0"	2,275	1,250	3,525
7000			wide stile	hdwre.	3'-0" x 10'-0"	1,575	860	2,435

EXTERIOR CLOSURE

4

For expanded coverage of these items see *Means Assemblies Cost Data 1998*

	MATERIAL	TYPE	DOORS	SPECIFICATION	OPENING	COST PER OPNG.		
						MAT.	INST.	TOTAL
7050	Alum. & glass	w/transom		dbl. door, Hdwre.	6'-0" x 10'-0"	2,625	1,500	4,125
7100			full vision	hdwre.	3'-0" x 10'-0"	1,725	960	2,685
7150				dbl. door, Hdwre.	6'-0" x 10'-0"	2,875	1,650	4,525
7200			non-standard	hdwre.	3'-0" x 10'-0"	1,550	665	2,215
7250				dbl. door, Hdwre.	6'-0" x 10'-0"	3,100	1,325	4,425
7300			bronze fin.	hdwre.	3'-0" x 10'-0"	1,575	665	2,240
7350				dbl. door, Hdwre.	6'-0" x 10'-0"	3,150	1,325	4,475
7400			black fin.	hdwre.	3'-0" x 10'-0"	1,625	665	2,290
7450				dbl. door, Hdwre.	6'-0" x 10'-0"	3,250	1,325	4,575
7500		revolving	stock design	minimum	6'-10" x 7'-0"	14,100	2,325	16,425
7550				average	6'-0" x 7'-0"	17,600	2,900	20,500
7600				maximum	6'-10" x 7'-0"	23,000	3,875	26,875
7650				min., automatic	6'-10" x 7'-0"	25,100	2,700	27,800
7700				avg., automatic	6'-10" x 7'-0"	28,600	3,275	31,875
7750				max., automatic	6'-10" x 7'-0"	34,000	4,250	38,250
7800		balanced	standard	economy	3'-0" x 7'-0"	3,600	970	4,570
7850				premium	3'-0" x 7'-0"	4,975	1,250	6,225
7900		mall front	sliding panels	alum. fin.	16'-0" x 9'-0"	2,200	485	2,685
7950					24'-0" x 9'-0"	3,200	900	4,100
8000				bronze fin.	16'-0" x 9'-0"	2,575	565	3,140
8050					24'-0" x 9'-0"	3,725	1,050	4,775
8100			fixed panels	alum. fin.	48'-0" x 9'-0"	5,950	700	6,650
8150				bronze fin.	48'-0" x 9'-0"	6,925	815	7,740
8200		sliding entrance	5' x 7' door	electric oper.	12'-0" x 7'-6"	5,600	900	6,500
8250		sliding patio	temp. glass	economy	6'-0" x 7'-0"	695	167	862
8300			temp. glass	economy	12'-0" x 7'-0"	2,300	222	2,522
8350				premium	6'-0" x 7'-0"	1,050	251	1,301
8400					12'-0" x 7'-0"	3,450	335	3,785

Important: See the Reference Section for critical supporting data - Location Factors & Historical Cost Indexes

The table below lists window systems by material, type and size. Prices between sizes listed can be interpolated with reasonable accuracy. Prices include frame, hardware, and casing as illustrated in the component block below.

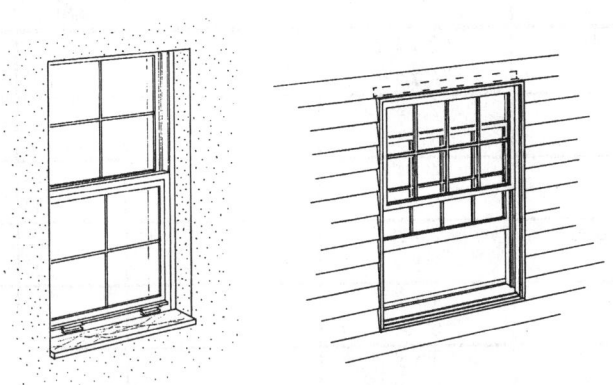

EXTERIOR CLOSURE

4

4.7-110			Wood, Steel & Aluminum					
	MATERIAL	**TYPE**	**GLAZING**	**SIZE**	**DETAIL**	**COST PER UNIT**		
						MAT.	**INST.**	**TOTAL**
3000	Wood	double hung	std. glass	2'-8" x 4'-6"		152	145	297
3050				3'-0" x 5'-6"		177	167	344
3100			insul. glass	2'-8" x 4'-6"		184	145	329
3150				3'-0" x 5'-6"		230	167	397
3200		sliding	std. glass	3'-4" x 2'-7"		157	122	279
3250				4'-4" x 3'-3"		220	133	353
3300				5'-4" x 6'-0"		288	160	448
3350			insul. glass	3'-4" x 2'-7"		190	142	332
3400				4'-4" x 3'-3"		267	154	421
3450				5'-4" x 6'-0"		350	182	532
3500		awning	std. glass	2'-10" x 1'-9"		190	73	263
3600				4'-4" x 2'-8"		227	71	298
3700			insul. glass	2'-10" x 1'-9"		233	84	317
3800				4'-4" x 2'-8"		277	77.50	354.50
3900		casement	std. glass	1'-10" x 3'-2"	1 lite	234	96	330
3950				4'-2" x 4'-2"	2 lite	365	127	492
4000				5'-11" x 5'-2"	3 lite	615	164	779
4050				7'-11" x 6'-3"	4 lite	780	197	977
4100				9'-11" x 6'-3"	5 lite	980	229	1,209
4150			insul. glass	1'-10" x 3'-2"	1 lite	264	96	360
4200				4'-2" x 4'-2"	2 lite	465	127	592
4250				5'-11" x 5'-2"	3 lite	810	164	974
4300				7'-11" x 6'-3"	4 lite	1,025	197	1,222
4350				9'-11" x 6'-3"	5 lite	1,300	229	1,529
4400		picture	std. glass	4'-6" x 4'-6"		223	164	387
4450				5'-8" x 4'-6"		273	183	456
4500		picture	insul. glass	4'-6" x 4'-6"		270	191	461
4550				5'-8" x 4'-6"		330	212	542
4600		fixed bay	std. glass	8' x 5'		640	300	940
4650				9'-9" x 5'-4"		750	425	1,175
4700			insul. glass	8' x 5'		795	300	1,095
4750				9'-9" x 5'-4"		960	425	1,385
4800		casement bay	std. glass	8' x 5'		675	345	1,020
4850			insul. glass	8' x 5'		950	345	1,295
4900		vert. bay	std. glass	8' x 5'		1,200	345	1,545
4950			insul. glass	8' x 5'		1,450	345	1,795
5000	Steel	double hung	1/4" tempered	2'-8" x 4'-6"		520	115	635
5050				3'-4" x 5'-6"		795	176	971

	MATERIAL	TYPE	GLAZING	SIZE	DETAIL	COST PER UNIT		
						MAT.	INST.	TOTAL
5100	Steel		insul. glass	2'-8" x 4'-6"		535	132	667
5150				3'-4" x 5'-6"		820	201	1,021
5202		horiz. pivoted	std. glass	2' x 2'		124	38.50	162.50
5250				3' x 3'		278	86.50	364.50
5300				4' x 4'		495	154	649
5350				6' x 4'		740	231	971
5400			insul. glass	2' x 2'		129	44	173
5450				3' x 3'		290	98.50	388.50
5500				4' x 4'		515	175	690
5550				6' x 4'		775	263	1,038
5600		picture window	std. glass	3' x 3'		171	86.50	257.50
5650				6' x 4'		455	231	686
5700			insul. glass	3' x 3'		183	98.50	281.50
5750				6' x 4'		490	263	753
5800		industrial security	std. glass	2'-9" x 4'-1"		505	108	613
5850				4'-1" x 5'-5"		995	212	1,207
5900			insul. glass	2'-9" x 4'-1"		520	123	643
5950				4'-1" x 5'-5"		1,025	242	1,267
6000		comm. projected	std. glass	3'-9" x 5'-5"		740	195	935
6050				6'-9" x 4'-1"		1,000	265	1,265
6100			insul. glass	3'-9" x 5'-5"		765	223	988
6150				6'-9" x 4'-1"		1,050	300	1,350
6200		casement	std. glass	4'-2" x 4'-2"	2 lite	605	167	772
6250			insul. glass	4'-2" x 4'-2"		630	190	820
6300			std. glass	5'-11" x 5'-2"	3 lite	1,225	294	1,519
6350			insul. glass	5'-11" x 5'-2"		1,275	335	1,610
6400	Aluminum	projecting	std. glass	3'-1" x 3'-2"		195	87	282
6450				4'-5" x 5'-3"		275	109	384
6500			insul. glass	3'-1" x 3'-2"		234	104	338
6550				4'-5" x 5'-3"		330	131	461
6600		sliding	std. glass	3' x 2'		147	87	234
6650				5' x 3'		187	97	284
6700				8' x 4'		270	145	415
6750				9' x 5'		410	218	628
6800			insul. glass	3' x 2'		163	87	250
6850				5' x 3'		262	97	359
6900				8' x 4'		435	145	580
6950				9' x 5'		650	218	868
7000		single hung	std. glass	2' x 3'		131	87	218
7050				2'-8" x 6'-8"		278	109	387
7100				3'-4" x 5'0"		179	97	276
7150			insul. glass	2' x 3'		159	87	246
7200				2'-8" x 6'-8"		360	109	469
7250				3'-4" x 5'		252	97	349
7300		double hung	std. glass	2' x 3'		188	57.50	245.50
7350				2'-8" x 6'-8"		560	171	731
7400				3'-4" x 5'		525	160	685
7450			insul. glass	2' x 3'		196	66	262
7500				2'-8" x 6'-8"		580	195	775
7550				3'-4" x 5'-0"		545	182	727
7600		casements	std. glass	3'-1" x 3'-2"		143	93.50	236.50
7650				4'-5" x 5'-3"		345	227	572
7700			insul. glass	3'-1" x 3'-2"		156	107	263
7750				4'-5" x 5'-3"		375	259	634

Important: See the Reference Section for critical supporting data - Location Factors & Historical Cost Indexes

	MATERIAL	TYPE	GLAZING	SIZE	DETAIL	COST PER UNIT		
						MAT.	INST.	TOTAL
7800	Aluminum	hinged swing	std. glass	3' x 4'		335	115	450
7850				4' x 5'		560	192	752
7900			insul. glass	3' x 4'		350	132	482
7950				4' x 5'		585	219	804
8002		folding type	std. glass	3'-0" x 4'-0"		197	94.50	291.50
8050				4'-0" x 5'-0"		275	94.50	369.50
8100			insul. glass	3'-0" x 4'-0"		247	94.50	341.50
8150				4'-0" x 5'-0"		360	94.50	454.50
8200		picture unit	std. glass	2'-0" x 3'-0"		97	57.50	154.50
8250				2'-8" x 6'-8"		287	171	458
8300				3'-4" x 5'-0"		269	160	429
8350			insul. glass	2'-0" x 3'-0"		105	66	171
8400				2'-8" x 6'-8"		310	195	505
8450				3'-4" x 5'-0"		291	182	473
8500		awning type	std. glass	3'-0" x 3'-0"	2 lite	320	62	382
8550				3'-0" x 4'-0"	3 lite	375	87	462
8600				3'-0" x 5'-4"	4 lite	455	87	542
8650				4'-0" x 5'-4"	4 lite	500	97	597
8700			insul. glass	3'-0" x 3'-0"	2 lite	345	62	407
8750				3'-0" x 4'-0"	3 lite	435	87	522
8800				3'-0" x 5'-4"	4 lite	535	87	622
8850				4'-0" x 5'-4"	4 lite	595	97	692
8900		jalousie type	std. glass	1'-7" x 3'-2"		118	87	205
8950				2'-3" x 4'-0"		169	87	256
9000				3'-1" x 2'-0"		116	87	203
9050				3'-1" x 5'-3"		241	87	328
9051								
9052								

4.7-582	Tubular Aluminum Framing	COST/S.F. OPNG.		
		MAT.	INST.	TOTAL
1100	Alum flush tube frame, for 1/4"glass,1-3/4"x4",5'x6'opng, no inter horizntls	8.50	7.75	16.25
1150	One intermediate horizontal	11.55	9.10	20.65
1200	Two intermediate horizontals	14.60	10.40	25
1250	5' x 20' opening, three intermediate horizontals	7.65	6.45	14.10
1400	1-3/4" x 4-1/2", 5' x 6' opening, no intermediate horizontals	9.65	7.75	17.40
1450	One intermediate horizontal	12.90	9.10	22
1500	Two intermediate horizontals	16.05	10.40	26.45
1550	5' x 20' opening, three intermediate horizontals	8.60	6.45	15.05
1700	For insulating glass, 2"x4-1/2", 5'x6' opening, no intermediate horizontals	10.85	8.10	18.95
1750	One intermediate horizontal	14.20	9.55	23.75
1800	Two intermediate horizontals	17.50	10.95	28.45
1850	5' x 20' opening, three intermediate horizontals	9.50	6.80	16.30
2000	Thermal break frame, 2-1/4"x4-1/2", 5'x6'opng, no intermediate horizontals	11.65	8.20	19.85
2050	One intermediate horizontal	15.65	9.90	25.55
2100	Two intermediate horizontals	19.60	11.55	31.15
2150	5' x 20' opening, three intermediate horizontals	10.65	7.05	17.70

EXTERIOR CLOSURE

4

The table below lists costs of curtain wall and spandrel panels per S.F. Costs do not include structural framing used to hang the panels.

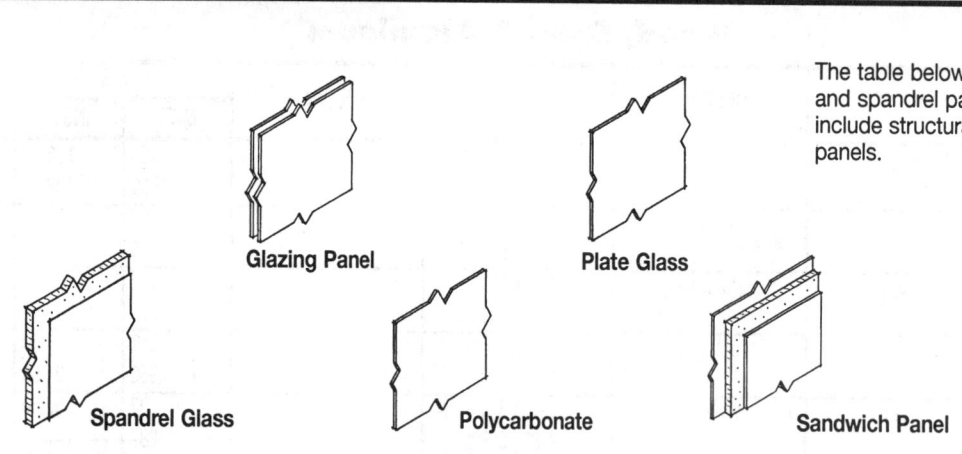

Glazing Panel   Plate Glass

Spandrel Glass   Polycarbonate   Sandwich Panel

4.7-584	Curtain Wall Panels	COST PER S.F.		
		MAT.	INST.	TOTAL
1000	Glazing panel, insulating, 1/2" thick, 2 lites 1/8" float, clear	5.70	6.60	12.30
1100	Tinted	8.50	6.60	15.10
1200	5/8" thick units, 2 lites 3/16" float, clear	6.45	7	13.45
1300	Tinted	7.20	7	14.20
1400	1" thick units, 2 lites, 1/4" float, clear	9.90	8.40	18.30
1500	Tinted	9.15	8.40	17.55
1600	Heat reflective film inside	13.90	7.40	21.30
1700	Light and heat reflective glass, tinted	15.20	7.40	22.60
2000	Plate glass, 1/4" thick, clear	3.99	5.25	9.24
2050	Tempered	4.39	5.25	9.64
2100	Tinted	4.08	5.25	9.33
2200	3/8" thick, clear	6.70	8.40	15.10
2250	Tempered	9.15	8.40	17.55
2300	Tinted	7.90	8.40	16.30
2400	1/2" thick, clear	12.70	11.45	24.15
2450	Tempered	14.65	11.45	26.10
2500	Tinted	13.55	11.45	25
2600	3/4" thick, clear	17.55	17.95	35.50
2650	Tempered	20.50	17.95	38.45
3000	Spandrel glass, panels, 1/4" plate glass insul w/fiberglass, 1" thick	8.45	5.25	13.70
3100	2" thick	9.85	5.25	15.10
3200	Galvanized steel backing, add	2.92		2.92
3300	3/8" plate glass, 1" thick	14.15	5.25	19.40
3400	2" thick	15.55	5.25	20.80
4000	Polycarbonate, masked, clear or colored, 1/8" thick	5.20	3.70	8.90
4100	3/16" thick	6.05	3.81	9.86
4200	1/4" thick	6.90	4.06	10.96
4300	3/8" thick	11.80	4.19	15.99
5000	Facing panel, textured al, 4' x 8' x 5/16" plywood backing, sgl face	2.40	2.02	4.42
5100	Double face	3.61	2.02	5.63
5200	4' x 10' x 5/16" plywood backing, single face	2.29	2.02	4.31
5300	Double face	3.47	2.02	5.49
5400	4' x 12' x 5/16" plywood backing, single face	3.29	2.02	5.31
5500	Sandwich panel, 22 Ga. galv., both sides 2" insulation, enamel exterior	7.20	3.55	10.75
5600	Polyvinylidene floride exterior finish	7.60	3.55	11.15
5700	26 Ga., galv. both sides, 1" insulation, colored 1 side	3.72	3.36	7.08
5800	Colored 2 sides	4.85	3.36	8.21

For information about Means Estimating Seminars, see yellow pages 11 and 12 in back of book

Important: See the Reference Section for critical supporting data - Location Factors & Historical Cost Indexes

Division 5
Roofing

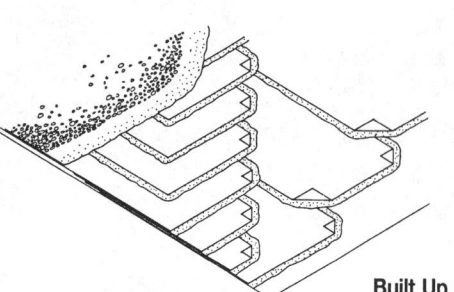

Multiple ply roofing is the most popular covering for minimum pitch roofs.

Built Up Ply

ROOFING 5

5.1-103	Built-Up	COST PER S.F.		
		MAT.	INST.	TOTAL
1200	Asphalt flood coat w/gravel; not incl. insul, flash., nailers			
1300				
1400	Asphalt base sheets & 3 plies #15 asphalt felt, mopped	.38	1.17	1.55
1500	On nailable deck	.43	1.22	1.65
1600	4 plies #15 asphalt felt, mopped	.54	1.28	1.82
1700	On nailable deck	.49	1.35	1.84
1800	Coated glass base sheet, 2 plies glass (type IV), mopped	.41	1.17	1.58
1900	For 3 plies	.49	1.28	1.77
2000	On nailable deck	.47	1.35	1.82
2300	4 plies glass fiber felt (type IV), mopped	.59	1.28	1.87
2400	On nailable deck	.55	1.35	1.90
2500	Organic base sheet & 3 plies #15 organic felt, mopped	.41	1.28	1.69
2600	On nailable deck	.43	1.35	1.78
2700	4 plies #15 organic felt, mopped	.52	1.17	1.69
2750				
2800	Asphalt flood coat, smooth surface			
2900	Asphalt base sheet & 3 plies #15 asphalt felt, mopped	.39	1.07	1.46
3000	On nailable deck	.37	1.12	1.49
3100	Coated glass fiber base sheet & 2 plies glass fiber felt, mopped	.36	1.03	1.39
3200	On nailable deck	.34	1.07	1.41
3300	For 3 plies, mopped	.43	1.12	1.55
3400	On nailable deck	.41	1.17	1.58
3700	4 plies glass fiber felt (type IV), mopped	.51	1.12	1.63
3800	On nailable deck	.49	1.17	1.66
3900	Organic base sheet & 3 plies #15 organic felt, mopped	.39	1.07	1.46
4000	On nailable decks	.37	1.12	1.49
4100	4 plies #15 organic felt, mopped	.46	1.17	1.63
4200	Coal tar pitch with gravel surfacing			
4300	4 plies #15 tarred felt, mopped	1.07	1.22	2.29
4400	3 plies glass fiber felt (type IV), mopped	.88	1.35	2.23
4500	Coated glass fiber base sheets 2 plies glass fiber felt, mopped	.88	1.35	2.23
4600	On nailable decks	.78	1.43	2.21
4800	3 plies glass fiber felt (type IV), mopped	1.20	1.22	2.42
4900	On nailable decks	1.10	1.28	2.38

Important: See the Reference Section for critical supporting data - Location Factors & Historical Cost Indexes

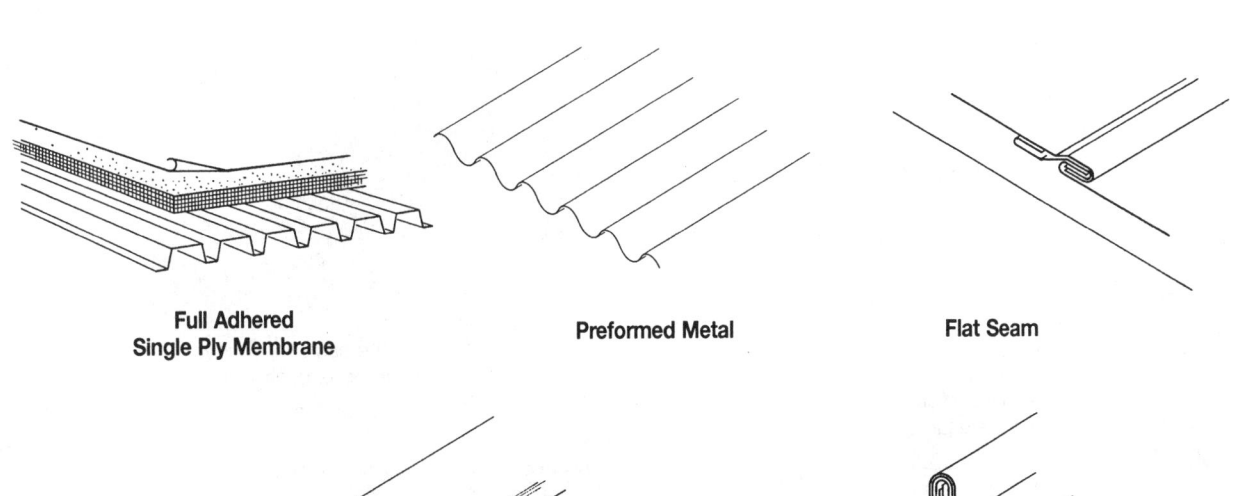

Full Adhered Single Ply Membrane

Preformed Metal

Flat Seam

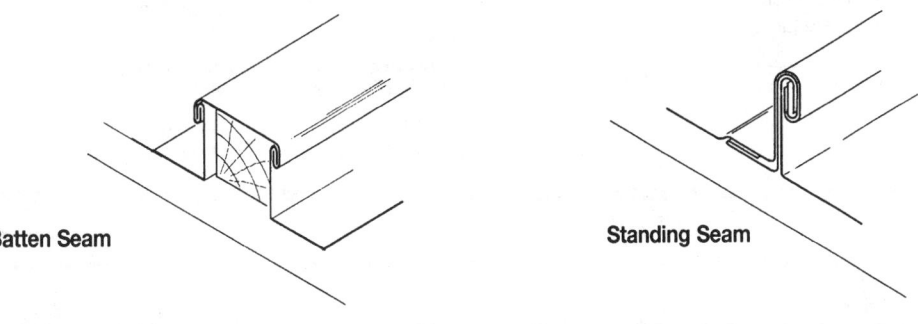

Batten Seam

Standing Seam

5.1-220	Single Ply Membrane	COST PER S.F.		
		MAT.	INST.	TOTAL
1000	CSPE (Chlorosulfonated polyethylene), 35 mils, fully adhered	1.34	.64	1.98
2000	EPDM (Ethylene propylene diene monomer), 45 mils, fully adhered	.83	.64	1.47
3000	55 mils, fully adhered	.97	.64	1.61
4000	Modified bit., SBS modified, granule surface cap sheet, mopped, 150 mils	.42	1.28	1.70
4500	APP modified, granule surface cap sheet, torched, 180 mils	.42	.83	1.25
6000	Reinforced PVC, 48 mils, loose laid and ballasted with stone	.95	.33	1.28
6200	Fully adhered with adhesive	1.22	.64	1.86

5.1-310	Preformed Metal Roofing	COST PER S.F.		
		MAT.	INST.	TOTAL
0200	Corrugated roofing, aluminum, mill finish, .0175" thick, .272 P.S.F.	.73	1.06	1.79
0250	.0215 thick, .334 P.S.F.	.92	1.06	1.98

5.1-330	Formed Metal	COST PER S.F.		
		MAT.	INST.	TOTAL
1000	Batten seam, formed copper roofing, 3"min slope, 16 oz., 1.2 P.S.F.	4.86	3.51	8.37
1100	18 oz., 1.35 P.S.F.	5.40	3.86	9.26
2000	Zinc copper alloy, 3" min slope, .020" thick, .88 P.S.F.	5.85	3.21	9.06
3000	Flat seam, copper, 1/4" min. slope, 16 oz., 1.2 P.S.F.	4.31	3.21	7.52
3100	18 oz., 1.35 P.S.F.	4.86	3.36	8.22
5000	Standing seam, copper, 2-1/2" min. slope, 16 oz., 1.25 P.S.F.	4.66	2.97	7.63
5100	18 oz., 1.40 P.S.F.	5.20	3.21	8.41
6000	Zinc copper alloy, 2-1/2" min. slope, .020" thick, .87 P.S.F.	5.70	3.21	8.91
6100	.032" thick, 1.39 P.S.F.	7.75	3.51	11.26

5 ROOFING

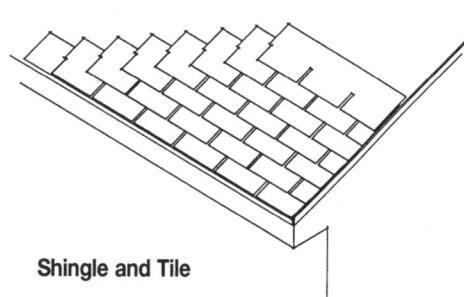

Shingle and Tile

Shingles and tiles are practical in applications where the roof slope is more than 3-1/2" per foot of rise. Lines 1100 through 6000 list the various materials and the weight per square foot.

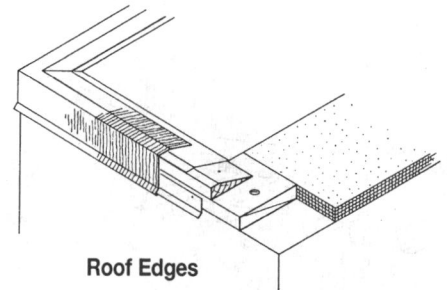

Roof Edges

The table below lists the costs for various types of roof perimeter edge treatments.

Roof edge systems include the cost per L.F. for a 2" x 8" treated wood nailer fastened at 4'-0" O.C. and a diagonally cut 4" x 6" treated wood cant.

Roof edge and base flashing are assumed to be made from the same material.

5.1-410	Shingle & Tile	COST PER S.F.		
		MAT.	INST.	TOTAL
1095	Asphalt roofing			
1100	Strip shingles, 4" slope, inorganic class A 210-235 lb/Sq.	.32	.68	1
1150	Organic, class C, 235-240 lb./sq.	.42	.74	1.16
1200	Premium laminated multi-layered, class A, 260-300 lb./Sq.	.55	1.02	1.57
1545	Metal roofing			
1550	Alum., shingles, colors, 3" min slope, .019" thick, 0.4 PSF	1.88	.73	2.61
1850	Steel, colors, 3" min slope, 26 gauge, 1.0 PSF	2.15	1.54	3.69
2795	Slate roofing			
2800	4" min. slope, shingles, 3/16" thick, 8.0 PSF	6.10	1.92	8.02
3495	Wood roofing			
3500	4" min slope, cedar shingles, 16" x 5", 5" exposure 1.6 PSF	1.82	1.43	3.25
4000	Shakes, 18", 8-1/2" exposure, 2.8 PSF	1.64	1.77	3.41
5095	Tile roofing			
5100	Aluminum, mission, 3" min slope, .019" thick, 0.65 PSF	3.81	1.39	5.20
6000	Clay, Americana, 3" minimum slope, 8 PSF	5.45	2.04	7.49

5.1-520	Roof Edges							
	EDGE TYPE	DESCRIPTION	SPECIFICATION	FACE HEIGHT		COST PER L.F.		
						MAT.	INST.	TOTAL
1000	Aluminum	mill finish	.050" thick	4"		7.05	6.45	13.50
1100				6"		7.60	6.65	14.25
1300		duranodic	.050" thick	4"		8.10	6.45	14.55
1400				6"		8.85	6.65	15.50
1600		painted	.050" thick	4"		8.75	6.45	15.20
1700				6"		9.60	6.65	16.25
2000	Copper	plain	16 oz.	4"		5.70	6.45	12.15
2100				6"		6.60	6.65	13.25
2300			20 oz.	4"		6.25	7.30	13.55
2400				6"		6.95	7.05	14
2700	Sheet Metal	galvanized	20 Ga.	4"		6.40	7.55	13.95
2800				6"		7.65	7.55	15.20
3000			24 Ga.	4"		5.70	6.45	12.15
3100				6"		6.55	6.45	13

ROOFING 5

5.1-620 — Flashing

	MATERIAL	BACKING	SIDES	SPECIFICATION	QUANTITY	COST PER S.F.		
						MAT.	INST.	TOTAL
0040	Aluminum	none		.019″		.87	2.61	3.48
0050				.032″		1.05	2.61	3.66
0300		fabric	2	.004″		.94	1.15	2.09
0400		mastic		.004″		.89	1.15	2.04
0700	Copper	none		16 oz.	<500 lbs.	3.68	3.29	6.97
0800				24 oz.	<500 lbs.	5.50	3.61	9.11
2000	Copper lead	fabric	1	2 oz.		1.54	1.15	2.69
3500	PVC black	none		.010″		.14	1.14	1.28
3700				.030″		.33	1.14	1.47
4200	Neoprene			1/16″		1.62	1.14	2.76
4500	Stainless steel	none		.015″	<500 lbs.	3.30	3.29	6.59
4600	Copper clad				>2000 lbs.	3.28	2.44	5.72
5000	Plain			32 ga.		2.37	2.44	4.81

For expanded coverage of these items see *Means Assemblies Cost Data 1998*

ROOFING

5

5.7-101	Roof Deck Rigid Insulation	COST PER S.F.		
		MAT.	INST.	TOTAL
0100	Fiberboard low density, 1/2" thick, R1.39			
0150	1" thick R2.78	.35	.41	.76
0300	1 1/2" thick R4.17	.53	.41	.94
0350	2" thick R5.56	.70	.41	1.11
0370	Fiberboard high density, 1/2" thick R1.3	.20	.32	.52
0380	1" thick R2.5	.37	.41	.78
0390	1 1/2" thick R3.8	.61	.41	1.02
0410	Fiberglass, 3/4" thick R2.78	.43	.32	.75
0450	15/16" thick R3.70	.55	.32	.87
0550	1-5/16" thick R5.26	.96	.32	1.28
0650	2 7/16" thick R10	1.15	.41	1.56
1510	Polyisocyanurate 2#/CF density, 1" thick R7.14	.34	.23	.57
1550	1 1/2" thick R10.87	.37	.26	.63
1600	2" thick R14.29	.48	.30	.78
1650	2 1/2" thick R16.67	.55	.31	.86
1700	3" thick R21.74	.69	.32	1.01
1750	3 1/2" thick R25	.80	.32	1.12
1800	Tapered for drainage	.41	.23	.64
1810	Expanded polystyrene, 1#/CF density, 3/4" thick R2.89	.22	.22	.44
1820	2" thick R7.69	.35	.26	.61
1830	Extruded polystyrene, 15 PSI compressive strength, 1" thick R5	.26	.22	.48
1835	2" thick R10	.35	.26	.61
1840	3" thick R15	.80	.32	1.12
2550	40 PSI compressive strength, 1" thick R5	.37	.22	.59
2600	2" thick R10	.73	.26	.99
2650	3" thick R15	1.09	.32	1.41
2700	4" thick R20	1.45	.32	1.77
2750	Tapered for drainage	.52	.23	.75
2810	60 PSI compressive strength, 1" thick R5	.44	.22	.66
2850	2" thick R10	.81	.27	1.08
2900	Tapered for drainage	.63	.23	.86
2910	115 PSI compressive strength, 1" thick R5	.95	.23	1.18
2950	2" thick R10	1.89	.28	2.17
3000	Tapered for drainage	1.06	.23	1.29
3050	Composites with 2" EPS			
3060	1" Fiberboard	.79	.34	1.13
3070	7/16" oriented strand board	.94	.41	1.35
3080	1/2" plywood	1	.41	1.41
3090	1" perlite	.84	.41	1.25
4000	Composites with 1-1/2" polyisocyanurate			
4010	1" fiberboard	.85	.41	1.26
4020	1" perlite	.87	.38	1.25
4030	7/16" oriented strand board	.99	.41	1.40

Important: See the Reference Section for critical supporting data - Location Factors & Historical Cost Indexes

Roof Hatch

Smoke Hatch

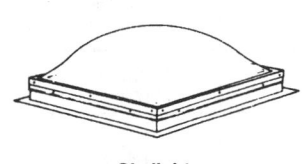

Skylight

5.8-100	Hatches	COST PER OPNG.		
		MAT.	INST.	TOTAL
0200	Roof hatches, with curb, and 1" fiberglass insulation, 2'-6"x3'-0",aluminum	445	128	573
0300	Galvanized steel 165 lbs.	375	128	503
0400	Primed steel 164 lbs.	330	128	458
0500	2'-6"x4'-6" aluminum curb and cover, 150 lbs.	615	142	757
0600	Galvanized steel 220 lbs.	525	142	667
0650	Primed steel 218 lbs.	505	142	647
0800	2'x6"x8'-0" aluminum curb and cover, 260 lbs.	1,100	194	1,294
0900	Galvanized steel, 360 lbs.	1,000	194	1,194
0950	Primed steel 358 lbs.	995	194	1,189
1200	For plexiglass panels, add to the above	380		380
2100	Smoke hatches, unlabeled not incl. hand winch operator, 2'-6"x3', galv	455	155	610
2200	Plain steel, 160 lbs.	400	155	555
2400	2'-6"x8'-0",galvanized steel, 360 lbs.	1,100	212	1,312
2500	Plain steel, 350 lbs.	1,100	212	1,312
3000	4'-0"x8'-0", double leaf low profile, aluminum cover, 359 lb.	1,575	160	1,735
3100	Galvanized steel 475 lbs.	1,375	160	1,535
3200	High profile, aluminum cover, galvanized curb, 361 lbs.	1,425	160	1,585

5.8-100	Skylights	COST PER S.F.		
		MAT.	INST.	TOTAL
5100	Skylights, plastic domes, insul curbs, nom. size to 10 S.F., single glaze	19.55	10.85	30.40
5200	Double glazing	22	12	34
5300	10 S.F. to 20 S.F., single glazing	24.50	13.35	37.85
5400	Double glazing	17.95	4.95	22.90
5500	20 S.F. to 30 S.F., single glazing	17.30	3.73	21.03
5600	Double glazing	18.35	3.94	22.29
5700	30 S.F. to 65 S.F., single glazing	17.75	2.83	20.58
5800	Double glazing	19.10	3.36	22.46
6000	Sandwich panels fiberglass, 9-1/16"thick, 2 S.F. to 10 S.F.	14.90	6.40	21.30
6100	10 S.F. to 18 S.F.	13.35	4.82	18.17
6200	2-3/4" thick, 25 S.F. to 40 S.F.	21.50	4.33	25.83
6300	40 S.F. to 70 S.F.	17.60	3.87	21.47
6301				

5 ROOFING

5.8-400 — Gutters

	SECTION	MATERIAL	THICKNESS	SIZE	FINISH	COST PER L.F. MAT.	INST.	TOTAL
0050	Box	aluminum	.027"	5"	enameled	1.15	3.16	4.31
0100					mill	.95	3.16	4.11
0200			.032"	5"	enameled	1.44	3.16	4.60
0500		copper	16 Oz.	4"	lead coated	7.65	3.16	10.81
0600					mill	3.91	3.16	7.07
1000		steel galv.	28 Ga.	5"	enameled	.92	3.16	4.08
1200			26 Ga.	5"	mill	.90	3.16	4.06
1800		vinyl		4"	colors	.94	3.03	3.97
1900				5"	colors	1.10	3.03	4.13
2300		hemlock or fir		4"x5"	treated	8	3.33	11.33
3000	Half round	copper	16 Oz.	4"	lead coated	5.35	3.16	8.51
3100					mill	3.55	3.16	6.71
3600		steel galv.	28 Ga.	5"	enameled	.92	3.16	4.08
4102		stainless steel		5"	mill	5.50	3.16	8.66
5000		vinyl		4"	white	.75	3.03	3.78

5.8-500 — Downspouts

	MATERIALS	SECTION	SIZE	FINISH	THICKNESS	COST PER V.L.F. MAT.	INST.	TOTAL
0100	Aluminum	rectangular	2"x3"	embossed mill	.020"	.77	1.99	2.76
0150				enameled	.020"	.71	1.99	2.70
0250			3"x4"	enameled	.024"	1.49	2.71	4.20
0300		round corrugated	3"	enameled	.020"	.94	1.99	2.93
0350			4"	enameled	.025"	1.42	2.71	4.13
0500	Copper	rectangular corr.	2"x3"	mill	16 Oz.	4.42	1.99	6.41
0600		smooth		mill	16 Oz.	5.20	1.99	7.19
0700		rectangular corr.	3"x4"	mill	16 Oz.	5.90	2.61	8.51
1300	Steel	rectangular corr.	2"x3"	galvanized	28 Ga.	.58	1.99	2.57
1350				epoxy coated	24 Ga.	1.11	1.99	3.10
1400		smooth		galvanized	28 Ga.	.87	1.99	2.86
1450		rectangular corr.	3"x4"	galvanized	28 Ga.	1.63	2.61	4.24
1500				epoxy coated	24 Ga.	1.88	2.61	4.49
1550		smooth		galvanized	28 Ga.	1.33	2.61	3.94
1652		round corrugated	3"	galvanized	28 Ga.	.78	1.99	2.77
1700			4"	galvanized	28 Ga.	1.01	2.61	3.62
1750			5"	galvanized	28 Ga.	1.35	2.91	4.26
2000	Steel pipe	round	4"	black	X.H.	6.70	18.95	25.65
2552	S.S. tubing sch.5	rectangular	3"x4"	mill		17.95	2.61	20.56

5.8-500 — Gravel Stop

	MATERIALS	SECTION	SIZE	FINISH	THICKNESS	COST PER L.F. MAT.	INST.	TOTAL
5100	Aluminum	extruded	4"	mill	.050"	2.99	2.61	5.60
5200			4"	duranodic	.050"	4.06	2.61	6.67
5300			8"	mill	.050"	4.37	3.03	7.40
5400			8"	duranodic	.050"	5.50	3.03	8.53
6000			12"-2 pc.	duranodic	.050"	6.90	3.79	10.69
6100	Stainless	formed	6"	mill	24 Ga.	7.85	2.81	10.66

For information about Means Estimating Seminars, see yellow pages 11 and 12 in back of book

ROOFING 5

Division 6
Interior Construction

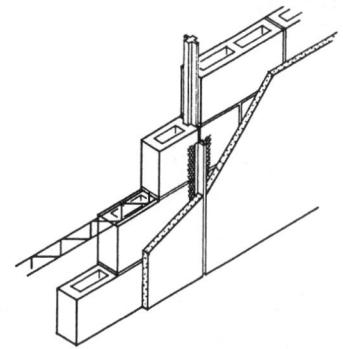

The Concrete Block Partition Systems are defined by weight and type of block, thickness, type of finish and number of sides finished. System components include joint reinforcing on alternate courses and vertical control joints.

6.1-210 Concrete Block Partitions - Regular Weight

	TYPE	THICKNESS (IN.)	TYPE FINISH	SIDES FINISHED		MAT.	INST.	TOTAL
1000	Hollow	4	none	0		.92	3.69	4.61
1010			gyp. plaster 2 coat	1		1.23	5.35	6.58
1020				2		1.53	7	8.53
1200			portland - 3 coat	1		1.13	5.60	6.73
1400			5/8" drywall	1		1.27	4.91	6.18
1500		6	none	0		1.12	3.96	5.08
1510			gyp. plaster 2 coat	1		1.43	5.60	7.03
1520				2		1.73	7.25	8.98
1700			portland - 3 coat	1		1.33	5.85	7.18
1900			5/8" drywall	1		1.47	5.20	6.67
1910				2		1.82	6.40	8.22
2000		8	none	0		1.60	4.23	5.83
2010			gyp. plaster 2 coat	1		1.91	5.90	7.81
2020			gyp. plaster 2 coat	2		2.21	7.55	9.76
2200			portland - 3 coat	1		1.81	6.10	7.91
2400			5/8" drywall	1		1.95	5.45	7.40
2410				2		2.30	6.65	8.95
2500		10	none	0		1.94	4.42	6.36
2510			gyp. plaster 2 coat	1		2.25	6.05	8.30
2520				2		2.55	7.70	10.25
2700			portland - 3 coat	1		2.15	6.30	8.45
2900			5/8" drywall	1		2.29	5.65	7.94
2910				2		2.64	6.85	9.49
3000	Solid	2	none	0		1.16	3.65	4.81
3010			gyp. plaster	1		1.47	5.30	6.77
3020				2		1.77	6.95	8.72
3200			portland - 3 coat	1		1.37	5.55	6.92
3400			5/8" drywall	1		1.51	4.87	6.38
3410				2		1.86	6.10	7.96
3500		4	none	0		1.35	3.82	5.17
3510			gyp. plaster	1		1.73	5.50	7.23
3520				2		1.96	7.10	9.06
3700			portland - 3 coat	1		1.56	5.70	7.26
3900			5/8" drywall	1		1.70	5.05	6.75
3910				2		2.05	6.25	8.30

COST PER S.F.

6.1-210 — Concrete Block Partitions - Regular Weight

	TYPE	THICKNESS (IN.)	TYPE FINISH	SIDES FINISHED		COST PER S.F. MAT.	COST PER S.F. INST.	COST PER S.F. TOTAL
4002	Solid	6	none	0		1.82	4.10	5.92
4010			gyp. plaster	1		2.13	5.75	7.88
4020				2		2.43	7.40	9.83
4200			portland - 3 coat	1		2.03	6	8.03
4400			5/8" drywall	1		2.17	5.30	7.47
4410				2		2.52	6.55	9.07

6.1-210 — Concrete Block Partitions - Lightweight

	TYPE	THICKNESS (IN.)	TYPE FINISH	SIDES FINISHED		COST PER S.F. MAT.	COST PER S.F. INST.	COST PER S.F. TOTAL
5000	Hollow	4	none	0		1.03	3.61	4.64
5010			gyp. plaster	1		1.34	5.25	6.59
5020				2		1.64	6.90	8.54
5200			portland - 3 coat	1		1.24	5.50	6.74
5400			5/8" drywall	1		1.38	4.83	6.21
5410				2		1.73	6.05	7.78
5500		6	none	0		1.28	3.86	5.14
5510			gyp. plaster	1		1.59	5.50	7.09
5520			gyp. plaster	2		1.89	7.15	9.04
5700			portland - 3 coat	1		1.49	5.75	7.24
5900			5/8" drywall	1		1.63	5.10	6.73
5910				2		1.98	6.30	8.28
6000		8	none	0		1.56	4.12	5.68
6010			gyp. plaster	1		1.87	5.75	7.62
6020				2		2.17	7.40	9.57
6200			portland - 3 coat	1		1.77	6	7.77
6400			5/8" drywall	1		1.91	5.35	7.26
6410				2		2.26	6.55	8.81
6500		10	none	0		2.08	4.31	6.39
6510			gyp. plaster	1		2.39	5.95	8.34
6520				2		2.69	7.60	10.29
6700			portland - 3 coat	1		2.29	6.20	8.49
6900			5/8" drywall	1		2.43	5.55	7.98
6910				2		2.78	6.75	9.53
7000	Solid	4	none	0		1.39	3.77	5.16
7010			gyp. plaster	1		1.70	5.40	7.10
7020				2		2	7.05	9.05
7200			portland - 3 coat	1		1.60	5.65	7.25
7400			5/8" drywall	1		1.74	4.99	6.73
7410				2		2.09	6.20	8.29
7500		6	none	0		2	4.05	6.05
7510			gyp. plaster	1		2.38	5.75	8.13
7520				2		2.61	7.35	9.96
7700			portland - 3 coat	1		2.21	5.95	8.16
7900			5/8" drywall	1		2.35	5.25	7.60
7910				2		2.70	6.50	9.20
8000		8	none	0		2.30	4.34	6.64
8010			gyp. plaster	1		2.61	6	8.61
8020				2		2.91	7.65	10.56
8200			portland - 3 coat	1		2.51	6.25	8.76
8400			5/8" drywall	1		2.65	5.55	8.20
8410				2		3	6.80	9.80

INTERIOR CONSTRUCTION

6

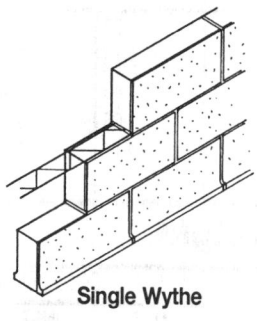

Structural facing tile
8W series
8" x 16"

Single Wythe

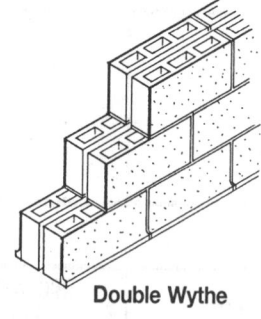

Double Wythe

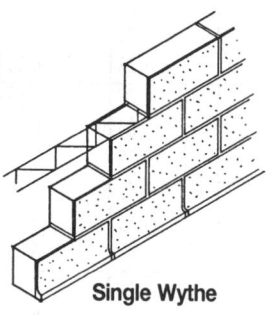

Structural facing tile
6T series
5-1/3" x 12"

Single Wythe

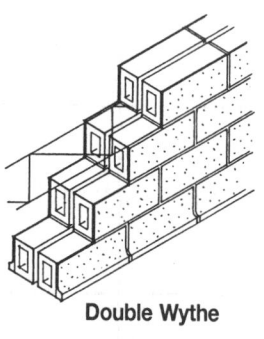

Double Wythe

INTERIOR CONSTRUCTION 6

6.1-270	Tile Partitions	COST PER S.F.		
		MAT.	**INST.**	**TOTAL**
1000	8W series 8"x16", 4" thick wall, reinf every 2 courses, glazed 1 side	6.85	4.43	11.28
1100	Glazed 2 sides	11	4.71	15.71
1200	Glazed 2 sides, using 2 wythes of 2" thick tile	12.70	8.50	21.20
1300	6" thick wall, horizontal reinf every 2 courses, glazed 1 side	9.75	4.64	14.39
1400	Glazed 2 sides, each face different color, 2" and 4" tile	13.20	8.70	21.90
1500	8" thick wall, glazed 2 sides using 2 wythes of 4" thick tile	13.70	8.85	22.55
1600	10" thick wall, glazed 2 sides using 1 wythe of 4" tile & 1 wythe of 6" tile	16.60	9.05	25.65
1700	Glazed 2 sides cavity wall, using 2 wythes of 4" thick tile	13.70	8.85	22.55
1800	12" thick wall, glazed 2 sides using 2 wythes of 6" thick tile	19.50	9.30	28.80
1900	Glazed 2 sides cavity wall, using 2 wythes of 4" thick tile	13.70	8.85	22.55
2100	6T series 5-1/3"x12" tile, 4" thick, non load bearing glazed one side,	7.55	6.95	14.50
2200	Glazed two sides	12.05	7.85	19.90
2300	Glazed two sides, using two wythes of 2" thick tile	12.50	13.60	26.10
2400	6" thick, glazed one side	11.20	7.30	18.50
2500	Glazed two sides	22.50	8.25	30.75
2600	Glazed two sides using 2" thick tile and 4" thick tile	13.80	13.75	27.55
2700	8" thick, glazed one side	14.10	8.50	22.60
2800	Glazed two sides using two wythes of 4" thick tile	15.10	13.90	29
2900	Glazed two sides using 6" thick tile and 2" thick tile	17.45	14.10	31.55
3000	10" thick cavity wall, glazed two sides using two wythes of 4" tile	15.10	13.90	29
3100	12" thick, glazed two sides using 4" thick tile and 8" thick tile	21.50	15.45	36.95
3200	2" thick facing tile, glazed one side, on 6" concrete block	7.30	8.10	15.40
3300	On 8" concrete block	7.80	8.35	16.15
3400	On 10" concrete block	8.10	8.50	16.60

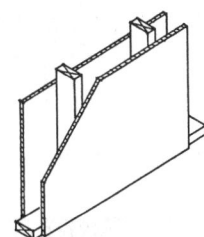

Gypsum board, single layer each side on wood studs.

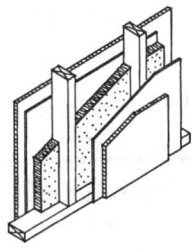

Gypsum board, sound deadening board each side, with 1-1/2" insulation on wood studs.

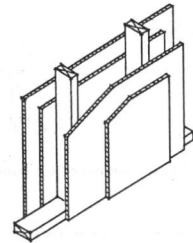

Gypsum board, two layers each side on wood studs.

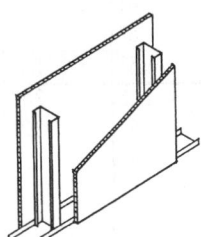

Gypsum board, single layer each side on metal studs.

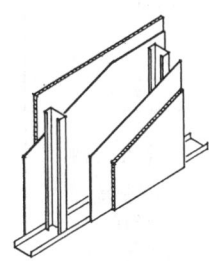

Gypsum board, sound deadening board each side on metal studs.

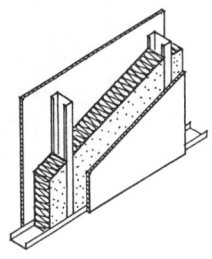

Gypsum board two layers one side, single layer opposite side, with 3-1/2" insulation on metal studs.

6.1-510 — Drywall Partitions/Wood Stud Framing

	FACE LAYER	BASE LAYER	FRAMING	OPPOSITE FACE	INSULATION	COST PER S.F. MAT.	INST.	TOTAL
1200	5/8" FR drywall	none	2 x 4, @ 16" O.C.	same	0	1.10	1.99	3.09
1250				5/8" reg. drywall	0	1.09	1.99	3.08
1300				nothing	0	.77	1.33	2.10
1400		1/4" SD gypsum	2 x 4 @ 16" O.C.	same	1-1/2" fiberglass	1.77	3.06	4.83
1450				5/8" FR drywall	1-1/2" fiberglass	1.60	2.69	4.29
1500				nothing	1-1/2" fiberglass	1.27	2.03	3.30
1600		resil. channels	2 x 4 @ 16", O.C.	same	1-1/2" fiberglass	1.68	3.89	5.57
1650				5/8" FR drywall	1-1/2" fiberglass	1.55	3.11	4.66
1700				nothing	1-1/2" fiberglass	1.22	2.45	3.67
1800		5/8" FR drywall	2 x 4 @ 24" O.C.	same	0	1.46	2.52	3.98
1850				5/8" FR drywall	0	1.23	2.19	3.42
1900				nothing	0	.90	1.53	2.43
2200		5/8" FR drywall	2 rows-2 x 4	same	2" fiberglass	2.40	3.64	6.04
2250			16"O.C.	5/8" FR drywall	2" fiberglass	2.17	3.31	5.48
2300				nothing	2" fiberglass	1.84	2.65	4.49
2400	5/8" WR drywall	none	2 x 4, @ 16" O.C.	same	0	1.20	1.99	3.19
2450				5/8" FR drywall	0	1.15	1.99	3.14
2500				nothing	0	.82	1.33	2.15
2600		5/8" FR drywall	2 x 4, @ 24" O.C.	same	0	1.56	2.52	4.08
2650				5/8" FR drywall	0	1.28	2.19	3.47
2700				nothing	0	.95	1.53	2.48
2800	5/8 VF drywall	none	2 x 4, @ 16" O.C.	same	0	1.80	2.15	3.95
2850				5/8" FR drywall	0	1.45	2.07	3.52
2900				nothing	0	1.12	1.41	2.53
3000		5/8" FR drywall	2 x 4 , 24" O.C.	same	0	2.16	2.68	4.84
3050				5/8" FR drywall	0	1.58	2.27	3.85
3100				nothing	0	1.25	1.61	2.86

INTERIOR CONSTRUCTION

6

6.1-510 — Drywall Partitions/Wood Stud Framing

	FACE LAYER	BASE LAYER	FRAMING	OPPOSITE FACE	INSULATION	COST PER S.F.		
						MAT.	INST.	TOTAL
3200	1/2" reg drywall	3/8" reg drywall	2 x 4, @ 16" O.C.	same	0	1.30	2.65	3.95
3252	1/2" reg. drywall	3/8" reg. drywall	2x4, @ 16"O.C.	5/8"FR drywall	0	1.20	2.32	3.52
3300				nothing	0	.87	1.66	2.53

6.1-510 — Drywall Partitions/Metal Stud Framing

	FACE LAYER	BASE LAYER	FRAMING	OPPOSITE FACE	INSULATION	COST PER S.F.		
						MAT.	INST.	TOTAL
5200	5/8" FR drywall	none	1-5/8" @ 24" O.C.	same	0	.82	1.96	2.78
5250				5/8" reg. drywall	0	.81	1.96	2.77
5300				nothing	0	.49	1.30	1.79
5400			3-5/8" @ 24" O.C.	same	0	.88	1.99	2.87
5450				5/8" reg. drywall	0	.87	1.99	2.86
5500				nothing	0	.55	1.33	1.88
5600		1/4" SD gypsum	1-5/8" @ 24" O.C.	same	0	1.16	2.70	3.86
5650				5/8" FR drywall	0	.99	2.33	3.32
5700				nothing	0	.66	1.67	2.33
5800			2-1/2" @ 24" O.C.	same	0	1.18	2.71	3.89
5850				5/8" FR drywall	0	1.01	2.34	3.35
5900				nothing	0	.68	1.68	2.36
6000		5/8" FR drywall	2-1/2" @ 16" O.C.	same	0	1.65	2.74	4.39
6050				5/8" FR drywall	0	1.42	2.41	3.83
6100				nothing	0	1.09	1.75	2.84
6200			3-5/8" @ 24" O.C.	same	0	1.34	2.65	3.99
6250				5/8"FR drywall	3-1/2" fiberglass	1.39	2.53	3.92
6300				nothing	0	.78	1.66	2.44
6400	5/8" WR drywall	none	1-5/8" @ 24" O.C.	same	0	.92	1.96	2.88
6450				5/8" FR drywall	0	.87	1.96	2.83
6500				nothing	0	.54	1.30	1.84
6600			3-5/8" @ 24" O.C.	same	0	.98	1.99	2.97
6650				5/8" FR drywall	0	.93	1.99	2.92
6700				nothing	0	.60	1.33	1.93
6800		5/8" FR drywall	2-1/2" @ 16" O.C.	same	0	1.75	2.74	4.49
6850				5/8" FR drywall	0	1.47	2.41	3.88
6900				nothing	0	1.14	1.75	2.89
7000			3-5/8" @ 24" O.C.	same	0	1.44	2.65	4.09
7050				5/8"FR drywall	3-1/2" fiberglass	1.44	2.53	3.97
7100				nothing	0	.83	1.66	2.49
7200	5/8" VF drywall	none	1-5/8" @ 24" O.C.	same	0	1.52	2.12	3.64
7250				5/8" FR drywall	0	1.17	2.04	3.21
7300				nothing	0	.84	1.38	2.22
7400			3-5/8" @ 24" O.C.	same	0	1.58	2.15	3.73
7450				5/8" FR drywall	0	1.23	2.07	3.30
7500				nothing	0	.90	1.41	2.31
7600		5/8" FR drywall	2-1/2" @ 16" O.C.	same	0	2.35	2.90	5.25
7650				5/8" FR drywall	0	1.77	2.49	4.26
7700				nothing	0	1.44	1.83	3.27
7800			3-5/8" @ 24" O.C.	same	0	2.04	2.81	4.85
7850				5/8"FR drywall	3-1/2" fiberglass	1.74	2.61	4.35
7900				nothing	0	1.13	1.74	2.87

Important: See the Reference Section for critical supporting data - Location Factors & Historical Cost Indexes

6.1-580	Drywall Components	COST PER S.F.		
		MAT.	INST.	TOTAL
0060	Metal studs, 24" O.C. including track, load bearing, 20 gage, 2-1/2"	.52	.93	1.45
0080	3-5/8"	.60	.97	1.57
0100	4"	.63	1.01	1.64
0120	6"	.77	1.06	1.83
0140	Metal studs, 24" O.C. including track, load bearing, 18 gage, 2-1/2"	.52	.93	1.45
0160	3-5/8"	.60	.97	1.57
0180	4"	.63	1.01	1.64
0200	6"	.77	1.06	1.83
0220	16 gage, 2-1/2"	.62	1.01	1.63
0240	3-5/8"	.85	1.06	1.91
0260	4"	.86	1.11	1.97
0280	6"	1.05	1.48	2.53
0300	Non load bearing, 25 gage, 1-5/8"	.16	.64	.80
0340	3-5/8"	.22	.67	.89
0360	4"	.23	.68	.91
0380	6"	.29	.69	.98
0400	20 gage, 2-1/2"	.42	.65	1.07
0420	3-5/8"	.51	.67	1.18
0440	4"	.54	.68	1.22
0460	6"	.69	.69	1.38
0540	Wood studs including blocking, shoe and double top plate, 2"x4", 12"O.C.	.55	.84	1.39
0560	16" O.C.	.44	.67	1.11
0580	24" O.C.	.34	.54	.88
0600	2"x6", 12" O.C.	.78	.95	1.73
0620	16" O.C.	.63	.74	1.37
0640	24" O.C.	.48	.58	1.06
0642	Furring one side only, steel channels, 3/4", 12" O.C.	.19	1.33	1.52
0644	16" O.C.	.17	1.18	1.35
0646	24" O.C.	.12	.89	1.01
0647	1-1/2" , 12" O.C.	.27	1.49	1.76
0648	16" O.C.	.25	1.31	1.56
0649	24" O.C.	.16	1.03	1.19
0650	Wood strips, 1" x 3", on wood, 12" O.C.	.31	.61	.92
0651	16" O.C.	.23	.46	.69
0652	On masonry, 12" O.C.	.31	.67	.98
0653	16" O.C.	.23	.50	.73
0654	On concrete, 12" O.C.	.31	1.28	1.59
0655	16" O.C.	.23	.96	1.19
0665	Gypsum board, one face only, exterior sheathing, 1/2"	.19	.59	.78
0680	Fire resistant, 1/2"	.23	.33	.56
0700	5/8"	.23	.33	.56
0720	Sound deadening board 1/4"	.17	.37	.54
0740	Standard drywall 3/8"	.17	.33	.50
0760	1/2"	.16	.33	.49
0780	5/8"	.22	.33	.55
0800	Tongue & groove coreboard 1"	.57	1.39	1.96
0820	Water resistant, 1/2"	.25	.33	.58
0840	5/8"	.28	.33	.61
0860	Add for the following:, foil backing	.08		.08
0880	Fiberglass insulation, 3-1/2"	.28	.21	.49
0900	6"	.33	.25	.58
0920	Rigid insulation 1"	.35	.33	.68
0940	Resilient furring @ 16" O.C.	.16	1.05	1.21
0960	Taping and finishing	.10	.33	.43
0980	Texture spray	.13	.39	.52
1000	Thin coat plaster	.13	.41	.54
1040	2"x4" staggered studs 2"x6" plates & blocking	.63	.73	1.36

INTERIOR CONSTRUCTION

6

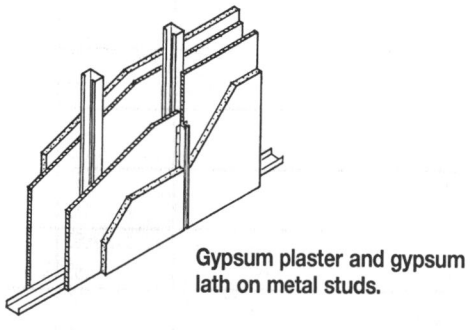

Gypsum plaster and gypsum lath on metal studs.

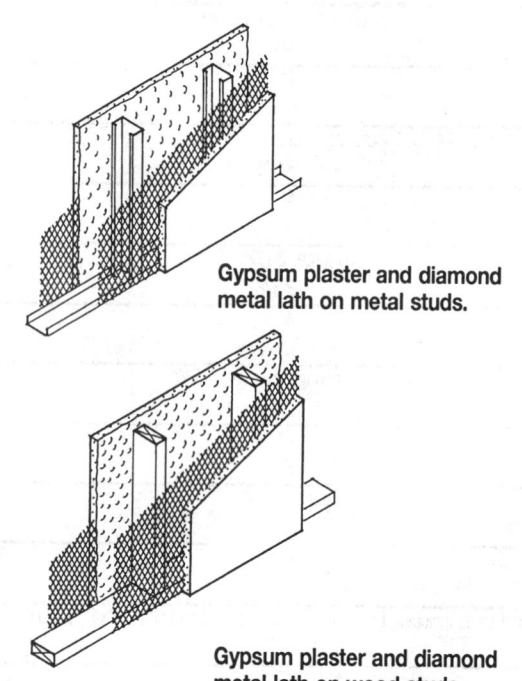

Gypsum plaster and diamond metal lath on metal studs.

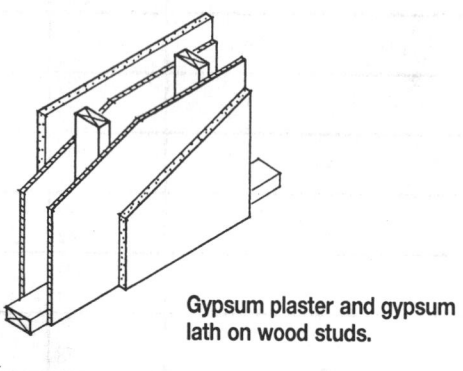

Gypsum plaster and gypsum lath on wood studs.

Gypsum plaster and diamond metal lath on wood studs.

6.1-610   Plaster Partitions/Metal Stud Framing

	TYPE	FRAMING	LATH	OPPOSITE FACE		COST PER S.F.		
						MAT.	INST.	TOTAL
1000	2 coat gypsum	2-1/2" @ 16"O.C.	3/8" gypsum	same		2	4.95	6.95
1010				nothing		1.15	2.91	4.06
1100		3-1/4" @ 24"O.C.	1/2" gypsum	same		2.04	4.92	6.96
1110				nothing		1.16	2.86	4.02
1500	2 coat vermiculite	2-1/2" @ 16"O.C.	3/8" gypsum	same		2.05	5.40	7.45
1510				nothing		1.17	3.13	4.30
1600		3-1/4" @ 24"O.C.	1/2" gypsum	same		2.09	5.35	7.44
1610				nothing		1.18	3.08	4.26
2000	3 coat gypsum	2-1/2" @ 16"O.C.	3/8" gypsum	same		1.93	5.60	7.53
2010				nothing		1.11	3.24	4.35
2020			3.4lb. diamond	same		1.53	5.60	7.13
2030				nothing		.91	3.24	4.15
2040			2.75lb. ribbed	same		1.44	5.60	7.04
2050				nothing		.87	3.24	4.11
2100		3-1/4" @ 24"O.C.	1/2" gypsum	same		1.97	5.55	7.52
2110				nothing		1.12	3.19	4.31
2120			3.4lb. ribbed	same		1.80	5.55	7.35
2130				nothing		1.04	3.19	4.23
3500	3 coat gypsum W/med. Keenes	2-1/2" @ 16"O.C.	3/8" gypsum	same		2.36	7.10	9.46
3510				nothing		1.33	4	5.33
3520			3.4lb. diamond	same		1.96	7.10	9.06
3530				nothing		1.13	4	5.13
3540			2.75lb. ribbed	same		1.87	7.10	8.97
3550				nothing		1.09	4	5.09
3600		3-1/4" @ 24"O.C.	1/2" gypsum	same		2.40	7.10	9.50
3610				nothing		1.34	3.95	5.29
3620			3.4lb. ribbed	same		2.23	7.10	9.33
3630				nothing		1.26	3.95	5.21

6.1-610 — Plaster Partitions/Metal Stud Framing

	TYPE	FRAMING	LATH	OPPOSITE FACE		COST PER S.F. MAT.	INST.	TOTAL
4000	3 coat gypsum	2-1/2" @ 16"O.C.	3/8" gypsum	same		2.37	7.90	10.27
4010	W/hard Keenes			nothing		1.33	4.38	5.71
4022	3 coat gypsum	2-1/2" @ 16"O.C.	3.4 lb. diamond	same		1.97	7.90	9.87
4032	W/hard Keenes			nothing		1.13	4.38	5.51
4040			2.75lb. ribbed	same		1.88	7.90	9.78
4050				nothing		1.09	4.38	5.47
4100		3-1/4" @ 24"O.C.	1/2" gypsum	same		2.41	7.85	10.26
4110				nothing		1.34	4.33	5.67
4120			3.4lb. ribbed	same		2.24	7.85	10.09
4130				nothing		1.26	4.33	5.59

6.1-610 — Plaster Partitions/Wood Stud Framing

	TYPE	FRAMING	LATH	OPPOSITE FACE		COST PER S.F. MAT.	INST.	TOTAL
5000	2 coat gypsum	2"x4" @ 16"O.C.	3/8" gypsum	same		2.10	4.75	6.85
5010				nothing		1.31	2.77	4.08
5100		2"x4" @ 24"O.C.	1/2" gypsum	same		2.16	4.67	6.83
5110				nothing		1.28	2.67	3.95
5500	2 coat vermiculite	2"x4" @ 16"O.C.	3/8" gypsum	same		2.15	5.20	7.35
5510				nothing		1.33	2.99	4.32
5600		2"x4" @ 24"O.C.	1/2" gypsum	same		2.21	5.10	7.31
5610				nothing		1.30	2.89	4.19
6000	3 coat gypsum	2"x4" @ 16"O.C.	3/8" gypsum	same		2.03	5.40	7.43
6010				nothing		1.27	3.10	4.37
6020			3.4lb. diamond	same		1.74	5.45	7.19
6030				nothing		1.12	3.13	4.25
6040			2.75lb. ribbed	same		1.64	5.45	7.09
6050				nothing		1.07	3.14	4.21
6100		2"x4" @ 24"O.C.	1/2" gypsum	same		2.09	5.30	7.39
6110				nothing		1.24	3	4.24
6120			3.4lb. ribbed	same		1.55	5.35	6.90
6130				nothing		.98	3.02	4
7500	3 coat gypsum	2"x4" @ 16"O.C.	3/8" gypsum	same		2.46	6.90	9.36
7510	W/med Keenes			nothing		1.49	3.86	5.35
7520			3.4lb. diamond	same		2.17	6.95	9.12
7530				nothing		1.34	3.89	5.23
7540			2.75lb. ribbed	same		2.07	7	9.07
7550				nothing		1.29	3.90	5.19
7600		2"x4" @ 24"O.C.	1/2" gypsum	same		2.52	6.85	9.37
7610				nothing		1.46	3.76	5.22
7620			3.4lb. ribbed	same		2.35	6.95	9.30
7630				nothing		1.38	3.82	5.20
8000	3 coat gypsum	2"x4" @ 16"O.C.	3/8" gypsum	same		2.47	7.70	10.17
8010	W/hard Keenes			nothing		1.49	4.24	5.73
8020			3.4lb. diamond	same		2.18	7.75	9.93
8030				nothing		1.34	4.27	5.61
8040			2.75lb. ribbed	same		2.08	7.80	9.88
8050				nothing		1.29	4.28	5.57
8100		2"x4" @ 24"O.C.	1/2" gypsum	same		2.53	7.60	10.13
8110				nothing		1.46	4.14	5.60
8120			3.4lb. ribbed	same		2.36	7.75	10.11
8130				nothing		1.38	4.20	5.58

6 INTERIOR CONSTRUCTION

6.1-680	Plaster Partition Components	COST PER S.F.		
		MAT.	INST.	TOTAL
0060	Metal studs, 16" O.C., including track, non load bearing, 25 gage, 1-5/8"	.23	.76	.99
0080	2-1/2"	.23	.76	.99
0100	3-1/4"	.26	.77	1.03
0120	3-5/8"	.26	.77	1.03
0140	4"	.29	.79	1.08
0160	6"	.36	.81	1.17
0180	Load bearing, 20 gage, 2-1/2"	.65	1.11	1.76
0200	3-5/8"	.74	1.17	1.91
0220	4"	.79	1.23	2.02
0240	6"	.96	1.31	2.27
0260	16 gage 2-1/2"	.77	1.23	2
0280	3-5/8"	1	1.31	2.31
0300	4"	1.08	1.39	2.47
0320	6"	1.32	1.75	3.07
0340	Wood studs, including blocking, shoe and double plate, 2"x4", 12" O.C.	.55	.84	1.39
0360	16" O.C.	.44	.67	1.11
0380	24" O.C.	.34	.54	.88
0400	2"x6", 12" O.C.	.78	.95	1.73
0420	16" O.C.	.63	.74	1.37
0440	24" O.C.	.48	.58	1.06
0460	Furring one face only, steel channels, 3/4", 12" O.C.	.19	1.33	1.52
0480	16" O.C.	.17	1.18	1.35
0500	24" O.C.	.12	.89	1.01
0520	1-1/2", 12" O.C.	.27	1.49	1.76
0540	16" O.C.	.25	1.31	1.56
0560	24"O.C.	.16	1.03	1.19
0580	Wood strips 1"x3", on wood., 12" O.C.	.31	.61	.92
0600	16"O.C.	.23	.46	.69
0620	On masonry, 12" O.C.	.31	.67	.98
0640	16" O.C.	.23	.50	.73
0660	On concrete, 12" O.C.	.31	1.28	1.59
0680	16" O.C.	.23	.96	1.19
0700	Gypsum lath. plain or perforated, nailed to studs, 3/8" thick	.40	.41	.81
0720	1/2" thick	.47	.44	.91
0740	Clipped to studs, 3/8" thick	.45	.46	.91
0760	1/2" thick	.47	.50	.97
0780	Metal lath, diamond painted, nailed to wood studs, 2.5 lb.	.22	.41	.63
0800	3.4 lb.	.25	.44	.69
0820	Screwed to steel studs, 2.5 lb.	.22	.44	.66
0840	3.4 lb.	.26	.46	.72
0860	Rib painted, wired to steel, 2.75 lb	.21	.46	.67
0880	3.4 lb	.39	.50	.89
0900	4.0 lb	.34	.54	.88
0910				
0920	Gypsum plaster, 2 coats	.35	1.57	1.92
0940	3 coats	.49	1.90	2.39
0960	Perlite or vermiculite plaster, 2 coats	.43	1.79	2.22
0980	3 coats	.68	2.23	2.91
1000	Stucco, 3 coats, 1" thick, on wood framing	.69	1.48	2.17
1020	On masonry	.24	.83	1.07
1100	Metal base galvanized and painted 2-1/2" high	.47	1.31	1.78

Important: See the Reference Section for critical supporting data - Location Factors & Historical Cost Indexes

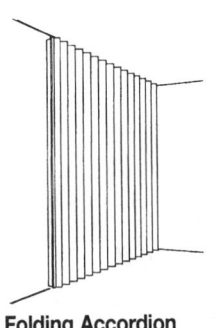

Folding Accordion

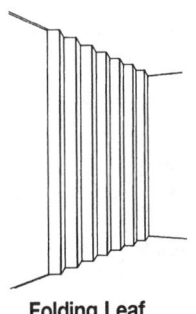

Folding Leaf

Movable and Borrow Lites

6.1-820	Partitions	MAT.	INST.	TOTAL
		COST PER S.F.		
0360	Folding accordion, vinyl covered, acoustical, 3 lb. S.F., 17 ft max. hgt	19.25	6.65	25.90
0380	5 lb. per S.F. 27 ft max height	28.50	7	35.50
0400	5.5 lb. per S.F., 17 ft. max height	32	7.40	39.40
0420	Commercial, 1.75 lb per S.F., 8 ft. max height	12.75	2.96	15.71
0440	2.0 Lb per S.F., 17 ft. max height	15.50	4.44	19.94
0460	Industrial, 4.0 lb. per S.F. 27 ft. max height	26	8.90	34.90
0480	Vinyl clad wood or steel, electric operation 6 psf	32	4.17	36.17
0500	Wood, non acoustic, birch or mahogany	17.10	2.22	19.32
0560	Folding leaf, aluminum framed acoustical 12 ft.high.,5.5 lb per S.F.,min.	27	11.10	38.10
0580	Maximum	32.50	22	54.50
0600	6.5 lb. per S.F., minimum	28.50	11.10	39.60
0620	Maximum	35	22	57
0640	Steel acoustical, 7.5 per S.F., vinyl faced, minimum	40	11.10	51.10
0660	Maximum	49	22	71
0680	Wood acoustic type, vinyl faced to 18' high 6 psf, minimum	38	11.10	49.10
0700	Average	45	14.80	59.80
0720	Maximum	58.50	22	80.50
0740	Formica or hardwood faced, minimum	39	11.10	50.10
0760	Maximum	41.50	22	63.50
0780	Wood, low acoustical type to 12 ft. high 4.5 psf	28.50	13.35	41.85
0840	Demountable, trackless wall, cork finish, semi acous, 1-5/8"th, min	19.70	2.05	21.75
0860	Maximum	24	3.51	27.51
0880	Acoustic, 2" thick, minimum	20.50	2.18	22.68
0900	Maximum	30	2.96	32.96
0920	In-plant modular office system, w/prehung steel door			
0940	3" thick honeycomb core panels			
0960	12' x 12', 2 wall	10.30	.41	10.71
0970	4 wall	10.60	.55	11.15
0980	16' x 16', 2 wall	10.25	.29	10.54
0990	4 wall	7.35	.29	7.64
1000	Gypsum, demountable, 3" to 3-3/4" thick x 9' high, vinyl clad	2.72	1.54	4.26
1020	Fabric clad	8.45	1.68	10.13
1040	1.75 system, vinyl clad hardboard, paper honeycomb core panel			
1060	1-3/4" to 2-1/2" thick x 9' high	5.70	1.54	7.24
1080	Unitized gypsum panel system, 2" to 2-1/2" thick x 9' high			
1100	Vinyl clad gypsum	7.70	1.54	9.24
1120	Fabric clad gypsum	13.90	1.68	15.58
1140	Movable steel walls, modular system			
1160	Unitized panels, 48" wide x 9' high			
1180	Baked enamel, pre-finished	9	1.23	10.23
1200	Fabric clad	13.75	1.32	15.07

INTERIOR CONSTRUCTION

6

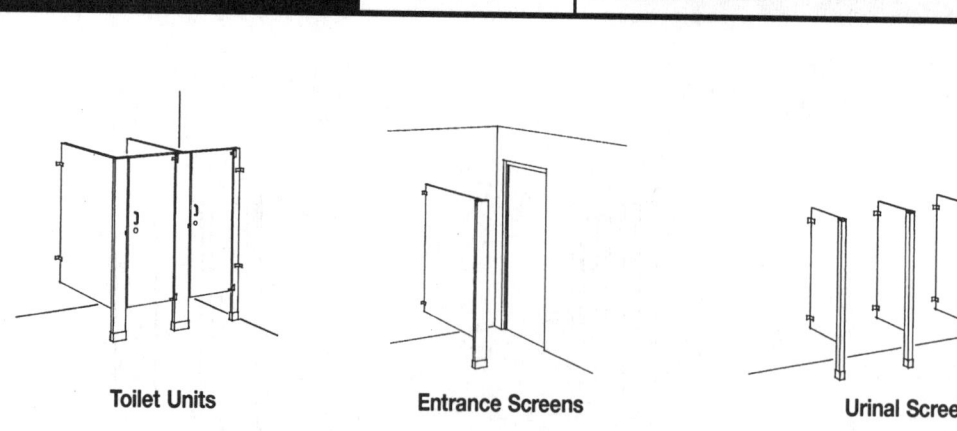

Toilet Units   **Entrance Screens**   **Urinal Screens**

6.1-870	Toilet Partitions	COST PER UNIT		
		MAT.	INST.	TOTAL
0380	Toilet partitions, cubicles, ceiling hung, marble	1,375	330	1,705
0400	Painted metal	420	167	587
0420	Plastic laminate	565	167	732
0440				
0460	Stainless steel	1,025	167	1,192
0480	Handicap addition	268		268
0520	Floor and ceiling anchored, marble	1,500	266	1,766
0540	Painted metal	425	133	558
0560	Plastic laminate	565	133	698
0580				
0600	Stainless steel	1,200	133	1,333
0620	Handicap addition	268		268
0660	Floor mounted marble	890	222	1,112
0680	Painted metal	385	95	480
0700	Plastic laminate	565	95	660
0720				
0740	Stainless steel	1,150	95	1,245
0760	Handicap addition	268		268
0780	Juvenile deduction	38.50		38.50
0820	Floor mounted with handrail marble	1,025	222	1,247
0840	Painted metal	395	111	506
0860	Plastic laminate	550	111	661
0880				
0900	Stainless steel	1,175	111	1,286
0920	Handicap addition	268		268
0960	Wall hung, painted metal	485	95	580
1000				
1020	Stainless steel	1,150	95	1,245
1040	Handicap addition	268		268
1080	Entrance screens, floor mounted, 54" high, marble	565	74	639
1100	Painted metal	177	44.50	221.50
1120				
1140	Stainless steel	640	44.50	684.50
1300	Urinal screens, floor mounted, 24" wide, laminated plastic	256	83.50	339.50
1320	Marble	530	99.50	629.50
1340	Painted metal	183	83.50	266.50
1360				
1380	Stainless steel	530	83.50	613.50
1428	Wall mounted wedge type, painted metal	207	66.50	273.50
1440				
1460	Stainless steel	425	66.50	491.50

INTERIOR CONSTRUCTION 6

Important: See the Reference Section for critical supporting data - Location Factors & Historical Cost Indexes

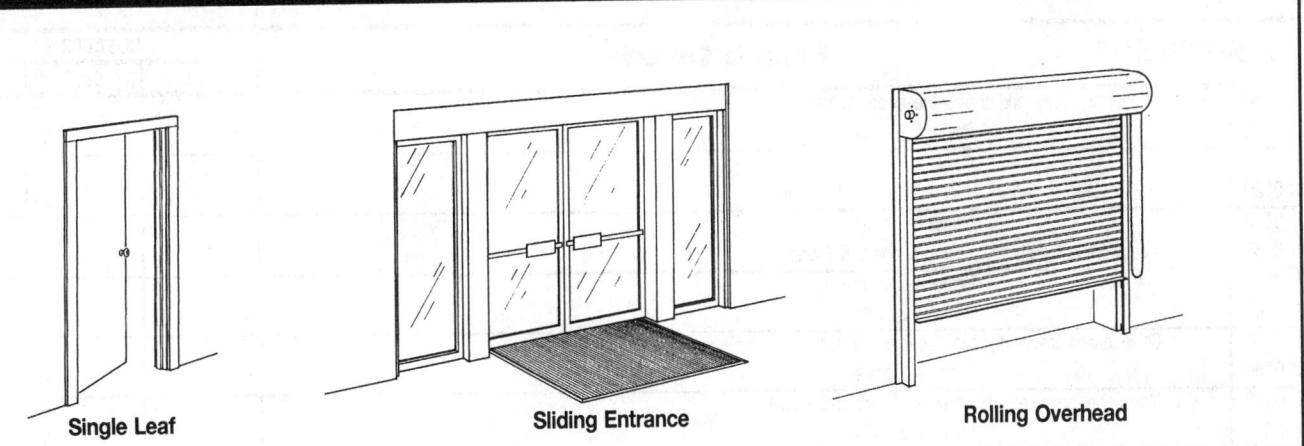

Single Leaf **Sliding Entrance** **Rolling Overhead**

6.4-100	Special Doors	COST PER OPNG.		
		MAT.	INST.	TOTAL
2500	Single leaf, wood, 3'-0"x7'-0"x1 3/8", birch, solid core	282	126	408
2510	Hollow core	242	120	362
2530	Hollow core, lauan	221	120	341
2540	Louvered pine	335	120	455
2550	Paneled pine	330	120	450
2600	Hollow metal, comm. quality, flush, 3'-0"x7'-0"x1-3/8"	395	131	526
2650	3'-0"x10'-0" openings with panel	630	170	800
2700	Metal fire, comm. quality, 3'-0"x7'-0"x1-3/8"	445	137	582
2800	Kalamein fire, comm. quality, 3'-0"x7'-0"x1-3/4"	380	250	630
3200	Double leaf, wood, hollow core, 2 - 3'-0"x7'-0"x1-3/8"	380	230	610
3300	Hollow metal, comm. quality, B label, 2'-3'-0"x7'-0"x1-3/8"	840	288	1,128
3400	6'-0"x10'-0" opening, with panel	1,200	355	1,555
3500	Double swing door system, 12'-0"x7'-0", mill finish	5,250	1,900	7,150
3700	Black finish	5,750	1,950	7,700
3800	Sliding entrance door and system mill finish	6,250	1,600	7,850
3900	Bronze finish	6,875	1,725	8,600
4000	Black finish	7,175	1,775	8,950
4100	Sliding panel mall front, 16'x9' opening, mill finish	2,200	485	2,685
4200	Bronze finish	2,850	630	3,480
4300	Black finish	3,525	775	4,300
4400	24'x9' opening mill finish	3,200	900	4,100
4500	Bronze finish	4,150	1,175	5,325
4600	Black finish	5,125	1,450	6,575
4700	48'x9' opening mill finish	5,950	700	6,650
4800	Bronze finish	7,725	910	8,635
4900	Black finish	9,525	1,125	10,650
5000	Rolling overhead steel door, manual, 8' x 8' high	710	545	1,255
5100	10' x 10' high	950	620	1,570
5200	20' x 10' high	2,425	870	3,295
5300	12' x 12' high	1,250	725	1,975
5400	Motor operated, 8' x 8' high	1,650	720	2,370
5500	10' x 10' high	1,900	795	2,695
5600	20' x 10' high	3,375	1,050	4,425
5700	12' x 12' high	2,200	900	3,100
5800	Roll up grille, aluminum, manual, 10' x 10' high, mill finish	1,800	1,050	2,850
5900	Bronze anodized	2,850	1,050	3,900
6000	Motor operated, 10' x 10' high, mill finish	2,600	1,225	3,825
6100	Bronze anodized	3,650	1,225	4,875
6200	Steel, manual, 10' x 10' high	1,600	870	2,470
6300	15' x 8' high	1,875	1,100	2,975
6400	Motor operated, 10' x 10' high	2,400	1,050	3,450
6500	15' x 8' high	2,675	1,275	3,950

5 INTERIOR CONSTRUCTION

For expanded coverage of these items see *Means Assemblies Cost Data 1998*

6.5-100	Paint & Covering	COST PER S.F.		
		MAT.	**INST.**	**TOTAL**
0060	Painting, interior on plaster and drywall, brushwork, primer & 1 coat	.09	.46	.55
0080	Primer & 2 coats	.13	.60	.73
0100	Primer & 3 coats	.17	.74	.91
0120	Walls & ceilings, roller work, primer & 1 coat	.09	.30	.39
0140	Primer & 2 coats	.13	.39	.52
0160	Woodwork incl. puttying, brushwork, primer & 1 coat	.09	.66	.75
0180	Primer & 2 coats	.13	.87	1
0200	Primer & 3 coats	.17	1.19	1.36
0260	Cabinets and casework, enamel, primer & 1 coat	.09	.74	.83
0280	Primer & 2 coats	.13	.91	1.04
0300	Masonry or concrete, latex, brushwork, primer & 1 coat	.16	.62	.78
0320	Primer & 2 coats	.21	.88	1.09
0340	Addition for block filler	.10	.76	.86
0380	Fireproof paints, intumescent, 1/8" thick 3/4 hour	1.67	.60	2.27
0400	3/16" thick 1 hour	3.72	.91	4.63
0420	7/16" thick 2 hour	4.79	2.12	6.91
0440	1-1/16" thick 3 hour	7.80	4.23	12.03
0480	Miscellaneous metal brushwork, exposed metal, primer & 1 coat	.06	.74	.80
0500	Gratings, primer & 1 coat	.15	.93	1.08
0600	Pipes over 12" diameter	.40	2.96	3.36
0700	Structural steel, brushwork, light framing 300-500 S.F./Ton	.05	1.21	1.26
0720	Heavy framing 50-100 S.F./Ton	.05	.60	.65
0740	Spraywork, light framing 300-500 S.F./Ton	.06	.27	.33
0760	Heavy framing 50-100 S.F./Ton	.06	.30	.36
0800	Varnish, interior wood trim, no sanding sealer & 1 coat	.07	.74	.81
0820	Hardwood floor, no sanding 2 coats	.12	.16	.28
0840	Wall coatings, acrylic glazed coatings, minimum	.24	.56	.80
0860	Maximum	.52	.97	1.49
0880	Epoxy coatings, minimum	.32	.56	.88
0900	Maximum	.98	1.74	2.72
0940	Exposed epoxy aggregate, troweled on, 1/16" to 1/4" aggregate, minimum	.50	1.26	1.76
0960	Maximum	1.06	2.28	3.34
0980	1/2" to 5/8" aggregate, minimum	.97	2.28	3.25
1000	Maximum	1.66	3.70	5.36
1020	1" aggregate, minimum	1.69	3.29	4.98
1040	Maximum	2.60	5.40	8
1060	Sprayed on, minimum	.46	1	1.46
1080	Maximum	.85	2.04	2.89
1100	High build epoxy 50 mil, minimum	.53	.76	1.29
1120	Maximum	.92	3.12	4.04
1140	Laminated epoxy with fiberglass minimum	.58	1	1.58
1160	Maximum	1.05	2.04	3.09
1180	Sprayed perlite or vermiculite 1/16" thick, minimum	.20	.10	.30
1200	Maximum	.59	.46	1.05
1260	Wall coatings, vinyl plastic, minimum	.26	.40	.66
1280	Maximum	.66	1.23	1.89
1300	Urethane on smooth surface, 2 coats, minimum	.19	.26	.45
1320	Maximum	.45	.45	.90
1340	3 coats, minimum	.26	.35	.61
1360	Maximum	.59	.63	1.22
1380	Ceramic-like glazed coating, cementitious, minimum	.40	.67	1.07
1400	Maximum	.65	.86	1.51
1420	Resin base, minimum	.26	.46	.72
1440	Maximum	.44	.90	1.34
1460	Wall coverings, aluminum foil	.83	1.09	1.92
1480	Copper sheets, .025" thick, phenolic backing	5.65	1.25	6.90
1500	Vinyl backing	4.37	1.25	5.62
1520	Cork tiles, 12"x12", light or dark, 3/16" thick	2.80	1.25	4.05
1540	5/16" thick	2.92	1.28	4.20
1560	Basketweave, 1/4" thick	4.48	1.25	5.73

Important: See the Reference Section for critical supporting data - Location Factors & Historical Cost Indexes

INTERIOR CONSTRUCTION 6

6.5-100	Paint & Covering	COST PER S.F.		
		MAT.	INST.	TOTAL
1580	Natural, non-directional, 1/2" thick	6.45	1.25	7.70
1600	12"x36", granular, 3/16" thick	.97	.78	1.75
1620	1" thick	1.25	.81	2.06
1640	12"x12", polyurethane coated, 3/16" thick	3.03	1.25	4.28
1660	5/16" thick	4.29	1.28	5.57
1661	Paneling, prefinished plywood, birch	1.13	1.59	2.72
1662	Mahogany, African	2.10	1.67	3.77
1663	Philippine (lauan)	.90	1.33	2.23
1664	Oak or cherry	2.95	1.67	4.62
1665	Rosewood	4.19	2.08	6.27
1666	Teak	2.95	1.67	4.62
1667	Chestnut	4.37	1.78	6.15
1668	Pecan	1.87	1.67	3.54
1669	Walnut	4.76	1.67	6.43
1670	Wood board, knotty pine, finished	1.60	2.78	4.38
1671	Rough sawn cedar	2	2.78	4.78
1672	Redwood	4.37	2.78	7.15
1673	Aromatic cedar	3.35	2.98	6.33
1680	Cork wallpaper, paper backed, natural	1.72	.63	2.35
1700	Color	2.12	.63	2.75
1720	Gypsum based, fabric backed, minimum	.68	.38	1.06
1740	Average	.95	.42	1.37
1760	Maximum	1.06	.47	1.53
1780	Vinyl wall covering, fabric back, light weight	.55	.47	1.02
1800	Medium weight	.69	.63	1.32
1820	Heavy weight	1.24	.69	1.93
1840	Wall paper, double roll, solid pattern, avg. workmanship	.29	.47	.76
1860	Basic pattern, avg. workmanship	.51	.56	1.07
1880	Basic pattern, quality workmanship	.97	.69	1.66
1900	Grass cloths with lining paper, minimum	.62	.75	1.37
1920	Maximum	1.98	.86	2.84
1940	Ceramic tile, thin set, 4-1/4" x 4-1/4"	2.23	2.93	5.16
1942				

6.5-100	Paint Trim	COST PER L.F.		
		MAT.	INST.	TOTAL
2040	Painting, wood trim, to 6" wide, enamel, primer & 1 coat	.09	.37	.46
2060	Primer & 2 coats	.13	.47	.60
2080	Misc. metal brushwork, ladders	.29	3.70	3.99
2100	Pipes, to 4" dia.	.05	.78	.83
2120	6" to 8" dia.	.11	1.56	1.67
2140	10" to 12" dia.	.32	2.33	2.65
2160	Railings, 2" pipe	.10	1.85	1.95
2180	Handrail, single	.08	.74	.82

6 INTERIOR CONSTRUCTION

6.6-100	Tile & Covering	MAT.	INST.	TOTAL
0060	Carpet tile, nylon, fusion bonded, 18" x 18" or 24" x 24", 24 oz.	3	.43	3.43
0080	35 oz.	3.72	.43	4.15
0100	42 oz.	4.33	.43	4.76
0140	Carpet, tufted, nylon, roll goods, 12' wide, 26 oz.	2.83	.49	3.32
0160	36 oz.	4.50	.49	4.99
0180	Woven, wool, 36 oz.	7.10	.49	7.59
0200	42 oz.	7.25	.49	7.74
0220	Padding, add to above, minimum	.38	.23	.61
0240	Maximum	.65	.23	.88
0260	Composition flooring, acrylic, 1/4" thick	1.21	3.29	4.50
0280	3/8" thick	1.60	3.80	5.40
0300	Epoxy, minimum	2.15	2.53	4.68
0320	Maximum	2.58	3.49	6.07
0340	Epoxy terrazzo, minimum	4.46	3.41	7.87
0360	Maximum	7.45	4.54	11.99
0380	Mastic, hot laid, 1-1/2" thick, minimum	3.14	2.48	5.62
0400	Maximum	4.02	3.29	7.31
0420	Neoprene 1/4" thick, minimum	3.08	3.14	6.22
0440	Maximum	4.23	3.98	8.21
0460	Polyacrylate with ground granite 1/4", minimum	2.37	2.33	4.70
0480	Maximum	4.84	3.56	8.40
0500	Polyester with colored quart 2 chips 1/16", minimum	2.15	1.61	3.76
0520	Maximum	3.85	2.53	6.38
0540	Polyurethane with vinyl chips, minimum	6	1.61	7.61
0560	Maximum	8.65	1.99	10.64
0600	Concrete topping, granolithic concrete, 1/2" thick	.17	1.47	1.64
0620	1" thick	.35	1.51	1.86
0640	2" thick	.69	1.73	2.42
0660	Heavy duty 3/4" thick, minimum	.23	2.28	2.51
0680	Maximum	.23	2.71	2.94
0700	For colors, add to above, minimum	.46	.53	.99
0720	Maximum	2.05	.58	2.63
0740	Exposed aggregate finish, minimum	.40	.49	.89
0760	Maximum	1.25	.65	1.90
0780	Abrasives, .25 P.S.F. add to above, minimum	.16	.36	.52
0800	Maximum	.19	.36	.55
0820	Dust on coloring, add, minimum	.62	.23	.85
0840	Maximum	2.16	.49	2.65
0860	1/2" integral, minimum	.19	1.47	1.66
0880	Maximum	.19	1.47	1.66
0900	Dustproofing, add, minimum	.07	.16	.23
0920	Maximum	.12	.23	.35
0930	Paint	.21	.88	1.09
0940	Hardeners, metallic add, minimum	.22	.36	.58
0960	Maximum	.49	.53	1.02
0980	Non-metallic, minimum	.13	.36	.49
1000	Maximum	.37	.53	.90
1020	Integral topping, 3/16" thick	.06	.87	.93
1040	1/2" thick	.16	.91	1.07
1060	3/4" thick	.25	1.02	1.27
1080	1" thick	.32	1.16	1.48
1100	Terrazzo, minimum	2.42	5.25	7.67
1120	Maximum	4.51	11.35	15.86
1280	Resilient, asphalt tile, 1/8" thick on concrete, minimum	1.03	.77	1.80
1300	Maximum	1.13	.77	1.90
1320	On wood, add for felt underlay	.20		.20
1340	Cork tile, minimum	2.66	.98	3.64
1360	Maximum	10.10	.98	11.08
1380	Polyethylene, in rolls, minimum	2.41	1.12	3.53
1400	Maximum	4.68	1.12	5.80

COST PER S.F.

Important: See the Reference Section for critical supporting data - Location Factors & Historical Cost Indexes

INTERIOR CONSTRUCTION 6

6.6-100	Tile & Covering	COST PER S.F.		
		MAT.	INST.	TOTAL
1420	Polyurethane, thermoset, minimum	3.71	3.09	6.80
1440	Maximum	4.40	6.20	10.60
1460	Rubber, sheet goods, minimum	3.31	2.57	5.88
1480	Maximum	5.45	3.43	8.88
1500	Tile, minimum	3.33	.77	4.10
1520	Maximum	6.45	1.12	7.57
1580	Vinyl, composition tile, minimum	.75	.62	1.37
1600	Maximum	2.04	.62	2.66
1620	Tile, minimum	1.69	.62	2.31
1640	Maximum	4.92	.62	5.54
1660	Sheet goods, minimum	1.38	1.24	2.62
1680	Maximum	3.93	1.54	5.47
1720	Tile, ceramic natural clay	3.85	3.05	6.90
1730	Marble, synthetic 12"x12"x5/8"	7.40	9.30	16.70
1740	Porcelain type, minimum	4.15	3.05	7.20
1760	Maximum	4.38	2.93	7.31
1800	Quarry tile, mud set, minimum	3.20	3.98	7.18
1820	Maximum	4.97	5.05	10.02
1840	Thin set, deduct		.80	.80
1860	Terrazzo precast, minimum	7.20	4.12	11.32
1880	Maximum	8.05	4.12	12.17
1900	Non-slip, minimum	16.35	20.50	36.85
1920	Maximum	17.60	28.50	46.10
1960	Wood, block, end grain factory type, creosoted, 2" thick	2.92	1.13	4.05
1980	2-1/2" thick	3.03	2.67	5.70
2000	3" thick	2.97	2.67	5.64
2020	Natural finish, 2" thick	3.03	2.67	5.70
2040	Fir, vertical grain, 1"x4", no finish, minimum	2.38	1.31	3.69
2060	Maximum	2.53	1.31	3.84
2080	Prefinished white oak, prime grade, 2-1/4" wide	6.75	1.96	8.71
2100	3-1/4" wide	8.35	1.80	10.15
2120	Maple strip, sanded and finished, minimum	3.62	2.84	6.46
2140	Maximum	4.28	2.84	7.12
2160	Oak strip, sanded and finished, minimum	4.38	2.84	7.22
2180	Maximum	3.96	2.84	6.80
2200	Parquetry, sanded and finished, minimum	2.40	2.96	5.36
2220	Maximum	6.40	4.21	10.61
2260	Add for sleepers on concrete, treated, 24" O.C., 1"x2"	1.10	1.70	2.80
2280	1"x3"	1.10	1.33	2.43
2300	2"x4"	.87	.68	1.55
2340	Underlayment, plywood, 3/8" thick	.50	.44	.94
2350	1/2" thick	.58	.46	1.04
2360	5/8" thick	.69	.48	1.17
2370	3/4" thick	.83	.51	1.34
2380	Particle board, 3/8" thick	.34	.44	.78
2390	1/2" thick	.36	.46	.82
2400	5/8" thick	.39	.48	.87
2410	3/4" thick	.45	.51	.96
2420	Hardboard, 4' x 4', .215" thick	.45	.44	.89

6 INTERIOR CONSTRUCTION

Plaster Partitions are defined as follows: type of plaster, type and spacing of framing, type of lath and treatment of opposite face.

Included in the system components are expansion joints. Metal studs are assumed to be nonload bearing.

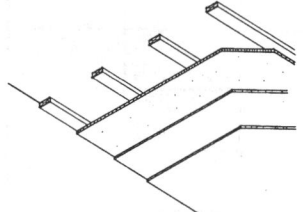

2 Coats of Plaster on Gypsum Lath on Wood Furring

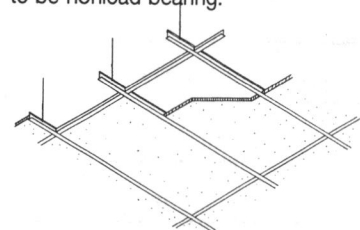

Fiberglass Board on Exposed Suspended Grid System

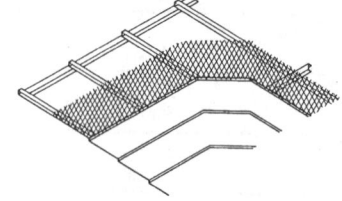

Plaster and Metal Lath on Metal Furring

6.7-100 — Plaster Ceilings

	TYPE	LATH	FURRING	SUPPORT		COST PER S.F. MAT.	INST.	TOTAL
2400	2 coat gypsum	3/8" gypsum	1"x3" wood, 16" O.C.	wood		1.07	3.37	4.44
2500	Painted			masonry		1.07	3.44	4.51
2600				concrete		1.07	3.85	4.92
2700	3 coat gypsum	3.4# metal	1"x3" wood, 16" O.C.	wood		1.06	3.62	4.68
2800	Painted			masonry		1.06	3.69	4.75
2900				concrete		1.06	4.10	5.16
3000	2 coat perlite	3/8" gypsum	1"x3" wood, 16" O.C.	wood		1.15	3.50	4.65
3100	Painted			masonry		1.15	3.57	4.72
3200				concrete		1.15	3.98	5.13
3300	3 coat perlite	3.4# metal	1"x3" wood, 16" O.C.	wood		1	3.59	4.59
3400	Painted			masonry		1	3.66	4.66
3500				concrete		1	4.07	5.07
3600	2 coat gypsum	3/8" gypsum	3/4" CRC, 12" O.C.	1-1/2" CRC, 48"O.C.		1.03	4.15	5.18
3700	Painted		3/4" CRC, 16" O.C.	1-1/2" CRC, 48"O.C.		1.01	3.74	4.75
3800			3/4" CRC, 24" O.C.	1-1/2" CRC, 48"O.C.		.96	3.41	4.37
3900	2 coat perlite	3/8" gypsum	3/4" CRC, 12" O.C.	1-1/2" CRC, 48"O.C		1.11	4.44	5.55
4000	Painted		3/4" CRC, 16" O.C.	1-1/2" CRC, 48"O.C.		1.09	4.03	5.12
4100			3/4" CRC, 24" O.C.	1-1/2" CRC, 48"O.C.		1.04	3.70	4.74
4200	3 coat gypsum	3.4# metal	3/4" CRC, 12" O.C.	1-1/2" CRC, 36" O.C.		1.28	5.40	6.68
4300	Painted		3/4" CRC, 16" O.C.	1-1/2" CRC, 36" O.C.		1.26	4.98	6.24
4400			3/4" CRC, 24" O.C.	1-1/2" CRC, 36" O.C.		1.21	4.65	5.86
4500	3 coat perlite	3.4# metal	3/4" CRC, 12" O.C.	1-1/2" CRC,36" O.C.		1.47	5.95	7.42
4600	Painted		3/4" CRC, 16" O.C.	1-1/2" CRC, 36" O.C.		1.45	5.50	6.95
4700			3/4" CRC, 24" O.C.	1-1/2" CRC, 36" O.C.		1.40	5.20	6.60

6.7-100 — Drywall Ceilings

	TYPE	FINISH	FURRING	SUPPORT		COST PER S.F. MAT.	INST.	TOTAL
4800	1/2" F.R. drywall	painted and textured	1"x3" wood, 16" O.C.	wood		.78	2.09	2.87
4900				masonry		.78	2.16	2.94
5000				concrete		.78	2.57	3.35
5100	5/8" F.R. drywall	painted and textured	1"x3" wood, 16" O.C.	wood		.78	2.09	2.87
5200				masonry		.78	2.16	2.94
5300				concrete		.78	2.57	3.35
5400	1/2" F.R. drywall	painted and textured	7/8"resil. channels	24" O.C.		.65	2.04	2.69
5500			1"x2" wood	stud clips		.78	1.97	2.75
5602	1/2" F.R. drywall	painted	1-5/8" metal studs	24" O.C.		.71	2.02	2.73

Important: See the Reference Section for critical supporting data - Location Factors & Historical Cost Indexes

INTERIOR CONSTRUCTION 6

6.7-100 Drywall Ceilings

	TYPE	FINISH	FURRING	SUPPORT		COST PER S.F.		
						MAT.	INST.	TOTAL
5700 5702	5/8" F.R. drywall	painted and textured	1-5/8"metal studs	24" O.C.		.71	2.02	2.73

6.7-100 Acoustical Ceilings

	TYPE	TILE	GRID	SUPPORT		COST PER S.F.		
						MAT.	INST.	TOTAL
5800	5/8" fiberglass board	24" x 48"	tee	suspended		.84	.96	1.80
5900		24" x 24"	tee	suspended		.92	1.05	1.97
6000	3/4" fiberglass board	24" x 48"	tee	suspended		1.38	.99	2.37
6100		24" x 24"	tee	suspended		1.46	1.08	2.54
6500	5/8" mineral fiber	12" x 12"	1"x3" wood, 12" O.C.	wood		.87	1.28	2.15
6600				masonry		.87	1.37	2.24
6700				concrete		.87	1.92	2.79
6800	3/4" mineral fiber	12" x 12"	1"x3" wood, 12" O.C.	wood		1.54	1.28	2.82
6900				masonry		1.54	1.28	2.82
7000				concrete		1.54	1.28	2.82
7100 7102	3/4"mineral fiber on 5/8" F.R. drywall	12" x 12"	25 ga. channels	runners		1.71	1.76	3.47
7200 7201 7202	5/8" plastic coated Mineral fiber	12" x 12"		adhesive backed		.87	.33	1.20
7300 7301 7302	3/4" plastic coated Mineral fiber	12" x 12"		adhesive backed		1.54	.33	1.87
7400 7401 7402	3/4" mineral fiber	12" x 12"	conceal 2" bar & channels	suspended		1.38	1.69	3.07

6.7-810 Acoustical Ceilings

		COST PER S.F.		
		MAT.	INST.	TOTAL
2480	Ceiling boards, eggcrate, acrylic, 1/2" x 1/2" x 1/2" cubes	1.33	.67	2
2500	Polystyrene, 3/8" x 3/8" x 1/2" cubes	1.11	.65	1.76
2520	1/2" x 1/2" x 1/2" cubes	1.50	.67	2.17
2540	Fiberglass boards, plain, 5/8" thick	.50	.53	1.03
2560	3/4" thick	1.04	.56	1.60
2580	Grass cloth faced, 3/4" thick	1.62	.67	2.29
2600	1" thick	1.83	.69	2.52
2620	Luminous panels, prismatic, acrylic	1.61	.83	2.44
2640	Polystyrene	.83	.83	1.66
2660	Flat or ribbed, acrylic	2.82	.83	3.65
2680	Polystyrene	1.91	.83	2.74
2700	Drop pan, white, acrylic	4.11	.83	4.94
2720	Polystyrene	3.44	.83	4.27
2740	Mineral fiber boards, 5/8" thick, standard	.57	.49	1.06
2760	Plastic faced	.89	.83	1.72
2780	2 hour rating	.79	.49	1.28
2800	Perforated aluminum sheets, .024 thick, corrugated painted	1.64	.68	2.32
2820	Plain			
3080	Mineral fiber, plastic coated, 12" x 12" or 12" x 24", 5/8" thick	.56	.33	.89
3100	3/4" thick	1.23	.33	1.56
3120	Fire rated, 3/4" thick, plain faced	.83	.33	1.16
3140	Mylar faced	.96	.33	1.29
3160	Add for ceiling primer	.12		.12
3180	Add for ceiling cement	.31		.31
3240	Suspension system, furring, 1" x 3" wood 12" O.C.	.31	.95	1.26
3260	T bar suspension system, 2' x 4' grid	.33	.42	.75

INTERIOR CONSTRUCTION

6

6.7-810	Acoustical Ceilings	COST PER S.F.		
		MAT.	INST.	TOTAL
3280	2' x 2' grid	.41	.51	.92
3300	Concealed Z bar suspension system 12" module	.36	.64	1
3320	Add to above for 1-1/2" carrier channels 4' O.C.	.18	.71	.89
3340	Add to above for carrier channels for recessed lighting	.35	.72	1.07

6.7-820	Plaster Ceilings	COST PER S.F.		
		MAT.	INST.	TOTAL
0060	Plaster, gypsum incl. finish, 2 coats			
0080	3 coats	.49	2.11	2.60
0100	Perlite, incl. finish, 2 coats	.43	2.08	2.51
0120	3 coats	.68	2.63	3.31
0140	Thin coat on drywall	.13	.41	.54
0200	Lath, gypsum, 3/8" thick	.40	.57	.97
0220	1/2" thick	.47	.60	1.07
0240	5/8" thick	.48	.70	1.18
0260	Metal, diamond, 2.5 lb.	.22	.46	.68
0280	3.4 lb.	.25	.50	.75
0300	Flat rib, 2.75 lb.	.21	.46	.67
0320	3.4 lb.	.39	.50	.89
0440	Furring, steel channels, 3/4" galvanized , 12" O.C.	.19	1.49	1.68
0460	16" O.C.	.17	1.08	1.25
0480	24" O.C.	.12	.75	.87
0500	1-1/2" galvanized , 12" O.C.	.27	1.65	1.92
0520	16" O.C.	.25	1.20	1.45
0540	24" O.C.	.16	.80	.96
0560	Wood strips, 1"x3", on wood, 12" O.C.	.31	.95	1.26
0580	16" O.C.	.23	.71	.94
0600	24" O.C.	.16	.48	.64
0620	On masonry, 12" O.C.	.31	1.04	1.35
0640	16" O.C.	.23	.78	1.01
0660	24" O.C.	.16	.52	.68
0680	On concrete, 12" O.C.	.31	1.59	1.90
0700	16" O.C.	.23	1.19	1.42
0720	24" O.C.	.16	.80	.96
0940	Paint on plaster or drywall, roller work, primer + 1 coat	.09	.30	.39
0960	Primer + 2 coats	.13	.39	.52

For information about Means Estimating Seminars, see yellow pages 11 and 12 in back of book

INTERIOR CONSTRUCTION 6

Important: See the Reference Section for critical supporting data - Location Factors & Historical Cost Indexes

Division 7
Conveying Systems

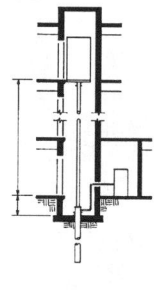

Hydraulic

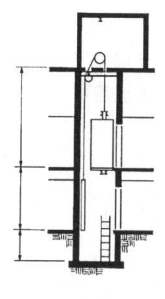

Traction Geared

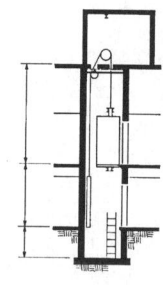

Traction Gearless

7.1-100	Hydraulic	COST EACH		
		MAT.	INST.	TOTAL
1300	Pass. elev., 1500 lb., 2 Floors, 100 FPM	32,100	7,900	40,000
1400	5 Floors, 100 FPM	43,500	21,000	64,500
1600	2000 lb., 2 Floors, 100 FPM	33,000	7,900	40,900
1700	5 floors, 100 FPM	44,300	21,000	65,300
1900	2500 lb., 2 Floors, 100 FPM	33,400	7,900	41,300
2000	5 floors, 100 FPM	44,800	21,000	65,800
2200	3000 lb., 2 Floors, 100 FPM	34,900	7,900	42,800
2300	5 floors, 100 FPM	46,200	21,000	67,200
2500	3500 lb., 2 Floors, 100 FPM	37,000	7,900	44,900
2600	5 floors, 100 FPM	48,400	21,000	69,400
2800	4000 lb., 2 Floors, 100 FPM	38,000	7,900	45,900
2900	5 floors, 100 FPM	49,400	21,000	70,400
3100	4500 lb., 2 Floors, 100 FPM	38,900	7,900	46,800
3200	5 floors, 100 FPM	50,500	21,000	71,500
4000	Hospital elevators, 3500 lb., 2 Floors, 100 FPM	46,400	7,900	54,300
4100	5 floors, 100 FPM	70,000	21,000	91,000
4300	4000 lb., 2 Floors, 100 FPM	46,400	7,900	54,300
4400	5 floors, 100 FPM	70,000	21,000	91,000
4600	4500 lb., 2 Floors, 100 FPM	52,000	7,900	59,900
4800	5 floors, 100 FPM	75,500	21,000	96,500
4900	5000 lb., 2 Floors, 100 FPM	54,500	7,900	62,400
5000	5 floors, 100 FPM	78,000	21,000	99,000
6700	Freight elevators (Class "B"), 3000 lb., 2 Floors, 50 FPM	44,600	10,100	54,700
6800	5 floors, 100 FPM	64,000	26,400	90,400
7000	4000 lb., 2 Floors, 50 FPM	47,500	10,100	57,600
7100	5 floors, 100 FPM	67,000	26,400	93,400
7500	10,000 lb., 2 Floors, 50 FPM	59,000	10,100	69,100
7600	5 floors, 100 FPM	78,500	26,400	104,900
8100	20,000 lb., 2 Floors, 50 FPM	71,500	10,100	81,600
8200	5 Floors, 100 FPM	90,500	26,400	116,900

7.1-200	Traction Geared Elevators	COST EACH		
		MAT.	INST.	TOTAL
1300	Passenger, 2000 Lb., 5 floors, 200 FPM	71,500	19,800	91,300
1500	15 floors, 350 FPM	114,000	60,000	174,000
1600	2500 Lb., 5 floors, 200 FPM	74,500	19,800	94,300
1800	15 floors, 350 FPM	117,000	60,000	177,000
1900	3000 Lb., 5 floors, 200 FPM	75,500	19,800	95,300
2100	15 floors, 350 FPM	118,000	60,000	178,000
2200	3500 Lb., 5 floors, 200 FPM	77,500	19,800	97,300
2400	15 floors, 350 FPM	119,500	60,000	179,500
2500	4000 Lb., 5 floors, 200 FPM	77,500	19,800	97,300
2700	15 floors, 350 FPM	119,500	60,000	179,500
2800	4500 Lb., 5 floors, 200 FPM	79,500	19,800	99,300
3000	15 floors, 350 FPM	121,500	60,000	181,500

Important: See the Reference Section for critical supporting data - Location Factors & Historical Cost Indexes

CONVEYING SYSTEMS 7

7.1-200 — Traction Geared Elevators

		MAT.	INST.	TOTAL
			COST EACH	
3100	5000 Lb., 5 floors, 200 FPM	81,000	19,800	100,800
3300	15 floors, 350 FPM	123,500	60,000	183,500
4000	Hospital, 3500 Lb., 5 floors, 200 FPM	88,500	19,800	108,300
4200	15 floors, 350 FPM	151,000	60,000	211,000
4300	4000 Lb., 5 floors, 200 FPM	88,500	19,800	108,300
4500	15 floors, 350 FPM	151,000	60,000	211,000
4600	4500 Lb., 5 floors, 200 FPM	93,000	19,800	112,800
4800	15 floors, 350 FPM	155,500	60,000	215,500
4900	5000 Lb., 5 floors, 200 FPM	94,500	19,800	114,300
5100	15 floors, 350 FPM	157,000	60,000	217,000
6000	Freight, 4000 Lb., 5 floors, 50 FPM class 'B'	73,000	20,500	93,500
6200	15 floors, 200 FPM class 'B'	131,500	71,500	203,000
6300	8000 Lb., 5 floors, 50 FPM class 'B'	87,500	20,500	108,000
6500	15 floors, 200 FPM class 'B'	146,000	71,500	217,500
7000	10,000 Lb., 5 floors, 50 FPM class 'B'	103,000	20,500	123,500
7200	15 floors, 200 FPM class 'B'	286,000	71,500	357,500
8000	20,000 Lb., 5 floors, 50 FPM class 'B'	113,500	20,500	134,000
8200	15 floors, 200 FPM class 'B'	296,500	71,500	368,000

7.1-300 — Traction Gearless Elevators

		MAT.	INST.	TOTAL
			COST EACH	
1700	Passenger, 2500 Lb., 10 floors, 200 FPM	146,500	51,500	198,000
1900	30 floors, 600 FPM	255,500	131,500	387,000
2000	3000 Lb., 10 floors, 200 FPM	147,000	51,500	198,500
2200	30 floors, 600 FPM	256,500	131,500	388,000
2300	3500 Lb., 10 floors, 200 FPM	149,000	51,500	200,500
2500	30 floors, 600 FPM	258,000	131,500	389,500
2700	50 floors, 800 FPM	339,000	212,000	551,000
2800	4000 lb., 10 floors, 200 FPM	149,000	51,500	200,500
3000	30 floors, 600 FPM	258,000	131,500	389,500
3200	50 floors, 800 FPM	339,000	212,000	551,000
3300	4500 lb., 10 floors, 200 FPM	151,000	51,500	202,500
3500	30 floors, 600 FPM	260,000	131,500	391,500
3700	50 floors, 800 FPM	341,000	212,000	553,000
3800	5000 lb., 10 floors, 200 FPM	152,500	51,500	204,000
4000	30 floors, 600 FPM	262,000	131,500	393,500
4200	50 floors, 800 FPM	343,000	212,000	555,000
6000	Hospital, 3500 Lb., 10 floors, 200 FPM	170,000	51,500	221,500
6200	30 floors, 600 FPM	320,000	131,500	451,500
6400	4000 Lb., 10 floors, 200 FPM	170,000	51,500	221,500
6600	30 floors, 600 FPM	320,000	131,500	451,500
6800	4500 Lb., 10 floors, 200 FPM	174,500	51,500	226,000
7000	30 floors, 600 FPM	324,500	131,500	456,000
7200	5000 Lb., 10 floors, 200 FPM	176,000	51,500	227,500
7400	30 floors, 600 FPM	326,000	131,500	457,500

For information about Means Estimating Seminars, see yellow pages 11 and 12 in back of book

CONVEYING SYSTEMS

7

Division 8
Mechanical

8.1-040	Piping - Installed - Unit Costs	COST PER L.F.		
		MAT.	INST.	TOTAL
0840	Cast iron, soil, B & S, service weight, 2" diameter	4.72	10.95	15.67
0860	3" diameter	6.50	11.50	18
0880	4" diameter	9.10	12.55	21.65
0900	5" diameter	11.15	14.15	25.30
0920	6" diameter	13.60	14.70	28.30
0940	8" diameter	21.50	25	46.50
0960	10" diameter	34.50	27	61.50
0980	12" diameter	48	30.50	78.50
1040	No hub, 1-1/2" diameter	5.60	9.75	15.35
1060	2" diameter	5.60	10.30	15.90
1080	3" diameter	7.40	10.80	18.20
1100	4" diameter	9.40	11.90	21.30
1120	5" diameter	13.40	12.95	26.35
1140	6" diameter	17.10	13.60	30.70
1160	8" diameter	26	21	47
1180	10" diameter	42	24	66
1220	Copper tubing, hard temper, solder, type K, 1/2" diameter	1.93	4.92	6.85
1260	3/4" diameter	3.07	5.20	8.27
1280	1" diameter	3.93	5.80	9.73
1300	1-1/4" diameter	4.87	6.85	11.72
1320	1-1/2" diameter	7	7.65	14.65
1340	2" diameter	9.45	9.60	19.05
1360	2-1/2" diameter	14	11.50	25.50
1380	3" diameter	19.30	12.80	32.10
1400	4" diameter	30.50	18.20	48.70
1480	5" diameter	57.50	21.50	79
1500	6" diameter	83.50	28.50	112
1520	8" diameter	144	31.50	175.50
1560	Type L, 1/2" diameter	1.66	4.74	6.40
1600	3/4" diameter	2.28	5.05	7.33
1620	1" diameter	2.49	5.65	8.14
1640	1-1/4" diameter	4.21	6.60	10.81
1660	1-1/2" diameter	5.25	7.40	12.65
1680	2" diameter	8	9.15	17.15
1700	2-1/2" diameter	11.95	11.15	23.10
1720	3" diameter	16.15	12.35	28.50
1740	4" diameter	26	17.70	43.70
1760	5" diameter	52.50	20.50	73
1780	6" diameter	67	27	94
1800	8" diameter	112	30	142
1840	Type M, 1/2" diameter	1.34	4.57	5.91
1880	3/4" diameter	1.81	4.92	6.73
1900	1" diameter	2.49	5.50	7.99
1920	1-1/4" diameter	3.51	6.40	9.91
1940	1-1/2" diameter	4.69	7.10	11.79
1960	2" diameter	7.45	8.70	16.15
1980	2-1/2" diameter	10.55	10.80	21.35
2000	3" diameter	13.60	11.90	25.50
2020	4" diameter	23.50	17.25	40.75
2060	6" diameter	65.50	25.50	91
2080	8" diameter	105	28.50	133.50
2120	Type DWV, 1-1/4" diameter	3.44	6.40	9.84
2160	1-1/2" diameter	4.21	7.10	11.31
2180	2" diameter	5.55	8.70	14.25
2200	3" diameter	9.60	11.90	21.50
2220	4" diameter	17.10	17.25	34.35
2240	5" diameter	37	19.20	56.20
2260	6" diameter	53	25.50	78.50

Important: See the Reference Section for critical supporting data - Location Factors & Historical Cost Indexes

8.1-040	Piping - Installed - Unit Costs	COST PER L.F.		
		MAT.	INST.	TOTAL
2280	8″ diameter	119	28.50	147.50
2800	Plastic,PVC, DWV, schedule 40, 1-1/4″ diameter	2.19	9.15	11.34
2820	1-1/2″ diameter	2.20	10.65	12.85
2830	2″ diameter	2.44	11.70	14.14
2840	3″ diameter	3.95	13.05	17
2850	4″ diameter	6	14.40	20.40
2890	6″ diameter	9.20	17.70	26.90
3010	Pressure pipe 200 PSI, 1/2″ diameter	1.84	7.10	8.94
3030	3/4″ diameter	1.89	7.50	9.39
3040	1″ diameter	2.02	8.35	10.37
3050	1-1/4″ diameter	2.23	9.15	11.38
3060	1-1/2″ diameter	2.36	10.65	13.01
3070	2″ diameter	2.63	11.70	14.33
3080	2-1/2″ diameter	3.92	12.35	16.27
3090	3″ diameter	4.40	13.05	17.45
3100	4″ diameter	5.80	14.40	20.20
3110	6″ diameter	10.25	17.70	27.95
3120	8″ diameter	17.25	22.50	39.75
4000	Steel, schedule 40, threaded, black, 1/2″ diameter	1.65	6.10	7.75
4020	3/4″ diameter	1.83	6.30	8.13
4030	1″ diameter	2.38	7.25	9.63
4040	1-1/4″ diameter	3.07	7.75	10.82
4050	1-1/2″ diameter	3.36	8.65	12.01
4060	2″ diameter	4.53	10.80	15.33
4070	2-1/2″ diameter	6.60	13.80	20.40
4080	3″ diameter	8.05	16.05	24.10
4090	4″ diameter	11.60	19.20	30.80
4100	5″ diameter	18.30	26.50	44.80
4110	6″ diameter	23.50	34.50	58
4120	8″ diameter	32.50	40	72.50
4130	10″ diameter	48.50	46.50	95
4140	12″ diameter	64	59.50	123.50
4200	Galvanized, 1/2″ diameter	1.49	6.10	7.59
4220	3/4″ diameter	1.72	6.30	8.02
4230	1″ diameter	2.03	7.25	9.28
4240	1-1/4″ diameter	2.51	7.75	10.26
4250	1-1/2″ diameter	3.44	8.65	12.09
4260	2″ diameter	3.61	10.80	14.41
4270	2-1/2″ diameter	5.25	13.80	19.05
4280	3″ diameter	6.55	16.05	22.60
4290	4″ diameter	9.10	19.20	28.30
4300	5″ diameter	16.90	26.50	43.40
4310	6″ diameter	22	34.50	56.50
4320	8″ diameter	32	40	72
4330	10″ diameter	47	46.50	93.50
4340	12″ diameter	62.50	59.50	122
5010	Flanged, black, 1″ diameter	6.75	10.60	17.35
5020	1-1/4″ diameter	7.45	11.65	19.10
5030	1-1/2″ diameter	7.70	12.80	20.50
5040	2″ diameter	8.75	16.55	25.30
5050	2-1/2″ diameter	11.15	20.50	31.65
5060	3″ diameter	12	23	35
5070	4″ diameter	15.80	28.50	44.30
5080	5″ diameter	25.50	35.50	61
5090	6″ diameter	29.50	45	74.50
5100	8″ diameter	44.50	59.50	104
5110	10″ diameter	76.50	70.50	147
5120	12″ diameter	102	81	183
5720	Grooved joints, black, 3/4″ diameter	1.98	5.40	7.38
5730	1″ diameter	2.17	6.10	8.27

8.1-040	Piping - Installed - Unit Costs	COST PER L.F.		
		MAT.	INST.	TOTAL
5740	1-1/4" diameter	2.72	6.60	9.32
5750	1-1/2" diameter	3.08	7.50	10.58
5760	2" diameter	3.69	9.60	13.29
5770	2-1/2" diameter	5.30	12.10	17.40
5900	3" diameter	6.55	13.80	20.35
5910	4" diameter	9.15	15.35	24.50
5920	5" diameter	15.10	18.65	33.75
5930	6" diameter	19.20	25.50	44.70
5940	8" diameter	28.50	29	57.50
5950	10" diameter	43	34.50	77.50
5960	12" diameter	55	40	95

Important: See the Reference Section for critical supporting data - Location Factors & Historical Cost Indexes

Installation includes piping and fittings
within 10′ of heater. Gas heaters require
vent piping (not included with these units).

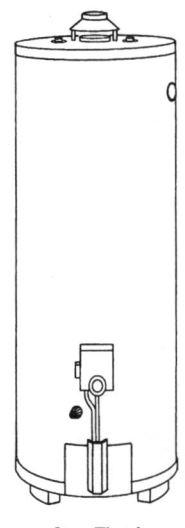

Gas Fired

Oil Fired

8.1-120	Gas Fired Water Heaters - Residential Systems	COST EACH		
		MAT.	INST.	TOTAL
2200	Gas fired water heater, residential, 100° F rise			
2220	20 gallon tank, 25 GPH	540	775	1,315
2260	30 gallon tank, 32 GPH	545	790	1,335
2300	40 gallon tank, 32 GPH	620	885	1,505
2340	50 gallon tank, 63 GPH	700	890	1,590
2380	75 gallon tank, 63 GPH	1,075	995	2,070
2420	100 gallon tank, 63 GPH	1,525	1,050	2,575
2422				

8.1-130	Oil Fired Water Heaters - Residential Systems	COST EACH		
		MAT.	INST.	TOTAL
2200	Oil fired water heater, residential, 100° F rise			
2220	30 gallon tank, 103 GPH	1,025	735	1,760
2260	50 gallon tank, 145 GPH	1,425	820	2,245
2300	70 gallon tank, 164 GPH	1,625	925	2,550
2340	85 gallon tank, 181 GPH	2,200	950	3,150

MECHANICAL

8

Systems below include piping and fittings within 10' of heater. Electric water heaters do not require venting. Gas fired heaters require vent piping (not included in these prices).

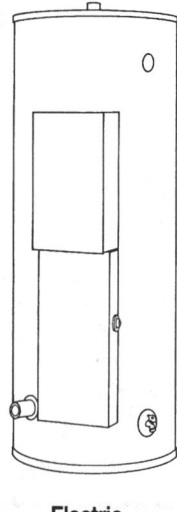

Electric

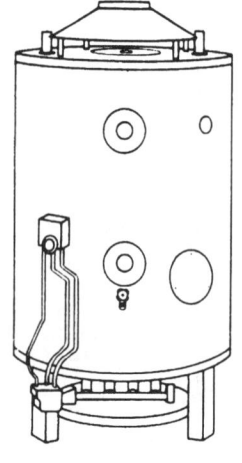

Gas Fired

8.1-160	Electric Water Heaters - Commercial Systems	COST EACH		
		MAT.	INST.	TOTAL
1800	Electric water heater, commercial, 100° F rise			
1820	50 gallon tank, 9 KW 37 GPH	2,925	660	3,585
1860	80 gal, 12 KW 49 GPH	3,725	820	4,545
1900	36 KW 147 GPH	5,075	885	5,960
1940	120 gal, 36 KW 147 GPH	5,475	955	6,430
1980	150 gal, 120 KW 490 GPH	18,700	1,025	19,725
2020	200 gal, 120 KW 490 GPH	19,600	1,050	20,650
2060	250 gal, 150 KW 615 GPH	21,800	1,225	23,025
2100	300 gal, 180 KW 738 GPH	33,800	1,300	35,100
2140	350 gal, 30 KW 123 GPH	16,700	1,400	18,100
2180	180 KW 738 GPH	24,400	1,400	25,800
2220	500 gal, 30 KW 123 GPH	21,500	1,650	23,150
2260	240 KW 984 GPH	32,100	1,650	33,750
2300	700 gal, 30 KW 123 GPH	26,100	1,875	27,975
2340	300 KW 1230 GPH	41,500	1,875	43,375
2380	1000 gal, 60 KW 245 GPH	31,700	2,600	34,300
2420	480 KW 1970 GPH	56,000	2,625	58,625
2460	1500 gal, 60 KW 245 GPH	46,900	3,250	50,150
2500	480 KW 1970 GPH	66,500	3,250	69,750

8.1-170	Gas Fired Water Heaters - Commercial Systems	COST EACH		
		MAT.	INST.	TOTAL
1760	Gas fired water heater, commercial, 100° F rise			
1780	75.5 MBH input, 63 GPH	1,600	1,025	2,625
1820	95 MBH input, 86 GPH	2,475	1,025	3,500
1860	100 MBH input, 91 GPH	2,675	1,075	3,750
1900	115 MBH input, 110 GPH	2,075	1,100	3,175
1980	155 MBH input, 150 GPH	2,400	1,250	3,650
2020	175 MBH input, 168 GPH	2,725	1,325	4,050
2060	200 MBH input, 192 GPH	3,200	1,500	4,700
2100	240 MBH input, 230 GPH	3,775	1,625	5,400
2140	300 MBH input, 278 GPH	4,825	1,875	6,700
2180	390 MBH input, 374 GPH	5,425	1,875	7,300
2220	500 MBH input, 480 GPH	7,150	2,025	9,175

8.1-170	Gas Fired Water Heaters - Commercial Systems	COST EACH		
		MAT.	INST.	TOTAL
2260	600 MBH input, 576 GPH	8,100	2,200	10,300
2300	800 MBH input, 768 GPH	9,775	2,425	12,200
2340	1000 MBH input, 960 GPH	11,500	2,475	13,975
2380	1200 MBH input, 1150 GPH	14,400	3,075	17,475
2420	1500 MBH input, 1440 GPH	17,500	3,150	20,650
2460	1800 MBH input, 1730 GPH	19,600	3,500	23,100
2500	2450 MBH input, 2350 GPH	25,700	4,050	29,750
2540	3000 MBH input, 2880 GPH	31,000	4,950	35,950
2580	3750 MBH input, 3600 GPH	38,700	5,100	43,800

MECHANICAL

8

Oil fired water heater systems include piping and fittings within 10' of heater. Oil fired heaters require vent piping (not included in these systems).

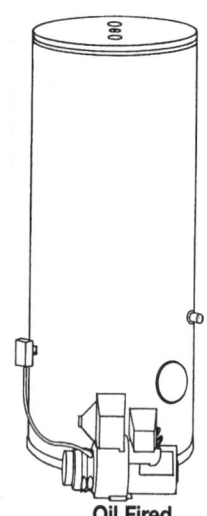

Oil Fired

8.1-180	Oil Fired Water Heaters - Commercial Systems	COST EACH		
		MAT.	INST.	TOTAL
1800	Oil fired water heater, commercial, 100° F rise			
1820	103 MBH output, 116 GPH	2,100	895	2,995
1900	134 MBH output, 161 GPH	2,400	1,150	3,550
1940	161 MBH output, 192 GPH	2,850	1,300	4,150
1980	187 MBH output, 224 GPH	3,100	1,525	4,625
2060	262 MBH output, 315 GPH	3,850	1,725	5,575
2100	341 MBH output, 409 GPH	5,100	1,850	6,950
2140	420 MBH output, 504 GPH	5,775	2,025	7,800
2180	525 MBH output, 630 GPH	7,325	2,275	9,600
2220	630 MBH output, 756 GPH	8,100	2,275	10,375
2260	735 MBH output, 880 GPH	9,975	2,625	12,600
2300	840 MBH output, 1000 GPH	11,300	2,625	13,925
2340	1050 MBH output, 1260 GPH	13,600	2,725	16,325
2380	1365 MBH output, 1640 GPH	17,100	3,075	20,175
2420	1680 MBH output, 2000 GPH	20,400	3,850	24,250
2460	2310 MBH output, 2780 GPH	25,900	4,750	30,650
2500	2835 MBH output, 3400 GPH	31,900	4,750	36,650
2540	3150 MBH output, 3780 GPH	35,700	6,550	42,250

Important: See the Reference Section for critical supporting data - Location Factors & Historical Cost Indexes

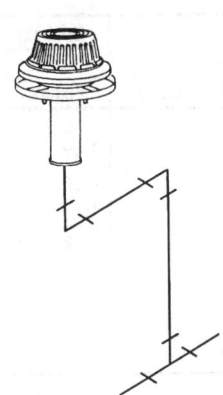

Design Assumptions: Vertical conductor size is based on a maximum rate of rainfall of 4″ per hour. To convert roof area to other rates multiply "Max. S.F. Roof Area" shown by four and divide the result by desired local rate. The answer is the local roof area that may be handled by the indicated pipe diameter.

Basic cost is for roof drain, 10′ of vertical leader and 10′ of horizontal, plus connection to the main.

Pipe Dia.	Max. S.F. Roof Area	Gallons per Min.
2″	544	23
3″	1610	67
4″	3460	144
5″	6280	261
6″	10,200	424
8″	22,000	913

8.1-310	Roof Drain Systems	COST PER SYSTEM		
		MAT.	INST.	TOTAL
1880	Roof drain, DWV PVC, 2″ diam., piping, 10′ high	110	400	510
1920	For each additional foot add	2.44	11.70	14.14
1960	3″ diam., 10′ high	147	485	632
2000	For each additional foot add	3.95	13.05	17
2040	4″ diam., 10′ high	202	545	747
2080	For each additional foot add	6	14.40	20.40
2120	5″ diam., 10′ high	375	585	960
2160	For each additional foot add	7.65	16.05	23.70
2200	6″ diam., 10′ high	510	695	1,205
2240	For each additional foot add	9.20	17.70	26.90
2280	8″ diam., 10′ high	995	1,100	2,095
2320	For each additional foot add	14.15	22.50	36.65
3940	C.I., soil, single hub, service wt., 2″ diam. piping, 10′ high	190	420	610
3980	For each additional foot add	4.72	10.95	15.67
4120	3″ diam., 10′ high	246	455	701
4160	For each additional foot add	6.50	11.50	18
4200	4″ diam., 10′ high	320	495	815
4240	For each additional foot add	9.10	12.55	21.65
4280	5″ diam., 10′ high	440	550	990
4320	For each additional foot add	11.15	14.15	25.30
4360	6″ diam., 10′ high	520	585	1,105
4400	For each additional foot add	13.60	14.70	28.30
4440	8″ diam., 10′ high	1,075	1,200	2,275
4480	For each additional foot add	21.50	25	46.50
6040	Steel galv. sch 40 threaded, 2″ diam. piping, 10′ high	197	410	607
6080	For each additional foot add	3.61	10.80	14.41
6120	3″ diam., 10′ high	360	585	945
6160	For each additional foot add	6.55	16.05	22.60
6200	4″ diam., 10′ high	530	755	1,285
6240	For each additional foot add	9.10	19.20	28.30
6280	5″ diam., 10′ high	755	955	1,710
6320	For each additional foot add	16.90	26.50	43.40
6360	6″ diam, 10′ high	1,175	1,175	2,350
6400	For each additional foot add	22	34.50	56.50
6440	8″ diam., 10′ high	2,225	1,600	3,825
6480	For each additional foot add	32	40	72

MECHANICAL

8

Reference—Minimum Plumbing Fixture Requirements

Type of Building/Use	Water Closets		Urinals		Lavatories		Bathtubs or Showers		Drinking Fountain	Other
	Persons	Fixtures	Persons	Fixtures	Persons	Fixtures	Persons	Fixtures	Fixtures	Fixtures
Assembly Halls										
Auditoriums	1-100	1	1-200	1	1-200	1			1 for each 1000 persons	1 service sink
Theater	101-200	2	201-400	2	201-400	2				
Public assembly	201-400	3	401-600	3	401-750	3				
	Over 400 add 1 fixt. for ea. 500 men; 1 fixt. for ea. 300 women		Over 600 add 1 fixture for each 300 men		Over 750 add 1 fixture for each 500 persons					
Assembly										
Public Worship	300 men	1	300 men	1	men	1			1	
	150 women	1			women	1				
Dormitories	Men: 1 for each 10 persons Women: 1 for each 8 persons		1 for each 25 men; over 150 add 1 fixture for each 50 men		1 for ea. 12 persons 1 separate dental lav. for each 50 persons recom.		1 for ea. 8 persons For women add 1 additional for each 30. Over 150 persons add 1 for each 20.		1 for each 75 persons	Laundry trays 1 for each 50 serv. sink 1 for ea. 100
Dwellings										
Apartments and homes	1 fixture for each unit				1 fixture for each unit		1 fixture for each unit			
Hospitals										
Indiv. Room		1				1		1		1 service sink per floor
Ward	8 persons	1			10 persons	1	20 persons	1	1 for 100 patients	
Waiting room		1				1				
Industrial										
Mfg. plants	1-10	1	0-30	1	1-100	1 for ea. 10	1 Shower for each 15 persons subject to excessive heat or occupational hazard		1 for each 75 persons	
Warehouses	11-25	2	31-80	2						
	26-50	3	81-160	3						
	51-75	4	161-240	4	over 100	1 for ea. 15				
	76-100	5								
	1 fixture for each additional 30 persons									
Public Buildings										
Businesses	1-15	1	Urinals may be provided in place of water closets but may not replace more than 1/3 required number of men's water closets		1-15	1			1 for each 75 persons	1 service sink per floor
Offices	16-35	2			16-35	2				
	36-55	3			36-60	3				
	56-80	4			61-90	4				
	81-110	5			91-125	5				
	111-150	6								
	1 fixture for ea. additional 40 persons				1 fixture for ea. additional 45 persons					
Schools										
Elementary	1 for ea. 30 boys 1 for ea. 25 girls		1 for ea. 25 boys		1 for ea. 35 boys 1 for ea. 35 girls		For gym or pool shower room 1/5 of a class		1 for each 40 pupils	
Schools										
Secondary	1 for ea. 40 boys 1 for ea. 30 girls		1 for ea. 25 boys		1 for ea. 40 boys 1 for ea. 40 girls		For gym or pool shower room 1/5 of a class		1 for each 50 pupils	

Important: See the Reference Section for critical supporting data - Location Factors & Historical Cost Indexes

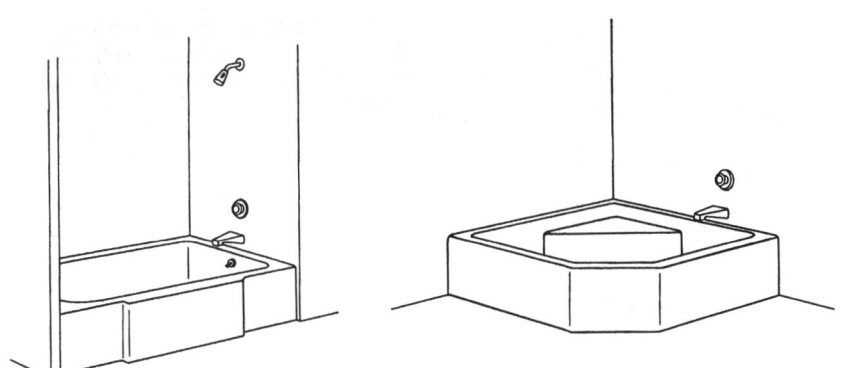

Recessed Bathtub Corner Bathtub

Systems are complete with trim and rough-in (supply, waste and vent) to connect to supply branches and waste mains.

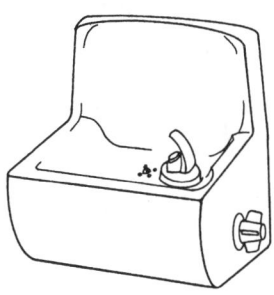

Wall Mounted, Low Back

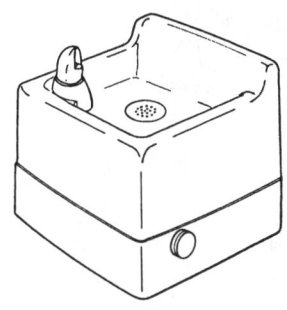

Wall Mounted, No Back

8.1-410	Bathtub Systems	COST EACH		
		MAT.	INST.	TOTAL
1960	Bathtub, recessed, P.E. on Cl., 42" x 37"	970	470	1,440
2000	48" x 42"	1,425	505	1,930
2040	72" x 36"	1,500	565	2,065
2080	Mat bottom, 5' long	565	490	1,055
2120	5'-6" long	1,175	505	1,680
2160	Corner, 48" x 42"	1,600	490	2,090
2200	Formed steel, enameled, 4'-6" long	445	455	900
2240	5' long	425	460	885

8.1-420	Drinking Fountain Systems	COST EACH		
		MAT.	INST.	TOTAL
1740	Drinking fountain, one bubbler, wall mounted			
1760	Non recessed			
1800	Bronze, no back	1,025	269	1,294
1840	Cast iron, enameled, low back	495	269	764
1880	Fiberglass, 12" back	565	269	834
1920	Stainless steel, no back	905	269	1,174
1960	Semi-recessed, poly marble	670	269	939
2040	Stainless steel	705	269	974
2080	Vitreous china	520	269	789
2120	Full recessed, poly marble	755	269	1,024
2200	Stainless steel	650	269	919
2240	Floor mounted, pedestal type, aluminum	485	365	850
2320	Bronze	1,125	365	1,490
2360	Stainless steel	1,125	365	1,490

MECHANICAL

8

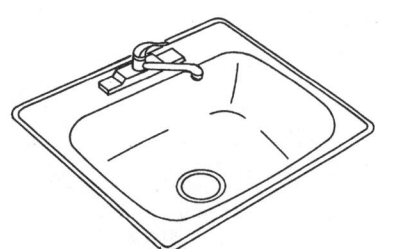

**Counter Top
Single Bowl**

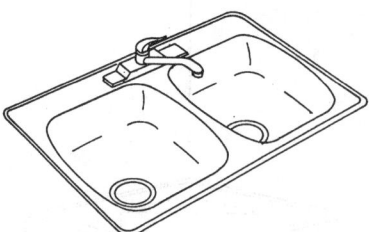

**Counter Top
Double Bowl**

Systems are complete with trim and
rough-in (supply, waste and vent) to
connect to supply branches and waste
mains.

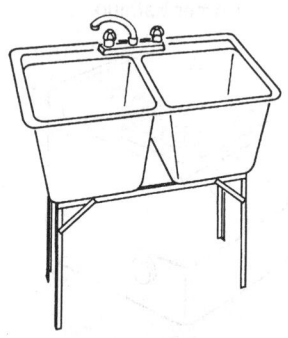

Single Compartment Sink

Double Compartment Sink

8.1-431	Kitchen Sink Systems	COST EACH		
		MAT.	INST.	TOTAL
1720	Kitchen sink w/trim, countertop, PE on CI, 24"x21", single bowl	350	445	795
1760	30" x 21" single bowl	425	445	870
1800	32" x 21" double bowl	470	480	950
1840	42" x 21" double bowl	630	490	1,120
1880	Stainless steel, 19" x 18" single bowl	390	445	835
1920	25" x 22" single bowl	415	445	860
1960	33" x 22" double bowl	560	480	1,040
2000	43" x 22" double bowl	630	490	1,120
2040	44" x 22" triple bowl	810	510	1,320
2080	44" x 24" corner double bowl	575	490	1,065
2120	Steel, enameled, 24" x 21" single bowl	229	445	674
2160	32" x 21" double bowl	255	480	735
2240	Raised deck, PE on CI, 32" x 21", dual level, double bowl	470	610	1,080
2280	42" x 21" dual level, triple bowl	740	665	1,405

8.1-432	Laundry Sink Systems	COST EACH		
		MAT.	INST.	TOTAL
1740	Laundry sink w/trim, PE on CI, black iron frame			
1760	24" x 20", single compartment	440	440	880
1840	48" x 21" double compartment	920	475	1,395
1920	Molded stone, on wall, 22" x 21" single compartment	242	440	682
1960	45"x 21" double compartment	350	475	825
2040	Plastic, on wall or legs, 18" x 23" single compartment	203	430	633
2080	20" x 24" single compartment	232	430	662
2120	36" x 23" double compartment	261	465	726
2160	40" x 24" double compartment	325	465	790
2200	Stainless steel, countertop, 22" x 17" single compartment	380	440	820
2240	19" x 22" single compartment	450	440	890
2280	33" x 22" double compartment	435	475	910

Systems are complete with trim and rough-in (supply, waste and vent) to connect to supply branches and waste mains.

Vanity Top

Wall Hung

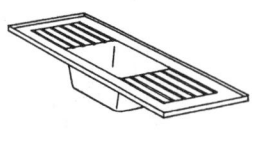

Laboratory Sink

Service Sink

8.1-433	Lavatory Systems	MAT.	INST.	TOTAL
		COST EACH		
1560	Lavatory w/trim, vanity top, PE on CI, 20" x 18", Vanity top by others.	296	410	706
1600	19" x 16" oval	289	410	699
1640	18" round	300	410	710
1680	Cultured marble, 19" x 17"	209	410	619
1720	25" x 19"	239	410	649
1760	Stainless, self-rimming, 25" x 22"	265	410	675
1800	17" x 22"	261	410	671
1840	Steel enameled, 20" x 17"	198	420	618
1880	19" round	196	420	616
1920	Vitreous china, 20" x 16"	300	430	730
1960	19" x 16"	300	430	730
2000	22" x 13"	310	430	740
2040	Wall hung, PE on CI, 18" x 15"	595	450	1,045
2080	19" x 17"	570	450	1,020
2120	20" x 18"	420	450	870
2160	Vitreous china, 18" x 15"	415	465	880
2200	19" x 17"	350	465	815
2240	24" x 20"	485	465	950
2241				
2243				

8.1-434	Laboratory Sink Systems	MAT.	INST.	TOTAL
		COST EACH		
1580	Laboratory sink w/trim, polyethylene, single bowl,			
1600	Double drainboard, 54" x 24" O.D.	910	570	1,480
1640	Single drainboard, 47" x 24"O.D.	945	570	1,515
1680	70" x 24" O.D.	965	570	1,535
1760	Flanged, 14-1/2" x 14-1/2" O.D.	350	515	865
1800	18-1/2" x 18-1/2" O.D.	315	515	830
1840	23-1/2" x 20-1/2" O.D.	365	515	880
1920	Polypropylene, cup sink, oval, 7" x 4" O.D.	163	455	618
1960	10" x 4-1/2" O.D.	174	455	629
1961				

8.1-434	Service Sink Systems	MAT.	INST.	TOTAL
		COST EACH		
4260	Service sink w/trim, PE on CI, corner floor, 28" x 28", w/rim guard	870	575	1,445
4300	Wall hung w/rim guard, 22" x 18"	780	670	1,450
4340	24" x 20"	820	670	1,490
4380	Vitreous china, wall hung 22" x 20"	835	670	1,505

MECHANICAL

8

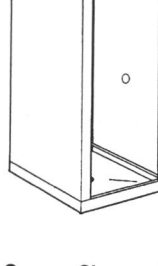
Square Shower Stall

Corner Angle Shower Stall

Systems are complete with trim and rough-in (supply, waste and vent) to connect to supply branches and waste mains.

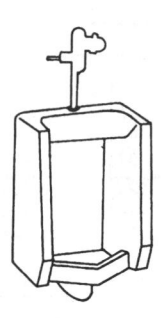
Wall Hung Urinal

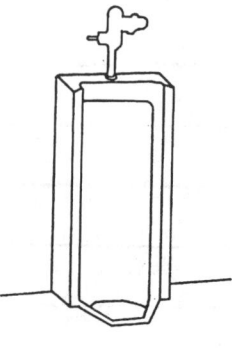
Stall Type Urinal

Wall Hung Water Cooler

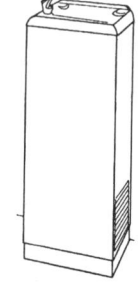
Floor Mounted Water Cooler

8.1-440	Shower Systems	COST EACH		
		MAT.	INST.	TOTAL
1560	Shower, stall, baked enamel, molded stone receptor, 30" square	475	685	1,160
1600	32" square	495	685	1,180
1640	Terrazzo receptor, 32" square	930	685	1,615
1680	36" square	990	725	1,715
1720	36" corner angle	910	725	1,635
1800	Fiberglass one piece, three walls, 32" square	530	625	1,155
1840	36" square	585	625	1,210
1880	Polypropylene, molded stone receptor, 30" square	485	685	1,170
1920	32" square	495	685	1,180
1960	Built-in head, arm, bypass, stops and handles	96	177	273

8.1-450	Urinal Systems	COST EACH		
		MAT.	INST.	TOTAL
2000	Urinal, vitreous china, wall hung	460	495	955
2040	Stall type	780	545	1,325

8.1-460	Water Cooler Systems	COST EACH		
		MAT.	INST.	TOTAL
1840	Water cooler, electric, wall hung, 8.2 GPH	595	345	940
1880	Dual height, 14.3 GPH	840	355	1,195
1920	Wheelchair type, 7.5 G.P.H.	1,300	345	1,645
1960	Semi recessed, 8.1 G.P.H.	790	345	1,135
2000	Full recessed, 8 G.P.H.	1,200	370	1,570
2040	Floor mounted, 14.3 G.P.H.	625	300	925
2080	Dual height, 14.3 G.P.H.	860	365	1,225
2120	Refrigerated compartment type, 1.5 G.P.H.	1,100	300	1,400

MECHANICAL 8

Important: See the Reference Section for critical supporting data - Location Factors & Historical Cost Indexes

One Piece Wall Hung Water Closet

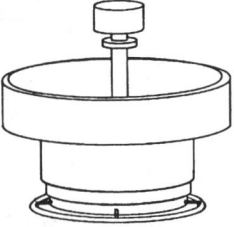

Circular Wash Fountain

Floor Mount Water Closet

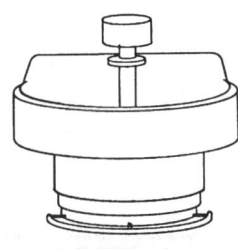

Semi-Circular Wash Fountain

Side By Side Water Closet Group

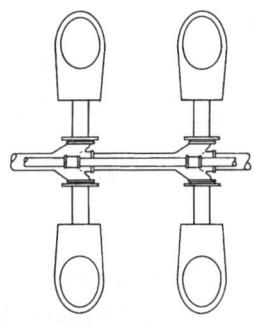

Back to Back Water Closet Group

8.1-470	Water Closet Systems	COST EACH		
		MAT.	INST.	TOTAL
1800	Water closet, vitreous china, elongated			
1840	Tank type, wall hung, one piece	900	400	1,300
1880	Close coupled two piece	570	400	970
1920	Floor mount, one piece	630	435	1,065
1960	One piece low profile	1,025	435	1,460
2000	Two piece close coupled	310	435	745
2040	Bowl only with flush valve			
2080	Wall hung	640	455	1,095
2120	Floor mount	505	440	945
2122				

8.1-510	Water Closets, Group	COST EACH		
		MAT.	INST.	TOTAL
1760	Water closets, battery mount, wall hung, side by side, first closet	670	470	1,140
1800	Each additional water closet, add	640	440	1,080
3000	Back to back, first pair of closets	1,225	630	1,855
3100	Each additional pair of closets, back to back	1,200	620	1,820

8.1-560	Group Wash Fountain Systems	COST EACH		
		MAT.	INST.	TOTAL
1740	Group wash fountain, precast terrazzo			
1760	Circular, 36" diameter	2,150	740	2,890
1800	54" diameter	2,625	810	3,435
1840	Semi-circular, 36" diameter	1,975	740	2,715
1880	54" diameter	2,375	810	3,185
1960	Stainless steel, circular, 36" diameter	2,475	685	3,160
2000	54" diameter	3,150	765	3,915
2040	Semi-circular, 36" diameter	2,175	685	2,860
2080	54" diameter	2,750	765	3,515
2160	Thermoplastic, circular, 36" diameter	1,900	560	2,460
2200	54" diameter	2,175	650	2,825
2240	Semi-circular, 36" diameter	1,750	560	2,310
2280	54" diameter	2,125	650	2,775

MECHANICAL

8

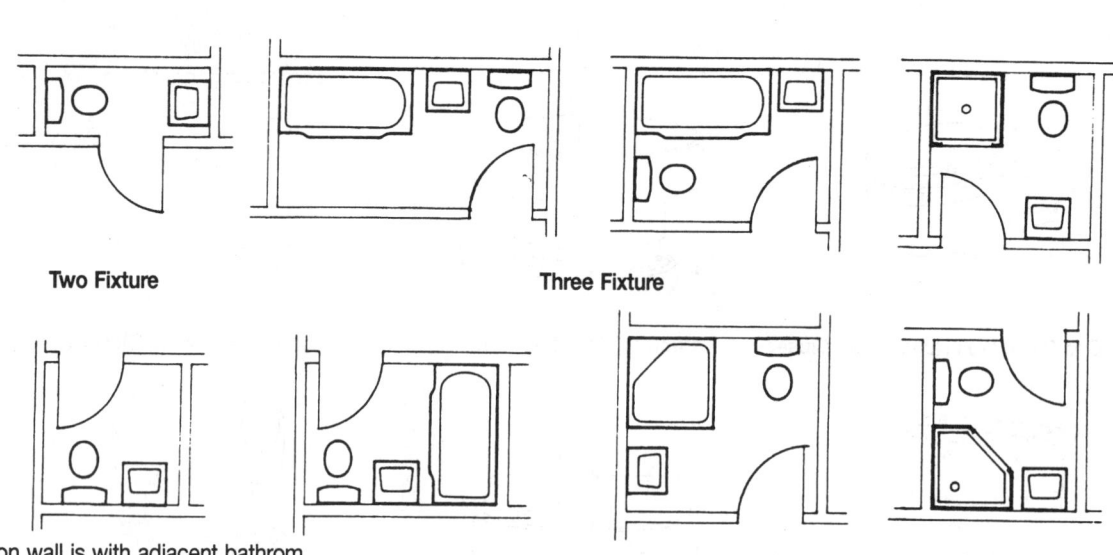

Two Fixture   Three Fixture

*Common wall is with adjacent bathrom

8.1-620	Two Fixture Bathroom, Two Wall Plumbing	COST EACH		
		MAT.	INST.	TOTAL
1180	Bathroom, lavatory & water closet, 2 wall plumbing, stand alone	840	1,125	1,965
1200	Share common plumbing wall*	785	970	1,755

8.1-620	Two Fixture Bathroom, One Wall Plumbing	COST EACH		
		MAT.	INST.	TOTAL
2220	Bathroom, lavatory & water closet, one wall plumbing, stand alone	800	1,000	1,800
2240	Share common plumbing wall*	715	860	1,575

8.1-630	Three Fixture Bathroom, One Wall Plumbing	COST EACH		
		MAT.	INST.	TOTAL
1150	Bathroom, three fixture, one wall plumbing			
1160	Lavatory, water closet & bathtub			
1170	Stand alone	1,250	1,325	2,575
1180	Share common plumbing wall *	1,075	950	2,025

8.1-630	Three Fixture Bathroom, Two Wall Plumbing	COST EACH		
		MAT.	INST.	TOTAL
2130	Bathroom, three fixture, two wall plumbing			
2140	Lavatory, water closet & bathtub			
2160	Stand alone	1,275	1,325	2,600
2180	Long plumbing wall common *	1,150	1,075	2,225
3610	Lavatory, bathtub & water closet			
3620	Stand alone	1,350	1,525	2,875
3640	Long plumbing wall common *	1,275	1,375	2,650
4660	Water closet, corner bathtub & lavatory			
4680	Stand alone	2,325	1,350	3,675
4700	Long plumbing wall common *	2,175	1,025	3,200
6100	Water closet, stall shower & lavatory			
6120	Stand alone	1,300	1,725	3,025
6140	Long plumbing wall common *	1,250	1,600	2,850
7060	Lavatory, corner stall shower & water closet			
7080	Stand alone	1,625	1,575	3,200
7100	Short plumbing wall common *	1,475	1,150	2,625

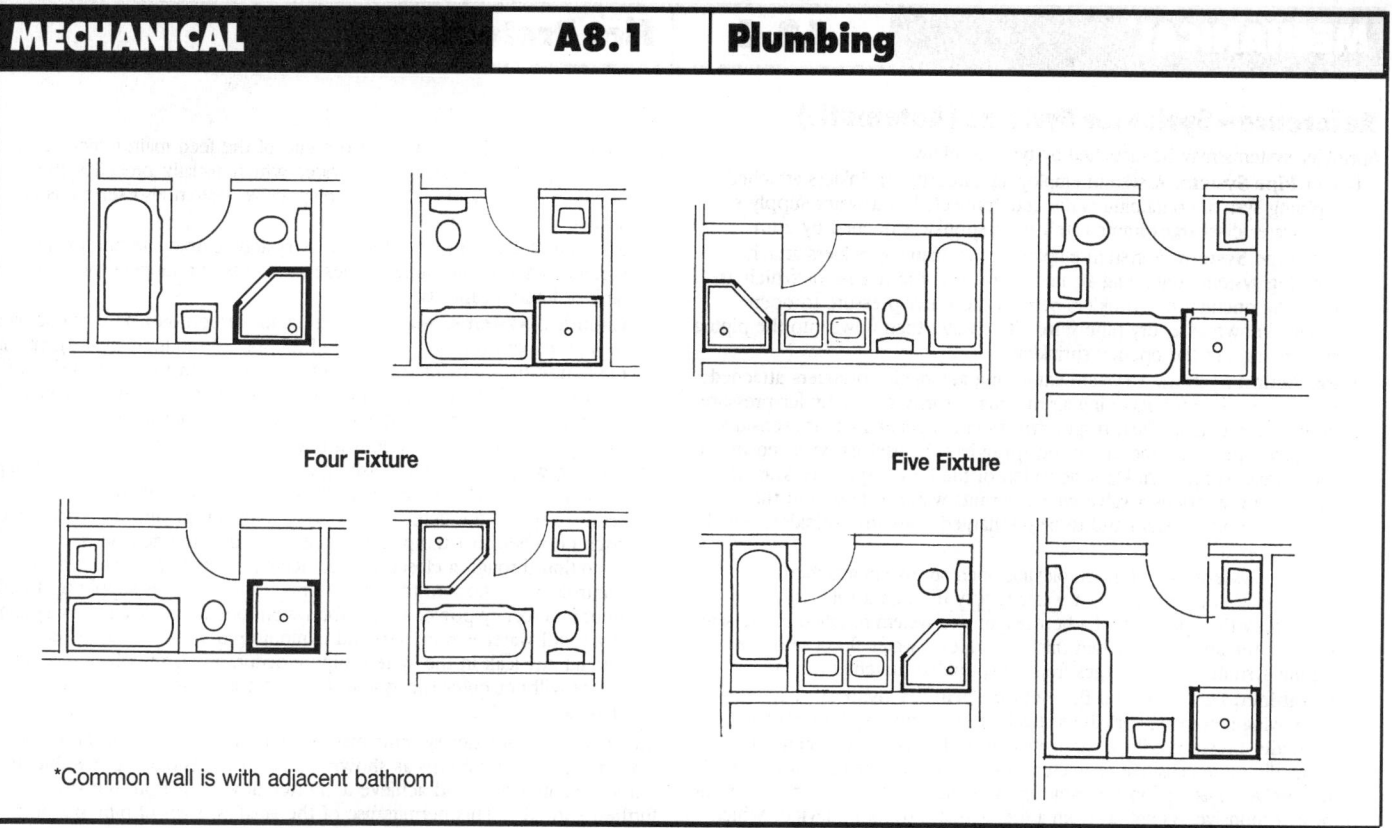

Four Fixture

Five Fixture

*Common wall is with adjacent bathrom

8.1-640	Four Fixture Bathroom, Two Wall Plumbing	COST EACH		
		MAT.	INST.	TOTAL
1140	Bathroom, four fixture, two wall plumbing			
1150	Bathtub, water closet, stall shower & lavatory			
1160	Stand alone	1,600	1,650	3,250
1180	Long plumbing wall common *	1,450	1,325	2,775
2260	Bathtub, lavatory, corner stall shower & water closet			
2280	Stand alone	2,000	1,700	3,700
2320	Long plumbing wall common *	1,850	1,350	3,200
3620	Bathtub, stall shower, lavatory & water closet			
3640	Stand alone	1,775	2,025	3,800
3660	Long plumbing wall (opp. door) common *	1,625	1,700	3,325
4680	Bathroom, four fixture, three wall plumbing			
4700	Bathtub, stall shower, lavatory & water closet			
4720	Stand alone	2,250	2,250	4,500
4761	Long plumbing wall (opposite door) common*	2,200	2,100	4,300

8.1-650	Five Fixture Bathroom, Two Wall Plumbing	COST EACH		
		MAT.	INST.	TOTAL
1320	Bathroom, five fixture, two wall plumbing			
1340	Bathtub, water closet, stall shower & two lavatories			
1360	Stand alone	2,200	2,525	4,725
1400	One short plumbing wall common *	2,050	2,200	4,250
2360	Bathroom, five fixture, three wall plumbing			
2380	Water closet, bathtub, two lavatories & stall shower			
2400	Stand alone	2,600	2,575	5,175
2440	One short plumbing wall common *	2,450	2,225	4,675
4080	Bathroom, five fixture, one wall plumbing			
4100	Bathtub, two lavatories, corner stall shower & water closet			
4120	Stand alone	2,500	2,325	4,825
4160	Share common wall *	2,125	1,600	3,725

MECHANICAL

8

Reference—Sprinkler Systems (Automatic)

Sprinkler systems may be classified by type as follows:

1. **Wet Pipe System.** A system employing automatic sprinklers attached to a piping system containing water and connected to a water supply so that water discharges immediately from sprinklers opened by a fire.

2. **Dry Pipe System.** A system employing automatic sprinklers attached to a piping system containing air under pressure, the release of which as from the opening of sprinklers permits the water pressure to open a valve known as a "dry pipe valve". The water then flows into the piping system and out the opened sprinklers.

3. **Pre-Action System.** A system employing automatic sprinklers attached to a piping system containing air that may or may not be under pressure, with a supplemental heat responsive system of generally more sensitive characteristics than the automatic sprinklers themselves, installed in the same areas as the sprinklers; actuation of the heat responsive system, as from a fire, opens a valve which permits water to flow into the sprinkler piping system and to be discharged from any sprinklers which may be open.

4. **Deluge System.** A system employing open sprinklers attached to a piping system connected to a water supply through a valve which is opened by the operation of a heat responsive system installed in the same areas as the sprinklers. When this valve opens, water flows into the piping system and discharges from all sprinklers attached thereto.

5. **Combined Dry Pipe and Pre-Action Sprinkler System.** A system employing automatic sprinklers attached to a piping system containing air under pressure with a supplemental heat responsive system of generally more sensitive characteristics than the automatic sprinklers themselves, installed in the same areas as the sprinklers; operation of the heat responsive system, as from a fire, actuates tripping devices which open dry pipe valves simultaneously and without loss of air pressure in the system. Operation of the heat responsive system also opens

approved air exhaust valves at the end of the feed main which facilita the filling of the system with water which usually precedes the opening of sprinklers. The heat responsive system also serves as an automatic fire alarm system.

6. **Limited Water Supply System.** A system employing automatic sprinklers and conforming to these standards but supplied by a press tank of limited capacity.

7. **Chemical Systems.** Systems using halon, carbon dioxide, dry chemic or high expansion foam as selected for special requirements. Agent m extinguish flames by chemically inhibiting flame propagation, suffocate flames by excluding oxygen, interrupting chemical action of oxygen uniting with fuel or sealing and cooling the combustion center.

8. **Firecycle System.** Firecycle is a fixed fire protection sprinkler system utilizing water as its extinguishing agent. It is a time delayed, recyclin preaction type which automatically shuts the water off when heat is reduced below the detector operating temperature and turns the wate back on when that temperature is exceeded. The sytem senses a fire condition through a closed circuit electrical detector system which controls water flow to the fire automatically. Batteries supply up to 90 hour emergency power supply for system operation. The piping system dry (until water is required) and is monitored with pressurized air. Should any leak in the system piping occur, an alarm will sound, but water will not enter the system until heat is sensed by a firecycle detector.

Area coverage sprinkler systems may be laid out and fed from the supply i any one of several patterns as shown below. It is desirable, if possible, to utilize a central feed and achieve a shorter flow path from the riser to th furthest sprinkler. This permits use of the smallest sizes of pipe possible with resulting savings.

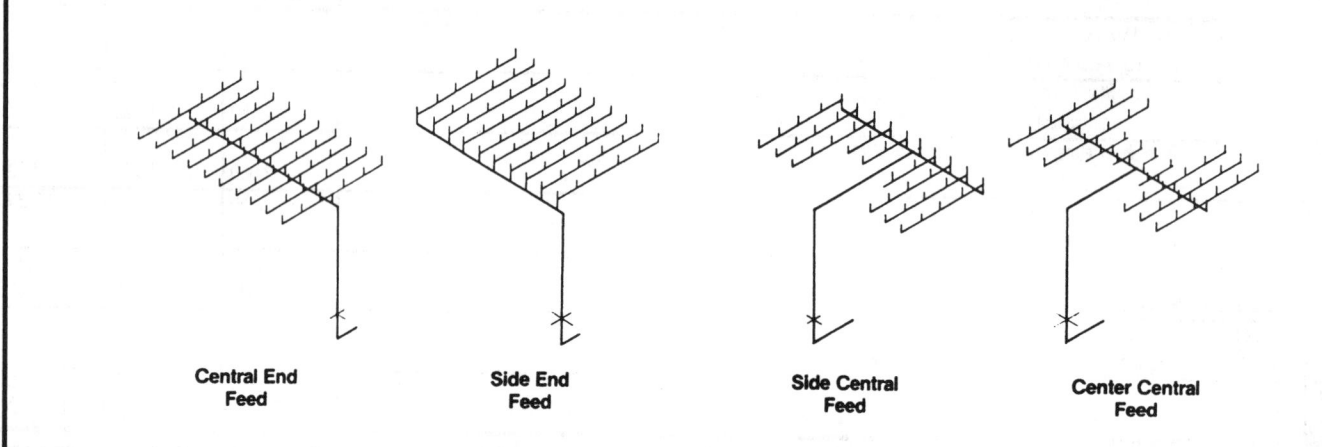

| Central End Feed | Side End Feed | Side Central Feed | Center Central Feed |

Reference—System Classification

System Classification

Rules for installation of sprinkler systems vary depending on the classification of occupancy falling into one of three categories as follows:

Light Hazard Occupancy

The protection area allotted per sprinkler should not exceed 200 S.F. with the maximum distance between lines and sprinklers on lines being 15'. The sprinklers do not need to be staggered. Branch lines should not exceed eight sprinklers on either side of a cross main. Each large area requiring more than 100 sprinklers and without a sub-dividing partition should be supplied by feed mains or risers sized for ordinary hazard occupancy.

Included in this group are:

Auditoriums	Museums
Churches	Nursing Homes
Clubs	Offices
Educational	Residential
Hospitals	Restaurants
Institutional	Schools
Libraries	Theaters
(except large stack rooms)	

Ordinary Hazard Occupancy

The protection area allotted per sprinkler shall not exceed 130 S.F. of noncombustible ceiling and 120 S.F. of combustible ceiling. The maximum allowable distance between sprinkler lines and sprinklers on line is 15'. Sprinklers shall be staggered if the distance between heads exceeds 12'. Branch lines should not exceed eight sprinklers on either side of a cross main.

Included in this group are:

Automotive garages	Electric generating stations
Bakeries	Feed mills
Beverage manufacturing	Grain elevators
Bleacheries	Ice manufacturing
Boiler houses	Laundries
Canneries	Machine shops
Cement plants	Mercantiles
Clothing factories	Paper mills
Cold storage warehouses	Printing and Publishing
Dairy products manufacturing	Shoe factories
Distilleries	Warehouses
Dry cleaning	Wood product assembly

Extra Hazard Occupancy

The protection area allotted per sprinkler shall not exceed 90 S.F. of noncombustible ceiling and 80 S.F. of combustible ceiling. The maximum allowable distance between lines and between sprinklers on lines is 12'. Sprinklers on alternate lines shall be staggered if the distance between sprinklers on lines exceeds 8'. Branch lines should not exceed six sprinklers on either side of a cross main.

Included in this group are:

Aircraft hangars	Paint shops
Chemical works	Shade cloth manufacturing
Explosives manufacturing	Solvent extracting
Linoleum manufacturing	Varnish works
Linseed oil mills	Volatile flammable
Oil refineries	liquid manufacturing & use

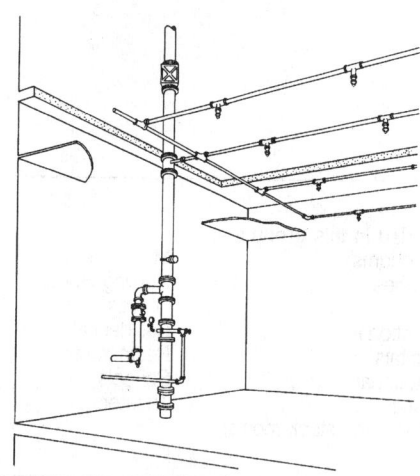

Wet Pipe System. A system employing automatic sprinklers attached to a piping system containing water and connected to a water supply so that water discharges immediately from sprinklers opened by heat from a fire.

All areas are assumed to be open.

8.2-110	Wet Pipe Sprinkler Systems	COST PER S.F.		
		MAT.	INST.	TOTAL
0520	Wet pipe sprinkler systems, steel, black, sch. 40 pipe			
0530	Light hazard, one floor, 500 S.F.	1.11	1.75	2.86
0560	1000 S.F.	1.53	1.82	3.35
0580	2000 S.F.	1.56	1.84	3.40
0600	5000 S.F.	.80	1.29	2.09
0620	10,000 S.F.	.52	1.08	1.60
0640	50,000 S.F.	.40	1.01	1.41
0660	Each additional floor, 500 S.F.	.58	1.49	2.07
0680	1000 S.F.	.49	1.39	1.88
0700	2000 S.F.	.45	1.25	1.70
0720	5000 S.F.	.35	1.08	1.43
0740	10,000 S.F.	.32	.98	1.30
0760	50,000 S.F.	.28	.79	1.07
1000	Ordinary hazard, one floor, 500 S.F.	1.23	1.88	3.11
1020	1000 S.F.	1.50	1.79	3.29
1040	2000 S.F.	1.63	1.94	3.57
1060	5000 S.F.	.89	1.38	2.27
1080	10,000 S.F.	.68	1.45	2.13
1100	50,000 S.F.	.62	1.42	2.04
1140	Each additional floor, 500 S.F.	.70	1.69	2.39
1160	1000 S.F.	.46	1.37	1.83
1180	2000 S.F.	.53	1.38	1.91
1200	5000 S.F.	.54	1.32	1.86
1220	10,000 S.F.	.49	1.35	1.84
1240	50,000 S.F.	.48	1.25	1.73
1500	Extra hazard, one floor, 500 S.F.	2.86	2.92	5.78
1520	1000 S.F.	1.83	2.53	4.36
1540	2000 S.F.	1.74	2.61	4.35
1560	5000 S.F.	1.17	2.24	3.41
1580	10,000 S.F.	1.13	2.15	3.28
1600	50,000 S.F.	1.21	2.01	3.22
1660	Each additional floor, 500 S.F.	.78	2.09	2.87
1680	1000 S.F.	.74	1.98	2.72
1700	2000 S.F.	.67	1.98	2.65
1720	5000 S.F.	.60	1.75	2.35
1740	10,000 S.F.	.71	1.61	2.32
1760	50,000 S.F.	.72	1.49	2.21
2020	Grooved steel, black sch. 40 pipe, light hazard, one floor, 2000 S.F.	1.56	1.58	3.14
2060	10,000 S.F.	.60	.97	1.57
2100	Each additional floor, 2000 S.F.	.46	1.02	1.48
2150	10,000 S.F.	.29	.83	1.12
2200	Ordinary hazard, one floor, 2000 S.F.	1.57	1.69	3.26

Important: See the Reference Section for critical supporting data - Location Factors & Historical Cost Indexes

8.2-110	**Wet Pipe Sprinkler Systems**	COST PER S.F.		
		MAT.	INST.	TOTAL
2250	10,000 S.F.	.59	1.20	1.79
2300	Each additional floor, 2000 S.F.	.47	1.13	1.60
2350	10,000 S.F.	.40	1.10	1.50
2400	Extra hazard, one floor, 2000 S.F.	1.72	2.15	3.87
2450	10,000 S.F.	.84	1.58	2.42
2500	Each additional floor, 2000 S.F.	.67	1.64	2.31
2550	10,000 S.F.	.58	1.39	1.97
3050	Grooved steel black sch. 10 pipe, light hazard, one floor, 2000 S.F.	1.53	1.57	3.10
3100	10,000 S.F.	.47	.91	1.38
3150	Each additional floor, 2000 S.F.	.43	1.01	1.44
3200	10,000 S.F.	.27	.81	1.08
3250	Ordinary hazard, one floor, 2000 S.F.	1.54	1.67	3.21
3300	10,000 S.F.	.56	1.19	1.75
3350	Each additional floor, 2000 S.F.	.44	1.11	1.55
3400	10,000 S.F.	.37	1.09	1.46
3450	Extra hazard, one floor, 2000 S.F.	1.70	2.12	3.82
3500	10,000 S.F.	.79	1.56	2.35
3550	Each additional floor, 2000 S.F.	.65	1.61	2.26
3600	10,000 S.F.	.57	1.38	1.95
4050	Copper tubing, type M, light hazard, one floor, 2000 S.F.	1.66	1.54	3.20
4100	10,000 S.F.	.66	.94	1.60
4150	Each additional floor, 2000 S.F.	.57	1	1.57
4200	10,000 S.F.	.46	.84	1.30
4250	Ordinary hazard, one floor, 2000 S.F.	1.74	1.73	3.47
4300	10,000 S.F.	.77	1.11	1.88
4350	Each additional floor, 2000 S.F.	.67	1.12	1.79
4400	10,000 S.F.	.56	.99	1.55
4450	Extra hazard, one floor, 2000 S.F.	1.95	2.13	4.08
4500	10,000 S.F.	1.44	1.71	3.15
4550	Each additional floor, 2000 S.F.	.90	1.62	2.52
4600	10,000 S.F.	1	1.51	2.51
5050	Copper tubing, type M, T-drill system, light hazard, one floor			
5060	2000 S.F.	1.67	1.43	3.10
5100	10,000 S.F.	.61	.77	1.38
5150	Each additional floor, 2000 S.F.	.58	.89	1.47
5200	10,000 S.F.	.41	.67	1.08
5250	Ordinary hazard, one floor, 2000 S.F.	1.67	1.47	3.14
5300	10,000 S.F.	.73	.99	1.72
5350	Each additional floor, 2000 S.F.	.57	.91	1.48
5400	10,000 S.F.	.54	.89	1.43
5450	Extra hazard, one floor, 2000 S.F.	1.75	1.75	3.50
5500	10,000 S.F.	1.12	1.23	2.35
5550	Each additional floor, 2000 S.F.	.73	1.27	2
5600	10,000 S.F.	.68	1.03	1.71

MECHANICAL

8

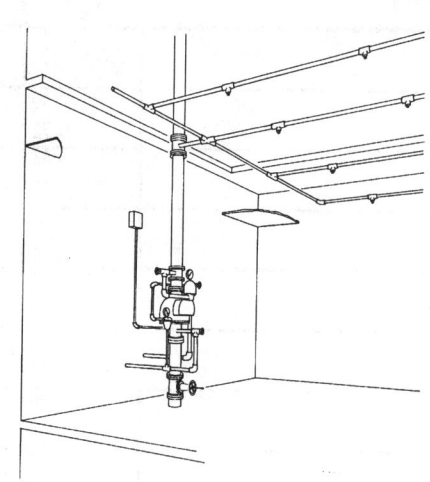

Dry Pipe System: A system employing automatic sprinklers attached to a piping system containing air under pressure, the release of which as from the opening of sprinklers permits the water pressure to open a valve known as a "dry pipe valve". The water then flows into the piping system and out the opened sprinklers.

All areas are assumed to be open.

8.2-120	Dry Pipe Sprinkler Systems	COST PER S.F.		
		MAT.	INST.	TOTAL
0520	Dry pipe sprinkler systems, steel, black, sch. 40 pipe			
0530	Light hazard, one floor, 500 S.F.	4.15	3.47	7.62
0560	1000 S.F.	2.23	2.01	4.24
0580	2000 S.F.	1.90	2.04	3.94
0600	5000 S.F.	.99	1.36	2.35
0620	10,000 S.F.	.67	1.13	1.80
0640	50,000 S.F.	.51	1.02	1.53
0660	Each additional floor, 500 S.F.	.81	1.67	2.48
0680	1000 S.F.	.60	1.37	1.97
0700	2000 S.F.	.57	1.27	1.84
0720	5000 S.F.	.47	1.09	1.56
0740	10,000 S.F.	.43	1	1.43
0760	50,000 S.F.	.40	.91	1.31
1000	Ordinary hazard, one floor, 500 S.F.	4.20	3.51	7.71
1020	1000 S.F.	2.25	2.03	4.28
1040	2000 S.F.	1.97	2.13	4.10
1060	5000 S.F.	1.12	1.46	2.58
1080	10,000 S.F.	.88	1.50	2.38
1100	50,000 S.F.	.80	1.47	2.27
1140	Each additional floor, 500 S.F.	.86	1.71	2.57
1160	1000 S.F.	.68	1.53	2.21
1180	2000 S.F.	.69	1.40	2.09
1200	5000 S.F.	.63	1.21	1.84
1220	10,000 S.F.	.57	1.19	1.76
1240	50,000 S.F.	.56	1.08	1.64
1500	Extra hazard, one floor, 500 S.F.	5.45	4.37	9.82
1520	1000 S.F.	3.12	3.13	6.25
1540	2000 S.F.	2.14	2.73	4.87
1560	5000 S.F.	1.28	2.03	3.31
1580	10,000 S.F.	1.31	1.96	3.27
1600	50,000 S.F.	1.38	1.81	3.19
1660	Each additional floor, 500 S.F.	1.01	2.12	3.13
1680	1000 S.F.	.97	2.01	2.98
1700	2000 S.F.	.90	2.01	2.91
1720	5000 S.F.	.79	1.75	2.54
1740	10,000 S.F.	.93	1.60	2.53
1760	50,000 S.F.	.95	1.50	2.45
2020	Grooved steel, black, sch. 40 pipe, light hazard, one floor, 2000 S.F.	1.86	1.77	3.63
2060	10,000 S.F.	.64	.98	1.62
2100	Each additional floor, 2000 S.F.	.58	1.04	1.62
2150	10,000 S.F.	.40	.85	1.25
2200	Ordinary hazard, one floor, 2000 S.F.	1.91	1.88	3.79

Important: See the Reference Section for critical supporting data - Location Factors & Historical Cost Indexes

8.2-120	Dry Pipe Sprinkler Systems	COST PER S.F.		
		MAT.	INST.	TOTAL
2250	10,000 S.F.	.79	1.25	2.04
2300	Each additional floor, 2000 S.F.	.63	1.15	1.78
2350	10,000 S.F.	.56	1.12	1.68
2400	Extra hazard, one floor, 2000 S.F.	2.13	2.35	4.48
2450	10,000 S.F.	1.12	1.64	2.76
2500	Each additional floor, 2000 S.F.	.90	1.67	2.57
2550	10,000 S.F.	.82	1.42	2.24
3050	Grooved steel black sch. 10 pipe, light hazard, one floor, 2000 S.F.	1.83	1.76	3.59
3100	10,000 S.F.	.62	.96	1.58
3150	Each additional floor, 2000 S.F.	.55	1.03	1.58
3200	10,000 S.F.	.38	.83	1.21
3250	Ordinary hazard, one floor, 2000 S.F.	1.88	1.86	3.74
3300	10,000 S.F.	.76	1.24	2
3350	Each additional floor, 2000 S.F.	.60	1.13	1.73
3400	10,000 S.F.	.53	1.11	1.64
3450	Extra hazard, one floor, 2000 S.F.	2.11	2.32	4.43
3500	10,000 S.F.	1.07	1.62	2.69
3550	Each additional floor, 2000 S.F.	.88	1.64	2.52
3600	10,000 S.F.	.81	1.41	2.22
4050	Copper tubing, type M, light hazard, one floor, 2000 S.F.	1.96	1.73	3.69
4100	10,000 S.F.	.81	.99	1.80
4150	Each additional floor, 2000 S.F.	.69	1.02	1.71
4200	10,000 S.F.	.57	.86	1.43
4250	Ordinary hazard, one floor, 2000 S.F.	2.08	1.92	4
4300	10,000 S.F.	.97	1.16	2.13
4350	Each additional floor, 2000 S.F.	.91	1.18	2.09
4400	10,000 S.F.	.72	1.01	1.73
4450	Extra hazard, one floor, 2000 S.F.	2.36	2.33	4.69
4500	10,000 S.F.	1.74	1.78	3.52
4550	Each additional floor, 2000 S.F.	1.13	1.65	2.78
4600	10,000 S.F.	1.24	1.54	2.78
5050	Copper tubing, type M, T-drill system, light hazard, one floor			
5060	2000 S.F.	1.97	1.62	3.59
5100	10,000 S.F.	.76	.82	1.58
5150	Each additional floor, 2000 S.F.	.70	.91	1.61
5200	10,000 S.F.	.52	.69	1.21
5250	Ordinary hazard, one floor, 2000 S.F.	2.01	1.66	3.67
5300	10,000 S.F.	.93	1.04	1.97
5350	Each additional floor, 2000 S.F.	.73	.93	1.66
5400	10,000 S.F.	.66	.87	1.53
5450	Extra hazard, one floor, 2000 S.F.	2.16	1.95	4.11
5500	10,000 S.F.	1.42	1.30	2.72
5550	Each additional floor, 2000 S.F.	.93	1.27	2.20
5600	10,000 S.F.	.92	1.06	1.98

MECHANICAL

8

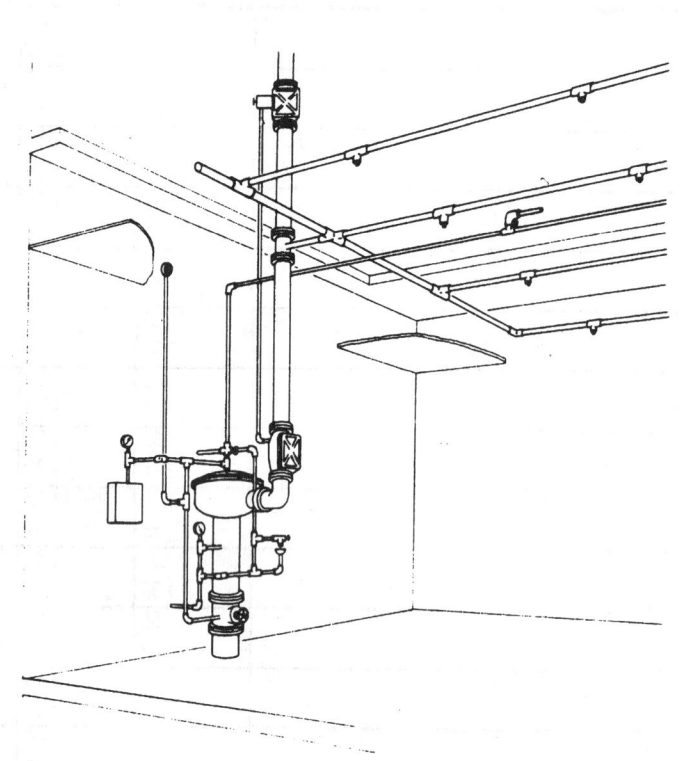

Preaction System: A system employing automatic sprinklers attached to a piping system containing air that may or may not be under pressure, with a supplemental heat responsive system of generally more sensitive characteristics than the automatic sprinklers themselves, installed in the same areas as the sprinklers. Actuation of the heat responsive system, as from a fire, opens a valve which permits water to flow into the sprinkler piping system and to be discharged from those sprinklers which were opened by heat from the fire.

All areas are assumed to be opened.

8.2-130	Preaction Sprinkler Systems	COST PER S.F.		
		MAT.	INST.	TOTAL
0520	Preaction sprinkler systems, steel, black, sch. 40 pipe			
0530	Light hazard, one floor, 500 S.F.	4.05	2.75	6.80
0560	1000 S.F.	2.27	2.05	4.32
0580	2000 S.F.	1.90	2.03	3.93
0600	5000 S.F.	.99	1.36	2.35
0620	10,000 S.F.	.67	1.12	1.79
0640	50,000 S.F.	.51	1.02	1.53
0660	Each additional floor, 500 S.F.	.87	1.48	2.35
0680	1000 S.F.	.63	1.36	1.99
0700	2000 S.F.	.60	1.26	1.86
0720	5000 S.F.	.47	1.09	1.56
0740	10,000 S.F.	.43	.99	1.42
0760	50,000 S.F.	.44	.94	1.38
1000	Ordinary hazard, one floor, 500 S.F.	1.54	2.02	3.56
1020	1000 S.F.	2.24	2.02	4.26
1040	2000 S.F.	2.04	2.13	4.17
1060	5000 S.F.	1.08	1.45	2.53
1080	10,000 S.F.	.83	1.49	2.32
1100	50,000 S.F.	.74	1.45	2.19
1140	Each additional floor, 500 S.F.	1	1.72	2.72
1160	1000 S.F.	.61	1.38	1.99
1180	2000 S.F.	.61	1.39	2
1200	5000 S.F.	.65	1.31	1.96
1220	10,000 S.F.	.60	1.36	1.96
1240	50,000 S.F.	.58	1.26	1.84
1500	Extra hazard, one floor, 500 S.F.	5.40	3.84	9.24
1520	1000 S.F.	2.96	2.84	5.80
1540	2000 S.F.	2.06	2.71	4.77
1560	5000 S.F.	1.29	2.19	3.48
1580	10,000 S.F.	1.27	2.17	3.44

8.2-130	Preaction Sprinkler Systems	COST PER S.F.		
		MAT.	**INST.**	**TOTAL**
1600	50,000 S.F.	1.32	2.03	3.35
1660	Each additional floor, 500 S.F.	1.08	2.12	3.20
1680	1000 S.F.	.89	1.99	2.88
1700	2000 S.F.	.82	1.99	2.81
1720	5000 S.F.	.72	1.76	2.48
1740	10,000 S.F.	.82	1.62	2.44
1760	50,000 S.F.	.80	1.48	2.28
2020	Grooved steel, black, sch. 40 pipe, light hazard, one floor, 2000 S.F.	1.89	1.76	3.65
2060	10,000 S.F.	.64	.97	1.61
2100	Each additional floor of 2000 S.F.	.61	1.03	1.64
2150	10,000 S.F.	.40	.84	1.24
2200	Ordinary hazard, one floor, 2000 S.F.	1.90	1.87	3.77
2250	10,000 S.F.	.74	1.24	1.98
2300	Each additional floor, 2000 S.F.	.62	1.14	1.76
2350	10,000 S.F.	.51	1.11	1.62
2400	Extra hazard, one floor, 2000 S.F.	2.05	2.33	4.38
2450	10,000 S.F.	1.01	1.63	2.64
2500	Each additional floor, 2000 S.F.	.82	1.65	2.47
2550	10,000 S.F.	.69	1.40	2.09
3050	Grooved steel, black, sch. 10 pipe light hazard, one floor, 2000 S.F.	1.86	1.75	3.61
3100	10,000 S.F.	.62	.95	1.57
3150	Each additional floor, 2000 S.F.	.58	1.02	1.60
3200	10,000 S.F.	.38	.82	1.20
3250	Ordinary hazard, one floor, 2000 S.F.	1.85	1.73	3.58
3300	10,000 S.F.	.63	1.22	1.85
3350	Each additional floor, 2000 S.F.	.59	1.12	1.71
3400	10,000 S.F.	.48	1.10	1.58
3450	Extra hazard, one floor, 2000 S.F.	2.03	2.30	4.33
3500	10,000 S.F.	.94	1.60	2.54
3550	Each additional floor, 2000 S.F.	.80	1.62	2.42
3600	10,000 S.F.	.68	1.39	2.07
4050	Copper tubing, type M, light hazard, one floor, 2000 S.F.	1.99	1.72	3.71
4100	10,000 S.F.	.81	.98	1.79
4150	Each additional floor, 2000 S.F.	.72	1.01	1.73
4200	10,000 S.F.	.49	.84	1.33
4250	Ordinary hazard, one floor, 2000 S.F.	2.07	1.91	3.98
4300	10,000 S.F.	.92	1.15	2.07
4350	Each additional floor, 2000 S.F.	.73	1.04	1.77
4400	10,000 S.F.	.61	.92	1.53
4450	Extra hazard, one floor, 2000 S.F.	2.28	2.31	4.59
4500	10,000 S.F.	1.59	1.75	3.34
4550	Each additional floor, 2000 S.F.	1.05	1.63	2.68
4600	10,000 S.F.	1.11	1.52	2.63
5050	Copper tubing, type M, T-drill system, light hazard, one floor			
5060	2000 S.F.	2	1.61	3.61
5100	10,000 S.F.	.76	.81	1.57
5150	Each additional floor, 2000 S.F.	.73	.90	1.63
5200	10,000 S.F.	.52	.68	1.20
5250	Ordinary hazard, one floor, 2000 S.F.	2	1.65	3.65
5300	10,000 S.F.	.88	1.03	1.91
5350	Each additional floor, 2000 S.F.	.72	.93	1.65
5400	10,000 S.F.	.65	.90	1.55
5450	Extra hazard, one floor, 2000 S.F.	2.08	1.93	4.01
5500	10,000 S.F.	1.27	1.27	2.54
5550	Each additional floor, 2000 S.F.	.85	1.25	2.10
5600	10,000 S.F.	.79	1.04	1.83

MECHANICAL

8

Deluge System: A system employing open sprinklers attached to a piping system connected to a water supply through a valve which is opened by the operation of a heat responsive system installed in the same areas as the sprinklers. When this valve opens, water flows into the piping system and discharges from all sprinklers attached thereto.

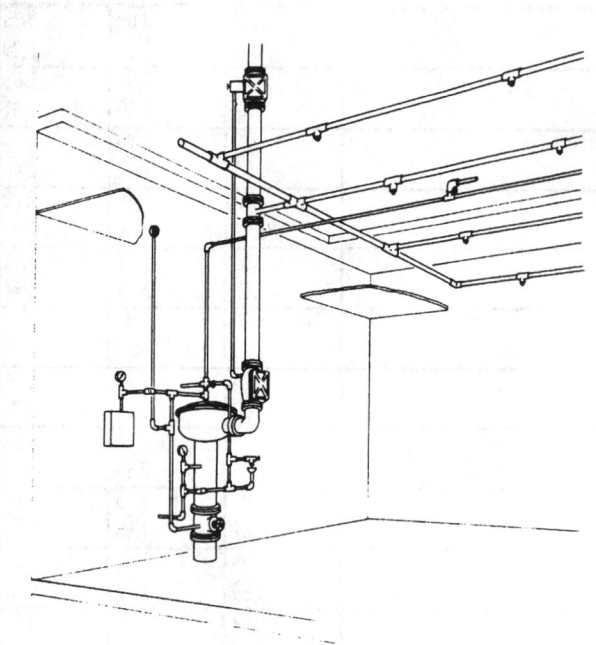

8.2-140	Deluge Sprinkler Systems	COST PER S.F.		
		MAT.	INST.	TOTAL
0520	Deluge sprinkler systems, steel, black, sch. 40 pipe			
0530	Light hazard, one floor, 500 S.F.	5.70	2.78	8.48
0560	1000 S.F.	3.07	1.98	5.05
0580	2000 S.F.	2.42	2.04	4.46
0600	5000 S.F.	1.21	1.36	2.57
0620	10,000 S.F.	.77	1.12	1.89
0640	50,000 S.F.	.53	1.02	1.55
0660	Each additional floor, 500 S.F.	.87	1.48	2.35
0680	1000 S.F.	.63	1.36	1.99
0700	2000 S.F.	.60	1.26	1.86
0720	5000 S.F.	.47	1.09	1.56
0740	10,000 S.F.	.43	.99	1.42
0760	50,000 S.F.	.44	.94	1.38
1000	Ordinary hazard, one floor, 500 S.F.	6.10	3.20	9.30
1020	1000 S.F.	3.06	2.04	5.10
1040	2000 S.F.	2.57	2.14	4.71
1060	5000 S.F.	1.30	1.45	2.75
1080	10,000 S.F.	.93	1.49	2.42
1100	50,000 S.F.	.80	1.47	2.27
1140	Each additional floor, 500 S.F.	1	1.72	2.72
1160	1000 S.F.	.61	1.38	1.99
1180	2000 S.F.	.61	1.39	2
1200	5000 S.F.	.59	1.20	1.79
1220	10,000 S.F.	.56	1.21	1.77
1240	50,000 S.F.	.54	1.16	1.70
1500	Extra hazard, one floor, 500 S.F.	7.05	3.87	10.92
1520	1000 S.F.	3.96	2.96	6.92
1540	2000 S.F.	2.59	2.72	5.31
1560	5000 S.F.	1.41	2.01	3.42
1580	10,000 S.F.	1.30	1.98	3.28
1600	50,000 S.F.	1.34	1.86	3.20
1660	Each additional floor, 500 S.F.	1.08	2.12	3.20
1680	1000 S.F.	.89	1.99	2.88
1700	2000 S.F.	.82	1.99	2.81

8.2-140	Deluge Sprinkler Systems	COST PER S.F.		
		MAT.	INST.	TOTAL
1720	5000 S.F.	.72	1.76	2.48
1740	10,000 S.F.	.84	1.68	2.52
1760	50,000 S.F.	.86	1.57	2.43
2000	Grooved steel, black, sch. 40 pipe, light hazard, one floor			
2020	2000 S.F.	2.43	1.77	4.20
2060	10,000 S.F.	.75	.98	1.73
2100	Each additional floor, 2,000 S.F.	.61	1.03	1.64
2150	10,000 S.F.	.40	.84	1.24
2200	Ordinary hazard, one floor, 2000 S.F.	1.90	1.87	3.77
2250	10,000 S.F.	.84	1.24	2.08
2300	Each additional floor, 2000 S.F.	.62	1.14	1.76
2350	10,000 S.F.	.51	1.11	1.62
2400	Extra hazard, one floor, 2000 S.F.	2.58	2.34	4.92
2450	10,000 S.F.	1.12	1.63	2.75
2500	Each additional floor, 2000 S.F.	.82	1.65	2.47
2550	10,000 S.F.	.69	1.40	2.09
3000	Grooved steel, black, sch. 10 pipe, light hazard, one floor			
3050	2000 S.F.	2.16	1.68	3.84
3100	10,000 S.F.	.72	.95	1.67
3150	Each additional floor, 2000 S.F.	.58	1.02	1.60
3200	10,000 S.F.	.38	.82	1.20
3250	Ordinary hazard, one floor, 2000 S.F.	2.40	1.86	4.26
3300	10,000 S.F.	.73	1.22	1.95
3350	Each additional floor, 2000 S.F.	.59	1.12	1.71
3400	10,000 S.F.	.48	1.10	1.58
3450	Extra hazard, one floor, 2000 S.F.	2.56	2.31	4.87
3500	10,000 S.F.	1.04	1.60	2.64
3550	Each additional floor, 2000 S.F.	.80	1.62	2.42
3600	10,000 S.F.	.68	1.39	2.07
4000	Copper tubing, type M, light hazard, one floor			
4050	2000 S.F.	2.52	1.73	4.25
4100	10,000 S.F.	.91	.98	1.89
4150	Each additional floor, 2000 S.F.	.72	1.01	1.73
4200	10,000 S.F.	.49	.84	1.33
4250	Ordinary hazard, one floor, 2000 S.F.	2.60	1.92	4.52
4300	10,000 S.F.	1.02	1.15	2.17
4350	Each additional floor, 2000 S.F.	.73	1.04	1.77
4400	10,000 S.F.	.61	.92	1.53
4450	Extra hazard, one floor, 2000 S.F.	2.81	2.32	5.13
4500	10,000 S.F.	1.72	1.76	3.48
4550	Each additional floor, 2000 S.F.	1.05	1.63	2.68
4600	10,000 S.F.	1.11	1.52	2.63
5000	Copper tubing, type M, T-drill system, light hazard, one floor			
5050	2000 S.F.	2.53	1.62	4.15
5100	10,000 S.F.	.86	.81	1.67
5150	Each additional floor, 2000 S.F.	.71	.91	1.62
5200	10,000 S.F.	.52	.68	1.20
5250	Ordinary hazard, one floor, 2000 S.F.	2.53	1.66	4.19
5300	10,000 S.F.	.98	1.03	2.01
5350	Each additional floor, 2000 S.F.	.72	.92	1.64
5400	10,000 S.F.	.65	.90	1.55
5450	Extra hazard, one floor, 2000 S.F.	2.61	1.94	4.55
5500	10,000 S.F.	1.37	1.27	2.64
5550	Each additional floor, 2000 S.F.	.85	1.25	2.10
5600	10,000 S.F.	.79	1.04	1.83

MECHANICAL

8

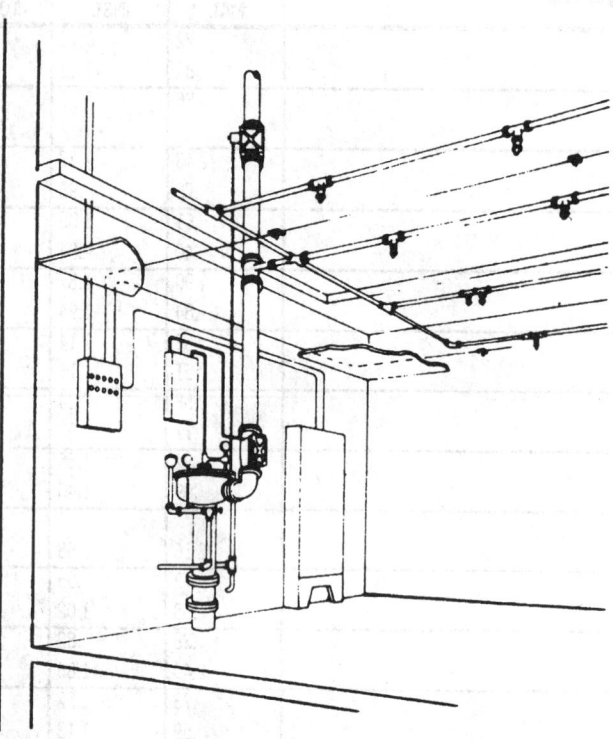

Firecycle is a fixed fire protection sprinkler system utilizing water as its extinguishing agent. It is a time delayed, recycling, preaction type which automatically shuts the water off when heat is reduced below the detector operating temperature and turns the water back on when that temperature is exceeded.

The system senses a fire condition through a closed circuit electrical detector system which controls water flow to the fire automatically. Batteries supply up to 90 hours emergency power supply for system operation. The piping system is dry (until water is required) and is monitored with pressurized air. Should any leak in the system piping occur, an alarm will sound, but water will not enter the system until heat is sensed by a Firecycle detector.

8.2-150	Firecycle Sprinkler Systems	COST PER S.F.		
		MAT.	**INST.**	**TOTAL**
0520	Firecycle sprinkler systems, steel black sch. 40 pipe			
0530	Light hazard, one floor, 500 S.F.	16.20	5.45	21.65
0560	1000 S.F.	8.30	3.41	11.71
0580	2000 S.F.	4.92	2.50	7.42
0600	5000 S.F.	2.22	1.55	3.77
0620	10,000 S.F.	1.30	1.21	2.51
0640	50,000 S.F.	.67	1.03	1.70
0660	Each additional floor of 500 S.F.	.95	1.49	2.44
0680	1000 S.F.	.67	1.37	2.04
0700	2000 S.F.	.55	1.26	1.81
0720	5000 S.F.	.50	1.09	1.59
0740	10,000 S.F.	.47	.99	1.46
0760	50,000 S.F.	.48	.94	1.42
1000	Ordinary hazard, one floor, 500 S.F.	16.35	5.70	22.05
1020	1000 S.F.	8.30	3.38	11.68
1040	2000 S.F.	4.98	2.59	7.57
1060	5000 S.F.	2.31	1.64	3.95
1080	10,000 S.F.	1.46	1.58	3.04
1100	50,000 S.F.	1.03	1.64	2.67
1140	Each additional floor, 500 S.F.	1.08	1.73	2.81
1160	1000 S.F.	.65	1.39	2.04
1180	2000 S.F.	.68	1.28	1.96
1200	5000 S.F.	.62	1.20	1.82
1220	10,000 S.F.	.56	1.18	1.74
1240	50,000 S.F.	.56	1.10	1.66
1500	Extra hazard, one floor, 500 S.F.	17.50	6.55	24.05
1520	1000 S.F.	9	4.20	13.20
1540	2000 S.F.	5.10	3.18	8.28
1560	5000 S.F.	2.42	2.20	4.62
1580	10,000 S.F.	1.87	2.24	4.11
1600	50,000 S.F.	1.61	2.41	4.02
1660	Each additional floor, 500 S.F.	1.16	2.13	3.29

MECHANICAL 8

		COST PER S.F.		
		MAT.	INST.	TOTAL
1680	1000 S.F.	.93	2	2.93
1700	2000 S.F.	.86	2	2.86
1720	5000 S.F.	.75	1.76	2.51
1740	10,000 S.F.	.86	1.62	2.48
1760	50,000 S.F.	.87	1.51	2.38
2020	Grooved steel, black, sch. 40 pipe, light hazard, one floor			
2030	2000 S.F.	4.91	2.23	7.14
2060	10,000 S.F.	1.38	1.49	2.87
2100	Each additional floor, 2000 S.F.	.65	1.04	1.69
2150	10,000 S.F.	.44	.84	1.28
2200	Ordinary hazard, one floor, 2000 S.F.	4.92	2.34	7.26
2250	10,000 S.F.	1.47	1.40	2.87
2300	Each additional floor, 2000 S.F.	.66	1.15	1.81
2350	10,000 S.F.	.55	1.11	1.66
2400	Extra hazard, one floor, 2000 S.F.	5.05	2.80	7.85
2450	10,000 S.F.	1.61	1.70	3.31
2500	Each additional floor, 2000 S.F.	.86	1.66	2.52
2550	10,000 S.F.	.73	1.40	2.13
3050	Grooved steel, black, sch. 10 pipe light hazard, one floor,			
3060	2000 S.F.	4.88	2.22	7.10
3100	10,000 S.F.	1.25	1.04	2.29
3150	Each additional floor, 2000 S.F.	.62	1.03	1.65
3200	10,000 S.F.	.42	.82	1.24
3250	Ordinary hazard, one floor, 2000 S.F.	4.89	2.32	7.21
3300	10,000 S.F.	1.34	1.32	2.66
3350	Each additional floor, 2000 S.F.	.63	1.13	1.76
3400	10,000 S.F.	.52	1.10	1.62
3450	Extra hazard, one floor, 2000 S.F.	5.05	2.77	7.82
3500	10,000 S.F.	1.56	1.68	3.24
3550	Each additional floor, 2000 S.F.	.84	1.63	2.47
3600	10,000 S.F.	.72	1.39	2.11
4060	Copper tubing, type M, light hazard, one floor, 2000 S.F.	5	2.19	7.19
4100	10,000 S.F.	1.44	1.07	2.51
4150	Each additional floor, 2000 S.F.	.76	1.02	1.78
4200	10,000 S.F.	.61	.85	1.46
4250	Ordinary hazard, one floor, 2000 S.F.	5.10	2.38	7.48
4300	10,000 S.F.	1.55	1.24	2.79
4350	Each additional floor, 2000 S.F.	.77	1.05	1.82
4400	10,000 S.F.	.63	.91	1.54
4450	Extra hazard, one floor, 2000 S.F.	5.30	2.78	8.08
4500	10,000 S.F.	2.26	1.86	4.12
4550	Each additional floor, 2000 S.F.	1.09	1.64	2.73
4600	10,000 S.F.	1.15	1.52	2.67
5060	Copper tubing, type M, T-drill system, light hazard, one floor 2000 S.F.	5	2.08	7.08
5100	10,000 S.F.	1.39	.90	2.29
5150	Each additional floor, 2000 S.F.	.82	.96	1.78
5200	10,000 S.F.	.56	.68	1.24
5250	Ordinary hazard, one floor, 2000 S.F.	5	2.12	7.12
5300	10,000 S.F.	1.51	1.12	2.63
5350	Each additional floor, 2000 S.F.	.76	.93	1.69
5400	10,000 S.F.	.69	.90	1.59
5450	Extra hazard, one floor, 2000 S.F.	5.10	2.40	7.50
5500	10,000 S.F.	1.89	1.35	3.24
5550	Each additional floor, 2000 S.F.	.89	1.26	2.15
5600	10,000 S.F.	.83	1.04	1.87

MECHANICAL

8

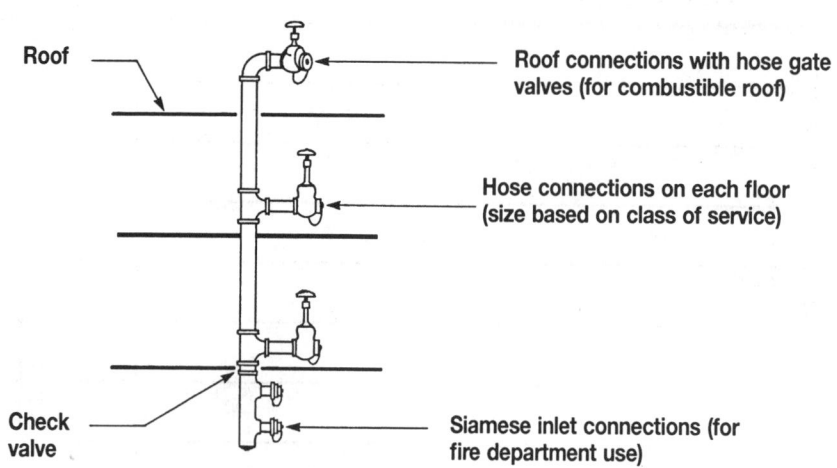

Roof — Roof connections with hose gate valves (for combustible roof)

Hose connections on each floor (size based on class of service)

Check valve — Siamese inlet connections (for fire department use)

8.2-310	Wet Standpipe Risers, Class I	COST PER FLOOR		
		MAT.	INST.	TOTAL
0550	Wet standpipe risers, Class I, steel black sch. 40, 10' height			
0560	4" diameter pipe, one floor	1,900	1,950	3,850
0580	Additional floors	555	620	1,175
0600	6" diameter pipe, one floor	3,350	3,400	6,750
0620	Additional floors	945	965	1,910
0640	8" diameter pipe, one floor	4,875	4,100	8,975
0660	Additional floors	1,250	1,175	2,425

8.2-310	Wet Standpipe Risers, Class II	COST PER FLOOR		
		MAT.	INST.	TOTAL
1030	Wet standpipe risers, Class II, steel black sch. 40, 10' height			
1040	2" diameter pipe, one floor	800	695	1,495
1060	Additional floors	284	269	553
1080	2-1/2" diameter pipe, one floor	1,025	1,025	2,050
1100	Additional floors	310	315	625

8.2-310	Wet Standpipe Risers, Class III	COST PER FLOOR		
		MAT.	INST.	TOTAL
1530	Wet standpipe risers, Class III, steel black sch. 40, 10' height			
1540	4" diameter pipe, one floor	1,975	1,950	3,925
1560	Additional floors	450	520	970
1580	6" diameter pipe, one floor	3,425	3,400	6,825
1600	Additional floors	980	965	1,945
1620	8" diameter pipe, one floor	4,950	4,100	9,050
1640	Additional floors	1,275	1,175	2,450

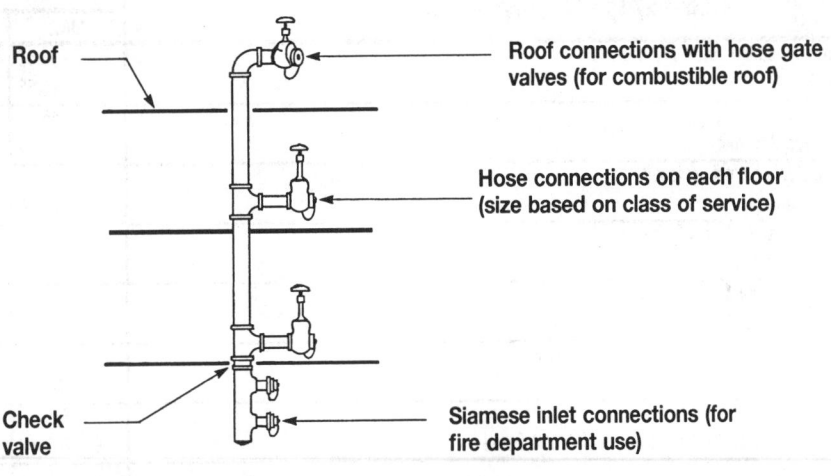

- Roof
- Roof connections with hose gate valves (for combustible roof)
- Hose connections on each floor (size based on class of service)
- Check valve
- Siamese inlet connections (for fire department use)

8.2-320	Dry Standpipe Risers, Class I	COST PER FLOOR		
		MAT.	INST.	TOTAL
0530	Dry standpipe riser, Class I, steel black sch. 40, 10' height			
0540	4" diameter pipe, one floor	1,275	1,550	2,825
0560	Additional floors	455	560	1,015
0580	6" diameter pipe, one floor	2,600	2,700	5,300
0600	Additional floors	850	910	1,760
0620	8" diameter pipe, one floor	3,800	3,250	7,050
0640	Additional floors	1,150	1,100	2,250

8.2-320	Dry Standpipe Risers, Class II	COST PER FLOOR		
		MAT.	INST.	TOTAL
1030	Dry standpipe risers, Class II, steel black sch. 40, 10' height			
1040	2" diameter pipe, one floor	760	725	1,485
1060	Additional floor	251	236	487
1080	2-1/2" diameter pipe, one floor	845	845	1,690
1100	Additional floors	278	280	558

8.2-320	Dry Standpipe Risers, Class III	COST PER FLOOR		
		MAT.	INST.	TOTAL
1530	Dry standpipe risers, Class III, steel black sch. 40, 10' height			
1540	4" diameter pipe, one floor	1,300	1,550	2,850
1560	Additional floors	365	505	870
1580	6" diameter pipe, one floor	2,625	2,700	5,325
1600	Additional floors	885	910	1,795
1620	8" diameter pipe, one floor	3,850	3,250	7,100
1640	Additional floor	1,200	1,100	2,300

MECHANICAL

8

8.2-390	Standpipe Equipment	COST EACH		
		MAT.	INST.	TOTAL
0100	Adapters, reducing, 1 piece, FxM, hexagon, cast brass, 2-1/2" x 1-1/2"	36.50		36.5
0200	Pin lug, 1-1/2" x 1"	11.55		11.5
0250	3" x 2-1/2"	45		45
0300	For polished chrome, add 75% mat.			
0400	Cabinets, D.S. glass in door, recessed, steel box, not equipped			
0500	Single extinguisher, steel door & frame	72.50	87.50	160
0550	Stainless steel door & frame	169	87.50	256.5
0600	Valve, 2-1/2" angle, steel door & frame	80	58.50	138.5
0650	Aluminum door & frame	123	58.50	181.5
0700	Stainless steel door & frame	172	58.50	230.5
0750	Hose rack assy, 2-1/2" x 1-1/2" valve & 100' hose, steel door & frame	162	117	279
0800	Aluminum door & frame	280	117	397
0850	Stainless steel door & frame	385	117	502
0900	Hose rack assy,& extinguisher,2-1/2"x1-1/2" valve & hose,steel door & frame	203	140	343
0950	Aluminum	355	140	495
1000	Stainless steel	450	140	590
1550	Compressor, air, dry pipe system, automatic, 200 gal., 1/3 H.P.	770	299	1,069
1600	520 gal., 1 H.P.	805	299	1,104
1650	Alarm, electric pressure switch (circuit closer)	120	14.95	134.95
2500	Couplings, hose, rocker lug, cast brass, 1-1/2"	28		28
2550	2-1/2"	58.50		58.50
3000	Escutcheon plate, for angle valves, polished brass, 1-1/2"	13.75		13.75
3050	2-1/2"	34		34
3500	Fire pump, electric, w/controller, fittings, relief valve			
3550	4" pump, 30 H.P., 500 G.P.M.	13,900	2,175	16,075
3600	5" pump, 40 H.P., 1000 G.P.M.	20,300	2,475	22,775
3650	5" pump, 100 H.P., 1000 G.P.M.	22,800	2,750	25,550
3700	For jockey pump system, add	2,500	350	2,850
5000	Hose, per linear foot, synthetic jacket, lined,			
5100	300 lb. test, 1-1/2" diameter	2.11		2.11
5150	2-1/2" diameter	3.41		3.41
5200	500 lb. test, 1-1/2" diameter	2.50		2.50
5250	2-1/2" diameter	4.30		4.30
5500	Nozzle, plain stream, polished brass, 1-1/2" x 10"	32		32
5550	2-1/2" x 15" x 13/16" or 1-1/2"	116		116
5600	Heavy duty combination adjustable fog and straight stream w/handle 1-1/2"	253		253
5650	2-1/2" direct connection	360		360
6000	Rack, for 1-1/2" diameter hose 100 ft. long, steel	32.50	35	67.50
6050	Brass	51.50	35	86.50
6500	Reel, steel, for 50 ft. long 1-1/2" diameter hose	87	50	137
6550	For 75 ft. long 2-1/2" diameter hose	141	50	191
7050	Siamese, w/plugs & chains, polished brass, sidewalk, 4" x 2-1/2" x 2-1/2"	299	280	579
7100	6" x 2-1/2" x 2-1/2"	455	350	805
7200	Wall type, flush, 4" x 2-1/2" x 2-1/2"	345	140	485
7250	6" x 2-1/2" x 2-1/2"	470	152	622
7300	Projecting, 4" x 2-1/2" x 2-1/2"	315	140	455
7350	6" x 2-1/2" x 2-1/2"	515	152	667
7400	For chrome plate, add 15% mat.			
8000	Valves, angle, wheel handle, 300 Lb., rough brass, 1-1/2"	35	32.50	67.50
8050	2-1/2"	65.50	55.50	121
8100	Combination pressure restricting, 1-1/2"	54.50	32.50	87
8150	2-1/2"	116	55.50	171.50
8200	Pressure restricting, adjustable, satin brass, 1-1/2"	63	32.50	95.50
8250	2-1/2"	105	55.50	160.50
8300	Hydrolator, vent and drain, rough brass, 1-1/2"	33.50	32.50	66
8350	2-1/2"	96.50	55.50	152
8400	Cabinet assy, incls. 2-1/2" valve, adapter, rack, hose, nozzle & hydrolator	670	263	933

MECHANICAL 8

Important: See the Reference Section for critical supporting data - Location Factors & Historical Cost Indexes

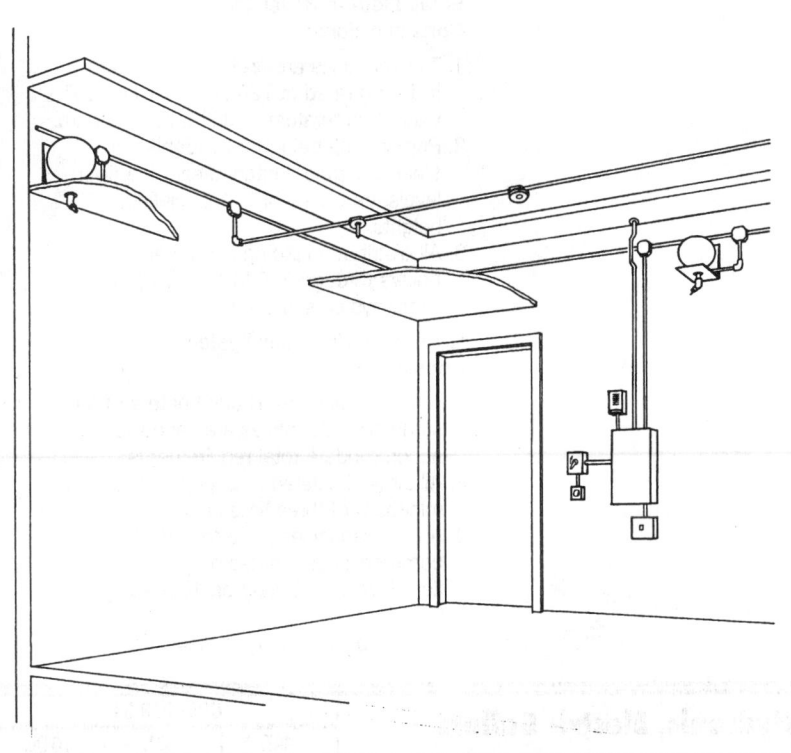

General: Automatic fire protection (suppression) systems other than water sprinklers may be desired for special environments, high risk areas, isolated locations or unusual hazards. Some typical applications would include:

1. Paint dip tanks
2. Securities vaults
3. Electronic data processing
4. Tape and data storage
5. Transformer rooms
6. Spray booths
7. Petroleum storage
8. High rack storage

Piping and wiring costs are highly variable and are not included.

8.2-810	FM200 Fire Suppression	COST EACH		
		MAT.	INST.	TOTAL
0020	Detectors with brackets			
0040	Fixed temperature heat detector	31	46	77
0060	Rate of temperature rise detector	37	46	83
0080	Ion detector (smoke) detector	82.50	59.50	142
0200	Extinguisher agent			
0240	200 lb FM200, container	6,375	173	6,548
0280	75 lb carbon dioxide cylinder	1,050	115	1,165
0320	Dispersion nozzle			
0340	FM200 1-1/2" dispersion nozzle	55	27.50	82.50
0380	Carbon dioxide 3" x 5" dispersion nozzle	55	21.50	76.50
0420	Control station			
0440	Single zone control station with batteries	1,525	370	1,895
0470	Multizone (4) control station with batteries	2,700	735	3,435
0500	Electric mechanical release	127	188	315
0550	Manual pull station	49.50	64	113.50
0640	Battery standby power 10" x 10" x 17"	760	92	852
0740	Bell signalling device	54.50	46	100.50

8.2-810	FM200 Systems	COST PER C.F.		
		MAT.	INST.	TOTAL
0820	Average FM200 system, minimum			1.38
0840	Maximum			2.75

MECHANICAL

8

Small Electric Boiler Systems Considerations:

1. Terminal units are fin tube baseboard radiation rated at 720 BTU/hr with 200° water temperature or 820 BTU/hr steam.
2. Primary use being for residential or smaller supplementary areas, the floor levels are based on 7-1/2″ ceiling heights.
3. All distribution piping is copper for boilers through 205 MBH. All piping for larger systems is steel pipe.

Large Electric Boiler System Considerations:

1. Terminal units are all unit heaters of the same size. Quantities are varied to accommodate total requirements.
2. All air is circulated through the heaters a minimum of three times per hour.
3. As the capacities are adequate for commercial use, gross output rating by floor levels are based on 10′ ceiling height.
4. All distribution piping is black steel pipe.

8.3-110	Small Heating Systems, Hydronic, Electric Boilers	COST PER S.F.		
		MAT.	INST.	TOTAL
1100	Small heating systems, hydronic, electric boilers			
1120	Steam, 1 floor, 1480 S.F., 61 M.B.H.	8.65	4.79	13.44
1160	3,000 S.F., 123 M.B.H.	5.20	4.19	9.39
1200	5,000 S.F., 205 M.B.H.	3.99	3.87	7.86
1240	2 floors, 12,400 S.F., 512 M.B.H.	2.90	3.82	6.72
1280	3 floors, 24,800 S.F., 1023 M.B.H.	2.56	3.80	6.36
1320	34,750 S.F., 1,433 M.B.H.	2.29	3.68	5.97
1360	Hot water, 1 floor, 1,000 S.F., 41 M.B.H.	7.80	2.67	10.47
1400	2,500 S.F., 103 M.B.H.	5	4.79	9.79
1440	2 floors, 4,850 S.F., 205 M.B.H.	4.69	5.75	10.44
1480	3 floors, 9,700 S.F., 410 M.B.H.	4.67	5.95	10.62

8.3-120	Large Heating Systems, Hydronic, Electric Boilers	COST PER S.F.		
		MAT.	INST.	TOTAL
1230	Large heating systems, hydronic, electric boilers			
1240	9,280 S.F., 150 K.W., 510 M.B.H., 1 floor	3.22	1.79	5.01
1280	14,900 S.F., 240 K.W., 820 M.B.H., 2 floors	3.36	2.91	6.27
1320	18,600 S.F., 300 K.W., 1,024 M.B.H., 3 floors	3.53	3.21	6.74
1360	26,100 S.F., 420 K.W., 1,432 M.B.H., 4 floors	3.38	3.13	6.51
1400	39,100 S.F., 630 K.W., 2,148 M.B.H., 4 floors	2.98	2.60	5.58
1440	57,700 S.F., 900 K.W., 3,071 M.B.H., 5 floors	2.84	2.55	5.39
1480	111,700 S.F., 1,800 K.W., 6,148 M.B.H., 6 floors	2.54	2.21	4.75
1520	149,000 S.F., 2,400 K.W., 8,191 M.B.H., 8 floors	2.48	2.20	4.68
1560	223,300 S.F., 3,600 K.W., 12,283 M.B.H., 14 floors	2.51	2.50	5.01

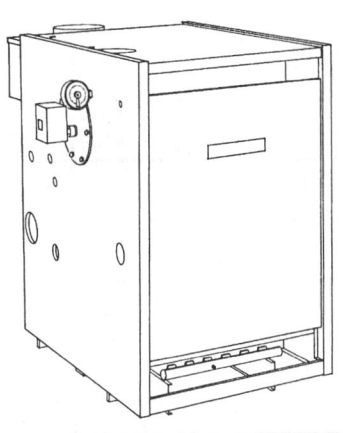

Boiler Selection: The maximum allowable working pressures are limited by ASME "Code for Heating Boilers" to 15 PSI for steam and 160 PSI for hot water heating boilers, with a maximum temperature limitation of 250° F. Hot water boilers are generally rated for a working pressure of 30 PSI. High pressure boilers are governed by the ASME "Code for Power Boilers" which is used almost universally for boilers operating over 15 PSIG. High pressure boilers used for a combination of heating/process loads are usually designed for 150 PSIG.

Boiler ratings are usually indicated as either Gross or Net Output. The Gross Load is equal to the Net Load plus a piping and pickup allowance. When this allowance cannot be determined, divide the gross output rating by 1.25 for a value equal to or greater than the next heat loss requirement of the building.

Table below lists installed cost per boiler and includes insulating jacket, standard controls, burner and safety controls. Costs do not include piping or boiler base pad. Outputs are Gross.

8.3-130	Boilers, Hot Water & Steam	COST EACH		
		MAT.	INST.	TOTAL
0600	Boiler, electric, steel, hot water, 12 K.W., 41 M.B.H.	3,650	820	4,470
0620	30 K.W., 103 M.B.H.	4,000	885	4,885
0640	60 K.W., 205 M.B.H.	5,100	965	6,065
0660	120 K.W., 410 M.B.H.	6,500	1,175	7,675
0680	210 K.W., 716 M.B.H.	10,800	1,775	12,575
0700	510 K.W., 1,739 M.B.H.	20,400	3,300	23,700
0720	720 K.W., 2,452 M.B.H.	25,600	3,725	29,325
0740	1,200 K.W., 4,095 M.B.H.	36,300	4,275	40,575
0760	2,100 K.W., 7,167 M.B.H.	57,000	5,375	62,375
0780	3,600 K.W., 12,283 M.B.H.	78,000	9,075	87,075
0820	Steam, 6 K.W., 20.5 M.B.H.	8,700	885	9,585
0840	24 K.W., 81.8 M.B.H.	9,075	965	10,040
0860	60 K.W., 205 M.B.H.	10,500	1,075	11,575
0880	150 K.W., 512 M.B.H.	13,600	1,625	15,225
0900	510 K.W., 1,740 M.B.H.	22,200	4,025	26,225
0920	1,080 K.W., 3,685 M.B.H.	34,000	5,800	39,800
0940	2,340 K.W., 7,984 M.B.H.	64,500	9,075	73,575
0980	Gas, cast iron, hot water, 80 M.B.H.	1,175	1,000	2,175
1000	100 M.B.H.	1,325	1,100	2,425
1020	163 M.B.H.	1,775	1,475	3,250
1040	280 M.B.H.	2,500	1,650	4,150
1060	544 M.B.H.	3,900	2,450	6,350
1080	1,088 M.B.H.	6,575	2,950	9,525
1100	2,000 M.B.H.	11,300	3,875	15,175
1120	2,856 M.B.H.	16,000	4,475	20,475
1140	4,720 M.B.H.	45,800	8,200	54,000
1160	6,970 M.B.H.	77,000	18,400	95,400
1180	For steam systems under 2,856 M.B.H., add 8%			
1240	Steel, hot water, 72 M.B.H.	1,925	540	2,465
1260	101 M.B.H.	2,200	600	2,800
1280	132 M.B.H.	2,500	635	3,135
1300	150 M.B.H.	2,900	720	3,620
1320	240 M.B.H.	4,450	835	5,285
1340	400 M.B.H.	6,350	1,350	7,700
1360	640 M.B.H.	8,625	1,800	10,425
1380	800 M.B.H.	10,100	2,175	12,275
1400	960 M.B.H.	12,500	2,400	14,900
1420	1,440 M.B.H.	17,400	3,100	20,500
1440	2,400 M.B.H.	28,100	5,400	33,500
1460	3,000 M.B.H.	34,700	7,225	41,925

MECHANICAL

8

8.3-130	Boilers, Hot Water & Steam	COST EACH		
		MAT.	INST.	TOTAL
1520	Oil, cast iron, hot water, 109 M.B.H.	1,300	1,225	2,525
1540	173 M.B.H.	2,200	1,475	3,675
1560	236 M.B.H.	2,550	1,725	4,275
1580	1,084 M.B.H.	9,125	3,500	12,625
1600	1,600 M.B.H.	12,100	4,750	16,850
1620	2,480 M.B.H.	17,100	5,900	23,000
1640	3,550 M.B.H.	16,900	6,700	23,600
1660	Steam systems same price as hot water			
1700	Steel, hot water, 103 M.B.H.	2,275	570	2,845
1720	137 M.B.H.	2,425	675	3,100
1740	225 M.B.H.	3,425	775	4,200
1760	315 M.B.H.	5,050	985	6,035
1780	420 M.B.H.	5,675	1,350	7,025
1800	630 M.B.H.	7,325	1,800	9,125
1820	735 M.B.H.	9,775	1,975	11,750
1840	1,050 M.B.H.	13,500	2,575	16,075
1860	1,365 M.B.H.	17,300	2,925	20,225
1880	1,680 M.B.H.	20,300	3,275	23,575
1900	2,310 M.B.H.	26,200	4,500	30,700
1920	2,835 M.B.H.	32,500	6,375	38,875
1940	3,150 M.B.H.	36,500	8,325	44,825

Important: See the Reference Section for critical supporting data - Location Factors & Historical Cost Indexes

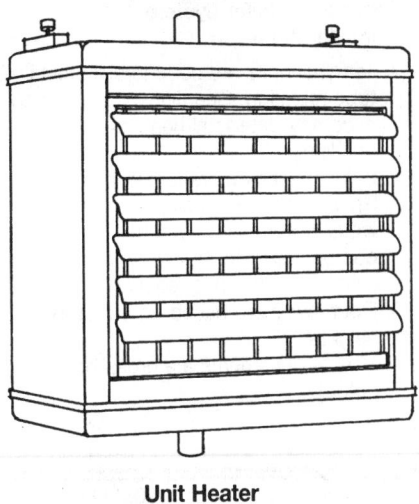

Unit Heater

Fossil Fuel Boiler System Considerations:

1. Terminal units are horizontal unit heaters. Quantities are varied to accommodate total heat loss per building.
2. Unit heater selection was determined by their capacity to circulate the building volume a minimum of three times per hour in addition to the BTU output.
3. Systems shown are forced hot water. Steam boilers cost slightly more than hot water boilers. However, this is compensated for by the smaller size or fewer terminal units required with steam.
4. Floor levels are based on 10' story heights.
5. MBH requirements are gross boiler output.

8.3-141	Heating Systems, Unit Heaters	COST PER S.F.		
		MAT.	INST.	TOTAL
1260	Heating systems, hydronic, fossil fuel, terminal unit heaters,			
1280	Cast iron boiler, gas, 80 M.B.H., 1,070 S.F. bldg.	6.55	5.70	12.25
1320	163 M.B.H., 2,140 S.F. bldg.	4.46	3.85	8.31
1360	544 M.B.H., 7,250 S.F. bldg.	2.93	2.68	5.61
1400	1,088 M.B.H., 14,500 S.F. bldg.	2.60	2.54	5.14
1440	3,264 M.B.H., 43,500 S.F. bldg.	2.19	1.88	4.07
1480	5,032 M.B.H., 67,100 S.F. bldg.	2.56	1.98	4.54
1520	Oil, 109 M.B.H., 1,420 S.F. bldg.	6.25	5.05	11.30
1560	235 M.B.H., 3,150 S.F. bldg.	4.30	3.63	7.93
1600	940 M.B.H., 12,500 S.F. bldg.	3.29	2.33	5.62
1640	1,600 M.B.H., 21,300 S.F. bldg.	3.15	2.22	5.37
1680	2,480 M.B.H., 33,100 S.F. bldg.	3.12	1.99	5.11
1720	3,350 M.B.H., 44,500 S.F. bldg.	2.61	2.04	4.65
1760	Coal, 148 M.B.H., 1,975 S.F. bldg.	5.05	3.31	8.36
1800	300 M.B.H., 4,000 S.F. bldg.	3.98	2.59	6.57
1840	2,360 M.B.H., 31,500 S.F. bldg.	2.81	2.08	4.89
1880	Steel boiler, gas, 72 M.B.H., 1,020 S.F. bldg.	6.15	4.22	10.37
1920	240 M.B.H., 3,200 S.F. bldg.	4.25	3.32	7.57
1960	480 M.B.H., 6,400 S.F. bldg.	3.52	2.47	5.99
2000	800 M.B.H., 10,700 S.F. bldg.	3.10	2.19	5.29
2040	1,960 M.B.H., 26,100 S.F. bldg.	2.77	1.96	4.73
2080	3,000 M.B.H., 40,000 S.F. bldg.	2.72	2.02	4.74
2120	Oil, 97 M.B.H., 1,300 S.F. bldg.	6.95	4.67	11.62
2160	315 M.B.H., 4,550 S.F. bldg.	4.13	2.40	6.53
2200	525 M.B.H., 7,000 S.F. bldg.	4.14	2.43	6.57
2240	1,050 M.B.H., 14,000 S.F. bldg.	3.62	2.44	6.06
2280	2,310 M.B.H. 30,800 S.F. bldg.	3.43	2.08	5.51
2320	3,150 M.B.H., 42,000 S.F. bldg.	3.26	2.16	5.42

MECHANICAL

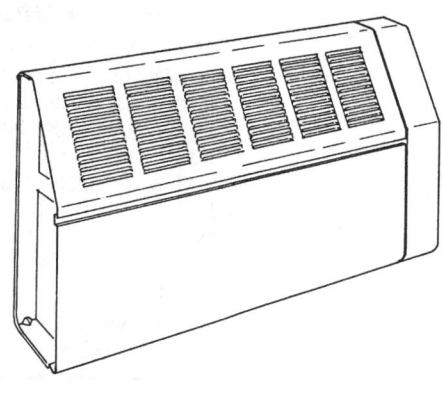

Fossil Fuel Boiler System Considerations:

1. Terminal units are commercial steel fin tube radiation. Quantities are varied to accommodate total heat loss per building.
2. Systems shown are forced hot water. Steam boilers cost slightly more than hot water boilers. However, this is compensated for by the smaller size or fewer terminal units required with steam.
3. Floor levels are based on 10' story heights.
4. MBH requirements are gross boiler output.

Fin Tube Radiator

8.3-142	Heating System, Fin Tube Radiation	COST PER S.F.		
		MAT.	INST.	TOTAL
3230	Heating systems, hydronic, fossil fuel, fin tube radiation			
3240	Cast iron boiler, gas, 80 MBH, 1,070 S.F. bldg.	7.60	8.60	16.20
3280	169 M.B.H., 2,140 S.F. bldg.	4.76	5.45	10.21
3320	544 M.B.H., 7,250 S.F. bldg.	3.79	4.60	8.39
3360	1,088 M.B.H., 14,500 S.F. bldg.	3.55	4.52	8.07
3400	3,264 M.B.H., 43,500 S.F. bldg.	3.27	3.95	7.22
3440	5,032 M.B.H., 67,100 S.F. bldg.	3.66	4.06	7.72
3480	Oil, 109 M.B.H., 1,420 S.F. bldg.	8.60	9.35	17.95
3520	235 M.B.H., 3,150 S.F. bldg.	5.15	5.50	10.65
3560	940 M.B.H., 12,500 S.F. bldg.	4.27	4.34	8.61
3600	1,600 M.B.H., 21,300 S.F. bldg.	4.23	4.31	8.54
3640	2,480 M.B.H., 33,100 S.F. bldg.	4.21	4.08	8.29
3680	3,350 M.B.H., 44,500 S.F. bldg.	3.68	4.12	7.80
3720	Coal, 148 M.B.H., 1,975 S.F. bldg.	5.85	5.20	11.05
3760	300 M.B.H., 4,000 S.F. bldg.	4.77	4.46	9.23
3800	2,360 M.B.H., 31,500 S.F. bldg.	3.84	4.13	7.97
3840	Steel boiler, gas, 72 M.B.H., 1,020 S.F. bldg.	8.55	8.15	16.70
3880	240 M.B.H., 3,200 S.F. bldg.	5.15	5.30	10.45
3920	480 M.B.H., 6,400 S.F. bldg.	4.42	4.42	8.84
3960	800 M.B.H., 10,700 S.F. bldg.	4.42	4.68	9.10
4000	1,960 M.B.H., 26,100 S.F. bldg.	3.80	4.01	7.81
4040	3,000 M.B.H., 40,000 S.F. bldg.	3.74	4.05	7.79
4080	Oil, 97 M.B.H., 1,300 S.F. bldg.	7.30	7.60	14.90
4120	315 M.B.H., 4,550 S.F. bldg.	4.94	4.17	9.11
4160	525 M.B.H., 7,000 S.F. bldg.	5.15	4.46	9.61
4200	1,050 M.B.H., 14,000 S.F. bldg.	4.61	4.47	9.08
4240	2,310 M.B.H., 30,800 S.F. bldg.	4.45	4.12	8.57
4280	3,150 M.B.H., 42,000 S.F. bldg.	4.30	4.21	8.51

Important: See the Reference Section for critical supporting data - Location Factors & Historical Cost Indexes

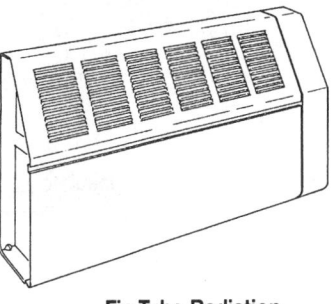

Fin Tube Radiation

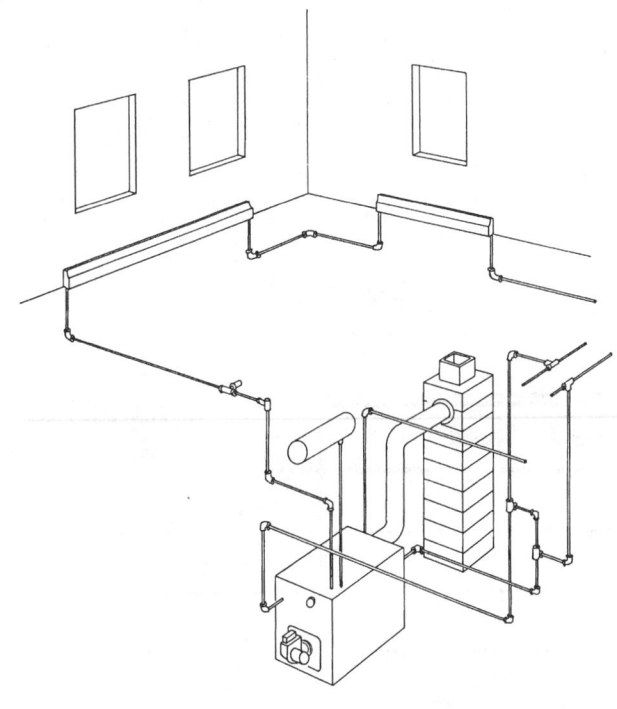

Forced Hot Water Heating System

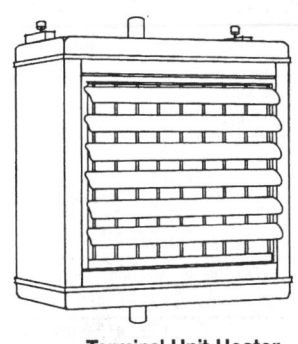

Terminal Unit Heater

8.3-151	Apartment Building Heating - Fin Tube Radiation	COST PER S.F.		
		MAT.	INST.	TOTAL
1740	Heating systems, fin tube radiation, forced hot water			
1760	1,000 S.F. area, 10,000 C.F. volume	3.87	3.72	7.59
1800	10,000 S.F. area, 100,000 C.F. volume	1.56	2.24	3.80
1840	20,000 S.F. area, 200,000 C.F. volume	1.65	2.51	4.16
1880	30,000 S.F. area, 300,000 C.F. volume	1.56	2.45	4.01

8.3-161	Commercial Building Heating - Fin Tube Radiation	COST PER S.F.		
		MAT.	INST.	TOTAL
1940	Heating systems, fin tube radiation, forced hot water			
1960	1,000 S.F. bldg, one floor	8.75	8.40	17.15
2000	10,000 S.F., 100,000 C.F., total two floors	2.53	3.23	5.76
2040	100,000 S.F., 1,000,000 C.F., total three floors	1.16	1.43	2.59
2080	1,000,000 S.F., 10,000,000 C.F., total five floors	.52	.77	1.29

8.3-162	Commercial Bldg. Heating - Terminal Unit Heaters	COST PER S.F.		
		MAT.	INST.	TOTAL
1860	Heating systems, terminal unit heaters, forced hot water			
1880	1,000 S.F. bldg., one floor	9.15	7.70	16.85
1920	10,000 S.F. bldg., 100,000 C.F. total two floors	2.36	2.67	5.03
1960	100,000 S.F. bldg., 1,000,000 C.F. total three floors	1.29	1.30	2.59
2000	1,000,000 S.F. bldg., 10,000,000 C.F. total five floors	.82	.85	1.67

MECHANICAL

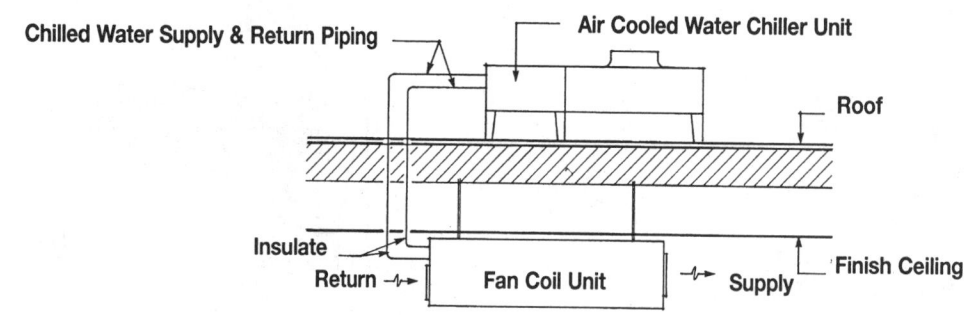

Chilled Water Supply & Return Piping

Air Cooled Water Chiller Unit

Roof

Insulate

Return

Fan Coil Unit

Supply

Finish Ceiling

*Cooling requirements would lead to a
choice of multiple chillers.

8.4-110	Chilled Water, Air Cooled Condenser Systems	COST PER S.F.		
		MAT.	INST.	TOTAL
1180	Packaged chiller, air cooled, with fan coil unit			
1200	Apartment corridors, 3,000 S.F., 5.50 ton	4.55	4.74	9.29
1240	6,000 S.F., 11.00 ton	3.57	3.81	7.38
1280	10,000 S.F., 18.33 ton	3.13	2.88	6.01
1320	20,000 S.F., 36.66 ton	2.53	2.15	4.68
1360	40,000 S.F., 73.33 ton	3	2.25	5.25
1440	Banks and libraries, 3,000 S.F., 12.50 ton	6.75	5.55	12.30
1480	6,000 S.F., 25.00 ton	6.25	4.52	10.77
1520	10,000 S.F., 41.66 ton	5.10	3.30	8.40
1560	20,000 S.F., 83.33 ton	5.50	3.18	8.68
1600	40,000 S.F., 167 ton*			
1680	Bars and taverns, 3,000 S.F., 33.25 ton	12.65	6.95	19.60
1720	6,000 S.F., 66.50 ton	11.90	6	17.90
1760	10,000 S.F., 110.83 ton	9.60	2.73	12.33
1800	20,000 S.F., 220 ton*			
1840	40,000 S.F., 440 ton*			
1920	Bowling alleys, 3,000 S.F., 17.00 ton	8.55	6.25	14.80
1960	6,000 S.F., 34.00 ton	6.90	4.58	11.48
2000	10,000 S.F., 56.66 ton	6.30	3.65	9.95
2040	20,000 S.F., 113.33 ton	6.10	3.34	9.44
2080	40,000 S.F., 227 ton*			
2160	Department stores, 3,000 S.F., 8.75 ton	6.45	5.25	11.70
2200	6,000 S.F., 17.50 ton	4.74	4.09	8.83
2240	10,000 S.F., 29.17 ton	3.96	3.05	7.01
2280	20,000 S.F., 58.33 ton	3.52	2.41	5.93
2320	40,000 S.F., 116.66 ton	3.82	2.48	6.30
2400	Drug stores, 3,000 S.F., 20.00 ton	9.80	6.45	16.25
2440	6,000 S.F., 40.00 ton	8.15	5	13.15
2480	10,000 S.F., 66.66 ton	8.45	4.64	13.09
2520	20,000 S.F., 133.33 ton	7.40	3.76	11.16
2560	40,000 S.F., 267 ton*			
2640	Factories, 2,000 S.F., 10.00 ton	5.85	5.35	11.20
2680	6,000 S.F., 20.00 ton	5.35	4.30	9.65
2720	10,000 S.F., 33.33 ton	4.34	3.13	7.47
2760	20,000 S.F., 66.66 ton	4.69	2.98	7.67
2800	40,000 S.F., 133.33 ton	4.20	2.58	6.78
2880	Food supermarkets, 3,000 S.F., 8.50 ton	6.30	5.25	11.55
2920	6,000 S.F., 17.00 ton	4.65	4.07	8.72
2960	10,000 S.F., 28.33 ton	3.78	2.97	6.75
3000	20,000 S.F., 56.66 ton	3.40	2.36	5.76
3040	40,000 S.F., 113.33 ton	3.69	2.45	6.14
3120	Medical centers, 3,000 S.F., 7.00 ton	5.60	5.10	10.70
3160	6,000 S.F., 14.00 ton	4.11	3.93	8.04
3200	10,000 S.F., 23.33 ton	3.72	2.99	6.71

Important: See the Reference Section for critical supporting data - Location Factors & Historical Cost Indexes

8.4-110	Chilled Water, Air Cooled Condenser Systems	COST PER S.F.		
		MAT.	INST.	TOTAL
3240	20,000 S.F., 46.66 ton	3.06	2.27	5.33
3280	40,000 S.F., 93.33 ton	3.30	2.32	5.62
3360	Offices, 3,000 S.F., 9.50 ton	5.45	5.20	10.65
3400	6,000 S.F., 19.00 ton	5.15	4.26	9.41
3440	10,000 S.F., 31.66 ton	4.19	3.10	7.29
3480	20,000 S.F., 63.33 ton	4.64	3.01	7.65
3520	40,000 S.F., 126.66 ton	4.11	2.59	6.70
3600	Restaurants, 3,000 S.F., 15.00 ton	7.65	5.80	13.45
3640	6,000 S.F., 30.00 ton	6.50	4.57	11.07
3680	10,000 S.F., 50.00 ton	5.95	3.65	9.60
3720	20,000 S.F., 100.00 ton	6.10	3.47	9.57
3760	40,000 S.F., 200 ton*			
3840	Schools and colleges, 3,000 S.F., 11.50 ton	6.40	5.45	11.85
3880	6,000 S.F., 23.00 ton	5.90	4.43	10.33
3920	10,000 S.F., 38.33 ton	4.80	3.23	8.03
3960	20,000 S.F., 76.66 ton	5.30	3.18	8.48
4000	40,000 S.F., 153 ton*			

MECHANICAL

8

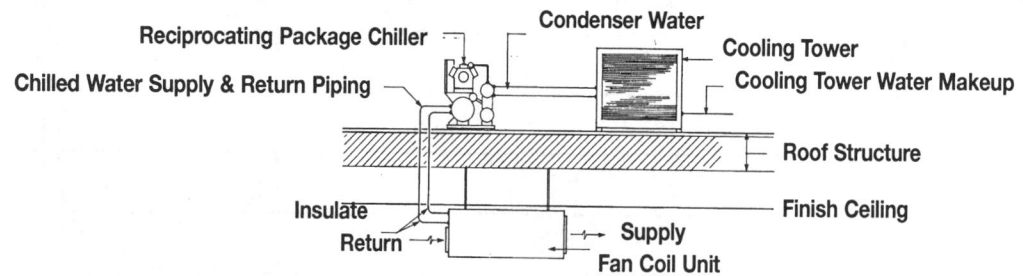

Reciprocating Package Chiller
Chilled Water Supply & Return Piping
Condenser Water
Cooling Tower
Cooling Tower Water Makeup
Roof Structure
Finish Ceiling
Insulate
Return — Supply
Fan Coil Unit

*Cooling requirements would lead to a choice of multiple chillers

8.4-120	Chilled Water, Cooling Tower Systems	COST PER S.F.		
		MAT.	INST.	TOTAL
1300	Packaged chiller, water cooled, with fan coil unit			
1320	Apartment corridors, 4,000 S.F., 7.33 ton	5.50	4.72	10.22
1360	6,000 S.F., 11.00 ton	4.30	4	8.30
1400	10,000 S.F., 18.33 ton	3.62	3.01	6.63
1440	20,000 S.F., 26.66 ton	2.85	2.25	5.10
1480	40,000 S.F., 73.33 ton	3.20	2.37	5.57
1520	60,000 S.F., 110.00 ton	3.14	2.44	5.58
1600	Banks and libraries, 4,000 S.F., 16.66 ton	8.20	5.25	13.45
1640	6,000 S.F., 25.00 ton	6.60	4.58	11.18
1680	10,000 S.F., 41.66 ton	5.25	3.47	8.72
1720	20,000 S.F., 83.33 ton	5.90	3.45	9.35
1760	40,000 S.F., 166.66 ton	5.70	4.18	9.88
1800	60,000 S.F., 250.00 ton	5.60	4.53	10.13
1880	Bars and taverns, 4,000 S.F., 44.33 ton	12.65	6.75	19.40
1920	6,000 S.F., 66.50 ton	12.90	7	19.90
1960	10,000 S.F., 110.83 ton	12.75	6	18.75
2000	20,000 S.F., 221.66 ton	11.90	6.55	18.45
2040	40,000 S.F., 440 ton*			
2080	60,000 S.F., 660 ton*			
2160	Bowling alleys, 4,000 S.F., 22.66 ton	9.55	5.75	15.30
2200	6,000 S.F., 34.00 ton	8	5.10	13.10
2240	10,000 S.F., 56.66 ton	6.50	3.83	10.33
2280	20,000 S.F., 113.33 ton	6.95	3.75	10.70
2320	40,000 S.F., 226.66 ton	6.70	4.47	11.17
2360	60,000 S.F., 340 ton			
2440	Department stores, 4,000 S.F., 11.66 ton	6.10	5.05	11.15
2480	6,000 S.F., 17.50 ton	6	4.29	10.29
2520	10,000 S.F., 29.17 ton	4.34	3.18	7.52
2560	20,000 S.F., 58.33 ton	3.34	2.40	5.74
2600	40,000 S.F., 116.66 ton	4.02	2.59	6.61
2640	60,000 S.F., 175.00 ton	4.59	4.05	8.64
2720	Drug stores, 4,000 S.F., 26.66 ton	9.95	5.90	15.85
2760	6,000 S.F., 40.00 ton	8.30	5.15	13.45
2800	10,000 S.F., 66.66 ton	8.15	4.69	12.84
2840	20,000 S.F., 133.33 ton	7.95	4.05	12
2880	40,000 S.F., 266.67 ton	7.75	5.10	12.85
2920	60,000 S.F., 400 ton*			
3000	Factories, 4,000 S.F., 13.33 ton	6.95	4.99	11.94
3040	6,000 S.F., 20.00 ton	6	4.27	10.27
3080	10,000 S.F., 33.33 ton	4.79	3.29	8.08
3120	20,000 S.F., 66.66 ton	4.48	2.97	7.45
3160	40,000 S.F., 133.33 ton	4.47	2.72	7.19
3200	60,000 S.F., 200.00 ton	4.94	4.25	9.19
3280	Food supermarkets, 4,000 S.F., 11.33 ton	6	5	11
3320	6,000 S.F., 17.00 ton	5.25	4.16	9.41
3360	10,000 S.F., 28.33 ton	4.24	3.16	7.40

MECHANICAL 8

8.4-120	Chilled Water, Cooling Tower Systems	COST PER S.F.		
		MAT.	INST.	TOTAL
3400	20,000 S.F., 56.66 ton	3.43	2.41	5.84
3440	40,000 S.F., 113.33 ton	3.97	2.57	6.54
3480	60,000 S.F., 170.00 ton	4.52	4.04	8.56
3560	Medical centers, 4.000 S.F., 9.33 ton	5.15	4.59	9.74
3600	6,000 S.F., 14.00 ton	5.10	4.10	9.20
3640	10,000 S.F., 23.33 ton	3.91	3.03	6.94
3680	20,000 S.F., 46.66 ton	3.03	2.34	5.37
3720	40,000 S.F., 93.33 ton	3.63	2.47	6.10
3760	60,000 S.F., 140.00 ton	4.16	3.96	8.12
3840	Offices, 4,000 S.F., 12.66 ton	6.70	4.94	11.64
3880	6,000 S.F., 19.00 ton	5.85	4.37	10.22
3920	10,000 S.F., 31.66 ton	4.72	3.32	8.04
3960	20,000 S.F., 63.33 ton	4.40	2.97	7.37
4000	40,000 S.F., 126.66 ton	4.79	3.94	8.73
4040	60,000 S.F., 190.00 ton	4.78	4.19	8.97
4120	Restaurants, 4,000 S.F., 20.00 ton	8.55	5.35	13.90
4160	6,000 S.F., 30.00 ton	7.25	4.76	12.01
4200	10,000 S.F., 50.00 ton	5.95	3.65	9.60
4240	20,000 S.F., 100.00 ton	6.70	3.67	10.37
4280	40,000 S.F., 200.00 ton	6	4.24	10.24
4320	60,000 S.F., 300.00 ton	6.35	4.77	11.12
4400	Schools and colleges, 4,000 S.F., 15.33 ton	7.70	5.15	12.85
4440	6,000 S.F., 23.00 ton	6.25	4.49	10.74
4480	10,000 S.F., 38.33 ton	4.92	3.39	8.31
4520	20,000 S.F., 76.66 ton	5.60	3.38	8.98
4560	40,000 S.F., 153.33 ton	5.35	4.07	9.42
4600	60,000 S.F., 230.00 ton	5.20	4.35	9.55

MECHANICAL

8

For expanded coverage of these items see *Means Mechanical or Plumbing Cost Data 1998*

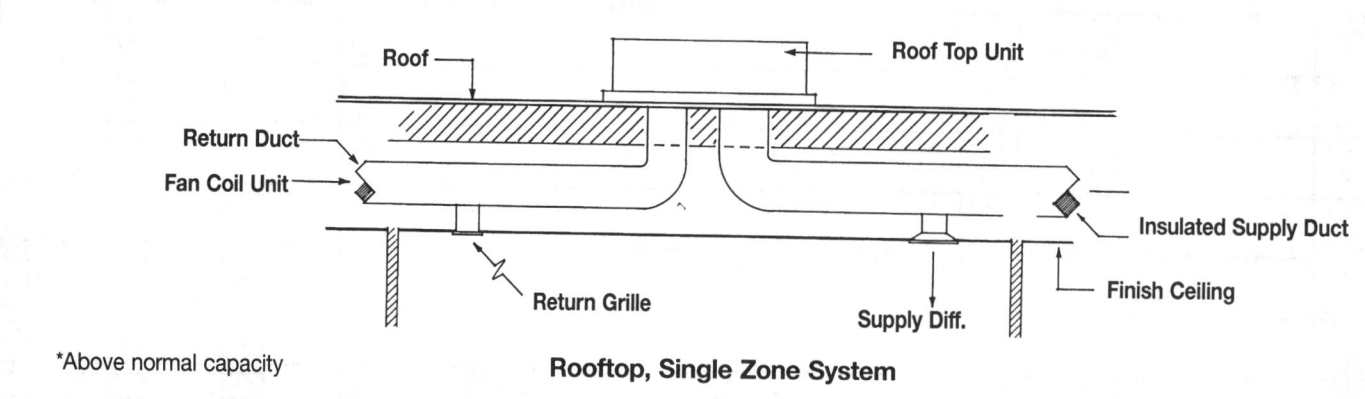

Roof

Roof Top Unit

Return Duct

Fan Coil Unit

Insulated Supply Duct

Return Grille

Supply Diff.

Finish Ceiling

*Above normal capacity

Rooftop, Single Zone System

8.4-210	Rooftop Single Zone Unit Systems	COST PER S.F.		
		MAT.	INST.	TOTAL
1260	Rooftop, single zone, air conditioner			
1280	Apartment corridors, 500 S.F., .92 ton	4.62	1.80	6.42
1320	1,000 S.F., 1.83 ton	4.59	1.79	6.38
1360	1500 S.F., 2.75 ton	2.66	1.59	4.25
1400	3,000 S.F., 5.50 ton	2.28	1.76	4.04
1440	5,000 S.F., 9.17 ton	2.30	1.73	4.03
1480	10,000 S.F., 18.33 ton	2.47	1.72	4.19
1560	Banks or libraries, 500 S.F., 2.08 ton	10.45	4.06	14.51
1600	1,000 S.F., 4.17 ton	6.05	3.62	9.67
1640	1,500 S.F., 6.25 ton	5.20	4	9.20
1680	3,000 S.F., 12.50 ton	5.25	3.94	9.19
1720	5,000 S.F., 20.80 ton	5.60	3.91	9.51
1760	10,000 S.F., 41.67 ton	5.25	3.92	9.17
1840	Bars and taverns, 500 S.F. 5.54 ton	12.50	6.70	19.20
1880	1,000 S.F., 11.08 ton	12.60	6.55	19.15
1920	1,500 S.F., 16.62 ton	11.30	6.50	17.80
1960	3,000 S.F., 33.25 ton	12.90	6.40	19.30
2000	5,000 S.F., 55.42 ton	12.30	6.45	18.75
2040	10,000 S.F., 110.83 ton*			
2080	Bowling alleys, 500 S.F., 2.83 ton	8.25	4.92	13.17
2120	1,000 S.F., 5.67 ton	7.05	5.45	12.50
2160	1,500 S.F., 8.50 ton	7.10	5.35	12.45
2200	3,000 S.F., 17.00 ton	6.45	5.35	11.80
2240	5,000 S.F., 28.33 ton	7.25	5.30	12.55
2280	10,000 S.F., 56.67 ton	6.95	5.30	12.25
2360	Department stores, 500 S.F., 1.46 ton	7.30	2.85	10.15
2400	1,000 S.F., 2.92 ton	4.24	2.53	6.77
2480	3,000 S.F., 8.75 ton	3.67	2.76	6.43
2520	5,000 S.F., 14.58 ton	3.31	2.75	6.06
2560	10,000 S.F., 29.17 ton	3.74	2.72	6.46
2640	Drug stores, 500 S.F., 3.33 ton	9.70	5.80	15.50
2680	1,000 S.F., 6.67 ton	8.30	6.40	14.70
2720	1,500 S.F., 10.00 ton	8.35	6.30	14.65
2760	3,000 S.F., 20.00 ton	8.95	6.25	15.20
2800	5,000 S.F., 33.33 ton	8.55	6.20	14.75
2840	10,000 S.F., 66.67 ton	8.20	6.25	14.45
2920	Factories, 500 S.F., 1.67 ton	8.35	3.26	11.61
3000	1,500 S.F., 5.00 ton	4.16	3.20	7.36
3040	3,000 S.F., 10.00 ton	4.19	3.15	7.34
3080	5,000 S.F., 16.67 ton	3.79	3.15	6.94
3120	10,000 S.F., 33.33 ton	4.28	3.11	7.39
3200	Food supermarkets, 500 S.F., 1.42 ton	7.10	2.78	9.88
3240	1,000 S.F., 2.83 ton	4.08	2.45	6.53
3280	1,500 S.F., 4.25 ton	3.53	2.72	6.25

Important: See the Reference Section for critical supporting data - Location Factors & Historical Cost Indexes

8.4-210	Rooftop Single Zone Unit Systems	COST PER S.F.		
		MAT.	INST.	TOTAL
3320	3,000 S.F., 8.50 ton	3.56	2.68	6.24
3360	5,000 S.F., 14.17 ton	3.22	2.67	5.89
3400	10,000 S.F., 28.33 ton	3.63	2.64	6.27
3480	Medical centers, 500 S.F., 1.17 ton	5.85	2.28	8.13
3520	1,000 S.F., 2.33 ton	5.85	2.27	8.12
3560	1,500 S.F., 3.50 ton	3.39	2.03	5.42
3640	5,000 S.F., 11.67 ton	2.93	2.21	5.14
3680	10,000 S.F., 23.33 ton	3.14	2.20	5.34
3760	Offices, 500 S.F., 1.58 ton	7.95	3.09	11.04
3800	1,000 S.F., 3.17 ton	4.61	2.75	7.36
3840	1,500 S.F., 4.75 ton	3.95	3.04	6.99
3880	3,000 S.F., 9.50 ton	3.98	2.99	6.97
3920	5,000 S.F., 15.83 ton	3.60	2.99	6.59
3960	10,000 S.F., 31.67 ton	4.07	2.96	7.03
4000	Restaurants, 500 S.F., 2.50 ton	12.55	4.89	17.44
4040	1,000 S.F., 5.00 ton	6.25	4.80	11.05
4080	1,500 S.F., 7.50 ton	6.30	4.73	11.03
4120	3,000 S.F., 15.00 ton	5.70	4.72	10.42
4160	5,000 S.F., 25.00 ton	6.75	4.70	11.45
4200	10,000 S.F., 50.00 ton	6.15	4.68	10.83
4240	Schools and colleges, 500 S.F., 1.92 ton	9.65	3.76	13.41
4280	1,000 S.F., 3.83 ton	5.55	3.33	8.88
4360	3,000 S.F., 11.50 ton	4.81	3.62	8.43
4400	5,000 S.F., 19.17 ton	5.15	3.60	8.75

MECHANICAL

8

Roof — / Rooftop Unit

Return Ducts →

Insulated Supply Ducts →

Finish Ceiling

Return Grille

Supply Diffusers

*Note A: Small single zone unit recommended.

*Note B: A combination of multizone units recommended.

8.4-220	Rooftop Multizone Unit Systems	COST PER S.F.		
		MAT.	INST.	TOTAL
1240	Rooftop, multizone, air conditioner			
1260	Apartment corridors, 1,500 S.F., 2.75 ton. See Note A.			
1280	3,000 S.F., 5.50 ton	6.95	3.50	10.45
1320	10,000 S.F., 18.30 ton	5.10	3.16	8.26
1360	15,000 S.F., 27.50 ton	4.29	3.12	7.41
1400	20,000 S.F., 36.70 ton	4.38	3.12	7.50
1440	25,000 S.F., 45.80 ton	4.19	3.13	7.32
1520	Banks or libraries, 1,500 S.F., 6.25 ton	15.80	7.95	23.75
1560	3,000 S.F., 12.50 ton	13.40	7.60	21
1600	10,000 S.F., 41.67 ton	9.55	7.10	16.65
1640	15,000 S.F., 62.50 ton	8.30	6.70	15
1680	20,000 S.F., 83.33 ton	8.30	6.70	15
1720	25,000 S.F., 104.00 ton	8.30	6.70	15
1800	Bars and taverns, 1,500 S.F., 16.62 ton	33.50	12.55	46.05
1840	3,000 S.F., 33.24 ton	24.50	11.25	35.75
1880	10,000 S.F., 110.83 ton	20	10.20	30.20
1920	15,000 S.F., 165 ton, See Note B			
1960	20,000 S.F., 220 ton, See Note B			
2000	25,000 S.F., 275 ton, See Note B			
2080	Bowling alleys, 1,500 S.F., 8.50 ton	21.50	10.85	32.35
2120	3,000 S.F., 17.00 ton	18.20	10.30	28.50
2160	10,000 S.F., 56.70 ton	12.95	9.70	22.65
2200	15,000 S.F., 85.00 ton	11.30	9.10	20.40
2240	20,000 S.F., 113.00 ton	11.25	9.10	20.35
2280	25,000 S.F., 140.00 ton see Note B			
2360	Department stores, 1,500 S.F., 4.37 ton, See Note A.			
2400	3,000 S.F., 8.75 ton	11.05	5.55	16.60
2440	10,000 S.F., 29.17 ton	6.95	4.96	11.91
2520	20,000 S.F., 58.33 ton	5.80	4.69	10.49
2560	25,000 S.F., 72.92 ton	5.80	4.68	10.48
2640	Drug stores, 1,500 S.F., 10.00 ton	25.50	12.75	38.25
2680	3,000 S.F., 20.00 ton	18.65	11.50	30.15
2720	10,000 S.F., 66.66 ton	13.25	10.75	24
2760	15,000 S.F., 100.00 ton	13.30	10.75	24.05
2800	20,000 S.F., 135 ton, See Note B			
2840	25,000 S.F., 165 ton, See Note B			
2920	Factories, 1,500 S.F., 5 ton, See Note A			
3000	10,000 S.F., 33.33 ton	7.95	5.65	13.60
3040	15,000 S.F., 50.00 ton	7.60	5.70	13.30
3080	20,000 S.F., 66.66 ton	6.65	5.35	12
3120	25,000 S.F., 83.33 ton	6.65	5.35	12
3200	Food supermarkets, 1,500 S.F., 4.25 ton, See Note A			
3240	3,000 S.F., 8.50 ton	10.75	5.40	16.15
3280	10,000 S.F., 28.33 ton	6.65	4.83	11.48

Important: See the Reference Section for critical supporting data - Location Factors & Historical Cost Indexes

8.4-220	Rooftop Multizone Unit Systems	COST PER S.F.		
		MAT.	INST.	TOTAL
3320	15,000 S.F., 42.50 ton	6.50	4.84	11.34
3360	20,000 S.F., 56.67 ton	5.65	4.56	10.21
3400	25,000 S.F., 70.83 ton	5.65	4.56	10.21
3480	Medical centers, 1,500 S.F., 3.5 ton, See Note A			
3520	3,000 S.F., 7.00 ton	8.85	4.46	13.31
3560	10,000 S.F., 23.33 ton	5.70	3.98	9.68
3600	15,000 S.F., 35.00 ton	5.55	3.97	9.52
3640	20,000 S.F., 46.66 ton	5.35	3.99	9.34
3680	25,000 S.F., 58.33 ton	4.64	3.76	8.40
3760	Offices, 1,500 S.F., 4.75 ton, See Note A			
3800	3,000 S.F., 9.50 ton	12	6.05	18.05
3840	10,000 S.F., 31.66 ton	7.55	5.40	12.95
3920	20,000 S.F., 63.33 ton	6.30	5.10	11.40
3960	25,000 S.F., 79.16 ton	6.30	5.10	11.40
4000	Restaurants, 1,500 S.F., 7.50 ton	18.95	9.55	28.50
4040	3,000 S.F., 15.00 ton	16.05	9.10	25.15
4080	10,000 S.F., 50.00 ton	11.45	8.55	20
4120	15,000 S.F., 75.00 ton	9.95	8.05	18
4160	20,000 S.F., 100.00 ton	9.95	8.05	18
4200	25,000 S.F., 125 ton, See Note B			
4240	Schools and colleges, 1,500 S.F., 5.75 ton	14.50	7.30	21.80
4320	10,000 S.F., 38.33 ton	8.75	6.55	15.30
4360	15,000 S.F., 57.50 ton	7.60	6.15	13.75
4400	20,000 S.F., 76.66 ton	7.60	6.15	13.75
4440	25,000 S.F., 95.83 ton	7.65	6.15	13.80

MECHANICAL

8

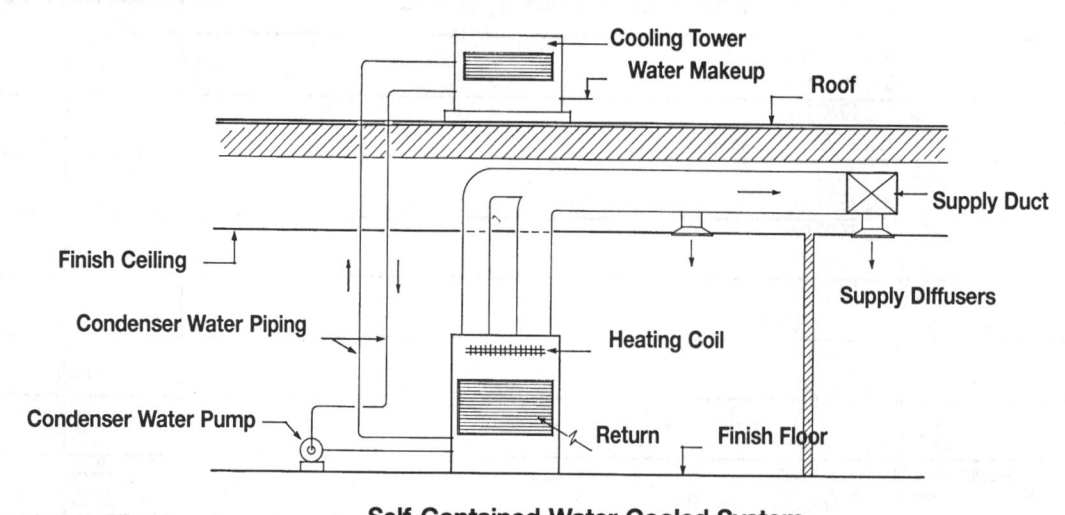

Self-Contained Water Cooled System

8.4-230	Self-contained, Water Cooled Unit Systems	COST PER S.F.		
		MAT.	INST.	TOTAL
1280	Self-contained, water cooled unit	2.54	1.78	4.32
1300	Apartment corridors, 500 S.F., .92 ton	2.54	1.79	4.33
1320	1,000 S.F., 1.83 ton	2.53	1.78	4.31
1360	3,000 S.F., 5.50 ton	1.90	1.49	3.39
1400	5,000 S.F., 9.17 ton	1.93	1.39	3.32
1440	10,000 S.F., 18.33 ton	1.76	1.25	3.01
1520	Banks or libraries, 500 S.F., 2.08 ton	5.45	1.81	7.26
1560	1,000 S.F., 4.17 ton	4.34	3.40	7.74
1600	3,000 S.F., 12.50 ton	4.41	3.18	7.59
1640	5,000 S.F., 20.80 ton	4.01	2.84	6.85
1680	10,000 S.F., 41.66 ton	3.51	2.84	6.35
1760	Bars and taverns, 500 S.F., 5.54 ton	10.65	3.05	13.70
1800	1,000 S.F., 11.08 ton	11.25	5.45	16.70
1840	3,000 S.F., 33.25 ton	9.25	4.29	13.54
1880	5,000 S.F., 55.42 ton	8.70	4.48	13.18
1920	10,000 S.F., 110.00 ton	8.50	4.42	12.92
2000	Bowling alleys, 500 S.F., 2.83 ton	7.40	2.46	9.86
2040	1,000 S.F., 5.66 ton	5.90	4.63	10.53
2080	3,000 S.F., 17.00 ton	5.45	3.87	9.32
2120	5,000 S.F., 28.33 ton	4.97	3.73	8.70
2160	10,000 S.F., 56.66 ton	4.67	3.82	8.49
2200	Department stores, 500 S.F., 1.46 ton	3.83	1.27	5.10
2240	1,000 S.F., 2.92 ton	3.04	2.39	5.43
2280	3,000 S.F., 8.75 ton	3.09	2.23	5.32
2320	5,000 S.F., 14.58 ton	2.82	1.99	4.81
2360	10,000 S.F., 29.17 ton	2.56	1.91	4.47
2440	Drug stores, 500 S.F., 3.33 ton	8.75	2.90	11.65
2480	1,000 S.F., 6.66 ton	6.95	5.45	12.40
2520	3,000 S.F., 20.00 ton	6.45	4.55	11
2560	5,000 S.F., 33.33 ton	6.75	4.49	11.24
2600	10,000 S.F., 66.66 ton	5.50	4.49	9.99
2680	Factories, 500 S.F., 1.66 ton	4.36	1.44	5.80
2720	1,000 S.F. 3.37 ton	3.51	2.76	6.27
2760	3,000 S.F., 10.00 ton	3.53	2.55	6.08
2800	5,000 S.F., 16.66 ton	3.22	2.27	5.49
2840	10,000 S.F., 33.33 ton	2.92	2.19	5.11
2920	Food supermarkets, 500 S.F., 1.42 ton	3.70	1.23	4.93
2960	1,000 S.F., 2.83 ton	3.92	2.76	6.68
3000	3,000 S.F., 8.50 ton	2.95	2.31	5.26
3040	5,000 S.F., 14.17 ton	3	2.15	5.15

8.4-230	Self-contained, Water Cooled Unit Systems	COST PER S.F.		
		MAT.	INST.	TOTAL
3080	10,000 S.F., 28.33 ton	2.48	1.86	4.34
3160	Medical centers, 500 S.F., 1.17 ton	3.05	1.02	4.07
3200	1,000 S.F., 2.33 ton	3.23	2.28	5.51
3240	3,000 S.F., 7.00 ton	2.43	1.91	4.34
3280	5,000 S.F., 11.66 ton	2.46	1.78	4.24
3320	10,000 S.F., 23.33 ton	2.25	1.60	3.85
3400	Offices, 500 S.F., 1.58 ton	4.15	1.38	5.53
3440	1,000 S.F., 3.17 ton	4.39	3.09	7.48
3480	3,000 S.F., 9.50 ton	3.35	2.41	5.76
3520	5,000 S.F., 15.83 ton	3.06	2.16	5.22
3560	10,000 S.F., 31.67 ton	2.78	2.08	4.86
3640	Restaurants, 500 S.F., 2.50 ton	6.55	2.17	8.72
3680	1,000 S.F., 5.00 ton	5.25	4.08	9.33
3720	3,000 S.F., 15.00 ton	5.30	3.81	9.11
3760	5,000 S.F., 25.00 ton	4.84	3.41	8.25
3800	10,000 S.F., 50.00 ton	3.04	3.13	6.17
3880	Schools and colleges, 500 S.F., 1.92 ton	5	1.67	6.67
3920	1,000 S.F., 3.83 ton	4	3.14	7.14
3960	3,000 S.F., 11.50 ton	4.06	2.93	6.99
4000	5,000 S.F., 19.17 ton	3.70	2.62	6.32
4040	10,000 S.F., 38.33 ton	3.24	2.62	5.86

MECHANICAL

8

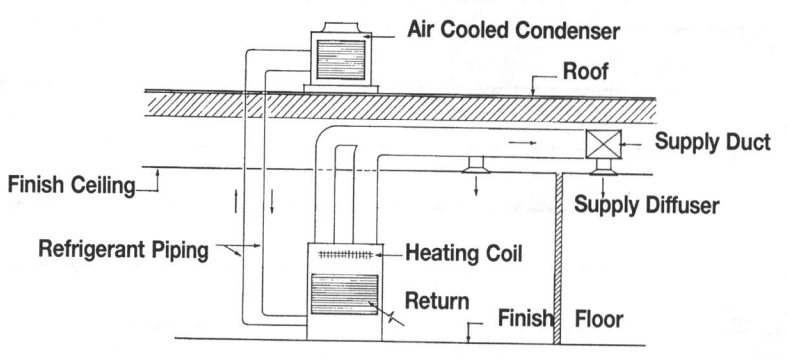

Air Cooled Condenser

Roof

Finish Ceiling

Supply Duct

Supply Diffuser

Refrigerant Piping

Heating Coil

Return

Finish Floor

Self-Contained Air Cooled System

8.4-240	Self-contained, Air Cooled Unit Systems	COST PER S.F.		
		MAT.	INST.	TOTAL
1300	Self-contained, air cooled unit			
1320	Apartment corridors, 500 S.F., .92 ton	4.12	2.29	6.41
1360	1,000 S.F., 1.83 ton	4.06	2.28	6.34
1400	3,000 S.F., 5.50 ton	3.27	2.12	5.39
1440	5,000 S.F., 9.17 ton	2.82	2.02	4.84
1480	10,000 S.F., 18.33 ton	2.40	1.86	4.26
1560	Banks or libraries, 500 S.F., 2.08 ton	8.90	2.91	11.81
1600	1,000 S.F., 4.17 ton	7.40	4.82	12.22
1640	3,000 S.F., 12.50 ton	6.40	4.60	11
1680	5,000 S.F., 20.80 ton	5.50	4.23	9.73
1720	10,000 S.F., 41.66 ton	5	4.19	9.19
1800	Bars and taverns, 500 S.F., 5.54 ton	16.05	6.35	22.40
1840	1,000 S.F., 11.08 ton	16.60	9.20	25.80
1880	3,000 S.F., 33.25 ton	13.10	8.05	21.15
1920	5,000 S.F., 55.42 ton	12.60	8.25	20.85
1960	10,000 S.F., 110.00 ton	12.75	8.15	20.90
2040	Bowling alleys, 500 S.F., 2.83 ton	12.15	3.97	16.12
2080	1,000 S.F., 5.66 ton	10.10	6.55	16.65
2120	3,000 S.F., 17.00 ton	7.45	5.75	13.20
2160	5,000 S.F., 28.33 ton	6.90	5.65	12.55
2200	10,000 S.F., 56.66 ton	6.65	5.70	12.35
2240	Department stores, 500 S.F., 1.46 ton	6.25	2.04	8.29
2280	1,000 S.F., 2.92 ton	5.20	3.38	8.58
2320	3,000 S.F., 8.75 ton	4.48	3.21	7.69
2360	5,000 S.F., 14.58 ton	4.48	3.20	7.68
2400	10,000 S.F., 29.17 ton	3.56	2.90	6.46
2480	Drug stores, 500 S.F., 3.33 ton	14.30	4.67	18.97
2520	1,000 S.F., 6.66 ton	11.85	7.70	19.55
2560	3,000 S.F., 20.00 ton	8.80	6.75	15.55
2600	5,000 S.F., 33.33 ton	8.15	6.65	14.80
2640	10,000 S.F., 66.66 ton	7.90	6.75	14.65
2720	Factories, 500 S.F., 1.66 ton	7.25	2.36	9.61
2760	1,000 S.F., 3.33 ton	6	3.87	9.87
2800	3,000 S.F., 10.00 ton	5.10	3.68	8.78
2840	5,000 S.F., 16.66 ton	4.34	3.38	7.72
2880	10,000 S.F., 33.33 ton	4.06	3.33	7.39
2960	Food supermarkets, 500 S.F., 1.42 ton	6.05	1.98	8.03
3000	1,000 S.F., 2.83 ton	6.25	3.51	9.76
3040	3,000 S.F., 8.50 ton	5.05	3.28	8.33
3080	5,000 S.F., 14.17 ton	4.35	3.12	7.47
3120	10,000 S.F., 28.33 ton	3.46	2.82	6.28
3200	Medical centers, 500 S.F., 1.17 ton	5	1.65	6.65
3240	1,000 S.F., 2.33 ton	5.20	2.91	8.11
3280	3,000 S.F., 7.00 ton	4.16	2.70	6.86

Important: See the Reference Section for critical supporting data - Location Factors & Historical Cost Indexes

8.4-240	Self-contained, Air Cooled Unit Systems	COST PER S.F.		
		MAT.	INST.	TOTAL
3320	5,000 S.F., 16.66 ton	3.60	2.58	6.18
3360	10,000 S.F., 23.33 ton	3.06	2.37	5.43
3440	Offices, 500 S.F., 1.58 ton	6.80	2.23	9.03
3480	1,000 S.F., 3.16 ton	7.10	3.94	11.04
3520	3,000 S.F., 9.50 ton	4.88	3.50	8.38
3560	5,000 S.F., 15.83 ton	4.17	3.22	7.39
3600	10,000 S.F., 31.66 ton	3.87	3.15	7.02
3680	Restaurants, 500 S.F., 2.50 ton	10.65	3.49	14.14
3720	1,000 S.F., 5.00 ton	8.95	5.80	14.75
3760	3,000 S.F., 15.00 ton	7.70	5.50	13.20
3800	5,000 S.F., 25.00 ton	6.05	4.98	11.03
3840	10,000 S.F., 50.00 ton	6.10	5.05	11.15
3920	Schools and colleges, 500 S.F., 1.92 ton	8.20	2.68	10.88
3960	1,000 S.F., 3.83 ton	6.90	4.45	11.35
4000	3,000 S.F., 11.50 ton	5.90	4.22	10.12
4040	5,000 S.F., 19.17 ton	5.05	3.89	8.94
4080	10,000 S.F., 38.33 ton	4.62	3.86	8.48

For expanded coverage of these items see *Means Mechanical or Plumbing Cost Data 1998*

MECHANICAL

8

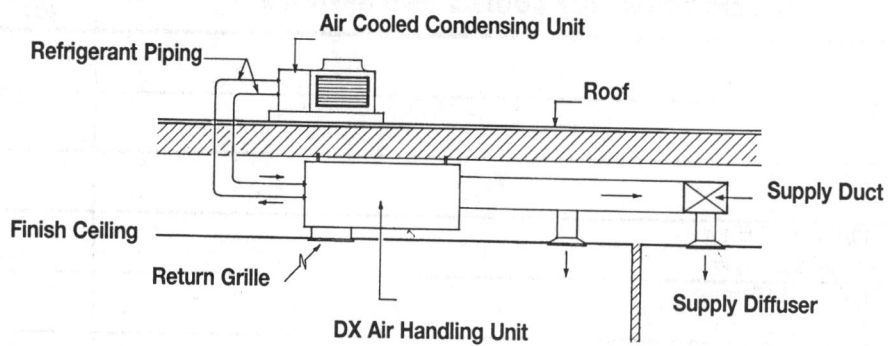

Air Cooled Condensing Unit

Refrigerant Piping

Roof

Supply Duct

Finish Ceiling

Supply Diffuser

Return Grille

DX Air Handling Unit

*Cooling requirements would lead to more than one system.

8.4-250	Split Systems With Air Cooled Condensing Units	COST PER S.F.		
		MAT.	INST.	TOTAL
1260	Split system, air cooled condensing unit			
1280	Apartment corridors, 1,000 S.F., 1.83 ton	1.65	1.43	3.08
1320	2,000 S.F., 3.66 ton	1.45	1.43	2.88
1360	5,000 S.F., 9.17 ton	1.58	1.81	3.39
1400	10,000 S.F., 18.33 ton	1.61	1.97	3.58
1440	20,000 S.F., 36.66 ton	1.64	2	3.64
1520	Banks and libraries, 1,000 S.F., 4.17 ton	3.31	3.28	6.59
1560	2,000 S.F., 8.33 ton	3.62	4.11	7.73
1600	5,000 S.F., 20.80 ton	3.67	4.48	8.15
1640	10,000 S.F., 41.66 ton	3.75	4.55	8.30
1680	20,000 S.F., 83.32 ton	4.55	4.73	9.28
1760	Bars and taverns, 1,000 S.F., 11.08 ton	8.80	6.45	15.25
1800	2,000 S.F., 22.16 ton	10.75	7.75	18.50
1840	5,000 S.F., 55.42 ton	8.80	7.25	16.05
1880	10,000 S.F., 110.84 ton	11.80	7.55	19.35
1920	20,000 S.F., 220 ton*			
2000	Bowling alleys, 1,000 S.F., 5.66 ton	4.62	6.10	10.72
2040	2,000 S.F., 11.33 ton	4.91	5.60	10.51
2080	5,000 S.F., 28.33 ton	4.99	6.10	11.09
2120	10,000 S.F., 56.66 ton	5.10	6.20	11.30
2160	20,000 S.F., 113.32 ton	7	6.70	13.70
2320	Department stores, 1,000 S.F., 2.92 ton	2.33	2.26	4.59
2360	2,000 S.F., 5.83 ton	2.38	3.16	5.54
2400	5,000 S.F., 14.58 ton	2.54	2.88	5.42
2440	10,000 S.F., 29.17 ton	2.57	3.14	5.71
2480	20,000 S.F., 58.33 ton	2.63	3.18	5.81
2560	Drug stores, 1,000 S.F., 6.66 ton	5.45	7.20	12.65
2600	2,000 S.F., 13.32 ton	5.80	6.55	12.35
2640	5,000 S.F., 33.33 ton	6	7.30	13.30
2680	10,000 S.F., 66.66 ton	7.40	7.60	15
2720	20,000 S.F., 133.32 ton*			
2800	Factories, 1,000 S.F., 3.33 ton	2.65	2.56	5.21
2840	2,000 S.F., 6.66 ton	2.72	3.60	6.32
2880	5,000 S.F., 16.66 ton	2.93	3.60	6.53
2920	10,000 S.F., 33.33 ton	2.99	3.65	6.64
2960	20,000 S.F., 66.66 ton	3.70	3.80	7.50
3040	Food supermarkets, 1,000 S.F., 2.83 ton	2.25	2.19	4.44
3080	2,000 S.F., 5.66 ton	2.31	3.06	5.37
3120	5,000 S.F., 14.66 ton	2.47	2.80	5.27
3160	10,000 S.F., 28.33 ton	2.50	3.05	5.55
3200	20,000 S.F., 56.66 ton	2.56	3.09	5.65
3280	Medical centers, 1,000 S.F., 2.33 ton	1.88	1.77	3.65
3320	2,000 S.F., 4.66 ton	1.90	2.53	4.43
3360	5,000 S.F., 11.66 ton	2.03	2.31	4.34

Important: See the Reference Section for critical supporting data - Location Factors & Historical Cost Indexes

8.4-250	Split Systems With Air Cooled Condensing Units	COST PER S.F.		
		MAT.	INST.	TOTAL
3400	10,000 S.F., 23.33 ton	2.06	2.52	4.58
3440	20,000 S.F., 46.66 ton	2.10	2.55	4.65
3520	Offices, 1,000 S.F., 3.17 ton	2.52	2.45	4.97
3560	2,000 S.F., 6.33 ton	2.58	3.43	6.01
3600	5,000 S.F., 15.83 ton	2.75	3.12	5.87
3640	10,000 S.F., 31.66 ton	2.79	3.42	6.21
3680	20,000 S.F., 63.32 ton	3.52	3.62	7.14
3760	Restaurants, 1,000 S.F., 5.00 ton	4.08	5.40	9.48
3800	2,000 S.F., 10.00 ton	4.34	4.94	9.28
3840	5,000 S.F., 25.00 ton	4.41	5.40	9.81
3880	10,000 S.F., 50.00 ton	4.50	5.50	10
3920	20,000 S.F., 100.00 ton	6.15	5.90	12.05
4000	Schools and colleges, 1,000 S.F., 3.83 ton	3.03	3.02	6.05
4040	2,000 S.F., 7.66 ton	3.33	3.78	7.11
4080	5,000 S.F., 19.17 ton	3.37	4.13	7.50
4120	10,000 S.F., 38.33 ton	3.44	4.19	7.63

For expanded coverage of these items see *Means Mechanical or Plumbing Cost Data 1998*

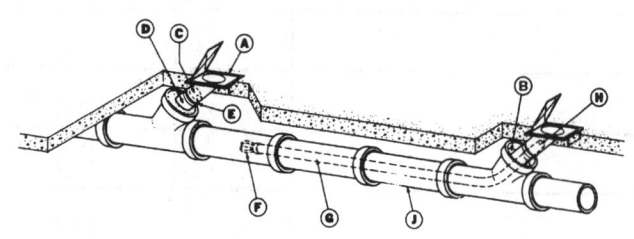

Vitrified Clay Garage Exhaust System

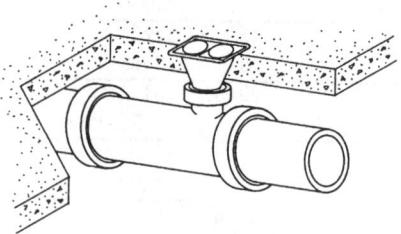

Dual Exhaust System

8.5-110	Garage Exhaust Systems	COST PER BAY		
		MAT.	INST.	TOTAL
1040	Garage, single exhaust, 3" outlet, cars & light trucks, one bay	1,675	760	2,435
1060	Additional bays up to seven bays	305	113	418
1500	4" outlet, trucks, one bay	1,675	760	2,435
1520	Additional bays up to six bays	315	113	428
1600	5" outlet, diesel trucks, one bay	1,725	760	2,485
1650	Additional single bays up to six	375	131	506
1700	Two adjoining bays	1,725	760	2,485
2000	Dual exhaust, 3" outlets, pair of adjoining bays	1,975	850	2,825
2100	Additional pairs of adjoining bays	570	131	701

For information about Means Estimating Seminars, see yellow pages 11 and 12 in back of book

Division 9
Electrical

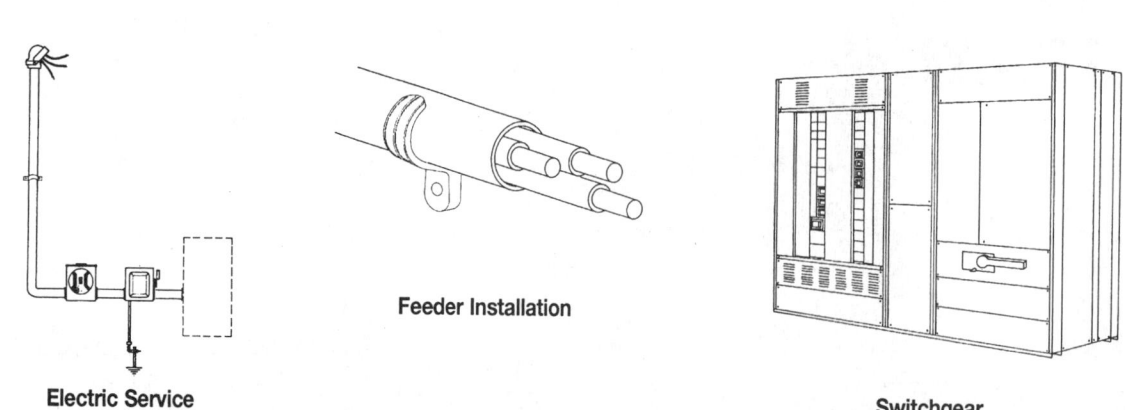

Electric Service

Feeder Installation

Switchgear

9.1-210	Electric Service, 3 Phase - 4 Wire	COST EACH		
		MAT.	INST.	TOTAL
0200	Service installation, includes breakers, metering, 20' conduit & wire			
0220	3 phase, 4 wire, 120/208 volts, 60 amp	500	570	1,070
0240	100 amps	620	685	1,305
0280	200 amps	920	1,050	1,970
0320	400 amps	2,150	1,925	4,075
0360	600 amps	4,100	2,625	6,725
0400	800 amps	5,375	3,150	8,525
0440	1000 amps	6,875	3,600	10,475
0480	1200 amps	8,200	3,700	11,900
0520	1600 amps	16,800	5,325	22,125
0560	2000 amps	18,600	6,050	24,650
0570	Add 25% for 277/480 volt			

9.1-310	Feeder Installation	COST PER L.F.		
		MAT.	INST.	TOTAL
0200	Feeder installation 600 volt, including conduit and wire, 60 amperes	3.14	6.85	9.99
0240	100 amperes	5.70	9.10	14.80
0280	200 amperes	12.30	14.10	26.40
0320	400 amperes	24.50	28	52.50
0360	600 amperes	51.50	45.50	97
0400	800 amperes	66	55	121
0440	1000 amperes	84.50	70	154.50
0480	1200 amperes	94	71.50	165.50
0520	1600 amperes	132	110	242
0560	2000 amperes	169	140	309

9.1-410	Switchgear	COST EACH		
		MAT.	INST.	TOTAL
0200	Switchgear inst., incl. swbd., panels & circ bkr, 400 amps, 120/208volt	3,375	2,300	5,675
0240	600 amperes	7,875	3,150	11,025
0280	800 amperes	10,100	4,450	14,550
0320	1200 amperes	13,600	6,800	20,400
0360	1600 amperes	18,900	9,550	28,450
0400	2000 amperes	23,600	12,100	35,700
0410	Add 20% for 277/480 volt			

Important: See the Reference Section for critical supporting data - Location Factors & Historical Cost Indexes

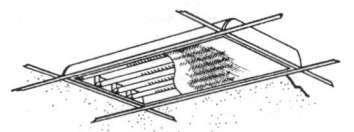

Fluorescent Fixture

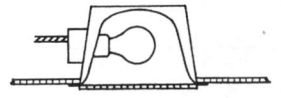

Incandescent Fixture

9.2-213	Fluorescent Fixtures (by Wattage)	COST PER S.F.		
		MAT.	INST.	TOTAL
0190	Fluorescent fixtures recess mounted in ceiling			
0200	1 watt per S.F., 20 FC, 5 fixtures per 1000 S.F.	.53	1.07	1.60
0240	2 watts per S.F., 40 FC, 10 fixtures per 1000 S.F.	1.06	2.13	3.19
0280	3 watts per S.F., 60 FC, 15 fixtures per 1000 S.F	1.59	3.21	4.80
0320	4 watts per S.F., 80 FC, 20 fixtures per 1000 S.F.	2.11	4.25	6.36
0400	5 watts per S.F., 100 FC, 25 fixtures per 1000 S.F.	2.64	5.35	7.99

9.2-223	Incandescent Fixture (by Wattage)	COST PER S.F.		
		MAT.	INST.	TOTAL
0190	Incandescent fixture recess mounted, type A			
0200	1 watt per S.F., 8 FC, 6 fixtures per 1000 S.F.	.51	.87	1.38
0240	2 watt per S.F., 16 FC, 12 fixtures per 1000 S.F.	1.05	1.74	2.79
0280	3 watt per S.F., 24 FC, 18 fixtures, per 1000 S.F.	1.57	2.57	4.14
0320	4 watt per S.F., 32 FC, 24 fixtures per 1000 S.F.	2.09	3.44	5.53
0400	5 watt per S.F., 40 FC, 30 fixtures per 1000 S.F.	2.60	4.31	6.91

9.2-235	H.I.D. Fixture, High Bay (by Wattage)	COST PER S.F.		
		MAT.	INST.	TOTAL
0190	High intensity discharge fixture, 16' above work plane			
0200	1 watt/S.F., type D, 23 FC, 1 fixture/1000 S.F.	.65	.87	1.52
0240	Type E, 42 FC, 1 fixture/1000 S.F.	.77	.90	1.67
0280	Type G, 52 FC, 1 fixture/1000 S.F.	.77	.90	1.67
0320	Type C, 54 FC, 2 fixture/1000 S.F.	.90	1.02	1.92
0400	2 watt/S.F., type D, 45 FC, 2 fixture/1000 S.F.	1.30	1.75	3.05
0440	Type E, 84 FC, 2 fixture/1000 S.F.	1.55	1.86	3.41
0480	Type G, 105 FC, 2 fixture/1000 S.F.	1.55	1.86	3.41
0520	Type C, 108 FC, 4 fixture/1000 S.F.	1.82	2.02	3.84
0600	3 watt/S.F., type D, 68 FC, 3 fixture/1000 S.F.	1.95	2.58	4.53
0640	Type E, 126 FC, 3 fixture/1000 S.F.	2.33	2.76	5.09
0680	Type G, 157 FC, 3 fixture/1000 S.F.	2.33	2.76	5.09
0720	Type C, 162 FC, 6 fixture/1000 S.F.	2.72	3.04	5.76
0800	4 watt/ S.F., type D, 91 FC, 4 fixture/1000 S.F.	2.80	4.08	6.88
0840	Type E, 168 FC, 4 fixture/1000 S.F.	3.10	3.71	6.81
0880	Type G, 210 FC, 4 fixture/1000 S.F.	3.10	3.71	6.81
0920	Type C, 243 FC, 9 fixture/1000 S.F.	3.98	4.24	8.22
1000	5 watt/S.F., type D, 113 FC, 5 fixture/1000 S.F.	3.26	4.31	7.57
1040	Type E, 210 FC, 5 fixture/1000 S.F.	3.89	4.60	8.49
1080	Type G, 262 FC, 5 fixture/1000 S.F.	3.89	4.60	8.49
1120	Type C, 297 FC, 11 fixture/1000 S.F.	4.89	5.25	10.14

ELECTRICAL

9

9.2-239	H.I.D. Fixture, High Bay (by Wattage)	COST PER S.F.		
		MAT.	INST.	TOTAL
0190	High intensity discharge fixture, 30' above work plane			
0200	1 watt/S.F., type D, 23 FC, 1 fixture/1000 S.F.	.74	1.10	1.84
0240	Type E, 37 FC, 1 fixture/1000 S.F.	.86	1.14	2
0280	Type G, 45 FC., 1 fixture/1000 S.F.	.86	1.14	2
0320	Type F, 50 FC, 1 fixture/1000 S.F.	.72	.89	1.61
0400	2 watt/S.F., type D, 40 FC, 2 fixtures/1000 S.F.	1.44	2.20	3.64
0440	Type E, 74 FC, 2 fixtures/1000 S.F.	1.68	2.30	3.98
0480	Type G, 92 FC, 2 fixtures/1000 S.F.	1.68	2.30	3.98
0520	Type F, 100 FC, 2 fixtures/1000 S.F.	1.45	1.84	3.29
0600	3 watt/S.F., type D, 60 FC, 3 fixtures/1000 S.F.	2.19	3.30	5.49
0640	Type E, 110 FC, 3 fixtures/1000 S.F.	2.56	3.48	6.04
0680	Type G, 138FC, 3 fixtures/1000 S.F.	2.56	3.48	6.04
0720	Type F, 150 FC, 3 fixtures/1000 S.F.	2.18	2.73	4.91
0800	4 watt/ S.F., type D, 80 FC, 4 fixtures/1000 S.F.	2.90	4.38	7.28
0840	Type E, 148 FC, 4 fixtures/1000 S.F.	3.40	4.65	8.05
0880	Type G, 185 FC, 4 fixtures/1000 S.F.	3.40	4.65	8.05
0920	Type F, 200 FC, 4 fixtures/1000 S.F.	2.92	3.68	6.60
1000	5 watt/ S.F., type D, 100 FC 5 fixtures/1000 S.F.	3.64	5.50	9.14
1040	Type E, 185 FC, 5 fixtures/1000 S.F.	4.26	5.80	10.06
1080	Type G, 230 FC, 5 fixtures/1000 S.F.	4.26	5.80	10.06
1120	Type F, 250 FC, 5 fixtures/1000 S.F.	3.64	4.56	8.20

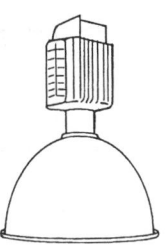

High Bay Fixture

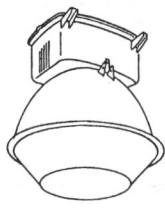

Low Bay Fixture

9.2-242	H.I.D. Fixture, Low Bay (by Wattage)	COST PER S.F.		
		MAT.	INST.	TOTAL
0190	High intensity discharge fixture, 8'-10' above work plane			
0200	1 watt/S.F., type H, 19 FC, 4 fixtures/1000 S.F.	1.70	1.86	3.56
0240	Type J, 30 FC, 4 fixtures/1000 S.F.	1.91	1.88	3.79
0280	Type K, 29 FC, 5 fixtures/1000 S.F.	1.83	1.68	3.51
0360	2 watt/S.F. type H, 33 FC, 7 fixtures/1000 S.F.	3.08	3.58	6.66
0400	Type J, 52 FC, 7 fixtures/1000 S.F.	3.39	3.44	6.83
0440	Type K, 63 FC, 11 fixtures/1000 S.F.	3.97	3.55	7.52
0520	3 watt/S.F., type H, 51 FC, 11 fixtures/1000 S.F.	4.78	5.45	10.23
0560	Type J, 81 FC, 11 fixtures/1000 S.F.	5.25	5.20	10.45
0600	Type K, 92 FC, 16 fixtures/1000 S.F.	5.80	5.20	11
0680	4 watt/S.F., type H, 65 FC, 14 fixtures/1000 S.F.	6.15	7.20	13.35
0720	Type J, 103 FC, 14 fixtures/1000 S.F.	6.75	6.85	13.60
0760	Type K, 127 FC, 22 fixtures/1000 S.F.	7.95	7.10	15.05
0840	5 watt/S.F., type H, 84 FC, 18 fixtures/1000 S.F.	7.85	9.05	16.90
0880	Type J, 133 FC, 18 fixtures/1000 S.F.	8.65	8.65	17.30
0920	Type K, 155 FC, 27 fixtures/1000 S.F.	9.75	8.75	18.50

ELECTRICAL

9

9.2-244	H.I.D. Fixture, Low Bay (by Wattage)	COST PER S.F.		
		MAT.	INST.	TOTAL
0190	High intensity discharge fixture, mounted 16' above work plane			
0200	1 watt/S.F., type H, 19 FC, 4 fixtures/1000 S.F.	1.78	2.11	3.89
0240	Type J, 28 FC, 4 fixt./1000 S.F.	1.99	2.13	4.12
0280	Type K, 27 FC, 5 fixt./1000 S.F.	2.05	2.39	4.44
0360	2 watts/S.F., type H, 30 FC, 7 fixt/1000 S.F.	3.24	4.09	7.33
0400	Type J, 48 FC, 7 fixt/1000 S.F.	3.60	4.12	7.72
0440	Type K, 58 FC, 11 fixt/1000 S.F.	4.40	4.88	9.28
0520	3 watts/S.F., type H, 47 FC, 11 fixt/1000 S.F.	5.05	6.20	11.25
0560	Type J, 75 FC, 11 fixt/1000 S.F.	5.60	6.25	11.85
0600	Type K, 85 FC, 16 fixt/1000 S.F.	6.45	7.25	13.70
0680	4 watts/S.F., type H, 60 FC, 14 fixt/1000 S.F.	6.50	8.20	14.70
0720	Type J, 95 FC, 14 fixt/1000 S.F.	7.20	8.25	15.45
0760	Type K, 117 FC, 22 fixt/1000 S.F.	8.80	9.75	18.55
0840	5 watts/S.F., type H, 77 FC, 18 fixt/1000 S.F.	8.35	10.60	18.95
0880	Type J, 122 FC, 18 fixt/1000 S.F.	9.20	10.35	19.55
0920	Type K, 143 FC, 27 fixt/1000 S.F.	10.85	12.15	23

Light Pole

9.2-252	Light Pole (Installed)	COST EACH		
		MAT.	INST.	TOTAL
0200	Light pole, aluminum, 20' high, 1 arm bracket	780	675	1,455
0240	2 arm brackets	865	675	1,540
0280	3 arm brackets	945	700	1,645
0320	4 arm brackets	1,025	700	1,725
0360	30' high, 1 arm bracket	1,375	845	2,220
0400	2 arm brackets	1,450	845	2,295
0440	3 arm brackets	1,525	870	2,395
0480	4 arm brackets	1,600	870	2,470
0680	40' high, 1 arm bracket	1,650	1,125	2,775
0720	2 arm brackets	1,725	1,125	2,850
0760	3 arm brackets	1,825	1,150	2,975
0800	4 arm brackets	1,900	1,150	3,050
0840	Steel, 20' high, 1 arm bracket	980	715	1,695
0880	2 arm brackets	1,050	715	1,765
0920	3 arm brackets	1,075	740	1,815
0960	4 arm brackets	1,150	740	1,890
1000	30' high, 1 arm bracket	1,125	905	2,030
1040	2 arm brackets	1,200	905	2,105
1080	3 arm brackets	1,225	930	2,155
1120	4 arm brackets	1,300	930	2,230
1320	40' high, 1 arm bracket	1,500	1,225	2,725
1360	2 arm brackets	1,575	1,225	2,800
1400	3 arm brackets	1,575	1,250	2,825
1440	4 arm brackets	1,675	1,250	2,925

ELECTRICAL

9

Duplex Wall Receptacle

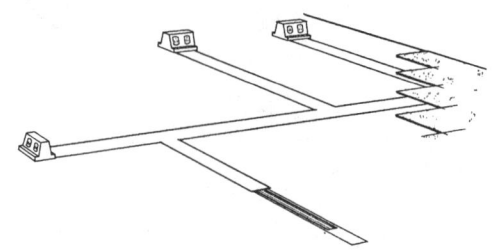

Undercarpet Receptacle System

9.2-522	Receptacle (by Wattage)	COST PER S.F.		
		MAT.	INST.	TOTAL
0190	Receptacles include plate, box, conduit, wire & transformer when required			
0200	2.5 per 1000 S.F., .3 watts per S.F.	.24	.81	1.05
0240	With transformer	.27	.85	1.12
0280	4 per 1000 S.F., .5 watts per S.F.	.28	.95	1.23
0320	With transformer	.33	1.01	1.34
0360	5 per 1000 S.F., .6 watts per S.F.	.32	1.13	1.45
0400	With transformer	.38	1.21	1.59
0440	8 per 1000 S.F., .9 watts per S.F.	.35	1.26	1.61
0480	With transformer	.44	1.37	1.81
0520	10 per 1000 S.F., 1.2 watts per S.F.	.36	1.36	1.72
0560	With transformer	.51	1.54	2.05
0600	16.5 per 1000 S.F., 2.0 watts per S.F.	.44	1.70	2.14
0640	With transformer	.69	2.02	2.71
0680	20 per 1000 S.F., 2.4 watts per S.F.	.47	1.85	2.32
0720	With transformer	.76	2.22	2.98

9.2-524	Receptacles	COST PER S.F.		
		MAT.	INST.	TOTAL
0200	Receptacle systems, underfloor duct, 5' on center, low density	3	1.81	4.81
0240	High density	3.24	2.33	5.57
0280	7' on center, low density	2.41	1.55	3.96
0320	High density	2.65	2.07	4.72
0400	Poke thru fittings, low density	.87	.84	1.71
0440	High density	1.74	1.63	3.37
0520	Telepoles, using Romex, low density	.64	.55	1.19
0560	High density	1.28	1.09	2.37
0600	Using EMT, low density	.66	.71	1.37
0640	High density	1.35	1.46	2.81
0720	Conduit system with floor boxes, low density	.73	.62	1.35
0760	High density	1.47	1.25	2.72
0840	Undercarpet power system, 3 conductor with 5 conductor feeder, low density	1.03	.24	1.27
0880	High density	2.04	.45	2.49

Important: See the Reference Section for critical supporting data - Location Factors & Historical Cost Indexes

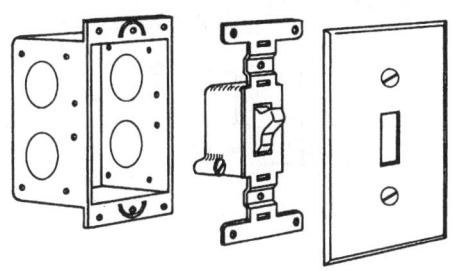

Description: Table 9.2-542 includes the cost for switch, plate, box, conduit in slab or EMT exposed and copper wire. Add 20% for exposed conduit.

No power required for switches.

Federal energy guidelines recommend the maximum lighting area controlled per switch shall not exceed 1000 S.F. and that areas over 500 S.F. shall be so controlled that total illumination can be reduced by at least 50%.

9.2-542	Wall Switch by Sq. Ft.	COST PER S.F.		
		MAT.	INST.	TOTAL
0200	Wall switches, 1.0 per 1000 S.F.	.04	.13	.17
0240	1.2 per 1000 S.F.	.04	.15	.19
0280	2.0 per 1000 S.F.	.05	.21	.26
0320	2.5 per 1000 S.F.	.08	.26	.34
0360	5.0 per 1000 S.F.	.16	.57	.73
0400	10.0 per 1000 S.F.	.34	1.16	1.50

9.2-582	Miscellaneous Power	COST PER S.F.		
		MAT.	INST.	TOTAL
0200	Miscellaneous power, to .5 watts	.02	.07	.09
0240	.8 watts	.03	.10	.13
0280	1 watt	.04	.12	.16
0320	1.2 watts	.05	.14	.19
0360	1.5 watts	.06	.17	.23
0400	1.8 watts	.07	.20	.27
0440	2 watts	.08	.24	.32
0480	2.5 watts	.09	.29	.38
0520	3 watts	.11	.34	.45

9.2-610	Central A. C. Power (by Wattage)	COST PER S.F.		
		MAT.	INST.	TOTAL
0200	Central air conditioning power, 1 watt	.05	.14	.19
0220	2 watts	.05	.16	.21
0240	3 watts	.07	.19	.26
0280	4 watts	.10	.25	.35
0320	6 watts	.18	.34	.52
0360	8 watts	.21	.36	.57
0400	10 watts	.28	.42	.70

ELECTRICAL

9

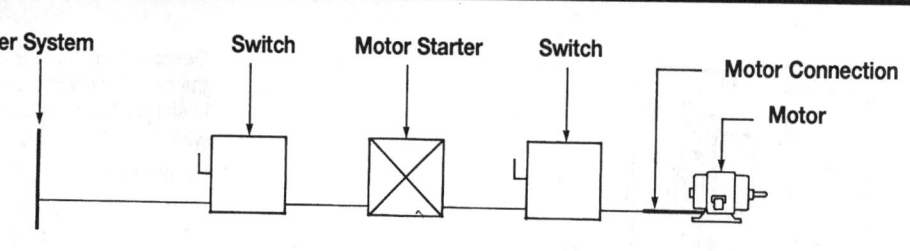

Motor Installation

9.2-710	Motor Installation	COST EACH		
		MAT.	INST.	TOTAL
0200	Motor installation, single phase, 115V, to and including 1/3 HP motor size	385	560	94
0240	To and incl. 1 HP motor size	400	560	96
0280	To and incl. 2 HP motor size	435	595	1,03
0320	To and incl. 3 HP motor size	490	605	1,09
0360	230V, to and including 1 HP motor size	385	565	95
0400	To and incl. 2 HP motor size	405	565	97
0440	To and incl. 3 HP motor size	455	610	1,06
0520	Three phase, 200V, to and including 1-1/2 HP motor size	440	625	1,06
0560	To and incl. 3 HP motor size	475	680	1,15
0600	To and incl. 5 HP motor size	525	755	1,28
0640	To and incl. 7-1/2 HP motor size	540	770	1,31
0680	To and incl. 10 HP motor size	880	965	1,845
0720	To and incl. 15 HP motor size	1,200	1,075	2,275
0760	To and incl. 20 HP motor size	1,475	1,250	2,725
0800	To and incl. 25 HP motor size	1,500	1,250	2,750
0840	To and incl. 30 HP motor size	2,400	1,475	3,875
0880	To and incl. 40 HP motor size	2,850	1,725	4,575
0920	To and incl. 50 HP motor size	5,200	2,025	7,225
0960	To and incl. 60 HP motor size	5,325	2,125	7,450
1000	To and incl. 75 HP motor size	6,700	2,450	9,150
1040	To and incl. 100 HP motor size	13,700	2,875	16,575
1080	To and incl. 125 HP motor size	14,000	3,150	17,150
1120	To and incl. 150 HP motor size	16,600	3,700	20,300
1160	To and incl. 200 HP motor size	20,000	4,525	24,525
1240	230V, to and including 1-1/2 HP motor size	420	615	1,035
1280	To and incl. 3 HP motor size	455	670	1,125
1320	To and incl. 5 HP motor size	505	745	1,250
1360	To and incl. 7-1/2 HP motor size	505	745	1,250
1400	To and incl. 10 HP motor size	805	915	1,720
1440	To and incl. 15 HP motor size	905	1,000	1,905
1480	To and incl. 20 HP motor size	1,400	1,200	2,600
1520	To and incl. 25 HP motor size	1,475	1,250	2,725
1560	To and incl. 30 HP motor size	1,500	1,250	2,750
1600	To and incl. 40 HP motor size	2,800	1,700	4,500
1640	To and incl. 50 HP motor size	2,925	1,775	4,700
1680	To and incl. 60 HP motor size	5,200	2,025	7,225
1720	To and incl. 75 HP motor size	6,175	2,300	8,475
1760	To and incl. 100 HP motor size	6,950	2,550	9,500
1800	To and incl. 125 HP motor size	14,000	2,975	16,975
1840	To and incl. 150 HP motor size	14,900	3,375	18,275
1880	To and incl. 200 HP motor size	16,000	3,750	19,750
1960	460V, to and including 2 HP motor size	515	620	1,135
2000	To and incl. 5 HP motor size	555	675	1,230
2040	To and incl. 10 HP motor size	590	745	1,335
2080	To and incl. 15 HP motor size	805	860	1,665
2120	To and incl. 20 HP motor size	825	920	1,745
2160	To and incl. 25 HP motor size	890	965	1,855
2200	To and incl. 30 HP motor size	1,175	1,050	2,225

Important: See the Reference Section for critical supporting data - Location Factors & Historical Cost Indexes

ELECTRICAL 9

9.2-710 Motor Installation

		COST EACH		
		MAT.	INST.	TOTAL
2240	To and incl. 40 HP motor size	1,475	1,125	2,600
2280	To and incl. 50 HP motor size	1,625	1,250	2,875
2320	To and incl. 60 HP motor size	2,525	1,450	3,975
2360	To and incl. 75 HP motor size	2,850	1,600	4,450
2400	To and incl. 100 HP motor size	3,100	1,800	4,900
2440	To and incl. 125 HP motor size	5,375	2,025	7,400
2480	To and incl. 150 HP motor size	6,525	2,275	8,800
2520	To and incl. 200 HP motor size	7,375	2,575	9,950
2600	575V, to and including 2 HP motor size	515	620	1,135
2640	To and incl. 5 HP motor size	555	675	1,230
2680	To and incl. 10 HP motor size	590	745	1,335
2720	To and incl. 20 HP motor size	805	860	1,665
2760	To and incl. 25 HP motor size	825	920	1,745
2800	To and incl. 30 HP motor size	1,175	1,050	2,225
2840	To and incl. 50 HP motor size	1,250	1,075	2,325
2880	To and incl. 60 HP motor size	2,500	1,450	3,950
2920	To and incl. 75 HP motor size	2,525	1,450	3,975
2960	To and incl. 100 HP motor size	2,850	1,600	4,450
3000	To and incl. 125 HP motor size	5,300	2,000	7,300
3040	To and incl. 150 HP motor size	5,375	2,025	7,400
3080	To and incl. 200 HP motor size	6,625	2,300	8,925

9.2-720 Motor Feeder

		COST PER L.F.		
		MAT.	INST.	TOTAL
0200	Motor feeder systems, single phase, feed up to 115V 1HP or 230V 2 HP	1.35	4.35	5.70
0240	115V 2HP, 230V 3HP	1.43	4.42	5.85
0280	115V 3HP	1.62	4.60	6.22
0360	Three phase, feed to 200V 3HP, 230V 5HP, 460V 10HP, 575V 10HP	1.42	4.69	6.11
0440	200V 5HP, 230V 7.5HP, 460V 15HP, 575V 20HP	1.54	4.79	6.33
0520	200V 10HP, 230V 10HP, 460V 30HP, 575V 30HP	1.83	5.05	6.88
0600	200V 15HP, 230V 15HP, 460V 40HP, 575V 50HP	2.36	5.80	8.16
0680	200V 20HP, 230V 25HP, 460V 50HP, 575V 60HP	3.39	7.35	10.74
0760	200V 25HP, 230V 30HP, 460V 60HP, 575V 75HP	3.78	7.45	11.23
0840	200V 30HP	4.20	7.70	11.90
0920	230V 40HP, 460V 75HP, 575V 100HP	5.30	8.40	13.70
1000	200V 40HP	6.10	9	15.10
1080	230V 50HP, 460V 100HP, 575V 125HP	7.35	9.95	17.30
1160	200V 50HP, 230V 60HP, 460V 125HP, 575V 150HP	8.60	10.55	19.15
1240	200V 60HP, 460V 150HP	10.45	12.35	22.80
1320	230V 75HP, 575V 200HP	12.10	12.85	24.95
1400	200V 75HP	13.80	13.15	26.95
1480	230V 100HP, 460V 200HP	17.75	15.30	33.05
1560	200V 100HP	24.50	19.15	43.65
1640	230V 125HP	28.50	20.50	49
1720	200V 125HP, 230V 150HP	29	24	53
1800	200V 150HP	34	26	60
1880	200V 200HP	49	38.50	87.50
1960	230V 200HP	45.50	28.50	74

9 ELECTRICAL

For expanded coverage of these items see *Means Electrical Cost Data 1998*

9.4-100	Communication & Alarm Systems	COST EACH		
		MAT.	INST.	TOTAL
0200	Communication & alarm systems, includes outlets, boxes, conduit & wire			
0210	Sound system, 6 outlets	4,050	4,650	8,70
0220	12 outlets	5,600	7,450	13,05
0240	30 outlets	9,675	14,100	23,77
0280	100 outlets	29,000	47,100	76,10
0320	Fire detection systems, 12 detectors	2,150	3,875	6,02
0360	25 detectors	3,775	6,550	10,32
0400	50 detectors	7,250	12,800	20,05
0440	100 detectors	13,200	23,100	36,30
0480	Intercom systems, 6 stations	2,350	3,175	5,52
0560	25 stations	7,975	12,200	20,17
0640	100 stations	30,500	44,700	75,20
0680	Master clock systems, 6 rooms	3,675	5,200	8,87
0720	12 rooms	5,450	8,875	14,32
0760	20 rooms	7,300	12,600	19,90
0800	30 rooms	11,600	23,100	34,700
0840	50 rooms	18,400	39,000	57,40
0920	Master TV antenna systems, 6 outlets	2,050	3,275	5,32
0960	12 outlets	3,825	6,125	9,95
1000	30 outlets	7,525	14,200	21,725
1040	100 outlets	24,900	46,200	71,10

Important: See the Reference Section for critical supporting data - Location Factors & Historical Cost Indexes

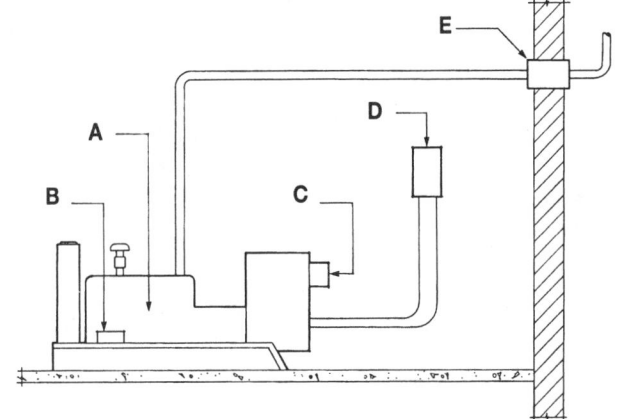

Generator System

A: Engine
B: Battery
C: Charger
D: Transfer Switch
E: Muffler

9.4-310	Generators (by KW)	COST PER KW		
		MAT.	INST.	TOTAL
0190	Generator sets, include battery, charger, muffler & transfer switch			
0200	Gas/gasoline operated, 3 phase, 4 wire, 277/480 volt, 7.5 KW	880	171	1,051
0240	10 KW	905	151	1,056
0280	15 KW	735	112	847
0320	30 KW	440	65	505
0360	70 KW	325	38.50	363.50
0400	85 KW	292	38	330
0440	115 KW	390	33	423
0480	170 KW	395	25	420
0560	Diesel engine with fuel tank, 30 KW	585	65	650
0600	50 KW	430	50.50	480.50
0640	75 KW	375	40.50	415.50
0680	100 KW	315	34.50	349.50
0760	150 KW	257	27.50	284.50
0840	200 KW	209	22	231
0880	250 KW	183	18.50	201.50
0960	350 KW	169	15.15	184.15
1040	500 KW	167	11.85	178.85

For information about Means Estimating Seminars, see yellow pages 11 and 12 in back of book

ELECTRICAL

9

Division 11
Special Construction

11.1-100	Architectural Specialties/Each	COST EACH		
		MAT.	INST.	TOTAL
1000	Specialties, bathroom accessories, st. steel, curtain rod, 5' long, 1" diam	30	25.50	55.5
1100	Dispenser, towel, surface mounted	45	21	66
1200	Grab bar, 1-1/4" diam., 12" long	39	13.90	52.9
1300	Mirror, framed with shelf, 18" x 24"	99.50	16.65	116.1
1340	72" x 24"	680	55.50	735.5
1400	Toilet tissue dispenser, surface mounted, single roll	10.55	11.10	21.6
1500	Towel bar, 18" long	30	14.50	44.5
1600	Canopies, wall hung, prefinished aluminum, 8' x 10'	1,250	1,050	2,300
1700	Chutes, linen or refuse incl. sprinklers, galv. steel, 18" diam., per floor	800	216	1,016
1740	Aluminized steel, 36" diam., per floor	1,075	271	1,346
1800	Mail, 8-3/4" x 3-1/2", aluminum & glass, per floor	570	152	722
1840	Bronze or stainless, per floor	885	168	1,053
1900	Control boards, magnetic, porcelain finish, framed, 24" x 18"	106	83.50	189.5
1940	96" x 48"	480	133	613
2000	Directory boards, outdoor, black plastic, 36" x 24"	605	335	940
2040	36" x 36"	695	445	1,140
2100	Indoor, economy, open faced, 18" x 24"	86.50	95	181.5
2300	Disappearing stairways, folding, pine, 8'-6" ceiling	91.50	83.50	175
2400	Fire escape, galvanized steel, 8'-0" to 10'-6" ceiling	1,225	665	1,890
2500	Automatic electric, wood, 8' to 9' ceiling	6,075	665	6,740
2600	Display cases, freestanding, glass and aluminum, 3'-6" x 3' x 1'-0" deep	490	83.50	573.5
2700	Wall mounted, glass and aluminum, 3' x 4' x 1'-4" deep	940	133	1,073
2900	Fireplace prefabricated, freestanding or wall hung, economy	995	256	1,251
2940	Deluxe	2,875	370	3,245
3000	Woodburning stoves, cast iron, economy	730	515	1,245
3040	Deluxe	2,650	835	3,485
3100	Flagpoles, on grade, aluminum, tapered, 20' high	710	395	1,105
3140	70' high	6,125	985	7,110
3240	59' high	6,150	875	7,025
3300	Concrete, internal halyard, 20' high	1,325	315	1,640
3340	100' high	13,400	785	14,185
3400	Lockers, steel, single tier, 5' to 6' high, per opening, min.	108	27	135
3440	Maximum	184	31.50	215.50
3700	Mail boxes, horizontal, rear loaded, aluminum, 5" x 6" x 15" deep	37	9.80	46.80
3740	Front loaded, aluminum, 10" x 12" x 15" deep	141	16.65	157.65
3800	Vertical, front loaded, aluminum, 15" x 5" x 6" deep	28.50	9.80	38.30
3840	Bronze, duranodic finish	53.50	9.80	63.30
3900	Letter slot, post office	208	41.50	249.50
4000	Mail counter, window, post office, with grille	560	167	727
4100	Medicine cabinets, sliding mirror doors, 20" x 16" x 4-3/4", unlighted	88.50	47.50	136
4140	24" x 19" x 8-1/2", lighted	149	66.50	215.50
4400	Partitions, shower stall, single wall, painted steel, 2'-8" x 2'-8"	625	152	777
4440	Fiberglass, 2'-8" x 2'-8"	510	168	678
4500	Double wall, enameled steel, 2'-8" x 2'-8"	635	152	787
4540	Stainless steel, 2'-8" x 2'-8"	1,150	152	1,302
4700	Tub enclosure, sliding panels, tempered glass, aluminum frame	275	189	464
4740	Chrome/brass frame-deluxe	785	253	1,038
5400	Projection screens, wall hung, manual operation, 50 S.F., economy	293	66.50	359.50
5440	Electric operation, 100 S.F., deluxe	1,575	335	1,910
5500	Scales, dial type, built in floor, 5 ton capacity, 8' x 6' platform	5,050	2,000	7,050
5540	10 ton capacity, 9' x 7' platform	9,525	2,850	12,375
5600	Truck (including weigh bridge), 20 ton capacity, 24' x 10'	9,600	3,325	12,925
5900	Security gates-scissors type, painted steel, single, 6' high, 5-1/2' wide	125	218	343
5940	Double gate, 7-1/2' high, 14' wide	400	435	835
6000	Shelving, metal industrial, braced, 3' wide, 1' deep	19.95	7.45	27.40
6040	2' deep	28.50	7.90	36.40
6200	Signs, interior electric exit sign, wall mounted, 6"	41.50	46	87.50
6240	Street, reflective alum., dbl. face, 4 way, w/bracket	79.50	22	101.50

11.1-100	Architectural Specialties/Each	COST EACH		
		MAT.	INST.	TOTAL
6300	Letters, cast aluminum, 1/2" deep, 4" high	15	18.50	33.50
6340	1" deep, 10" high	45.50	18.50	64
6400	Plaques, cast aluminum, 20" x 30"	890	167	1,057
6440	Cast bronze, 36" x 48"	3,500	335	3,835
6700	Turnstiles, one way, 4' arm, 46" diam., manual	268	133	401
6740	Electric	945	555	1,500
6800	3 arm, 5'-5' diam. & 7' high, manual	2,475	665	3,140
6840	Electric	3,075	1,100	4,175

11.1-100	Architectural Specialties/S.F.	COST PER S.F.		
		MAT.	INST.	TOTAL
7100	Bulletin board, cork sheets, no frame, 1/4" thick	2.68	2.30	4.98
7200	Aluminum frame, 1/4" thick, 3' x 5'	6.95	2.78	9.73
7500	Chalkboards, wall hung, alum, frame & chalktrough	9.60	1.47	11.07
7540	Wood frame & chalktrough	8.85	1.60	10.45
7600	Sliding board, one board with back panel	31	1.42	32.42
7640	Two boards with back panel	47	1.42	48.42
7700	Liquid chalk type, alum. frame & chalktrough	10.25	1.47	11.72
7740	Wood frame & chalktrough	17.50	1.47	18.97

11.1-100	Architectural Specialties/L.F.	COST PER L.F.		
		MAT.	INST.	TOTAL
8600	Partitions, hospital curtain, ceiling hung, polyester oxford cloth	12.05	3.25	15.30
8640	Designer oxford cloth	23.50	4.11	27.61

11.1-200	Architectural Equipment/Each	COST EACH		
		MAT.	INST.	TOTAL
0200	Arch. equip., appliances, range, cook top, 4 burner, economy	199	61.50	260.50
0240	Built in, single oven 30" wide, economy	455	13	468
0300	Standing, single oven-21" wide, economy	330	52	382
0340	Double oven-30" wide, deluxe	1,650	52	1,702
0400	Compactor, residential, economy	330	66.50	396.50
0440	Deluxe	460	111	571
0500	Dish washer, built-in, 2 cycles, economy	239	188	427
0540	4 or more cycles, deluxe	710	375	1,085
0600	Garbage disposer, sink type, economy	40.50	75	115.50
0640	Deluxe	169	75	244
0700	Refrigerator, no frost, 10 to 12 C.F., economy	455	52	507
0740	21 to 29 C.F., deluxe	2,500	173	2,673
0800	Automotive equipment, compressors, electric, 1-1/2 H.P., std. controls	2,700	630	3,330
0840	5 H.P., dual controls	3,800	945	4,745
0900	Hoists, single post, 4 ton capacity, swivel arms	3,850	2,350	6,200
0940	Dual post, 12 ton capacity, adjustable frame	6,600	6,300	12,900
1000	Lube equipment, 3 reel type, with pumps	6,600	1,900	8,500
1040	Product dispenser, 6 nozzles, w/vapor recovery, not incl. piping, installed	16,500		16,500
1100	Bank equipment, drive up window, drawer & mike, no glazing, economy	5,250	870	6,120
1140	Deluxe	7,625	1,750	9,375
1200	Night depository, economy	6,800	870	7,670
1240	Deluxe	10,800	1,750	12,550
1300	Pneumatic tube systems, 2 station, standard	20,800	2,825	23,625
1340	Teller, automated, 24 hour, single unit	39,200	2,825	42,025
1400	Teller window, bullet proof glazing, 44" x 60"	3,475	525	4,000
1440	Pass through, painted steel, 72" x 40"	2,425	1,100	3,525
1500	Barber equipment, chair, hydraulic, economy	440	13.90	453.90
1540	Deluxe	2,975	21	2,996
1700	Church equipment, altar, wood, custom, plain	1,375	238	1,613
1740	Granite, custom, deluxe	11,400	3,325	14,725
1800	Baptistry, fiberglass, economy	1,550	860	2,410
1840	Bells & carillons, keyboard operation	9,900	5,300	15,200
1900	Confessional, wood, single, economy	1,750	555	2,305
1940	Double, deluxe	16,400	1,675	18,075

11 **SPECIAL CONSTRUCTION**

11.1-200	Architectural Equipment/Each	COST EACH		
		MAT.	INST.	TOTAL
2000	Steeples, translucent fiberglas, 30" square, 15' high	1,875	1,075	2,950
2100	Checkout counter, single belt	2,125	52	2,177
2140	Double belt, power take-away	3,625	58	3,683
2600	Dental equipment, central suction system, economy	2,425	320	2,745
2640	Compressor-air, deluxe	6,825	690	7,515
2700	Chair, hydraulic, economy	3,325	690	4,015
2740	Deluxe	7,675	1,375	9,050
2800	Drill console with accessories, economy	4,175	216	4,391
2840	Deluxe	10,900	216	11,116
2900	X-ray unit, portable	4,700	86.50	4,786.50
2940	Panoramic unit	12,700	575	13,275
3000	Detention equipment, cell front rolling door, 7/8" bars, 5' x 7' high	4,475	930	5,405
3040	Cells, prefab., including front, 5' x 7' x 7' deep	6,975	1,225	8,200
3100	Doors and frames, 3' x 7', single plate	3,325	465	3,790
3140	Double plate	4,100	465	4,565
3200	Toilet apparatus, incl wash basin	2,175	570	2,745
3240	Visitor cubicle, vision panel, no intercom	2,525	930	3,455
3300	Dock bumpers, rubber blocks, 4-1/2" thick, 10" high, 14" long	42.50	12.80	55.30
3340	6" thick, 20" high, 11" long	102	25.50	127.50
3400	Dock boards, H.D., 5' x 5', aluminum, 5000 lb. capacity	1,125		1,125
3440	16,000 lb. capacity	1,550		1,550
3500	Dock levelers, hydraulic, 7' x 8', 10 ton capacity	6,125	915	7,040
3540	Dock lifters, platform, 6' x 6', portable, 3000 lb. capacity	6,000		6,000
3600	Dock shelters, truck, scissor arms, economy	1,025	335	1,360
3640	Deluxe	1,350	665	2,015
3700	Kitchen equipment, bake oven, single deck	3,475	86.50	3,561.50
3740	Broiler, without oven	3,350	86.50	3,436.50
3800	Commercial dish washer, semiautomatic, 50 racks/hr.	5,000	530	5,530
3840	Automatic, 275 racks/hr.	20,600	2,150	22,750
3900	Cooler, beverage, reach-in, 6 ft. long	2,600	115	2,715
3940	Food warmer, counter, 1.65 kw	745		745
4000	Fryers, with submerger, single	3,525	98.50	3,623.50
4040	Double	4,525	138	4,663
4100	Kettles, steam jacketed, 20 gallons	4,575	159	4,734
4140	Range, restaurant type, burners, 2 ovens and 24" griddle	4,150	115	4,265
4200	Range hood, incl. carbon dioxide system, economy	2,700	230	2,930
4240	Deluxe	19,400	690	20,090
4300	Laboratory equipment, glassware washer, distilled water, economy	4,850	420	5,270
4340	Deluxe	18,800	750	19,550
4440	Radio isotope	12,300		12,300
4600	Laundry equipment, dryers, gas fired, residential, 16 lb. capacity	550	128	678
4640	Commercial, 30 lb. capacity, single	2,425	128	2,553
4700	Dry cleaners, electric, 20 lb. capacity	30,300	3,750	34,050
4740	30 lb. capacity	41,500	5,000	46,500
4800	Ironers, commercial, 120" with canopy, 8 roll	146,000	10,700	156,700
4840	Institutional, 110", single roll	25,800	1,850	27,650
4900	Washers, residential, 4 cycle	630	128	758
4940	Commercial, coin operated, deluxe	2,625	128	2,753
5000	Library equipment, carrels, metal, economy	195	66.50	261.50
5040	Hardwood, deluxe	770	83.50	853.50
5100	Medical equipment, autopsy table, standard	6,600	385	6,985
5140	Deluxe	8,250	640	8,890
5200	Incubators, economy	2,875		2,875
5240	Deluxe	15,000		15,000
5300	Station, scrub-surgical, single, economy	3,900		3,900
5340	Dietary, medium, with ice	11,800		11,800
5400	Sterilizers, general purpose, single door, 20" x 20" x 28"	11,600		11,600
5440	Floor loading, double door, 28" x 67" x 52"	154,000		154,000
5500	Surgery tables, standard	10,500	620	11,120
5540	Deluxe	23,800	870	24,670

Important: See the Reference Section for critical supporting data - Location Factors & Historical Cost Indexes

11.1-200	Architectural Equipment/Each	COST EACH		
		MAT.	INST.	TOTAL
5600	Tables, standard, with base cabinets, economy	985	222	1,207
5640	Deluxe	2,925	335	3,260
5700	X-ray, mobile, economy	10,100		10,100
5740	Stationary, deluxe	164,500		164,500
5800	Movie equipment, changeover, economy	385		385
5840	Film transport, incl. platters and autowind, economy	7,825		7,825
5900	Lamphouses, incl. rectifiers, xenon, 1000W	6,475	184	6,659
5940	4000W	10,600	245	10,845
6000	Projector mechanisms, 35 mm, economy	10,900		10,900
6040	Deluxe	14,500		14,500
6100	Sound systems, incl. amplifier, single, economy	2,025	410	2,435
6140	Dual, Dolby/super sound	13,900	920	14,820
6200	Seating, painted steel, upholstered, economy	101	19.05	120.05
6240	Deluxe	260	24	284
6300	Parking equipment, automatic gates, 8 ft. arm, one way	2,775	670	3,445
6340	Traffic detectors, single treadle	2,050	305	2,355
6400	Booth for attendant, economy	4,400		4,400
6440	Deluxe	22,000		22,000
6500	Ticket printer/dispenser, rate computing	6,425	525	6,950
6540	Key station on pedestal	360	180	540
6700	Frozen food, chest type, 12 ft. long	4,575	211	4,786
7000	Safe, office type, 1 hr. rating, 34" x 20" x 20"	1,900		1,900
7040	4 hr. rating, 62" x 33" x 20"	7,575		7,575
7100	Data storage, 4 hr. rating, 63" x 44" x 16"	12,200		12,200
7140	Jewelers, 63" x 44" x 16"	20,900		20,900
7200	Money, "B" label, 9" x 14" x 14"	480		480
7240	Tool and torch resistive, 24" x 24" x 20"	8,250		8,250
7300	Sauna, prefabricated, incl. heater and controls, 7' high, 6' x 4'	3,225	505	3,730
7340	10' x 12'	7,800	1,100	8,900
7400	Heaters, wall mounted, to 200 C.F.	550		550
7440	Floor standing, to 1000 C.F., 12500 W	1,725	123	1,848
7500	School equipment, basketball backstops, wall mounted, wood, fixed	540	585	1,125
7540	Suspended type, electrically operated	2,650	1,125	3,775
7600	Bleachers-telescoping, manual operation, 15 tier, economy (per seat)	55	21	76
7640	Power operation, 30 tier, deluxe (per seat)	209	37	246
7700	Weight lifting gym, universal, economy	3,675	520	4,195
7740	Deluxe	13,200	1,050	14,250
7800	Scoreboards, basketball, 1 side, economy	2,525	515	3,040
7840	4 sides, deluxe	27,500	7,075	34,575
8300	Vocational shop equipment, benches, metal	248	133	381
8340	Wood	425	133	558
8400	Dust collector, not incl. ductwork, 6' diam.	2,350	345	2,695
8440	Planer, 13" x 6"	1,475	167	1,642
8500	Waste handling, compactors, single bag, 250 lbs./hr., hand fed	6,250	395	6,645
8540	Heavy duty industrial, 5 C.Y. capacity	16,300	1,900	18,200
8600	Incinerator, electric, 100 lbs./hr., economy	9,900	1,875	11,775
8640	Gas, 2000 lbs./hr., deluxe	171,000	14,600	185,600
8700	Shredder, no baling, 35 tons/hr.	211,000		211,000
8740	Incl. baling, 50 tons/day	422,000		422,000

SPECIAL CONSTRUCTION

11

11.1-200 Architectural Equipment/S.F.

		MAT.	INST.	TOTAL
8910	Arch. equip., lab equip., counter tops, acid proof, economy	15.40	8.15	23.55
8940	Stainless steel	65.50	8.15	73.65
9000	Movie equipment, projection screens, rigid in wall, acrylic, 1/4" thick	35	3.22	38.22
9040	1/2" thick	40.50	4.84	45.34
9100	School equipment, gym mats, naugahyde cover, 2" thick	4.19		4.19
9140	Wrestling, 1" thick, heavy duty	4.76		4.76
9200	Stage equipment, curtains, velour, medium weight	12.10	1.11	13.21
9240	Silica based yarn, fireproof	27.50	13.35	40.85
9300	Stages, portable with steps, folding legs, 8" high	7.65		7.65
9340	Telescoping platforms, aluminum, deluxe	44	17.30	61.30

11.1-200 Architectural Equipment/L.F.

		MAT.	INST.	TOTAL
9410	Arch. equip., church equip. pews, bench type, hardwood, economy	49	16.65	65.65
9440	Deluxe	79	22	101
9500	Laboratory equipment, cabinets, wall, open	70.50	33.50	104
9540	Base, drawer units	230	37	267
9600	Fume hoods, not incl. HVAC, economy	630	123	753
9640	Deluxe incl. fixtures	1,250	278	1,528
9700	Library equipment, book shelf, metal, single face, 90" high x 10" shelf	68	28	96
9740	Double face, 90" high x 10" shelf	194	60.50	254.50
9800	Charging desk, built-in, with counter, plastic laminate	415	47.50	462.50
9900	Stage equipment, curtain track, heavy duty	26	37	63
9940	Lights, border, quartz, colored	132	18.40	150.40

11.1-500 Furnishings/Each

		MAT.	INST.	TOTAL
1000	Furnishings, blinds, exterior, aluminum, louvered, 1'-4" wide x 3'-0" long	32.50	33.50	66
1040	1'-4" wide x 6'-8" long	52.50	37	89.50
1100	Hemlock, solid raised, 1-'4" wide x 3'-0" long	56	33.50	89.50
1140	1'-4" wide x 6'-9" long	105	37	142
1200	Polystyrene, louvered, 1'-3" wide x 3'-3" long	43	33.50	76.50
1240	1'-3" wide x 6'-8" long	93.50	37	130.50
1300	Interior, wood folding panels, louvered, 7" x 20" (per pair)	25	19.60	44.60
1340	18" x 40" (per pair)	79	19.60	98.60
1800	Hospital furniture, beds, manual, economy	865		865
1840	Deluxe	1,500		1,500
1900	All electric, economy	1,450		1,450
1940	Deluxe	3,575		3,575
2000	Patient wall systems, no utilities, economy, per room	870		870
2040	Deluxe, per room	1,600		1,600
2200	Hotel furnishings, standard room set, economy, per room	1,625		1,625
2240	Deluxe, per room	7,850		7,850
2400	Office furniture, standard employee set, economy, per person	360		360
2440	Deluxe, per person	2,025		2,025
2800	Posts, portable, pedestrian traffic control, economy	90		90
2840	Deluxe	315		315
3000	Restaurant furniture, booth, molded plastic, stub wall and 2 seats, economy	370	167	537
3040	Deluxe	790	222	1,012
3100	Upholstered seats, foursome, single-economy	530	66.50	596.50
3140	Foursome, double-deluxe	2,850	111	2,961
3300	Seating, lecture hall, pedestal type, economy	108	19.05	127.05
3340	Deluxe	335	33.50	368.50
3400	Auditorium chair, veneer construction	108	19.05	127.05
3440	Fully upholstered, spring seat	159	19.05	178.05

11.1-500 Furnishings/S.F.

		MAT.	INST.	TOTAL
4010	Furnishings, blinds-interior, venetian-aluminum, stock, 2" slats, economy	1.35	.56	1.91
4040	Custom, 1" slats, deluxe	8.10	.76	8.86

Important: See the Reference Section for critical supporting data - Location Factors & Historical Cost Indexes

11.1-500	**Furnishings/S.F.**	COST PER S.F.		
		MAT.	INST.	TOTAL
4100	Vertical, PVC or cloth, T&B track, economy	7.15	.72	7.87
4140	Deluxe	11.95	.83	12.78
4300	Draperies, unlined, economy	2.06		2.06
4440	Lightproof, deluxe	6.20		6.20
4700	Floor mats, recessed, inlaid black rubber, 3/8" thick, solid	21	1.68	22.68
4740	Colors, 1/2" thick, perforated	33	1.68	34.68
4800	Link-including nosings, steel-galvanized, 3/8" thick	5.70	1.68	7.38
4840	Vinyl, in colors	17.80	1.68	19.48
5000	Shades, mylar, wood roller, single layer, non-reflective	5.70	.49	6.19
5040	Metal roller, triple layer, heat reflective	10.05	.49	10.54
5100	Vinyl, light weight, 4 ga.	.41	.49	.90
5140	Heavyweight, 6 ga.	1.27	.49	1.76
5200	Vinyl coated cotton, lightproof decorator shades	1.27	.49	1.76
5300	Woven aluminum, 3/8" thick, light and fireproof	4.05	.95	5

11.1-500	**Furnishings/L.F.**	COST PER L.F.		
		MAT.	INST.	TOTAL
5510	Furnishings, cabinets, hospital, base, laminated plastic	188	66.50	254.50
5540	Stainless steel	235	66.50	301.50
5600	Counter top, laminated plastic, no backsplash	32.50	16.65	49.15
5640	Stainless steel	90	16.65	106.65
5700	Nurses station, door type, laminated plastic	218	66.50	284.50
5740	Stainless steel	251	66.50	317.50
5900	Household, base, hardwood, one top drawer & one door below x 12" wide	122	27	149
5940	Four drawer x 24" wide	190	30	220
6000	Wall, hardwood, 30" high with one door x 12" wide	91	30.50	121.50
6040	Two doors x 48" wide	188	36	224
6100	Counter top-laminated plastic, stock, economy	5.40	11.10	16.50
6140	Custom-square edge, 7/8" thick	11.35	25	36.35
6300	School, counter, wood, 32" high	149	33.50	182.50
6340	Metal, 84" high	230	44.50	274.50
6500	Dormitory furniture, desk top (built-in),laminated plastc, 24"deep, economy	24	13.35	37.35
6540	30" deep, deluxe	105	16.65	121.65
6600	Dressing unit, built-in, economy	177	55.50	232.50
6640	Deluxe	525	83.50	608.50
7000	Restaurant furniture, bars, built-in, back bar	142	66.50	208.50
7040	Front bar	195	66.50	261.50
7200	Wardrobes & coatracks, standing, steel, single pedestal, 30" x 18" x 63"	77.50		77.50
7300	Double face rack, 39" x 26" x 70"	131		131
7340	Wall mounted rack, steel frame & shelves, 12" x 15" x 26"	55	4.73	59.73
7400	12" x 15" x 50"	43.50	2.48	45.98

SPECIAL CONSTRUCTION

11

11.1-700	Special Construction/Each	COST EACH		
		MAT.	INST.	TOTAL
1000	Special construction, bowling alley incl. pinsetter, scorer etc., economy	35,500	6,675	42,175
1040	Deluxe	40,400	7,400	47,800
1100	For automatic scorer, economy, add	6,600		6,600
1140	Deluxe, add	7,475		7,475
1200	Chimney, metal, high temp. steel jacket, factory lining, 24″ diameter	7,750	3,050	10,800
1240	60″ diameter	58,000	13,300	71,300
1300	Poured concrete, brick lining, 10′ diam., 200′ high			1,120,000
1340	20′ diam., 500′ high			5,000,000
1400	Radial brick, 3′-6″ I.D., 75′ high			184,000
1440	7′ I.D., 175′ high			555,500
1500	Control tower, modular, 12′ x 10′, incl. instrumentation, economy			450,000
1540	Deluxe			600,000
1600	Garage costs, residential, prefab, wood, single car economy	2,800	665	3,465
1640	Two car deluxe	8,275	1,325	9,600
1700	Grandstands, permanent, closed deck, steel, economy (per seat)			25
1740	Deluxe (per seat)			80
1800	Composite design, economy (per seat)			30
1840	Deluxe (per seat)			90
1900	Hangars, prefab, galv. steel, bottom rolling doors, economy (per plane)	8,950	3,175	12,125
1940	Electrical bi-folding doors, deluxe (per plane)	12,500	4,375	16,875
2000	Ice skating rink, 85′ x 200′, 55° system, 5 mos., 100 ton			325,000
2040	90° system, 12 mos., 135 ton			350,000
2100	Dash boards, acrylic screens, polyethylene coated plywood	126,500	22,600	149,100
2140	Fiberglass and aluminum construction	137,500	22,600	160,100
2200	Integrated ceilings, radiant electric, 2′ x 4′ panel, manila finish	44.50	14.70	59.20
2240	ABS plastic finish	81.50	28.50	110
2300	Kiosks, round, 5′ diam., 1/4″ fiberglass wall, 8′ high	6,050		6,050
2340	Rectangular, 5′ x 9′, 1″ insulated dbl. wall fiberglass, 7′-6″ high	10,500		10,500
2400	Portable booth, acoustical, 27 db 1000 hz., 15 S.F. floor	3,175		3,175
2440	55 S.F. flr.	6,500		6,500
2500	Radio towers, guyed, 40 lb. section, 50′ high, 70 MPH basic wind speed	1,725	870	2,595
2540	90 lb. section, 400′ high, wind load 70 MPH basic wind speed	22,400	9,675	32,075
2600	Self supporting, 60′ high, 70 MPH basic wind speed	3,600	1,700	5,300
2640	190′ high, wind load 90 MPH basic wind speed	21,400	6,750	28,150
2700	Shelters, aluminum frame, acrylic glazing, 8′ high, 3′ x 9′	2,800	765	3,565
2740	9′ x 12′	4,300	1,200	5,500
2800	Shielding, lead x-ray protection, radiography room, 1/16″ lead, economy	5,050	2,500	7,550
2840	Deluxe	6,600	4,175	10,775
2900	Deep therapy x-ray, 1/4″ lead, economy	17,100	7,825	24,925
2940	Deluxe	23,100	10,400	33,500
3000	Shooting range incl. bullet traps, controls, separators, ceilings, economy	11,600	4,400	16,000
3040	Deluxe	22,100	7,400	29,500
3100	Silos, concrete stave, industrial, 12′ diam., 35′ high	16,500	13,900	30,400
3140	25′ diam., 75′ high	54,000	30,600	84,600
3200	Steel prefab, 30,000 gal., painted, economy	12,000	3,600	15,600
3240	Epoxy-lined, deluxe	24,000	7,150	31,150
3300	Sport court, squash, regulation, in existing building, economy			14,500
3340	Deluxe			27,100
3400	Racketball, regulation, in existing building, economy	1,925	6,300	8,225
3440	Deluxe	25,800	12,600	38,400
3500	Swimming pool equipment, diving stand, stainless steel, 1 meter	4,625	247	4,872
3540	3 meter	5,450	1,675	7,125
3600	Diving boards, 16 ft. long, aluminum	2,050	247	2,297
3640	Fiberglass	1,525	247	1,772
3700	Filter system, sand, incl. pump, 6000 gal./hr.	1,100	425	1,525
3800	Lights, underwater, 12 volt with transformer, 300W	190	920	1,110
3900	Slides, fiberglass with aluminum handrails & ladder, 6′ high, straight	4,550	415	4,965
3940	12′ high, straight with platform	14,100	555	14,655
4000	Tanks, steel, ground level, 100,000 gal.			96,000
4040	10,000,000 gal.			2,200,000

Important: See the Reference Section for critical supporting data - Location Factors & Historical Cost Indexes

11.1-700	Special Construction/Each	COST EACH		
		MAT.	INST.	TOTAL
4100	Elevated water, 50,000 gal.			163,000
4140	1,000,000 gal.			797,000
4200	Cypress wood, ground level, 3,000 gal.	5,500	6,625	12,125
4240	Redwood, ground level, 45,000 gal.	27,500	18,000	45,500

11.1-700	Special Construction/S.F.	COST PER S.F.		
		MAT.	INST.	TOTAL
4500	Special construction, acoustical, enclosure, 4″ thick, 8 psf panels	26.50	13.90	40.40
4540	Reverb chamber, 4″ thick, parallel walls	32	16.65	48.65
4600	Sound absorbing panels, 2′-6″ x 8′, painted metal	9.55	4.65	14.20
4640	Vinyl faced	7.75	4.17	11.92
4700	Flexible transparent curtain, clear	6	5.30	11.30
4740	With absorbing foam, 75% coverage	8.35	5.30	13.65
4800	Strip entrance, 2/3 overlap	3.91	8.40	12.31
4840	Full overlap	4.83	9.90	14.73
4900	Audio masking system, plenum mounted, over 10,000 S.F.	.57	.17	.74
4940	Ceiling mounted, under 5,000 S.F.	1.02	.31	1.33
5000	Air supported struc., polyester vinyl fabric, 24oz., warehouse, 5000 S.F.	16.95	.21	17.16
5040	50,000 S.F.	6.80	.17	6.97
5100	Tennis, 7,200 S.F.	14.30	.17	14.47
5140	24,000 S.F.	9.35	.17	9.52
5200	Woven polyethylene, 6 oz., shelter, 3,000 S.F.	6.50	.35	6.85
5240	24,000 S.F.	2.96	.17	3.13
5300	Teflon coated fiberglass, stadium cover, economy	38.50	.09	38.59
5340	Deluxe	44	.12	44.12
5400	Air supported storage tank covers, reinf. vinyl fabric, 12 oz., 400 S.F.	13.20	.29	13.49
5440	18,000 S.F.	3.91	.27	4.18
5500	Anechoic chambers, 7′ high, 100 cps cutoff, 25 S.F.			1,925
5540	200 cps cutoff, 100 S.F.			1,000
5600	Audiometric rooms, under 500 S.F.	42	13.60	55.60
5640	Over 500 S.F.	39.50	11.10	50.60
5700	Comfort stations, prefab, mobile on steel frame, economy	41		41
5740	Permanent on concrete slab, deluxe	220	31	251
5800	Darkrooms, shell, not including door, 240 S.F., 8′ high	35	5.55	40.55
5840	64 S.F., 12′ high	82.50	10.40	92.90
5900	Domes, bulk storage, wood framing, wood decking, 50′ diam.	27.50	1.08	28.58
5940	116′ diam.	16.50	1.25	17.75
6000	Steel framing, metal decking, 150′ diam.	29.50	7.20	36.70
6040	400′ diam.	24	5.50	29.50
6100	Geodesic, wood framing, wood panels, 30′ diam.	15.55	1.21	16.76
6140	60′ diam.	12.90	.76	13.66
6200	Aluminum framing, acrylic panels, 40′ diam.			75
6240	Aluminum panels, 400′ diam.			25
6300	Garden house, prefab, wood, shell only, 48 S.F.	14.80	3.33	18.13
6340	200 S.F.	27	13.90	40.90
6400	Greenhouse, shell-stock, residential, lean-to, 8′-6″ long x 3′-10″ wide	34	19.60	53.60
6440	Freestanding, 8′-6″ long x 13′-6″ wide	26.50	6.15	32.65
6500	Commercial-truss frame, under 2000 S.F., deluxe			38
6540	Over 5,000 S.F., economy			23
6600	Institutional-rigid frame, under 500 S.F., deluxe			99
6640	Over 2,000 S.F., economy			39
6700	Hangar, prefab, galv. roof and walls, bottom rolling doors, economy	8.55	2.87	11.42
6740	Electric bifolding doors, deluxe	10.10	3.74	13.84
6800	Integrated ceilings, Luminaire, suspended, 5′ x 5′ modules, 50% lighted	3.88	7.85	11.73
6840	100% lighted	5.80	14.15	19.95
6900	Dimensionaire, 2′ x 4′ module tile system, no air bar	2.35	1.33	3.68
6940	With air bar, deluxe	3.40	1.33	4.73
7000	Music practice room, modular, perforated steel, under 500 S.F.	23	9.50	32.50
7040	Over 500 S.F.	17.45	8.35	25.80
7100	Pedestal access floors, stl pnls, no stringers, vinyl covering, >6000 S.F.	11.40	1.48	12.88
7140	With stringers, high pressure laminate covering, under 6000 S.F.	10.20	1.67	11.87

For expanded coverage of these items see Means Assemblies Cost Data 1998

SPECIAL CONSTRUCTION

11

11.1-700 — Special Construction/S.F.

		MAT.	INST.	TOTAL
7200	Carpet covering, under 6000 S.F.	11.85	1.67	13.5
7240	Aluminum panels, no stringers, no covering	22.50	1.33	23.8
7300	Refrigerators, prefab, walk-in, 7'-6" high, 6' x 6'	105	12.15	117.1
7340	12' x 20'	66	6.10	72.1
7400	Shielding, lead, gypsum board, 5/8" thick, 1/16" lead	5.45	3.93	9.3
7440	1/8" lead	11.55	4.49	16.0
7500	Lath, 1/16" thick	5.20	4.64	9.8
7540	1/8" thick	11	5.20	16.2
7600	Radio frequency, copper, prefab screen type, economy	26.50	3.70	30.20
7640	Deluxe	34	4.60	38.60
7700	Swimming pool enclosure, transluscent, freestanding, economy	11	3.33	14.3
7740	Deluxe	275	9.50	284.50
7800	Swimming pools, residential, vinyl liner, metal sides	9.25	4.58	13.8
7840	Concrete sides	11.05	8.70	19.7
7900	Gunite shell, plaster finish, 350 S.F.	18.30	18	36.30
7940	800 S.F.	14.75	10.45	25.20
8000	Motel, gunite shell, plaster finish	22.50	22.50	45
8040	Municipal, gunite shell, tile finish, formed gutters	94	39.50	133.50
8100	Tension structures, steel frame, polyester vinyl fabric, 12,000 S.F.	9.10	1.54	10.64
8140	20,800 S.F.	9.80	1.39	11.19

11.1-700 — Special Construction/L.F.

		COST PER L.F.		
		MAT.	INST.	TOTAL
8500	Spec. const., air curtains, shipping & receiving, 8'high x 5'wide, economy	141	152	293
8540	20' high x 8' wide, heated, deluxe	1,250	291	1,541
8600	Customer entrance, 10' high x 5' wide, economy	208	152	360
8640	12' high x 4' wide, heated, deluxe	370	271	641

For information about Means Estimating Seminars, see yellow pages 11 and 12 in back of book

Important: See the Reference Section for critical supporting data - Location Factors & Historical Cost Indexes

SPECIAL CONSTRUCTION   11

Division 12
Site Work

12.3-110	Trenching	COST PER L.F.		
		MAT.	INST.	TOTAL
1310	Trenching, backhoe, 0 to 1 slope, 2' wide, 2' deep, 3/8 C.Y. bucket		2.37	2.
1330	4' deep, 3/8 C.Y. bucket		4.11	4.1
1360	10' deep, 1 C.Y. bucket		8.30	8.3
1400	4' wide, 2' deep, 3/8 C.Y. bucket		4.85	4.8
1420	4' deep, 1/2 C.Y. bucket		7.08	7.0
1450	10' deep, 1 C.Y. bucket		16.80	16.8
1480	18' deep, 2-1/2 C.Y. bucket		26.05	26.0
1520	6' wide, 6' deep, 5/8 C.Y. bucket		16.20	16.2
1540	10' deep, 1 C.Y. bucket		22.70	22.7
1570	20' deep, 3-1/2 C.Y. bucket		35.95	35.9
1640	8' wide, 12' deep, 1-1/4 C.Y. bucket		27.75	27.7
1680	24' deep, 3-1/2 C.Y. bucket		80.50	80.5
1730	10' wide, 20' deep, 3-1/2 C.Y. bucket		63	63
1740	24' deep, 3-1/2 C.Y. bucket		101.50	101.5
3500	1 to 1 slope, 2' wide, 2' deep, 3/8 C.Y. bucket		3.62	3.6
3540	4' deep, 3/8 C.Y. bucket		8.47	8.4
3600	10' deep, 1 C.Y. bucket		36.35	36.3
3800	4' wide, 2' deep, 3/8 C.Y. bucket		6.16	6.1
3840	4' deep, 1/2 C.Y. bucket		10.31	10.3
3900	10' deep, 1 C.Y. bucket		38.35	38.3
4030	6' wide, 6' deep, 5/8 C.Y. bucket		22.05	22.0
4050	10' deep, 1 C.Y. bucket		45.60	45.6
4080	20' deep, 3-1/2 C.Y. bucket		118	118
4500	8' wide, 12' deep, 1-1/4 C.Y. bucket		52.50	52.5
4650	24' deep, 3-1/2 C.Y. bucket		245	245
4800	10' wide, 20' deep, 3-1/2 C.Y. bucket		142.50	142.5
4850	24' deep, 3-1/2 C.Y. bucket		262	262

12.3-310	Pipe Bedding	COST PER L.F.		
		MAT.	INST.	TOTAL
1440	Pipe bedding, side slope 0 to 1, 1' wide, pipe size 6" diameter	.26	.61	.87
1460	2' wide, pipe size 8" diameter	.58	1.35	1.93
1500	Pipe size 12" diameter	.61	1.42	2.03
1600	4' wide, pipe size 20" diameter	1.53	3.51	5.04
1660	Pipe size 30" diameter	1.62	3.73	5.35
1680	6' wide, pipe size 32" diameter	2.83	6.50	9.33
1740	8' wide, pipe size 60" diameter	4.72	10.85	15.57
1780	12' wide, pipe size 84" diameter	9.25	21.50	30.75

12.3-710	Manholes & Catch Basins	COST PER EACH		
		MAT.	INST.	TOTAL
1920	Manhole/catch basin, brick, 4' I.D. riser, 4' deep	655	1,050	1,705
1980	10' deep	1,375	2,475	3,850
3200	Block, 4' I.D. riser, 4' deep	700	850	1,550
3260	10' deep	1,350	2,050	3,400
4620	Concrete, cast-in-place, 4' I.D. riser, 4' deep	725	1,475	2,200
4680	10' deep	1,550	3,525	5,075
5820	Concrete, precast, 4' I.D. riser, 4' deep	715	780	1,495
5880	10' deep	1,325	1,850	3,175
6200	6' I.D. riser, 4' deep	1,275	1,125	2,400
6260	10' deep	2,500	2,675	5,175

12.5-110	Roadway Pavement	COST PER L.F.		
		MAT.	INST.	TOTAL
1500	Roadways, bituminous conc. paving, 2-1/2" thick, 20' wide	34	35.50	69.50
1580	30' wide	40.50	41	81.50
1800	3" thick, 20' wide	36	34.50	70.50
1880	30' wide	43	41.50	84.50
2100	4" thick, 20' wide	39	34.50	73.50
2180	30' wide	47.50	41	88.50
2400	5" thick, 20' wide	45.50	35	80.50
2480	30' wide	57	42	99
3000	6" thick, 20' wide	50.50	33.50	84
3080	30' wide	65	40	105
3300	8" thick 20' wide	58	33.50	91.50
3380	30' wide	76	40.50	116.50
3600	12" thick 20' wide	75	34.50	109.50
3700	32' wide	106	43	149

12.5-510	Parking Lots	COST PER CAR		
		MAT.	INST.	TOTAL
1500	Parking lot, 90° angle parking, 3" bituminous paving, 6" gravel base	242	204	446
1540	10" gravel base	262	238	500
1560	4" bituminous paving, 6" gravel base	287	203	490
1600	10" gravel base	305	238	543
1620	6" bituminous paving, 6" gravel base	385	212	597
1660	10" gravel base	405	246	651
1800	60° angle parking, 3" bituminous paving, 6" gravel base	278	223	501
1840	10" gravel base	300	264	564
1860	4" bituminous paving, 6" gravel base	330	223	553
1900	10" gravel base	355	262	617
1920	6" bituminous paving, 6" gravel base	445	232	677
1960	10" gravel base	470	272	742
2200	45° angle parking, 3" bituminous paving, 6" gravel base	315	243	558
2240	10" gravel base	340	289	629
2260	4" bituminous paving, 6" gravel base	375	241	616
2300	10" gravel base	400	287	687
2320	6" bituminous paving, 6" gravel base	505	253	758
2360	10" gravel base	530	298	828

For information about Means Estimating Seminars, see yellow pages 11 and 12 in back of book

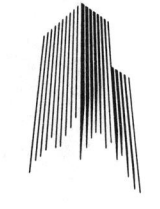

Reference Section

All the reference information is in one section making it easy to find what you need to know . . . and easy to use the book on a daily basis. This section is visually identified by a vertical gray bar on the edge of pages.

In this Reference Section, we've included General Conditions; Historical Cost Indexes for cost comparisons over time; Location Factors; a Glossary; and an explanation of all abbreviations used in the book.

Table of Contents

General Conditions (Overhead & Profit)

Total building costs in the Commercial/Industrial/Institutional section include a 15% allowance for general conditions. This allowance provides for the general contractor's overhead & profit and contingencies.

General contractor overhead includes indirect costs such as permits, Workers' Compensation, insurances, supervision and bonding fees. Overhead will vary with the size of project, the contractor's operating procedures and location. Profits will vary with economic activity and local conditions.

Contingencies provide for unforeseen construction difficulties which include material shortages and weather. In all situations, the appraiser should give consideration to possible adjustment of the 15% factor used in developing the Commercial/Industrial/Institutional models.

Architectural Fees

Tabulated below are typical percentage fees by project size, for good professional architectural service. Fees may vary from those listed depending upon degree of design difficulty and economic conditions in any particular area.

Rates can be interpolated horizontally and vertically. Various portions of the same project requiring different rates should be adjusted proportionately. For alterations, add 50% to the fee for the first $500,000 of project cost and add 25% to the fee for project cost over $500,000.

Architectural fees tabulated below include Engineering fees.

Insurance Exclusions

Many insurance companies exclude from coverage such items as architect's fees, excavation, foundations below grade, underground piping and site preparation. Since exclusions vary among insurance companies, it is recommended that for greatest accuracy each exclusion be priced separately using the unit-in-place section.

As a rule of thumb, exclusions can be calculated at 9% of total building cost plus the appropriate allowance for architect's fees.

Building Types	Total Project Size in Thousands of Dollars						
	100	250	500	1,000	5,000	10,000	50,000
Factories, garages, warehouses, repetitive housing	9.0%	8.0%	7.0%	6.2%	5.3%	4.9%	4.5%
Apartments, banks, schools, libraries, offices, municipal buildings	11.7	10.8	8.5	7.3	6.4	6.0	5.6
Churches, hospitals, homes, laboratories, museums, research	14.0	12.8	11.9	10.9	8.5	7.8	7.2
Memorials, monumental work, decorative furnishings	—	16.0	14.5	13.1	10.0	9.0	8.3

Location Factors

osts shown in *Means cost data publications* are based on National verages for materials and installation. To adjust these costs to a pecific location, simply multiply the base cost by the factor for that city. The data is arranged alphabetically by state and postal zip code numbers. For a city not listed, use the factor for a nearby city with similar economic characteristics.

STATE/ZIP	CITY	Residential	Commercial
ALABAMA			
350-352	Birmingham	.84	.85
354	Tuscaloosa	.81	.79
355	Jasper	.76	.77
356	Decatur	.82	.83
357-358	Huntsville	.82	.83
359	Gadsden	.81	.82
360-361	Montgomery	.82	.80
362	Anniston	.74	.75
363	Dothan	.81	.79
364	Evergreen	.82	.80
365-366	Mobile	.83	.84
367	Selma	.81	.79
368	Phenix City	.81	.79
369	Butler	.81	.79
ALASKA			
995-996	Anchorage	1.27	1.26
997	Fairbanks	1.27	1.26
998	Juneau	1.26	1.25
999	Ketchikan	1.31	1.30
ARIZONA			
850,853	Phoenix	.93	.90
852	Mesa/Tempe	.88	.86
855	Globe	.91	.88
856-857	Tucson	.91	.88
859	Show Low	.92	.88
860	Flagstaff	.95	.91
863	Prescott	.92	.88
864	Kingman	.91	.88
865	Chambers	.91	.88
ARKANSAS			
716	Pine Bluff	.80	.80
717	Camden	.72	.72
718	Texarkana	.77	.76
719	Hot Springs	.71	.71
720-722	Little Rock	.81	.81
723	West Memphis	.81	.81
724	Jonesboro	.81	.81
725	Batesville	.78	.78
726	Harrison	.79	.79
727	Fayetteville	.71	.68
728	Russellville	.80	.77
729	Fort Smith	.83	.80
CALIFORNIA			
900-902	Los Angeles	1.11	1.11
903-905	Inglewood	1.09	1.09
906-908	Long Beach	1.10	1.10
910-912	Pasadena	1.08	1.08
913-916	Van Nuys	1.10	1.10
917-918	Alhambra	1.10	1.10
919-921	San Diego	1.11	1.07
922	Palm Springs	1.12	1.08
923-924	San Bernardino	1.11	1.07
925	Riverside	1.13	1.09
926-927	Santa Ana	1.11	1.08
928	Anaheim	1.13	1.11
930	Oxnard	1.15	1.10
931	Santa Barbara	1.12	1.09
932-933	Bakersfield	1.11	1.06
934	San Luis Obispo	1.22	1.10
935	Mojave	1.11	1.07
936-938	Fresno	1.13	1.09
939	Salinas	1.12	1.12
940-941	San Francisco	1.21	1.24
942,956-958	Sacramento	1.12	1.11
943	Palo Alto	1.14	1.17
944	San Mateo	1.15	1.18
945	Vallejo	1.14	1.17
946	Oakland	1.16	1.19
947	Berkeley	1.29	1.33
948	Richmond	1.15	1.17
949	San Rafael	1.25	1.19
950	Santa Cruz	1.17	1.15
951	San Jose	1.22	1.20
952	Stockton	1.14	1.10
953	Modesto	1.14	1.10

STATE/ZIP	CITY	Residential	Commercial
CALIFORNIA (CONT'D)			
954	Santa Rosa	1.15	1.19
955	Eureka	1.12	1.11
959	Marysville	1.11	1.10
960	Redding	1.10	1.09
961	Susanville	1.10	1.09
COLORADO			
800-802	Denver	.98	.94
803	Boulder	.90	.86
804	Golden	.95	.91
805	Fort Collins	.98	.92
806	Greeley	.92	.86
807	Fort Morgan	.96	.90
808-809	Colorado Springs	.93	.91
810	Pueblo	.93	.91
811	Alamosa	.91	.89
812	Salida	.91	.89
813	Durango	.89	.87
814	Montrose	.87	.85
815	Grand Junction	.91	.87
816	Glenwood Springs	.96	.91
CONNECTICUT			
060	New Britain	1.05	1.06
061	Hartford	1.05	1.06
062	Willimantic	1.05	1.06
063	New London	1.06	1.05
064	Meriden	1.04	1.05
065	New Haven	1.05	1.06
066	Bridgeport	1.02	1.05
067	Waterbury	1.06	1.06
068	Norwalk	1.01	1.05
069	Stamford	1.03	1.07
D.C.			
200-205	Washington	.94	.95
DELAWARE			
197	Newark	.99	1.00
198	Wilmington	.98	.99
199	Dover	.99	1.00
FLORIDA			
320,322	Jacksonville	.86	.85
321	Daytona Beach	.90	.89
323	Tallahassee	.78	.80
324	Panama City	.72	.74
325	Pensacola	.87	.85
326	Gainesville	.87	.84
327-328,347	Orlando	.89	.87
329	Melbourne	.90	.89
330-332,340	Miami	.86	.88
333	Fort Lauderdale	.86	.88
334,349	West Palm Beach	.87	.84
335-336,346	Tampa	.83	.85
337	St. Petersburg	.84	.86
338	Lakeland	.82	.84
339	Fort Myers	.83	.83
342	Sarasota	.82	.84
GEORGIA			
300-303,399	Atlanta	.84	.89
304	Statesboro	.67	.69
305	Gainesville	.71	.76
306	Athens	.77	.81
307	Dalton	.68	.68
308-309	Augusta	.79	.81
310-312	Macon	.82	.82
313-314	Savannah	.82	.83
315	Waycross	.76	.76
316	Valdosta	.78	.78
317	Albany	.79	.81
318-319	Columbus	.80	.80
HAWAII			
967	Hilo	1.27	1.23
968	Honolulu	1.27	1.23

LOCATION FACTORS

STATE/ZIP	CITY	Residential	Commercial
STATES & POSS.			
969	Guam	.85	.82
IDAHO			
832	Pocatello	.95	.94
833	Twin Falls	.82	.81
834	Idaho Falls	.86	.85
835	Lewiston	1.11	1.02
836-837	Boise	.95	.94
838	Coeur d'Alene	1.00	.92
ILLINOIS			
600-603	North Suburban	1.08	1.07
604	Joliet	1.07	1.06
605	South Suburban	1.08	1.07
606	Chicago	1.12	1.11
609	Kankakee	1.00	1.00
610-611	Rockford	1.03	1.02
612	Rock Island	1.05	.96
613	La Salle	1.07	.99
614	Galesburg	1.04	.97
615-616	Peoria	1.07	1.00
617	Bloomington	1.03	.99
618-619	Champaign	1.03	1.00
620-622	East St. Louis	.99	.99
623	Quincy	.97	.94
624	Effingham	.99	.96
625	Decatur	1.00	.97
626-627	Springfield	1.01	.98
628	Centralia	.98	.98
629	Carbondale	.96	.96
INDIANA			
460	Anderson	.93	.92
461-462	Indianapolis	.97	.95
463-464	Gary	1.00	.98
465-466	South Bend	.92	.90
467-468	Fort Wayne	.90	.91
469	Kokomo	.91	.90
470	Lawrenceburg	.91	.88
471	New Albany	.92	.88
472	Columbus	.92	.89
473	Muncie	.91	.90
474	Bloomington	.93	.90
475	Washington	.93	.93
476-477	Evansville	.93	.93
478	Terre Haute	.95	.94
479	Lafayette	.91	.91
IOWA			
500-503,509	Des Moines	.96	.92
504	Mason City	.88	.82
505	Fort Dodge	.85	.80
506-507	Waterloo	.90	.84
508	Creston	.91	.86
510-511	Sioux City	.90	.84
512	Sibley	.81	.79
513	Spencer	.83	.81
514	Carroll	.87	.83
515	Council Bluffs	.94	.88
516	Shenandoah	.83	.77
520	Dubuque	.97	.87
521	Decorah	.92	.82
522-524	Cedar Rapids	.99	.91
525	Ottumwa	.95	.86
526	Burlington	.85	.80
527-528	Davenport	.97	.94
KANSAS			
660-662	Kansas City	.95	.93
664-666	Topeka	.87	.86
667	Fort Scott	.88	.86
668	Emporia	.83	.83
669	Belleville	.90	.84
670-672	Wichita	.89	.86
673	Independence	.84	.81
674	Salina	.87	.83
675	Hutchinson	.82	.78
676	Hays	.88	.84
677	Colby	.88	.84
678	Dodge City	.88	.85
679	Liberal	.81	.78
KENTUCKY			
400-402	Louisville	.93	.90
403-405	Lexington	.89	.86

STATE/ZIP	CITY	Residential	Commercial
KENTUCKY (CONT'D)			
406	Frankfort	.95	.89
407-409	Corbin	.80	.75
410	Covington	.96	.93
411-412	Ashland	.95	.96
413-414	Campton	.79	.75
415-416	Pikeville	.84	.84
417-418	Hazard	.79	.75
420	Paducah	.95	.90
421-422	Bowling Green	.94	.89
423	Owensboro	.92	.90
424	Henderson	.93	.91
425-426	Somerset	.77	.74
427	Elizabethtown	.91	.88
LOUISIANA			
700-701	New Orleans	.87	.86
703	Thibodaux	.86	.86
704	Hammond	.85	.84
705	Lafayette	.86	.83
706	Lake Charles	.85	.85
707-708	Baton Rouge	.85	.84
710-711	Shreveport	.81	.81
712	Monroe	.80	.80
713-714	Alexandria	.79	.79
MAINE			
039	Kittery	.80	.82
040-041	Portland	.89	.91
042	Lewiston	.90	.91
043	Augusta	.81	.81
044	Bangor	.93	.93
045	Bath	.80	.80
046	Machias	.85	.85
047	Houlton	.82	.82
048	Rockland	.85	.85
049	Waterville	.81	.80
MARYLAND			
206	Waldorf	.88	.88
207-208	College Park	.90	.90
209	Silver Spring	.89	.89
210-212	Baltimore	.92	.92
214	Annapolis	.90	.91
215	Cumberland	.87	.88
216	Easton	.70	.70
217	Hagerstown	.90	.89
218	Salisbury	.79	.79
219	Elkton	.85	.86
MASSACHUSETTS			
010-011	Springfield	1.06	1.04
012	Pittsfield	1.01	1.01
013	Greenfield	1.03	1.01
014	Fitchburg	1.11	1.06
015-016	Worcester	1.12	1.08
017	Framingham	1.09	1.10
018	Lowell	1.11	1.11
019	Lawrence	1.10	1.10
020-022	Boston	1.16	1.17
023-024	Brockton	1.08	1.10
025	Buzzards Bay	1.05	1.07
026	Hyannis	1.07	1.08
027	New Bedford	1.09	1.10
MICHIGAN			
480,483	Royal Oak	1.00	.99
481	Ann Arbor	1.02	1.01
482	Detroit	1.06	1.05
484-485	Flint	.98	.99
486	Saginaw	.95	.96
487	Bay City	.95	.96
488-489	Lansing	1.00	.97
490	Battle Creek	.99	.93
491	Kalamazoo	.99	.93
492	Jackson	.97	.94
493,495	Grand Rapids	.91	.88
494	Muskegan	.95	.92
496	Traverse City	.91	.88
497	Gaylord	.91	.92
498-499	Iron mountain	.97	.94
MINNESOTA			
550-551	Saint Paul	1.11	1.08
553-554	Minneapolis	1.14	1.10

STATE/ZIP	CITY	Residential	Commercial
MINNESOTA (CONT'D)			
556-558	Duluth	1.03	1.04
559	Rochester	1.06	1.02
560	Mankato	.99	.98
561	Windom	.89	.88
562	Willmar	.89	.88
563	St. Cloud	1.09	1.01
564	Brainerd	1.06	.99
565	Detroit Lakes	.89	.96
566	Bemidji	.91	.98
567	Thief River Falls	.88	.94
MISSISSIPPI			
386	Clarksdale	.71	.68
387	Greenville	.83	.79
388	Tupelo	.72	.73
389	Greenwood	.73	.70
390-392	Jackson	.83	.79
393	Meridian	.78	.77
394	Laurel	.74	.70
395	Biloxi	.85	.81
396	Mccomb	.70	.68
397	Columbus	.72	.73
MISSOURI			
630-631	St. Louis	.99	1.02
633	Bowling Green	.93	.96
634	Hannibal	1.02	.95
635	Kirksville	.86	.90
636	Flat River	.96	.99
637	Cape Girardeau	.95	.98
638	Sikeston	.91	.93
639	Poplar Bluff	.91	.92
640-641	Kansas City	.98	.95
644-645	St. Joseph	.88	.92
646	Chillicothe	.81	.84
647	Harrisonville	.95	.93
648	Joplin	.85	.87
650-651	Jefferson City	.98	.92
652	Columbia	.96	.89
653	Sedalia	.96	.90
654-655	Rolla	.89	.83
656-658	Springfield	.85	.87
MONTANA			
590-591	Billings	.98	.96
592	Wolf Point	.97	.95
593	Miles City	.98	.96
594	Great Falls	.97	.96
595	Havre	.95	.94
596	Helena	.97	.96
597	Butte	.95	.94
598	Missoula	.95	.94
599	Kalispell	.94	.93
NEBRASKA			
680-681	Omaha	.90	.89
683-685	Lincoln	.89	.84
686	Columbus	.76	.75
687	Norfolk	.86	.85
688	Grand Island	.88	.84
689	Hastings	.85	.81
690	Mccook	.76	.72
691	North Platte	.85	.81
692	Valentine	.80	.76
693	Alliance	.77	.73
NEVADA			
889-891	Las Vegas	1.04	1.03
893	Ely	.96	.97
894-895	Reno	.94	.98
897	Carson City	.95	.98
898	Elko	.93	.96
NEW HAMPSHIRE			
030	Nashua	.96	.97
031	Manchester	.96	.97
032-033	Concord	.95	.96
034	Keene	.82	.83
035	Littleton	.84	.85
036	Charleston	.80	.81
037	Claremont	.80	.81
038	Portsmouth	.95	.94

STATE/ZIP	CITY	Residential	Commercial
NEW JERSEY			
070-071	Newark	1.15	1.13
072	Elizabeth	1.11	1.09
073	Jersey City	1.12	1.12
074-075	Paterson	1.13	1.13
076	Hackensack	1.10	1.10
077	Long Branch	1.11	1.09
078	Dover	1.13	1.11
079	Summit	1.10	1.08
080,083	Vineland	1.11	1.07
081	Camden	1.12	1.09
082,084	Atlantic City	1.11	1.08
085-086	Trenton	1.14	1.12
087	Point Pleasant	1.12	1.10
088-089	New Brunswick	1.12	1.10
NEW MEXICO			
870-872	Albuquerque	.89	.91
873	Gallup	.90	.92
874	Farmington	.90	.92
875	Santa Fe	.89	.91
877	Las Vegas	.89	.91
878	Socorro	.89	.91
879	Truth/Consequences	.88	.88
880	Las Cruces	.85	.85
881	Clovis	.91	.91
882	Roswell	.91	.91
883	Carrizozo	.92	.92
884	Tucumcari	.91	.91
NEW YORK			
100-102	New York	1.34	1.34
103	Staten Island	1.29	1.29
104	Bronx	1.28	1.28
105	Mount Vernon	1.21	1.21
106	White Plains	1.20	1.20
107	Yonkers	1.24	1.24
108	New Rochelle	1.22	1.22
109	Suffern	1.15	1.15
110	Queens	1.28	1.28
111	Long Island City	1.29	1.29
112	Brooklyn	1.28	1.28
113	Flushing	1.30	1.30
114	Jamaica	1.28	1.28
115,117,118	Hicksville	1.26	1.26
116	Far Rockaway	1.29	1.29
119	Riverhead	1.27	1.27
120-122	Albany	.98	.98
123	Schenectady	.99	.99
124	Kingston	1.13	1.11
125-126	Poughkeepsie	1.15	1.13
127	Monticello	1.12	1.10
128	Glens Falls	.96	.94
129	Plattsburgh	.97	.95
130-132	Syracuse	1.01	.99
133-135	Utica	.93	.96
136	Watertown	.95	.98
137-139	Binghamton	.96	.96
140-142	Buffalo	1.08	1.04
143	Niagara Falls	1.08	1.04
144-146	Rochester	1.02	1.03
147	Jamestown	.96	.93
148-149	Elmira	.97	.95
NORTH CAROLINA			
270,272-274	Greensboro	.78	.79
271	Winston-Salem	.77	.78
275-276	Raleigh	.79	.79
277	Durham	.78	.79
278	Rocky Mount	.67	.67
279	Elizabeth City	.67	.67
280	Gastonia	.77	.77
281-282	Charlotte	.77	.78
283	Fayetteville	.78	.78
284	Wilmington	.75	.77
285	Kinston	.66	.66
286	Hickory	.64	.65
287-288	Asheville	.76	.78
289	Murphy	.66	.67
NORTH DAKOTA			
580-581	Fargo	.79	.84
582	Grand Forks	.79	.84
583	Devils Lake	.79	.84
584	Jamestown	.79	.84
585	Bismarck	.81	.85

STATE/ZIP	CITY	Residential	Commercial
NORTH DAKOTA (CONT'D)			
586	Dickinson	.80	.84
587	Minot	.81	.85
588	Williston	.80	.83
OHIO			
430-432	Columbus	.96	.94
433	Marion	.91	.92
434-436	Toledo	.98	.97
437-438	Zanesville	.91	.90
439	Steubenville	.96	.96
440	Lorain	1.02	.96
441	Cleveland	1.08	1.01
442-443	Akron	1.00	.99
444-445	Youngstown	.99	.96
446-447	Canton	.96	.95
448-449	Mansfield	.95	.93
450	Hamilton	.98	.93
451-452	Cincinnati	.98	.92
453-454	Dayton	.93	.92
455	Springfield	.94	.92
456	Chillicothe	1.00	.94
457	Athens	.89	.88
458	Lima	.93	.92
OKLAHOMA			
730-731	Oklahoma City	.81	.83
734	Ardmore	.83	.82
735	Lawton	.84	.83
736	Clinton	.80	.82
737	Enid	.83	.82
738	Woodward	.82	.81
739	Guymon	.70	.70
740-741	Tulsa	.86	.83
743	Miami	.85	.82
744	Muskogee	.77	.75
745	Mcalester	.76	.78
746	Ponca City	.83	.82
747	Durant	.79	.81
748	Shawnee	.79	.80
749	Poteau	.86	.82
OREGON			
970-972	Portland	1.09	1.07
973	Salem	1.07	1.06
974	Eugene	1.06	1.05
975	Medford	1.06	1.05
976	Klamath Falls	1.06	1.05
977	Bend	1.07	1.05
978	Pendleton	1.04	1.02
979	Vale	.99	.97
PENNSYLVANIA			
150-152	Pittsburgh	1.05	1.03
153	Washington	1.03	1.01
154	Uniontown	1.02	1.00
155	Bedford	1.04	.97
156	Greensburg	1.03	1.01
157	Indiana	1.05	.99
158	Dubois	1.04	.97
159	Johnstown	1.05	.98
160	Butler	1.01	.98
161	New Castle	1.01	.98
162	Kittanning	1.03	1.00
163	Oil City	.90	.95
164-165	Erie	.98	.97
166	Altoona	1.05	.97
167	Bradford	.99	.97
168	State College	.98	.98
169	Wellsboro	.95	.96
170-171	Harrisburg	.99	.98
172	Chambersburg	.97	.96
173-174	York	.99	.97
175-176	Lancaster	.98	.96
177	Williamsport	.94	.93
178	Sunbury	.97	.96
179	Pottsville	.97	.96
180	Lehigh Valley	1.03	1.03
181	Allentown	1.03	1.02
182	Hazleton	.98	.97
183	Stroudsburg	.98	.97
184-185	Scranton	.97	1.01
186-187	Wilkes-Barre	.94	.97
188	Montrose	.95	.98
189	Doylestown	.94	1.06

STATE/ZIP	CITY	Residential	Commercial
PENNSYLVANIA (CONT'D)			
190-191	Philadelphia	1.12	1.10
193	Westchester	1.06	1.05
194	Norristown	1.09	1.07
195-196	Reading	.98	.99
RHODE ISLAND			
028	Newport	1.04	1.06
029	Providence	1.04	1.06
SOUTH CAROLINA			
290-292	Columbia	.74	.77
293	Spartanburg	.73	.76
294	Charleston	.76	.78
295	Florence	.73	.75
296	Greenville	.73	.76
297	Rock Hill	.66	.69
298	Aiken	.67	.70
299	Beaufort	.70	.72
SOUTH DAKOTA			
570-571	Sioux Falls	.89	.83
572	Watertown	.87	.81
573	Mitchell	.86	.80
574	Aberdeen	.88	.82
575	Pierre	.87	.81
576	Mobridge	.88	.81
577	Rapid City	.86	.80
TENNESSEE			
370-372	Nashville	.85	.85
373-374	Chattanooga	.85	.84
375,380-381	Memphis	.86	.86
376	Johnson City	.82	.81
377-379	Knoxville	.81	.81
382	Mckenzie	.71	.71
383	Jackson	.70	.77
384	Columbia	.78	.78
385	Cookeville	.70	.70
TEXAS			
750	Mckinney	.91	.84
751	Waxahackie	.84	.84
752-753	Dallas	.91	.86
754	Greenville	.81	.75
755	Texarkana	.91	.80
756	Longview	.86	.76
757	Tyler	.92	.81
758	Palestine	.76	.76
759	Lufkin	.79	.79
760-761	Fort Worth	.84	.83
762	Denton	.90	.81
763	Wichita Falls	.82	.82
764	Eastland	.77	.76
765	Temple	.79	.79
766-767	Waco	.83	.82
768	Brownwood	.75	.74
769	San Angelo	.81	.77
770-772	Houston	.89	.90
773	Huntsville	.76	.76
774	Wharton	.78	.79
775	Galveston	.87	.88
776-777	Beaumont	.85	.87
778	Bryan	.82	.83
779	Victoria	.82	.82
780	Laredo	.79	.80
781-782	San Antonio	.83	.84
783-784	Corpus Christi	.82	.81
785	Mc Allen	.81	.79
786-787	Austin	.80	.83
788	Del Rio	.70	.70
789	Giddings	.76	.75
790-791	Amarillo	.82	.82
792	Childress	.77	.80
793-794	Lubbock	.80	.82
795-796	Abilene	.79	.79
797	Midland	.81	.82
798-799,885	El Paso	.79	.78
UTAH			
840-841	Salt Lake City	.88	.87
842,844	Ogden	.88	.86
843	Logan	.89	.87
845	Price	.83	.82
846-847	Provo	.89	.88

STATE/ZIP	CITY	Residential	Commercial
VERMONT			
050	White River Jct.	.75	.75
051	Bellows Falls	.76	.75
052	Bennington	.72	.71
053	Brattleboro	.76	.76
054	Burlington	.85	.86
056	Montpelier	.84	.85
057	Rutland	.87	.86
058	St. Johnsbury	.77	.78
059	Guildhall	.76	.77
VIRGINIA			
220-221	Fairfax	.89	.90
222	Arlington	.89	.90
223	Alexandria	.90	.91
224-225	Fredericksburg	.85	.86
226	Winchester	.80	.81
227	Culpeper	.79	.80
228	Harrisonburg	.76	.76
229	Charlottesville	.84	.82
230-232	Richmond	.85	.83
233-235	Norfolk	.82	.82
236	Newport News	.83	.82
237	Portsmouth	.81	.81
238	Petersburg	.85	.83
239	Farmville	.77	.75
240-241	Roanoke	.79	.78
242	Bristol	.80	.76
243	Pulaski	.72	.71
244	Staunton	.74	.72
245	Lynchburg	.82	.78
246	Grundy	.71	.71
WASHINGTON			
980-981,987	Seattle	1.00	1.05
982	Everett	.98	1.04
983-984	Tacoma	1.07	1.05
985	Olympia	1.07	1.04
986	Vancouver	1.10	1.04
988	Wenatchee	.97	1.01
989	Yakima	1.04	1.02
990-992	Spokane	1.01	1.00
993	Richland	1.02	1.01
994	Clarkston	1.00	.99
WEST VIRGINIA			
247-248	Bluefield	.86	.86
249	Lewisburg	.90	.90
250-253	Charleston	.94	.94
254	Martinsburg	.78	.79
255-257	Huntington	.94	.96
258-259	Beckley	.92	.92
260	Wheeling	.93	.95
261	Parkersburg	.93	.94
262	Buckhannon	.96	.94
263-264	Clarksburg	.97	.94
265	Morgantown	.97	.94
266	Gassaway	.94	.94
267	Romney	.92	.92
268	Petersburg	.95	.92
WISCONSIN			
530,532	Milwaukee	1.01	1.00
531	Kenosha	1.00	.98
534	Racine	1.04	.99
535	Beloit	.99	.97
537	Madison	.97	.95
538	Lancaster	.92	.90
539	Portage	.96	.94
540	New Richmond	1.00	.93
541-543	Green Bay	.99	.96
544	Wausau	.95	.91
545	Rhinelander	.96	.92
546	La Crosse	.96	.93
547	Eau Claire	1.02	.94
548	Superior	1.00	.94
549	Oshkosh	.95	.92
WYOMING			
820	Cheyenne	.88	.84
821	Yellowstone Nat. Pk.	.84	.81
822	Wheatland	.85	.80
823	Rawlins	.85	.80
824	Worland	.81	.79
825	Riverton	.85	.81
826	Casper	.88	.84

STATE/ZIP	CITY	Residential	Commercial
WYOMING (CONT'D)			
827	Newcastle	.82	.78
828	Sheridan	.86	.84
829-831	Rock Springs	.86	.81
CANADIAN FACTORS (reflect Canadian currency)			
ALBERTA			
	Calgary	1.02	.99
	Edmonton	1.02	.99
BRITISH COLUMBIA			
	Vancouver	1.08	1.09
	Victoria	1.08	1.09
MANITOBA			
	Winnipeg	1.01	1.00
NEW BRUNSWICK			
	Moncton	.96	.94
	Saint John	.99	.97
NEWFOUNDLAND			
	St. John's	.97	.96
NOVA SCOTIA			
	Halifax	.99	.98
ONTARIO			
	Barrie	1.12	1.10
	Brantford	1.14	1.12
	Cornwall	1.12	1.10
	Hamilton	1.15	1.11
	Kingston	1.12	1.10
	Kitchener	1.07	1.05
	London	1.11	1.09
	North Bay	1.10	1.09
	Oshawa	1.12	1.10
	Ottawa	1.12	1.10
	Owen Sound	1.11	1.09
	Peterborough	1.12	1.10
	Sarnia	1.12	1.10
	St. Catharines	1.07	1.05
	Sudbury	1.07	1.05
	Thunder Bay	1.08	1.06
	Toronto	1.14	1.13
	Windsor	1.09	1.07
PRINCE EDWARD ISLAND			
	Charlottetown	.95	.93
QUEBEC			
	Chicoutimi	1.03	1.02
	Montreal	1.10	1.03
	Quebec	1.11	1.03
SASKATCHEWAN			
	Regina	.92	.92
	Saskatoon	.92	.92

The following tables are the revised Historical Cost Indexes based on a 30-city national average with a base of 100 on January 1, 1993.

The indexes may be used to:

1. Estimate and compare construction costs for different years in the same city.

2. Estimate and compare construction costs in different cities for the same year.

3. Estimate and compare construction costs in different cities for different years.

4. Compare construction trends in any city with the national average.

EXAMPLES

1. Estimate and compare construction costs for different years in the same city.

   A. To estimate the construction cost of a building in Lexington, KY in 1970, knowing that it cost $900,000 in 1998.

Index Lexington, KY in 1970 =	26.9
Index Lexington, KY in 1998 =	98.2

   $$\frac{\text{Index 1970}}{\text{Index 1998}} \times \text{Cost 1998} = \text{Cost 1970}$$

   $$\frac{26.9}{98.2} \times \$900,000 = \$246,500$$

   Construction Cost in Lexington, KY in 1970 = $246,500

   B. To estimate the current construction cost of a building in Boston, MA that was built in 1978 for $900,000.

Index Boston, MA in 1978 =	54.0
Index Boston, MA in 1998 =	133.1

   $$\frac{\text{Index 1998}}{\text{Index 1978}} \times \text{Cost 1978} = \text{Cost 1998}$$

   $$\frac{133.1}{54.0} \times \$900,000 = \$2,218,500$$

   Construction Cost in Boston in 1998 = $2,218,500

2. Estimate and compare construction costs in different cities for the same year.

   To compare the construction cost of a building in Topeka, KS in 1998 with the known cost of $800,000 in Baltimore, MD in 1998

Index Topeka, KS in 1998 =	97.9
Index Baltimore, MD in 1998 =	104.8

   $$\frac{\text{Index Topeka}}{\text{Index Baltimore}} \times \text{Cost Baltimore} = \text{Cost Topeka}$$

   $$\frac{97.9}{104.8} \times \$800,000 = \$747,500$$

   Construction Cost in Topeka in 1998 = $747,500

3. Estimate and compare construction costs in different cities for different years.

   To compare the construction cost of a building in Detroit, MI in 1998 with the known construction cost of $4,000,000 for the same building San Francisco, CA in 1978.

Index Detroit, MI in 1998 =	119.9
Index San Francisco, CA in 1978 =	62.6

   $$\frac{\text{Index Detroit 1998}}{\text{Index San Francisco 1978}} \times \text{Cost San Francisco 1978} = \text{Cost Detroit 1998}$$

   $$\frac{119.9}{62.6} \times \$4,000,000 = \$7,661,500$$

   Construction Cost in Detroit in 1998 = $7,661,500

4. Compare construction trends in any city with the national average.

   To compare the construction cost in Las Vegas, NV from 1974 to 1998 with the increase in the National Average during the same time period.

Index Las Vegas, NV for 1974 =	40.0	For 1998 =	116.7
Index 30 City Average for 1974 =	39.4	For 1998 =	114.4

   A. National Average increase
      From 1974 to 1998

   $$= \frac{\text{Index 30 City 1998}}{\text{Index 30 City 1974}}$$

   $$= \frac{114.4}{39.4}$$

   National Average increase
   From 1974 to 1998 = 2.90 or 290%

   B. Increase for Las Vegas, NV
      From 1974 to 1998

   $$= \frac{\text{Index Las Vegas, NV 1998}}{\text{Index Las Vegas, NV 1974}}$$

   $$= \frac{116.7}{40.0}$$

   Las Vegas increase 1974 — 1998 = 2.92 or 292%

   Conclusion: Construction costs in Las Vegas are higher than National average costs and increased at a greater rate from 1974 to 1998 than the National Average.

Historical Cost Indexes

Year	National 30 City Average	Alabama Birming-ham	Hunts-ville	Mobile	Mont-gomery	Tusca-loosa	Alaska Anchor-age	Arizona Phoenix	Tuscon	Arkansas Fort Smith	Little Rock	California Anaheim	Bakers-field	Fresno	Los Angeles	Oxnard
Jn 1998	114.4E	97.0E	94.7E	95.5E	91.0E	90.5E	144.5E	102.3E	101.2E	91.2E	92.0E	125.7E	120.9E	124.3E	126.4E	125.3E
1997	111.5	94.6	92.4	93.3	88.8	88.3	142.0	101.8	100.7	89.3	90.1	124.0	119.1	122.3	124.6	123.5
1996	108.9	90.9	91.4	92.3	87.5	86.5	140.4	98.3	97.4	86.5	86.7	121.7	117.2	120.2	122.4	121.5
1995	105.6	87.8	88.0	88.8	84.5	83.5	138.0	96.1	95.5	85.0	85.6	120.1	115.7	117.5	120.9	120.0
1994	103.0	85.9	86.1	86.8	82.6	81.7	134.0	93.7	93.1	82.4	83.9	118.1	113.6	114.5	119.0	117.5
1993	100.0	82.8	82.7	86.1	82.1	78.7	132.0	90.9	90.9	80.9	82.4	115.0	111.3	112.9	115.6	115.4
1992	97.9	81.7	81.5	84.9	80.9	77.6	128.6	88.8	89.4	79.7	81.2	113.5	108.1	110.3	113.7	113.7
1991	95.7	80.5	78.9	83.9	79.8	76.6	127.4	88.1	88.5	78.3	80.0	111.0	105.8	108.2	110.9	111.4
1990	93.2	79.4	77.6	82.7	78.4	75.2	125.8	86.4	87.0	77.1	77.9	107.7	102.7	103.3	107.5	107.4
1989	91.0	77.5	76.1	81.1	76.8	73.7	123.4	85.3	85.6	75.6	76.4	105.6	101.0	101.7	105.3	105.4
1988	88.5	75.7	74.7	79.7	75.4	72.3	121.4	84.4	84.3	74.2	75.0	102.7	97.6	99.2	102.6	102.3
1987	85.7	74.0	73.8	77.9	74.5	71.7	119.3	79.7	80.6	72.2	72.8	100.9	96.1	98.0	100.0	101.1
1986	83.7	72.8	71.7	76.7	72.3	69.4	117.3	78.9	78.8	70.7	72.3	99.2	94.2	93.8	97.8	100.4
1985	81.8	71.1	70.5	72.7	70.9	67.9	116.0	78.1	77.7	69.6	71.0	95.4	92.2	92.6	94.6	96.6
1984	80.6	69.7	69.0	75.2	69.4	66.3	113.8	79.4	79.2	69.1	70.3	93.4	89.8	90.8	92.1	94.0
1983	78.2	67.4	68.8	72.9	69.4	65.5	104.5	72.3	77.9	66.8	68.7	90.9	88.8	88.3	89.7	91.8
1982	72.1	63.5	63.1	67.3	64.4	61.8	96.1	72.3	71.7	62.2	64.6	83.1	83.0	82.5	80.9	83.4
1981	66.1	60.0	58.1	62.2	61.3	58.1	91.5	69.0	67.5	57.7	60.2	75.7	76.5	75.2	74.8	77.1
1980	60.7	55.2	54.1	56.8	56.8	54.0	91.4	63.7	62.4	53.1	55.8	68.7	69.4	68.7	67.4	69.9
1979	54.9	49.9	48.9	52.5	50.8	48.7	82.5	55.9	56.5	47.8	49.5	62.9	61.9	62.5	61.7	62.9
1978	51.3	46.7	45.5	48.8	46.9	45.0	75.3	51.9	52.3	44.8	45.4	57.4	57.1	57.7	57.0	58.6
1977	47.9	43.1	43.2	45.1	42.4	40.5	70.6	48.7	48.5	41.6	42.2	53.2	52.8	53.2	53.7	53.4
1976	45.3	40.8	40.7	42.3	40.3	38.6	64.0	46.5	46.4	39.2	39.4	49.6	49.3	50.4	50.2	48.2
1975	43.7	40.0	40.9	41.8	40.1	37.8	57.3	44.5	45.2	39.0	38.7	47.6	46.6	47.7	48.3	47.0
1974	39.4	34.8	35.6	36.6	36.0	34.4	55.9	40.1	40.5	34.6	32.4	44.1	43.7	44.2	42.2	44.1
1970	27.8	24.1	25.1	25.8	25.4	24.2	43.0	27.2	28.5	24.3	22.3	31.0	30.8	31.1	29.0	31.0
1965	21.5	19.6	19.3	19.6	19.5	18.7	34.9	21.8	22.0	18.7	18.5	23.9	23.7	24.0	22.7	23.9
1960	19.5	17.9	17.5	17.8	17.7	16.9	31.7	19.9	20.0	17.0	16.8	21.7	21.5	21.8	20.6	21.7
1955	16.3	14.8	14.7	14.9	14.9	14.2	26.6	16.7	16.7	14.3	14.5	18.2	18.1	18.3	17.3	18.2
1950	13.5	12.2	12.1	12.3	12.3	11.7	21.9	13.8	13.8	11.8	11.6	15.1	15.0	15.1	14.3	15.1
1945	8.6	7.8	7.7	7.8	7.8	7.5	14.0	8.8	8.8	7.5	7.4	9.6	9.5	9.6	9.1	9.6
1940	6.6	6.0	6.0	6.1	6.0	5.8	10.8	6.8	6.8	5.8	5.7	7.4	7.4	7.4	7.0	7.4

Year	National 30 City Average	California River-side	Sacra-mento	San Diego	San Francisco	Santa Barbara	Stockton	Vallejo	Colorado Colorado Springs	Denver	Pueblo	Connecticut Bridge-Port	Bristol	Hartford	New Britain	New Haven
Jan 1998	114.4E	124.3E	126.7E	121.9E	141.5E	124.5E	125.4E	132.8E	103.3E	106.8E	104.1E	120.0E	120.0E	120.4E	120.6E	120.6E
1997	111.5	122.6	124.7	120.3	139.2	122.4	123.3	130.6	101.1	104.4	102.0	119.2	119.5	119.9	119.7	120.0
1996	108.9	120.6	122.4	118.4	136.8	120.5	121.4	128.1	98.5	101.4	99.7	117.4	117.6	117.9	117.8	118.1
1995	105.6	119.2	119.5	115.4	133.8	119.0	119.0	122.5	96.1	98.9	96.8	116.0	116.5	116.9	116.3	116.5
1994	103.0	116.6	114.8	113.4	131.4	116.4	115.5	119.7	94.2	95.9	94.6	114.3	115.0	115.3	114.7	114.8
1993	100.0	114.3	112.5	111.3	129.6	114.3	115.2	117.5	92.1	93.8	92.6	108.7	107.8	108.4	106.3	106.1
1992	97.9	111.8	110.8	109.5	127.9	112.2	112.7	115.2	90.6	91.5	90.9	106.7	106.0	106.6	104.5	104.4
1991	95.7	109.4	108.5	107.7	125.8	110.0	111.0	113.4	88.9	90.4	89.4	97.6	97.3	98.0	97.3	97.9
1990	93.2	107.0	104.9	105.6	121.8	106.4	105.4	111.4	86.9	88.8	88.8	96.3	95.9	96.6	95.9	96.5
1989	91.0	105.2	103.1	103.6	119.2	104.6	103.7	109.8	85.6	88.7	87.4	94.5	94.2	94.9	94.2	94.6
1988	88.5	102.6	100.5	101.0	115.2	101.4	101.2	107.2	83.8	87.0	85.6	92.8	92.6	93.3	92.6	93.0
1987	85.7	100.3	98.7	99.0	111.0	99.3	99.6	103.2	81.6	84.8	83.8	92.4	93.0	93.3	93.0	92.6
1986	83.7	98.6	94.4	96.9	108.0	97.0	96.1	99.1	81.9	83.1	82.7	88.5	88.7	89.4	88.6	88.8
1985	81.8	94.8	92.2	94.2	106.2	93.2	93.7	97.0	79.6	81.0	80.4	86.1	86.2	87.2	86.1	86.3
1984	80.6	91.8	93.8	92.0	102.8	90.6	91.9	95.6	79.8	84.1	80.5	83.5	83.4	84.4	83.0	83.2
1983	78.2	89.4	92.0	90.3	101.0	88.8	91.0	93.0	77.3	80.5	77.4	79.6	79.9	81.1	79.5	79.3
1982	72.1	82.1	84.9	83.8	91.9	81.8	84.9	85.7	71.2	73.4	75.0	71.9	71.2	73.2	71.4	72.0
1981	66.1	75.5	77.7	74.5	82.8	75.9	77.4	78.5	65.5	66.3	65.4	67.0	66.3	67.3	66.3	67.0
1980	60.7	68.7	71.3	68.1	75.2	71.1	71.2	71.9	60.7	60.9	59.5	61.5	60.7	61.9	60.7	61.4
1979	54.9	62.2	64.0	61.6	67.5	65.0	64.2	65.2	54.9	56.0	54.0	56.0	55.3	56.4	55.4	55.8
1978	51.3	57.7	60.0	58.3	62.6	59.2	59.5	59.9	51.1	52.0	50.6	52.5	50.9	51.9	51.8	52.0
1977	47.9	53.3	55.7	53.4	58.1	53.0	54.8	54.2	47.2	48.3	47.0	49.3	48.6	48.6	48.3	49.0
1976	45.3	49.7	51.5	50.1	53.9	49.0	51.1	50.6	44.5	45.6	44.4	47.1	46.6	46.8	46.6	47.8
1975	43.7	47.3	49.1	47.7	49.8	46.1	47.6	46.5	43.1	42.7	42.5	45.2	45.5	46.0	45.2	46.0
1974	39.4	44.0	45.7	43.1	44.7	44.5	45.2	45.1	39.2	37.3	38.8	41.7	40.1	41.8	40.1	41.8
1970	27.8	31.0	32.1	30.7	31.6	31.3	31.8	31.8	27.6	26.1	27.3	29.2	28.2	29.6	28.2	29.3
1965	21.5	23.9	24.8	24.0	23.7	24.1	24.5	24.5	21.3	20.9	21.0	22.4	21.7	22.6	21.7	23.2
1960	19.5	21.7	22.5	21.7	21.5	21.9	22.3	22.2	19.3	19.0	19.1	20.0	19.8	20.0	19.8	20.0
1955	16.3	18.2	18.9	18.2	18.0	18.4	18.7	18.6	16.2	15.9	16.0	16.8	16.6	16.8	16.6	16.8
1950	13.5	15.0	15.6	15.0	14.9	15.2	15.4	15.4	13.4	13.2	13.2	13.9	13.7	13.8	13.7	13.9
1945	8.6	9.6	10.0	9.6	9.5	9.7	9.9	9.8	8.5	8.4	8.4	8.8	8.7	8.8	8.7	8.9
1940	6.6	7.4	7.7	7.4	7.3	7.5	7.6	7.6	6.6	6.5	6.5	6.8	6.8	6.8	6.8	6.8

Historical Cost Indexes

Year	National 30 City Average	Connecticut Norwalk	Stamford	Water-bury	Delaware Wilming-ton	D.C. Washing-ton	Florida Fort Lau-derdale	Jackson-ville	Miami	Orlando	Talla-hassee	Tampa	Georgia Albany	Atlanta	Colum-bus	M
Jan 1998	114.4E	119.9E	121.5E	121.0E	113.5E	108.9E	100.8E	96.9E	99.8E	99.4E	91.7E	97.1E	92.3E	101.4E	91.0E	
1997	111.5	118.9	120.9	120.2	110.7	106.4	98.6	95.4	98.2	97.2	89.9	95.5	89.9	98.5	88.7	
1996	108.9	117.0	119.0	118.4	108.6	105.4	95.9	92.6	95.9	95.1	88.0	93.6	86.6	94.3	83.9	
1995	105.6	115.5	117.9	117.3	106.1	102.3	94.0	90.9	93.7	93.5	86.5	92.2	85.0	92.0	82.6	
1994	103.0	113.9	116.4	115.8	105.0	99.6	92.2	88.9	91.8	91.5	84.5	90.2	82.3	89.6	80.6	
1993	100.0	108.8	110.6	104.8	101.5	96.3	87.4	86.1	87.1	88.5	82.1	87.7	79.5	85.7	77.8	
1992	97.9	107.2	109.0	103.1	100.3	94.7	85.7	84.0	85.3	87.1	80.8	86.2	78.2	84.3	76.5	
1991	95.7	100.6	103.2	96.5	94.5	92.9	85.1	82.8	85.2	85.5	79.7	86.3	76.3	82.6	75.4	
1990	93.2	96.3	98.9	95.1	92.5	90.4	83.9	81.1	84.0	82.9	78.4	85.0	75.0	80.4	74.0	
1989	91.0	94.4	97.0	93.4	89.5	87.5	82.4	79.7	82.5	81.6	76.9	83.5	73.4	78.6	72.4	
1988	88.5	92.3	94.9	91.8	87.7	85.0	80.9	78.0	81.0	80.1	75.4	81.9	71.8	76.9	70.8	
1987	85.7	92.0	92.7	92.5	85.1	82.1	78.8	76.0	77.9	76.6	74.4	79.4	72.2	73.6	70.3	
1986	83.7	88.2	89.7	87.8	83.8	80.8	78.9	75.0	79.9	76.2	72.3	79.1	68.5	72.0	67.5	
1985	81.8	85.3	86.8	85.6	81.1	78.8	76.7	73.4	78.3	73.9	70.6	77.3	66.9	70.3	66.1	
1984	80.6	82.6	83.7	82.5	79.7	79.1	73.8	72.6	75.6	73.0	69.6	76.5	65.9	68.6	65.2	
1983	78.2	78.5	79.8	78.7	76.3	76.0	71.5	70.4	72.9	71.1	67.6	73.6	65.6	68.6	64.7	
1982	72.1	71.7	72.0	71.8	69.1	69.4	65.4	65.2	65.7	66.0	62.8	67.7	60.1	61.9	59.3	
1981	66.1	66.2	66.2	67.3	63.4	64.9	60.3	61.0	60.7	62.0	58.6	62.1	56.2	58.3	55.7	
1980	60.7	60.7	60.9	62.3	58.5	59.6	55.3	55.8	56.5	56.7	53.5	57.2	51.7	54.0	51.1	
1979	54.9	55.2	55.6	57.0	53.6	53.9	50.0	51.7	50.8	51.8	48.9	51.8	46.5	48.3	46.1	
1978	51.3	51.5	51.6	52.9	50.0	51.5	47.0	47.7	47.6	48.0	45.4	48.4	42.7	45.0	42.5	
1977	47.9	48.1	48.3	49.0	47.1	48.4	43.7	43.3	45.7	44.9	42.0	45.9	41.6	42.1	37.7	
1976	45.3	46.3	45.9	47.4	43.3	45.7	42.2	42.1	43.9	42.8	40.1	43.0	39.9	39.6	35.8	
1975	43.7	44.7	45.0	46.3	42.9	43.7	42.1	40.3	43.2	41.5	38.1	41.3	37.5	38.4	36.2	
1974	39.4	39.9	40.0	41.0	38.4	38.5	36.5	34.2	39.6	37.2	34.9	35.8	33.8	34.6	32.2	
1970	27.8	28.1	28.2	28.9	27.0	26.3	25.7	22.8	27.0	26.2	24.5	24.2	23.8	25.2	22.8	
1965	21.5	21.6	21.7	22.2	20.9	21.8	19.8	17.4	19.3	20.2	18.9	18.6	18.3	19.8	17.6	
1960	19.5	19.7	19.7	20.2	18.9	19.4	18.0	15.8	17.6	18.3	17.2	16.9	16.7	17.1	16.0	
1955	16.3	16.5	16.5	17.0	15.9	16.3	15.1	13.2	14.7	15.4	14.4	14.1	14.0	14.4	13.4	
1950	13.5	13.6	13.7	14.0	13.1	13.4	12.5	11.0	12.2	12.7	11.9	11.7	11.5	11.9	11.0	
1945	8.6	8.7	8.7	8.9	8.4	8.6	7.9	7.0	7.8	8.1	7.6	7.5	7.4	7.6	7.0	
1940	6.6	6.7	6.7	6.9	6.4	6.6	6.1	5.4	6.0	6.2	5.8	5.7	5.7	5.8	5.5	

Year	National 30 City Average	Georgia Savan-nah	Hawaii Hono-lulu	Idaho Boise	Poca-tello	Illinois Chicago	Decatur	Joliet	Peoria	Rock-ford	Spring-field	Indiana Ander-son	Evans-ville	Fort Wayne	Gary	Indi-ap
Jan 1998	114.4E	94.2E	140.6E	106.7E	106.8E	125.8E	110.5E	120.5E	114.3E	116.0E	111.7E	104.3E	106.1E	103.4E	112.0E	10
1997	111.5	92.0	139.8	104.6	104.7	121.3	107.8	117.5	111.5	113.1	108.9	101.9	104.4	101.8	110.3	10
1996	108.9	88.6	134.5	102.2	102.1	118.8	106.6	116.2	109.3	111.5	106.5	100.0	102.1	99.9	107.5	10
1995	105.6	87.4	130.3	99.5	98.2	114.2	98.5	110.5	102.3	103.6	98.1	96.4	97.2	95.0	100.7	10
1994	103.0	85.3	124.0	94.8	95.0	111.3	97.3	108.9	100.9	102.2	97.0	93.6	95.8	93.6	99.1	9
1993	100.0	82.0	122.0	92.2	92.1	107.6	95.6	106.8	98.9	99.6	95.2	91.2	94.3	91.5	96.7	9
1992	97.9	80.8	120.0	91.0	91.0	104.3	94.4	104.2	97.3	98.2	94.0	89.5	92.9	89.9	95.0	8
1991	95.7	79.5	106.1	89.5	89.4	100.9	92.3	100.0	95.9	95.8	91.5	87.8	91.4	88.3	93.3	8
1990	93.2	77.9	104.7	88.2	88.1	98.4	90.9	98.4	93.7	94.0	90.1	84.6	89.3	83.4	88.4	8
1989	91.0	76.0	102.8	86.6	86.5	93.7	89.4	92.8	91.6	92.1	88.6	82.3	87.8	81.7	86.5	8
1988	88.5	74.5	101.1	83.9	83.7	90.6	87.5	89.8	88.5	87.9	86.8	80.8	85.5	80.0	84.6	8
1987	85.7	72.0	99.1	81.4	81.5	86.6	84.4	86.7	86.3	85.2	85.0	78.8	82.4	78.1	81.7	80
1986	83.7	70.8	97.5	80.6	80.3	84.4	83.5	85.3	85.1	84.5	83.4	77.0	80.9	76.5	79.6	78
1985	81.8	68.9	94.7	78.0	78.0	82.4	81.9	83.4	83.7	83.0	81.5	75.2	79.8	75.0	77.8	7
1984	80.6	67.9	90.7	76.6	76.8	80.2	80.4	82.2	83.6	80.6	80.1	73.5	77.4	73.5	77.3	7
1983	78.2	66.3	87.2	76.0	75.8	79.0	78.8	80.6	81.5	78.9	78.8	70.9	75.0	71.2	75.2	74
1982	72.1	61.1	79.3	71.0	70.1	75.0	72.7	74.4	75.3	72.6	72.7	66.4	69.8	67.1	70.5	68
1981	66.1	57.1	73.4	65.4	64.8	68.6	67.2	68.8	70.3	67.3	67.0	61.5	64.3	61.5	65.3	62
1980	60.7	52.2	68.9	60.3	59.5	62.8	62.3	63.4	64.5	61.6	61.1	56.5	59.0	56.7	59.8	57
1979	54.9	47.2	63.0	54.4	53.7	56.5	56.0	57.0	58.4	54.8	55.6	50.8	53.3	51.2	54.0	5
1978	51.3	43.6	58.9	50.0	49.9	52.9	51.9	52.7	53.8	50.4	51.5	46.7	49.1	46.9	49.4	48
1977	47.9	40.2	52.0	45.1	45.2	49.0	47.7	49.1	49.4	46.8	47.5	43.2	45.7	43.0	46.7	45
1976	45.3	38.3	45.7	41.9	41.9	47.3	45.0	46.0	46.5	44.1	44.1	40.6	42.8	41.4	43.9	4
1975	43.7	36.9	44.6	40.8	40.5	45.7	43.1	44.5	44.7	42.7	42.4	39.5	41.7	39.9	41.9	4
1974	39.4	31.0	40.6	37.9	37.7	41.7	39.7	40.6	41.2	39.0	39.2	36.0	36.5	36.3	38.4	36
1970	27.8	21.0	30.4	26.7	26.6	29.1	28.0	28.6	29.0	27.5	27.6	25.4	26.4	25.5	27.1	26
1965	21.5	16.4	21.8	20.6	20.5	22.7	21.5	22.1	22.4	21.2	21.3	19.5	20.4	19.7	20.8	20
1960	19.5	14.9	19.8	18.7	18.6	20.2	19.6	20.0	20.3	19.2	19.3	17.7	18.7	17.9	18.9	20
1955	16.3	12.5	16.6	15.7	15.6	16.9	16.4	16.8	17.0	16.1	16.2	14.9	15.7	15.0	15.9	15
1950	13.5	10.3	13.7	13.0	12.9	14.0	13.6	13.9	14.0	13.3	13.4	12.3	12.9	12.4	13.1	12
1945	8.6	6.6	8.8	8.3	8.2	8.9	8.6	8.9	9.0	8.5	8.6	7.8	8.3	7.9	8.4	7
1940	6.6	5.1	6.8	6.4	6.4	6.9	6.7	6.8	6.9	6.5	6.6	6.0	6.4	6.1	6.5	

Year	National 30 City Average	Indiana Muncie	South Bend	Terre Haute	Iowa Cedar Rapids	Daven-port	Des Moines	Sioux City	Water-loo	Kansas Topeka	Wichita	Kentucky Lexing-ton	Louis-ville	Louisiana Baton Rouge	Lake Charles	New Orleans
Jan 1998	114.4E	103.2E	103.2E	107.0E	103.4E	107.5E	104.5E	95.9E	95.6E	97.9E	97.8E	98.2E	102.2E	95.2E	97.5E	98.0E
1997	111.5	101.3	101.2	104.6	101.2	105.6	102.5	93.8	93.5	96.3	96.2	96.1	99.9	93.3	96.8	96.2
1996	108.9	99.4	99.1	102.1	99.1	102.0	99.1	90.9	90.7	93.6	93.8	94.4	98.3	91.5	94.9	94.3
1995	105.6	95.6	94.6	97.2	95.4	96.6	94.4	88.2	88.9	91.1	91.0	91.6	94.6	89.6	92.7	91.6
1994	103.0	93.2	93.0	95.8	93.0	92.3	92.4	85.9	86.8	89.4	88.1	89.6	92.2	87.7	89.9	89.5
1993	100.0	91.0	90.7	94.4	91.1	90.5	90.7	84.1	85.2	87.4	86.2	87.3	89.4	86.4	88.5	87.8
1992	97.9	89.4	89.3	93.1	89.8	89.3	89.4	82.9	83.8	86.2	85.0	85.8	88.0	85.2	87.3	86.6
1991	95.7	87.7	87.5	91.7	88.5	88.0	88.4	81.8	81.6	84.4	83.9	84.1	84.4	83.2	85.3	85.8
1990	93.2	83.9	85.1	89.1	87.1	86.5	86.9	80.4	80.3	83.2	82.6	82.8	82.6	82.0	84.1	84.5
1989	91.0	81.9	83.5	86.7	85.3	84.9	85.3	79.0	78.8	81.6	81.2	81.3	80.1	80.6	82.7	83.4
1988	88.5	80.4	82.1	84.9	83.4	83.1	83.6	76.8	77.2	80.1	79.5	79.4	78.2	79.0	81.1	82.0
1987	85.7	78.5	79.6	83.0	82.4	81.1	81.2	75.4	75.6	78.9	78.4	77.3	76.3	77.1	78.0	81.3
1986	83.7	76.9	77.5	81.1	79.5	80.0	79.8	73.8	73.7	76.9	76.2	76.5	75.3	76.1	78.1	79.2
1985	81.8	75.2	76.0	79.2	75.9	77.6	77.1	72.1	72.3	75.0	74.7	75.5	74.5	74.9	78.5	78.2
1984	80.6	73.6	75.1	78.0	80.3	77.8	76.6	74.4	73.7	75.1	74.6	75.6	73.7	77.3	78.3	77.5
1983	78.2	71.2	74.5	76.1	79.6	77.0	75.5	74.6	73.0	73.2	72.4	74.9	74.1	75.1	76.6	74.7
1982	72.1	66.0	68.3	69.9	73.5	71.7	71.1	70.1	67.7	68.4	66.1	69.4	69.2	69.3	70.5	69.2
1981	66.1	61.0	63.1	65.1	68.6	66.7	67.1	65.5	63.5	63.0	62.2	64.7	65.1	64.0	64.8	62.7
1980	60.7	56.1	58.3	59.7	62.7	59.6	61.8	59.3	57.7	58.9	58.0	59.3	59.8	59.1	60.0	57.2
1979	54.9	50.7	52.6	54.0	56.6	54.1	55.6	53.6	52.0	53.5	52.9	53.2	54.5	53.5	54.5	52.6
1978	51.3	46.8	48.2	49.9	52.3	50.0	51.1	49.2	48.1	50.0	48.9	49.2	50.0	49.2	50.2	48.5
1977	47.9	43.4	44.7	46.4	48.1	45.7	47.6	45.3	44.2	46.2	45.8	45.8	46.1	45.2	45.2	46.0
1976	45.3	40.9	42.2	43.9	45.8	43.9	45.1	42.3	42.0	43.7	43.3	43.5	43.7	41.7	42.3	43.5
1975	43.7	39.2	40.3	41.9	43.1	41.0	43.1	41.0	40.0	42.0	43.1	42.7	42.5	40.4	40.6	41.5
1974	39.4	36.0	37.2	38.5	40.1	38.2	38.1	37.8	36.9	38.1	37.4	38.1	37.9	35.3	37.9	37.4
1970	27.8	25.3	26.2	27.0	28.2	26.9	27.6	26.6	25.9	27.0	25.5	26.9	25.9	25.0	26.7	27.2
1965	21.5	19.5	20.2	20.9	21.7	20.8	21.7	20.5	20.0	20.8	19.6	20.7	20.3	19.4	20.6	20.4
1960	19.5	17.8	18.3	18.9	19.7	18.8	19.5	18.6	18.2	18.9	17.8	18.8	18.4	17.6	18.7	18.5
1955	16.3	14.9	15.4	15.9	16.6	15.8	16.3	15.6	15.2	15.8	15.0	15.8	15.4	14.8	15.7	15.6
1950	13.5	12.3	12.7	13.1	13.7	13.1	13.5	12.9	12.6	13.1	12.4	13.0	12.8	12.2	13.0	12.8
1945	8.6	7.8	8.1	8.4	8.7	8.3	8.6	8.2	8.0	8.3	7.9	8.3	8.1	7.8	8.3	8.2
1940	6.6	6.0	6.3	6.5	6.7	6.4	6.6	6.3	6.2	6.4	6.1	6.4	6.3	6.0	6.4	6.3

Year	National 30 City Average	Louisiana Shreve-port	Maine Lewis-ton	Portland	Maryland Balti-more	Massachusetts Boston	Brockton	Fall River	Law-rence	Lowell	New Bedford	Pitts-field	Spring-field	Wor-cester	Michigan Ann Arbor	Dear-born
Jan 1998	114.4E	92.6E	103.7E	103.4E	104.8E	133.1E	125.4E	125.6E	126.0E	126.4E	125.5E	115.4E	118.3E	123.7E	115.1E	120.1E
1997	111.5	90.1	101.7	101.4	102.2	132.1	124.4	124.7	125.1	125.6	124.7	114.6	117.4	122.9	113.5	117.9
1996	108.9	88.5	99.5	99.2	99.6	128.7	121.6	122.2	121.7	122.1	122.2	112.7	115.0	120.2	112.9	116.2
1995	105.6	86.7	96.8	96.5	96.1	128.6	119.6	117.7	120.5	119.6	117.2	110.7	112.7	114.8	106.1	110.5
1994	103.0	85.0	95.2	94.9	94.4	124.9	114.2	113.7	116.8	116.9	111.6	109.2	111.1	112.9	105.0	109.0
1993	100.0	83.6	93.0	93.0	93.1	121.1	111.7	111.5	114.9	114.3	109.4	107.1	108.9	110.3	102.8	105.8
1992	97.9	82.3	91.7	91.7	90.9	118.0	110.0	109.8	110.9	110.2	107.7	105.7	107.2	108.6	101.5	103.2
1991	95.7	81.6	89.8	89.9	89.1	115.2	105.9	104.7	107.3	105.5	105.1	100.6	103.2	105.7	95.1	98.3
1990	93.2	80.3	88.5	88.6	85.6	110.9	103.7	102.9	105.2	102.8	102.6	98.7	101.4	103.2	93.2	96.5
1989	91.0	78.8	86.7	86.7	83.5	107.2	100.4	99.7	102.9	99.7	99.8	94.7	96.4	99.1	89.6	94.6
1988	88.5	77.0	84.0	84.1	81.2	102.5	96.7	96.4	97.7	96.6	96.4	91.9	92.7	94.4	85.8	91.4
1987	85.7	74.5	81.3	81.4	78.6	97.3	94.5	94.8	94.9	94.8	95.0	89.4	90.5	92.2	87.4	88.7
1986	83.7	74.1	79.0	79.0	75.7	95.1	91.2	90.8	92.5	90.8	90.7	87.1	87.8	89.8	81.7	85.3
1985	81.8	73.6	76.7	77.0	72.7	92.8	88.8	88.7	89.4	88.2	88.6	85.0	85.6	86.7	80.1	82.7
1984	80.6	73.3	75.0	75.1	72.4	88.1	85.5	85.8	86.8	84.1	84.8	82.8	83.5	83.6	78.9	81.1
1983	78.2	72.2	73.1	72.9	70.6	84.0	82.5	82.2	82.7	80.9	81.3	79.5	80.6	81.3	77.7	79.4
1982	72.1	67.0	67.4	67.4	64.7	76.9	75.7	75.0	74.5	73.6	74.2	72.3	73.7	73.1	73.7	75.8
1981	66.1	62.5	62.2	63.1	59.0	67.8	68.7	69.2	68.4	67.1	68.7	66.3	67.0	66.7	68.6	70.1
1980	60.7	58.7	57.3	58.5	53.6	64.0	63.7	64.1	63.4	62.7	63.1	61.8	62.0	62.3	62.9	64.0
1979	54.9	53.1	52.3	52.6	48.2	57.9	58.0	57.5	57.3	57.0	57.4	56.2	55.9	56.0	57.0	57.8
1978	51.3	49.2	48.6	48.8	45.9	54.0	53.4	53.3	53.5	53.1	53.2	52.6	52.3	52.2	52.5	54.1
1977	47.9	44.8	46.2	45.2	44.4	49.3	50.4	50.1	50.5	49.5	49.9	49.2	50.0	49.4	48.6	48.6
1976	45.3	42.0	43.2	42.5	41.5	47.7	46.9	47.8	47.2	46.9	47.3	46.3	47.6	47.1	46.5	46.4
1975	43.7	40.5	41.9	42.1	39.8	46.6	45.8	45.7	46.2	45.7	46.1	45.5	45.8	46.0	44.1	44.9
1974	39.4	37.3	37.5	35.6	37.4	41.7	41.3	41.3	41.3	40.9	41.1	40.4	41.0	40.6	40.4	40.9
1970	27.8	26.4	26.5	25.8	25.1	29.2	29.1	29.2	29.1	28.9	29.0	28.6	28.5	28.7	28.5	28.9
1965	21.5	20.3	20.4	19.4	20.2	23.0	22.5	22.5	22.5	22.2	22.4	22.0	22.4	22.1	22.0	22.3
1960	19.5	18.5	18.6	17.6	17.5	20.5	20.4	20.4	20.4	20.2	20.3	20.0	20.1	20.1	20.0	20.2
1955	16.3	15.5	15.6	14.7	14.7	17.2	17.1	17.1	17.1	16.9	17.1	16.8	16.9	16.8	16.7	16.9
1950	13.5	12.8	12.9	12.2	12.1	14.2	14.1	14.1	14.1	14.0	14.1	13.8	14.0	13.9	13.8	14.0
1945	8.6	8.1	8.2	7.8	7.7	9.1	9.0	9.0	9.0	8.9	9.0	8.9	8.9	8.9	8.8	8.9
1940	6.6	6.3	6.3	6.0	6.0	7.0	6.9	7.0	7.0	6.9	6.9	6.8	6.9	6.8	6.8	6.9

HISTORICAL COST INDEXES

Historical Cost Indexes

Year	National 30 City Average	Michigan Detroit	Flint	Grand Rapids	Kala-mazoo	Lansing	Sagi-naw	Minnesota Duluth	Minne-apolis	Roches-ter	Mississippi Biloxi	Jackson	Missouri Kansas City	St. Joseph	St. Louis	Sp...
Jan 1998	114.4E	119.9E	112.3E	100.3E	105.5E	110.4E	109.0E	118.5E	125.3E	116.3E	92.8E	90.2E	108.6E	104.8E	116.4E	9
1997	111.5	117.6	110.8	98.9	104.2	108.7	107.5	115.9	121.9	114.1	90.9	88.3	106.4	102.4	113.2	9
1996	108.9	116.0	109.7	93.9	103.5	107.8	106.9	115.0	120.4	113.5	87.1	85.4	103.2	99.9	110.1	9
1995	105.6	110.1	104.1	91.1	96.4	98.3	102.3	100.3	111.9	102.5	84.2	83.5	99.7	96.1	106.3	8
1994	103.0	108.5	102.9	89.7	95.0	97.1	101.1	99.1	109.3	101.1	82.4	81.7	97.3	92.8	103.2	8
1993	100.0	105.4	100.7	87.7	92.7	95.3	98.8	99.8	106.7	99.9	80.0	79.4	94.5	90.9	99.9	8
1992	97.9	102.9	99.4	86.4	91.4	94.0	97.4	98.5	105.3	98.6	78.8	78.2	92.9	89.5	98.6	8
1991	95.7	97.6	93.6	85.3	90.0	92.1	90.3	96.0	102.9	97.3	77.6	76.9	90.9	88.1	96.4	8
1990	93.2	96.0	91.7	84.0	87.4	90.2	88.9	94.3	100.5	95.6	76.4	75.6	89.5	86.8	94.0	8
1989	91.0	94.0	90.2	82.4	85.8	88.6	87.4	92.5	97.4	93.7	75.0	74.1	87.3	85.1	91.8	8
1988	88.5	90.8	87.8	80.7	84.1	86.3	85.6	90.5	94.6	91.9	73.5	72.9	84.9	82.9	89.0	7
1987	85.7	87.8	85.4	79.7	82.6	85.2	84.5	88.5	92.0	90.6	72.4	71.3	81.7	82.7	85.5	7
1986	83.7	84.3	82.8	77.5	80.9	82.7	81.5	87.2	89.7	88.4	70.6	69.9	79.9	79.4	83.5	7
1985	81.8	81.6	80.7	75.8	79.4	80.0	80.5	85.2	87.9	86.5	69.4	68.5	78.2	77.3	80.8	7
1984	80.6	79.4	79.5	74.3	78.7	79.0	79.9	84.9	86.9	86.0	68.6	67.5	77.7	76.7	77.7	7
1983	78.2	77.0	77.4	74.1	76.2	77.1	78.0	81.6	81.4	82.6	68.0	67.3	76.2	76.8	76.2	7
1982	72.1	74.8	73.5	69.9	72.3	73.2	73.7	74.6	75.2	75.6	63.1	62.6	70.2	70.3	69.2	66
1981	66.1	68.6	68.2	64.6	67.1	67.5	68.5	68.8	69.6	69.8	58.7	59.1	63.5	65.7	64.5	6
1980	60.7	62.5	63.1	59.8	61.7	60.3	63.3	63.4	64.1	64.8	54.3	54.3	59.1	60.4	59.7	56
1979	54.9	56.2	56.4	53.7	55.5	55.8	56.8	59.2	58.5	58.7	49.3	49.3	54.4	54.6	54.9	51
1978	51.3	52.6	52.0	49.1	51.7	51.4	53.2	53.1	54.1	53.3	45.0	45.8	50.1	51.4	51.1	47
1977	47.9	48.7	48.4	44.9	48.4	47.4	49.1	49.5	50.0	48.9	41.2	42.3	46.8	48.8	48.2	44
1976	45.3	47.0	46.4	42.7	46.2	45.0	46.5	45.8	47.0	45.1	39.2	39.3	43.5	47.5	44.9	44
1975	43.7	45.8	44.9	41.8	43.7	44.1	44.8	45.2	46.0	45.3	37.9	37.7	42.7	45.4	44.8	41
1974	39.4	41.8	40.3	37.8	39.8	39.4	40.6	41.0	42.4	40.9	34.5	30.8	37.7	40.0	39.7	36
1970	27.8	29.7	28.5	26.7	28.1	27.8	28.7	28.9	29.5	28.9	24.4	21.3	25.3	28.3	28.4	26
1965	21.5	22.1	21.9	20.6	21.7	21.5	22.2	22.3	23.4	22.3	18.8	16.4	20.6	21.8	21.8	20
1960	19.5	20.1	20.0	18.7	19.7	19.5	20.1	20.3	20.5	20.2	17.1	14.9	19.0	19.8	19.5	18
1955	16.3	16.8	16.7	15.7	16.5	16.3	16.9	17.0	17.2	17.0	14.3	12.5	16.0	16.6	16.3	15
1950	13.5	13.9	13.8	13.0	13.6	13.5	13.9	14.0	14.2	14.0	11.8	10.3	13.2	13.7	13.5	12
1945	8.6	8.9	8.8	8.3	8.7	8.6	8.9	8.9	9.0	8.9	7.6	6.6	8.4	8.8	8.6	12
1940	6.6	6.8	6.8	6.4	6.7	6.6	6.9	6.9	7.0	6.9	5.8	5.1	6.5	6.7	6.7	6

Year	National 30 City Average	Montana Billings	Great Falls	Nebraska Lincoln	Omaha	Nevada Las Vegas	Reno	New Hampshire Man-chester	Nashua	New Jersey Camden	Jersey City	Newark	Pater-son	Trenton	NM Albu-querque	Alb...
Jan 1998	114.4E	109.5E	109.7E	95.4E	101.8E	116.7E	111.9E	110.3E	110.3E	123.6E	126.8E	128.2E	128.1E	127.0E	103.9E	112
1997	111.5	107.4	107.6	93.3	99.5	114.6	109.8	108.6	108.6	121.9	125.1	126.4	126.4	125.2	100.8	110
1996	108.9	108.0	107.5	91.4	97.4	111.8	108.9	106.5	106.4	119.1	122.5	122.5	123.8	121.8	98.6	108
1995	105.6	104.7	104.9	85.6	93.4	108.5	105.0	100.9	100.8	107.0	112.2	111.9	112.1	111.2	96.3	103
1994	103.0	100.2	100.7	84.3	91.0	105.3	102.2	97.8	97.7	105.4	110.6	110.1	110.6	108.2	93.4	102
1993	100.0	97.9	97.8	82.3	88.7	102.8	99.9	95.4	95.4	103.7	109.0	108.2	109.0	106.4	90.1	99
1992	97.9	95.5	96.4	81.1	87.4	99.4	98.4	90.4	90.4	102.0	107.5	107.0	107.8	103.9	87.5	98
1991	95.7	94.2	95.1	80.1	86.4	97.5	95.8	87.8	87.8	95.0	98.5	95.6	100.0	96.6	86.2	96
1990	93.2	92.9	93.8	78.8	85.0	96.3	94.5	86.3	86.3	93.0	93.5	93.8	97.2	94.5	84.9	93
1989	91.0	91.3	92.2	77.3	83.6	94.8	92.3	84.8	84.8	90.4	91.4	91.8	95.5	91.9	83.3	88
1988	88.5	89.5	90.4	75.8	82.0	93.0	90.5	83.2	83.2	87.7	89.5	89.7	92.5	89.5	81.7	86
1987	85.7	85.4	87.3	75.3	79.3	91.5	88.3	82.4	82.5	86.1	88.3	88.7	90.3	87.9	78.1	84
1986	83.7	86.2	87.3	72.6	78.8	90.1	87.2	79.7	79.7	84.6	86.1	86.7	87.8	86.1	78.9	81
1985	81.8	83.9	84.3	71.5	77.5	87.6	85.0	78.1	78.1	81.3	83.3	83.5	84.5	82.8	76.7	79
1984	80.6	83.2	82.9	70.9	77.6	85.9	83.3	76.4	75.5	78.2	80.3	80.1	81.3	78.5	75.9	77
1983	78.2	80.3	80.0	72.1	77.3	83.0	81.9	73.0	72.2	74.6	76.5	76.7	76.5	74.2	73.1	74
1982	72.1	74.3	74.1	67.8	72.5	76.9	75.8	67.0	66.7	67.9	69.8	70.4	70.1	69.1	67.1	69
1981	66.1	69.5	70.3	64.1	68.2	70.2	68.5	61.5	61.1	62.5	65.0	65.1	64.8	63.4	63.3	64
1980	60.7	63.9	64.6	58.5	63.5	64.6	63.0	56.6	56.0	58.6	60.6	60.1	60.0	58.9	59.0	59
1979	54.9	57.5	58.9	52.9	56.2	58.4	56.7	51.2	50.8	53.5	55.6	55.3	54.7	54.0	54.3	54
1978	51.3	53.4	54.0	49.3	52.2	53.9	51.9	47.5	47.1	50.1	51.7	51.4	51.7	50.8	49.8	50
1977	47.9	48.9	49.1	44.6	48.0	49.7	48.1	43.9	43.6	47.2	47.4	47.2	47.7	47.0	45.1	47
1976	45.3	45.4	46.2	42.1	45.2	46.3	45.1	40.8	41.1	44.6	44.8	45.4	45.3	45.2	41.9	45
1975	43.7	43.1	43.8	40.9	43.2	42.8	41.9	41.3	40.8	42.3	43.4	44.2	43.9	43.8	40.3	43
1974	39.4	40.4	41.0	37.3	37.7	40.0	39.7	36.3	36.3	38.4	39.3	40.5	39.4	39.4	35.5	40
1970	27.8	28.5	28.9	26.4	26.8	29.4	28.0	26.2	25.6	27.2	27.8	29.0	27.8	27.4	26.4	28
1965	21.5	22.0	22.3	20.3	20.6	22.4	21.6	20.6	19.7	20.9	21.4	23.8	21.4	21.3	20.6	22
1960	19.5	20.0	20.3	18.5	18.7	20.2	19.6	18.0	17.9	19.0	19.4	19.4	19.4	19.2	18.5	19
1955	16.3	16.7	17.0	15.5	15.7	16.9	16.4	15.1	15.0	16.0	16.3	16.3	16.3	16.1	15.6	16
1950	13.5	13.9	14.0	12.8	13.0	14.0	13.6	12.5	12.4	13.2	13.5	13.5	13.5	13.3	12.9	16
1945	8.6	8.8	9.0	3.1	8.3	8.9	8.7	8.0	7.9	8.4	8.6	8.6	8.6	8.5	8.2	13
1940	6.6	6.8	6.9	6.3	6.4	6.9	6.7	6.2	6.1	6.5	6.6	6.6	6.6	6.6	6.3	8

New York / North Carolina / N. Dakota / Ohio

Year	National 30 City Average	New York — Binghamton	Buffalo	New York	Rochester	Schenectady	Syracuse	Utica	Yonkers	N. Carolina — Charlotte	Durham	Greensboro	Raleigh	Winston-Salem	N. Dakota — Fargo	Ohio — Akron
Jan 1998	114.4E	109.3E	118.5E	152.4E	117.4E	113.1E	112.7E	108.9E	140.5E	88.7E	89.6E	89.7E	89.9E	89.5E	95.7E	112.5E
1997	111.5	107.2	115.7	150.3	115.2	111.2	110.6	107.0	138.8	86.8	87.7	87.8	87.9	87.6	93.5	110.6
1996	108.9	105.5	114.0	148.0	113.3	109.5	108.2	105.2	137.2	85.0	85.9	86.0	86.1	85.8	91.7	107.9
1995	105.6	99.3	110.1	140.7	106.6	104.6	104.0	97.7	129.4	81.8	82.6	82.6	82.7	82.6	88.4	103.4
1994	103.0	98.1	107.2	137.1	105.2	103.5	102.3	96.5	128.1	80.3	81.1	81.0	81.2	81.1	87.0	102.2
1993	100.0	95.7	102.2	133.3	102.0	100.8	99.6	94.1	126.3	78.2	78.9	78.9	78.9	78.8	85.8	100.3
1992	97.9	93.7	100.1	128.2	99.3	99.3	98.1	90.5	123.9	77.1	77.7	77.8	77.8	77.6	83.1	98.0
1991	95.7	89.5	96.8	124.4	96.0	96.7	95.5	88.7	121.5	76.0	76.6	76.8	76.7	76.6	82.6	96.8
1990	93.2	87.0	94.1	118.1	94.6	93.9	91.0	85.6	111.4	74.8	75.4	75.5	75.5	75.3	81.3	94.6
1989	91.0	85.4	91.8	114.8	92.6	89.6	88.4	84.1	102.1	73.3	74.0	74.0	74.0	73.8	79.7	92.9
1988	88.5	83.6	89.4	106.5	87.9	87.7	86.5	82.4	99.9	71.5	72.2	72.2	72.2	72.0	78.2	91.1
1987	85.7	82.4	85.8	102.6	85.8	85.3	85.1	82.1	98.9	70.6	71.2	71.2	71.1	71.0	77.4	90.1
1986	83.7	80.2	84.4	99.8	83.7	82.4	83.3	79.2	96.6	68.4	69.1	69.1	69.1	68.9	75.4	87.7
1985	81.8	77.5	83.2	94.9	81.6	80.1	81.0	77.0	92.8	66.9	67.6	67.7	67.6	67.5	73.4	86.8
1984	80.6	75.4	81.3	91.1	80.6	78.5	79.0	76.1	89.5	66.3	66.6	66.8	66.2	66.0	72.2	82.8
1983	78.2	73.2	78.2	86.5	77.1	75.3	76.5	74.5	85.4	64.6	65.1	65.7	64.9	64.6	70.5	79.4
1982	72.1	67.5	70.3	78.3	71.3	69.8	70.2	68.4	77.8	58.7	60.0	60.5	59.5	59.4	66.1	73.2
1981	66.1	62.6	65.2	71.6	65.6	64.7	64.6	63.5	71.0	55.3	56.6	56.7	55.9	56.3	61.7	67.7
1980	60.7	58.0	60.6	66.0	60.5	60.3	61.6	58.5	65.9	51.1	52.2	52.5	51.7	50.8	57.4	62.3
1979	54.9	52.8	56.0	60.2	55.0	54.9	56.2	53.2	60.0	45.9	46.9	47.1	46.4	45.9	52.2	56.3
1978	51.3	50.0	52.6	56.4	51.7	50.8	52.2	49.1	54.7	41.9	43.0	42.9	42.4	41.9	47.8	52.1
1977	47.9	46.5	49.0	52.9	48.3	48.5	49.1	46.0	51.8	40.2	41.5	40.7	40.0	39.6	44.2	48.5
1976	45.3	44.1	46.2	50.6	45.6	45.3	46.6	43.8	48.5	37.5	37.9	38.1	37.6	37.0	41.3	45.8
1975	43.7	42.5	45.1	49.5	44.8	43.7	44.8	42.7	47.1	36.1	37.0	37.0	37.3	36.1	39.3	44.7
1974	39.4	38.2	41.8	45.4	41.5	39.3	40.3	38.1	42.5	30.3	33.5	33.5	33.1	32.6	36.5	40.1
1970	27.8	27.0	28.9	32.8	29.2	27.8	28.5	26.9	30.0	20.9	23.7	23.6	23.4	23.0	25.8	28.3
1965	21.5	20.8	22.2	25.5	22.8	21.4	21.9	20.7	23.1	16.0	18.2	18.2	18.0	17.7	19.9	21.8
1960	19.5	18.9	19.9	21.5	19.7	19.5	19.9	18.8	21.0	14.4	16.6	16.6	16.4	16.1	18.1	19.9
1955	16.3	15.8	16.7	18.1	16.5	16.3	16.7	15.8	17.6	12.1	13.9	13.9	13.7	13.5	15.2	16.6
1950	13.5	13.1	13.8	14.9	13.6	13.5	13.8	13.0	14.5	10.0	11.5	11.5	11.4	11.2	12.5	13.7
1945	8.6	8.3	8.8	9.5	8.7	8.6	8.8	8.3	9.3	6.4	7.3	7.3	7.2	7.1	8.0	8.8
1940	6.6	6.4	6.8	7.4	6.7	6.7	6.8	6.4	7.2	4.9	5.6	5.6	5.6	5.5	6.2	6.8

Ohio / Oklahoma / Oregon / PA

Year	National 30 City Average	Ohio — Canton	Cincinnati	Cleveland	Columbus	Dayton	Lorain	Springfield	Toledo	Youngstown	Oklahoma — Lawton	Oklahoma City	Tulsa	Oregon — Eugene	Portland	PA — Allentown
Jan 1998	114.4E	108.6E	105.5E	114.8E	107.3E	104.5E	109.4E	104.3E	110.5E	109.1E	94.9E	95.3E	95.0E	119.9E	121.2E	115.7E
1997	111.5	106.4	102.9	112.9	104.7	102.6	107.4	102.2	108.5	107.2	92.9	93.2	93.0	118.1	119.6	113.9
1996	108.9	103.2	100.2	110.1	101.1	98.8	103.8	98.4	105.1	104.1	90.9	91.3	91.2	114.4	116.1	112.0
1995	105.6	98.8	97.1	106.4	99.1	94.6	97.1	92.0	100.6	100.1	85.3	88.0	89.0	112.2	114.3	108.3
1994	103.0	97.7	95.0	104.8	95.4	93.0	96.0	90.1	99.3	98.9	84.0	86.5	87.6	107.6	109.2	106.4
1993	100.0	95.9	92.3	101.9	93.9	90.6	93.9	87.7	98.3	97.2	81.3	83.9	84.8	107.2	108.8	103.6
1992	97.9	94.5	90.6	98.7	92.6	89.2	92.2	86.4	97.1	96.0	80.1	82.7	83.4	101.1	102.6	101.6
1991	95.7	93.4	88.6	97.2	90.6	87.7	91.1	84.9	94.2	92.7	78.3	80.7	81.4	99.5	101.1	98.9
1990	93.2	92.1	86.7	95.1	88.1	85.9	89.4	83.4	92.9	91.8	77.1	79.9	80.0	98.0	99.4	95.0
1989	91.0	90.3	84.6	93.6	85.8	82.6	87.9	81.0	91.4	88.4	75.8	78.5	78.4	95.2	97.0	92.4
1988	88.5	89.1	82.9	92.1	83.9	81.0	85.2	79.5	89.6	86.9	74.3	77.0	76.8	93.5	95.3	89.8
1987	85.7	87.7	80.9	89.6	81.5	79.5	86.9	78.0	86.1	85.4	72.6	74.4	74.7	91.4	91.9	86.9
1986	83.7	84.9	78.8	87.3	79.7	78.0	81.6	76.1	84.6	83.8	71.6	74.1	73.9	90.0	91.7	84.7
1985	81.8	83.7	78.2	86.2	77.7	76.3	80.7	74.3	84.7	82.7	71.2	73.9	73.8	88.7	90.1	81.9
1984	80.6	81.3	77.4	82.5	76.1	75.9	79.5	73.9	83.5	80.3	70.6	72.9	72.8	89.5	89.3	78.4
1983	78.2	76.9	74.9	77.9	74.5	71.6	77.2	71.9	80.4	77.3	69.1	71.5	72.2	89.0	89.2	75.0
1982	72.1	70.8	70.1	71.6	68.6	65.2	70.8	66.5	74.6	71.7	63.5	65.2	66.2	82.3	83.6	68.2
1981	66.1	65.8	65.3	66.0	63.9	60.6	65.7	61.3	69.4	67.2	57.0	60.3	61.5	74.2	74.5	62.7
1980	60.7	60.6	59.8	61.0	58.6	56.6	60.8	56.2	64.0	61.5	52.2	55.5	57.2	68.9	68.4	58.7
1979	54.9	55.2	54.3	54.8	52.7	51.2	55.2	51.0	57.9	56.0	47.7	49.8	51.4	61.9	60.6	54.1
1978	51.3	51.1	50.4	51.3	49.3	47.2	50.3	46.9	54.0	52.2	44.5	46.3	48.0	57.2	56.0	50.4
1977	47.9	47.7	46.9	48.2	45.3	44.3	46.7	42.9	50.2	48.7	40.9	43.0	44.8	51.5	51.7	46.7
1976	45.3	45.4	44.8	45.9	43.7	42.3	44.3	40.7	47.4	45.8	39.0	40.6	41.7	48.7	48.2	45.1
1975	43.7	44.0	43.7	44.6	42.2	40.5	42.7	40.0	44.9	44.5	38.6	38.6	39.2	44.6	44.9	42.3
1974	39.4	39.4	39.9	41.0	38.4	37.6	38.9	35.9	41.9	40.3	34.1	33.3	34.1	43.0	40.8	38.6
1970	27.8	27.8	28.2	29.6	26.7	27.0	27.5	25.4	29.1	29.6	24.1	23.4	24.1	30.4	30.0	27.3
1965	21.5	21.5	20.7	21.1	20.2	20.0	21.1	19.6	21.3	21.0	18.5	17.8	20.6	23.4	22.6	21.0
1960	19.5	19.5	19.3	19.6	18.7	18.1	19.2	17.8	19.4	19.1	16.9	16.1	18.1	17.8	17.7	16.0
1955	16.3	16.3	16.2	16.5	15.7	15.2	16.1	14.9	16.2	16.0	14.1	13.5	15.1	14.7	14.6	13.2
1950	13.5	13.5	13.3	13.6	13.0	12.5	13.3	12.3	13.4	13.2	11.7	11.2	12.5	12.9	12.8	10.7
1945	8.6	8.6	8.5	8.7	8.3	8.0	8.5	7.8	8.6	8.4	7.4	7.1	8.0	9.4	9.3	8.4
1940	6.6	6.7	6.5	6.7	6.4	6.2	6.5	6.1	6.6	6.5	5.7	5.5	6.1	7.2	7.2	6.5

HISTORICAL COST INDEXES

Historical Cost Indexes

Year	National 30 City Average	Erie	Harris-burg	Phila-delphia	Pitts-burgh	Reading	Scranton	Provi-dence	Charles-ton	Colum-bia	Rapid City	Sioux Falls	Chatta-nooga	Knox-ville	Memphis	Na...
Jan 1998	114.4E	110.5E	111.2E	126.0E	117.6E	112.7E	114.4E	120.9E	88.5E	87.7E	91.4E	94.3E	95.7E	92.7E	97.9E	92
1997	111.5	108.5	108.8	123.3	113.9	110.8	112.3	118.9	86.5	85.6	89.3	92.1	93.6	90.8	96.1	94
1996	108.9	106.8	105.7	120.3	110.8	108.7	109.9	117.2	84.9	84.1	87.2	90.0	91.6	88.8	94.0	91
1995	105.6	99.8	100.6	117.1	106.3	103.6	103.8	111.1	82.7	82.2	84.0	84.7	89.2	86.0	91.2	8.
1994	103.0	98.2	99.4	115.2	103.7	102.2	102.6	109.6	81.2	80.6	82.7	83.1	87.4	84.1	89.0	84
1993	100.0	94.9	96.6	107.4	99.0	98.6	99.8	108.2	78.4	77.9	81.2	81.9	85.1	82.3	86.8	81
1992	97.9	92.1	94.8	105.3	96.5	96.7	97.8	106.9	77.1	76.8	80.0	80.7	83.6	77.7	85.4	80
1991	95.7	90.5	92.5	101.7	93.2	94.1	94.8	96.1	75.0	75.8	78.7	79.7	81.9	76.6	83.0	79
1990	93.2	88.5	89.5	98.5	91.2	90.8	91.3	94.1	73.7	74.5	77.2	78.4	79.9	75.1	81.3	77
1989	91.0	86.6	86.5	94.2	89.4	87.9	88.8	92.4	72.1	72.8	75.7	77.0	78.5	73.7	81.2	74
1988	88.5	84.9	84.0	89.8	87.7	85.3	86.4	89.1	70.5	71.3	74.2	75.5	77.0	72.1	79.7	73
1987	85.7	82.6	81.8	86.8	85.6	82.9	83.3	86.3	70.2	70.2	73.6	74.9	74.4	70.2	77.5	71
1986	83.7	81.5	79.6	84.9	83.6	79.7	81.3	85.2	67.4	68.1	71.0	72.3	74.2	69.2	75.3	68
1985	81.8	79.6	77.2	82.2	81.5	77.7	80.0	83.0	65.9	66.9	69.7	71.2	72.5	67.7	74.3	66
1984	80.6	78.3	75.4	79.0	78.9	75.9	78.8	80.3	64.3	65.5	68.7	70.4	71.4	66.4	73.3	66
1983	78.2	76.3	72.2	74.4	75.5	74.0	75.0	76.3	65.3	65.2	67.7	69.7	69.3	64.1	72.8	65
1982	72.1	71.0	66.3	69.0	70.9	67.7	68.4	70.0	59.9	60.0	64.4	67.6	64.5	59.8	66.5	62
1981	66.1	64.6	61.7	63.4	66.3	62.5	62.3	64.3	56.3	55.9	59.7	63.6	59.9	55.9	60.3	57
1980	60.7	59.4	56.9	58.7	61.3	57.7	58.4	59.2	50.6	51.1	54.8	57.1	55.0	51.6	55.9	53
1979	54.9	54.0	52.6	54.2	55.6	53.8	53.5	53.3	45.7	44.4	49.7	51.5	49.8	46.4	51.0	47
1978	51.3	50.2	49.6	51.3	51.4	49.7	50.0	50.2	42.1	40.7	46.0	47.4	45.9	42.8	47.4	43
1977	47.9	46.6	45.2	48.1	48.3	46.5	46.9	47.0	39.2	39.4	42.5	44.3	42.5	40.2	45.0	40
1976	45.3	44.7	43.7	45.8	45.6	44.2	44.4	44.2	36.6	37.9	40.1	42.0	39.7	38.3	42.0	38
1975	43.7	43.4	42.2	44.5	44.5	43.6	41.9	42.7	35.0	36.4	37.7	39.6	39.1	37.3	40.7	37
1974	39.4	39.2	36.7	39.6	39.9	38.4	38.2	39.7	32.3	32.4	35.1	36.5	35.2	31.5	33.9	32
1970	27.8	27.5	25.8	27.5	28.7	27.1	27.0	27.3	22.8	22.9	24.8	25.8	24.9	22.0	23.0	22
1965	21.5	21.1	20.1	21.7	22.4	20.9	20.8	21.9	17.6	17.6	19.1	19.9	19.2	17.1	18.3	17
1960	19.5	19.1	18.6	19.4	19.7	19.0	18.9	19.1	16.0	16.0	17.3	18.1	17.4	15.5	16.6	15
1955	16.3	16.1	15.6	16.3	16.5	15.9	15.9	16.0	13.4	13.4	14.5	15.1	14.6	13.0	13.9	13.
1950	13.5	13.3	12.9	13.5	13.6	13.2	13.1	13.2	11.1	11.1	12.0	12.5	12.1	10.7	11.5	10
1945	8.6	8.5	8.2	8.6	8.7	8.4	8.3	8.4	7.1	7.1	7.7	8.0	7.7	6.8	7.3	7.
1940	6.6	6.5	6.3	6.6	6.7	6.4	6.5	6.5	5.4	5.5	5.9	6.2	6.0	5.3	5.6	5.

Year	National 30 City Average	Abi-lene	Ama-rillo	Austin	Beau-mont	Corpus Christi	Dallas	El Paso	Fort Worth	Houston	Lubbock	Odessa	San Antonio	Waco	Wichita Falls	Og...
Jan 1998	114.4E	90.4E	93.4E	93.8E	98.6E	92.4E	98.5E	88.9E	95.1E	102.0E	93.9E	90.7E	95.4E	93.2E	93.6E	98.
1997	111.5	88.4	91.3	92.8	96.8	90.3	96.1	87.0	93.3	100.1	91.9	88.8	93.4	91.4	91.5	96.
1996	108.9	86.8	89.6	90.9	95.3	88.5	94.1	86.7	91.5	97.9	90.2	87.1	92.3	89.7	89.9	94.
1995	105.6	85.2	87.4	89.3	93.7	87.4	91.4	85.2	89.5	95.9	88.4	85.6	88.9	86.4	86.8	92.
1994	103.0	83.3	85.5	87.0	91.8	84.6	89.6	82.2	87.4	93.4	87.0	83.9	87.0	84.9	85.4	89.
1993	100.0	81.4	83.4	84.9	90.0	82.8	87.8	80.2	85.5	91.1	84.8	81.9	85.0	83.0	83.5	87.
1992	97.9	80.3	82.3	83.8	88.9	81.6	86.2	79.0	84.1	89.8	83.6	80.7	83.9	81.8	82.4	85.
1991	95.7	79.2	81.1	82.8	87.8	80.6	85.9	78.0	83.3	87.9	82.7	79.8	83.3	80.6	81.5	84.
1990	93.2	78.0	80.1	81.3	86.5	79.3	84.5	76.7	82.1	85.4	81.5	78.6	80.7	79.6	80.3	83.
1989	91.0	76.6	78.7	79.7	85.1	77.8	82.5	75.2	80.6	84.0	80.0	77.0	78.7	78.1	78.7	81.
1988	88.5	75.0	77.2	78.3	83.9	76.3	81.1	73.6	79.8	82.5	78.4	75.5	77.1	76.5	77.4	80.
1987	85.7	75.9	76.4	76.0	81.3	75.6	79.9	72.2	77.7	81.3	77.3	74.6	75.7	75.2	77.4	79.
1986	83.7	72.5	74.1	75.3	80.5	73.5	78.9	70.7	76.7	80.3	75.4	72.4	74.3	72.8	74.5	77.
1985	81.8	71.1	72.5	74.5	79.3	72.3	77.6	69.4	75.1	79.6	74.0	71.2	73.9	71.7	73.3	77.
1984	80.6	70.5	71.6	72.8	78.7	71.6	79.8	68.3	77.1	80.7	72.7	69.8	73.1	71.3	72.4	75.
1983	78.2	68.3	70.6	71.0	75.0	70.2	77.8	66.3	74.9	79.8	71.0	69.0	71.8	69.4	70.1	76.
1982	72.1	62.7	65.5	64.9	65.7	64.7	70.7	62.6	68.2	72.1	65.6	62.5	66.0	64.3	64.8	69.
1981	66.1	57.9	60.1	59.3	62.4	59.6	63.1	58.6	61.0	65.1	61.0	58.3	60.5	59.2	60.0	65.
1980	60.7	53.4	55.2	54.5	57.6	54.5	57.9	53.1	57.0	59.4	55.6	57.2	55.0	54.9	55.4	62.
1979	54.9	48.6	49.8	49.0	52.3	48.7	51.5	48.5	51.4	53.3	50.3	51.1	49.2	49.7	49.4	53.
1978	51.3	44.7	45.8	46.2	48.0	44.9	48.2	45.3	47.5	50.2	46.5	43.9	46.0	46.4	47.1	49.
1977	47.9	42.9	42.5	43.1	43.9	42.2	44.9	41.7	44.7	47.6	42.9	41.5	42.5	42.1	41.1	45.
1976	45.3	40.4	39.7	40.3	40.7	39.6	41.2	39.0	41.5	43.8	40.3	38.6	40.8	39.7	37.7	43.
1975	43.7	37.6	39.0	39.0	39.6	38.1	40.7	38.0	40.4	41.2	38.9	37.9	39.0	38.6	38.0	40.
1974	39.4	34.7	35.2	35.3	36.5	34.7	35.8	32.3	36.6	36.1	35.6	34.9	33.8	35.2	34.7	37.
1970	27.8	24.5	24.9	24.9	25.7	24.5	25.5	23.7	25.9	25.4	25.1	24.6	23.3	24.8	24.5	26.
1965	21.5	18.9	19.2	19.2	19.9	18.9	19.9	19.0	19.9	20.0	19.4	19.0	18.5	19.2	18.9	20.
1960	19.5	17.1	17.4	17.4	18.1	17.1	18.2	17.0	18.1	18.2	17.6	17.3	16.8	17.4	17.2	18.
1955	16.3	14.4	14.6	14.6	15.1	14.4	15.3	14.3	15.2	15.2	14.8	14.5	14.1	14.6	14.4	15.
1950	13.5	11.9	12.1	12.1	12.5	11.9	12.6	11.8	12.5	12.6	12.2	12.0	11.6	12.1	11.9	13.
1945	8.6	7.6	7.7	7.7	8.0	7.6	8.0	7.5	8.0	8.0	7.8	7.6	7.4	7.7	7.6	8.
1940	6.6	5.9	5.9	5.9	6.1	5.8	6.2	5.8	6.2	6.2	6.0	5.9	5.7	5.9	5.8	6.

HISTORICAL COST INDEXES

Year	National 30 City Average	Utah Salt Lake City	Vermont Burlington	Vermont Rutland	Virginia Alexandria	Virginia Newport News	Virginia Norfolk	Virginia Richmond	Virginia Roanoke	Washington Seattle	Washington Spokane	Washington Tacoma	West Virginia Charleston	West Virginia Huntington	Wisconsin Green Bay	Wisconsin Kenosha
Jan 1998	114.4E	99.2E	98.7E	98.0E	103.6E	93.5E	93.7E	95.0E	88.8E	120.0E	113.3E	119.0E	107.2E	109.6E	109.1E	112.1E
1997	111.5	97.2	96.6	96.3	101.2	91.6	91.7	92.9	86.9	118.1	111.7	117.2	105.3	107.7	105.6	109.1
1996	108.9	94.9	95.1	94.8	99.7	90.2	90.4	91.6	85.5	115.2	109.2	114.3	103.1	104.8	103.8	106.4
1995	105.6	93.1	91.1	90.8	96.3	86.0	86.4	87.8	82.8	113.7	107.4	112.8	95.8	97.2	97.6	97.9
1994	103.0	90.2	89.5	89.3	93.9	84.6	84.8	86.3	81.4	109.9	104.0	108.3	94.3	95.3	96.3	96.2
1993	100.0	87.9	87.6	87.6	91.6	82.9	83.0	84.3	79.5	107.3	103.9	106.7	92.6	93.5	94.0	94.3
1992	97.9	86.0	86.1	86.1	90.1	81.0	81.6	82.0	78.3	105.1	101.4	103.7	91.4	92.3	92.0	92.1
1991	95.7	84.9	84.2	84.2	88.2	77.6	77.9	79.8	77.3	102.2	100.0	102.2	89.7	88.6	88.6	89.8
1990	93.2	84.3	83.0	82.9	86.1	76.3	76.7	77.6	76.1	100.1	98.5	100.5	86.1	86.8	86.7	87.8
1989	91.0	82.8	81.4	81.3	83.4	74.9	75.2	76.0	74.2	96.1	96.8	98.4	84.5	84.8	84.3	85.6
1988	88.5	81.3	79.7	79.7	81.0	73.4	73.7	74.1	72.3	94.2	95.0	96.6	82.8	83.1	82.0	83.0
1987	85.7	79.8	79.0	79.0	77.9	71.0	71.7	72.7	70.2	91.9	92.4	92.9	81.1	81.3	80.2	81.7
1986	83.7	78.1	76.4	76.4	76.8	70.3	70.5	71.0	67.8	90.5	91.9	92.8	79.9	80.2	77.8	79.3
1985	81.8	75.9	74.8	74.9	75.1	68.7	68.8	69.5	67.2	88.3	89.0	91.2	77.7	77.7	76.7	77.4
1984	80.6	75.2	73.7	73.7	75.0	67.9	68.3	68.8	66.2	89.2	88.1	90.0	75.4	75.8	75.0	75.1
1983	78.2	74.5	70.9	71.5	72.9	65.9	66.5	68.9	66.0	87.6	86.9	88.8	73.8	74.4	73.4	73.2
1982	72.1	68.1	64.9	65.4	67.0	60.6	60.9	63.7	61.0	80.7	80.6	81.7	67.6	68.4	67.5	69.1
1981	66.1	61.9	59.3	59.7	61.6	56.0	56.8	58.2	55.6	75.7	72.9	73.8	62.6	63.3	63.6	63.5
1980	60.7	57.0	55.3	58.3	57.3	52.5	52.4	54.3	51.3	67.9	66.3	66.7	57.7	58.3	58.6	58.3
1979	54.9	52.5	49.8	51.2	51.7	47.6	47.6	49.1	46.8	61.1	60.1	60.0	51.8	52.5	52.7	53.2
1978	51.3	49.0	46.1	47.2	48.9	44.0	44.9	46.0	43.6	56.7	55.4	55.6	48.2	48.5	48.7	49.5
1977	47.9	45.0	44.5	46.3	44.6	41.2	41.5	42.5	41.2	51.8	51.1	50.7	44.7	43.2	44.8	45.5
1976	45.3	42.8	42.3	45.0	42.5	38.4	38.7	39.9	38.6	48.5	47.5	47.1	42.3	40.3	42.6	42.9
1975	43.7	40.1	41.8	43.9	41.7	37.2	36.9	37.1	37.1	44.9	44.4	44.5	41.0	40.0	40.9	40.5
1974	39.4	36.1	35.7	38.0	37.2	33.8	31.4	31.1	33.6	39.8	40.1	42.0	37.0	36.5	37.3	37.5
1970	27.8	26.1	25.4	26.8	26.2	23.9	21.5	22.0	23.7	28.8	29.3	29.6	26.1	25.8	26.4	26.5
1965	21.5	20.0	19.8	20.6	20.2	18.4	17.1	17.2	18.3	22.4	22.5	22.8	20.1	19.9	20.3	20.4
1960	19.5	18.4	18.0	18.8	18.4	16.7	15.4	15.6	16.6	20.4	20.8	20.8	18.3	18.1	18.4	18.6
1955	16.3	15.4	15.1	15.7	15.4	14.0	12.9	13.1	13.9	17.1	17.4	17.4	15.4	15.2	15.5	15.6
1950	13.5	12.7	12.4	13.0	12.7	11.6	10.7	10.8	11.5	14.1	14.4	14.4	12.7	12.5	12.8	12.9
1945	8.6	8.1	7.9	8.3	8.1	7.4	6.8	6.9	7.3	9.0	9.2	9.2	8.1	8.0	8.1	8.2
1940	6.6	6.3	6.1	6.4	6.2	5.7	5.3	5.3	5.7	7.0	7.1	7.1	6.2	6.2	6.3	6.3

Year	National 30 City Average	Wisconsin Madison	Wisconsin Milwaukee	Wisconsin Racine	Wyoming Cheyenne	Canada Calgary	Canada Edmonton	Canada Hamilton	Canada London	Canada Montreal	Canada Ottawa	Canada Quebec	Canada Toronto	Canada Vancouver	Canada Winnipeg
Jan 1998	114.4E	108.4E	113.5E	112.2E	95.2E	112.8E	112.7E	126.5E	124.1E	116.7E	124.9E	117.3E	128.0E	124.3E	113.4E
1997	111.5	106.1	110.1	109.3	93.2	110.7	110.6	124.4	121.9	114.6	122.8	115.4	125.7	121.9	111.4
1996	108.9	104.4	107.1	106.5	91.1	109.1	109.0	122.6	120.2	112.9	121.0	113.6	123.9	119.0	109.7
1995	105.6	96.5	103.9	97.8	87.6	107.4	107.4	119.9	117.5	110.8	118.2	111.5	121.6	116.2	107.6
1994	103.0	94.5	100.6	96.1	85.4	106.7	106.6	116.5	114.2	109.5	115.0	110.2	117.8	115.1	105.5
1993	100.0	91.3	96.7	93.8	82.9	104.7	104.5	113.7	111.9	106.9	112.0	107.0	114.9	109.2	102.8
1992	97.9	89.2	93.9	91.7	81.7	103.4	103.2	112.4	110.7	104.2	110.8	103.6	113.7	108.0	101.6
1991	95.7	86.2	91.6	89.3	80.3	102.1	102.0	108.2	106.7	101.8	106.9	100.4	109.0	106.8	98.6
1990	93.2	84.3	88.9	87.3	79.1	98.0	97.1	103.8	101.4	99.0	102.7	96.8	104.6	103.2	95.1
1989	91.0	81.8	86.4	85.0	77.8	95.6	94.7	98.2	96.5	94.9	98.5	92.5	98.5	97.6	92.9
1988	88.5	79.9	84.1	82.4	76.3	93.9	93.0	95.4	93.2	90.6	94.0	88.3	94.8	95.9	90.0
1987	85.7	78.1	81.1	81.1	76.7	90.2	89.3	89.7	90.1	87.2	89.6	85.0	90.4	94.2	87.1
1986	83.7	76.1	78.9	78.6	73.4	91.2	90.1	88.9	87.9	84.8	87.7	82.4	89.3	93.2	85.3
1985	81.8	74.3	77.4	77.0	72.3	90.2	89.1	85.8	84.9	82.4	83.9	79.9	86.5	89.8	83.4
1984	80.6	72.5	76.3	74.7	73.8	89.6	88.1	85.1	84.1	81.6	82.9	80.0	85.2	89.2	82.5
1983	78.2	71.9	74.0	73.0	73.8	84.6	82.8	80.1	79.1	76.4	77.6	75.8	79.9	83.3	77.9
1982	72.1	65.9	70.5	68.6	68.3	76.2	74.1	73.7	71.8	70.2	72.3	69.6	71.8	76.4	70.7
1981	66.1	61.4	64.9	63.3	62.7	70.6	68.1	68.8	67.1	64.6	65.8	64.3	67.5	70.3	65.4
1980	60.7	56.8	58.8	58.1	56.9	64.9	63.3	63.7	61.9	59.2	60.9	59.3	60.9	65.0	61.7
1979	54.9	51.4	53.0	52.8	51.2	58.9	58.5	58.1	56.6	54.4	55.5	53.9	56.0	59.2	56.1
1978	51.3	47.3	49.9	49.2	47.5	54.9	54.4	54.0	52.6	50.4	51.7	49.7	51.9	55.1	51.7
1977	47.9	44.0	46.9	45.7	44.8	49.9	49.9	48.8	48.2	43.7	48.0	43.0	48.2	50.2	47.8
1976	45.3	41.9	44.2	43.5	42.2	45.6	45.4	44.5	43.3	41.7	43.8	40.6	44.1	45.9	42.1
1975	43.7	40.7	43.3	40.7	40.6	42.2	41.6	42.9	41.6	39.7	41.5	39.0	42.2	42.4	39.2
1974	39.4	37.6	40.9	37.6	36.7	40.9	40.5	40.3	39.4	34.7	39.0	36.8	36.4	37.1	33.9
1970	27.8	26.5	29.4	26.5	26.0	28.9	28.6	28.5	27.8	25.6	27.6	26.0	25.6	26.0	23.1
1965	21.5	20.6	21.8	20.4	20.0	22.3	22.0	22.0	21.4	18.7	21.2	20.1	19.4	20.5	17.5
1960	19.5	18.1	19.0	18.6	18.2	20.2	20.0	20.0	19.5	17.0	19.3	18.2	17.6	18.6	15.8
1955	16.3	15.2	15.9	15.6	15.2	17.0	16.8	16.7	16.3	14.3	16.2	15.3	14.8	15.5	13.3
1950	13.5	12.5	13.2	12.9	12.6	14.0	13.9	13.8	13.5	11.8	13.4	12.6	12.2	12.8	10.9
1945	8.6	8.0	8.4	8.2	8.0	9.0	8.8	8.8	8.6	7.5	8.5	8.0	7.8	8.2	7.0
1940	6.6	6.2	6.5	6.3	6.2	6.9	6.8	6.8	6.6	5.8	6.6	6.2	6.0	6.3	5.4

Accent lighting–Fixtures or directional beams of light arranged to bring attention to an object or area.

Acoustical material–A material fabricated for the sole purpose of absorbing sound.

Addition–An expansion to an existing structure generally in the form of a room, floor(s) or wing(s). An increase in the floor area or volume of a structure.

Aggregate–Materials such as sand, gravel, stone, vermiculite, perlite and fly ash (slag) which are essential components in the production of concrete, mortar or plaster.

Air conditioning system–An air treatment to control the temperature, humidity and cleanliness of air and to provide for its distribution throughout the structure.

Air curtain–A stream of air directed downward to prevent the loss of hot or cool air from the structure and inhibit the entrance of dust and insects. Generally installed at loading platforms.

Anodized aluminum–Aluminum treated by an electrolytic process to produce an oxide film that is corrosion resistant.

Arcade–A covered passageway between buildings often with shops and offices on one or both sides.

Ashlar–A square-cut building stone or a wall constructed of cut stone.

Asphalt paper–A sheet paper material either coated or saturated with asphalt to increase its strength and resistance to water.

Asphalt shingles–Shingles manufactured from saturated roofing felts, coated with asphalt and topped with mineral granules to prevent weathering.

Assessment ratio–The ratio between the market value and assessed valuation of a property.

Back-up–That portion of a masonry wall behind the exterior facing-usually load bearing.

Balcony–A platform projecting from a building either supported from below or cantilevered. Usually protected by railing.

Bay–Structural component consisting of beams and columns occurring consistently throughout the structure.

Beam–A horizontal structural framing member that transfers loads to vertical members (columns) or bearing walls.

Bid–An offer to perform work described in a contract at a specified price.

Board foot–A unit or measure equal in volume to a board one foot long, one foot wide and one inch thick.

Booster pump–A supplemental pump installed to increase or maintain adequate pressure in the system.

Bowstring roof–A roof supported by trusses fabricated in the shape of a bow and tied together by a straight member.

Brick veneer–A facing of brick laid against, but not bonded to, a wall.

Bridging–A method of bracing joists for stiffness, stability and load distribution.

Broom finish–A method of finishing concrete by lightly dragging a broom over freshly placed concrete.

Built-up roofing–Installed on flat or extremely low-pitched roofs, a roof covering composed of plies or laminations of saturated roofing felts alternated with layers of coal tar pitch or asphalt and surfaced with a layer of gravel or slag in a thick coat of asphalt and finished with a capping sheet.

Caisson–A watertight chamber to expedite work on foundations or structure below water level.

Cantilever–A structural member supported only at one end.

Carport–An automobile shelter having one or more sides open to the weather.

Casement window–A vertical opening window having one side fixed.

Caulking–A resilient compound generally having a silicone or rubber base used to prevent infiltration of water or outside air.

Cavity wall–An exterior wall usually of masonry having an inner and outer wall separated by a continuous air space for thermal insulation.

Cellular concrete–A lightweight concrete consisting of cement mixed with gas-producing materials which, in turn, forms bubbles resulting in good insulating qualities.

Chattel mortgage–A secured interest in a property as collateral for payment of a note.

Clapboard–A wood siding used as exterior covering in frame construction. It is applied horizontally with grain running lengthwise. The thickest section of the board is on the bottom.

Clean room–An environmentally controlled room usually found in medical facilities or precision manufacturing spaces where it is essential to eliminate dust, lint or pathogens.

Close studding–A method of construction whereby studs are spaced close together and the intervening spaces are plastered.

Cluster housing–A closely grouped series of houses resulting in a high density land use.

Cofferdam–A watertight enclosure used for foundation construction in waterfront areas. Water is pumped out of the cofferdam allowing free access to the work area.

Collar joint–Joint between the collar beam and roof rafters.

Column–A vertical structural member supporting horizontal members (beams) along the direction of its longitudinal axis.

Combination door (windows)–Door (windows) having interchangeable screens and glass for seasonal use.

Common area–Spaces either inside or outside the building designated for use by the occupant of the building but not the general public.

Common wall–A wall used jointly by two dwelling units.

Compound wall–A wall constructed of more than one material.

Concrete–A composite material consisting of sand, coarse aggregate (gravel, stone or slag), cement and water that when mixed and allowed harden forms a hard stone-like material.

Concrete floor hardener–A mixture of chemicals applied to the surface of concrete to produce a dense, wear-resistant bearing surface.

Conduit–A tube or pipe used to protect electric wiring.

Coping–The top cover or capping on a wall.

Craneway–Steel or concrete column and beam supports and rails on which a crane travels.

Curtain wall–A non-bearing exterior wall not supported by beams or girders of the steel frame.

Dampproofing–Coating of a surface to prevent the passage of water or moisture.

Dead load–Total weight of all structural components plus permanently attached fixtures and equipment.

Decibel–Unit of acoustical measurement.

Decorative block–A concrete masonry unit having a special treatment on its face.

Deed restriction–A restriction on the use of a property as set forth in the deed.

Depreciation–A loss in property value caused by physical aging, functional or economic obsolescence.

Direct heating–Heating of spaces by means of exposed heated surfaces (stove, fire, radiators, etc.).

Distributed load–A load that acts evenly over a structural member.

Dock bumper–A resilient material attached to a loading dock to absorb the impact of trucks backing in.

Dome–A curved roof shape spanning an area.

Dormer–A structure projecting from a sloping roof.

Double-hung window–A window having two vertical sliding sashes-one covering the upper section and one covering the lower.

Downspout–A vertical pipe for carrying rain water from the roof to the ground.

Glossary

ain tile–A hollow tile used to drain water-soaked soil.

essed size–In lumber about 3/8″ - 1/2″ less than nominal size after ning and sawing.

op panel–The depressed surface on the bottom side of a flat concrete which surrounds a column.

y pipe sprinkler system–A sprinkler system that is activated with ter only in case of fire. Used in areas susceptible to freezing or to id the hazards of leaking pipes.

ywall–An interior wall constructed of gypsum board, plywood or od paneling. No water is required for application.

ct–Pipe for transmitting warm or cold air. A pipe containing electrical le or wires.

velling–A structure designed as living quarters for one or more ilies.

sement–A right of way or free access over land owned by another.

ve–The lower edge of a sloping roof that overhangs the side wall of structure.

onomic rent–That rent on a property sufficient to pay all operating ts exclusive of services and utilities.

onomic life–The term during which a structure is expected to be ofitable. Generally shorter than the physical life of the structure.

onomic obsolescence–Loss in value due to unfavorable economic uences occurring from outside the structure itself.

fective age–The age a structure appears to be based on observed ysical condition determined by degree of maintenance and repair.

evator–A mechanism for vertical transport of personnel or freight uipped with car or platform.

l–A secondary wing or addition to a structure at right angles to the in structure.

ninent domain–The right of the state to take private property for blic use.

velope–The shape of a building indicating volume.

quity–Value of an owner's interest calculated by subtracting outstanding ortgages and expenses from the value of the property.

cheat–The assumption of ownership by the state, of property whose ner cannot be determined.

cade–The exterior face of a building sometimes decorated with borate detail.

ce brick–Brick manufactured to present an attractive appearance.

cing block–A concrete masonry unit having a decorative exterior ish.

asibility study–A detailed evaluation of a proposed project to termine its financial potential.

lt paper–Paper sheathing used on walls as an insulator and to prevent filtration of moisture.

nestration–Pertaining to the density and arrangement of windows in a ructure.

berboard–A building material composed of wood fiber compressed ith a binding agent, produced in sheet form.

berglass–Fine spun filaments of glass processed into various densities produce thermal and acoustical insulation.

eld house–A long structure used for indoor athletic events.

nish floor–The top or wearing surface of the floor system.

reproofing–The use of fire-resistant materials for the protection of ructural members to ensure structural integrity in the event of fire.

re stop–A material or member used to seal an opening to prevent the read of fire.

ashing–A thin, impervious material such as copper or sheet metal used prevent air or water penetration.

oat finish–A concrete finish accomplished by using a flat tool with a ndle on the back.

Floor drain–An opening installed in a floor for removing excess water into a plumbing system.

Floor load–The live load the floor system has been designed for and which may be applied safely.

Flue–A heat-resistant enclosed passage in a chimney to remove gaseous products of combustion from a fireplace or boiler.

Footing–The lowest portion of the foundation wall that transmits the load directly to the soil.

Foundation–Below grade wall system that supports the structure.

Foyer–A portion of the structure that serves as a transitional space between the interior and exterior.

Frontage–That portion of a lot that is placed facing a street, public way or body of water.

Frost action–The effects of freezing and thawing on materials and the resultant structural damage.

Functional obsolescence–An inadequacy caused by outmoded design, dated construction materials or over or undersized areas, all of which cause excessive operating costs.

Furring strips–Wood or metal channels used for attaching gypsum or metal lath to masonry walls as a finish or for leveling purposes.

Gambrel roof (mansard)–A roof system having two pitches on each side.

Garden apartment–Two or three story apartment building with common outside areas.

General contractor–The prime contractor responsible for work on the construction site.

Girder–A principal horizontal supporting member.

Girt–A horizontal framing member for adding rigid support to columns which also acts as a support for sheathing or siding.

Grade beam–That portion of the foundation system that directly supports the exterior walls of the building.

Gross area–The total enclosed floor area of a building.

Ground area–The area computed by the exterior dimensions of the structure.

Ground floor–The floor of a building in closest proximity to the ground.

Grout–A mortar containing a high water content used to fill joints and cavities in masonry work.

Gutter–A wood or sheet metal conduit set along the building eaves used to channel rainwater to leaders or downspouts.

Gunite–A concrete mix placed pneumatically.

Gypsum–Hydrated calcium sulphate used as both a retarder in Portland cement and as a major ingredient in plaster of Paris. Used in sheets as a substitute for plaster.

Hall–A passageway providing access to various parts of a building–a large room for entertainment or assembly.

Hangar–A structure for the storage or repair of aircraft.

Head room–The distance between the top of the finished floor to the bottom of the finished ceiling.

Hearth–The floor of a fireplace and the adjacent area of fireproof material.

Heat pump–A mechanical device for providing either heating or air conditioning.

Hip roof–A roof whose four sides meet at a common point with no gabled ends.

I-beam–A structural member having a cross-section resembling the letter "I".

Insulation–Material used to reduce the effects of heat, cold or sound.

Jack rafter–An unusually short rafter generally found in hip roofs.

Jalousie–Adjustable glass louvers which pivot simultaneously in a common frame.

Jamb–The vertical member on either side of a door frame or window frame.

Joist–Parallel beams of timber, concrete or steel used to support floor and ceiling.

Junction box–A box that protects splices or joints in electrical wiring.

Kalamein door–Solid core wood doors clad with galvanized sheet metal.

Kip–A unit of weight equal to 1,000 pounds.

Lally column–A concrete-filled steel pipe used as a vertical support.

Laminated beam–A beam built up by gluing together several pieces of timber.

Landing–A platform between flights of stairs.

Lath–Strips of wood or metal used as a base for plaster.

Lead-lined door, sheetrock–Doors or sheetrock internally lined with sheet lead to provide protection from X-ray radiation.

Lean-to–A small shed or building addition with a single pitched roof attached to the exterior wall of the main building.

Lintel–A horizontal framing member used to carry a load over a wall opening.

Live load–The moving or movable load on a structure composed of furnishings, equipment or personnel weight.

Load bearing partition–A partition that supports a load in addition to its own weight.

Loading dock leveler–An adjustable platform for handling off-on loading of trucks.

Loft building–A commercial/industrial type building containing large, open unpartitioned floor areas.

Louver window–A window composed of a series of sloping, overlapping blades or slats that may be adjusted to admit varying degrees of air or light.

Main beam–The principal load bearing beam used to transmit loads directly to columns.

Mansard roof–A roof having a double pitch on all four sides, the lower level having the steeper pitch.

Mansion–An extremely large and imposing residence.

Masonry–The utilization of brick, stone, concrete or block for walls and other building components.

Mastic–A sealant or adhesive compound generally used as either a binding agent or floor finish.

Membrane fireproofing–A coating of metal lath and plaster to provide resistance to fire and heat.

Mesh reinforcement–An arrangement of wire tied or welded at their intersection to provide strength and resistance to cracking.

Metal lath–A diamond-shaped metallic base for plaster.

Mezzanine–A low story situated between two main floors.

Mill construction–A heavy timber construction that achieves fire resistance by using large wood structural members, noncombustible bearing and non-bearing walls and omitting concealed spaces under floors and roof.

Mixed occupancy–Two or more classes of occupancy in a single structure.

Molded brick–A specially shaped brick used for decorative purposes.

Monolithic concrete–Concrete poured in a continuous process so there are no joints.

Movable partition–A non-load bearing demountable partition that can be relocated and can be either ceiling height or partial height.

Moving walk–A continually moving horizontal passenger carrying device.

Multi-zone system–An air conditioning system that is capable of handling several individual zones simultaneously.

Net floor area–The usable, occupied area of a structure excluding stairwells, elevator shafts and wall thicknesses.

Non-combustible construction–Construction in which the walls, partitions and the structural members are of non-combustible materials.

Nurses call system–An electrically operated system for use by patient or personnel for summoning a nurse.

Occupancy rate–The number of persons per room, per dwelling unit etc.

One-way joist connection–A framing system for floors and roofs in concrete building consisting of a series of parallel joists supported by girders between columns.

Open web joist–A lightweight prefabricated metal chord truss.

Parapet–A low guarding wall at the point of a drop. That portion of t exterior wall that extends above the roof line.

Parging–A thin coat of mortar applied to a masonry wall used primar for waterproofing purposes.

Parquet–Inlaid wood flooring set in a simple design.

Peaked roof–A roof of two or more slopes that rises to a peak.

Penthouse–A structure occupying usually half the roof area and used house elevator, HVAC equipment, or other mechanical or electrical systems.

Pier–A column designed to support a concentrated load.

Pilaster–A column usually formed of the same material and integral w but projecting from, a wall.

Pile–A concrete, wood or steel column usually less than 2′ in diamete which is driven or otherwise introduced into the soil to carry a vertica load or provide vertical support.

Pitched roof–A roof having one or more surfaces with a pitch greate than 10°.

Plenum–The space between the suspended ceiling and the floor abov

Plywood–Structural wood made of three or more layers of veneer and bonded together with glue.

Porch–A structure attached to a building to provide shelter for an entranceway or to serve as a semi-enclosed space.

Post and beam framing–A structural framing system where beams re on posts rather than bearing walls.

Post-tensioned concrete–Concrete that has the reinforcing tendons tensioned after the concrete has set.

Pre-cast concrete–Concrete structural components fabricated at a location other than in-place.

Pre-stressed concrete–Concrete that has the reinforcing tendons tensioned prior to the concrete setting.

Purlins–A horizontal structural member supporting the roof deck and resting on the trusses, girders, beams or rafters.

Rafters–Structural members supporting the roof deck and covering.

Resilient flooring–A manufactured interior floor covering material in either sheet or tile form, that is resilient.

Ridge–The horizontal line at the junction of two sloping roof edges.

Rigid frame–A structural framing system in which all columns and beams are rigidly connected; there are no hinged joints.

Roof drain–A drain designed to accept rainwater and discharge it into leader or downspout.

Rough floor–A layer of boards or plywood nailed to the floor joists th serves as a base for the finished floor.

Sanitary sewer–A sewer line designed to carry only liquid or waterborne waste from the structure to a central treatment plant.

Sawtooth roof–Roof shape found primarily in industrial roofs that creates the appearance of the teeth of a saw.

Semi-rigid frame–A structural system wherein the columns and beam are so attached that there is some flexibility at the joints.

Septic tank–A covered tank in which waste matter is decomposed by natural bacterial action.

Service elevator–A combination passenger and freight elevator.

Sheathing–The first covering of exterior studs by boards, plywood or particle board.

...ring–Temporary bracing for structural components during ...struction.

...n–The horizontal distance between supports.

...cification–A detailed list of materials and requirements for ...struction of a building.

...rm sewer–A system of pipes used to carry rainwater or surface ...ers.

...cco–A cement plaster used to cover exterior wall surfaces usually ...lied over a wood or metal lath base.

...d–A vertical wooden structural component.

...floor–A system of boards, plywood or particle board laid over the ...r joists to form a base.

...erstructure–That portion of the structure above the foundation or ...und level.

...razzo–A durable floor finish made of small chips of colored stone or ...ble, embedded in concrete, then polished to a high sheen.

...e–A thin piece of fired clay, stone or concrete used for floor, roof or ...l finishes.

Unit cost–The cost per unit of measurement.

Vault–A room especially designed for storage.

Veneer–A thin surface layer covering a base of common material.

Vent–An opening serving as an outlet for air.

Wainscot–The lower portion of an interior wall whose surface differs from that of the upper wall.

Wall bearing construction–A structural system where the weight of the floors and roof are carried directly by the masonry walls rather than the structural framing system.

Waterproofing–Any of a number of materials applied to various surfaces to prevent the infiltration of water.

Weatherstrip–A thin strip of metal, wood or felt used to cover the joint between the door or window sash and the jamb, casing or sill to keep out air, dust, rain, etc.

Wing–A building section or addition projecting from the main structure.

X-ray protection–Lead encased in sheetrock or plaster to prevent the escape of radiation.

A	Area Square Feet; Ampere
ABS	Acrylonitrile Butadiene Stryrene; Asbestos Bonded Steel
A.C.	Alternating Current; Air-Conditioning; Asbestos Cement; Plywood Grade A & C
A.C.I.	American Concrete Institute
AD	Plywood, Grade A & D
Addit.	Additional
Adj.	Adjustable
af	Audio-frequency
A.G.A.	American Gas Association
Agg.	Aggregate
A.H.	Ampere Hours
A hr.	Ampere-hour
A.H.U.	Air Handling Unit
A.I.A.	American Institute of Architects
AIC	Ampere Interrupting Capacity
Allow.	Allowance
alt.	Altitude
Alum.	Aluminum
a.m.	Ante Meridiem
Amp.	Ampere
Anod.	Anodized
Approx.	Approximate
Apt.	Apartment
Asb.	Asbestos
A.S.B.C.	American Standard Building Code
Asbe.	Asbestos Worker
A.S.H.R.A.E.	American Society of Heating, Refrig. & AC Engineers
A.S.M.E.	American Society of Mechanical Engineers
A.S.T.M.	American Society for Testing and Materials
Attchmt.	Attachment
Avg.	Average
A.W.G.	American Wire Gauge
Bbl.	Barrel
B. & B.	Grade B and Better; Balled & Burlapped
B. & S.	Bell and Spigot
B. & W.	Black and White
b.c.c.	Body-centered Cubic
B.C.Y.	Bank Cubic Yards
BE	Bevel End
B.F.	Board Feet
Bg. cem.	Bag of Cement
BHP	Boiler Horsepower; Brake Horsepower
B.I.	Black Iron
Bit.; Bitum.	Bituminous
Bk.	Backed
Bkrs.	Breakers
Bldg.	Building
Blk.	Block
Bm.	Beam
Boil.	Boilermaker
B.P.M.	Blows per Minute
BR	Bedroom
Brg.	Bearing
Brhe.	Bricklayer Helper
Bric.	Bricklayer
Brk.	Brick
Brng.	Bearing
Brs.	Brass
Brz.	Bronze
Bsn.	Basin
Btr.	Better
BTU	British Thermal Unit
BTUH	BTU per Hour
B.U.R.	Built-up Roofing
BX	Interlocked Armored Cable
c	Conductivity, Copper Sweat
C	Hundred; Centigrade
C/C	Center to Center, Cedar on Cedar

Cab.	Cabinet
Cair.	Air Tool Laborer
Calc	Calculated
Cap.	Capacity
Carp.	Carpenter
C.B.	Circuit Breaker
C.C.A.	Chromate Copper Arsenate
C.C.F.	Hundred Cubic Feet
cd	Candela
cd/sf	Candela per Square Foot
CD	Grade of Plywood Face & Back
CDX	Plywood, Grade C & D, exterior glue
Cefi.	Cement Finisher
Cem.	Cement
CF	Hundred Feet
C.F.	Cubic Feet
CFM	Cubic Feet per Minute
c.g.	Center of Gravity
CHW	Chilled Water; Commercial Hot Water
C.I.	Cast Iron
C.I.P.	Cast in Place
Circ.	Circuit
C.L.	Carload Lot
Clab.	Common Laborer
C.L.F.	Hundred Linear Feet
CLF	Current Limiting Fuse
CLP	Cross Linked Polyethylene
cm	Centimeter
CMP	Corr. Metal Pipe
C.M.U.	Concrete Masonry Unit
CN	Change Notice
Col.	Column
CO_2	Carbon Dioxide
Comb.	Combination
Compr.	Compressor
Conc.	Concrete
Cont.	Continuous; Continued
Corr.	Corrugated
Cos	Cosine
Cot	Cotangent
Cov.	Cover
C/P	Cedar on Paneling
CPA	Control Point Adjustment
Cplg.	Coupling
C.P.M.	Critical Path Method
CPVC	Chlorinated Polyvinyl Chloride
C.Pr.	Hundred Pair
CRC	Cold Rolled Channel
Creos.	Creosote
Crpt.	Carpet & Linoleum Layer
CRT	Cathode-ray Tube
CS	Carbon Steel, Constant Shear Bar Joist
Csc	Cosecant
C.S.F.	Hundred Square Feet
CSI	Construction Specifications Institute
C.T.	Current Transformer
CTS	Copper Tube Size
Cu	Copper, Cubic
Cu. Ft.	Cubic Foot
cw	Continuous Wave
C.W.	Cool White; Cold Water
Cwt.	100 Pounds
C.W.X.	Cool White Deluxe
C.Y.	Cubic Yard (27 cubic feet)
C.Y./Hr.	Cubic Yard per Hour
Cyl.	Cylinder
d	Penny (nail size)
D	Deep; Depth; Discharge
Dis.;Disch.	Discharge
Db.	Decibel
Dbl.	Double
DC	Direct Current
Demob.	Demobilization
d.f.u.	Drainage Fixture Units

D.H.	Double Hung
DHW	Domestic Hot Water
Diag.	Diagonal
Diam.	Diameter
Distrib.	Distribution
Dk.	Deck
D.L.	Dead Load; Diesel
DLH	Deep Long Span Bar Joist
Do.	Ditto
Dp.	Depth
D.P.S.T.	Double Pole, Single Throw
Dr.	Driver
Drink.	Drinking
D.S.	Double Strength
D.S.A.	Double Strength A Grade
D.S.B.	Double Strength B Grade
Dty.	Duty
DWV	Drain Waste Vent
DX	Deluxe White, Direct Expansion
dyn	Dyne
e	Eccentricity
E	Equipment Only; East
Ea.	Each
E.B.	Encased Burial
Econ.	Economy
EDP	Electronic Data Processing
EIFS	Exterior Insulation Finish System
E.D.R.	Equiv. Direct Radiation
Eq.	Equation
Elec.	Electrician; Electrical
Elev.	Elevator; Elevating
EMT	Electrical Metallic Conduit; Thin Wall Conduit
Eng.	Engine, Engineered
EPDM	Ethylene Propylene Diene Monomer
EPS	Expanded Polystyrene
Eqhv.	Equip. Oper., Heavy
Eqlt.	Equip. Oper., Light
Eqmd.	Equip. Oper., Medium
Eqmm.	Equip. Oper., Master Mechanic
Eqol.	Equip. Oper., Oilers
Equip.	Equipment
ERW	Electric Resistance Welded
E.S.	Energy Saver
Est.	Estimated
esu	Electrostatic Units
E.W.	Each Way
EWT	Entering Water Temperature
Excav.	Excavation
Exp.	Expansion, Exposure
Ext.	Exterior
Extru.	Extrusion
f.	Fiber stress
F	Fahrenheit; Female; Fill
Fab.	Fabricated
FBGS	Fiberglass
F.C.	Footcandles
f.c.c.	Face-centered Cubic
f'c.	Compressive Stress in Concrete; Extreme Compressive Stress
F.E.	Front End
FEP	Fluorinated Ethylene Propylene (Teflon)
F.G.	Flat Grain
F.H.A.	Federal Housing Administration
Fig.	Figure
Fin.	Finished
Fixt.	Fixture
Fl. Oz.	Fluid Ounces
Flr.	Floor
F.M.	Frequency Modulation; Factory Mutual
Fmg.	Framing
Fndtn.	Foundation
Fori.	Foreman, Inside
Foro.	Foreman, Outside

nt.	Fountain
	Feet per Minute
	Female Pipe Thread
	Frame
	Fire Rating
	Foil Reinforced Kraft
	Fiberglass Reinforced Plastic
	Forged Steel
	Cast Body; Cast Switch Box
	Foot; Feet
g.	Fitting
	Footing
Lb.	Foot Pound
rn.	Furniture
NR	Full Voltage Non-Reversing
M	Female by Male
	Minimum Yield Stress of Steel
	Gram
	Gauss
	Gauge
	Gallon
/Min.	Gallon per Minute
lv.	Galvanized
n.	General
EI.	Ground Fault Interrupter
az.	Glazier
D	Gallons per Day
H	Gallons per Hour
M	Gallons per Minute
	Grade
an.	Granular
nd.	Ground
	High; High Strength Bar Joist; Henry
C.	High Capacity
D.	Heavy Duty; High Density
D.O.	High Density Overlaid
dr.	Header
dwe.	Hardware
elp.	Helpers Average
EPA	High Efficiency Particulate Air Filter
	Mercury
IC	High Interrupting Capacity
M	Hollow Metal
O.	High Output
oriz.	Horizontal
P.	Horsepower; High Pressure
P.F.	High Power Factor
r.	Hour
rs./Day	Hours per Day
SC	High Short Circuit
t.	Height
tg.	Heating
trs.	Heaters
VAC	Heating, Ventilation & Air-Conditioning
vy.	Heavy
W	Hot Water
yd.;Hydr.	Hydraulic
z.	Hertz (cycles)
	Moment of Inertia
C.	Interrupting Capacity
D	Inside Diameter
D.	Inside Dimension; Identification
F.	Inside Frosted
M.C.	Intermediate Metal Conduit
n.	Inch
ncan.	Incandescent
ncl.	Included; Including
nt.	Interior
nst.	Installation
nsul.	Insulation/Insulated
P	Iron Pipe
PS.	Iron Pipe Size
PT.	Iron Pipe Threaded
W.	Indirect Waste
	Joule
J.I.C.	Joint Industrial Council
K	Thousand; Thousand Pounds; Heavy Wall Copper Tubing
K.A.H.	Thousand Amp. Hours
KCMIL	Thousand Circular Mils
KD	Knock Down
K.D.A.T.	Kiln Dried After Treatment
kg	Kilogram
kG	Kilogauss
kgf	Kilogram Force
kHz	Kilohertz
Kip.	1000 Pounds
KJ	Kiljoule
K.L.	Effective Length Factor
K.L.F.	Kips per Linear Foot
Km	Kilometer
K.S.F.	Kips per Square Foot
K.S.I.	Kips per Square Inch
K.V.	Kilovolt
K.V.A.	Kilovolt Ampere
K.V.A.R.	Kilovar (Reactance)
KW	Kilowatt
KWh	Kilowatt-hour
L	Labor Only; Length; Long; Medium Wall Copper Tubing
Lab.	Labor
lat	Latitude
Lath.	Lather
Lav.	Lavatory
lb.; #	Pound
L.B.	Load Bearing; L Conduit Body
L. & E.	Labor & Equipment
lb./hr.	Pounds per Hour
lb./L.F.	Pounds per Linear Foot
lbf/sq.in.	Pound-force per Square Inch
L.C.L.	Less than Carload Lot
Ld.	Load
LE	Lead Equivalent
LED	Light Emitting Diode
L.F.	Linear Foot
Lg.	Long; Length; Large
L & H	Light and Heat
LH	Long Span Bar Joist, Labor Hours
L.L.	Live Load
L.L.D.	Lamp Lumen Depreciation
L-O-L	Lateralolet
lm	Lumen
lm/sf	Lumen per Square Foot
lm/W	Lumen per Watt
L.O.A.	Length Over All
log	Logarithm
L.P.	Liquefied Petroleum; Low Pressure
L.P.F.	Low Power Factor
LR	Long Radius
L.S.	Lump Sum
Lt.	Light
Lt. Ga.	Light Gauge
L.T.L.	Less than Truckload Lot
Lt. Wt.	Lightweight
L.V.	Low Voltage
M	Thousand; Material; Male; Light Wall Copper Tubing
m/hr; M.H.	Man-hour
mA	Milliampere
Mach.	Machine
Mag. Str.	Magnetic Starter
Maint.	Maintenance
Marb.	Marble Setter
Mat; Mat'l.	Material
Max.	Maximum
MBF	Thousand Board Feet
MBH	Thousand BTU's per hr.
MC	Metal Clad Cable
M.C.F.	Thousand Cubic Feet
M.C.F.M.	Thousand Cubic Feet per Minute
M.C.M.	Thousand Circular Mils
M.C.P.	Motor Circuit Protector
MD	Medium Duty
M.D.O.	Medium Density Overlaid
Med.	Medium
MF	Thousand Feet
M.F.B.M.	Thousand Feet Board Measure
Mfg.	Manufacturing
Mfrs.	Manufacturers
mg	Milligram
MGD	Million Gallons per Day
MGPH	Thousand Gallons per Hour
MH, M.H.	Manhole; Metal Halide; Man-Hour
MHz	Megahertz
Mi.	Mile
MI	Malleable Iron; Mineral Insulated
mm	Millimeter
Mill.	Millwright
Min., min.	Minimum, minute
Misc.	Miscellaneous
ml	Milliliter
M.L.F.	Thousand Linear Feet
Mo.	Month
Mobil.	Mobilization
Mog.	Mogul Base
MPH	Miles per Hour
MPT	Male Pipe Thread
MRT	Mile Round Trip
ms	Millisecond
M.S.F.	Thousand Square Feet
Mstz.	Mosaic & Terrazzo Worker
M.S.Y.	Thousand Square Yards
Mtd.	Mounted
Mthe.	Mosaic & Terrazzo Helper
Mtng.	Mounting
Mult.	Multi; Multiply
M.V.A.	Million Volt Amperes
M.V.A.R.	Million Volt Amperes Reactance
MV	Megavolt
MW	Megawatt
MXM	Male by Male
MYD	Thousand Yards
N	Natural; North
nA	Nanoampere
NA	Not Available; Not Applicable
N.B.C.	National Building Code
NC	Normally Closed
N.E.M.A.	National Electrical Manufacturers Assoc.
NEHB	Bolted Circuit Breaker to 600V.
N.L.B.	Non-Load-Bearing
NM	Non-Metallic Cable
nm	Nanometer
No.	Number
NO	Normally Open
N.O.C.	Not Otherwise Classified
Nose.	Nosing
N.P.T.	National Pipe Thread
NQOD	Combination Plug-on/Bolt on Circuit Breaker to 240V.
N.R.C.	Noise Reduction Coefficient
N.R.S.	Non Rising Stem
ns	Nanosecond
nW	Nanowatt
OB	Opposing Blade
OC	On Center
OD	Outside Diameter
O.D.	Outside Dimension
ODS	Overhead Distribution System
O.G.	Ogee
O.H.	Overhead
O & P	Overhead and Profit
Oper.	Operator
Opng.	Opening
Orna.	Ornamental
OSB	Oriented Strand Board
O. S. & Y.	Outside Screw and Yoke
Ovhd.	Overhead
OWG	Oil, Water or Gas
Oz.	Ounce

P.	Pole; Applied Load; Projection
p.	Page
Pape.	Paperhanger
P.A.P.R.	Powered Air Purifying Respirator
PAR	Weatherproof Reflector
Pc., Pcs.	Piece, Pieces
P.C.	Portland Cement; Power Connector
P.C.F.	Pounds per Cubic Foot
P.C.M.	Phase Contract Microscopy
P.E.	Professional Engineer; Porcelain Enamel; Polyethylene; Plain End
Perf.	Perforated
Ph.	Phase
P.I.	Pressure Injected
Pile.	Pile Driver
Pkg.	Package
Pl.	Plate
Plah.	Plasterer Helper
Plas.	Plasterer
Pluh.	Plumbers Helper
Plum.	Plumber
Ply.	Plywood
p.m.	Post Meridiem
Pntd.	Painted
Pord.	Painter, Ordinary
pp	Pages
PP; PPL	Polypropylene
P.P.M.	Parts per Million
Pr.	Pair
P.E.S.B.	Pre-engineered Steel Building
Prefab.	Prefabricated
Prefin.	Prefinished
Prop.	Propelled
PSF; psf	Pounds per Square Foot
PSI; psi	Pounds per Square Inch
PSIG	Pounds per Square Inch Gauge
PSP	Plastic Sewer Pipe
Pspr.	Painter, Spray
Psst.	Painter, Structural Steel
P.T.	Potential Transformer
P. & T.	Pressure & Temperature
Ptd.	Painted
Ptns.	Partitions
Pu	Ultimate Load
PVC	Polyvinyl Chloride
Pvmt.	Pavement
Pwr.	Power
Q	Quantity Heat Flow
Quan.;Qty.	Quantity
Q.C.	Quick Coupling
r	Radius of Gyration
R	Resistance
R.C.P.	Reinforced Concrete Pipe
Rect.	Rectangle
Reg.	Regular
Reinf.	Reinforced
Req'd.	Required
Res.	Resistant
Resi.	Residential
Rgh.	Rough
R.H.W.	Rubber, Heat & Water Resistant; Residential Hot Water
rms	Root Mean Square
Rnd.	Round
Rodm.	Rodman
Rofc.	Roofer, Composition
Rofp.	Roofer, Precast
Rohe.	Roofer Helpers (Composition)
Rots.	Roofer, Tile & Slate
R.O.W.	Right of Way
RPM	Revolutions per Minute
R.R.	Direct Burial Feeder Conduit
R.S.	Rapid Start
Rsr	Riser
RT	Round Trip
S.	Suction; Single Entrance; South
SCFM	Standard Cubic Feet per Minute
Scaf.	Scaffold
Sch.; Sched.	Schedule
S.C.R.	Modular Brick
S.D.	Sound Deadening
S.D.R.	Standard Dimension Ratio
S.E.	Surfaced Edge
Sel.	Select
S.E.R.;	Service Entrance Cable
S.E.U.	Service Entrance Cable
S.F.	Square Foot
S.F.C.A.	Square Foot Contact Area
S.F.G.	Square Foot of Ground
S.F. Hor.	Square Foot Horizontal
S.F.R.	Square Feet of Radiation
S.F. Shlf.	Square Foot of Shelf
S4S	Surface 4 Sides
Shee.	Sheet Metal Worker
Sin.	Sine
Skwk.	Skilled Worker
SL	Saran Lined
S.L.	Slimline
Sldr.	Solder
SLH	Super Long Span Bar Joist
S.N.	Solid Neutral
S-O-L	Socketolet
S.P.	Static Pressure; Single Pole; Self-Propelled
Spri.	Sprinkler Installer
Sq.	Square; 100 Square Feet
S.P.D.T.	Single Pole, Double Throw
SPF	Spruce Pine Fir
S.P.S.T.	Single Pole, Single Throw
SPT	Standard Pipe Thread
Sq. Hd.	Square Head
Sq. In.	Square Inch
S.S.	Single Strength; Stainless Steel
S.S.B.	Single Strength B Grade
Sswk.	Structural Steel Worker
Sswl.	Structural Steel Welder
St.;Stl.	Steel
S.T.C.	Sound Transmission Coefficient
Std.	Standard
STK	Select Tight Knot
STP	Standard Temperature & Pressure
Stpi.	Steamfitter, Pipefitter
Str.	Strength; Starter; Straight
Strd.	Stranded
Struct.	Structural
Sty.	Story
Subj.	Subject
Subs.	Subcontractors
Surf.	Surface
Sw.	Switch
Swbd.	Switchboard
S.Y.	Square Yard
Syn.	Synthetic
S.Y.P.	Southern Yellow Pine
Sys.	System
t.	Thickness
T	Temperature; Ton
Tan	Tangent
T.C.	Terra Cotta
T & C	Threaded and Coupled
T.D.	Temperature Difference
T.E.M.	Transmission Electron Microscopy
TFE	Tetrafluoroethylene (Teflon)
T. & G.	Tongue & Groove; Tar & Gravel
Th.;Thk.	Thick
Thn.	Thin
Thrded	Threaded
Tilf.	Tile Layer, Floor
Tilh.	Tile Layer, Helper
THHN	Nylon Jacketed Wire
THW.	Insulated Strand Wire
THWN;	Nylon Jacketed Wire
T.L.	Truckload
T.M.	Track Mounted
Tot.	Total
T-O-L	Threadolet
T.S.	Trigger Start
Tr.	Trade
Transf.	Transformer
Trhv.	Truck Driver, Heavy
Trlr	Trailer
Trlt.	Truck Driver, Light
TV	Television
T.W.	Thermoplastic Water Resistant Wire
UCI	Uniform Construction Index
UF	Underground Feeder
UGND	Underground Feeder
U.H.F.	Ultra High Frequency
U.L.	Underwriters Laboratory
Unfin.	Unfinished
URD	Underground Residential Distribution
US	United States
USP	United States Primed
UTP	Unshielded Twisted Pair
V	Volt
V.A.	Volt Amperes
V.C.T.	Vinyl Composition Tile
VAV	Variable Air Volume
VC	Veneer Core
Vent.	Ventilation
Vert.	Vertical
V.F.	Vinyl Faced
V.G.	Vertical Grain
V.H.F.	Very High Frequency
VHO	Very High Output
Vib.	Vibrating
V.L.F.	Vertical Linear Foot
Vol.	Volume
VRP	Vinyl Reinforced Polyester
W	Wire; Watt; Wide; West
w/	With
W.C.	Water Column; Water Closet
W.F.	Wide Flange
W.G.	Water Gauge
Wldg.	Welding
W. Mile	Wire Mile
W-O-L	Weldolet
W.R.	Water Resistant
Wrck.	Wrecker
W.S.P.	Water, Steam, Petroleum
WT.,Wt.	Weight
WWF	Welded Wire Fabric
XFER	Transfer
XFMR	Transformer
XHD	Extra Heavy Duty
XHHW;	Cross-Linked Polyethylene Wire Insulation
XLP	Cross-linked Polyethylene
XLPE	Cross-Linked Polyethylene Wire Insulation
Y	Wye
yd	Yard
yr	Year
Δ	Delta
%	Percent
~	Approximately
Ø	Phase
@	At
#	Pound; Number
<	Less Than
>	Greater Than

Means Forms

**RESIDENTIAL
COST ESTIMATE**

OWNER'S NAME: _____   APPRAISER: _____

RESIDENCE ADDRESS: _____   PROJECT: _____

CITY, STATE, ZIP CODE: _____   DATE: _____

CLASS OF CONSTRUCTION	RESIDENCE TYPE	CONFIGURATION	EXTERIOR WALL SYSTEM
☐ ECONOMY	☐ 1 STORY	☐ DETACHED	☐ WOOD SIDING - WOOD FRAME
☐ AVERAGE	☐ 1-1/2 STORY	☐ TOWN/ROW HOUSE	☐ BRICK VENEER - WOOD FRAME
☐ CUSTOM	☐ 2 STORY	☐ SEMI-DETACHED	☐ STUCCO ON WOOD FRAME
☐ LUXURY	☐ 2-1/2 STORY		☐ PAINTED CONCRETE BLOCK
	☐ 3 STORY	OCCUPANCY	☐ SOLID MASONRY (AVERAGE & CUSTOM)
	☐ BI-LEVEL	☐ ONE FAMILY	☐ STONE VENEER - WOOD FRAME
	☐ TRI-LEVEL	☐ TWO FAMILY	☐ SOLID BRICK (LUXURY)
		☐ THREE FAMILY	☐ SOLID STONE (LUXURY)
		☐ OTHER _____	

* LIVING AREA (Main Building)		* LIVING AREA (Wing or Ell) ( )		* LIVING AREA (WING or ELL) ( )	
First Level	_____ S.F.	First Level	_____ S.F.	First Level	_____ S.F.
Second level	_____ S.F.	Second level	_____ S.F.	Second level	_____ S.F.
Third Level	_____ S.F.	Third Level	_____ S.F.	Third Level	_____ S.F.
Total	_____ S.F.	Total	_____ S.F.	Total	_____ S.F.

* Basement Area is not part of living area.

MAIN BUILDING	COSTS PER S.F. LIVING AREA
Cost per Square Foot of Living Area, from Page _____	$
Basement Addition: _____ % Finished, _____ % Unfinished	+
Roof Cover Adjustment: _____ Type, Page _____ (Add or Deduct)	( )
Central Air Conditioning: ☐ Separate Ducts ☐ Heating Ducts, Page _____	+
Heating System Adjustment: _____ Type, Page _____ (Add or Deduct)	( )
Main Building: Adjusted Cost per S.F. of Living Area	$

MAIN BUILDING TOTAL COST $ _____ /S.F.   X   _____ S.F.   X   _____   = $ _____

Cost per S.F. Living Area   Living Area   Town/Row House Multiplier (Use 1 for Detached)   TOTAL COST

WING OR ELL ( ) _____ STORY	COSTS PER S.F. LIVING AREA
Cost per Square Foot of Living Area, from Page _____	$
Basement Addition: _____ % Finished, _____ % Unfinished	+
Roof Cover Adjustment: _____ Type, Page _____ (Add or Deduct)	( )
Central Air Conditioning: ☐ Separate Ducts ☐ Heating Ducts, Page _____	+
Heating System Adjustment: _____ Type, Page _____ (Add or Deduct)	( )
Wing or Ell ( ): Adjusted Cost per S.F. of Living Area	$

WING OR ELL ( ) TOTAL COST $ _____ /S.F.   X   _____ S.F.   X   = $ _____

Cost per S.F. Living Area   Living Area   TOTAL COST

WING OR ELL ( ) _____ STORY	COSTS PER S.F. LIVING AREA
Cost per Square Foot of Living Area, from Page _____	$
Basement Addition: _____ % Finished, _____ % Unfinished	+
Roof Cover Adjustment: _____ Type, Page _____ (Add or Deduct)	( )
Central Air Conditioning: ☐ Separate Ducts ☐ Heating Ducts, Page _____	+
Heating System Adjustment: _____ Type, Page _____ (Add or Deduct)	( )
Wing or Ell ( ): Adjusted Cost per S.F. of Living Area	$

WING OR ELL ( ) TOTAL COST $ _____ /S.F.   X   _____ S.F.   X   = $ _____

Cost per S.F. Living Area   Living Area   TOTAL COST

TOTAL THIS PAGE   [_____]

Page 1 of 2

Means Forms

RESIDENTIAL
COST ESTIMATE

Total Page 1				$
		QUANTITY	UNIT COST	
Additional Bathrooms: _____ Full _____ Half				
Finished Attic: _____ Ft. x _____ Ft.		S.F.		+
Breezeway: ☐ Open ☐ Enclosed _____ Ft. x _____ Ft.		S.F.		+
Covered Porch: ☐ Open ☐ Enclosed _____ Ft. x _____ Ft.		S.F.		+
Fireplace: ☐ Interior Chimney ☐ Exterior Chimney				
☐ No. of Flues ☐ Additional Fireplaces				+
Appliances:				+
Kitchen Cabinets Adjustments: (±)				
☐ Garage ☐ Carport: _____ Car(s) Description _____ (±)				
Miscellaneous:				+

ADJUSTED TOTAL BUILDING COST $ _____

REPLACEMENT COST	
ADJUSTED TOTAL BUILDING COST	$ _____
Site Improvements	
(A) Paving & Sidewalks	$ _____
(B) Landscaping	$ _____
(C) Fences	$ _____
(D) Swimming Pools	$ _____
(E) Miscellaneous	$ _____
TOTAL	$ _____
Location Factor	x _____
Location Replacement Cost	$ _____
Depreciation	- $ _____
LOCAL DEPRECIATED COST	$ _____

INSURANCE COST	
ADJUSTED TOTAL BUILDING COST	$ _____
Insurance Exclusions	
(A) Footings, Site work, Underground Piping	- $ _____
(B) Architects Fees	- $ _____
Total Building Cost Less Exclusion	$ _____
Location Factor	x _____
LOCAL INSURABLE REPLACEMENT COST	$ _____

SKETCH AND ADDITIONAL CALCULATIONS

CII APPRAISAL

5. DATE: _____

1. SUBJECT PROPERTY: _____
6. APPRAISER: _____

2. BUILDING: _____

3. ADDRESS: _____

4. BUILDING USE: _____
7. YEAR BUILT: _____

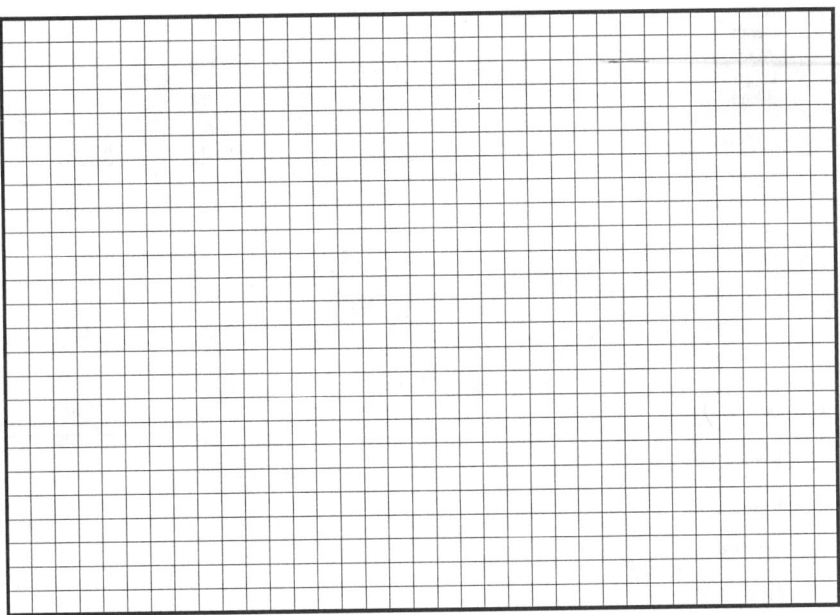

8. EXTERIOR WALL CONSTRUCTION: _____

9. FRAME: _____

10. GROUND FLOOR AREA: _____ S.F.   11. GROSS FLOOR AREA (EXCL. BASEMENT): _____ S.F.

12. NUMBER OF STORIES: _____   13. STORY HEIGHT: _____

14. PERIMETER: _____ L.F.   15. BASEMENT AREA: _____ S.F.

16. GENERAL COMMENTS: _____

_____

_____

_____

_____

_____

_____

_____

SQUARE FOOT AREA *(EXCLUDING BASEMENT)* FROM ITEM 11 _____ S.F. _____

PERIMETER FROM ITEM 14 _____ L.F. _____

ITEM 17 - MODEL SQUARE FOOT COSTS *(FROM MODEL SUB-TOTAL)* $ _____

Field Description & Calculation Section

NO.	SYSTEM/COMPONENT	DESCRIPTION	UNIT	UNIT COST	NEW S.F. COST	MODEL S.F. COST	+/- CHANGE
1.0	**FOUNDATION**						
.1	Bldg. Excavation	Depth					
		Area	S.F.				
.2	Trench Excavation	Depth Width Length	L.F.				
	Footings: Strip Width	L.F. Length	L.F.				
	Footings: Spread	Bay Size	S.F. Gnd.				
		Other					
.3	Foundation Wall	Material					
		Height Thickness	L.F. Wall				
.4	Slab on Grade	Material Thickness	S.F. Gnd.				
3.0	**SUPERSTRUCTURE**						
.1	Elevated Floors	Bay Size Material					
		Member Size	S.F. Floor				
.2	Roof Structure	Material					
		Bay Size					
		Member Size	S.F. Roof				
.3	Column/Bearing Wall						
	Column	Size	L.F. Col				
		Bay Size					
	Bearing Wall	Material					
		Thickness	L.F. Wall				
.4	Fireproofing	Material	S.F. Surf.				
4.0	**EXTERIOR WALLS**	Building Ht. _____ Ft.					
.1	Outer Wall	Material					
		Thickness	S.F. Wall				
.2	Interior Finish		S.F. Wall				
.3	Windows	Type					
		% of Wall	S.F. Wind.				
.4	Doors	Type					
		Density	S.F. Door				
5.0	**ROOFING**						
.1	Roof Covering	Material	S.F. Roof				

NO.	SYSTEM/COMPONENT		DESCRIPTION	UNIT	UNIT COST	NEW S.F. COST	MODEL S.F. COST	+/- CHANGE
6.0	INTERIOR CONSTRUCTION							
.1	Partitions	Material						
		Density		S.F. Part.				
.2	Doors	Type						
		Density		S.F. Door				
.3	Floor Finish	Material						
		% of Floor		S.F. Floor				
.4	Ceiling Finish	Material						
		% of Ceiling		S.F. Ceil.				
.5	Partition Finish	Material		S.F. Surf.				
7.0	CONVEYING SYSTEMS							
.1	Vertical	Type		Each				
		Capacity						
		Stops						
8.0	MECHANICAL SYSTEMS							
.1	Plumbing			Each				
.2	Fire Protection			S.F.				
.3	Heating	Type		S.F.				
.4	Air Conditioning			S.F.				
9.0	ELECTRICAL							
.1	Lighting			S.F.				
10.0	MISCELLANEOUS							
11.0	SPECIALTIES							

Total Change

ITEM		
18	Adjusted S.F. cost $ _____ item 17 +/- changes	
19	Building area - from item 11 _____ S.F. x adjusted S.F. cost	$ _____
20	Basement area - from item 15 _____ S.F. x S.F. cost $	$ _____
21	Base building sub-total - item 20 + item 19	$ _____
22	Miscellaneous addition (quality, etc.)	$ _____
23	Sub-total - item 22 + item 21	$ _____
24	General conditions - 15 % of item 23	$ _____
25	Sub-total - item 24 + item 23	$ _____
26	Architect's fees _____ % of item 25	$ _____
27	Sub-total - item 26 + item 25	$ _____
28	Location modifier	x _____
29	Local replacement cost - item 28 x item 27	$ _____
30	Depreciation _____ % of item 29	$ _____
31	Depreciated local replacement cost - item 29 less item 30	$ _____
32	Exclusions	$ _____
33	Net depreciated replacement cost - item 31 less item 32	$ _____

Base Cost per square foot floor area: *(from square foot table)*

Specify source:

Page: _____ , Model # _____ , Area _____ S.F.,
Exterior wall _____ , Frame _____

Adjustments for exterior wall variation:

Size Adjustment:
(Interpolate)

Height Adjustment: _____ + _____ = _____

Adjusted Base Cost per square foot:

Building Cost $ _____ x _____ = _____
                    Adjusted Base Cost   Floor Area
                    per square foot

Basement Cost $ _____ x _____ = _____
                    Basement Cost   Basement Area

Lump Sum Additions

TOTAL BUILDING COST *(Sum of above costs)*   _____

Modifications: *(complexity, workmanship, size)*   +/- _____ %   _____

Location Modifier: City _____ Date _____   x _____

Local cost of replacement   _____

Less depreciation: _____   - _____

Local cost of replacement less depreciation $ _____

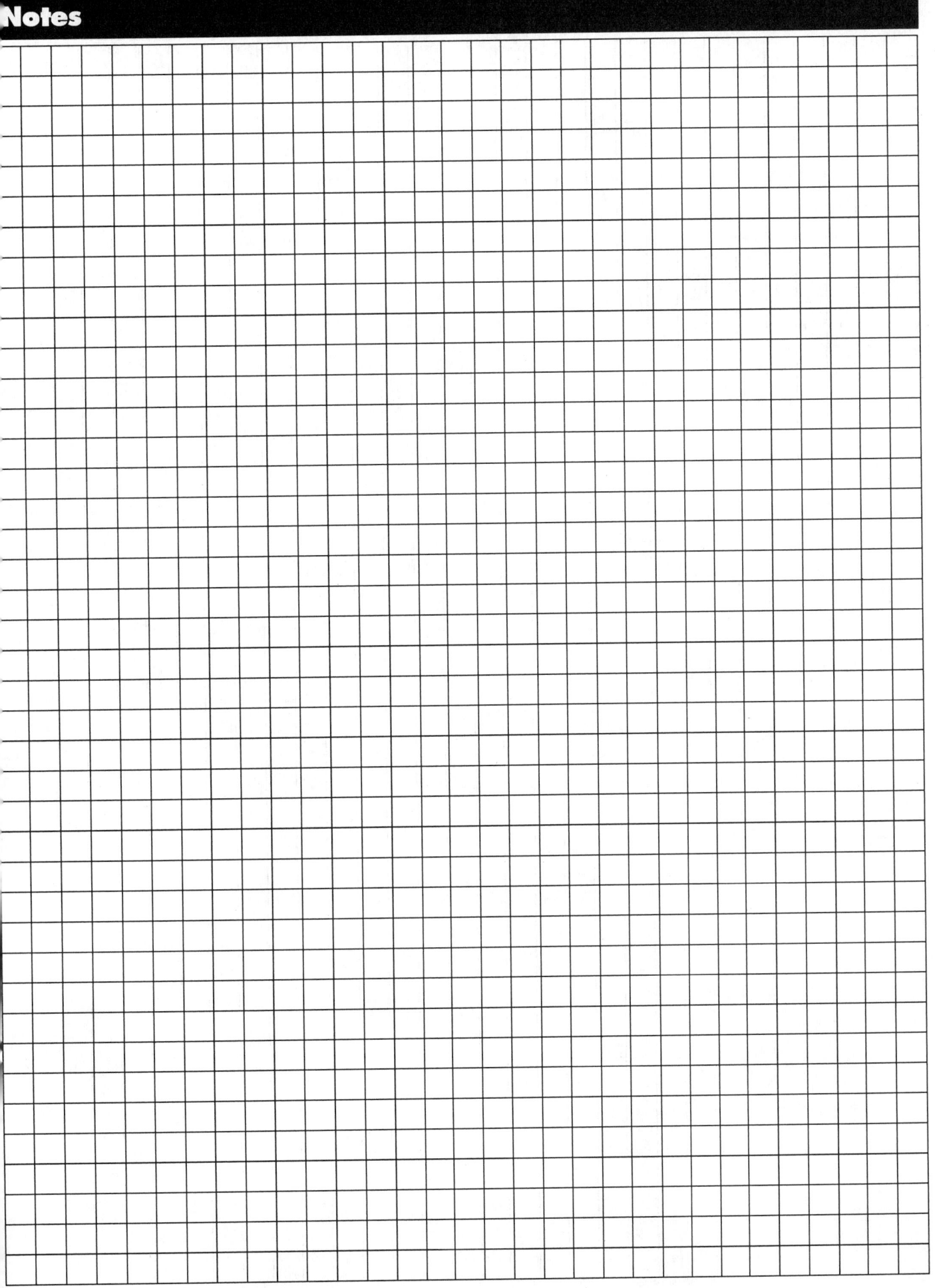

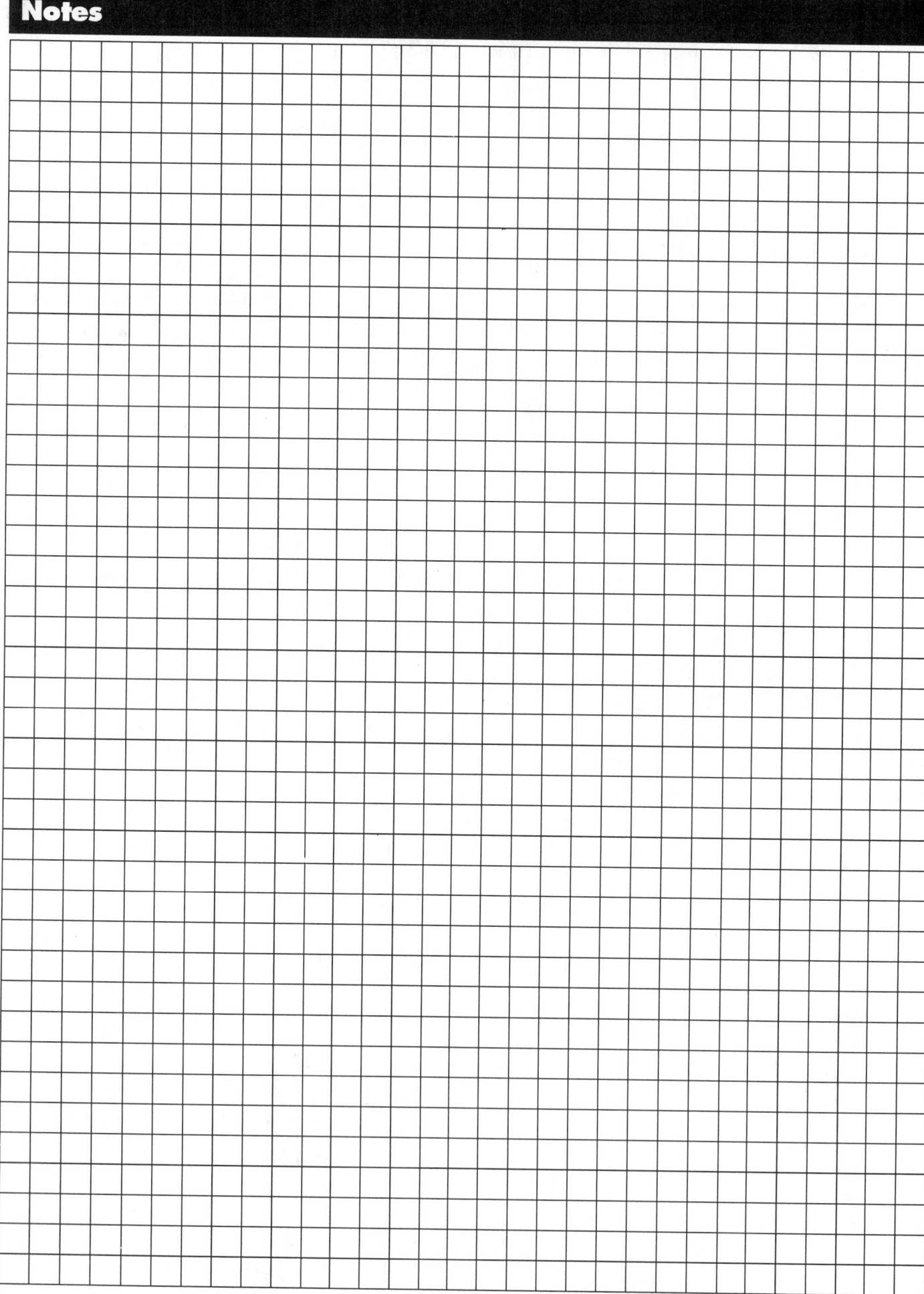

Means Project Cost Report

By filling out and returning the Project Description, you can receive a discount of $20.00 off any one of the Means products advertised in the following pages. The cost information required includes all items marked (✔) except those where no costs occurred. The sum of all major items should equal the Total Project Cost.

$20.00 Discount per product for each report you submit.

DISCOUNT PRODUCTS AVAILABLE—FOR U.S. CUSTOMERS ONLY—STRICTLY CONFIDENTIAL

Project Description (No remodeling projects, please.)

✔ Type Building _____

✔ Location _____

Capacity _____

✔ Frame _____

✔ Exterior _____

✔ Basement: full ☐   partial ☐   none ☐   crawl ☐

✔ Height in Stories _____

✔ Total Floor Area _____

Ground Floor Area _____

✔ Volume in C.F. _____

% Air Conditioned _____ Tons _____

Comments _____

Owner _____

Architect _____

General Contractor _____

✔ Bid Date _____

Typical Bay Size _____

✔ Labor Force: _____ % Union   _____ % Non-Union

✔ Project Description (Circle one number in each line)
1. Economy 2. Average   3. Custom 4. Luxury
1. Square   2. Rectangular 3. Irregular 4. Very Irregular

	✔ **Total Project Cost**	$			
A	✔ **General Conditions**	$			
B	✔ **Site Work**	$			
BS	Site Clearing & Improvement				
BE	Excavation	(	C.Y.)		
BF	Caissons & Piling	(	L.F.)		
BU	Site Utilities				
BP	Roads & Walks Exterior Paving	(	S.Y.)		
C	✔ **Concrete**	$			
C	Cast in Place	(	C.Y.)		
CP	Precast	(	S.F.)		
D	✔ **Masonry**	$			
DB	Brick	(	M)		
DC	Block	(	M)		
DT	Tile	(	S.F.)		
DS	Stone	(	S.F.)		
E	✔ **Metals**	$			
ES	Structural Steel	(	Tons)		
EM	Misc. & Ornamental Metals				
F	✔ **Wood & Plastics**	$			
FR	Rough Carpentry	(	MBF)		
FF	Finish Carpentry				
FM	Architectural Millwork				
G	✔ **Thermal & Moisture Protection**	$			
GW	Waterproofing-Dampproofing	(	S.F.)		
GN	Insulation	(	S.F.)		
GR	Roofing & Flashing	(	S.F.)		
GM	Metal Siding/Curtain Wall	(	S.F.)		
H	✔ **Doors and Windows**	$			
HD	Doors	(	Ea.)		
HW	Windows	(	S.F.)		
HH	Finish Hardware				
HG	Glass & Glazing	(	S.F.)		
HS	Storefronts	(	S.F.)		

J	✔ **Finishes**				$
JL	Lath & Plaster	(	S.Y.)		
JD	Drywall	(	S.F.)		
JM	Tile & Marble	(	S.F.)		
JT	Terrazzo	(	S.F.)		
JA	Acoustical Treatment	(	S.F.)		
JC	Carpet	(	S.Y.)		
JF	Hard Surface Flooring	(	S.F.)		
JP	Painting & Wall Covering	(	S.F.)		
K	✔ **Specialties**				$
KB	Bathroom Partitions & Access.	(	S.F.)		
KF	Other Partitions	(	S.F.)		
KL	Lockers	(	Ea.)		
L	✔ **Equipment**				$
LK	Kitchen				
LS	School				
LO	Other				
M	✔ **Furnishings**				$
MW	Window Treatment				
MS	Seating	(	Ea.)		
N	✔ **Special Construction**				$
NA	Acoustical	(	S.F.)		
NB	Prefab. Bldgs.	(	S.F.)		
NO	Other				
P	✔ **Conveying Systems**				$
PE	Elevators	(	Ea.)		
PS	Escalators	(	Ea.)		
PM	Material Handling				
Q	✔ **Mechanical**				$
QP	Plumbing	(No. of fixtures	)		
QS	Fire Protection (Sprinklers)				
QF	Fire Protection (Hose Standpipes)				
QB	Heating, Ventilating & A.C.				
QH	Heating & Ventilating	(BTU Output	)		
QA	Air Conditioning	(	Tons)		
R	✔ **Electrical**				$
RL	Lighting	(	S.F.)		
RP	Power Service				
RD	Power Distribution				
RA	Alarms				
RG	Special Systems				
S	✔ **Mech./Elec. Combined**				$

Product Name _____

Product Number _____

Your Name _____

Title _____

Company _____

☐ Company

☐ Home   Street Address _____

City, State, Zip _____

☐ Please send _____ forms.

Please specify the Means product you wish to receive. Complete the address information as requested and return this form with your check (product cost less $20.00) to address below.

R.S. Means Company, Inc.,
Square Foot Costs Department
100 Construction Plaza, P.O. Box 800
Kingston, MA 02364-9988

R.S. Means Company, Inc. . . .
a tradition of excellence in Construction Cost Information and Services since 1942.

Table of Contents

Book Selection Guide

The following table provides definitive information on the content of each cost data publication. The number of lines of data provided in each unit price or assemblies division, as well as the number of reference tables and crews is listed for each book. The presence of other elements such as an historical cost index, city cost indexes, square foot models or cross-referenced index is also indicated. You can use the table to help select the Means' book that has the quantity and type of information you most need in your work.

Cost Divisions	Building Construction Costs	Mechanical	Electrical	Repair & Remodel.	Square Foot	Site Work Landsc.	Assemblies	Interior	Concrete Masonry	Open Shop	Heavy Construc.	Residential	Light Commercial	Facil. Construc.	Plumbing	Western Construction Costs
1	1033	521	519	702		945		436	969	1032	1045	372	556	1480	592	1030
2	3366	1384	451	2158		5281		1134	1594	3315	5169	1131	1206	4925	1636	3346
3	1502	91	68	747		1321		188	1932	1478	1449	262	206	1342	43	1480
4	874	27	0	666		805		633	1227	856	709	317	402	1134	0	851
5	1284	170	165	461		787		457	653	1247	1087	267	293	1347	93	1264
6	1370	96	83	1275		475		1287	328	1339	593	1521	1429	1367	59	1727
7	1378	60	9	1304		484		539	458	1377	364	824	1014	1370	70	1377
8	1837	60	0	1861		339		1746	717	1812	9	1094	1151	1967	0	1837
9	1639	50	0	1484		186		1734	322	1637	114	1311	1380	1833	50	1629
10	973	56	30	546		206		804	195	973	0	244	446	971	249	973
11	1019	261	200	586		53		896	34	1002	16	115	234	1016	231	1001
12	353	0	0	47		227		1533	30	344	0	73	68	1536	0	344
13	881	413	201	154		464		356	105	879	325	120	133	966	239	930
14	365	43	0	262		36		330	0	365	40	9	18	363	17	363
15	2724	15477	703	2302		2022		1772	80	2643	2327	986	1613	13058	11712	2600
16	1547	798	10589	1121		1098		1346	62	1562	1034	704	1275	10170	705	1489
17	459	369	368	1		0		0	0	460	0	0	0	459	369	459
Totals	22604	19876	13386	15677		14729		15191	8706	22321	14281	9350	11424	45304	16065	22700

Assembly Divisions	Building Construction Costs	Mechanical	Electrical	Repair & Remodel.	Square Foot	Site Work Landsc.	Assemblies	Interior	Concrete Masonry	Open Shop	Heavy Construc.	Residential	Light Commercial	Facil. Construc.	Plumbing	Western Construction Costs
1		0	0	202	131	738	775	0	689		752	844	131	89	0	0
2		0	0	40	33	35	48	0	49		0	589	32	0	0	0
3		0	0	445	1114	0	3064	228	1245		0	1549	806	174	0	0
4		0	0	713	1353	0	3175	255	1273		0	2047	1171	26	0	0
5		0	0	289	231	0	465	0	0		0	1082	227	28	0	0
6		0	0	1006	857	0	1308	1736	152		0	1084	749	333	0	0
7		0	0	42	89	0	179	159	0		0	723	33	71	0	0
8		2255	149	998	1699	0	2725	925	0		0	1694	1197	1114	2091	0
9		0	1347	355	357	0	1284	307	0		0	242	358	319	0	0
10		0	0	0	0	0	0	0	0		0	0	0	0	0	0
11		0	0	365	465	0	724	174	0		0	0	466	222	0	0
12		501	160	539	84	2392	699	0	698		541	0	84	119	850	
Totals		2756	1656	4994	6413	3165	14446	3784	4106		1293	9854	5254	2495	2941	

Reference Section	Building Construction Costs	Mechanical	Electrical	Repair & Remodel.	Square Foot	Site Work Landsc.	Assemblies	Interior	Concrete Masonry	Open Shop	Heavy Construc.	Residential	Light Commercial	Facil. Construc.	Plumbing	Western Construction Costs
Tables	156	46	85	70	4	87	223	60	84	156	56	51	74	88	49	158
Models					102							32	43			
Crews	397	397	397	397		397		397	397	397	397	397	397	397	397	397
City Cost Indexes	yes	yes	yes	yes	yes	yes	yes	yes	yes	yes	yes	yes	yes	yes	yes	yes
Historical Cost Indexes	yes	yes	yes	yes	yes	yes	yes	yes	yes	yes	yes	no	yes	yes	yes	yes
Index	yes	yes	yes	yes	no	yes	yes	yes	yes	yes	yes	yes	yes	yes	yes	yes

1

Annual Cost Guides

Means Building Construction Cost Data 1998

Available in Both Softbound and Looseleaf Editions

The "Bible" of the industry comes in the standard softcover edition or the looseleaf edition.

Many customers enjoy the convenience and flexibility of the looseleaf binder, which increases the usefulness of *Means Building Construction Cost Data 1998* by making it easy to add and remove pages. You can insert your own cost information pages, so everything is in one place. Copying pages for faxing is easier also. Whichever edition you prefer, softbound or the convenient looseleaf edition, you'll get *The DI Quarterly/Change Notice* newsletter at no extra cost.

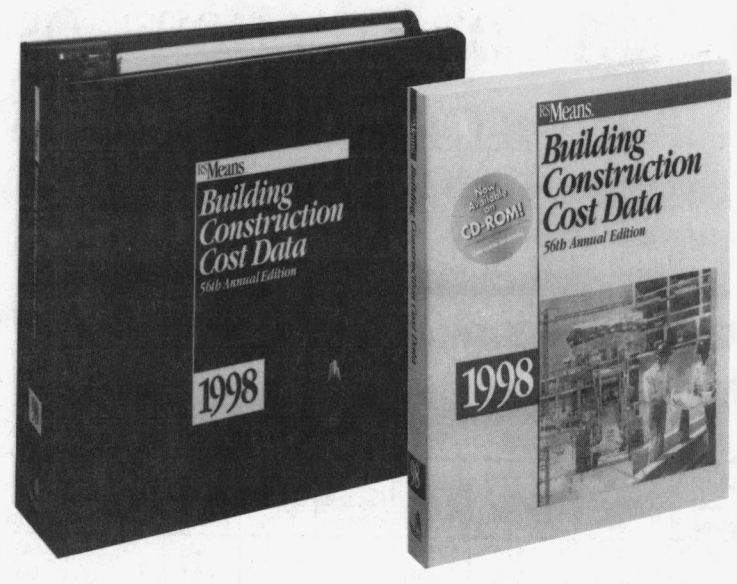

$84.95 per copy, Softbound
Catalog No. 60018

$109.95 per copy, Looseleaf
Catalog No. 61018

Means Building Construction Cost Data 1998

Offers you unchallenged unit price reliability in an easy-to-use arrangement. Whether used for complete, finished estimates or for periodic checks, it supplies more cost facts better and faster than any comparable source. Over 21,000 unit prices for 1998. The City Cost Indexes now cover over 930 areas, for indexing to any project location in North America. Order and get *The DI Quarterly/Change Notice* newsletter sent to you FREE. You'll have year-long access to the Means Estimating **Hotline** FREE with your subscription. Expert assistance when using Means data is just a phone call away.

$84.95 per copy
Over 650 pages, illustrated, available Oct. 1997
Catalog No. 60018

Means Building Construction Cost Data 1998

Metric Version

The Federal Government has stated that all federal construction projects must now use metric documentation. The *Metric Version* of *Means Building Construction Cost Data 1998* is presented in metric measurements covering all construction areas. Don't miss out on these billion dollar opportunities. Make the switch to metric today.

$89.95 per copy
Over 650 pages, illustrated, available Nov. 1997
Catalog No. 63018

Annual Cost Guides

eans Mechanical ost Data 1998

VAC · Controls

al unit and systems price guidance for
chanical construction . . . materials, parts,
ngs, and complete labor cost information.
udes prices for piping, heating, air conditioning,
tilation, and all related construction.

s new 1998 unit costs for:

ver 3000 installed HVAC/controls assemblies
On Site" Location Factors for close to 1,000
ties and towns in the U.S. and Canada
rews, labor and equipment

.95 per copy
er 600 pages, illustrated, available Oct. 1997
talog No. 60028

Means Plumbing Cost Data 1998

Comprehensive unit prices and assemblies for
plumbing, irrigation systems, commercial and
residential fire protection, point-of-use water
heaters, and the latest approved materials. This
publication and its companion, *Means Mechanical
Cost Data*, provide full-range cost estimating
coverage for all the mechanical trades.

$87.95 per copy
Over 500 pages, illustrated, available Oct. 1997
Catalog No. 60218

eans Electrical ost Data 1998

cing information for every part of electrical cost
nning: More than 17,000 unit and systems costs
h design tables; clear specifications and
wings; engineering guides and illustrated
imating procedures; complete labor-hour and
terials costs for better scheduling and
ocurement; the latest electrical products and
nstruction methods.

NEW FOR 1998) Underground Marking Tape
nd Polyethylene Pull Rope costs
Variety of Special Electrical Systems including
athodic Protection
osts for maintenance, demolition, HVAC/
echanical, specialties, equipment, and more

7.95 per copy
er 450 pages, illustrated, available Oct. 1997
talog No. 60038

Means Electrical Change Order Cost Data 1998

You are provided with electrical unit prices
exclusively for pricing change orders—based on the
recent, direct experience of contractors and
suppliers. Analyze and check your own change
order estimates against the experience others have
had doing the same work. It also covers
productivity analysis and change order cost
justifications. With useful information for
calculating the effects of change orders and dealing
with their administration.

$89.95 per copy
Over 440 pages, available Oct. 1997
Catalog No. 60238

eans Facilities aintenance & epair Cost Data 1998

blished in a looseleaf format, *Means Facilities
aintenance & Repair Cost Data* gives you a
mplete system to manage and plan your facility
pair and maintenance costs and budget efficiently.
idelines for auditing a facility and developing an
nual maintenance plan. Budgeting is included,
ng with reference tables on cost and
anagement and information on frequency and
oductivity of maintenance operations.

e only nationally recognized source of
aintenance and repair costs. Developed in
operation with the Army Corps of Engineers.

99.95 per copy
ver 600 pages, illustrated, available Dec. 1997
atalog No. 60308

Means Square Foot Costs 1998

It's Accurate and Easy To Use!

· NEW 1998 price information, based on
  nationwide figures from suppliers, estimators,
  labor experts and contractors.

· NEW "How-to-Use" Sections, with better,
  clearer examples of commercial, residential,
  industrial, and institutional structures.

· NEW, more realistic graphics, offering true-to-life
  illustrations of building projects.

· NEW, more extensive information on using
  square foot cost data, including sample estimates
  and alternate pricing methods.

$97.95 per copy
Over 460 pages, illustrated, available Nov. 1997
Catalog No. 60058

Annual Cost Guides

Means Repair & Remodeling Cost Data 1998

Commercial/Residential

You can use this valuable tool to estimate commercial and residential renovation and remodeling.

Includes: New 1998 costs for hundreds of unique methods, materials and conditions that only come up in repair and remodeling. PLUS:

- 1998 unit costs for over 15,000 construction components
- 1998 installed costs for over 4,000 assemblies.
- 1998 costs for 300+ construction crews
- Over 930 "On Site" localization factors for the U.S. and Canada.

$79.95 per copy
Over 600 pages, illustrated, available Oct. 1997
Catalog No. 60048

Means Facilities Construction Cost Data 1998

For the maintenance and construction of commercial, industrial, municipal, and institution properties. Costs are shown for new and remodeli construction and are broken down into materials, labor, equipment, overhead, and profit. Special emphasis is given to sections on mechanical, electrical, furnishings, site work, building maintenance, finish work, and demolition. More than 40,000 unit costs plus assemblies and reference sections are included.

$209.95 per copy
Over 1100 pages, illustrated, available Nov. 1997
Catalog No. 60208

Means Residential Cost Data 1998

Now contains square foot costs for 30 basic home models with the look of today—plus hundreds of custom additions and modifications you can quote right off the page. With costs for the 100 residential systems you're most likely to use in the year ahead. Complete with blank estimating forms, sample estimates and step-by-step instructions.

$74.95 per copy
Over 550 pages, illustrated, available Dec. 1997
Catalog No. 60178

Means Light Commercial Cost Data 1998

Specifically addresses the light commercial marke which is an increasingly specialized niche in the industry. Aids you, the owner/designer/contractor in preparing all types of estimates, from budgets t detailed bids. Includes new advances in methods and materials. Assemblies section allows you to evaluate alternatives in the early stages of design/planning.

Over 10,000 unit costs for 1998 ensure you have the prices you need... when you need them.

$76.95 per copy
Over 600 pages, illustrated, available Nov. 1997
Catalog No. 60188

Means Assemblies Cost Data 1998

Means Assemblies Cost Data 1998 takes the guesswork out of preliminary or conceptual estimates. Now you don't have to try to calculate the assembled cost by working up individual components costs. We've done all the work for you.

Presents detailed illustrations, descriptions, specifications and costs for every conceivable building assembly—240 types in all—arranged in the easy-to-use UniFormat system. Each illustrated "assembled" cost includes a complete grouping of materials and associated installation costs including the installing contractor's overhead and profit.

$139.95 per copy
Over 570 pages, illustrated, available Oct. 1997
Catalog No. 60068

Means Site Work & Landscape Cost Data 1998

Means Site Work & Landscape Cost Data 1998 is organized to assist you in all your estimating need Hundreds of fact-filled pages help you make accurate cost estimates efficiently.

New for 1998!

- New or expanded demolition features— ceilings, doors, electrical, flooring, HVAC, millwork, plumbing, roofing, walls and windows
- State-of-the-art segmental retaining walls
- Flywheel trenching costs and details
- Expanded Wells section
- Thousands of landscape materials, flowers, shrubs and trees

$89.95 per copy
Over 570 pages, illustrated, available Nov. 1997
Catalog No. 60288

Annual Cost Guides

Means Open Shop Building Construction Cost Data 1998

...latest costs for accurate budgeting and estimating of commercial and residential construction... ...vation work... change orders... cost engineering.
...ns Open Shop BCCD will assist you to...
...velop benchmark prices for change orders
...g gaps in preliminary estimates, budgets
...timate complex projects
...bstantiate invoices on contracts
...ce ADA-related renovations

...95 per copy
...er 650 pages, illustrated, available Nov. 1997
...alog No. 60158

Means Heavy Construction Cost Data 1998

A comprehensive guide to heavy construction costs. Includes costs for highly specialized projects such as tunnels, dams, highways, airports, and waterways. Information on different labor rates, equipment, and material costs is included. Has unit price costs, systems costs, and numerous reference tables for costs and design. Valuable not only to contractors and civil engineers, but also to government agencies and city/town engineers.

$89.95 per copy
Over 450 pages, illustrated, available Dec. 1997
Catalog No. 60168

Means Building Construction Cost Data 1998
Western Edition

...is regional edition provides more precise cost
...ormation for western North America. Labor rates are
...ed on union rates from 13 western states and
...stern Canada. Included are western practices and
...terials not found in our national edition: tilt-up
...ncrete walls, glu-lam structural systems, specialized
...ber construction, seismic restraints, landscape and
...gation systems.

...9.95 per copy
...er 600 pages, illustrated, available Nov. 1997
...atalog No. 60228

Means Heavy Construction Cost Data 1998
Metric Version

Make sure you have the Means industry standard metric costs for the federal, state, municipal and private marketplace. With thousands of up-to-date metric unit prices in tables by CSI standard divisions. Supplies you with assemblies costs using the metric standard for reliable cost projections in the design stage of your project. Helps you determine sizes, material amounts, and has tips for handling metric estimates.

$89.95 per copy
Over 450 pages, illustrated, available Dec. 1997
Catalog No. 63168

Means Construction Cost Indexes 1998

...ho knows what 1998 holds? What materials
...d labor costs will change unexpectedly?
...how much?
...reakdowns for 305 major cities.
...ational averages for 30 key cities.
...xpanded five major city indexes.
...istorical construction cost indexes.

...98.00 per year/$49.50 individual quarters
...atalog No. 60148

Means Interior Cost Data 1998

Provides you with prices and guidance needed to make accurate interior work estimates. Contains costs on materials, equipment, hardware, custom installations, furnishings, labor costs . . . every cost factor for new and remodel commercial and industrial interior construction, plus more than 50 reference tables. For contractors, facility managers, owners.

Newly expanded information on office furnishings.

$83.95 per copy
Over 550 pages, illustrated, available Nov. 1997
Catalog No. 60098

Means Concrete & Masonry Cost Data 1998

...rovides you with cost facts for virtually all
...oncrete/masonry estimating needs, from complicated
...rm work to various sizes and face finishes of brick and
...lock, all in great detail. The comprehensive unit cost
... section contains more than 15,000 selected entries. The
...ssemblies cost section is illustrated with isometric
...rawings. A detailed reference section supplements the
...ost data.

...77.95 per copy
...Over 520 pages, illustrated, available Dec. 1997
...Catalog No. 60118

Means Labor Rates for the Construction Industry 1998

Complete information for estimating labor costs, making comparisons and negotiating wage rates by trade for over 300 cities (United States and Canada), 46 construction trades listed by local union number in each city, historical wage rates included for comparison. No similar book is available through the trade.

New: Each city chart now lists the county and is alphabetically arranged with handy visual flip tabs for quick reference.

$189.95 per copy
Over 330 pages, available Dec. 1997
Catalog No. 60128

Reference Books

Means Environmental Remediation Estimating Methods

By Richard R. Rast **NEW TITLE**

The first-ever guide to estimating any size environmental remediation project... anywhere in the country.

Field-tested guidelines for estimating 50 standard remediation technologies. This resource will help you: prepare preliminary budgets, develop detailed estimates, compare costs, select solutions, estimate liability, review quotes, and negotiate settlements.

A valuable support tool for the *Environmental Restoration* cost data books.

$99.95 per copy
Over 600 pages, illustrated, Hardcover
Catalog No. 64777

Cyberplaces: The Internet Guide for Architects, Engineers & Contractors

By Paul Doherty **NEW TITL**

Internet applications for business and project management. Includes book, CD-ROM and Web site.

The CD-ROM offers built-in links to Web sites, tours of captured sites, FREE browser software and a document workshop, plus a test for Continuing Education Credits. **The Web Site** keeps the book current with updates on new technologies, interactive workshops, links to new tools and site. and reports from top firms.

See Page 14 for full-page description

$59.95 per copy
Over 700 pages, illustrated, Softcover
Catalog No. 67317

Value Engineering: Practical Applications

...For Design, Construction, Maintenance & Operations **NEW TITLE**

By Alphonse Dell'Isola, P.E., the recognized leading authority on Value Engineering

A tool for immediate application—for engineers, architects, facility managers, owners and contractors. Includes: Making the Case for VE—The Management Briefing, Integrating VE into Planning and Budgeting, Conducting Life Cycle Costing, Integrating VE into the Design Process, and Using VE Methodology in Design Review and Consultant Selection, Case Studies (corporate, commercial, hospital, industrial and civil), and an expertly organized VE Workbook.

$79.95 per copy
Over 450 pages, illustrated, Hardcover
Catalog No. 67319

HVAC: Design Criteria, Options, Selection
Expanded Second Edition

By William H. Rowe III, AIA, PE

Now including Indoor Air Quality, CFC Removal, Energy Efficient Systems and Special Systems by Building Type.

The book that helps you solve a wide range of HVAC system design and selection problems effectively and economically. Gives you explanations of the latest ASHRAE standards.

$84.95 per copy
Over 600 pages, illustrated, Hardcover
Catalog No. 67306

The ADA in Practice

(Revised, expanded edition of the award-winning* New ADA: Compliance & Costs)

By Deborah S. Kearney, PhD

Helps you meet and budget for the requirements of the Americans with Disabilities Act. Shows you how to do the job right, by understanding what the law actually requires and allows. Includes an objective "authoritative buyers guide" for 70 ADA-compliant products with specs and purchasing information. With sample evaluation forms and illustrations of ADA-compliant products.

*Winner of the "Distinguished Author of the Year" award from the International Facility Managers Association.

$72.95 per copy
Over 600 pages, illustrated, Softcover
Catalog No. 67147A

Means ADA Compliance Pricing Guide

Gives you detailed cost estimates for budgeting modification projects, including estimates for each of 260 alternates. You get the 75 most commonly needed modifications for ADA compliance, each an assembly estimate with detailed cost breakdown including materials, labor hours and contractor's overhead. With 3,000 additional ADA compliance-related unit cost line items and costs easily adjusted to 900 cities and towns.

$72.95 per copy
Over 350 pages, illustrated, Softcover
Catalog No. 67310

Reference Books

Cost Planning & Estimating for Facilities Maintenance

In this unique book, a team of facilities management authorities shares their expertise at:
- Evaluating and budgeting operations
- Maintaining & repairing key building components
- Applying *Means Facilities Maintenance & Repair Cost Data* to your estimating

With the special maintenance requirements of 10 different building types.

$82.95 per copy
Over 475 pages, Hardcover
Catalog No. 67314

Facilities Planning & Relocation

By David D. Owen

An A–Z working guide—complete with the checklists, schematic diagrams and questionnaires to ensure the success of every office relocation. Complete with a step-by-step manual, 100-page technical reference section, over 50 reproducible forms in hard copy and on computer diskettes.

Winner of the International Facility Managers Assoc. "Distinguished Author of the Year" award.

$109.95 per copy, Textbook-384 pages,
Forms Binder-137 pages, illustrated, Hardcover
Catalog No. 67301

Means Facilities Maintenance Standards

Unique features of this one-of-a-kind working guide for facilities maintenance

A working encyclopedia that points the way to solutions to every kind of maintenance and repair dilemma. With a labor-hours section to provide productivity figures for over 180 maintenance tasks. Included are ready-to-use forms, checklists, worksheets and comparisons, as well as analysis of materials systems and remedies for deterioration and wear.

$159.95 per copy, 600 pages, 205 tables,
checklists and diagrams, Hardcover
Catalog No. 67246

Maintenance Management Audit

At a time when many companies and institutions are reorganizing or downsizing, this annual audit program is essential for every organization in need of a proper assessment of its maintenance operation. The forms presented in this easy-to-use workbook allow managers to identify and correct problems, enhance productivity, and impact the bottom line.

New reduced price

Now $32.48 per copy, limited quantity
125 pages, spiral bound, illustrated, Hardcover
Catalog No. 67299

The Facilities Manager's Reference

By Harvey H. Kaiser, PhD

The tasks and tools the facility manager needs to accomplish the organization's objectives, and develop individual and staff skills. Includes Facilities and Property Management, Administrative Control, Planning and Operations, Support Services, and a complete building audit with forms and instructions, widely used by facilities managers nationwide.

$86.95 per copy, over 250 pages,
with prototype forms and graphics, Hardcover
Catalog No. 67264

Facilities Maintenance Management

By Gregory H. Magee, PE

Now you can get successful management methods and techniques for all aspects of facilities maintenance. This comprehensive reference explains and demonstrates successful management techniques for all aspects of maintenance, repair and improvements for buildings, machinery, equipment and grounds. Plus, guidance for outsourcing and managing internal staffs.

$86.95 per copy
Over 280 pages with illustrations, Hardcover
Catalog No. 67249

Understanding Building Automation Systems

- Direct Digital Control
- Energy Management
- Security/Access Control
- Life Safety
- Lighting

By Reinhold A. Carlson, PE & Robert Di Giandomenico

The authors, leading authorities on the design and installation of these systems, describe the major building systems in both an overview with estimating and selection criteria, and in system configuration-level detail.

$79.95 per copy
Over 200 pages, illustrated, Hardcover
Catalog No. 67284

Planning and Managing Interior Projects

By Carol E. Farren

Expert, up-to-date guidance for managing interior installation projects. Includes: project phases; winning client support; space planning and design; budgeting, bidding and purchasing, and more.

For interior designers, architects, facilities professionals.

$76.95 per copy
Over 330 pages, illustrated, Hardcover
Catalog No. 67245

Reference Books

Basics For Builders: Plan Reading & Material Takeoff
By Wayne J. DelPico

For Residential and Light Commercial Construction

A valuable tool for understanding plans and specs, and accurately calculating material quantities. Step-by-step instructions and takeoff procedures based on a full set of working drawings.

$35.95 per copy
Over 420 pages, Softcover
Catalog No. 67307

Means Illustrated Construction Dictionary Unabridged Edition

Written in contractor's language, the information is adaptable for report writing, specifications or just intelligent discussion. Contains over 17,000 construction terms, words, phrases, acronyms and abbreviations, slang, regional terminology, and hundreds of illustrations.

$99.95 per copy
Over 700 pages, Hardcover
Catalog No. 67292

Superintending for Contractors:
How to Bring Jobs in On-time, On-Budget
By Paul J. Cook

This book examines the complex role of the superintendent/field project manager, and provides guidelines for the efficient organization of this job. Includes administration of contracts, change orders, purchase orders, and more.

$35.95 per copy
Over 220 pages, illustrated, Softcover
Catalog No. 67233

Means Forms for Contractors
For general and specialty contractors

Means editors have created and collected the most needed forms as requested by contractors of various-sized firms and specialties. This book covers all project phases. Includes sample correspondence and personnel administration.
Full-size forms on durable paper for photocopying or reprinting.

$79.95 per copy
Over 400 pages, three-ring binder, more than 80 forms
Catalog No. 67288

Bidding for Contractors:
How to Make Bids that Make Money
By Paul J. Cook

The author shares the benefit of his more than 30 years of experience in construction project management, providing contractors with the tools they need to develop competitive bids.

New reduced price

Now $17.98 per copy, limited quantity
Over 225 pages with graphics, Softcover
Catalog No. 67180

Means Heavy Construction Handbook

Informed guidance for planning, estimating and performing today's heavy construction projects. Provides expert advice on every aspect of heavy construction work including hazardous waste remediation and estimating. To assist planning, estimating, performing, or overseeing work.

$74.95 per copy
Over 430 pages, illustrated, Hardcover
Catalog No. 67148

Estimating for Contractors:
How to Make Estimates that Win Jobs
By Paul J. Cook

Estimating for Contractors is a reference that will be used over and over, whether to check a specific estimating procedure, or to take a complete course in estimating.

$35.95 per copy
Over 225 pages, illustrated, Softcover
Catalog No. 67160

HVAC Systems Evaluation
By Harold R. Colen, PE

You get direct comparisons of how each type of system works, with the relative costs of installation, operation, maintenance and applications by type of building. With requirements for hooking up electrical power to HVAC components. Contains experienced advice for repairing operational problems in existing HVAC systems, ductwork, fans, cooling coils, and much more!

$84.95 per copy
Over 500 pages, illustrated, Hardcover
Catalog No. 67281

Quantity Takeoff for Contractors:
How to Get Accurate Material Counts
By Paul J. Cook

Contractors who are new to material takeoffs or want to be sure they are using the best techniques will find helpful information in this book, organized by CSI MasterFormat division.

New reduced price

Now $17.48 per copy, limited quantity
Over 250 pages, illustrated, Softcover
Catalog No. 67262

Roofing: Design Criteria, Options, Selection
By R.D. Herbert, III

This book is required reading for those who specify, install or have to maintain roofing systems. It covers all types of roofing technology and systems. You'll get the facts needed to intelligently evaluate and select both traditional and new roofing systems.

$62.95 per copy
Over 225 pages with illustrations, Hardcover
Catalog No. 67253